W9-CLD-590

Cosmology

Advances in science have greatly changed our ideas on the nature of the universe. *Cosmology: The Science of the Universe* is a broad and elementary introduction to cosmology that includes aspects of its history, theology, and philosophy. The book explores the realm of receding galaxies, the fascinating properties of space and time, the bizarre world of black holes, the astonishing expansion of the universe, the elegant simplicity of cosmic redshifts, and the momentous issues of inflation. Its subjects cover modern views on the origin of atoms, galaxies, life, and the universe itself; they range from the subatomic to the extragalactic, from the beginning to the end of time, and from terrestrial to extraterrestrial life. Old problems (e.g., the cosmic-edge) are revived and new perplexities (e.g., the containment riddle) are reviewed. In this unique book, Professor Harrison shows how in every age societies devise universes that make sense of the human experience. He explores the cosmic scenery of the Babylonian, Pythagorean, Aristotelian, Stoic, Epicurean, Medieval, Cartesian, and Newtonian world systems and shows how these and other systems laid the foundations of the modern physical universe.

The first edition of this best-selling book received world-wide acclaim for its far ranging treatment and clarity of explanation. This eagerly awaited second edition updates and extends the first edition. The additional chapters discuss *Early Scientific Cosmology, Cartesian and Newtonian World Systems, Cosmology After Newton and Before Einstein, Observational Cosmology, Inflation,* and *Creation of the Universe.*

EDWARD HARRISON, distinguished university professor emeritus of physics and astronomy at the University of Massachusetts, was born in London at the end of World War I. He studied at London University and served for several years in action with the British Army in World War II. He was a scientist at the Atomic Energy Research Establishment and the Rutherford High Energy Laboratory in England until 1966 when he became a Five College professor at the University of Massachusetts and taught at Amherst, Hampshire, Mount Holyoke, and Smith Colleges. Professor Harrison is author of *The Masks of the Universe* (which gained the Melcher Award), *Darkness at Night: A Riddle of the Universe,* and numerous scientific articles that have contributed to the advance of modern cosmology. He has also written many articles on the history and philosophy of early cosmology. He is married to Photeni, has two children, John-Peter and June Zöe, and is now adjunct professor at the Steward Observatory, University of Arizona.

Cosmology

THE SCIENCE OF THE UNIVERSE
SECOND EDITION

EDWARD HARRISON
Five College Astronomy Department, University of Massachusetts
Steward Observatory, University of Arizona

CAMBRIDGE
UNIVERSITY PRESS

CAMBRIDGE
UNIVERSITY PRESS

University Printing House, Cambridge CB2 8BS, United Kingdom

Cambridge University Press is part of the University of Cambridge.

It furthers the University's mission by disseminating knowledge in the pursuit of education, learning and research at the highest international levels of excellence.

www.cambridge.org
Information on this title: www.cambridge.org/9780521661485

© Cambridge University Press 1981, 2000

This publication is in copyright. Subject to statutory exception and to the provisions of relevant collective licensing agreements, no reproduction of any part may take place without the written permission of Cambridge University Press.

First published 1981
Reprinted 1985, 1986, 1988, 1989, 1991
Second edition 2000
8th printing 2013

A catalogue record for this publication is available from the British Library

Library of Congress Cataloguing in Publication data

Harrison, Edward Robert.
Cosmology: the science of the universe / Edward R. Harrison. —
2nd ed.
    p.   cm.
Includes bibliographical references and index.
ISBN 0 521 66148 X
1. Cosmology. I. Title. II. Title: Cosmology, the science of the universe.
QB981.H276   1999
523.1–dc21   99-10172 CIP

ISBN 978-0-521-66148-5 Hardback

Cambridge University Press has no responsibility for the persistence or accuracy of URLs for external or third-party internet websites referred to in this publication, and does not guarantee that any content on such websites is, or will remain, accurate or appropriate.

CONTENTS

PREFACE

This second edition of *Cosmology: The Science of the Universe* revises and extends the first edition published in 1981. Much has happened since the first edition; many developments have occurred, and cosmology has become a wider field of research.

As before, the treatment is elementary yet broad in scope, and the aim is to present an outline that appeals to the thoughtful person at a level not requiring an advanced knowledge in the natural sciences. Cosmology has many faces, scientific and nonscientific; in this work the primary emphasis is on cosmology as a science, but the important historical, philosophical, and theological aspects are not ignored. Mathematics is avoided except in a few places, mostly at the end of chapters, and the treatment is varied enough to meet the needs of both those who enjoy and do not enjoy mathematics.

At the end of each chapter are two sections entitled *Reflections* and *Projects*. The Reflections section presents topics for reflection and discussion. The Projects section raises questions and issues that a challenged reader might care to tackle. Cosmology impels us to ask deep questions, read widely, and think deeply. It is not the sort of subject that lends itself readily to simple yes and no answers. On most issues there are conflicting arguments to be investigated, weighed, rejected, accepted, or modified according to one's personal tastes and beliefs. Cosmology challenges the mind, shapes our way of thinking about the world in which we live, and leaves impressions and ideas that last a lifetime.

Many texts on cosmology and general relativity tend to be too technical for college students and nonspecialists. Numerous less-technical treatments now exist that are often too brief and of insufficient scope and depth for a course of study. At the end of each chapter are suggestions for further reading to help the reader explore alternative treatments (sometimes in greater depth and detail) of the subjects discussed in the chapter. Also provided is a list of sources containing references that are usually readable and not too technical; the few that are more technical are included for their historical interest.

The first edition of this book evolved from class notes used for teaching elementary cosmology in the Five College Astronomy Department of Amherst College, Hampshire College, Mount Holyoke College, Smith College, and the University of Massachusetts. At that time the method of grading consisted of brief weekly papers, mostly on topics (germane to the lectures) of each student's choice. It was evident that a text of broad scope was needed that might hold the attention of students of different backgrounds and interests, and provide the information needed for discussions and the preparation of papers. After the publication of the first edition, the method of grading changed and consisted of four equally spaced take-home examinations followed by an end-of-semester examination. Many questions included in the examinations did not require mathematical skills. Both methods of grading have their advantages and disadvantages. There must be a better way!

I am indebted to many persons for their comments and helpful suggestions, particularly Thomas Arny (University of Massachusetts, Amherst), Gregory Benford (University of California, Irvine), Robert Brandenberger (Brown University), Mario Bunge (McGill University), Thomas Dennis (Mount Holyoke College), James Ellern (University of Southern California, Los Angeles), George Ellis (University of Capetown), Stephen Gottesman (University of Florida, Gainsville), George Greenstein (Amherst College), Gary Hinshaw (NASA/Goddard Space Flight Center), Paul Hodge (University of Washington), Duane Howells (Hughs Research Laboratories), John Huchra (Harvard–Smithsonian Center for Astrophysics), John

Lathrop, Charles Leffert, William McCrea (University of Sussex), A. J. Meadows (Loughborough University of Technology), Heinz Pagels (University of California, Santa Cruz), Joel Primack (University of California, Santa Cruz), Martin Rees (Cambridge University), Joe Rosen (University of Central Arkansas), Rick Shafer (NASA/Goddard Space Flight Center), Stephen Schneider (University of Massachusetts, Amherst), Joseph Snider (Oberlin College), Joseph Tenn (Sonoma State University), Virginia Trimble (University of California, Irvine), David Van Blerkom (University of Massachusetts, Amherst), Gerard de Vaucouleurs (University of Texas, Austin), and Robert Wilson (Smithsonian Astrophysical Observatory).

I am particularly grateful to Fred Stevenson (University of Leeds) for his helpful comments and corrections.

EDWARD HARRISON
Mesilla, New Mexico, May 1998

INTRODUCTION

With equal passion I have sought knowledge. I have wished to understand the hearts of men, I have wished to know why the stars shine. And I have tried to apprehend the Pythagorean power by which number holds sway above the flux. A little of this, but not much, I have achieved.
Bertrand Russell (1872–1970), Autobiography, *Prologue*

PROLOGUE

Cosmology, the science of the universe, attracts and fascinates us all. In one sense, it is the science of the large-scale structure of the universe: of the realm of extragalactic nebulae, of distant and receding horizons, and of the dynamic curvature of cosmic space and time. In another sense, it seeks to assemble all knowledge into a unifying cosmic picture. Most sciences tear things apart into smaller and smaller constituents in order to examine the world in ever greater detail, whereas cosmology is the one science that puts the pieces together into a "mighty frame." In yet another sense, it is the history of mankind's search for understanding of the universe, a quest that began long ago at the dawn of the human race. We cannot study cosmology in the broadest sense without heeding the many cosmic pictures of the past that have shaped human history. We trace the rise of the scientific method and how it has increased our understanding of the physical universe. Which brings us to the major aim of this book: gaining an elementary understanding of the physical universe of modern times.

Cosmology compels us, willy-nilly, to examine our deepest and sometimes most cherished beliefs. It awakens an awareness of ancient vestigial paradigms that control our lives and direct the destiny of societies. A person who migrates to a new land, joins a revolution, goes to war, seeks political power, gains or loses a fortune, gets married, or does any momentous thing is influenced by cosmic beliefs.

A brief summary of the contents of this book serves as an introduction to the scope of cosmology as a modern science. In outline only, chapter by chapter, the subjects covered are as follows.

CHAPTER 1
WHAT IS COSMOLOGY?

The history of cosmology shows that in every age in all societies people believe that they have at last discovered the true nature of the Universe. But in each case they have devised a mask fitted on the face of the unknown Universe. In this book we use "universe" to denote "a model of the Universe" and avoid making claims to true and final knowledge of the Universe. Where there is a society of rational individuals, there we find a universe, and where there is a universe, there we find a society of rational individuals. Proud of their knowledge and confident of its final truth, the members of a society pity the ignorance of their ancestors and fail to foresee that their descendants will also pity them for their ignorance.

Cosmology is the study of universes, how they originate, how they evolve. Plausibly, hundreds of thousands of years ago, in an Age of Magic, the world was explained by the activity of ambient spirits. In an Age of Myths, tens of thousand years ago, lasting until recent times, the world was explained by the capricious acts of nature spirits and the will of remote gods and goddesses. In an Age of Science, we have abandoned much of our anthropocentric heritage, and have devised a series of mechanistic universes. The old historic universes (Sumerian, Egyptian, Judaic, Zoroastrian, Confucian, Taoist, Jainic, Buddhist, Aristotelian, Platonic, Stoic, Epicurean, Neoplatonic, Medieval, ...) dealt with cosmic themes that gave meaning to human life, themes that now fail to fit naturally into the current physical universe. This causes concern and prompts us all to think deeply. The last section in this chapter considers how cosmology relates to society and affects our everyday thoughts, actions, and beliefs.

CHAPTER 2
EARLY SCIENTIFIC COSMOLOGY

This chapter briefly reviews known early scientific cosmology with comments on the Babylonian, Pythagorean, and Platonic systems. Emphasis is placed on the three important and enduring Hellenistic world systems: Aristotelianism, Epicureanism, and Stoicism. The Aristotelian universe, finite in size, consisted of planetary spheres bounded by an outer sphere of stars; the Epicurean universe, infinite in extent, consisted of endless worlds composed of atoms; and the Stoic universe, finite in size, consisted of a cosmos of planets and stars surrounded by empty infinite space. These world systems have shaped the history of subsequent cosmology.

The Medieval universe, with Aristotelian foundations, reached its peak in the High Middle Ages. Because of the Condemnations by the bishop of Paris of Aristotelian cosmology in 1277, the Medieval universe evolved into a Stoic-like system, able to accommodate an omnipotent God of unlimited extent. The Copernican revolution overthrew geocentric astronomy in favor of heliocentric astronomy, which in turn was soon overthrown by the rise of the Cartesian and Newtonian world systems.

CHAPTER 3
CARTESIAN AND NEWTONIAN WORLD SYSTEMS

In the seventeenth century, the revolutionary Cartesian and Newtonian systems mathematized and mechanized the natural world. From medieval mathematics and dynamics, René Descartes fashioned a mechanized atomless Epicurean-like world of matter and motion operating in strict obedience to natural laws. The repercussions – scientific, philosophical, and theological – were, and still are, profound. The body–mind (or body–soul) duality became more sharply etched than ever before and haunts us to this day.

Isaac Newton reacted strongly against Cartesian materialism and at first believed in a finite Stoic cosmos surrounded by an infinite mysterious space. What Descartes had denied – the existence of atoms, the vacuum, and forces acting at a distance – Newton affirmed. Newton's laws of motion and the theory of universal gravity transformed astronomy. The atomic theory lost its atheistic associations and began to make sense of the properties of matter. Where there is no matter, declared Newton, space still exists by virtue of the presence of spirit. Bodies act upon one another across empty space by means of long-range gravity. The implications of universal gravity caused Newton later to change his mind and believe in an infinite Epicurean-like universe, endlessly populated with uniformly distributed stars.

CHAPTER 4
COSMOLOGY AFTER NEWTON AND BEFORE EINSTEIN

But even the naked eye sees that stars do not cover the sky uniformly. Thomas Wright in the eighteenth century proposed that the Milky Way is an enormous assembly of stars, and that possibly other milky ways exist far away. Immanuel Kant expanded on this idea and devised a hierarchical universe. The renowned astronomer William Herschel explored the heavens, surveyed the Galaxy, and formed the opinion that many of the small fuzzy patches of light (nebulae) not only are clusters of unresolved stars but also some are distant milky ways (galaxies) similar to our own Milky Way (Galaxy). The nebula hypothesis, the idea that the Sun and planets formed from a rotating and contracting cloud of interstellar gas, was suggested by Kant and later considered in more detail by Pierre Simon de Laplace. Thus began the riddle of the nebulae: are the nebulae distant milky ways in a many-island universe or are they solar systems in the process of formation in a one-island universe? Herschel later in life changed his mind and favored the one-island universe. In the nineteenth century, the spectroscopic analysis of starlight by William Huggins and other astronomers and the development of photography established the "new astronomy" that later became known as astrophysics. At last human beings knew the stars consist of chemical elements exactly the same as on Earth. And astronomers knew that many nebulae consist only of gas (tipping the balance in favor of the Kant–Laplace nebula hypothesis and against the Wright–Kant milky way hypothesis). Astronomers succeeded in measuring the radial velocities of stars by the Fizeau–Doppler displacement in spectral lines. The Victorian universe of the nineteenth century was a one-island universe. The Solar System occupied the center of the Galaxy, which existed in a void of infinite, mysterious space. The Darwinian theory of natural selection exacerbated the age-of-the-universe problem and brought fundamental cosmological issues into every home. The old conflict between the Stoic and Epicurean systems climaxed in the early years of the twentieth century and the many-island universe emerged triumphant. We now know that some of the fuzzy patches of light are unresolved star clusters, some are swirling gas clouds, and others are distant galaxies.

CHAPTERS 5 AND 6
STARS AND GALAXIES
These two chapters discuss stars and galaxies and their treatment is oriented toward cosmology. Readers familiar with elementary astronomy may wish to skip these two chapters and proceed immediately to the next two chapters that discuss the important subjects of location and containment.

CHAPTER 7
LOCATION AND THE COSMIC CENTER
Generally, the subject of location (Chapter 7) deals with the cosmic center, and the subject of containment (Chapter 8) deals with the cosmic edge. The location and containment principles, which seem deceptively simple, serve to guide us among the pitfalls that trapped earlier cosmologists and still trap students.

This chapter deals with the rise and fall of the geocentric and heliocentric universes, and the rise of the centerless universes. We live in an isotropic universe in which all directions in space are alike. The location principle states that "probably we do not occupy a cosmic center." The observed isotropy of the universe, coupled with the location principle, leads us to the conclusion that the universe is probably homogeneous.

The homogeneity of the universe, meaning that all places in space are alike at a common instant in time, is the essence of the cosmological principle. The perfect cosmological principle, which states that all places in both space and time are alike, applies not only to the Cartesian and Newtonian world systems, but also to the more recent expanding steady-state universe.

CHAPTER 8
CONTAINMENT AND THE COSMIC EDGE
Containment deals with the edge and contents of the physical universe. The containment principle states: "the physical universe contains only physical things." In modern physics, both space and time in the form of spacetime are physically real and therefore part of the physical universe. As cosmophysicists we deal with the physical universe. But the Universe contains also nonphysical things and this aspect of

containment has implications in the social and life sciences. Various topics are considered, such as cosmic design and the finely tuned fundamental physical constants, the theistic and anthropic principles, and the laws of nature.

A word of warning comes not amiss while on the subject of containment. Cosmology is incomplete in the fundamental sense that we do not know how to put ourselves, as cosmologists, into our world systems. The Universe is self-aware – it contains us who are conscious beings – but the physical universe is not self-aware and does not contain us as self-aware beings. We can put our physical bodies and biochemical brains into a physical universe, which is a model of the Universe, but we cannot put our minds (whatever that means) into a universe conceived and studied by our minds. When we try, we fall into an infinite regression: the cosmologist studies a universe, which contains the cosmologist studying that universe, which contains the cosmologist, ... and so on, indefinitely. For the same reason, painters in the act of painting landscapes leave themselves out of the landscapes they paint. Otherwise they would have to include themselves painting a picture that includes themselves painting a picture that includes.... This subject is referred to as the containment riddle: "Where in a universe is the cosmologist studying that universe?" The solution to the riddle requires that we distinguish between the inconceivable Universe (of which we are totally a part) and our conceived universes (of which we are not totally a part) that we create to make sense of our experiences.

CHAPTER 9
SPACE AND TIME

In more depth than usual in an elementary work, we consider the fascinating nature of space and time in pre-Newtonian and post-Newtonian universes. Some topics discussed are the arrow of time, the "now," time travel, Zeno's paradoxes, Parmenidean states of being, Heraclitean acts of becoming,

and conjugate time in the Islamic Kalam universe.

Our everyday understanding of time is a patchwork of primitive and sophisticated concepts. The time that is used in special relativity is not the same as that used in most other sciences, which is not the same as that in everyday speech, which in turn is not the same as the time we actually experience. Conflict and contrast abound whenever we discuss the nature of time. With not much hope of success, we try to clarify some of the issues involved in this perplexing subject; any fundamental change in our understanding of time will undoubtedly profoundly affect cosmology.

CHAPTER 10
CURVED SPACE

The development of non-Euclidean geometry in the nineteenth century forms an engrossing subject in the history of science and mathematics. Understanding curved space is not easy, even for people who live in curved spaces. Much of our attention in this chapter focuses on the three homogeneous and isotropic spaces that are of basic importance in modern cosmology.

CHAPTER 11
SPECIAL RELATIVITY

Special relativity, contrary to students' expectations, is easy to understand, and a true grasp of the essential ideas does not require mathematics. The secret lies in spacetime pictures and the realization that in spacetime the shortest distance is not a straight line. Space travel close to the speed of light provides interesting applications of relativity theory. The "twin paradox" is puzzling only when the most elementary aspects of the theory are not understood.

CHAPTER 12
GENERAL RELATIVITY

Special relativity and curved space lead us to general relativity and the labors of Albert Einstein. The first stepping-stone is the principle of equivalence. This is established by means of experiments in imaginary laboratories that move freely in space near to and far from stars. The second stepping-stone is the realization that this dynamic state of affairs is analogous in many respects to the geometric properties of curved space. In a flight of inspiration we are catapulted to the theory of general relativity and the Einstein master equation. Many ingenious tests of general relativity have been performed, successfully verifying the validity of the theory on astronomical (not cosmic) scales.

We consider the bootstrap ideas embodied in Mach's principle, a principle so-named by Einstein who found Mach's ideas inspiring. The old bootstrap theory, periodically revived, asserts that all things are immanent within one another, and the nature of any one thing is determined by the universe as a whole. So far, science has failed to make sense of the bootstrap theory. Mach's principle, a bootstrap theory, claims that the inertia of a body is determined by all the matter distributed in the universe. Many persons dislike an "undressed" space that exists in its own right, and with the ancients, Bishop Berkeley, and Ernst Mach, think that space cannot exist in a real sense unless decently dressed in a distribution of matter. Berkeley's ideas, revamped by Mach, played a historic role in the formulation of general relativity. But the idea: the materialization of space, championed at first by Einstein, was dropped when Einstein performed the converse: the geometrization of matter.

CHAPTER 13
BLACK HOLES

Although black holes were anticipated in the eighteenth century on the basis of Newtonian theory, the proper theory for their study is general relativity. Of spherical bodies of similar mass, black holes have the highest density and the strongest gravitational force at their surface. They are wrapped in their own curved spacetime. A black hole exists in a frozen state of permanent free-fall collapse. Owing to the extreme distortion of spacetime, an external observer sees the black hole in a frozen state, from which nothing (according to the classical

theory), not even light, can escape. We consider several topics of interest, such as nonrotating and rotating black holes, the energy liberated by accretion of matter, miniholes and superholes, the temperature of black holes, Hawking radiation, and violation of certain cherished laws of conservation.

CHAPTER 14
EXPANSION OF THE UNIVERSE

The expansion of the physical universe ranks as one of the greatest discoveries in the history of the human race. We invoke the expanding space paradigm and perform imaginary experiments with ERSU – Expanding Rubber Sheet Universe. To aid us in our investigations we use the two different observers introduced in Chapter 7: the ordinary stay-at-home "observer" who looks out at distant things in the same way that we do, and the imaginary gadabout "explorer" who rushes around at infinite speed and traverses the universe in zero cosmic time. The explorer in these experiments is really us looking down on ERSU as external observers. Our experiments shed light on many topics, such as homogeneous expansion, cosmic time, recession of the galaxies, the velocity–distance law of expansion, and the Hubble redshift–distance law. The experiments stress that the galaxies are not hurling away through space but are actually at rest in space that is expanding. This is why distant galaxies can recede from us faster than the speed of light. Recession velocity is unlike the ordinary velocity with which we are familiar. Measuring the expansion requires the introduction of comoving coordinates, coordinate distances, the scaling factor, the Hubble term, and the deceleration term, and we show how universes are classified according to the way the scaling factor changes in time.

CHAPTER 15
REDSHIFTS

Light rays from distant galaxies are redshifted because of the expansion of the universe. This cosmological redshift, which is distinct from the Doppler and gravitational redshifts, is produced by the stretching of wavelengths as radiation propagates through expanding space. Space expands, wavelengths are stretched, and the cosmic redshift is as simple as that. There are a few redshift curiosities. The oddest curiosity of all is the unwise practice in popular literature of failing to distinguish between cosmic and Doppler redshifts. The Doppler effect implies that galaxies are rushing away through space and that special relativity explains the universe. This is a dangerous interpretation and leads to endless confusion for those trying to understand modern cosmology. It restores the cosmic edge at which recession reaches the speed of light. Our treatment stresses two concepts: first, recession is the result of the expansion of space (and galaxies are more or less stationary in expanding space); and second, cosmic redshifts are the result of the stretching of wavelengths as light and other forms of radiation travel through expanding space. It is now clear why we have previously insisted that space and time are physically real (this is the essence of general relativity) and are contained in the universe; the universe is not expanding in space, but consists of expanding space.

CHAPTER 16
NEWTONIAN COSMOLOGY
Isaac Newton resolved the gravity paradox (or war of cosmic forces) by assuming that the universe is perfectly homogeneous. Under certain limiting conditions Newtonian theory gives the same results as general relativity. The dynamics of the universe showing how gravity and the lambda force determine expansion are discussed with the aid of Newtonian ideas. We try to explain why Newtonian theory under certain conditions yields the same results in cosmology as general relativity.

CHAPTER 17
THE COSMIC BOX
The principle that all places in the universe are alike at each moment in cosmic time has far-reaching consequences. Distant regions are in the same state as local regions when compared at the same time, and we can discover much about the universe by studying the history of only a sample region. This is the basic idea of the "universe in a nutshell." We suppose that a part of the universe is enclosed in an imaginary cosmic box that has perfectly reflecting walls and expands with the universe. What happens inside is exactly the same as what happens outside. The cosmic box is small on the cosmic scale, hence we assume Euclidean space and the ordinary laws of physics, as used in the laboratory, to study the various forms of cosmic phenomena. The enclosed cosmic box serves as a useful tool for tackling subjects that otherwise would be difficult, such as the entropy of the universe and the nonconservation of energy on the cosmic scale.

CHAPTER 18
THE MANY UNIVERSES
In the past, many cosmological theories, now mostly of historical interest, have been proposed. We look at various "mighty frames," or cosmic models, such as the Einstein, de Sitter, Friedmann, and Friedmann–Lemaître universes. These models may be classified as static, bang, whimper, or oscillating. Other methods of classifications are given. In this great gallery of universes, the lambda force, popularized by Einstein, adds much to the variety. From this cosmological supermarket we select and examine Milne's kinematic universe, steady-state universes, and universes in compression, tension, and convulsion. We consider also inflation, chaos, and antichaos. The "dream machine" of the scalar–tensor theory is discussed; by adjusting its control knobs the cosmologist converts a universe into any one of an infinite number of different universes.

CHAPTER 19
OBSERVATIONAL COSMOLOGY
We consider first local observations, then observations at intermediate distances, and finally observations at cosmically large

distances. The local observations are confined to the Solar System, Galaxy, and Local Group of galaxies and extend no farther than a few million light years. They determine the first steps in a distance ladder, the distribution and density of matter, the ages of stars and the age of the Galaxy (setting lower limits on the age of the universe), the abundance of the chemical elements, the cosmic background radiation that originated in the early universe (and reveals the peculiar motion of the Galaxy), and give cosmological information on topics such as the baryon density and properties of the cosmic background radiation.

Observations at intermediate distances are confined mainly to the Local (or Virgo) Supercluster and extend a few hundred million light years. They explore only the sub-Hubble sphere and do not extend into the full Hubble flow. They determine the structure, distribution, and motions of galaxies, extended distance scales, redshift–distance and velocity–distance laws in approximate form, and give information on topics such as the age of the universe, baryon density, the density parameters, and the approximate value of the Hubble term.

Observations at cosmically large distances extend deep into the Hubble flow where the redshift–distance relation ceases to be linear. We piece together evolutionary histories by comparing nearby and distant astronomical systems. What is seen in the world picture (on the observer's backward lightcone) must be projected forward onto the world map (in which the linear velocity–distance law holds). This mapping procedure greatly complicates the determination of the cosmological parameters and we are still far from a secure knowledge of the values of the Hubble term, the density parameters, the deceleration term, the cosmological term, and the curvature constant.

CHAPTER 20
THE EARLY UNIVERSE

The cosmic background radiation, discovered in 1965, provides unambiguous evidence of a big bang in the early history of the universe. We explore the big bang, not by traveling in space, but by remaining where we are and traveling far back in time. The big bang was everywhere. If the universe extends infinitely in space, then so also did the big bang. As we journey back in time, the cosmic density and temperature rise steadily and the universe at age a few hundred thousand years is filled with brilliant light. We stand at the threshold of the radiation era. From this epoch descends directly the cosmic background radiation, cooled by expansion, that we nowadays observe. When the universe is one second old, and the temperature is 10 billion kelvin and the density is one million times that of water, we quit the radiation era and enter the bizarre world of the lepton era. Hordes of electrons and muons, and their antiparticles, struggle to survive and from the lepton battlefields flee hosts of ghostly neutrinos condemned forever to wander unseen through the universe. We continue our journey back in time, traveling through the hadron era and its warring matter and antimatter ruled by the strong, electromagnetic, and weak forces. Eventually we enter the quark era ruled by the strong and electroweak forces. Finally phase transitions pass us through an inflation era into a world ruled by the hyperweak force in which matter and antimatter are indistinguishable. Our journey back in time ultimately comes to a halt at the impenetrable Planck barrier. At the Planck epoch the age of the universe is one billion-trillionth of a jiffy (a jiffy is one billion-trillionth of a second) and the density of the universe is 1 followed by 93 zeros times the density of water. Quantum fluctuations of space and time are now of cosmic magnitude and spacetime forms a foam of tangled discontinuities.

We return from the early universe and with our time machine turned to the future we journey to the end of time. We find that perhaps the universe ceases to expand, then collapses and terminates in a new big bang, or perhaps it expands forever and dies in a long drawn-out whimper. In the first possibility, during the collapse of the universe, galaxies are crushed together, and in the devastation that follows, dissolving stars zip through space at speeds close to that of light. The brilliance of the radiation era returns and the universe reverts to its original primordial state. In the second possibility, the galaxies continue to recede from one another and after hundreds of billions of years are dead and lifeless. In the enormous stretches of time that follow, star systems and galaxies contract to form black holes, and particles slowly decay and convert into radiation. After eons of time all black holes evaporate, mostly into low-temperature radiation, and the universe then contains almost no matter, only feeble radiation forever growing feebler.

CHAPTER 21
HORIZONS IN THE UNIVERSE
How far can we see in the universe? The answer depends on the things that we see, whether they are events or world lines. Event and particle (or world line) horizons are discussed, first in the static Newtonian universe to illustrate their nature, and then more generally in nonstatic universes. The horizon riddle, the horizon problem, the Hubble sphere, and other topics are discussed. Also discussed is the photon horizon, beyond which photons emitted in our direction actually recede.

CHAPTER 22
INFLATION
Possibly, the universe begins in a state of utmost symmetry, and progresses through a series of phase transitions to states of lower symmetry and richer diversity. Among the first-born in the very early universe are the magnetic monopole particles. These massive monopoles cannot decay and should still exist and be as abundant as the photons of the cosmic background radiation. An era of inflation explains why they have not been observed. During the grand-unified phase transition, in which the hyperweak force split into the electroweak and strong forces, the universe is thrown into a state of extreme tension. In this state, the universe expands (or inflates)

enormously at constant density. This inflation solves not only the monopole problem but also the flatness and horizon problems. But inflation exacts a price: it creates the problem of missing nonbaryonic matter.

CHAPTER 23
THE COSMIC NUMBERS
Cosmic numbers connect the subatomic and cosmic properties of the universe. These dimensionless numbers have intriguing coincidences. Discussed are the cluster hypothesis and Dirac's large-number hypothesis, and their connection with the anthropic principle. The art of cosmonumerology began long ago in the ancient world when Archimedes calculated in the *Sand Reckoner* the number of grains of sand needed to fill the universe.

CHAPTER 24
DARKNESS AT NIGHT
The dark night-sky riddle, known as Olbers' paradox, originated during the Copernican revolution in the sixteenth century. Why the sky at night is dark, and not ablaze with light from countless stars has puzzled many scientists, and played a conspicuous role in the history of cosmology. Many writers in recent times have said the night sky is dark because of the expansion of the universe. But this cannot be true because calculation shows that if our universe were static the sky at night would still be dark. The universe does not contain enough energy to create a bright-sky universe. The correct solution was anticipated by the poet Edgar Allan Poe and investigated in depth by Lord Kelvin. The sky at night is dark because the stars shine for too short a time to fill the universe with radiation in equilibrium with stars; equivalently, stars shine for too short a time for the universe to contain sufficient visible stars to cover the sky.

CHAPTER 25
CREATION OF THE UNIVERSE
We consider miscellaneous topics in cosmogeny, beginning with the creation myths of earlier societies. The Mosaic chronology

that fixes the date of creation of the universe to five or six thousand years before the present has been the cause of considerable conflict between science and religion. Creation and fitness of the universe are distinguished as separate subjects and examined in current theistic, anthropic, spontaneous, and natural selection theories. Eschatological myths and end-of-the-world theories are also briefly reviewed.

CHAPTER 26
LIFE IN THE UNIVERSE
In this last chapter we consider past and present theories of the origin of life and discuss aspects of evolution and natural selection. Understanding the nature of intelligence is vitally important in cosmology, and we consider how human beings might have acquired their large brains. As cosmologists, in our finely tuned universe, we feel impelled to believe that intelligent life must exist elsewhere in the multitudes of galaxies. But what is life? What is intelligence? Does intelligent life, technologically advanced, exist elsewhere in our Galaxy? Avenues of inquiry open up in pursuit of answers to this and other questions.

Part I

1 WHAT IS COSMOLOGY?

1

He has ventured far beyond the flaming ramparts of the world
and in mind and spirit traversed the boundless universe.
Lucretius (99–55 BC), The Nature of the Universe

THE UNIVERSE

From the outset we must decide whether to use *Universe* or *universe*. This is not so trivial a matter as it might seem. We know of only one planet called Earth; similarly, we know of only one Universe. Surely then the proper word is *Universe*?

The Universe is everything and includes us thinking about what to call it. But what is the Universe? Do we truly know? It has many faces and means many different things to different people. To religious people it is a theistically created world ruled by supernatural forces; to artists it is an exquisite world revealed by sensitive perceptions; to professional philosophers it is a logical world of analytic and synthetic structures; and to scientists it is a world of controlled observations elucidated by natural forces. Or it may be all these things at different times. Even more diverse are the worlds or cosmic pictures held by people of different societies, such as the Australian aboriginals, Chinese, Eskimos, Hindus, Hopi, Maoris, Navajo, Polynesians, Zulus. Cosmic pictures evolve because cultures influence one another, and because knowledge advances. Thus in Europe the medieval picture, influenced by the rise of Islam, evolved into the Cartesian, then Newtonian, Victorian, and finally Einsteinian pictures. The standard Western world picture of the late nineteenth century – the Victorian picture – was totally unlike the standard picture – the Einsteinian picture – of a hundred years later. Each society in each age constructs a different cosmic picture that is like a mask fitted on the face of the unknown Universe.

If the word "Universe" is used we must distinguish between the various "models of the Universe." Each model, religious, artistic, philosophical, or scientific, is one of many representations; and similarly with the models of different societies. Thus in the history of science we distinguish between the Pythagorean model, the Atomist model, the Aristotelian model, and so on. More precisely, we should say, the Pythagorean model of the Universe, the Atomist model of the Universe, the Aristotelian model of the Universe, and so on. Inevitably, the models receive the abbreviated titles: the Pythagorean Universe, the Atomist Universe, the Aristotelian Universe, ..., and we confuse ourselves by using the word Universe to mean "a model of the Universe."

The grandiose word Universe has a further major disadvantage. When used alone, without specification of the model we have in mind, it conveys the impression that we know the true nature of the Universe. We find ourselves, in the company of multitudes of others in the past, speaking of the Universe as if it were at last discovered and revealed. By referring to the contemporary model of the Universe as the "Universe," we forget that our contemporary model will undoubtedly suffer the same fate as its predecessors. Always, we mistake the mask for the face, the model universe for the actual Universe. Our ancestors made this

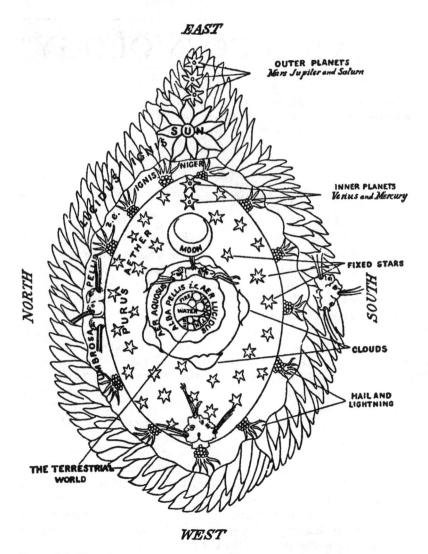

Figure 1.1. The universe according to Hildegaard of Bingen in Germany in the twelfth century. In her lifetime we see in her writings how the medieval picture evolved toward its climax in Dante's *Divine Comedy* (Figure 8.4). (Reproduced from the Wiesbaden Codex B as figure 2 in Charles Singer's "The scientific views and visions of Saint Hildegaard".)

mistake continually and most likely our descendants will look back and see us repeating the same mistake.

Because we cannot guess even in our wildest imaginings the true nature of the Universe, we may avoid referring to it directly by using the more modest word "universe." A universe is simply a model of the Universe (see Figure 1.3). Hence we may speak of the Pythagorean universe, the Atomist universe, Aristotelian universe, and so on, and each universe is a mask, a cosmic picture, a model that is invented, modified as knowledge advances, and finally discarded.

The word "universe," which we shall use, has the further advantage that it may be used freely and loosely without any need to

Figure 1.2. The Universe, *one and all-inclusive*, by Filippo Picinelli, 1694. In *The Cosmographical Glass: Renaissance Diagrams of the Universe* (1977), S. K. Heninger writes, "We might conjecture that the artist, not bound by the constraint of cosmological dogma, felt free to engage in cosmological speculations of his own sort. He assumed a license to create his own universe. The worlds of Hieronymus Bosch, of Leon Battista Alberti, and of John Milton, to name a few examples, are the result." (Courtesy of the Henry E. Huntingdon Library, San Marino, California.)

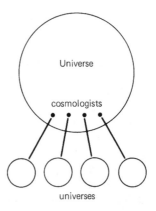

Figure 1.3. The Universe contains us who construct the many universes. Each universe is a model of the Universe. An intriguing thought is that each universe is the Universe attempting to understand itself.

remind ourselves constantly that the Universe is still mysterious and unknown. When the word "universe" is used alone, as in such phrases as "the vastness of the universe," it denotes our present universe as disclosed by modern science.

COSMOLOGY

We search the sky, the Earth, and within ourselves, and forever wonder about the mystery of the universe: What is it all about? Why did it all begin? How will it all end? And are these questions meaningful? Always we ask the burning question: What is the meaning of life? Each of us echoes the words of Erwin Schrödinger – "I know not whence I came nor whither I go nor who I am," and seeks the answer. The search is doomed to go astray from the beginning unless we familiarize ourselves with the universes of the past and particularly with the modern universe.

Cosmology is the study of universes. In the broadest sense it is a joint enterprise by science, philosophy, theology, and the arts that seeks to gain understanding of what unifies and is fundamental. As a science, which is the main concern in this book, it is the study of the large and small structures of the universe; it draws on knowledge from other sciences, such as physics and astronomy, and assembles a physically all-inclusive cosmic picture.

In our everyday life we deal with ordinary things, such as plants and flowerpots, and to understand these things of sensible size we explore the small-scale and large-scale realms of the universe. We delve deeply into the microscopic realms of cells, molecules, atoms, and subatomic particles, and reach far out into the macroscopic realms of planets, stars, galaxies, and the universe. We find that the very small and the very large are intimately related in cosmology.

Since the seventeenth century, knowledge has advanced rapidly and the number of sciences has grown enormously. Each science focuses on a domain of the universe and tends in the course of time to fragment into closely related new sciences of greater specialization. Originally, the characteristics of living and nonliving things defined the differences between the broad domains of biology and physics. Each of these basic sciences, as it advanced, branched into new sciences, which in turn branched into more specialized sciences. Physics – once known as natural philosophy – has grown and branched into high-energy subatomic particle physics, low-energy nuclear physics, atomic physics, chemical physics, condensed-matter physics, biophysics, geophysics, astrophysics, and so on, and each has its own theoreticians, experimenters, and technicians. Biology – once the subject of naturalists of broad interests – with associated sciences such as botany, zoology, entomology, ecology, and paleontology, and so on, has grown and branched into molecular biology, biochemistry, genetics, and so on. And astronomy – once the subject in which everybody had equal knowledge (but not computing skill) – has branched into planetary sciences, the study of stellar structure and atmospheres, interstellar media, galactic astronomy, extragalactic astronomy, and the separate fields of radio, infrared, optical, ultraviolet, x-ray, and gamma-ray astronomy.

It is evident that the sciences divide the universe in order that each can construct in detail a domain of special knowledge. Science tears things apart into constituents of greater and greater specialization – often into smaller and smaller pieces – and devotes closer and closer attention to detail. A person studying in depth a branch of science becomes a specialist, engrossed in a maze of detailed knowledge, who knows much about a small domain of the universe and is comparatively ignorant of all the rest.

Cosmology is the one science in which specialization is rather difficult. Its main aim is to assemble the cosmic jigsaw puzzle, not to study in detail any particular jigsaw piece. While other scientists are pulling the universe apart into progressively more detailed pieces, the cosmologists are endeavoring to put the pieces together to see the picture on the jigsaw puzzle. Unlike all other scientists, the cosmologists take a broad view; like the impressionist painters they stand well back from their canvases so as not to see too much distracting detail.

Introductory cosmology is not a branch of astronomy. It is a "cosmopedia," more than an inventory of the contents of the universe, and is not a "whole-universe catalogue" of descriptive astronomical data. Cosmology is the study of the primary cosmic constituents, such as the origin and history of the chemical elements, and of space and time that form the frame of the expanding universe. The primary things of importance are scattered over large regions of space and endure over long periods of time. The origin and evolution of stars and galaxies, even the origin of life and intelligence, are important cosmic subjects. Subatomic particles, the role they play during the earliest moments of the universe, their subsequent combination into atoms and molecules that form the complexity of the living cell and our surrounding world, are all of cosmic interest.

At each turn, the issues of cosmology cause us to pause and reflect. Many subjects of vital importance are still obscure and not understood: how human beings acquired speech and large brains; and how they developed the ability to create abstract mental structures and think quantitatively. What determines the way that human beings think also determines the design they perceive in their universes. Human beings form a vital part of cosmology and represent the Universe perceiving and thinking about itself.

Who are the cosmologists? Professional cosmologists are relatively few; they are well-versed in mathematics, physics, and astronomy, and they study the evolution and large-scale structure of the physical universe. In general, however, whenever a person seeks to understand the Universe, that person becomes a cosmologist. When we stand back from the study of a specialized area of knowledge, or just step aside from our everyday affairs, and reflect on things in general, and try to see the forest and not just the trees, the whole painting and not just the dabs of paint, the whole tapestry and not just the threads, we become cosmologists.

THE MAGIC UNIVERSE

Cosmology is as old as *Homo sapiens*. It goes back to a time when human beings, living in primitive social groups, developed language and made their first attempts to understand the world around them. Probably, hundreds of thousands of years ago, human beings explained their world by means of spirits. Spirits of all kinds, motivated by humanlike impulses and passions, activated everything. The early people projected their own inner thoughts and feelings into an outer animistic world, a world in which everything was alive. With supplications, prayers, sacrifices, and gifts to the spirits, human beings gained control of the phenomena of their world.

It was the Age of Magic, of benign and demonic spirits incarnate in plant, animal, and human form. Everything that happened was explained readily and easily by the passions, motives, and actions of ambient and indwelling spirits. It was an anthropomorphic world, of the living earth, water, wind, and fire, into which men and women projected their own emotions and motives as the guiding forces; the kind of world that children read about in fairy tales. From this "golden age" comes our primeval fear of the menace of darkness and the rage of storms, and our enchantment with the wizardry of sunrises, sunsets, and rainbows. For reasons not yet fully understood, human beings everywhere remained one species, and cultures (languages, social codes, belief systems, laws, technologies) interdiffused. Possibly, our moral codes of today, which regulate behavior in the family and society and determine in general what is ethically right and wrong, were naturally selected over long periods of time in primitive societies. Societies deficient in codes of mutual care and support among individuals had little chance of surviving.

THE MYTHIC UNIVERSE

At the dawn of history, ten or more thousand years ago, the early city-states attained more abstract concepts of the Universe. The magic universe evolved into the mythic universe. The long age of magic gave way to what might be called the Age of Theism. The spirits that had been everywhere, activating everything, amalgamated, retreated into remote mythic realms, and became powerful gods who personified abstractions of thought and language. James Frazer, in *The Golden Bough*, speculated on how magic among primitive people evolved into theism, and how the magic universe transformed into a variety of mythic universe:

> But with the growth of knowledge man learns to realize more clearly the vastness of nature and his own littleness and feebleness in the presence of it. The recognition of his helplessness does not, however, carry with it a corresponding belief in the impotence of those supernatural beings with which his imagination peoples the universe. On the contrary, it enhances his conception of their power.... If then he feels himself to be so frail and slight, how vast and powerful must he deem the beings who control the gigantic machinery of nature! ... Thus in the acuter minds magic is gradually superseded by religion, which explains the succession of natural phenomena as regulated by will, passion, or caprice of the spiritual beings like man in kind, though vastly superior to him in power.

Much of mythology consists of primitive cosmic imagery (Figure 1.4). The Sumerian, Assyro-Babylonian, Minoan, Greek, Chinese, Norse, Celtic, and Mayan mythologies, to name only a few, are of historical interest because they illustrate mankind's earlier views of the universe. The creation myths, often difficult to interpret, are of particular interest (see Chapter 25).

Human beings at the cosmic center

No matter how powerful and remote they became, the mythic gods continued to serve and protect human beings, and men and women everywhere remained secure and of central importance in an anthropocentric universe. The universe was assembled about a center and human beings were located prominently at the center.

Anthropocentricity formed the basis of the Greek cosmology of an Earth-centered

Figure 1.4. *The Ancient of Days* by William Blake (1757–1827). "When he sets a compass upon the face of the depths" (Proverbs 8:27).

universe. The universe of Aristotle in the fourth century BC was geocentric (or Earth centered); the spherical Earth rested at the center of the universe and the Moon, Sun, planets, and stars, fixed to translucent celestial spheres, revolved about the Earth. The innermost region of heaven – the sublunar sphere between the Earth and the Moon – contained earthly and tangible things in an ever-changing state, and the outer regions of heaven – the celestial spheres – contained ethereal and intangible things in a never-changing state. The subsequent elaborations of this system, bringing it into closer agreement with astronomical observations, culminated in the Ptolemaic system of AD 140.

The Middle Ages (fifth to fifteenth centuries) were not so terribly dark as was once supposed. The medieval universe from the thirteenth century to the sixteenth century was perhaps the most satisfying form of cosmology known in history. Christians, Jews, and Moslems were blessed with a finite universe in which they had utmost importance. By the Arab and European standards of those times it was a rational and well-organized universe that everybody could understand; it gave location and prominence to mankind's place in the scheme of things, it provided a secure foundation for religion and gave meaning and purpose to human life on Earth. Never before or since has cosmology served in so vivid a manner the everyday needs of ordinary people; it was simultaneously their religion, philosophy, and science.

The Copernican Revolution

The transition from the finite geocentric universe to the infinite and centerless universe is known as the Copernican Revolution. In the sixteenth century, Nicolaus Copernicus crystallized trends in astronomical thought that had originated in Greek science almost 2000 years before and proposed the heliocentric (or Sun-centered) universe. The Copernican heliocentric universe was soon transformed into the infinite and centerless Cartesian universe, which in turn was followed by the Newtonian universe. This revolution in outlook occupied the sixteenth and seventeenth centuries. The Copernican Revolution opened the way for modern cosmology.

But the spiritual universe, thought to be vastly more important than the physical universe, remained firmly anthropocentric. The spiritual universe was the "great chain of being," a chain of countless links that descended from human beings through all the lower forms of life to inanimate matter, and ascended from human beings through hierarchies of angelic beings to the throne of God. Mankind was the central link connecting the angelic and brute worlds. Even in an infinitely large physical universe, deprived first of the Earth and then of the Sun as its natural center, it was still possible to cling to old ideas that portrayed human beings as having central importance in the cosmic drama. The gods were ever mysterious and after the Copernican Revolution they became more mysterious than before.

The Darwinian revolution

In the middle of the nineteenth century came the most dreadful of all revolutions: the Darwinian Revolution. Human beings, hitherto the central figures in the cosmic drama, became akin to the beasts of the field. The gods who had attended and protected mankind for so long were cast out of the physical universe.

The anthropomorphic (magic) and anthropocentric (mythic) universes were wrong in almost every detail. The medieval universe has gone and with it has gone the great chain of being. Science at last is the victor, putting to flight the myths and superstitions of the past. We applaud the Renaissance (fifteenth to sixteenth centuries) with its revival of art and learning, we applaud the rise of the Cartesian and Newtonian world-systems in the seventeenth century, we applaud the Age of Reason (the Enlightenment of the eighteenth century) with its conviction in the power of human reason, and we applaud the Age of Science (seventeenth to twentieth centuries), and too easily forget the growing dismay of ordinary men and women in a universe that century by century progressively became more meaningless and senseless. With the decline and death of the old universes – anthropomorphic and anthropocentric – mankind was cast aimlessly adrift in an alien universe.

THE ANTHROPOMETRIC UNIVERSE

"Man is the measure of all things."
Protagoras (fifth century BC)

We believe that the universe is not anthropomorphic and not made in the image of human beings; it is not a magic realm alive with humanlike spirits. Also we believe that the universe is not anthropocentric with human beings occupying its center; we are not the central figures; and the world is not controlled by gods and goddesses.

Instead, as Protagoras said, we are the measure of the universe, and this means that the universe is anthropometric. Let us try to understand what this means.

We have minds, or as some would say, we have brains. For our purpose it is not necessary to inquire into the nature of the mind–brain and attempt to probe its mysteries. It does not matter if we think the mind is a nonphysical entity of psychic activity or is a physical brain throbbing with bioelectrochemical activity. We have mind–brains into which information pours via the sensory pathways and from this information we devise in our mind–brains the Aristotelian, Stoic, Epicurean, Zoroastrian, Neoplatonic, Medieval, Cartesian, Newtonian, and all the other universes that have dominated human thought in different ages. We observe plants and flowerpots and other things and devise grand theories that relate and explain them, and these theories reside not in the things themselves but in our mind–brains. At each step in the history of cosmology, different universes prevail, and every universe in every society is a grand mental edifice that makes sense of the human experience. Each universe is anthropometric because it consists of ideas devised by human beings seeking to understand the things they observe and experience.

For those lost in the vast and apparently meaningless modern universe there is comfort in the realization that all universes are anthropometric. The Medieval universe was made and measured by men and women, although the medievalists themselves would have hotly denied the thought. The modern universe with its bioelectrochemical brains pondering over it is also human-made. Like the Medieval universe it will inevitably fade away in time and be replaced by other universes. The universes of the future will almost certainly differ from our modern version; nevertheless, they will all be anthropometric because "man is the measure of all things" entertained by man. The Universe itself, of course, is not human-made, but we have no true conception of what it actually is. All we know is that it contains us – the dreamers of universes.

COSMOLOGY AND SOCIETY

Cosmology and society are intimately related. Where there is a society, there is a universe, and where there is a universe, there is a society of thinking individuals. Each universe shapes the history and directs the destiny of its society.

This intimate relationship is most obvious in primitive cosmology where mythology and society mirror each other and the ways of gods and goddesses are the ways of men and women. Cruel people create cruel gods who sanction cruel behavior, and peaceful people create peaceful gods who foster peaceful behavior. The interplay between cosmology and society in the modern world is as strong as ever, if not stronger, but often in less easily recognized forms.

Without doubt the most powerful and influential ideas in any society are those that relate to the universe. They shape histories, inspire civilizations, foment wars, create monarchies, launch empires, and establish political systems. One such idea was the principle of plenitude, which can be traced back to Plato and has been enormously influential since the fifteenth century.

The principle of plenitude originated in the anthropocentric belief system that the universe is created for mankind by an intelligible supreme being. In its simplest form it states that a beneficent Creator has given to human beings for their own use an Earth of unlimited bounty. The more formal argument is as follows. The supreme being is without limitation because limitation implies imperfection and imperfection is contrary to belief. The unlimited potential of the supreme being is made manifest in the unlimited actuality of the created world. The Earth necessarily displays every form of reality in inexhaustible abundance. This is the principle of plenitude that saturates Western culture.

In the Late Middle Ages, telescopes disclosed the richness of the heavens, microscopes disclosed a teeming world of micro-organic life, and the worldwide voyages by mariners opened up dazzling vistas of a vast and bountiful Earth. An unlimited abundance of every conceivable thing provided sufficient proof of the principle of plenitude. Europeans developed the principle, were guided by it, and have since exported it to the rest of the world.

Political ideologies were shaped by the principle of plenitude. The principle guaranteed endless untapped wealth and free enterprise flourished as never before. To offset depletion and escape population growth it was necessary only to push farther east and west to the glittering prizes of unravished lands. "The real price of anything is the toil and trouble of acquiring it" said Adam Smith. Go east! the streets are paved in gold. Go west! beyond the sunset lie lands of unharvested wealth. Husbandry of finite resources was not part of plenitude philosophy. People confidently believed that everything existed in unlimited abundance, and when anything became exhausted (such as the elimination of the bison herds, the extinction of the carrier pigeons and the great auks), they were taken by surprise and felt cheated.

The inevitable question followed, and has since echoed around the world: Why should inequality of wealth exist in a world of unlimited abundance? One answer came in the message from Karl Marx: in the *Communist Manifesto* we are told the less wealthy "have nothing to lose but their chains. They have a world to win." The principle of plenitude, which now lies buried deep in our cultural heritage and has been disseminated in various forms throughout the world, is unfortunately nothing but a cosmological myth.

Old ideas of cosmological breadth still dominate our everyday thoughts and many of these ideas are totally unsuitable in the modern world. We are, it seems, locked into the misguiding logic of obsolete universes that threaten to destroy us. We live in an age of crises – unchecked population growth, rapid depletion of resources, environmental and atmospheric pollution – and are mesmerized by prophecies of doom.

In 1776 the engineering firm of Boulton and Watt began to sell steam engines that, unlike previous steam devices, were powerful, quick-acting, and easily adapted for driving machinery of various kinds. This event more than any other ushered in the Industrial Revolution that has transformed our way of life. Many persons say that the ills of today are the direct consequence of the Industrial Revolution. But it is not the technologies that are to blame, but the ideas – the belief systems – that govern the use of the technology.

To make the point clear, let us imagine that space travelers encounter a planet that has been devastated by unbridled technology and become lifeless. In their investigations the space travelers cannot automatically assume that technology was the cause of the devastation. They must search for evidence indicating the nature of the beliefs of the vanished inhabitants. What inner mental world resulted in the outer ruined world? In their reports they will probably draw the conclusion that the ruined world is the result of an ancient cosmology, a cosmology founded on principles that in their saner moments the inhabitants had rejected and yet had driven them to their doom.

REFLECTIONS

1 *"I don't pretend to understand the Universe – it's a great deal bigger than I am." Attributed to William Allingham (1828–1889).*

• *The word* Universe *can be thought of as combining* Uni*ty and the di*verse. *The word* cosmos *means the harmonious whole of all reality. But what are the full meanings of unity, diversity, harmony, and reality?*

2 *In cosmology, there are two distinct languages: the first refers to* universes *and the second refers to* cosmologies. *In the first, cosmology is the study of many universes, and each universe is a model of the*

Universe. (*Naturally in any age cosmology tends to be the study of the contemporary universe.*) *In the second, the Universe is studied by many cosmologies, and each cosmology is peculiar to a particular society. We have either a single cosmology studying many universes or a single Universe studied by many cosmologies. The first refers repeatedly to universes and the second refers repeatedly to the Universe. In this book we adopt the first method because it avoids using the word "Universe," except occasionally to make a point clear, and does not foster the illusion that the Universe is a known or even knowable thing.*

3 *Homo sapiens has existed for about one million years. How did the early human beings view the world around them? "I shall invite my readers to step outside the closed study of the theorist into the open air of the anthropological field,"* wrote Bransilaw Malinowski *in his book on the Tobriand Islanders of Melanesia. Through his observations and those of many other anthropologists studying different societies we find not primitive but sophisticated cultures and intricate languages existing everywhere. Truly primitive human beings, offering us insight into how our remote ancestors thought and lived, most probably exist nowhere in the world today.*

The world of primitive people was "possessed, pervaded, and crowded with spiritual beings," according to the Victorian anthropologist Edward Tylor in his book Primitive Culture. *He advanced the theory of animism. The early human beings projected their own emotions and motives into the surrounding world, and the world, thus animated, was able to explain almost everything that needed explaining. In the course of time, with the growth in language and abstract thought, the ambient spirits amalgamated into powerful nature spirits, godlings, gods, and goddesses.*

"The conception of gods as superhuman beings endowed with the powers to which man possesses nothing comparable in degree and hardly in kind has been slowly evolved in the course of history," wrote James Frazer *in* The Golden Bough. *Frazer discussed the*

evolution of animism into theism, and of how the management of "the gigantic machinery of nature" was handed over to the gods. He assumed as a basic premise that religion was born with the rise of the gods.

4 *Religion in general is not easily defined. It seems to comprise emotions and ideas. The religious emotions experienced by individuals are much the same in all societies, whereas the religious ideas that evoke those emotions are peculiar to each society. Religious emotions are probably an integral part of human nature and essential in the survival of human societies. Theology is the study of religious ideas, and faith is the conviction in the absolute truth of those ideas. Invariably, the ideas have cosmological significance (see Chapters 2, 3, 4, 7, 8, 25, and 26). We note that everywhere in every age people in different societies have similar religious emotions, but have totally different religious ideas in whose absolute truth they have complete faith.*

Recognition of the universality of religious emotions and the diversity of religious ideas suggests that Frazer was wrong when he traced the roots of religion back to the birth of gods. Possibly religion is as old as Homo sapiens. *The error of confusing religious emotions with religious ideas seems quite common. When members of religious institutions insist on keeping their mythic beliefs, they unwittingly make the mistake of confusing theory with emotional experience and think that without primitive cosmology they cannot have religion. They fail to realize that scientific rejection of mythic cosmology does not bring science into conflict with religious experience. The modern theory of light as quanta of energy, for example, has not robbed us of the sensation of color and the emotional experience that accompanies color.*

Mythology is the study of myths. Myths apparently are ideas and stories that provide historical insights into the belief systems of other and often earlier cultures. Although credible in the belief systems in which they first originated, myths become incredible when transplanted into the belief systems of other cultures.

5 *Cosmological concepts have great influence for good and evil. Consider:* "Thou shalt not suffer a witch to live." *It is estimated that in the witch universe of the late Middle Ages (known as the Renaissance) and of the Age of Reason (known as the Enlightenment) about half a million men, women, and children confessed heresy and witchcraft under torture and were burned to death. It was said that heretics would burn forever in hell and the temporary anguish of fire on Earth was justified if they were saved from eternal fire of hell. Here is an instance of the maxim:* "cruel people create cruel gods who sanction cruel behavior."

• "And the awful fact was that whenever you found one witch and used the just and proper instruments of inquiry, you inevitably found many others. Their numbers multiplied and seemed without limit. Male and female witches and their evilly spawned children were consumed by fire in mounting numbers, and still they multiplied" (E. Harrison, Masks of the Universe).

"All Christianity, it seems, is at the mercy of these horrifying creatures. Countries in which they had previously been unknown are now suddenly found to be swarming with them, and the closer we look, the more of them we find. All contemporary observers agree that they are multiplying at an incredible rate. They have acquired powers hitherto unknown, a complex international organization and social habits of indecent sophistication. Some of the most powerful minds of the time turn from human sciences to explore this newly discovered continent, this America of the spiritual world" (Trevor-Roper, The European Witch Craze).

"The details they discovered are constantly and amply confirmed by other research workers – experimenters in confessional and torture chamber, theorists in library and cloister – leaving the facts still more securely established and the prospect even more alarming than before. Instead of being stamped out, the witches increased at a frightening rate, until the whole of Christendom seemed about to be overwhelmed by the marshaled forces of triumphant evil. To protest in any way against witch hunting as inhuman in a time of emergency was sheer lunacy, condemned by the popes as bewitchment and the result of consorting with devils" (E. Harrison, Masks of the Universe).

6 *Edward Milne in his last book* Modern Cosmology and the Christian Idea of God, *published posthumously in 1952, wrote:* "There is a remarkable difference between physics and philosophy. On the one hand, physicists agree with one another in general at any one time, yet the physical theories of any one decade differ profoundly from those of each succeeding decade – at any rate in the twentieth century. On the other hand, philosophers disagree with one another at any one time, yet the grand problems of philosophy remain the same from age to age.... The man of science should be essentially a rebel, a prophet rather than a priest, one who should not be ashamed of finding himself in opposition to the hierarchy.... The hard-baked or hardboiled scientist usually holds that science and religion, whilst on nodding terms, have no immediate bearing on one another. On the contrary, one cannot study cosmology without having a religious attitude to the universe. Cosmology assumes the rationality of the universe, but can give no reason for it short of a creator of the laws of nature being a rational creator."

7 "Whereas philosophers and theologians appear to possess an emotional attachment to their theories and ideas that requires them to believe in them, scientists tend to regard their ideas differently. They are interested in formulating many logically consistent possibilities, leaving any judgment regarding their truth to observation. Scientists feel no qualms about suggesting different but mutually exclusive explanations for the same phenomenon" (John Barrow and Frank Tipler, The Anthropic Cosmological Principle, 1986).

8 *The emergence of science, says Herbert Butterfield in* The Origins of Modern Science, "outshines everything since the rise of Christianity and reduces the Renaissance and Reformation to the rank of mere episodes," and "looms so large as the real origin both of the modern world and the modern mentality that our customary periodisation of European history has become an anachronism and an encumbrance." Butterfield argues that science saved Europe from the mad witch universe. Not the humanities, not religion, but the sciences ended the witch craze of the Renaissance. Science was reaching out to a new universe more capable of distinguishing between the supernatural and the natural and of defining the limits of human control over nature.

9 "Possibly the world of external facts is much more fertile and plastic than we have ventured to suppose: it may be that all these cosmologies and many more analyses and classifications are genuine ways of arranging what nature offers to our understanding, and that the main condition determining our selection between them is something in us rather than something in the external world" (Edwin Burtt, The Metaphysical Foundations of Modern Physical Science, 1932).

• "Natural science does not simply describe and explain nature; it is part of the interplay between nature and ourselves; it describes nature as exposed to our method of questioning" (Werner Heisenberg, Physics and Philosophy, 1958).

• In The Discarded Image (1967), C. S. Lewis writes: "The great masters do not take any Model quite so seriously as the rest of us. They know that it is, after all, only a model, possibly replaceable." Later he continues: "It is not impossible that our own Model will die a violent death, ruthlessly smashed by an unprovoked assault of new facts – unprovoked as the nova of 1572. But I think it is more likely to change when, and because, far-reaching changes in the mental temper of our descendants demand that it should. The new Model will not be set up without evidence, but the evidence will turn up when the inner need for it becomes sufficiently great. It will be true evidence. But nature gives most of her evidence in answer to the questions we ask her."

10 In The Great Chain of Being (1936) by Arthur Lovejoy, we read: "Next to the word 'nature,' the 'Great Chain of Being' was the sacred phrase of the eighteenth century,

playing a part somewhat analogous to that of the blessed word 'evolution' in the late nineteenth." The great chain inspired the notion of "missing links" long before Darwin. The great chain of being, according to Lovejoy, was intimately associated with the principle of plenitude. "Not so very long ago the world seemed almost infinite in its ability to provide for man's needs – and limitless as a receptacle for man's waste products. Those with an inclination to escape from worn-out farms or the clutter of urban life could always move out into a fresh, unspoiled environment. There were virgin forests, rich lodes waiting to be discovered, frontiers to push back, and large blank regions marked unexplored on the map.... It has, so far as I know, never been distinguished by an appropriate name; and for want of this, its identity in varying contexts and in different phrasings seems often to have escaped recognition by historians. I shall call it the principle of plenitude."

• Garrett Hardin in "The tragedy of the commons" (1968) discusses how old myths and cosmological beliefs affect the way we live. Individuals strive to maximize their share of a common resource in the belief that ownership is a natural and even divine right. When herdsmen graze their beasts on common land, each strives to increase the size of his herd. Disease and tribal warfare maintain a state of equilibrium by limiting the numbers of persons and beasts below the capacity of the land. Then comes a more orderly and civilized way of life that, with diminished war and disease, places an increased burden on the commons. A herdsman now thinks, "If I increase my herd, the loss owing to overgrazing will be shared by all, and my gain will exceed my loss." All herdsmen think this way and therein lies the tragedy. "Each person," states Hardin, "is locked into a system that compels him to increase his herd without limit – in a world that is limited.... Ruin is the destination to which all men rush." Unfortunately, most problems created by outdated cosmic myths (such as the Great Chain of Being, the principle of plenitude, and the freedom to reproduce without limit) do not have technical solutions.

"A technical solution may be defined as one that requires a change only in the techniques of the natural sciences, demanding little or nothing in the way of change in human values or ideas of morality." The "concern here is with that important concept of a class of human problems which can be called 'no technical solution problems.' . . . My thesis is that the 'population problem,' as conventionally conceived, is a member of this class. . . . It is fair to say that most people who anguish over the population problem . . . think that farming the seas or developing new strains of wheat will solve the problem – technically."

PROJECTS

1 Consider the old English prayer: "God help me in my search for truth, and protect me from those who believe they have found it."

• Consider also: In the *Memoirs of Zeus* by Maurice Druon, the goddess Mnemosyn declares "we would be better mirrors of the Universe if we were less concerned about our own image."

2 In the ancient world and in the Middle Ages astrology was the science of planets and stars, astrolatry was the worship of stars, and astromancy was the practice of soothsaying and divination by means of celestial configurations. We use the word biology for the science of living things and properly speaking we should use the word astrology for the science of stars. But astrology became corrupted and took the place of astrolatry and astromancy. Astrology now is the mythological belief that the affairs of human beings are influenced by the heavenly bodies.

Millions of people in America read the astrology (or rather the astromancy) columns in the daily newspapers; they find astromancy interesting and entertaining, for it is anthropocentric and connects human beings and the universe in ways that are meaningful to most people. Some persons take it seriously, and then, by modern standards, it becomes slightly ridiculous. But most people find it entertaining because it appeals to vestigial elements in our cultural heritage. Bart Bok, Lawrence Jerome, and 19 other leading scientists, in "Objections to astrology" (1975), vent their dismay: "Scientists in a variety of fields have become concerned about the increased acceptance of astrology in many parts of the world. . . . It should be apparent that those individuals who continue to have faith in astrology do so in spite of the fact that there is no verified scientific basis for their beliefs, and indeed that there is strong evidence to the contrary."

Discuss why astrology is still popular. Can it be that many persons find themselves in a largely meaningless universe from which their religions and philosophies have retreated? What can be done about this unhappy situation in which people find comfort in astromancy that science is resolved to eliminate? Sunday schools (in my day) did not arrest the flight from religion; will more introductory science courses arrest the flight from the scientific universe? Consider also Alfred Whitehead's statement in *Science and the Modern World*: "Nature is a dull affair, soundless, scentless, colourless; merely the hurrying of material, endlessly, meaninglessly."

3 Adam Smith's famous statement "The real price of anything is the toil and trouble of acquiring it" needs reexamining. In all undertakings with nature we should read the small print in the contract. This might disclose that the real price is paid by those who inherit the depletion and despoliation that follows. Are we already beginning to see the real price?

4 Give examples of problems that have no technical solution. Note that technical solutions, when they exist, often entail new problems. New drugs cure old diseases but add to the problem of population growth and may lead to greater suffering. Population growth has become a problem without technical solution, and requires, in Hardin's words, either a "change in human values or ideas of morality."

Do you think that colonizing space will technically solve the population problem? Sebastian von Hoerner, in "Population explosion and interstellar expansion" (1975), shows that this could solve the problem, with the present growth in birthrate, for at most only 500 years. The human space bubble, full of human beings, would expand faster and faster and in 500 years would expand at the speed of light. Each colonized planet would become more crowded and face the same problem that we now face on Earth. To what extent is the West with its technology, pharmacology, hygiene, and ideas of plenitude responsible for the alarming decrease in wild life and startling increase in human life?

5 Consider critically the syllogism:

> We are part of the Universe,
> we are alive,
> therefore the Universe is alive.

Consider also:

> The Universe contains us,
> we create universes,
> therefore no universe contains us.

6 Discuss the following examples of cosmic despair and hope:

"That man is the product of causes which had no prevision of the end they were achieving; that his origin, his growth, his hopes and fears, his loves and his beliefs, are but the outcome of accidental collocations of atoms; that no fire, no heroism, no intensity of thought or feeling, can preserve a life beyond the grave; that all the labors of the ages, all the devotion, all the inspiration, all the noonday brightness of human genius, are destined to extinction in the vast death of the solar system; and the whole temple of Man's achievement must inevitably be buried beneath the debris of a universe in ruins – all these things, if not quite beyond dispute, are yet so nearly certain, that no philosophy which rejects them can hope to stand. Only within the scaffolding of these truths, only on the firm foundation of unyielding despair, can the soul's habitation be safely built" (Bertrand Russell, *A Free Man's Worship*, 1923).

• "The same thrill, the same awe and mystery, come again and again when we look at any problem deeply enough. With more knowledge comes deeper, more wonderful mystery, luring one on to penetrate deeper still. Never concerned that the answer may prove disappointing, but with pleasure and confidence we turn over each new stone to find unimagined strangeness leading on to more wonderful questions and mysteries – certainly a grand adventure!" (Richard Feynman, "The value of science," 1958).

FURTHER READING

Blacker, C. and Loewe, M. Editors. *Ancient Cosmologies*. George Allen and Unwin, London, 1975.

Hamilton, E. *Mythology: Timeless Tales of Gods and Heroes*. Little, Brown, Boston, 1942.

Munitz, M. K. Editor. *Theories of the Universe: From Babylonian Myth to Modern Science*. Free Press, Glencoe, Illinois, 1957.

SOURCES

Barrow, J. D. and Tipler, F. J. *The Anthropic Cosmological Principle*. Oxford University Press, Oxford, 1986.

Bok, B. J., Jerome, L. E., Kurtz, P. et al. "Objections to astrology." *Humanist* 35, 4 (October 1975).

Burtt, E. *The Metaphysical Foundations of Modern Physical Science*. 1924. Revised edition: Humanities Press, New York, 1932. Reprint: Doubleday, Garden City, New York, 1954.

Butterfield, H. *The Origins of Modern Science, 1300–1800*. Bell and Sons, London, 1957. Revised edition: Free Press, New York, 1965.

Campbell, J. *The Masks of God: Primitive Mythology*. Viking Press, New York, 1959.

Campbell, J. *The Mythic Image*. Princeton University Press, Princeton, New Jersey, 1974.

Childe, V. G. *What Happened in History*. Penguin Books, London, 1942.

Feynman, R. "The value of science," in *Frontiers in Science: A Survey*. Editor E. Hutchings. Basic Books, New York, 1958.

Frankfort, H., Frankfort, H. A., Wilson, J. A., and Jacobsen, T. *Before Philosophy*. Penguin Books, London, 1949. First published as *The Intellectual Adventure of Ancient Man*. University of Chicago Press, Chicago, 1946.

Frazer, J. G. *The Golden Bough: A Study in Magic and Religion*. Abridged edition: Macmillan, London, 1922.

Greene, J. C. *Darwin and the Modern World View.* Louisiana State University Press, Baton Rouge, 1973.

Hardin, G. "The tragedy of the commons." *Science* 162, 1243 (1968).

Harrison, E. R. *Masks of the Universe.* Macmillan, New York, 1985.

Heisenberg, W. *Physics and Philosophy.* Harper and Row, New York, 1958.

Heninger, S. K. *The Cosmographical Glass: Renaissance Diagrams of the Universe.* Huntington Library, San Marino, 1977.

Hoerner, S. von. "Population explosion and interstellar expansion." *Journal of the British Interplanetary Society* 28, 691 (1975).

Hoyle, F. *Ten Faces of the Universe.* Freeman, San Francisco, 1977.

John, L. Editor. *Cosmology Now.* British Broadcasting Corporation, London, 1973.

Kruglak, H. and O'Bryan, M. "Astrology in the astronomy classroom." *Mercury* (November–December 1977).

Leach, M. *The Beginning: Creation Myths Around the World.* Funk and Wagnalls, New York, 1956.

Leslie, J. *Universes.* Routledge, New York, 1989.

Lewis, C. S. *The Abolition of Man.* Macmillan, New York, 1947.

Lewis, C. S. *The Discarded Image.* Cambridge University Press, Cambridge, 1967.

Lovejoy, A. O. *The Great Chain of Being: A Study of the History of an Idea.* Harvard University Press, Cambridge, Massachusetts, 1936.

Lucretius. *The Nature of the Universe.* Translated by R. E. Latham. Penguin, Harmondsworth, England, 1951.

Malinowski, B. *Magic, Science and Religion.* Free Press, New York, 1948.

Milne, E. A. *Modern Cosmology and the Christian Idea of God.* Oxford University Press, Oxford, 1952.

Neugebauer, O. *The Exact Sciences in Antiquity.* Brown University Press, Providence, Rhode Island, 1957.

Russell, B. *A Free Man's Worship.* Mosher, Portland, Maine, 1923.

Schrödinger, E. *Science and Humanism: Physics in Our Time.* Cambridge University Press, Cambridge, 1951.

Schrödinger, E. *Science, Theory and Man.* Dover Publications, New York, 1957.

Singer, C. "The scientific views and visions of Saint Hildegaard (1098–1180)." In *Studies in the History and Method of Science.* Editor C. Singer. Dawson, London, 1955.

Trevor-Roper, H. R. *The European Witch Craze.* Harper and Row, New York, 1969.

Tylor, E. B. *Primitive Culture.* London, 1871.

Whitehead, A. N. *Science and the Modern World.* Macmillan, London, 1925.

2 EARLY SCIENTIFIC COSMOLOGY

Philosophers of ancient times were diversely transported in the stream of their own opinions, both concerning the worlds originall and continuance: some determining that it once began; others imagining that it was without beginning, and that the circled orbs should spin out a thread as long as is eternite, before it found an ending.
John Swan, Speculum Mundi (1635)

THE BEGINNING OF WESTERN SCIENCE

Babylonic wizardry

Four thousand years ago Babylonian sky-gazers divided the sky into the constellations of the Zodiac, compiled star catalogs, and recorded the movements of planets. They invented multiplication tables, established rules of arithmetic, and were skilled in the arts of computation. By studying the rhythmic variations of the heavens, they predicted eclipses and prepared calendars forecasting the seasons and dates of full and new Moon. All this was done as religious worship (astrolatry) and religious divination (astromancy). The Babylonian wizards charted the heavens, guided by mythic principles, and failed to develop natural explanations of celestial movements.

Greek philosopher-scientists

From the time of Thales of Miletus (sixth century BC), the Greek philosopher-scientists sought to account for the complexity of the world by reducing it to an interplay of elements. The proposed elements or primary constituents were

water:	Thales (sixth century BC)
air:	Anaximenes (sixth century BC)
seeds:	Anaxagoras (sixth century BC)
atoms:	Leucippus (fifth century BC)
fire:	Heraclitus (fifth century BC)
earth:	Xenophanes (fifth century BC)
earth, water, air, fire:	Empedocles (fifth century BC)
earth, water, air, fire, ether:	Aristotle (fourth century BC)

"There exists an ultimate substance," said Thales, "from which all things come to be, it being conserved." Early scientists asked how things worked and used analogies drawn from the mechanical arts and crafts, such as pottery and metalwork. They believed in the conservation of elements that constantly combine and recombine to form objects of different shapes and colors. They rejected supernatural (mythological) explanations of natural phenomena, reduced everything to basic elements, and used conservation rules. They freely hypothesized and made experiments based on hypotheses. All this is the essence of the scientific method.

In the sixth century BC, starting with Pythagoras, Heraclitus, and Parmenides, we see the budding of novel and potent ideas. Pythagoras of Samos, who founded an influential school of advanced thought and may have been the first to use the word "philosopher," taught that the Earth is a sphere and the harmonies of the cosmos are governed by mathematical relations. Heraclitus of Ephesus, known as the "weeping philosopher" because of his pessimism, said everything changes, nothing endures, and the basic element is therefore fire. He conceived a universe of tempestuous flux governed by a conflict of opposing forces, and said wisdom consists of knowing how things change. Parmenides of Elia said nothing changes, everything endures, and wisdom consists of rejecting the sensory deceptions. He conceived an abstract universe of truth, beauty, and justice in which all change was an illusion of the senses. The Heraclitean flux and the timeless Parmenidean continuum are highly original concepts; the former anticipates the Newtonian universe of dynamic motions, and the latter anticipates the Riemannian spacetime of relativity theory.

Pythagorean harmony

The Babylonian astronomers were priests and prophets, whereas the Hellenic (or Greek) astronomers were philosophers and scientists. The Babylonians excelled in arithmetic and abstraction, whereas the Greeks excelled in geometry and metaphor.

Pythagoras visualized a universe of geometrical harmony and used spheres, circles, and vortical motions as basic forms in the design of the universe. "It was said," according to Diogenes Laertius (third century AD), "that Pythagoras was the first to call the heavens cosmos and the Earth a sphere." The Pythagoreans believed that the heavenly bodies were perfect spheres moving in perfect circles around a central cosmic fire that lay beyond the reach of mortal vision. The heavenly bodies emitted melodious notes and their celestial symphony or harmony of the spheres lay beyond the reach of mortal hearing.

PLATO'S UNIVERSE
Socrates

Socrates (about 470–399 BC) lived in Athens where he taught the liberal arts and stressed the importance of humanistic studies. Through his disciple Plato, he changed the course of philosophy. He wrote almost nothing that has survived and his teachings are known through the writings of Plato. Socrates believed that the immortal soul (or mind) inhabited a mortal house of clay. He stressed the importance of questions that commence with why, whereas the Greek scientists, with the aid of analogies from the mechanical arts, stressed the importance of how things worked. If atoms indeed explain how matter is constructed, what of it? Surely it is more important to know why atoms exist, and only pure reason

searching within the soul can determine their necessity and purpose. At the age of seventy he was condemned to death on a charge of misleading his students with heretical thoughts.

Plato

Plato (about 427–347 BC) established in Athens the first school of advanced learning, known as the Academy. Over the centuries his teachings have been more influential than those of any other philosopher. He interwove the Pythagorean and Socratic themes into a cosmology that stressed the wide difference between appearance and reality, between fugitive matter and concrete mind. The transitory phenomenal world is only a shadowy image of the eternal real world. Matter, which is innately disordered, incoherent, and discordant, is governed by Mind, which is the source of order, coherence, and harmony. In the parable of the cave, Plato imagines people chained in a cave and unable to move. The wall in front is illuminated by light shining from behind them that they cannot see. Behind them moving objects cast shadows on the illuminated wall. Throughout their life they are chained and know only of the shadows cast on the wall. "Surely," wrote Plato, "such persons would believe the shadows to be the only realities." According to this parable, we are prisoners of our senses, and the real world beyond the senses can be reached only by the intelligent mind.

THREE COSMIC SYSTEMS OF THE ANCIENT WORLD
Aristotelianism, Epicureanism, Stoicism

Three rational belief systems – Aristotelianism, Epicureanism, and Stoicism – dominated the Mediterranean world from the fourth century BC to the third century AD. Each was a universe combining philosophical, scientific, and ethical principles. Remnants of all three universes lie deep in the cultural heritage of the West.

The Aristotelian universe of celestial spheres originated in Athens, and centuries later was adopted by the Judaic–Christian–Islamic world. Outfitted with theistic additions of Babylonian and Zoroastrian origin, it evolved into the Medieval universe that survived until the sixteenth century.

The Epicurean universe of endless worlds originated in Athens in the late fourth century and enjoyed wide popularity among educated and middle class people for several centuries. It incorporated atomist ideas, emphasized the life sciences, rejected the gods as explicative agents in the natural world, and accepted the equality of men and women. Condemned later because of its atheism, it was suppressed everywhere in the Judaic–Christian–Islamic world.

The Stoic universe, founded in Athens in the third century, was popular throughout the Roman Empire. It consisted of two parts: a finite cosmos of stars, and a surrounding void extending to infinity. The Stoics stressed the importance not only of the sciences but also of ethical principles and duties.

THE ARISTOTELIAN UNIVERSE
The two-sphere universe
The Pythagorean two-sphere universe was popular at the time of Plato at the Academy in Athens in the fourth century BC. It consisted of a central sphere (the Earth) surrounded by an outer sphere (the stars). The planets moved in undetermined ways between the two spheres. "To all earnest students," Plato proposed the problem, "what are the uniform and ordered movements of the planets?"

The many-sphere universe
Eudoxus, a student of Plato, proposed that the universe can be represented by a central spherical Earth surrounded by concentric and rotating spheres. In Eudoxus's proposal, the outermost sphere rotated daily and supported the stars, as in the Pythagorean two-sphere model. In addition, intermediate spheres supported the planets (see Figure 2.2) and rotated at various rates about variously inclined axes. At first these geometrical contrivances were intended to be little

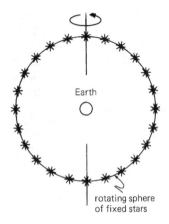

Figure 2.1. The two-sphere universe. The central sphere is the Earth. The outer sphere, studded with stars, rotates daily and the stars rise in the east and set in the west.

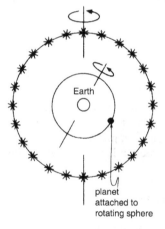

Figure 2.2. The many-sphere model proposed by Eudoxus. Additional intermediate spheres, rotating about inclined axes, support the planets. (Only one intermediate sphere is shown in the figure.)

more than analogies, much like the mechanical and optical contrivances of a modern planetarium. But Aristotle changed hypothesis into reality.

Aristotle
Aristotle (384–322 BC), born in Macedonia, studied at Plato's Academy. He founded his own school called the Lyceum, known also as the peripatetic school because Aristotle lectured while walking around with his students. His interests were universal

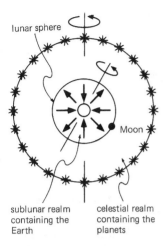

lunar sphere

Moon

sublunar realm
containing the
Earth

celestial realm
containing the
planets

Figure 2.3. The Aristotelian universe (planetary spheres not shown). The many-sphere model was invested with physical and ethereal reality. Physical things occupied the sublunar realm and ethereal things occupied the celestial realm.

and his lecture notes filled 150 volumes. He was a careful observer and his main field of interest tended to be the life sciences. He discussed the great chain of being and the evolution of various forms of life. When Alexander the Great, who was his student and then his patron, died, Aristotle feared accusations of impiety against the gods and fled from Athens in order that Athens should not "sin twice against philosophy."

Aristotle adopted the many-sphere model and invested the spheres with physical and etheric reality (see Figure 2.3). The planets – Moon, Mercury, Venus, Sun, Mars, Jupiter, and Saturn – were attached to the translucent spheres that rotated around the Earth. As many as 56 supporting spheres were needed to explain the motions of the planets and stars. The terrestrial elements of earth, water, air, and fire composed the Earth and inner sublunar sphere, and a fifth element, called ether (later also known as quintessence), composed the outer celestial spheres. The natural motion of the terrestrial elements was up and down whereby they sought to find their proper places according to weight, and the natural motion of the etheric element was endless revolution around the Earth.

In Aristotle's universe all motion required the continual application of force. Bodies moved because they were pushed and pulled by direct contact with other bodies that did the pushing and pulling. The void (or vacuum) could not exist, argued Aristotle, because isolated bodies would be motionless. Hence atoms (the smallest particles of matter) also could not exist, because atoms are isolated from one another by interatomic voids. Moreover, if atoms actually existed, as claimed by the atomist philosophers, then no force could have direct contact, and all bodies would be perpetually motionless.

The whole Aristotelian universe formed a cosmic sphere of finite size, necessarily finite according to Aristotle, because its outer boundary rotated about the Earth. "If the heaven be infinite, and revolve in a circle, it will traverse an infinite distance in a finite time ... this we know to be impossible." Straight lines could not be of infinite length because they would extend beyond the universe. Hence all straight lines were incomplete and therefore imperfect, whereas circles were complete and therefore perfect. It was fitting that the terrestrial elements of perishable form should have imperfect vertical motion, toward and away from the cosmic center, and this explained why the Earth did not rotate. It was also fitting that the etheric element composing the celestial spheres should have only perfect circular motion. "So the unceasing movement of the heavens is clearly understandable ... Everything ceases to move when it comes to its natural destination, but for the body whose natural path is a circle, every destination is a fresh starting point." Aristotle's system was a spatially finite universe in a steady state; it had existed unchanged through eternity, and its perfect motions had no beginning or end.

The rotating spheres of Eudoxus did not explain planetary retrogression satisfactorily (see Figure 2.4) and failed to explain why planets have increased brightness during their periods of retrograde motion. A step in the solution of this problem was the

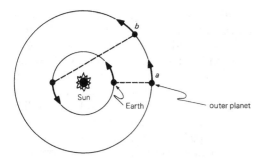

Figure 2.4. This figure shows why the outer planets, as seen from the Earth, appear to seesaw across the sky. The Earth and an outer planet, such as Mars, are shown revolving around the Sun. The Earth revolves faster than the outer planet. At position *a* the planet is closer (and brighter) and appears to move backward, but at position *b* the planet is farther away (and fainter) and appears to move forward.

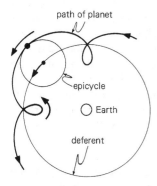

Figure 2.5. The motion of the planets according to the epicyclic theory. A planet revolves around a small circle (an epicycle) whose center moves around on a large circle (the deferent) that has the Earth at its center of the universe. The planet is thus seen to move backward and forward, and the backward (retrograde) motion occurs when the planet is closer to Earth and appears brighter, in agreement with observation.

idea of eccentric circles: the Earth remained at the center of the universe, but each celestial circle had its center displaced by a certain amount from the Earth. Another and more important step, taken in the third century BC, was the introduction of epicycles. The epicycles were additional circular motions, shown in Figures 2.5 and 2.6, that explained why the planets moved backward

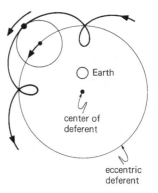

Figure 2.6. An epicycle on an eccentric deferent.

(retrogression) and forward (progression) across the sky and why they were brighter during the intervals of backward motion.

Ptolemy

Claudius Ptolemy (actual dates unknown), an astronomer and mathematician at the Museum of Alexandria in the second century AD, did for astronomy what Euclid (also at the Museum four centuries previously) had done for geometry. He brought together many of the ideas and observations made in previous centuries, and in his principal work, *Almagest* ("The Great System" in Arabic), he used not only eccentrics and epicycles, but also equants. The equant – or "equalizing point" – is an off-center point, shown in Figure 2.7, about which the epicycle moves at a uniform angular rate. With combinations of eccentrics,

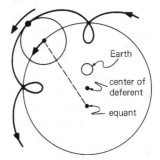

Figure 2.7. The equant introduced by Ptolemy. The equant, or "equalizing point," is a noncentral point about which the epicycle moves at constant angular rate.

epicycles, and equants, and by means of laborious calculations, it at last had become possible to mimic with precision the observed motions of the planets in a geocentric universe. The result was a geometric marvel. It endured for fourteen hundred years until overthrown by the revolutionary works of Copernicus, Kepler, and Galileo.

The final form of the Aristotelian universe, as presented by Ptolemy, failed to incorporate many developments in Greek science. It rejected the notion of atoms, the Democritean suggestion that the Milky Way is an agglomeration of stars, the proposal by Heracleides that the Earth rotates daily, and the theory by Aristarchus (accepted by Archimedes) that the Earth rotates daily and revolves annually about the Sun.

THE EPICUREAN UNIVERSE
Atomist origins
Emphasis by Parmenides on the unity of the One was countered by other philosophers who emphasized the plurality of the Many. The idea of a changeless, undifferentiated continuum was opposed by the idea of a void in which moved numberless discrete entities. Anaxagoras in the sixth century BC showed the way. He said the universe was infinite in extent and contained an infinite number of small seeds. These seeds – later called atoms (meaning indivisible) – had properties that impacted on our senses enabling us to perceive the world of matter. The universe was not ruled by gods but by a universal rational Mind, and the heavens and the Earth consisted of the same substances. Anaxagoras was accused of impiety and tried for heresy. Powerful friends defended him and although acquitted he fled the hostility of Athens.

Leucippus (fifth century BC) of Miletus, of whom very little is known, is credited with the invention of the atomic theory. He was the first to state clearly the principle of causality: all events are the effects of preceding causes. "Everything happens out of reason and necessity." Democritus of Abdera, a student of Leucippus, said, "Nothing can be created out of nothing, nor can it be destroyed and returned to nothing." Leucippus's atomist theory, elaborated by Democritus, was not accepted by the Athenian philosophers Socrates, Plato, and Aristotle, and we are indebted to the later teachings of the Epicureans for keeping alive the ideas of the Atomists.

The original atomist universe consisted of only atoms and the void. The atoms were infinite in number and the void was infinite in extent. Atoms consisted of the same substance, but differed in shape and size. They moved freely through the void, forever colliding and aggregating to form moons, planets, and stars, which slowly dissolved back into atoms.

Epicurus
Epicurus (341–270 BC) of Samos settled in Athens and founded the Epicurean school of philosophy – the first school to admit women students. The Epicureans adopted the atomist theory of numberless worlds formed by the aggregations of atoms. Epicureans believed that the human being is an evolved and superior animal, that the gods exist in ourselves and not the external world, and that the greatest pleasures in life stem from moderate and mutually supportive living. Their philosophy, based on atomist principles, flourished widely for six hundred years in the Greco-Roman world among thoughtful people and perished with the spread of Judaism, Christianity, and Islam.

The Nature of the Universe, written by the Roman poet Lucretius in the first century BC, is in praise of Epicureanism. "Bear this well in mind," wrote Lucretius in his epic poem, "that nature is free and uncontrolled by proud masters and runs the universe without the aid of gods. For who . . . can rule the sum total of the measureless? Who can hold in coercive hand the strong reins of the unfathomable? Who can spin all the firmaments alike and foment with the fires of ether all the fruitful earths? Who can be in all places at all times?" The answer was only nature itself. Religious institutions responded by suppressing Epicureanism wherever possible, and by making the gods even more remote and powerful than before. A surviving manuscript of the Lucretian poem was found in 1417 in an Eastern European monastery, and soon after became widely known through the invention of printing. Its effect, little documented by historians, was perhaps greater than the work of Copernicus.

THE STOIC UNIVERSE
Zeno of Citium

Zeno of Citium, born about 334 and died about 262 BC, founded in Athens the popular Stoic school of philosophy. He lectured in a roofed colonnade called a stoa, and his philosophy, which became known as stoicism, appealed to all classes from slaves to aristocrats. He exalted the ethical principles of duty and justice. We may imagine him calling to those strolling by: "Have fortitude in the face of adversity, for fate rules the world! Weep not for thou art strong. The gods exist in high places, in nature, and in ourselves, and the divine spirit throbs on Earth and in the heavens, swelling and subsiding from age to age, from cycle to cycle on the Wheel of Time. Gaze on it all, but be not amazed, for the soul has witnessed it many times before" (Edward Harrison, *Darkness at Night*). Stoic ethical values and codes of behavior, exemplified in the writings of Seneca and Marcus Aurelius, now permeate Western culture. Stoicism in the first century BC was much more popular than Epicureanism, particularly among the Romans.

The Stoics believed in fate, that all was predestinate, and that Mind, manifesting through the gods and mortals as divine spirit, governed the universe. They believed the stars were alive and the universe was a living organic whole. (Ge the Earth was also a living entity.) The Stoics believed that the starry cosmos was finite, and beyond the finite cosmos stretched an infinite mysterious void (Figure 2.8). Some Stoic schools believed the cosmos slowly pulsated in size and periodically passed

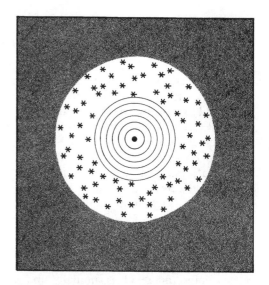

Figure 2.8. The Stoic universe consists of a finite cosmos of stars surrounded by a void of infinite extent. The Stoic system in various forms persisted until the early twentieth century.

through catastrophic upheavals. Two thousand years later the Stoic universe – a finite starry cosmos in an infinite void – formed the basis of the nineteenth century Victorian universe.

THE MYSTERY RELIGIONS

Science, the sort that analyzes and classifies, began in the Mediterranean world in the sixth century BC. At the same time intellectual activity quickened elsewhere in many lands. The teachings of Zoroaster (or Zarathustra) in Persia (Iran), Gautama the Buddha and Mahavira the Jain in India, and Confucius and Lao-tzu in China gave birth to ethical religious movements. The moral codes of civil behavior, previously unrelated to religion, were formulated in terms of religious doctrine. The spread of these enlightened teachings created an abhorrence of human sacrifice that previously had been a worldwide religious practice. Zoroaster in Persia was the first prophet to teach a doctrine of rewards and punishments in afterlife. His monotheism of a universe ruled by the lord of light – Ahura Mazda – and of an afterlife where the good go to heaven and the evil go to hell, was adopted by the Medes and the

Persians (with angelology and demonology additions), and enfolded in the derivative religions of Mithraism and Manichaeism.

THE MEDIEVAL UNIVERSE

From the sixth century BC to the second century AD, from Thales of Ionia to Ptolemy of Alexandria, Greek science flourished for 700 years. By the end of the second century AD the blaze of Greek inspiration had died to a feeble glow.

The Early Middle Ages

With the fall of the Roman Empire in the fifth century, cultural darkness descended on Europe and all intellectual pursuits languished under the rule of barbarians. In the ensuing "Dark Ages" (the Early Middle Ages) the universe reverted to a mythic polarization of heaven and hell, with the Earth in the form of a rectangular tabernacle surrounded by an abyss of water. A few scholars (such as Boethius and the Venerable Bede) were aware of Greek learning through the Latin commentaries of Cicero, Pliny, and others. While European learning was at its lowest ebb, remnants of ancient knowledge survived in Byzantium, Syria, and Persia.

The rise of Islam

In the seventh century the Arabs poured out of their deserts and created the great Islamic Empire that extended from the Atlantic to India. The crafts, arts, and sciences once again flourished; libraries of old forgotten manuscripts were assembled, and scholars migrated to Damascus, Baghdad, Cordoba, and other centers of the new civilization. Greek, Egyptian, Persian, Chinese, and Indian literature was translated into Syriac, later into Arabic, and synthesized into extensive commentaries. The new learning echoed throughout the known world, including Europe and China, and in the ninth century the Earth had regained its spherical form and the universe was once more spherically geocentric. We are all indebted to the Islamic Empire for its preservation and transmission of ancient knowledge that ultimately awakened Europe. Unfortunately, the Mongolian and Turkic invasions in the fourteenth century shattered the old Islamic and Chinese cultures.

The High Middle Ages

Exciting knowledge from Arab lands dispersed the darkness of the Early Middle Ages. New ideas, such as in order to believe it is necessary to understand, transformed old attitudes. Anselm, archbishop of Canterbury, introduced in the eleventh century the empyrean: a sphere of purest fire where God dwelt beyond the sphere of stars (Figure 2.9). Numerous industrious scholars translated the works of Plato, Aristotle, Euclid, Galen and other philosophers, mathematicians, and scientists of the ancient world into medieval Latin, first from Arabic, and then directly from the original Greek. The new knowledge exceeded the limits of the monastery and cathedral schools, and communities of translators and learned scholars founded the universities. Thomas Aquinas in the early thirteenth century showed how Christian doctrine could be accommodated in the Aristotelian universe with minor modifications. Human beings retained their immortality but the universe lost its eternity. The Medieval universe that followed attained its finest form in the fourteenth century; it was a grand unification of the material and spiritual worlds, sanctified by religion, sanctioned by philosophy, and rationalized by geocentric science.

But already in reawakened Europe of the thirteenth century there was much dissatisfaction with Aristotle's physics and Ptolemy's astronomy. Roger Bacon, a Franciscan monk, declared that the scientific method consisted of making observations, not reading old texts, using mathematics, and checking calculations with experiments.

The wholesale adoption of Aristotelian learning threatened Christian doctrine and ecclesiastical authorities grew alarmed. They conceded that the Earth is truly a sphere at the center of the universe, but denied the Aristotelian argument that God could not move the Earth if he willed, or

Figure 2.9. God creates and maintains the universe, and occupies the outermost sphere of purest fire, as suggested by Anselm, archbishop of Canterbury (eleventh century). (From Martin Luther's *Biblia*, published in Wittenberg by Hans Lufft in 1534.) The sphere of purest fire became known as the empyrean after the time of Milton. As the notion of God evolved, and God became infinite, the abode of God also became infinite in extent, and the medieval universe transformed into a Stoic universe.

could not create other worlds if he willed. Aristotelian constraints imposed on the power of God were totally rejected in the 219 Condemnations proclaimed in 1277 by Etienne Tempier, the bishop of Paris. God's power is without limit, said the bishop, and God, who is without limit, is everywhere and cannot be confined to any one place. The Condemnations stand as a landmark in the history of cosmology. The explosive idea of an unlimited God burst open the bounds of the Aristotelian universe. The new universe had to be capable of accommodating an unlimited God.

Thomas Bradwardine of Oxford, who became archbishop of Canterbury, said, "God is that whose power is not numbered and whose being is not enclosed." He echoed Empedocles of the fifth century BC who had written, "God is an infinite sphere whose center is everywhere and circumference nowhere." To this end, Bradwardine expanded Anselm's empyrean into an infinite extramundane void. Beyond the sphere of stars stretched a mysterious limitless realm where God dwelt. He transformed the bounded Aristotelian universe into an unbounded Stoic universe.

Nicolas Oresme in France, also in the fourteenth century, said, "motion can be perceived only when one body alters its position relative to another." He refuted old arguments claiming that the Earth could not rotate and pointed out that Heracleides' theory of a rotating Earth greatly simplified the structure of the heavens.

Undoubtedly, in the fifteenth century, the discovery of the poem *The Nature of the Universe* influenced the thoughts of many thinkers and made familiar the exciting idea of an infinite universe. Cardinal Nicholas of Cusa argued that because God created the universe, and God is boundless and without location, the universe must also be without edge and center. In his treatise *Of Learned Ignorance*, he made the famous statement, the universe "is a sphere of which the center is everywhere and the circumference nowhere." Thus the properties of the Creator were reflected in the created universe. It

is convenient, said Nicholas of Cusa, to regard the Earth as the center of the universe, although nothing in reality compels us to do so. An actual center need not exist and he could see no reason why the Earth should not move.

THE HELIOCENTRIC UNIVERSE

The Copernican Revolution began with the Pythagoreans and ended with the Cartesians and Newtonians. Its heroes were Aristarchus and Copernicus. It released human beings from their geocentric obsession and paved the way for the ultimate overthrow of the anthropocentric universe.

Copernicus

In the sixteenth century, Nicolaus Copernicus (1473–1543), a canon of the Catholic Church, demonstrated the feasibility of a heliocentric universe. As a student in Italy he had studied the Ptolemaic system and had been dismayed by its abdication of the Platonic ideal of perfect circular motion. This ideal, accepted by Aristotle, had been abandoned when Ptolemy introduced equants. Copernicus was also aware of the heliocentric theory proposed by Aristarchus. According to this theory, the Earth revolves about the Sun, thus causing the other planets to appear to move backward and forward across the sky. Copernicus thought perhaps in a heliocentric system equants could be discarded and the original ideal of perfect circular motion restored. He devoted his life to the construction and computation of heliocentric orbits.

Copernicus's great work, *Revolutions of the Celestial Spheres*, rivaling the *Almagest* in scope, appeared in print in 1543 shortly before his death. In an earlier work he had written, "All orbs revolve about the Sun, taken as their center point, and therefore the Sun is the center of the universe" (see Figure 2.10). In the *Revolutions of the Celestial Spheres* he wrote, "Why then do we hesitate to allow the Earth the mobility natural to its spherical shape, instead of proposing that the whole universe, whose boundaries are unknown and unknowable,

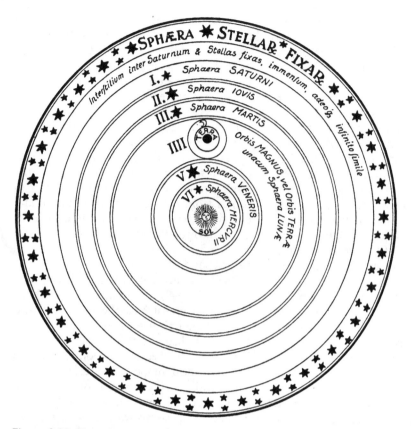

Figure 2.10. The universe according to Copernicus with the Sun occupying the center. The heliocentric universe originated in the third century BC and was proposed by Aristarchus of Samos who "brought out a book consisting of certain hypotheses in which the premises lead to the conclusion that the universe is many times greater than that now so called. His hypotheses are that the stars and the sun remain motionless, that the earth revolves about the sun in the circumference of a circle, the sun lying in the middle of the orbit" (Archimedes [about 287–212 BC], *The Sand Reckoner*. T. Heath, *Aristarchus of Samos*).

is in rotation?" The arguments he used to justify the Earth's rotation were similar to those proposed in the previous century by Oresme. From a rotating Earth it was a simple step to a moving Earth: "We therefore assert that the center of the Earth, carrying the Moon's orbit, passes in a great orbit among the other planets in an annual revolution around the Sun; that near the Sun is the center of the universe, and that whereas the Sun is at rest, any apparent motion of the Sun can be better explained by motion of the Earth."

Alas! the Copernican dream of a simpler universe was not fulfilled, and most astronomers at first were not convinced. The more Copernicus labored to bring the heliocentric system into conformity with observations, the larger became the required number of circles. By sacrificing equants, he required more circles than ever before and was still unable to match the precision achieved by Ptolemy.

THE INFINITE UNIVERSE
Thomas Digges

The astronomer and mathematician Thomas Digges (1543–1595) was born in the year that Copernicus died. In "A perfit description of the caelestiall orbes", published in

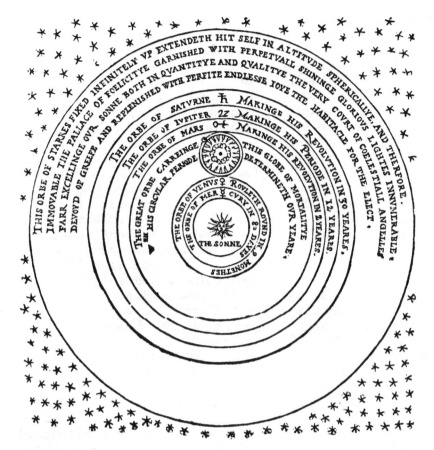

Figure 2.11. The infinite universe, as proposed by Thomas Digges in 1576. The legend on the diagram reads: "This orbe of starres fixed infinitely up extendeth hit self in altitude sphericallye, and therefore * immovable the pallace of foelicitye garnished with perpetuall shininge glorious lightes innumerable * farr excellinge our sonne both in quantitye and qualitye the very court of coelestiall angelles * devoyd of greefe and replenished with perfite endlesse joye the habitacle for the elect."

1576, he expounded on the Copernican system and introduced a major modification: the dispersal of the sphere of stars throughout unbounded space (see Figure 2.11). "This orbe of starres," wrote Digges, "fixed infinitely up, extendeth hit self in altitude sphericallye." Only 33 years after the publication of the *Revolutions of the Celestial Spheres* (and exactly two hundred years before the Industrial Revolution and American Declaration of Independence), the universe had its sphere of stars torn away. "The perfit description" passed through many editions in the latter part of the

sixteenth century and fostered thoughts that transformed the material universe.

Giordano Bruno

The fiery monk Giordano Bruno (1548–1600) lived in London while Digges's book was the talk of the town. He enthusiastically adopted the idea of an edgeless universe and stressed the logical conclusion, previously made by Nicholas of Cusa and others, that the universe must also be centerless: "In the universe no center and circumference exist, but the center is everywhere," he wrote. In his writings and travels, Bruno broadcast a Christianized version of the Epicurean universe (known from the poem by Lucretius): a universe of infinite extent, populated with numberless planetary systems teeming with life. "Thus is the excellence of God magnified and the greatness of his kingdom made manifest; he is glorified not in one but in countless Suns; not in a single Earth, but in a thousand, I say, in an infinity of worlds." Bruno was the revolutionary champion of the Copernican Revolution. His last seven years were spent in an ecclesiastical prison; tormented and tortured, he refused to recant and was burned at the stake in Rome in 1600.

Tycho Brahe

Tycho Brahe (1546–1601), a Danish nobleman, observed the planets using the utmost precision possible before the introduction of telescopes. He rejected the Copernican system for a simple reason. If the Earth moved in a great circle around the Sun once per year, he said, the stars would be seen in slightly different directions at different times of the year. His observations failed to detect this angular variation (parallax). Must the stars be banished to distances so great that parallax becomes unobservable? The stars, he argued, have a certain size as seen by the unaided eye and are not points of light of no size. If the stars were banished to distances so large that parallax became too small to be detected, they would have to be enormously larger than the Sun to have their observed size. Nowadays we know the apparent size of stars – caused by diffraction – is deceptive and that stars are even farther away than Tycho imagined.

Tycho constructed a compromise system. In the Tychonic system, the Earth is stationary, the Sun revolves around the Earth, and all the planets revolve around the Sun (see Figure 2.12).

Kepler

Johannes Kepler (1571–1630), an imaginative scientist who overcame ill health and became the imperial mathematician of the Holy Roman Empire (an "agglomeration" in the words of Voltaire that although it "was called and still calls itself the Holy Roman Empire was neither holy, nor Roman, nor an empire in any way"). He accepted the finite Copernican system with the Sun at its center and the sphere of fixed stars at its outer edge, and opposed the radical suggestion of an infinite and centerless universe. The thought of an infinite universe appalled him and in 1606 he wrote, "This very cogitation carries with it I don't know what secret, hidden horror; indeed one finds oneself wandering in this immensity to which are denied limits and center and also all determinate places." In his first book, *Cosmographical Mysteries*, published in 1596, Kepler sought to unravel the secrets of the cosmos. This work contained many germinal ideas that blossomed in Kepler's later research.

The previous imperial mathematician had been Tycho Brahe. Kepler inherited Tycho's careful and detailed observations of the planets, and for years struggled to explain their motions, particularly that of Mars. At last he triumphed and succeeded in freeing astronomy from the paradigm of epicyclic motion. His important three laws of elliptical planetary motion (Chapter 5) served as the foundation stones in the Newtonian world system.

Galileo

Galileo Galilei (1564–1642) was born the same year as Shakespeare and died in the year of Newton's birth. His great contribution to astronomy was the introduction of the telescope. What Galileo saw through his telescope was not in accord with Ptolemaic teaching. His ideas ran counter to the Aristotelian belief that the celestial realm is the abode of spirits. In his forthright manner he declared that the sublunar and celestial realms are physically alike, and to the observing eye and critical mind the Earth obviously rotates daily about an axis and revolves annually about the Sun.

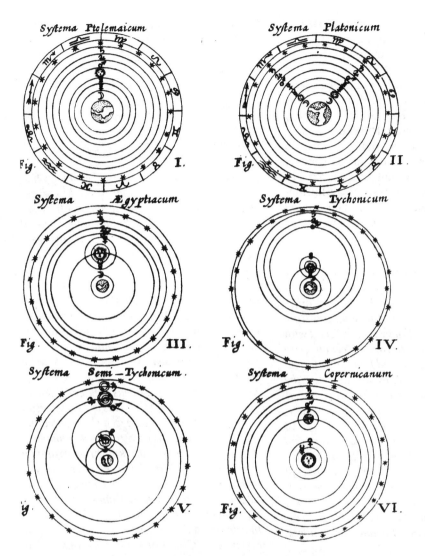

Figure 2.12. A gallery of universes from *The New Almagest* (1651) by Giovanni Riccioli. The Ptolemaic system (I), Tychonic system (IV), and Copernican system (VI) are discussed in the text. In the Platonic system (II), the Sun is interior to the orbits of Mercury and Venus (in the Ptolemaic system the Sun is exterior to these planetary orbits). In the Egyptian system (III), the inner planets Mercury and Venus revolve about the Sun, which revolves with the outer planets about the Earth. In the semi-Tychonic system (V), Mercury, Venus, and Mars revolve about the Sun, which revolves with Jupiter and Saturn about the Earth. (Courtesy of the Henry E. Huntington Library, San Marino, California.)

Galileo believed in the Copernican system but showed no great interest in Kepler's theories. His astronomical discoveries – mountains on the surface of the Moon, the satellites of Jupiter, and numerous hitherto unresolved stars of the Milky Way – were published in 1610 in *The Starry Messenger*. He gave an answer to Tycho's objection to the Copernican system. Even though Galileo's telescope magnified 30 times, stars remained the same apparent size as when seen with the unaided eye, and he argued correctly that their observed size is deceptive.

Galileo's most hostile opponents were academics and clerics steeped in the works of Aristotle. In *The Two Great Systems of the World*, he contrasted the geocentric and heliocentric systems and poured scorn on the physics of Aristotle and the astronomy of Ptolemy. This brought him in old age into conflict with ecclesiastical authorities and under the threat of torture he recanted and abjured the heliocentric system.

REFLECTIONS

1 *Pythagoras (about 582–497 BC), a Greek philosopher born on the Aegean island of Samos, traveled widely in Egypt and other lands. He founded a school in southern Italy and taught that the Earth is a sphere (with gravity acting always toward its center) and that all things are governed by mathematical regularities. The Pythagoreans held that the universe was constructed with numbers. A point was 1, a line 2, a surface 3, a solid 4, and the total number $1 + 2 + 3 + 4 = 10$ was sacred. Points had finite size. Lines and surfaces had finite thickness because they were constructed from points. The concept of geometric atomism, of a universe constructed from points, was one of the paths that led to the atomist theory. Many problems were analyzed in terms of triangular numbers, such as*

3   6   10

and square numbers, such as

4   9   16

Great importance was attached to the mean of any two numbers a and b:

$$\text{arithmetic mean} = \frac{a+b}{2},$$

$$\text{geometric mean} = \sqrt{(ab)},$$

$$\text{harmonic mean} = \frac{2}{(1/a + 1/b)}.$$

The harmonic mean in music and geometry had great importance. Note that the eight corners of a cube are the harmonic mean of its six faces and 12 edges. The Pythagoreans proved (but the Babylonians discovered) the geometric–algebraic rule: the square of the diagonal length of a right triangle equals the sum of the squares of its sides. Thus, 3, 4, 5 and 5, 12, 13 are examples of the magnitudes of the sides and diagonals of two right triangles. The Pythagoreans discovered that generally the sides and diagonal are incommensurable and cannot be expressed in integral or fractional numbers. For example, a right triangle of equal sides of unit length has a diagonal length $\sqrt{(1+1)} = 1.41421\ldots$. This irrational result ("devoid of logos") came as a shock, creating a Pythagorean crisis, and to this day numbers such as $\sqrt{2}$, $\sqrt{3}$, $\sqrt{5}$ are known as irrational numbers. (An "irrational number" is not expressible as a ratio of two integers.) The subsequent realization that lines are infinitely divisible meant that geometric points, basic to the design and construction of the universe, had no size. Critics of the Pythagorean school demanded to know how a universe can be composed of points of no size and yet be itself of finite size.

2 *Zeno of Elea (in southern Italy), a Greek philosopher in the fifth century BC denied that truth can be attained by the senses. The real world is absolute and timeless, the perceived world is illusory and transitory. He is best known for his paradoxes that sought to disprove the possibility of change as perceived by the senses. The paradoxes are all similar and the one most quoted is the race between Archilles and the tortoise. Suppose the tortoise has a 100-meter start and Archilles runs 100 times faster than the tortoise.*

While Archilles runs 100 meters, the tortoise moves 1 meter, while Archilles runs 1 meter, the tortoise moves 1 centimeter, and so on, in an infinite number of decreasing steps. Archilles therefore gets progressively closer but never overtakes the tortoise. It may be argued that the paradox is fallacious because the infinite number of steps occupy only a finite interval of time, and Archilles overtakes the tortoise, as observed by the senses. Philosophers, however, still debate the paradox; at issue are the assumptions that space and time are legitimately continuous and infinitely divisible.

3 "But multitudinous atoms, swept along in multitudinous courses through infinite time by mutual clashes and their own weight have come together in every possible way and realized everything that could be formed by their combinations. So it comes about that a voyage of immense duration, in which they have experienced every variety of movement and conjunction, has at length brought together those whose sudden encounter normally forms the starting-point of substantial fabrics – earth and sea and sky and the races of living creatures" (Lucretius, The Nature of the Universe).

4 The universal ideas (truth, beauty, justice, perfection, ...) that Plato had made divine, which impose form on the disorderly world of matter, were made secular by Aristotle and given an inseparable association with material things. In mythology the gods ruled, in Plato's universe the Mind ruled, and in Aristotle's universe the Forms ruled. In a sense, Aristotle restored the spirits of the age of magic, but with a difference: the spirits were now the hidden forces and innate properties of matter. In their new form they had become the souls of material things; they still exist, masquerading as forces, masses, momenta, energies, potentials, wave functions, and so forth. Once again, human beings live in a world of magic, but the capricious spirits are now the disciplined dancers in a ballet of weaving forces and waves.

5 In their exile in Babylon (sixth century BC), Jewish people encountered Zoroastrianism with its polarization of good and evil, and adapted much of its ethical idealism to their own brand of monotheism. Zoroastrianism inspired the Wisdom Literature of the Jews: in the books of Job ("Where was thou when I laid the foundations of the earth?"), Psalms ("Yea, though I walk through the valley of the shadow of death, I will fear no evil: for thou art with me, thy rod and thy staff they comfort me"), and the Song of Solomon ("Who is she that looketh forth as the morning, fair as the moon, clear as the sun, and terrible as an army with banners?").

"Zoroastrian idealism with its hereditary miscellany of angels and demons has greatly influenced Judaism, Christianity, and Islam. From the early fifth century BC, with the fall of Babylon, until the time of Saint Augustine of Hippo in the late fourth and early fifth centuries, the Mediterranean world was exposed to Zoroastrianism through its derivative religions of Mithraism and Manichaeism and by its infiltration of Greek philosophy, Jewish prophetic literature, and Gnostic and Neoplatonic theologies. Augustine, who molded Western Catholicism, was at first a Manichaean, and after his conversion he blended Zoroastrian ideals and Greek logic with Judaic scriptural history. In The Eternal City, Augustine compared the Heavenly and Earthly Cities and contrasted otherworldliness and the way of grace and salvation with worldliness and the way of evil and damnation" (Edward Harrison, Masks of the Universe).

6 "Human tides have washed across the globe, crushing nations and carving out empires, led by god-inspired men who sought to write their will across the sky in stars. One such leader was Alexander the Great, who crossed the Hellespont in the fourth century BC, subjugated Asia Minor and Egypt, vanquished the armies of the Persian Empire with his cohorts, quelled the turbulent forces of Afghanistan, crossed the Hindu Kush, invaded and defeated the nations of the Punjab. Eastward flowed Hellenic science and philosophy in the wake of Alexander's conquests; westward flowed oriental philosophy and religion. Westward into the Mediterranean world came the Zoroastrian lord of

light – the glorious *Ahura Mazda* – embattled with the lord of darkness, bringing the belief that the soul is divine, and the worship of gods other than the appointed one a sin. Westward into the Roman legions came the religion of the dying and resurrected martyred god – the triumphant *Mithra* – bringing the sacramental eating of the flesh of the god and the notion that evil is the privation of good. Westward came the Babylonian story of creation and the flood, and the Persian stories of heaven and hell, the last day of judgment, and the resurrection of the dead. All of which reshaped the theology of the Greco-Roman world in preparation for the rise of Christianity" *(Edward Harrison,* Masks of the Universe*). And we must not forget that westward came ethically inspiring cults such as the divine mother Isis holding her child.*

7 *"Tracing the development of ideas in the long Middle Ages leads the historian into a labyrinth of bewildering beliefs. The works of Jabir ibn Haiyan, court physician in the eighth century to Harun al Rashid (caliph of Baghdad famed in* The Thousand and One Nights*), became widely known for their medical lore and learned alchemy. Jabir was later latinized into Geber, and because of the rigmarole and obfuscation of the numerous works attributed to him, the word Geberish eventually became gibberish"* (Edward Harrison, Masks of the Universe). *In alchemy and astrology, the same principles applied to both the macrocosm (the universe) and the microcosm (the individual). This was the Great Analogy. The chemical elements and the parts of the human body had correspondence with the planets and other heavenly bodies. Silver was associated with the Moon, quicksilver with Mercury, copper with Venus, gold with the Sun, iron with Mars, tin with Jupiter, and lead with Saturn.*

8 *Roger Bacon (1220–1292) foresaw with prophetic vision the development of telescopes, submarines, steamships, automobiles, and flying machines. He wrote, "Machines for navigation can be made without rowers so that the largest ships on rivers or seas will be moved by a single man in charge with greater velocity than if they were full of men. Also cars can be made so that without animals they will move with unbelievable rapidity . . . . Also flying machines can be constructed so that a man sits in the midst of the machine revolving some engine by which artificial wings are made to beat the air like a flying bird." Europe was in the throes of a technology revolution – with the introduction of stirrups, heavy ploughs, water mills, windmills, textile mills, magnetic compass, gunpowder, pattern-welded steel, and papermaking. Bacon's extrapolations were not so fanciful in view of the contemporary technological developments occurring five centuries before the Industrial Revolution. It has not always been recognized that the Technology Revolution occurred not in the eighteenth, or nineteenth, or twentieth centuries but in the High Middle Ages of the thirteenth century. Many of the developments (stirrup, water mill, windmill, gunpowder, papermaking, and magnetic compass) came from outside Europe. But Europeans were the first to apply them widely and make technology a basic part of society. (See* Science in the Middle Ages, *David Lindberg, Editor.)*

9 *William of Ockham (about 1280–1349), an Oxford scholar at Merton College, was opposed to the Platonic philosophy that ideas are the true reality. Ideas are often nothing more than empty names, he said, and the true realities are the objects themselves. His viewpoint has been summarized in the words "entities must not be needlessly multiplied." This principle, known as Ockham's razor, is often interpreted to mean that the preferred theory has the fewest and simplest assumptions (or ideas). Ockham used this argument in his criticism of Aquinas's theological elaborations of the Aristotelian universe. Ockham's razor is useful in cosmology when comparing rival universes. All other things being equal, the preferred universe explains the observed world with the fewest and simplest ideas.*

10 *Dante Alighieri, in the early fourteenth century, while a political refugee from Florence, wrote the epic poem* The Comedy *(*Divine *was added later to the title in the* *sixteenth century). In this imaginative poem he placed the angelic spheres of the old Gnostic and Neoplatonic cosmologies in the empyrean, arranged in such a way that the realm of angelic spheres and the realm of celestial spheres mirrored each other. In his unified model, God became the center of a theocentric realm, the Earth remained the center of a geocentric realm, and God and Earth were the antipodes of each other. Dante's superb fusion of the material and spiritual realms came too late and never entered the mainstream of cosmology. Only a few decades previously, in 1277, the notion of God confined to a fixed place in space had been condemned as contrary to orthodox doctrine.*

11 *In* Two Great Systems of the World, *Galileo championed the virtues of the Copernican system in a fictional debate with an imaginary Aristotelian named Simplicius. Alfred Whitehead (*Science and the Modern World*) writes: "Galileo keeps harping on how things happen, whereas his adversaries had a complete theory as to why things happen. Unfortunately the two theories did not bring out the same results. Galileo insists upon 'irreducible and stubborn facts,' and Simplicius, his opponent, brings forward reasons, completely satisfactory, at least to himself. It is a great mistake to conceive this historical revolt as an appeal to reason. On the contrary, it was through and through an anti-intellectual movement. It was the return to the contemplation of brute fact; and it was based on a recoil from the inflexible rationality of medieval thought. In making this statement I am merely summarizing what at the time the adherents of the old regime themselves asserted."*

PROJECTS

1 Contrast the scientific aspects of the Aristotelian, Epicurean, and Stoic systems. (For more information see S. Sambursky, *The Physical World of the Greeks*.)

2 "The use of the sea and air is common to all; neither can a title to the ocean belong to any people or private persons, forasmuch as neither nature nor public use and custom permit any possessions thereof" (Elizabeth

I [1533–1603] to the Spanish ambassador in 1580). In some societies this Elizabethan principle applies also to the land. Why not in ours?

3 A simple model of Dante's unified universe can easily be constructed. On a white disk of cardboard, between 10 and 50 centimeters diameter, mark in the center a point representing the Earth. Draw around this point eight concentric circles of increasing radius to represent the celestial spheres of the Moon, Mercury, Venus, Sun, Mars, Jupiter, Saturn, and Stars. The rim of the disk is the primum mobile, an additional sphere introduced by the Arabs that moves all other spheres and is itself moved by God. On the other side of the disk mark in the center a point representing God. Again draw around this point eight concentric circles of increasing radius to represent the angelic spheres of the Seraphim, Cherubim, Thrones, Dominions, Virtues, Powers, Principalities, and Archangels. On this side of the disk the rim is the sphere of Angels. The disk is a model of the marvelously symmetric universe portrayed in *The Divine Comedy*. On one side can be seen the geocentric world of celestial spheres, on the other side can be seen the theocentric world of angelic spheres, and mediating between the two, at the rim, are the angels occupying the primum mobile. Suspend the disk for all to see as the most unified universe of all time. Make several and give them to friends as happy cosmos gifts.

4 Since the Enlightenment, with its reform bills and new constitutions, Western societies have stressed the rights and entitlements of individuals. Contrast this social change with the old Stoic practice of stressing the duties and obligations of individuals.

5 Cosmology in its broadest sense has few boundaries, and world systems of different religions legitimately fall within its scope. Some major unsolved theological riddles in cosmology are:

(a) If a supreme being of utmost power controls the universe, how can human beings have free will, and how can they be held responsible for their good and bad acts?

(b) If a supreme being of utmost goodness created the universe, why does the universe contain so much that is not good in the form of pain and evil?

Discuss these cosmological riddles. Discuss also the passage from *The Two Hands of God* by Alan Watts: "This, then, is the paradox that the greater the ethical idealism, the darker the shadow we cast, and that ethical monotheism became, in attitude if not in theory, the world's most startling dualism."

FURTHER READING

Allen, R. E. *Greek Philosophy: Thales to Aristotle.* Free Press, New York, 1966.

Copernicus, N. *Revolutions of the Heavenly Spheres.* In *Great Books of the Western World*, Vol. 16. Encyclopedia Britannica, Chicago, 1952.

Dampier, W. C. *A Shorter History of Science.* Cambridge University Press, Cambridge, 1944. Reprint: World Publishing Co., New York, 1957.

Dante. *The Comedy of Dante Alighieri.* Vol. 3. *Paradise.* Translated by D. Sayers and B. Reynolds. Penguin Books, Harmondsworth, Middlesex, 1962.

Drake, S. *Discoveries and Opinions of Galileo.* Doubleday, Garden City, New York, 1957.

Farrington, B. *Greek Science: Its Meaning For Us.* Vol. I, *Thales to Aristotle.* Vol. II, *Theothrastus to Galen.* Penguin Books, Harmondsworth, Middlesex, 1944.

Harrison, E. R. *Masks of the Universe.* Macmillan, New York, 1985.

Heninger, S. K. *The Cosmographical Glass: Renaissance Diagrams of the Universe.* Huntington Library, San Marino, Calif., 1977.

Koestler, A. *The Watershed: A Biography of Johannes Kepler.* Doubleday, Garden City, New Jersey, 1960.

Lerner, L. and Gosselin, E. "Giordano Bruno." *Scientific American* (April 1973).

Lewis, C. S. *The Discarded Image.* Cambridge University Press, Cambridge, 1967.

Lucretius. *The Nature of the Universe.* Translated in prose by R. E. Latham. Penguin Books, Harmondsworth, Middlesex, 1951.

Munitz, M. K. Editor. *Theories of the Universe. From Babylonian Myth to Modern Science.* Free Press, Glencoe, Illinois, 1957.

Murchie, G. *Music of the Spheres.* Vol. I, *The Macrocosm: Planets, Stars, Galaxies, Cosmology.* Vol. II, *The Microcosm: Matter, Atoms, Waves, Radiation, Relativity.* Dover Publications, New York, 1967.

Ravetz, J. "The origins of the Copernican Revolution." *Scientific American* (October 1966).

Reynolds, T. S. "Medieval roots of the industrial revolution." *Scientific American* (July 1984).

Tillyard, E. M. W. *The Elizabethan World Picture.* Macmillan, New York, 1944.

Toulmin, S. and Goodfield, J. *The Fabric of the Heavens: The Development of Astronomy and Dynamics.* Harper and Brothers, New York, 1961.

Van Melsen, A. G. *From Atomos to Atom: The History of the Concept Atom.* Duquesne University Press, Pittsburgh, 1951. Reprint: Harper and Row, New York, 1960.

Waerden, B. L. van der. *Science Awakening.* Noordhoff, Gronigen, Holland, 1954.

Wilson, C. A. "How did Kepler discover his first three laws?" *Scientific American* 226, 92 (March 1972).

SOURCES

Aristotle. *On the Heavens.* Translated by W. K. C. Guthrie. Harvard University Press, Cambridge, Massachusetts, 1939.

Aristotle. *On the Soul.* Translated by J. A. Smith. Clarendon Press, Oxford, 1908.

Bailey, C. *The Greek Atomists and Epicurus.* Clarendon Press, Oxford, 1928.

Brickman, R. "On physical space, Francesco Patrizi." *Journal of the History of Ideas* 4, 224 (1943).

Bruno, G. See Singer, D. W.

Burnet, J. *Early Greek Philosophy.* Meridian Books, New York, 1957.

Butterfield, H. *The Origins of Modern Science, 1300–1800.* Bell and Sons, London, 1957. Revised edition: Free Press, New York, 1965.

Caspar, M. *Kepler.* Translated and edited by C. D. Hellman. Abelard-Schuman, New York, 1959.

Clagett, M. *The Science of Mechanics in the Middle Ages*. University of Wisconsin Press, Madison, 1959.

Cornford, F. M. *Plato's Cosmology*. Routledge and Kegan Paul, London, 1937.

Crombie, A. C. *Robert Grosseteste and the Origin of Experimental Science 1180–1700*. Clarendon Press, Oxford, 1953.

Crombie, A. C. *Augustine to Galileo*. Vol. I, *Science in the Middle Ages*. Vol. II, *Science in the Later Middle Ages and Early Modern Times*. Mercury Books, London, 1961.

Cusa, N. *Of Learned Ignorance*. Translated by G. Heron. Yale University Press, New Haven, 1954.

Digges, T. "A perfit description of the caelestiall orbes." In *Prognostications Everlasting*. London, 1576. Reprinted in F. R. Johnson and S. V. Larkey, "Thomas Digges, the Copernican System, and the idea of the infinite universe in 1576." *Huntington Library Bulletin* No. 5, 69 (April 1934).

Diogenes Laertius. *Lives of Eminent Philosophers*. Translated by R. D. Hicke. 2 volumes. Loeb Classical Library, New York, 1935.

Dreyer, J. L. E. *A Shorter History of the Planetary Systems from Thales to Kepler*. Cambridge University Press, Cambridge, 1905. Reprint: Dover Publications, New York, 1953.

Duhem, P. *Medieval Cosmology: Theory of Infinity, Place, Time, Void, and the Plurality of Worlds*. Edited and translated by R. Ariew. Chicago University Press, Chicago, 1985.

Epicurus. *Epicurus: The Extant Remains*. Translated by C. Bailey. Clarendon Press, Oxford, 1926.

Galileo Galilei. *Two Chief World Systems – Ptolemaic and Copernican*. Translation by S. Drake. Foreword by A. Einstein. University of California Press, Berkeley, 1953.

Galileo Galilei. *The Starry Messenger*. In *Discoveries and Opinions of Galileo*. Translated by S. Drake. Doubleday, Garden City, New York, 1957.

Gershenson, D. E. and Greenberg, D. A. *Anaxagoras and the Birth of Physics*. Blaisdell, New York, 1964.

Grant, E. *Physical Sciences in the Middle Ages*. Wiley, New York, 1971.

Grant, E. *A Source Book in Medieval Science*. Harvard University Press, Cambridge, Massachusetts, 1974.

Harrison, E. R. *Darkness at Night: A Riddle of the Universe*. Harvard University Press, Cambridge, Massachusetts, 1987.

Haskins, C. H. *The Rise of Universities*. Cornell University Press, Ithaca, 1957.

Heath, T. *Aristarchus of Samos: The Ancient Copernicus*. Clarendon Press, Oxford, 1913.

Johnson. F. R. "Thomas Digges and the infinity of the universe." Reprinted in *Theories of the Universe: From Babylonian Myth to Modern Science*. Edited by M. K. Munitz. Free Press, Glencoe, Illinois, 1957.

Koyré, A. *From the Closed World to the Infinite Universe*. Johns Hopkins University Press, Baltimore, 1957. Reprint: Harper Torchbooks, New York, 1958.

Kuhn, T. S. *The Copernican Revolution: Planetary Astronomy in the Development of Western Thought*. Harvard University Press, Cambridge, Massachusetts, 1957.

Kuhn, T. S. *The Structure of Scientific Revolutions*. Chicago University Press, Chicago, 1970.

Lindberg, D. C. Editor. *Science in the Middle Ages*. University of Chicago Press, Chicago, 1978.

Lucretius. *Titi Lucruti Cari: De Rerum Natura*. Translated by V. Bailey. Clarendon Press, Oxford, 1947.

Maimonides, M. *The Guide for the Perplexed*. Translated by M. Friedlander. Dover Publications, New York, 1956.

Miller, D. G. "Ignored intellect: Pierre Duhem." *Physics Today* 19, 47 (1966).

Oates, W. J. *The Stoic and Epicurean Philosophies: The Complete Extant Writings of Epicurus, Epictetus, Lucretius, Marcus Aurelius*. Random House, New York, 1940.

Paterson, A. M. *The Infinite Worlds of Giordano Bruno*. Charles C. Thomas, Springfield, Illinois, 1970.

Rist, J. M. *Stoic Philosophy*. Cambridge University Press, Cambridge, 1969.

Sambursky, S. *The Physical World of the Greeks*. Routledge and Kegan Paul, London, 1956.

Santillana, G. de. *The Crime of Galileo*. University of Chicago Press, Chicago, 1955.

Saunders, J. L. *Greek and Roman Philosophy After Aristotle*. Free Press, New York, 1966.

Singer, D. W. *Giordano Bruno: His Life and Thoughts, With an Annotated Translation of His Work on The Infinite Universe and Worlds*. Schumann, New York, 1950.

Watts, A. W. *The Two Hands of God.* Collier Books, New York, 1969.

Wheelwright, P. *Aristotle: Containing Selections from Seven of the Most Important Books of Aristotle.* 3 volumes. Parker, London, 1835.

White, L. "Cultural climates and technological advance in the Middle Ages." *Viator* 2, 171 (1971).

Whitehead, A. N. *Science and the Modern World.* Macmillan, London, 1923.

3 CARTESIAN AND NEWTONIAN WORLD SYSTEMS

Awake! for Morning in the Bowl of Night
Has flung the Stone that puts the Stars to Flight.
Edward FitzGerald (1809–1883), The Rubáiyát of Omar Khayyám

THE DECLINE OF ARISTOTELIAN SCIENCE

Aristotle's law of motion

In Aristotle's day, ideas on space and time were vague and had yet to be sharpened into their modern forms. Space was associated with the distribution of things directly observed. Things distributed in time, however, were not directly observed, and generally intervals of time were not easily measured. How to define motion by combining intervals of space and time was not at all clear, and motion was poorly distinguished from other forms of change.

Aristotle's law of motion may be expressed by the relation

$$\text{applied force} = \text{resistance} \times \text{speed}. \quad [3.1]$$

But really he had no general formula, and no precise way of measuring force, resistance, and speed. He argued qualitatively, reasoning from the everyday experience that effort is needed to maintain a state of motion, and the faster the motion, the bigger the effort needed to maintain that motion. "A body will move through a given medium in a given time, and through the same distance in a thinner medium in a shorter time," said Aristotle, and "will move through air faster than through water by so much as air is thinner and less corporeal than water." Guided by this principle, it seemed natural to conclude that bodies of unequal weight fall through air at different speeds. Moreover, in the absence of all resistance, a body would move from place to place in no time at all, that is, at infinite speed.

Hence, argued the Aristotelians, a vacuum cannot exist in nature because a body in a vacuum would experience no resistance and every force would result in an infinite speed. Thus everywhere space was necessarily occupied by either solid, liquid, or gaseous substances that bestowed on space a substantial reality and moderated the motions of moving bodies. The Atomists (who later were the Epicureans) claimed that atoms move freely through the void at finite speed. If we grant that atoms exist, countered Aristotle, "what can be the cause of their motion?" The Atomists say there is always movement, "but why and what this movement is they do not say, nor, if the world moves in this way or that, do they tell us the cause of the motion." René Descartes later shared similar views and declared, "a vacuum is repugnant to reason." Aristotle's common-sense law of motion endured until finally eclipsed by the Newtonian law

$$\text{applied force} = \text{mass} \times \text{acceleration}, \quad [3.2]$$

that came two thousand years later.

Impetus (momentum)

We must not suppose the giant step from Aristotle to Newton was the result of the genius of Galileo alone. Impetus – now known as momentum – was discussed by Philoponus in the sixth century and can be traced back to Hipparchus in the second century BC. Scholars in the High and Late Middle Ages, such as Robert Grosseteste and Roger Bacon, developed the now familiar definitions that speed is the distance traveled in an interval of time, and acceleration is the change of speed in an interval of time. Bodies of unequal weight, they said, fall with equal acceleration. They used diagrams and graphs to illustrate the nature of motion.

In the fourteenth century, scholars of Merton College, Oxford, such as William Heytesbury, showed that if V represents speed at any moment, and a represents a constant positive acceleration, then in time interval t the speed attained is

$$V = at, \qquad [3.3]$$

and the distance S traveled is

$$S = \tfrac{1}{2}at^2. \qquad [3.4]$$

Their calculations by graphical methods anticipated the discovery of calculus by Newton and Gottfried Leibniz three hundred years later. William Ockham, also of Merton College, argued that forces can act at a distance without any need for direct contact between bodies.

Jean Buridan, a French philosopher who studied under Ockham and later taught Nicholas Oresme, conjectured that impetus was proportional to the speed of a body and to its quantity of matter. (We now refer to "quantity of matter" as mass and express impetus, or momentum, as mV, where m represents mass.) The planets, said Buridan, are not continually pushed along in their orbits, but move of their own accord because of their impetus. The impetus of a thrown stone keeps it moving and air resistance causes it to lose impetus slowly. Lacking the idea of impetus, Aristotelians had supposed that moving bodies were continually pushed. They thought the air displaced at the front of a moving body flowed to the rear and kept the body in motion by pushing from behind. Buridan and his successors argued that bodies falling vertically gain equal amounts of impetus in equal intervals of time; thus

Figure 3.1. Simon Stevinus (1548–1620) dropped bodies of unequal weight simultaneously and showed they reach the ground at about the same time. The following argument shows that this result is to be expected: A body of weight W is divided into two equal bodies of weight $\tfrac{1}{2}W$ and are separately dropped. Do the lesser weights $\tfrac{1}{2}W$ fall slower than the original weight W? If you think so, tie them together with a hair, and drop them again. How can the hair cause them to fall faster?

bodies of constant mass, falling freely, accelerate at a constant rate.

In the fifteenth century, Leonardo da Vinci, a famed Italian artist, scientist, and engineer wrote on subjects such as the rotation of the Earth, the origin and antiquity of fossils, and the impossibility of perpetual motion. He promoted the idea that bodies fall with an acceleration that is constant and independent of weight.

In the late sixteenth century, Simon Stevinus, a Dutch–Belgian scientist, performed the experiment later attributed to Galileo. He dropped bodies of unequal weight from a high building and found they reached the ground at approximately the same instant (see Figure 3.1).

In the early seventeenth century, Pierre Gassendi, a French philosopher, dropped stones from the top of the mast of a ship and showed that the stones always landed at the foot of the mast, even when the ship is in motion. Ptolemy had said that if the Earth rotated, a person jumping up from the ground would not return to the same place on the ground (see Figure 3.2). This meant a person jumping up from the deck of a moving ship would not return to the

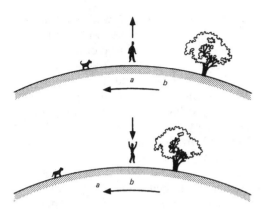

Figure 3.2. Ptolemy's "proof" that the Earth does not move or rotate. If a person on the Earth's surface at point *a* jumps up, and the Earth's surface moves, the person will fall back at point *b*. But observation shows that the person always falls back at the original point *a*. This proves, argued Ptolemy, that the Earth is stationary and hence the heavens must revolve around the Earth.

Figure 3.3. Pierre Gassendi (1592–1655) of France dropped stones from the mast of a moving ship. He found experimentally that the stones fall to the foot of the mast when the ship is in uniform motion, as when the ship is stationary, and thus showed that Ptolemy was wrong. Newton later said: "The motions of bodies enclosed in a given space are the same relatively to each other whether that space is at rest or moving uniformly in a straight line without circular motion."

same place on the deck. Gassendi's experiments showed that Ptolemy was wrong (see Figure 3.3).

Jeremiah Horrocks, a clergyman who died at age 22, showed that the Moon moves in a Keplerian elliptical orbit about the Earth. He argued that disturbances in the Moon's motion were due to the influence of the distant Sun, and anticipated the idea of universal gravity and suggested the planets perturb one another's orbits.

Galileo Galilei, in his lectures and publications, brought together (without acknowledgment) medieval discoveries concerning space, time, and motion. He showed that impetus is conserved in freely moving bodies, and with balls rolling down inclined planes showed that free-falling bodies in the Earth's gravity have constant acceleration. He demonstrated that the period of a pendulum depends on its length and not the weight of its bob. To measure time intervals he used either his pulse or the flow of a jet of water.

Galileo enthusiastically championed the Copernican heliocentric system, but ignored Kepler's work, and did not theorize about planetary motions. He failed to realize that a body following a circular orbit has constant acceleration toward the center of the

orbit, and thought a planet following a heliocentric circular orbit moved naturally in this manner without a force pulling it toward the Sun. Giovanni Borelli, an Italian contemporary, considered Kepler's work on planetary motions in the light of Galileo's developments and speculated on the nature of the Sun's gravity.

THE CARTESIAN WORLD SYSTEM

"Give me matter and motion and I will construct the universe."
René Descartes, Discourse on the Method *(1637)*

Leaders of the Protestant Reformation were at first hostile to Copernicus ("this fool," said Martin Luther, "wishes to reverse the entire history of astronomy"), but later relented, and science found sanctuary in Reformation countries from the rising hostility of the Counter Reformation. While Rome continued to cling to geocentrism for a further 200 years, the northwestern countries of Europe discarded geocentrism and soon abandoned even heliocentrism.

Then, in the seventeenth century, two immensely important world systems emerged that mathematized and mechanized the universe.

René Descartes (1596–1650)

The far-reaching thoughts of René Descartes, renowned French philosopher and mathematician, influenced all branches of learning. Guided by the Aristotelian method of "rightly conducting the reason" and the Platonic principle that "what is reasonable must be true," he created a grand and sweeping system of natural philosophy more innovative than any since antiquity. Its most important features were the mathematization of the physical sciences and the clear dichotomy of body and mind. He clarified ideas on the nature of space, time, and motion, and enunciated laws of motion that resembled in some respects the Newtonian laws of a few decades later.

Only God can be infinite, wrote Descartes in his *Principles of Philosophy* while living in the safety of Protestant Holland, and he would therefore refer to the spatial extent of material things as "indefinite rather than infinite in order to reserve to God alone the name of infinite." In the Aristotelian system, space had ended at the sphere of fixed stars; in the Cartesian system, developed by Descartes, it extended indefinitely and was strewn throughout with stars and their planets. In the Aristotelian system, matter and ether suffused every part of finite space; in the mechanistic Cartesian system, continuous matter of varying density suffused every part of indefinitely extended space.

Descartes's mechanized world of matter and motion obeyed natural laws: "we may well believe, without doing outrage to the miracle of creation, that by this means alone all things that are purely material might in course of time have become such as we observe them to be at present." All forces, he believed, acted by direct contact, that is, between bodies touching one another. Forces acting at a distance without direct material contact were contrary to reason, hence impossible, and reminiscent of the etheric forces of medieval mysticism. Real forces acting as pressures and tensions, which pushed and pulled, controlled all motions. In the Cartesian world system, planetary orbits resulted from the vortical motions of interplanetary matter (Figure 3.4), and gravity was the pressure exerted by swirling fluids.

Reason persuades us, said Descartes, that space by itself, being nothing, has no extension. How can space, which is empty and nothing, have length, breadth, and height? Only matter has the property of extension, and space does not exist where there is no matter. Matter of all kinds at all densities exists throughout all space, and the vacuum, "repugnant to reason," said Descartes, exists nowhere. The nonexistence of undressed space (the vacuum) was a cardinal concept in Cartesian science. From this concept sprang Descartes's firm belief in the impossibility of the atomist theory. The principle of action by direct contact required the absence of voids, and required that matter be continuous and infinitely divisible. Atoms by their nature, if they exist, would be separated by voids (how else could they be atoms?), and because voids are repugnant to reason, atoms are physically impossible.

In 1651, a year after the death of Descartes, Thomas Hobbes wrote in his book *Leviathan*, "for what is the heart but a spring, and the nerves but so many strings, and the joints but so many wheels giving motion to the whole body." Hobbes was one of the first to make clear the stunning implications of the Cartesian duality of body and mind. The human body with its physical brain was a part of the clockwork universe of matter and motion, and the mind was no more than a ghost haunting the machinery of body and brain. How can an immaterial mind control or influence a material body?

Thus began in clear-cut mechanistic terms the Cartesian duality of body and mind that to this day remains an unsolved problem (see Chapter 8).

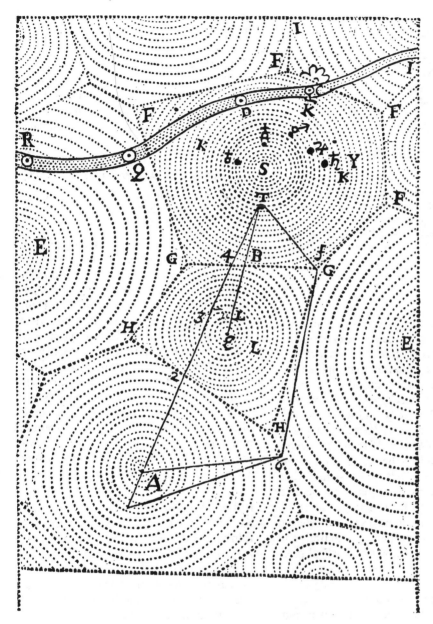

Figure 3.4. An illustration of the Cartesian system of vortical fluids and gyrating bodies reproduced from René Descartes's *The World* (1636). Each major vortex is a solar system in an endless expanse of solar systems. The centers (A, E, and S) of the vortices are stars made luminous by churning motions.

Christiaan Huygens (1629–1695)

A Dutch physicist and astronomer, a pillar of Cartesian science and philosophy, Christiaan Huygens improved the design of telescopes (he discovered the rings encircling Saturn) and is famed for the development of the pendulum clock. Hitherto, mechanical clocks, as used on church towers, had been no more accurate than candle-clocks and water clocks of

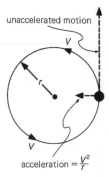

unaccelerated motion

V

r

V

acceleration $= \dfrac{V^2}{r}$

Figure 3.5. A body moves at constant speed V in a circular orbit of radius r. Its speed is constant but its velocity (which has also direction) continually changes, and the acceleration (rate of change of velocity) is V^2/r, directed toward the center. The body obviously accelerates toward the center, otherwise it would move away in a straight line.

the ancient world. Huygens's pendulum-regulated clocks made scientific experiments and marine navigation much more precise.

Descartes had realized what Galileo had failed to understand: motion not in a straight line is accelerated motion (see Figure 3.5). Huygens showed that in circular motion there is an acceleration of V^2/r toward the center of the circle, where V is the speed and r is the radius of the circle. A stone at the end of a length of string, when whirled around, is continually accelerated toward the hand by the string pulling on the stone. Huygens, like Descartes, believed that space was a property of matter and gravity was a force caused by vortical pressure.

Birth of the Age of Reason

The Cartesians believed in a universe of indefinite extent in which all things were pushed and pulled by forces acting in direct contact. Everything behaved in accordance with reason. Despite strong opposition from clerics steeped in biblical scripture and academics imprisoned in ancient doctrine, the liberating and exhilarating Cartesian philosophy spread rapidly, capturing the imagination of freethinkers everywhere. With the invention of the telescope,

microscope, thermometer, barometer, and the pendulum clock, and with Descartes doubting all except the irreducible, "I think, therefore I am," the Cartesian system signposted the way to the Age of Reason (the Enlightenment) of the eighteenth century, and triggered the explosion of thought in England that created the Newtonian system.

THE NEWTONIAN WORLD SYSTEM

"I do not define time, space, place and motion as being well known to all."
Isaac Newton, Principia *(1687)*

At first, English liberal theologians and philosophers viewed Descartes as a savior from medieval mysticism. But soon his philosophy received more criticism than praise. The Cambridge theologian Henry Moore, initially impressed with the vision of a universe of natural laws, in later years, aghast at the implications of Cartesian materialism, returned to the medieval idea that space exists without matter by virtue of the presence of ubiquitous spirit. Moore favored the idea, shared by many colleagues, of a finite Stoic cosmos of stars surrounded by an infinite mysterious void. This view of the universe (see Figures 3.6 and 3.7) was also shared by Isaac Newton.

Robert Hooke (1635–1703)

Robert Hooke, Christopher Wren, and Edmund Halley outlined qualitatively what Newton later explained quantitatively. A freely moving body, as explained by Descartes, travels at constant speed in a straight line when nothing forces it from that natural state. Because the planets move not in straight lines but in curved orbits about the Sun, they must be continually pulled by the Sun's gravitational force. Hooke demonstrated the idea with a conical pendulum (Figure 3.8) and said planetary motions can be understood by mechanical principles. At about the time when Newton was silently pondering these matters, Hooke, a scientific genius ("perhaps the most inventive man who ever lived," writes

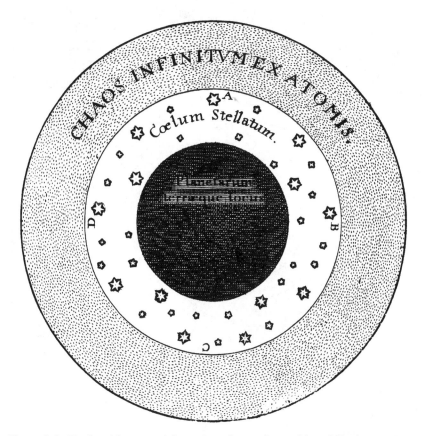

Figure 3.6. The "world system of the ancients," according to Edward Sherburne (1675). This illustration combines Stoic and Epicurean elements and plausibly represents Newton's initial view of the universe when he wrote *De Gravitatione* sometime between 1666 and 1668.

Edward Andrade), realized that the force controlling the Solar System, drawing the planets to the Sun and the Moon to the Earth, is the same as that which causes apples to fall from trees. "I shall explain," wrote Hooke, "a System of the World differing in many particulars from any yet known, answering in all things to the common rules of mechanical motions ... that all celestial bodies whatsoever have an attraction or gravitating power to their own centers, whereby they attract not only their own parts, and keep them from flying from them, as we may observe the Earth to do, but that they do also attract all other celestial bodies that are within the sphere of their activity." Gravity that in

Pythagoras's day made the sphericity of the Earth plausible (people on the other side could not fall off) had become a universal force in control of the heavens.

Isaac Newton (1642–1726)

Isaac Newton, the most illustrious of all scientists, gathered together the thoughts of many thinkers since the Middle Ages. He developed the dynamic theories of motion and universal gravity, and constructed a system that attained the power and elegance to which science had aspired from the beginning.

During his early years at Cambridge, in response to Descartes's *Principles of Philosophy*, Newton wrote in an unpublished

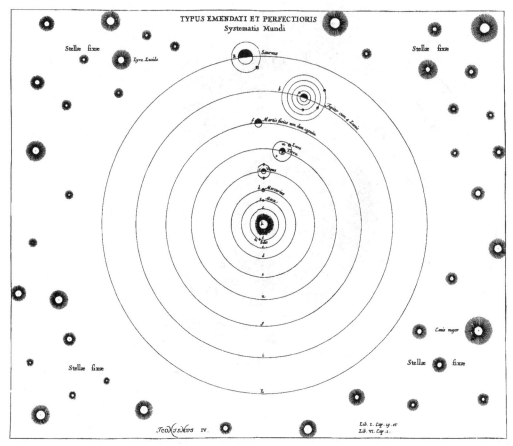

Figure 3.7. The system of the world according to Otto von Guericke in his *New Magdeburg Experiments on Void Space* (1672). Guericke, mayor of Magdeburg for 30 years, disagreed with Descartes and performed experiments demonstrating the properties of the vacuum. He believed in a finite starry cosmos surrounded by an infinite void, as in the Stoic system. He thought the sky is dark at night because we look between the stars and see a starless void beyond (Chapter 24).

manuscript, referred to by its opening words *De Gravitatione*), that an "infinite and eternal divine power" occupies all space and "extends infinitely in all directions." Descartes claimed that where there is no matter, there is no space; on the contrary, said Newton, where there is no matter, spirit alone endows space with extension. To say that space cannot exist where there is no matter, denies the presence of spirit, and hence the presence of God in the universe. Newton's ideas on the nature of space changed very little in his lifetime. Descartes claimed that matter extends indefinitely; on the contrary, said Newton, in infinite

space, God had created a material system of finite extent. Newton's Stoic picture of a finite cosmos of stars (Figure 3.6) changed abruptly 25 years later in response to questions by the theologian Richard Bentley.

Newton was appointed Lucasian Professor of Mathematics at Cambridge University at age 27, and resigned from this position 32 years later in 1701 after becoming master of the mint. Like Descartes, he remained single all his life. Most biographers have ignored Newton's interests in nonscientific subjects and have failed to mention that he spent many years engrossed in

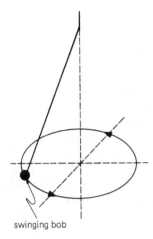

swinging bob

Figure 3.8. Robert Hooke said a conical pendulum illustrates the motion of a planet about the Sun. The bob of a conical pendulum follows an elliptical path. But, as shown by Newton, in a planetary system, the Sun is not at the center but at one of the two foci of the elliptical orbit.

alchemical pursuits and absorbed in scriptural studies.

Mathematical Principles of Natural Philosophy

Hooke's brilliant mechanistic vision foresaw the rise of the Newtonian mathematical universe. The astounding genius of Newton, meditating for many years on the natural philosophy of space, time, and motion, transformed all previous graphical descriptions into mathematical prescriptions. In his *Mathematical Principles of Natural Philosophy*, known as *Principia*, Newton said of space: "Absolute space, in its own nature, without relation to anything external, remains always similar and immovable." Of time, he said: "Absolute, true, and mathematical time, of itself, and from its own nature, flows equably without relation to anything external." We understand Newton's idea of absolute space, but not his idea of absolute time; what does the flow of time mean? (Chapter 9).

Newton's celebrated three laws of motion state:

1. A body continues in a state of rest, or of constant motion in a straight line, unless compelled to change that state by an applied force. We must note that velocity has magnitude (speed) and direction, and the momentum (mass × velocity) of a body is constant in the absence of an applied force.
2. The rate momentum changes in time equals the applied force and is in the direction of the force. If mass is constant, the rate of change of momentum equals mass × rate of change of velocity, and the law of motion becomes

applied force = mass × acceleration,

(Equation [3.2]), and acceleration is in the direction of the force.

3. To every force there exists at the same place and time an equal and opposite force. This important law states that the sum of all forces at any point is zero. For example, the weight of a person is balanced by an equal and opposite force in the ground pushing upward. The third law creates the concept of inertial force. Notice that the equation of motion can be written in the form

applied force − (mass × acceleration)

$$= 0. \qquad [3.5]$$

The first term on the left is the applied force and the second term is the negative of the inertial force, such that

applied force + inertial force = 0. [3.6]

Thus the sum of the forces acting on a freely moving body is zero.

The inertial force is the force experienced during acceleration and is opposite in direction to the acceleration. In circular motion, the acceleration of a body is toward the center, whereas the inertial force (in this case the centrifugal force) is directed away from the center. A stone whirled around at the end of a length of string has a centrifugal force that pulls outward and is equal and opposite to the tension in the string that pulls inward (Figure 3.9).

The third law explains why in a spaceship orbiting the Earth an astronaut is weightless

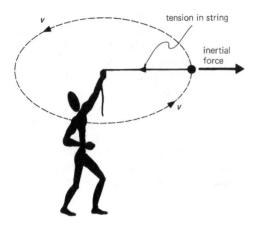

Figure 3.9. A body in circular motion is accelerated toward the center. The inertial force, which in this case is the centrifugal force, is in the opposite direction, and equals the mass of the body multiplied by its acceleration.

and does not feel the pull of the Earth's gravity: The applied force is gravity, and the inertial force caused by the motion of the spaceship exactly cancels gravity. The remarkable thing about the Newtonian laws is that a freely moving body follows a trajectory on which the inertial force cancels gravity and the body experiences no force at all. The Sun pulls on the Earth, and because of the Earth's orbital motion about the Sun, we on Earth cannot feel the Sun's pull. The third law is the principle of equivalence (Chapter 12) – of vital importance in the development of the theory of general relativity – according to which freely moving bodies follow trajectories that abolish gravity. Thus a person in free fall, such as an astronaut in a spaceship, experiences no gravitational force. But beware of tidal forces! (See Chapter 12.)

Relative and absolute motion

In the Newtonian world, motion is both relative and absolute. When the velocity of an automobile is constant, we feel no inertial force. Its constant velocity is measured relative to other moving vehicles or to things stationary at the side of the road. No matter what the relative velocity is, as long as it stays constant, no inertial force exists. A passenger with closed eyes and unable to hear cannot determine the automobile's velocity because the velocity is purely relative and has no absolute value. When the velocity changes, however, a force exists – the inertial force – that we feel during acceleration, and this force is not produced by motion relative to anything. Notice, in the Newtonian system, acceleration means change in velocity, either change in speed or in direction, or both. The passenger with closed eyes estimates the acceleration from the magnitude and direction of the inertial force that is experienced. When speed only changes, the inertial force is directed forward (if speed decreases) or backward (if speed increases). When only the direction changes, the inertial force is directed toward the left (when turning to the right) or toward the right (when turning to the left). The physical properties of Newtonian space are such that uniform velocities are relative, measured relative to one another, and changes in velocity (accelerations) are absolute, measured relative to nothing. The situation remains much the same in the modern world of relativity physics: uniform velocity is relative and acceleration is absolute.

Universal gravity

The gravitational attraction between any two bodies varies as the inverse square of their separating distance (Figure 3.10). The force pulling a planet to the Sun varies as the inverse square of the distance of the planet from the Sun. How was this discovered? Kepler's third law gave the clue (see p. 107 for Kepler's laws). Suppose that a planet moves about the Sun in a circular orbit of radius r at speed V. The acceleration toward the center of the orbit is V^2/r, as shown by Huygens; hence the inertial force directed away from the center is mass × acceleration, or mV^2/r, where m is the planet's mass. The gravitational pull of the Sun must be equal and opposite to the centrifugal force experienced by the planet:

$$\text{Sun's gravitational pull} = mV^2/r. \qquad [3.7]$$

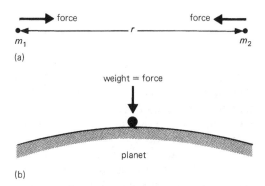

(a)

weight = force

planet

(b)

Figure 3.10. (a) Two bodies of mass m_1 and m_2, respectively, separated by a distance r, attract each other with a gravitational force Gm_1m_2/r^2, where G is the universal constant of gravity. (b) At the surface of a planet of mass M and radius R, a body of mass m has a weight $w = GMm/R^2 = mg$, where $g = GM/R^2$ is the acceleration produced by gravity at the Earth's surface. Weight on a planetary surface is equal to the gravitational attraction.

The circumference of the orbit is $2\pi r$, and because the period P is the time to revolve once about the center, we have $P = 2\pi r/V$. Kepler's third law for all planets states that P^2 is proportional to r^3. We have just seen that P varies as r/V, and therefore, according to Kepler's law, $(r/V)^2$ is proportional to r^3, and hence V^2 is proportional to $1/r$. With this result and Equation [3.7] we see that

$$\text{Sun's gravitational pull is proportional to } 1/r^2, \qquad [3.8]$$

and varies as the inverse square of distance. This important result was first derived by Robert Hooke.

Newton showed that a spherical body exerts a gravitational attraction as if all its mass were concentrated at the center of the body. He also showed that the natural orbits of planets are ellipses, and the orbits of all bodies freely moving in the Sun's gravitational field are either ellipses, parabolas, or hyperbolas. In Newton's *System of the World*, all bodies in the universe attract one another with gravitational forces proportional to their masses and the inverse square of their separating distances.

Isaac Newton, as professor of mathematics at Cambridge University, gave eight lectures a year, which few students attended. His great work in three volumes, the *Mathematical Principles of Natural Philosophy* (written in Latin and often referred to as the *Principia*), was written in less than two years and published in 1687 at Halley's encouragement and sold for seven shillings a set. With a few definitions and axioms, and an array of propositions, Newton proceeded to explain mathematically the twice-daily tides on Earth caused by the Sun and Moon, the flattening of the Earth at the poles owing to its daily rotation, the precession of the axis of the Earth's rotation once every 26 000 years due to the equatorial bulge of the Earth, the perturbations of the Moon's orbit, and the paths of the planets and comets (Figure 3.11). He, and Gottfried Leibniz, a German mathematician and philosopher, independently developed the mathematical principles of calculus.

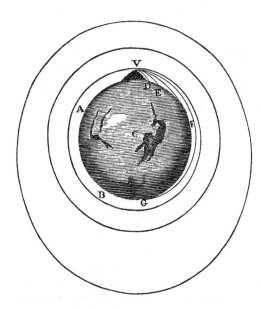

Figure 3.11. "The Stone that put the Stars to Flight" (*The Rubáiyát*). Newton wrote: "For a stone ... the greater the velocity with which it is projected, the farther it goes before if falls to Earth ... till at last, exceeding the limits of the Earth it will pass into space." Illustration and quotation from *The System of the World* by Isaac Newton.

NEWTON AND THE INFINITE UNIVERSE

The Bentley correspondence

Robert Boyle in his will left an endowment to support an annual lectureship to combat the atheism widely professed by wits in coffeehouses and taverns. In 1692, the young Richard Bentley, an erudite clergyman, was selected to give the first series of lectures. In his last two lectures, entitled *A Confutation of Atheism from the Origin and Frame of the World*, he showed how the marvels of the Newtonian system gave indisputable proof of the existence of a divine power. He argued that the laws of nature by themselves were insufficient to explain the wonders of the natural world and must be supplemented by acts of a divine power. Before publishing his lectures, Bentley took the precaution of consulting Newton on several technical points. His deep and disturbing questions jolted Newton into rethinking his cosmological ideas, and Newton's four letters to Bentley rank among the most important documents in the history of cosmology.

First letter (10 December 1692)

In the first letter, Newton responded to Bentley's query concerning the effect of gravity in a finite system of stars, and expressed the opinion that a universe composed of self-gravitating matter is necessarily unbounded, otherwise all matter would "fall down to the middle of the whole space and there compose one great spherical mass ... But if the matter was evenly diffused through an infinite space, it would never convene into one mass but some of it into one mass and some into another so as to make an infinite number of great masses scattered at great distances from one to another throughout all of infinite space. And thus might the Sun and fixt stars be formed." Newton thus abandoned the Stoic universe in favor of an Epicurean-like universe.

Second letter (17 January 1693)

"You argue," said Newton in his second letter, "that every particle of matter in an infinite space has an infinite quantity of matter on all sides and by consequence an infinite attraction everyway and therefore must rest in equilibrio because all infinities are equal." Newton had fully agreed with Bentley that gravity meant providence had created a universe of great precision. "And much harder it is to believe that all the particles in an infinite space should be so accurately poised one among another. For I reckon this as hard as to make not one needle only but an infinite number of them (so many as there are particles in an infinite space) stand accurately poised upon their points. Yet I grant it possible, at least by a divine power; and if they were once so placed I agree with you that they would continue in that position without motion forever, unless put into motion by the same power. When therefore I said that matter evenly spread through all spaces would convene by its gravity into one or more great masses, I understand it of matter not resting in accurate poise ... So then gravity may put the planets into motion but without the divine power it could never put them into such a circulating motion as they have about the Sun, and therefore for this as well as other reasons I am compelled to ascribe the frame of the system to an intelligent agent." (See Figure 3.12 and Chapter 16.)

Third letter (11 February 1693)

"The hypothesis of deriving the frame of the world by mechanical principles from matter evenly spread through the heavens being inconsistent with my system, I had considered it very little before your letters put me upon it, and therefore trouble you with a line or two more about it ..." Newton elaborated earlier arguments that a divine power was essential in the design of the initial conditions.

Fourth letter (25 February 1693)

Newton again assured Bentley that his mechanistic system of the world did not dispense with the necessity of a divine power to maintain it: "this frame of things

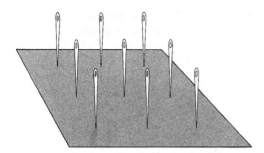

Figure 3.12. Newton agreed with Bentley that stars cannot form a finite and bounded system (as in the Stoic cosmos), for they would fall into the middle of such a system by reason of their gravitational attraction. They agreed that matter was uniformly distributed throughout infinite space, and realized that this was an unstable distribution. The particles of matter, wrote Newton, are like an array of needles standing upright on their points ready to fall one way or another, and "thus might the Sun and fixed stars be formed."

could not always subsist without a divine power to conserve it."

Universal gravity

The Newtonian theory of universal gravity, in which all bodies attract one another, reinforced the growing belief that the universe must be edgeless and therefore, according to Euclidean geometry, infinite. For a finite universe bounded by a cosmic edge would have a center of gravity, and the attraction between its parts, said Newton, would cause them to "fall down into the middle of the whole space, and there compose one great spherical mass." This argument caused him to abandon a finite Stoic cosmos in favor of an infinite Epicurean universe. In an infinite, uniform universe, no preferred direction exists in which gravity can pull and make matter fall into a single "middle." In the second edition of the *Principia*, published after the Bentley letters, Newton wrote, "The fixed stars, being equally spread out in all points of the heavens, cancel out their mutual pulls by opposite attractions." Each particle of matter, pulled equally in all directions, remains in equilibrium. (But an unstable equilibrium as Bentley pointed out.) The theory of universal gravity

supported the belief that the universe is static on the cosmic scale, and initiated the idea that on smaller scales gravity caused matter to condense and form astronomical bodies such as stars and planets.

THE ATOMIC THEORY

Atomism, an inspired theory, did not enter the mainstream of science until the seventeenth century. Pierre Gassendi, a French natural philosopher of that century, revived atomist theory despite lingering Aristotelian objections and its association with Epicurean atheism. The theory played a prominent role in the thinking of men like Robert Boyle, Otto von Guericke (mayor of Magdeburg), and Newton. In his book *Opticks*, Newton wrote: "It seems probable to me that God in the beginning formed matter in solid, massy, hard, impenetrable particles, of such size and figures, and with such other properties, in such proportions to space, as most conduced to the end for which he formed them; even so very hard as never to wear or break in pieces." Newton interwove the old atheistic atomic philosophy into the contemporary religious doctrine and triumphed by compromise. In prophetic words, he wrote: "There are therefore agents in nature able to make the particles of bodies stick together by very strong attractions. And it is the business of experimental philosophy to find them out." We now have high-energy particle accelerators for this purpose.

REFLECTIONS

1 *Three world systems survived from classical antiquity and formed the bases of European natural philosophy in the sixteenth and seventeenth centuries:*

The Aristotelian system of geocentric celestial spheres consisted of planetary orbits enclosed within a sphere of fixed stars. Medieval additions populated the celestial spheres with angelic creatures and surrounded the sphere of fixed stars with the empyrean where God dwelt. This was the orthodox Judaic–Christian–Islamic cosmology.

The Stoic system *consisted of a finite cosmos of stars in an infinite mysterious void where God dwelt. With the Epicurean addition that matter consists of atoms, this was the intellectual view, particularly in England, before the Newton–Bentley correspondence.*

The Epicurean system *consisted of an infinite void occupied by an infinity of uniformly distributed worlds composed of atoms and regulated by natural laws. Descartes modified this system by suffusing it with matter, and by denying the existence of atoms. After his correspondence with Bentley, Newton adopted this system by suffusing it with spirit and retaining its atomicity.*

2 *"Whirl is king" declared the Athenian playwright Aristophanes in the fourth century* BC. *Whirlwinds and whirlpools performed dramatic roles in mythology, and vortex scenarios dominated early science. Many philosopher-scientists thought the planets and stars were formed in a primordial vortex. Swirling fluids in large and small vortices dominated the Cartesian system. The Cartesian vortex theory was eventually abandoned after Newton showed that planetary orbits are explained by gravity and the laws of motion. The idea of swirling matter was later developed by Immanuel Kant and Pierre Simon de Laplace into the solar nebula hypothesis (Chapter 4). To this day whirl is king in our cosmogonic theories of star and galaxy formation.*

3 *"I don't say that matter and space are the same thing, I only say, there is no space where there is no matter; and that space in itself is not an absolute reality." Written by Gottfried Leibniz (1646–1716) in a letter to Samuel Clarke, who argued in defense of Newton's ideas and the reality of an absolute space that is independent of matter. Leibniz shared the Cartesian belief of many Continental philosophers that empty space is meaningless. He also shared the Cartesian belief that forces could not act at a distance unless conveyed by a material medium.*

4 *The universal gravitational constant (nowadays denoted by G) does not appear*

in Newton's Principia, *nor can I find it in any work during the next hundred years. In effect, G was replaced by* $(2\pi)^2/M_\odot$, *thus making* $GM_\odot/(2\pi)^2 = 1$ *in a system of units in which masses are measured in solar masses* $(M_\odot)$, *distances are measured in astronomical units (an astronomical unit is the distance from the Sun to the Earth), and time intervals are measured in years. Kepler's third law, for example, in these units reads* $P^2 = r^3$, *where P is the period and r the radius of an orbit.*

5 *In Book I of* De Rerum Natura, *Lucretius argued that space is infinite and unbounded: "If all the space in the whole universe were closed in on all sides with fixed boundaries, then all matter, because of its weight, would have flowed together from all sides and sunk to the bottom. Nothing could be carried on beneath the canopy of the sky; indeed, there could be no sky, nor light from the Sun, for all matter would be idle, piled together over limitless time. But no such rest has been granted the universe because all things have ceaseless movement and no bottom exists where matter can flow from all sides and settle" (translated from* T. Lucreti Cari: De Rerum Natura). *Lucretius's argument anticipated Newton's reasons for believing in infinite space. By drawing on the idea of "ceaseless movement," the Epicurean poet realized vaguely what Newton and Bentley omitted to discuss: the possibility of a finite system of moving stars in a state of stable equilibrium (as in a galaxy). They agreed that stars stretch away endlessly (as in the Epicurean universe), and believed, on the other hand, that if the material system were finite (as in the Stoic universe) the stars would fall into the middle of the system. But a static Stoic cosmos can exist in a state of equilibrium. Stars in a galaxy do not fall into the middle but move freely around the middle in various orbits. A cluster of stars in equilibrium obeys what is known as the virial theorem. According to this theorem, if V is the typical speed of the stars in a spherical cluster of radius R and mass M, then* $V = \sqrt{(aGM/R)}$, *where a is a numerical coefficient in the neighborhood of unity that* depends on the way mass is distributed in the cluster. (For a cluster of uniform density $a = 3/5$.) Stars move in various orbits about the center of mass of the cluster and need not "fall into the middle," as Newton supposed. The Victorian one-island universe (Chapter 4) – a single giant Galaxy in an infinite vacuum – obeyed the virial theorem.

6 In 1726, Voltaire fled from France to England for three years to escape his enemies (the victims of his uncompromising wit). His published "Letters from London on the English" contain comments contrasting the Cartesian and Newtonian systems. "A Frenchman coming to London," wrote Voltaire, "finds matters considerably changed, in philosophy as in everything else. He left the world filled, he finds it here empty. In Paris you see the universe consisting of vortices of a subtle matter; in London nothing is seen of this. With us it is the pressure of the Moon that causes the tides of the sea; with the English it is the sea that gravitates toward the Moon . . . . Moreover, you may perceive that the Sun, which in France is not at all involved in the affair, here has to contribute by nearly one quarter. With your Cartesians everything takes place through pressure, which is not easily comprehensible; with Monsieur Newton it takes place through attraction, the cause of which is not better known either."

7 Determining the distances to stars is not easy. Christiaan Huygens, a Dutch astronomer, in the late seventeenth century used a crude photometric method. He assumed that all stars are similar to the Sun. By observing the Sun in a dark room through a small hole in a screen covering the window, and by adjusting the size of the hole until the brightness of the hole looked like Sirius at night, he estimated that Sirius was at the distance 30 000 astronomical units, or roughly 0.5 light years. This method depends on judging in daytime how bright Sirius is at night. The young Scottish astronomer James Gregory, unknown to Huygens, had some years previously proposed a photometric method that did not depend on memory. Gregory assumed that the nearby bright stars are sunlike and

compared their brilliance with that of the outer planets Mars, Jupiter, and Saturn. Knowing the sizes of the planets and their distances from the Sun and Earth, and making allowance for the imperfect reflection of sunlight from their surfaces, he estimated the distance of the nearest stars as several hundred astronomical units. Isaac Newton referred to Gregory's method in his System of the World, and in unpublished work placed the brightest stars at 500 000 astronomical units, or roughly 8 light years. This estimate was a remarkable anticipation of more recent measurements (Table 5.5).

8 "This most beautiful System of the Sun, Planets, and Comets could only proceed from the counsel and dominion of an intelligent and powerful being. And if the fixed Stars are the centers of other like systems, these being form'd by the like wise counsel, must be all subject to the dominion of One . . ." (Isaac Newton, Principia, 2nd edition, translated by Andrew Motte, pages 389–390).

"Whence is it that Nature does nothing in vain and whence arises all the order and beauty in the world" (Isaac Newton, Opticks).

"Mortals! Rejoice at so great an ornament to the human race!" Words inscribed on Newton's tomb in Westminster Abbey.

PROJECTS

1 The "teacup effect" illustrates Descartes's idea of gravity. Stir water in a teacup that has a few tea leaves floating on the surface. Notice that the leaves tend to concentrate in the center as if attracted by gravity.

2 Show that the Cartesian system contained Aristotelian and Epicurean elements, and the Newtonian system, before the Bentley correspondence, contained Stoic and Epicurean elements.

3 Evangelista Torricelli, Galileo's companion during the last months of his life, in 1643 succeeded with surprising ease in creating a vacuum in a glass tube above a column of mercury of height one meter. He poured mercury into a glass tube that was longer

than one meter and sealed at one end. With his finger over the open end, he inverted the tube and dipped the open end into a bowl of mercury. Torricelli performed various tests and concluded that a vacuum exists in the tube above the mercury column. He noticed how the height of the mercury column varied from day to day because of atmospheric changes, and concluded that the height of the mercury column measures the pressure of the atmosphere. Torricelli was thus the inventor of the barometer. Discuss the vacuum in the barometer. Is it a perfect vacuum? (Actually the average density of the universe is much less.) Why, in wells deeper than thirty feet, is the water pump at the bottom and not the top of the well?

4 Weight ($W = mg$) is the gravitational force experienced by a stationary body of mass m that in free fall would have an acceleration g at that point. At the Earth's surface $g = 9.8$ meters per second per second. What is the weight in dynes of 1 gram at the Earth's surface?

FURTHER READING

Hall, A. R. *From Galileo to Newton.* Dover Publications, New York, 1981.

Halley, E. "Of the infinity of the sphere of fix'd stars." *Philosophical Transactions* 31, 22 (1720–1721). "Of the number, order, and light of the fix'd stars." *Philosophical Transactions* 31, 24 (1720–1721). Reproduced in Harrison, E. R. *Darkness at Night: A Riddle of the Universe.*

Harrison, E. R. *Masks of the Universe.* Macmillan, New York, 1985.

Harrison, E. R. "Newton and the infinite universe." *Physics Today* (February 1986).

Johnson, F. R. "Thomas Digges and the infinity of the universe." Reproduced in *Theories of the Universe: From Babylonian Myth to Modern Science.* Editor M. K. Munitz. Free Press, Glencoe, Illinois, 1957.

Jones, F. F. *Ancients and Moderns: A Study of the Rise of the Scientific Movement in Seventeenth-Century England.* Dover Publications, New York, 1982.

Kemble, E. C. *Structure and Development: From Geometric Astronomy to the Mechanical*

Theory of Heat. M.I.T. Press, Cambridge, Massachusetts, 1966.

Koyré, A. *Newtonian Studies.* Chapman and Hall, London, 1965.

Munitz, M. K. Editor. *Theories of the Universe: From Babylonian Myth to Modern Science.* Free Press, Glencoe, Illinois, 1957.

Toulmin, S. and Goodfield, J. *The Fabric of the Heavens: The Development of Astronomy and Dynamics.* Harper and Row, New York, 1961.

Whiteside, D. T. "Before the *Principia*: The maturing of Newton's thoughts on dynamical astronomy." *Journal for the History of Astronomy* 1, 5 (1970).

Wren, C. "The Life of Sir Christopher Wren," in *Parentalia: Or Memoirs of the Family of Wrens.* Stephen Wren, London, 1750. Reprinted: Gregg Press, Farnborough, 1965.

SOURCES

Alexander, H. G. Editor. *The Leibnitz–Clarke Correspondence, With Extracts From Newton's "Principia" and "Opticks."* University of Manchester, Manchester, 1956.

Andrade, E. N. Da C. "Robert Hooke." *Proceedings of the Royal Society* 201A, 439 (1950).

Bell, A. E. *Christiaan Huygens and the Development of Science in the Seventeenth Century.* Arnold, London, 1947.

Crombie, A. C. *Robert Grosseteste and the Origin of Experimental Science 1180–1700.* Clarendon Press, Oxford, 1953.

Descartes, R. *Discourse on Method of Rightly Conducting the Reasoning and Seeking for Truth in the Sciences* (1637). Translated by P. J. Olscamp. University of Indianapolis, Indianapolis, 1965.

Descartes, R. *The Philosophical Works of Descartes.* 2 volumes. Translated by E. S. Haldane and G. R. T. Ross. Cambridge University Press, Cambridge, 1911.

Dijksterhuis, E. J. *The Mechanization of the World Picture.* Translated by C. Dikshoom. Clarendon Press, Oxford, 1961.

Dyce, A. *The Works of Richard Bentley.* Volume 3, *Sermons Preached at Boyle's Lectures.* Macpherson, London, 1838.

Galileo Galilei. *Dialogue Concerning the Two Chief World Systems – Ptolemaic and Copernican.* Translated by S. Drake. University of California Press, Berkeley, 1953.

Grant, E. "Medieval and seventeenth century conceptions of an infinite void space beyond the cosmos." *Isis* 60, 39 (1969).

Grant, E. *Physical Science in the Middle Ages.* Wiley, New York, 1971.

Grant, E. *Much Ado About Nothing: Theories of Space and Vacuum from the Middle Ages to the Scientific Revolution.* Cambridge University Press, New York, 1981.

Grant, E. *The Foundations of Modern Science in the Middle Ages.* Cambridge University Press, Cambridge, 1996.

Guericke, O. von. *Experimenta Nova Magdeburgica de Vacuo Spatio.* Amsterdam 1672. Reprinted: Zellers Verlagsbuchhandlung, Aalen, 1962.

Guerlac, H. and Jacob, M. C. "Bentley, Newton, and providence." *Journal of the History of Ideas* 30, 307 (1969).

Hall, A. R. *From Galileo to Newton 1630–1720.* Harper and Row, New York, 1963.

Harrison, E. R. *Darkness at Night: A Riddle of the Universe.* Harvard University Press, Cambridge, Massachusetts, 1987.

Hooke, R. *The Posthumous Works of Robert Hooke, Containing his Cutlerian Lectures and Other Discourses.* Edited by R. Waller 1705. Revised by R. S. Westfall. Johnson Reprint Corporation, New York, 1969.

Huygens, C. *Cosmotheros* and *The Celestial Worlds Discover'd.* Reprinted: Frank Cass, London, 1968.

Lindberg, D. C. Editor. *Science in the Middle Ages.* University of Chicago Press, Chicago, 1978.

Lucretius. *Titi Lucruti Cari: De Rerum Natura.* With translation by V. Bailey. Clarendon Press, Oxford, 1947.

Lucretius. *The Nature of the Universe.* Translated in prose by R. E. Latham. Penguin Books, Harmondsworth, Middlesex, 1951.

Middleton, W. E. K. *The History of the Barometer.* Johns Hopkins University Press, Baltimore, 1964.

Miller, P. "Bentley and Newton," in *Isaac Newton's Papers & Letters on Natural Philosophy and Related Documents.* Edited by I. B. Cohen. Harvard University Press, Cambridge, Massachusetts, 1978.

Newton, I. "De Gravitatione," in *Unpublished Scientific Papers of Isaac Newton.* Edited by A. R. Hall and M. B. Hall. Cambridge University Press, New York, 1962.

Newton, I. *Sir Isaac Newton's Mathematical Principles of Natural Philosophy and his System of the World (1729).* Translated by A. Motte. Revised by F. Cajori. University of California Press, Berkeley, 1960.

Newton, I. *Opticks: A Treatise of the Reflections, Refractions, Inflections and Colours of Light.* 4th edition 1730. Dover Publications, 1952.

Newton, I. *A Treatise of the System of the World.* Translated by I. B. Cohen. Dawsons, London, 1969.

Pancheri, L. V. "Pierre Gassendi, a forgotten but important man in the history of physics." *American Journal of Physics* 46, 455 (1978).

Pedersen, O. and Pihl, M. *Early Physics and Astronomy: A Historical Introduction.* American Elsevier, New York, 1974.

Scott, J. F. *The Scientific Work of René Descartes.* Taylor and Francis, London, 1952.

Shapiro, H. *Motion, Time, and Place According to William Ockham.* Franciscan Institute, New York, 1957.

Tillyard, E. M. *The Elizabethan World Picture.* Macmillan, New York, 1944.

Van Helden, A. "The invention of the telescope." *Transactions of the American Philosophical Society* 64, part 4 (1977).

Van Helden, A. *Measuring the Universe: Cosmic Dimensions from Aristarchus to Halley.* Chicago University Press, Chicago, 1985.

Voltaire. *Letters Concerning the English Nation.* Davis and Lyon, London, 1733.

Vrooman, J. R. *René Descartes: A Biography.* Putnam's Sons, New York, 1970.

Webster, C. "Henry Moore and Descartes: Some new sources." *British Journal for the History of Science* 4, 359 (1969).

Westfall, R. S. *Force in Newton's Physics: The Science of Dynamics in the Seventeenth Century.* Elsevier, New York, 1971.

Westfall, R. S. *Never at Rest: A Biography of Isaac Newton.* Cambridge University Press, New York, 1980.

Wilson, C. A. *William Heytesbury.* University of Wisconsin Press, Madison, 1956.

Wolf, A. *A History of Science, Technology, and Philosophy in the Sixteenth and Seventeenth Centuries.* Allen and Unwin, London, 1935.

4 COSMOLOGY AFTER NEWTON AND BEFORE EINSTEIN

With what astonishment are we transported when we behold the multitude of worlds and systems that fill the extension of the Milky Way! But how this astonishment is increased when we become aware of the fact that all those immense orders of star-worlds again form but one of a number whose termination we do not know, and which perhaps, like the former, is a system inconceivably vast – and yet again but one member in a new combination of numbers!

Immanuel Kant, Universal Natural History and Theory of the Heavens (1755)

After Newton, astronomical advances in observation and theory were at first slow. Better telescopes had yet to be developed, photography and spectroscopy introduced into astronomy, and the chemical compositions and radial velocities of stars and nebulae determined. The puzzling nature of the nebulae had yet to be resolved, nebulae in the Galaxy to be distinguished from extragalactic nebulae, distance indicators to be found and calibrated, globular clusters to be identified as systems of stars lying in and on the outskirts of the Galaxy, and the confusing obscuration of starlight caused by interstellar absorption to be recognized. All this would be accomplished and accompanied by continual debate over controversial issues from the time of Newton to the time of Einstein during the eighteenth, nineteenth, and early twentieth centuries.

HIERARCHICAL UNIVERSES
The *via lactea* (Milky Way)

Those who live in deserts, or sail the seas, or live in out of-the-way places far from city lights understand why the night sky was so significant to the people of earlier times. On clear moonless nights the vault of heaven swarms with dazzling stars and nebulous lights, and the Milky Way – the *via lactea* – arches wraithlike across the sky.

At the beginning of the eighteenth century the centerless and edgeless Cartesian and Newtonian systems were uniformly strewn with stars. But as astronomers widened their horizons and developed better

telescopes it became increasingly difficult to ignore the obvious truth that stars are not scattered uniformly on the face of the sky.

Thomas Wright (1711–1786)

Thomas Wright of Durham, in the north of England, had novel ideas concerning the Milky Way that he presented in his book *An Original Theory of the Universe*, published in 1750. At first, he said, he had supposed that the stars were "promiscuously distributed through the mundane space," but later, because of the Milky Way, he realized that the stars were scattered "in some regular order" (Figure 4.1). A feature of Wright's universe was the existence of a supernatural galactic center, and at this "centre of creation," he "would willingly introduce a primitive fountain, perpetually overflowing with divine grace, from whence all the laws of nature have their origin." He proposed two possible constructions of the Milky Way system: either a ring-shaped distribution of stars encircling the center of the Milky Way, similar to the rings encircling Saturn; or a spherical shell of stars, concentric with the center, in which the Milky Way consists of the stars seen in a plane tangential to the shell (Figure 4.2).

Wright went further, and speculated on the possibility of many centers of creation. The distant nebulae, seen as faint and fuzzy lights in the sky, are perhaps other creations or "abodes of the blessed," similar to our Milky Way, and "the endless immensity is an unlimited plenum of creations not

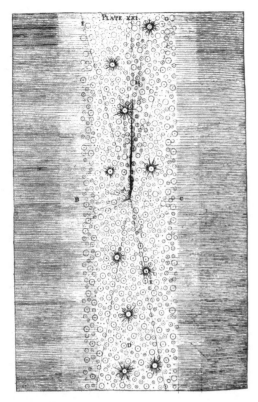

Figure 4.1. Thomas Wright's universe. At first he thought the stars were uniformly ("promiscuously") distributed, as Newton had supposed. But the Milky Way made him realize that stars are distributed in a disk, as shown by this illustration from Wright's book *An Original Theory of the Universe*.

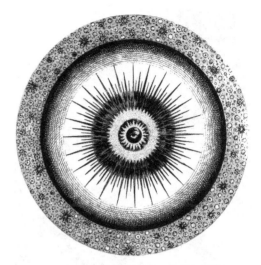

Figure 4.2. Wright considered two possibilities. First, the Milky Way is a disk composed of stars that rotate about a mysterious galactic center, and the universe is filled with similar disk-shaped milky ways. Second, as in this illustration, the stars are distributed in a spherical shell concentric with the galactic center (the Milky Way is seen in a plane tangential to the shell), and the universe in this case is filled with similar spherical milky ways.

unlike the known universe" (Figures 4.3 and 4.4).

Immanuel Kant (1724–1804)

Kant, a German philosopher and scientist, saw a review of Wright's work and adopted the idea that the Milky Way is a disk-shaped – or lens-shaped – distribution of stars, and the Milky Way is surrounded by distant similar milky-way systems. In his *Theory of the Heavens*, Kant presented in 1755 a scientific account of Wright's views. The stars of the Milky Way form a rotating disk held together by gravity, said Kant, and the fuzzy nebulae are similar rotating milky-way systems. "It is natural to assume that these nebulae are systems of numerous suns, which appear, because of their distance, crowded into a space so limited as to give a pale and uniform light. Their analogy with our own system of stars, their shape, which is just what it should be according to our theory; the faintness of their light, which denotes great distances, are in admirable agreement and lead us to consider these elliptical spots as systems of the same order as our own." These milky ways (now called galaxies), are perhaps themselves clustered together, forming vast systems of many galaxies, said Kant. He went on to conjecture that these vast systems are themselves clustered together to form even vaster systems, and these vaster systems are clustered together to form yet vaster systems, and so on, throughout infinite space (Figure 4.5). In such a cosmic hierarchy, each level consists of an infinite array of centers, and the centers at each level form clusters about the centers of the next higher level. At the highest level, of infinite order, was the ultimate center that dominated the structure

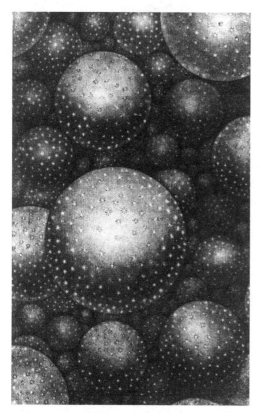

Figure 4.3. Wright's "endless immensity" of galaxies, as illustrated in *An Original Theory of the Universe.*

Figure 4.4. Thomas Wright's illustration of a universe of milky ways (galaxies) similar to our Milky Way (Galaxy).

of the universe. "We see the first members of a progressive relationship of worlds and systems: and the first part of this infinite progression enables us already to recognize what must be conjectured of the whole. There is here no end but an abyss of a real immensity, in the presence of which all the capability of human conception sinks exhausted."

Johann Lambert, a Swiss-German mathematician, entertained similar ideas. The main difference was Lambert's assumption that each center was occupied by a body that he called a "dark regent." In *Cosmological Letters*, Lambert wrote in 1761, "The eye, assisted by the telescope, may at length penetrate all the way to the centers of the milky ways, and why not even to the center of the universe?"

Island universes

von Humboldt, a restless man of broad scientific interests, introduced in 1855 in his book *Kosmos* the term "cosmical island" when he wrote "our cosmical island forms a lens-shaped system of stars." Authors soon popularized the term cosmical island in the form "island universe" and "island universes." Unfortunately, confusion creeps in when galaxies are referred to as universes. We shall instead refer to galaxies as islands; thus a Stoic-like system is a one-island universe and an Epicurean-like system is a many-island universe.

Fractal universe

When the arrangement in a hierarchy repeats itself on several levels the hierarchy is known as a fractal, a term introduced by the French scientist Benoit Mandelbrot.

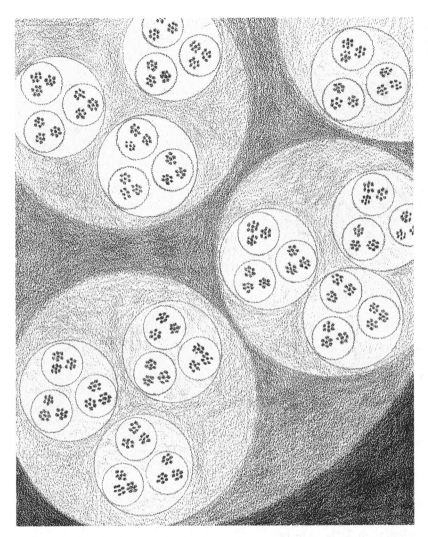

Figure 4.5. A polka-dot hierarchical universe of stars clustered into galaxies and of galaxies clustered into larger systems, which in turn are clustered into yet larger systems, and so on, indefinitely, as conceived by Immanuel Kant and Johann Lambert in the eighteenth century.

Irregularities of a coastline, for example, seen on the scale of tens of meters, then kilometers, then hundreds of kilometers may look much the same. A homemade cubic fractal of wooden blocks is shown in Figure 4.6. Kant's hierarchy possessed fractal properties: stars form clusters, each of N_1 stars, at the first level; star clusters form galaxies, each of N_2 star clusters, at the second level; galaxies form galaxy clusters, each of N_3 galaxies, at the third level; galaxy clusters form superclusters, each of N_4 galaxy clusters, at the fourth level; and so on. When $N_1 = N_2 = N_3 = N_4 = \cdots$, the fractal has a repetitive pattern and is said to be regular.

In a regular fractal, the total number N of arranged things is proportional to L^D, where L is the scale size and D the fractal dimension. If, at the first level, things have a size $L = 1$, then

$$N = L^D. \qquad [4.1]$$

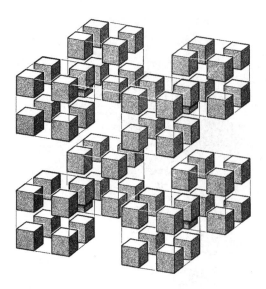

Figure 4.6. A homemade fractal of blocks of wood showing similarity on three levels. The first level consists of single blocks, each of unit size; the second level consists of groups of 8 blocks, each of size 3 units; and the third level consists of groups of 8 × 8 blocks, each of size 3 × 3 units. The fractal dimension of this arrangement is $D = \log 8 / \log 3 = 1.89$.

The average density of things per unit volume is NL^{-3}, and therefore

$$\text{density} = L^{D-3}. \qquad [4.2]$$

When things are uniformly distributed, $D = 3$, and the density is unity and independent of scale length L; when, however, the fractal dimension is less than 3, the density decreases as the scale length increases. In a regular fractal, the fractal dimension is

$$D = \frac{\text{logarithm of cluster number}}{\text{logarithm of cluster scale}}. \qquad [4.3]$$

The dimension of the cubic fractal shown in Figure 4.6 is therefore

$$D = \frac{\log 8}{\log 3} = \frac{\log 8 \times 8}{\log 3 \times 3}$$

$$= \frac{\log 8 \times 8 \times 8}{\log 3 \times 3 \times 3} = 1.89, \qquad [4.4]$$

and the density decreases with size as $L^{-1.11}$. Fractals are discussed further in Chapter 24. Suppose that 1000 galaxies, each of size 1,

form a large cluster of size 100, and 1000 similar large clusters form a supercluster of size 10 000, and 1000 similar superclusters form a higher order cluster of size 1 000 000. For these three levels we see from Equation [4.3], the fractal dimension is $D = 1.5$. As shown in Chapter 24, in an infinite universe containing an infinite number of galaxies hierarchically arranged, the galaxies do not cover the sky when D is less than 2.

THE NEBULA HYPOTHESIS
Pierre Simon de Laplace (1749–1827)
Laplace, a French mathematician and astronomer, demonstrated in his *Celestial Mechanics* the stability of the planetary system and thereby dispensed with periodic corrections by a divine power that Newton had invoked. In his nonmathematical and popular *System of the World* (first published in 1796), Laplace discussed the hypothesis that the Solar System had formed from a rotating and contracting cloud of gas. According to this idea, now accepted in modernized form (Chapter 5), the Sun and planets originally condensed from a large swirling cloud of interstellar gas. Possibly other solar systems had formed in a similar way. Kant earlier had advanced essentially the same idea in his book *The Theory of the Heavens*, but with less emphasis on the importance of rotation.

The Kant–Laplacian nebula hypothesis (nebula means cloud) caught the imagination of astronomers and natural philosophers. Every fuzzy patch of light in the night sky became, in the nineteenth century, a possible interstellar cloud of condensing gas. Even the Andromeda Nebula (a giant neighboring galaxy), and similar nebulae, became conjectural solar systems in the process of formation.

COSMICAL ISLANDS
The distant nebulae
The small, faint, and fuzzy patches of light in the night sky (Figure 4.8), according to the Wright–Kantian interpretation, were distant milky-way systems of stars, and

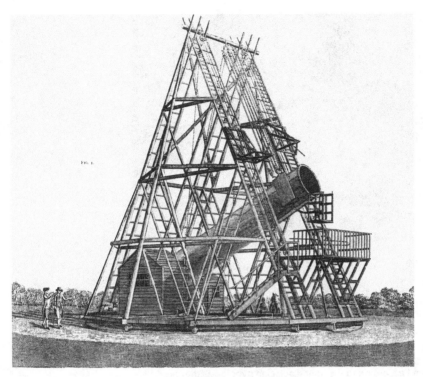

Figure 4.7. William Herschel's 40-foot telescope in 1795.

according to the Kant–Laplacian interpretation, were clouds of gas condensing into solar systems. The milky-way interpretation favored a many-island universe, whereas the solar-system interpretation favored a one-island universe. These opposing interpretations took center stage in a melodrama lasting until the early years of the twentieth century, in which astronomers see-sawed from one interpretation to the other.

William Herschel (1738–1822)

Born in Germany and a musician by profession, Herschel emigrated to England at age 19. Later, his sister Caroline joined him and both became ardently interested in astronomy. Using state-of-the-art telescopes, which they themselves made, they succeeded in resolving many nebulae into clusters of stars. Following his sensational discovery of a seventh planet beyond Saturn, later named Uranus, William became famous and was recognized as the leading astronomer of the eighteenth century.

The Herschels surveyed the heavens with telescopes of unrivaled precision and light-gathering power. In publications, William interpreted their results on the basis of three assumptions:

(i) Interstellar space is transparent to starlight.
(ii) All stars are similar to the Sun.
(iii) Stars are distributed uniformly in space.

(All three were later found to be in error.) The first and second assumptions meant the faintest stars were the farthest, and apparent brightness could be used as a measure of distance. The first and third assumptions meant the Milky Way extended the farthest where the sky appeared the most crowded with stars. On the basis of these assumptions, the Herschels charted the heavens and found the Milky Way to be a flattened system, as shown in Figure 4.9, with the Sun positioned near the center.

The galactocentric theory (the theory that the Sun is at the center of the Galaxy) was

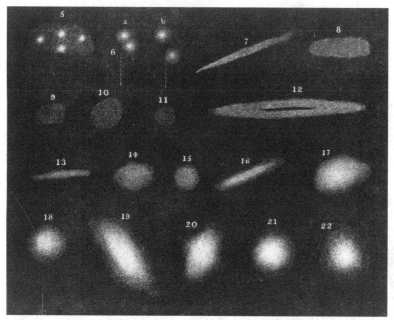

Figure 4.8. William Herschel's sketch of various nebulae in his paper "Astronomical observations relating to the construction of the heavens" (1811). According to the Wright–Kantian hypothesis the nebulae are distant milky ways like our Milky Way, and according to the Kant–Laplacian hypothesis they are swirling clouds of gas located in the Milky Way that are in the process of condensing to form new solar systems.

primarily the consequence of assumption (i). No astronomer at the time knew of the pronounced absorption of starlight caused by clouds of dusty gas drifting between the stars. (The center of the Galaxy is hidden from view in the constellation of Sagittarius.) William thought the dark and starless regions of the sky, now known to be caused by obscuring clouds of gas and dust, were "holes in the sky" through which we see the darkness beyond the Milky Way.

The Herschels succeeded in showing that the motions of double stars – two stars in orbit about each other – are in accord with Kepler's laws. In 1785, inspired by the Wright–Kantian hypothesis, William wrote in a paper entitled "On the construction of the heavens" that many nebulae may be very distant systems similar to our Milky Way. "For which reason they may also be called milky ways by way of distinction." In a letter he wrote he had "discovered 1500 universes! ... whole sidereal systems, some of which might well outvie our Milky Way in grandeur." For most of his life he supported the view that the universe is endlessly populated with galaxies much like the Milky Way. But the riddle of the nebulae grew more puzzling, and with the rise in popularity of the Kant–Laplacian nebula hypothesis, William grew less confident and eventually expressed the opinion that possibly most nebulae existed inside the Milky Way.

His observations of double stars led him to realize that stars in general can be greatly different in brightness, and the second assumption (all stars are alike) was untenable. So was the third assumption (stars are uniformly distributed in space), and he realized that stars actually are scattered with pronounced irregularity. Some nebulous regions, as in Orion, looked like "a shining fluid" and not at all like a collection of stars, thus raising disturbing doubts concerning the first assumption.

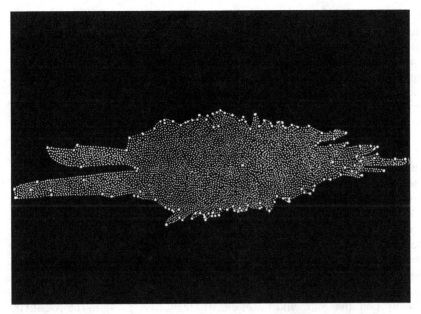

Figure 4.9. The Stellar System (Galaxy) according to William Herschel in 1785, in which the Sun was positioned close to the center. He assumed that stars were uniformly distributed in space, but because they do not uniformly cover the sky, the Stellar System extended the farthest where stars seem the faintest and are the most crowded in the sky.

William Herschel believed that the Moon and planets were inhabited by living creatures. It seems astonishing, however, that the foremost astronomer in the Age of Reason could believe that beneath the bright atmosphere of the Sun existed a cool surface also inhabited by living creatures. But this was before the marriage of physics and astronomy and the birth of the new astronomy.

THE NEW ASTRONOMY
The speed of light
Reason assured Descartes that light travels at infinite speed and we see the world as it is. How confused our reconstruction of the external world would be if the light rays composing an image and coming from different distances originated at different times! A finite speed meant that an object at the moment of observation was not what it seemed; it had moved elsewhere and changed its appearance. Objects at greater or lesser distances would have moved and changed by greater or lesser amounts. The idea that when we look out in space we look back in time seemed to Descartes and many Cartesians too incredible to be taken seriously.

But in 1676, the Danish astronomer Ole Roemer announced at a meeting of the Paris Academy of Sciences that the eclipse of Io (a moon of Jupiter) on November 9 would occur 10 minutes late. His prediction was based on earlier records. Roemer explained that the Earth was moving away from Jupiter and the delay would be caused by the extra distance that light had to travel to catch up with the Earth. This correct prediction established him as the discoverer of the finite speed of light. The results, updated by Edmund Halley, showed that light travels at the enormous speed of 300 000 kilometers a second and takes only 500 seconds to travel a Sun–Earth distance (known as an astronomical unit).

Noting that parallax of the stars, caused by the Earth's orbit about the Sun (Chapter

5), had not been detected and that stars were therefore farther away than a certain minimum distance, Francis Roberts in 1694, in a paper "Concerning the distance of the fixed stars," reckoned that "Light takes up more time Travelling from the Stars to us than we in making a West-India voyage (which is ordinarily performed in six weeks)." Light-travel time is important in cosmology and this was its first use as a measure of distance. But the parallax of even the nearest stars is far less than Roberts imagined, and their light takes years not weeks to reach us.

Opposition to the finite speed of light ended in 1729 when James Bradley discovered the aberration of light. He observed that stars move backward and forward by a small angle during the year because of the Earth's motion around the Sun, and the angular displacement, independent of the distance of stars (unlike parallax), was in agreement with Roemer's finite speed of light. When a person walks in vertically falling rain, the rain slants toward the person's face, and an umbrella is tilted forward. The slant of the rain is analogous to the aberration of light.

Agnes Clerke, a leading nineteenth-century historian of astronomy, in her *System of the Stars* (1890), wrote, "For our view of sidereal objects is not simultaneous. Communication with them by means of light takes time, and postdates the sensible impressions ... of their whereabouts in direct proportion of their distances. We see the stars not where they are – not even where they were at any one instant, but on a sliding scale of instants." A fact of paramount importance in cosmology is that when we look out in space we also look back in time, and the farther we look out in space, the farther we look back in time.

Rays of light

Although Newton spoke of "ether waves" that vibrate like sound waves, or are like the waves on the surface of water, he nonetheless believed that light is composed of particles, and asked, "Are not the rays of light very small bodies emitted from shining bodies?" Robert Hooke and Christiaan Huygens proposed wave theories of light that explained reflection and refraction (bending of light as in a prism). A century later, in the early years of the nineteenth century, Thomas Young, a physician, scientist, and authority on Egyptian hieroglyphics, showed how the wave theory explained interference (superposed waves of the same wavelength add and subtract) and diffraction (deflection according to wavelength of waves passing through an aperture). In the middle decades of the century, Michael Faraday, outstanding experimental physicist, and James Clerk Maxwell, outstanding theoretical physicist, explored and unified electricity, magnetism, and the properties of light, and developed the modern theory of electromagnetism. Early in the twentieth century, Max Planck and Albert Einstein showed that light has properties that are both wavelike and corpuscular, and particles of radiation are now called photons.

William Huggins (1824–1910)

The "new astronomy" of the nineteenth century, later known as "astrophysics," began when astronomers sought the aid of physics in the study of stars and nebulae. Literally, the laboratory moved into the observatory (Figure 4.10). A major contributor to the new science was William Huggins who pioneered the application of the new technologies of photography and spectroscopy to astronomy, and was later greatly aided by his wife Margaret. Spectroscopic analysis of sunlight had already shown that the Sun consisted of the same elements as the Earth, and Huggins resolved to do the same for the stars.

A principal problem in astronomy throughout most of the nineteenth century concerned the nature of the nebulae – the fuzzy patches of light in the night sky. Three possibilities confronted astronomers: the nebulae were

(a) distant galaxies like our Galaxy;

Figure 4.10. The birth of astrophysics: the laboratory invades the observatory. The Huggins observatory at Tulse Hill, outside London, showing photographic and spectroscopic equipment. (William and Margaret Huggins, in *Atlas of Representative Stellar Spectra*.)

(b) clouds of swirling gas condensing into stars and planets;

(c) clusters of unresolved stars in and on the outskirts of the Galaxy.

The first possibility (a) was the Wright–Kantian many-island universe, now less popular after losing the support of William Herschel; the second possibility (b) was the Kant–Laplacian nebular hypothesis, now gaining in popularity. And the third was that many nebulae were clusters of hitherto unresolved stars, a possibility of growing popularity as a consequence of the discoveries made by greatly improved telescopes.

Possibilities (b) and (c) were consistent with a one-island universe.

A spectrum of the light from a luminous source shows how the intensity varies with wavelength. A spectrum often displays bright and dark narrow regions called spectral lines. These are the emission (bright) and absorption (dark) lines at different wavelength that identify the emitting and absorbing atomic elements. Working in his home observatory, Huggins compared the spectra of light from stars, nebulae, and comets with the spectra of elements excited in spark-gap discharges. (Notice the battery, wires, and induction coil in Figure 4.10.) In 1863,

Huggins finally succeeded, and announced to the scientific world the news: the heavens are composed of the same elements as the Sun and Earth! This was the death blow to many old beliefs, and ordinary people began to turn from the churches to the observatories for knowledge on the nature of the starry heavens.

A year later, in 1864, William Huggins wrote, "I directed the telescope for the first time to a planetary nebula in Draco," and with hesitation "put my eye to the spectroscope. Was I not about to look into a secret place of creation?" What he saw was not what he expected. The spectroscope showed not the continuous spectrum characteristic of a hot star, but a spectrum of lines characteristic of an excited gas. "The riddle of the nebulae was solved. The answer, which came to us in the light itself, read: Not an aggregation of stars, but a luminous gas." This discovery swept away all doubts about the truth of the Kant–Laplacian nebular hypothesis. The puzzling elliptical and spiral nebulae were "not clusters of suns," as many had previously supposed, "but gaseous nebulae" that by their gradual loss of heat were contracting to form solar systems. Thus assumption (a), the Wright–Kant hypothesis, was dead.

Doppler effect

Huggins also was the first to measure the radial velocity (the velocity away from and toward us) of stars by observing the shift in their spectral lines. This shift, now known as the Doppler effect, was predicted and calculated in 1848 by the French scientist Armand Fizeau. (Fizeau in 1849 was also the first person to measure the speed of light by terrestrial methods.)

When a luminous source, such as a candle or a star, moves away from an observer, all wavelengths of its radiation are increased and all frequencies decreased; and when the source moves toward the observer, all wavelengths are decreased and all frequencies increased, as seen by the observer. "To a swimmer striking out from the shore each wave is shorter, and the number of

waves he goes through in a given time is greater than would be the case if he stood still in the water," wrote Huggins in "The new astronomy: A personal retrospect."

Christian Doppler, an Austrian physicist, showed in 1842 that a receding (or approaching) source of sound is heard with a lower (or higher) pitch than when stationary. In 1848, he suggested that stars have different colors because of their motions, and that stars in binary systems periodically change their color, becoming blue when approaching and red when receding. Fizeau made the correct calculation and showed this was not the case. The displacement of spectral lines caused by relative motion was known in the nineteenth century as the Fizeau–Doppler effect, and today, perhaps less justly, as the Doppler effect. Detection of the effect was difficult until photography advanced to the stage where dry and sensitive plates could be exposed for hours in clock-driven telescopes. Of the Fizeau–Doppler effect, William Huggins prophetically wrote in 1868: it would be scarcely possible "to sketch even in broad outline the many glorious achievements that doubtless lie before this method of research in the immediate future." (The cosmological achievements are discussed in Chapters 14 and 19.)

A working atomic theory

The modern atomic theory of matter began in 1803 with John Dalton's book *New System of Chemical Philosophy*, which became an immediate success. Dalton introduced the Greek word "atom" into chemistry and was the first scientist to make the atomic theory quantitative. From the known fact that elements combine in definite proportions by weight to form chemical compounds, he was able to show that matter is composed of atoms of different weights. Hydrogen was said to have an atomic weight 1, and on this scale carbon had an atomic weight 12, oxygen 16, and so on.

Joseph Thomson discovered the negatively charged electron in 1897; Ernest Rutherford discovered the positively charged atomic nucleus in 1911; Niels Bohr constructed in 1913 a mechanical model of the atom with electrons in orbits about the nucleus; and a few years later Erwin Schrödinger, Werner Heisenberg, and other physicists developed the modern wavemechanical model of the atom using quantum mechanics. "When it comes to atoms," said Bohr, "language can be used only as in poetry."

THE VICTORIAN UNIVERSE
The standard cosmological model of the nineteenth century

The Victorian universe was the standard model of the nineteenth century during the reign (1837–1901) of Queen Victoria of Great Britain. Propelled by the rise of the new astronomy, it reached final form toward the end of the century. It was a one-island universe. Providentially, the Earth and Sun were located at the center of the Galaxy – the mainland of the universe – which was surrounded by small islands. The Galaxy (Figure 4.12) consisted of approximately one billion stars, numerous clusters of stars (often unresolved and on the outskirts), and gaseous nebulae (mostly condensing solar systems and planetary nebulae), and beyond the Galaxy stretched an endless mysterious void more fit for contemplation by theologians than astronomers. The many wonders of the Victorian universe, glorified in hundreds of popular astronomy books and proclaimed from every pulpit, thrilled a large audience in Europe and North America.

Agnes Clerke, the leading late nineteenth-century historian of astronomy, summed up the consensus view in 1890 in her book *The System of the Stars*. Some nebulae, she wrote, were clouds of swirling gas, and some were clusters of unresolved stars, but none was a galaxy like our own Milky Way:

> No competent thinker, with the whole of the available evidence before him, can now, it is safe to say, maintain any single nebula to be a star system of coordinate rank with the Milky Way. A practical certainty has been attained that the entire contents, stellar and nebular, of the sphere belong to one mighty aggregation, and stand in

Figure 4.11. William Parsons (1800–1867), third Earl of Rosse, discovered that some nebulae have spiral structure. This 1845 sketch of M 51 is as seen in his 72-inch reflecting telescope. Such spiral patterns were attributed to the swirling motion of gas in conformity with the Kant–Laplace nebula hypothesis.

> ordered mutual relations within the limits of one all-embracing scheme – all-embracing that is to say, so far as our capacities of knowledge extend. With the infinite possibilities beyond, science has no concern.

This, more or less, was the standard model of the universe – the Victorian universe – inherited by the twentieth century. In some respects the Victorian universe was the old Stoic one-island universe, updated and refurbished with stars, planets, and gaseous nebulae made of atoms similar to those of the Solar System. The Stoic solution of the dark night-sky riddle (Chapter 24) lent further support to a one-island universe. Clerke wrote:

> The probability amounts almost to certainty that the stellar system is of measurable dimensions, otherwise darkness would be banished from our skies; and the "intense inane" glowing with the mingled beams of suns individually indistinguishable, would bewilder our feeble senses with its monotonous splendour.

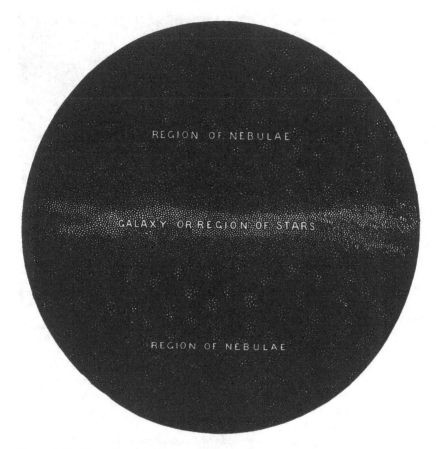

Figure 4.12. Simon Newcombe's "probable arrangement of the stars and nebulae" in his book *Popular Astronomy* (1878). According to Newcombe and other contemporary astronomers the Galaxy consisted of a billion stars and had a radius 1000 parsecs (1 parsec = 3.26 light years = 206 000 astronomical units; 1 astronomical unit = Sun–Earth distance).

In an infinite universe, endlessly populated with stars, every line of sight from the eye, extended out in space, must ultimately intercept the surface of a star. Stars must cover the entire sky. Hence the riddle, known as Olbers' paradox: why is the sky dark at night? The finite cosmos of stars discussed by Clerke, however, is only one of several possible solutions, and not the solution accepted one hundred years later.

THE AGE PROBLEM

Until the eighteenth century, Jews, Christians, and Moslems believed that the universe was only thousands of years old. Mounting evidence in geology and paleontology (the study of fossils) indicated a much greater age and brought scriptural records into conflict with science. Doctrines of compromise were developed in which the Earth had been periodically visited by catastrophes, such as life-destroying deluges, and supernatural and natural laws had alternated in their control of the Earth (see Chapter 25).

Far away and long ago

For more than two hundred years after Roemer's discovery, astronomers tended to ignore the significance of the finite speed of light concerning the age of the universe.

There were a few exceptions. William Herschel, bolder than most, occasionally remarked that we see the heavens as they were long ago. In 1802, he wrote "a telescope with a power of penetrating into space, like my 40-foot one (see Figure 4.7), has also, as it may be called, a power of penetrating into time past." In fact, the light rays from a certain very remote nebula must have been "almost two millions of years on their way; and consequently, so many years ago, this object must already have had an existence in the sidereal heavens, in order to send out those rays by which we now perceive it."

William Herschel's son, John, a more cautious man, wrote in 1830 in *A Treatise of Astronomy*, "Among the countless multitude of such stars visible in telescopes there must be many whose light has taken at least a thousand years to reach us; and that when we observe their places and note their changes we are, in fact, reading only their history of a thousand years anterior date, wonderfully recorded." John Herschel, a pillar of Victorian society, was not a Don Quixote to sally forth and tilt at solidly held religious beliefs concerning the age of the universe. But not von Humboldt, who wrote in his book *Kosmos*, "it still remains more than probable, from the knowledge that we possess of the velocity of transmission of luminous rays, that the light of remote heavenly bodies presents us with the most ancient perceptible evidence of the existence of matter." The fundamental fact that when we look out in space we also look back in time to the beginning was not openly explored in the Victorian universe. Lord Kelvin in 1901 (coincidentally, the year that Queen Victoria died) performed what no astronomer had dared. In an article "On ether and gravitational matter through infinite space," he made the relevant calculations on the connection between the age of the oldest luminous stars and the age and extent of the visible universe. But evidence from the heavens of cosmic age came too late; the cosmochronology of Genesis had already been controverted by geologists who showed that the age of the

universe must be measured not in thousands but in millions and perhaps even billions of years.

Geology

James Hutton, Scottish farmer and physician, proposed in 1785 that the formation and erosion of mountains are continuous processes that have acted over an indefinite period of time. From the evidence he found "no vestige of a beginning – no prospect of an end." This was the beginning of the steady-state uniformitarian principle, powerfully argued by Charles Lyell: the landscape is continually modified by mountain-uplift, erosion, and sedimentation, and the surface as a whole remains unchanged. Into this picture of an Earth of great, if not unlimited, age fitted Charles Darwin's theory of evolution, presented in 1859 in his book *On the Origin of Species by Natural Selection*. The catastrophists on the one hand believed in a world periodically visited by catastrophes, and the uniformitarians on the other hand believed in a steady-state world controlled by natural laws. The controversy between the two schools until the end of the nineteenth century was far more heated than the controversy between the big-bang and steady-state schools of the twentieth century.

In the second half of the nineteenth century the uniformitarians were attacked by physicists under the leadership of Lord Kelvin. Calculations (now known to be inapplicable) on tidal effects and terrestrial heat losses showed the Earth could not be as old as the geologists claimed. The calculated age of the Sun was the most decisive. Kelvin adopted Hermann von Helmholtz's idea that gravitational energy released by slow contraction of the Sun fueled the Sun's luminosity. He estimated an age of the Sun of 20 million years (see Chapter 5). Because a luminous Sun is essential for life on Earth, this implied that fossils could not be older than 20 million years. Insistence by the physicists on this short time span created dismay in the Earth and life sciences, and attempts were made to fit geological

in's
the
ned
eth-
e of
not
his
ium
dio-
ion
hus
ion
ime
for

FALL OF THE VICTORIAN UNIVERSE

The long competition between the Stoic and Epicurean systems and the alternation in their popularity, lasting for more than two thousand years, ended in the early twentieth century.

In the late Middle Ages the finite Medieval universe of celestial and angelic spheres evolved into a Stoic system of celestial spheres immersed in an infinite and mysterious void. Then the awesome poem *De Rerum Natura*, discovered in 1427, burst on the scene, opening the minds of Western Europeans to the dizzy prospect of a vast Epicurean system. Nicholas of Cusa, Thomas Digges, Giordano Bruno, William Gilbert, and numerous natural philosophers – Cartesian and Newtonian – spread the message. But in the eighteenth century the evidence was confusing. The pendulum eventually swung again in the nineteenth century and the standard model of the universe was a one-island Stoic system. In the early decades of the twentieth century, the old debate once more flared up: are we the inhabitants of a one-island or a many-island universe?

Harlow Shapley (1885–1972), an American astronomer, by studying the distribution of globular clusters (compact clusters of very old stars), in 1918 overthrew the Herschel galactocentric system. He found that these clusters form a spherical distribution whose center lies tens of thousands of light years away in the direction of the constellation Sagittarius. (A light year is the distance light travels in one year and equals 63 000 astronomical units.) The globular clusters observed by Shapley belong to our Galaxy, indicating that the center of their distribution is also the center of the Galaxy. Jan Oort, a Dutch astronomer, confirmed this result by showing that the stars of the Milky Way are orbiting about the distant center.

In the early 1920s, Shapley championed the one-island universe, and Heber Curtis, another American astronomer, championed the many-island universe. Because of insufficient allowance for absorption of starlight by interstellar dust, Shapley over-estimated the distances of the globular clusters and made the Galaxy much too large; Curtis underestimated the distances of the stars and made the Galaxy much too small. The Victorian vision of a giant Galaxy was defended by Shapley until 1930, and then finally abandoned. This tussle in ideas, referred to by some historians as the Great Debate, brought to an end the long struggle between the rival Stoic and Epicurean systems. The controversy ended in favor of an Epicurean-like system originally conceived by the Atomists of the ancient world.

A puzzle still remained. Our Galaxy seemed much larger than other galaxies. The puzzle was solved in 1952 by Walter Baade who distinguished between population I and II stars (Chapters 5 and 6), and between the cepheid variables of these two stellar populations. This had the effect of doubling the distances and sizes of other galaxies and the Milky Way no longer seemed disproportionately large.

REFLECTIONS

1 *Immanuel Kant of Königsberg, scientist and philosopher, published a deep-searching philosophy in 1781 under the title* Critique of Pure Reason. *According to Kant's philosophy, the world of sensations is organized into meaningful perceptions by an activity of a priori (subconscious) ideas of primitive*

origin. These primitive ideas, issuing from all past experience of the species, are essential for making sense of our fleeting and disjointed world of sensations. In Plato's philosophy, the ideas that organize our experiences belong to the universal Mind, in Kant's philosophy they belong to our own minds.

2   *William Herschel said that astronomy has much in common with botany. In a paper "On the construction of the heavens" he wrote: "This method of viewing the heavens seems to throw them into a new kind of light. They are now seen to resemble a luxuriant garden, which contains the greatest variety of productions in different flourishing beds: and one advantage we may at least reap from it is that we can, as it were, extend the range of our experience to an immense duration. For, to continue the simile borrowed from the vegetable kingdom, is it not always the same thing, whether we live successively to witness the germinations, blooming, foliage, fecundity, fading, withering, and corruption of a plant, or whether a vast number of specimens, selected from every stage through which the plant passes in the course of its existence, be brought at once to our view?" To learn how an oak tree grows, we do not study a single oak in isolation growing over a long period of time, but many oaks in a forest in different stages of growth. Similarly with stars.*

• *Richard Proctor in* Our Place Among the Infinities *(1876) wrote on William Herschel's change in cosmological ideas: "As the work progressed Sir William Herschel grew less confident. He began to recognize signs of a complexity of structure which set his method of star-gauging at defiance. It became more and more clear to him also, as he extended his survey, that the star-depths were in fact unfathomable."*

3   *The compilation of catalogs stating positions, descriptions, and spectral compositions is the main aim of observational astronomy. Charles Messier (1730–1817), French astronomer, and ardent comet hunter, compiled a catalog of 103 nebulae. He did this "so that astronomers would not confuse these same nebulae with comets just beginning*

to shine." The Messier nebulae are prefixed with letter M; thus M 1 is the Crab Nebula, described as "whitish light and spreading like a flame," and M 31 is the galaxy in Andromeda.

• *William Herschel published catalogs listing positions and descriptions of thousands of nebulae. John Herschel continued this work and published in 1864 the General Catalogue (referred to as GC), which was the first systematic survey of the entire sky and contained 5000 nebulae and star clusters. The GC was replaced in 1890 by the New General Catalogue (referred to as NGC) and subsequent supplements (Index Catalogues) were added. Thus the galaxy M 31 is also known as NGC 224.*

4   *We are told that Napoleon Buonaparte, emperor of the French, said to Laplace concerning his* Celestial Mechanics, *"You have written this huge book on the system of the world without once mentioning the author of the universe." Laplace replied, "Sire, I had no need of that hypothesis." Whether true or not, this story illustrates the difference between theism and deism. Theism is the ancient belief that God created and runs the universe; deism came in the Age of Reason and is the belief that God created a self-running universe. Newton was a theist, Laplace a deist.*

5   *Hierarchical astronomy became respectable as a serious possibility when John Herschel and Richard Proctor introduced it as a solution of the riddle of darkness at night. This idea was adopted by the Irish physicist Fournier d'Albe in England and the Swedish astronomer Carl Charlier in the early twentieth century. Charlier, whose work received wide publicity, showed that if the density of the clusters decreased sufficiently rapidly with increasing size (corresponding to a fractal dimension of 2 or less), stars do not cover the sky, and the sky at night is dark. Fournier d'Albe put forward the hierarchical idea that the visible universe is only one of a series of universes containing solar systems of increasing size, arranged such that the solar systems in one universe are the atoms in the next higher* universe, and so on. He showed that in universes made this way the sky at night is dark.

• The gathering of stars into galaxies, galaxies into clusters, clusters into superclusters constitutes what astronomers often call a hierarchy. Strictly speaking, this is a misuse of the word. A hierarchy is a "pecking-order" organization that usually, but not always, consists of human beings, in which each member bosses those below and obeys those above. Government agencies and large businesses are hierarchies. The medieval universe with its angelology and demonology was also hierarchical. The technically correct word for Kant's hierarchy is "multilevel." A complete multilevel universe (an infinite number of levels in an infinite universe), according to Kant and Lambert, has an ultimate cosmic center at the highest level. This center, however, is at infinite distance and its reality is open to debate. Fournier d'Albe and Charlier showed that in a multilevel universe the total amount of matter is much less than in a uniform universe. The mass and volume of the clusters steadily increase at higher and higher levels, and for clusters to stay separated from one another, their volume must increase faster than their mass. The average density (mass divided by volume) therefore steadily decreases on progressively larger scales. In an infinite multilevel universe, the density of matter averaged over an infinitely large scale is vanishingly small.

The clumpiness of the observed universe is nowadays best represented by a finite hierarchy, or a finite number of levels, consisting of stars, star clusters, galaxies, galaxy clusters, and superclusters. Possibly the universe is uniform on scales larger than superclusters, although this is not certain.

6 Our knowledge of the heavens comes from the study of whatever reaches Earth. In the old astronomy, observations were optical with color description. In the new astronomy, observations were optical with spectral decomposition. In modern astrophysics, observations include all electromagnetic radiation (radio, microwave, infrared, optical, ultraviolet, x-ray, gamma ray), particle radiation (cosmic rays, neutrinos), and meteorites. Before the new astronomy everybody believed that the chemical composition of stars was forever unknown. Auguste Comte, French mathematician, philosopher, and humanist, expressed this belief in his Course de Philosophie Positive (1830–1842): "Any research that cannot be reduced to actual visual observation is excluded where the stars are concerned.... We can see the possibility of determining their forms, their distances, their magnitudes, and their movements, but it is inconceivable that we should ever be able to study, by any means whatsoever, their chemical composition or mineralogical structure..."

7 "The observatory became a meeting place where terrestrial chemistry was brought into direct touch with celestial chemistry" (William and Margaret Huggins, Atlas of Representative Stellar Spectra).

• "An unsigned article in a magazine [Good Words] on how to make your own spectroscope launched Margaret Lindsay into the new science of spectroscopy. She was a keen photographer and also an observer of the heavens with instruments made by herself. By chance she met the author of the article – William Huggins – who was visiting her home city, Dublin, to inspect his new telescope manufactured by Howard Grubb. Spectroscopy sparked romance, and in 1875 they married. Working as a team – he with failing eyes and she keen-eyed – they made observations, using state-of-the-art photography and spectroscopy, that helped to launch the new science of astrophysics. Queen Victoria in 1897 knighted William by conferring the Order of the Bath for 'the great contributions which, with his gifted wife, he has made to the new science of astro-physics'" (Harrison, Darkness at Night).

8 For a long time the cosmic consequences of a finite speed of light received little attention. Occasionally astronomers pointed out that light travels a finite distance in a finite period of time. One or two expressed wonder that we see astronomical bodies as they were thousands and millions of years ago. The

implication that the universe is at least as old as the time taken by light to travel from its farthest visible regions was either overlooked or left aside in articles, books, and lectures, possibly because it ran counter to prevailing religious dogma. The embarrassing thought that when we look out in space to the limit of the visible universe we see things as they were in the beginning, when the universe was created, rarely surfaced in pre-relativity Victorian days. Many astronomers in Britain were ordained members of the Church of England, and astronomy, the most socially respectable and religiously correct of the sciences, was in the business of revealing the works of God and not of contradicting scriptural records.

PROJECTS

1 My first lesson in cosmology was as a child at Sunday school in 1924 where we were told that heaven is "up there," and the teacher pointed to the ceiling. Presumably, the teacher had in mind a Victorian universe, and "up there" was out in a mysterious extramundane space beyond the Galaxy. Can you remember your first lesson in cosmology?

2 The standard model of the universe at the end of the nineteenth century was unlike the standard model at the end of the twentieth century in almost every respect. This prompts the question: Is it possible that the standard model of the universe at the end of the twenty-first century will be totally unlike that at the end of the twentieth century? The Victorians were confident that they were close to the truth. What are we to make of the fact that today there is a similar attitude?

3 Thomas Huxley "is a great and even severe Agnostic, who goes about exhorting all men to know how little they know" (report in *The Spectator*, 1869). Theists and deists believe in the existence of God, atheists believe in the non-existence of God, and agnostics hold that we have no evidence for either belief. Can we apply the categories of theism, deism, atheism, and agnosticism to all religions?

• "In the clockwork universe, God appeared to be only the clockmaker, the Being who had shaped the atomic parts, established the laws of their motion, set them to work, and then left them to run themselves" (Thomas Kuhn, *The Copernican Revolution*, 1957). Consider, are you a theist or a deist? In other words, do you prefer a universe created and controlled by a divine power or a created self-controlled clockwork universe?

4 Draw a two-dimensional fractal showing three levels, and calculate the fractal dimension D. What does D less than 2 mean?

• Make a do-it-yourself fractal. Construct from simple materials (e.g., toothpicks and small rubber balls) a three-dimensional three-level fractal and calculate the fractal dimension.

• One hundred galaxies, each of size 1, form a cluster of size 10; 100 clusters form a supercluster of size 100, and so on; what is the fractal dimension?

• Kant imagined a multilevel universe covering infinite space and having an infinite number of levels. He said such a universe has an ultimate center. But this is puzzling. Is a center possible when on the largest scale the density is zero?

FURTHER READING

Andrade, E. N. da C. "Doppler and the Doppler effect." *Endeavour* p. 14 (January 1959).

Armitage, A. *William Herschel*. Thomas Nelson, 1962.

Barnard, E. E. "On the vacant regions of the Milky Way." *Popular Astronomy* 14, 579 (1906).

Berendzen, R., Hart, R., and Seeley, D. *Man Discovers the Galaxies*. Science History Publications, New York, 1976.

Gingerich, O. "Charles Messier and his catalog." *Sky and Telescope* 12, 255 and 288 (1953).

Haber, F. C. *The Age of the World: Moses to Darwin*. Johns Hopkins University Press, Baltimore, 1959.

Harrison, E. R. *Masks of the Universe*. Macmillan, New York, 1985.

Hoskin, M. A. *William Herschel and the Construction of the Heavens*. Science History Publications, New York, 1963.

King, H. C. *The History of the Telescope.* Charles Griffin, London, 1955.

Meadows, A. J. "The origins of astrophysics." *American Scientist* p. 269 (May–June 1984).

Munitz, M. K. Editor. *Theories of the Universe: From Babylonian Myth to Modern Science.* Free Press, Glencoe, Illinois, 1957.

Paneth, F. A. "Thomas Wright of Durham." *Endeavour* 9, 117 (1950).

Richardson, R. S. "Lady Huggins and Others." In *The Star Lovers.* Macmillan, New York, 1967.

Shapley, H. *Flights from Chaos: A Survey of Material Systems from Atoms to Galaxies.* McGraw-Hill, New York, 1930.

Shapley, H. *Beyond the Observatory.* Charles Scribner's Sons, New York, 1967.

Sharlin, H. I. and Sharlin, T. *Lord Kelvin: The Dynamic Victorian.* Pennsylvania State University Press, University Park, 1979.

Sidgwick, J. B. *William Herschel: Explorer of the Heavens.* Faber, London, 1953.

Toulmin, S. and Goodfield, J. *The Fabric of the Heavens: The Development of Astronomy and Dynamics.* Harper and Row, New York, 1961.

Waterfield, R. L. *A Hundred Years of Astronomy.* Macmillan, New York, 1938.

Whitney, C. A. *The Discovery of Our Galaxy.* Knopf, New York, 1971.

SOURCES

Aiton, E. J. *The Vortex Theory of Planetary Motions.* Macdonald, London, 1972.

Becker, C. L. *The Heavenly City of the Eighteenth-Century Philosophers.* Yale University Press, New Haven, 1968.

Burchfield, J. D. *Lord Kelvin and the Age of the Earth.* Science History Publications, New York, 1975.

Chambers, G. F. *Descriptive Astronomy.* Clarendon Press, Oxford, 1867.

Charlier, C. "How an infinite world may be built up." *Arkiv för Matematik, Astronomi och Fysik* 16, no. 22 (1922).

Clerke, A. M. *The System of the Stars.* Longmans, Green, London, 1890.

Clerke, A. M. *A Popular History of Astronomy During the Nineteenth Century.* Black, London, 1893.

Clerke, A. M. *The Herschels and Modern Astronomy.* Cassell, London, 1901.

Cohen, I. B. "Roemer and the first determination of the velocity of light (1676)." *Isis* 31, 327 (1940).

Crowe, M. J. *The Extraterrestrial Life Debate, 1750–1900.* Cambridge University Press, Cambridge, 1986.

Curtis, H. D. "Dimensions and structure of the Galaxy." *Bulletin of the Research Council of the National Academy of Sciences* 2, 194 (1921).

Darwin, C. *On the Origin of Species by Means of Natural Selection, or the Preservation of Favoured Races in the Struggle for Life.* Penguin Books, Harmondsworth, Middlesex, England, 1968. First published by John Murray, London, 1859.

Dick, S. J. *Plurality of Worlds: The Origin of the Extraterrestrial Life Debate from Democritus to Kant.* Cambridge University Press, New York, 1982.

Grant, R. *History of Physical Astronomy from the Earliest Ages to the Middle of the Nineteenth Century.* Baldwin, London, 1852.

Hall, A. R. *The Scientific Revolution 1500–1800. Formation of the Modern Scientific Attitude.* Longmans, Green, London, 1954.

Halley, E. "Of the infinity of the sphere of fix'd stars." *Philosophical Transactions* 31, 22 (1720–1721).

Halley, E. "Of the number, order, and light of the fix'd stars." *Philosophical Transactions* 31, 24 (1720–1721).

Harrison, E. R. "Newton and the infinite universe." *Physics Today* 39, 24 (1986).

Harrison, E. R. *Darkness at Night: A Riddle of the Universe.* Harvard University Press, Cambridge, Massachusetts, 1987.

Hearnshaw, J. B. *The Analysis of Starlight: One Hundred and Fifty Years of Astronomical Spectroscopy.* Cambridge University Press, Cambridge, 1986.

Helden, A. van. *Measuring the Universe: Cosmic Dimensions from Aristarchus to Halley.* Chicago University Press, Chicago, 1985.

Herrmann, D. B. *The History of Astronomy from Herschel to Hertzsprung.* Cambridge University Press, Cambridge, 1984.

Herschel, J. F. W. *A Treatise of Astronomy.* Longmans, London, 1830.

Herschel, J. F. W. *Outlines of Astronomy.* Longmans, Green, London, 1849.

Herschel, W. "On the construction of the heavens." *Philosophical Transactions* 75, 213 (1785).

Herschel, W. "Catalogue of 500 new nebulae, nebulous stars, planetary nebulae, and clusters of stars." *Philosophical Transactions* 92, 477 (1802).

Hoskin, M. A. *William Herschel and the Construction of the Heavens.* Science History Publications, New York, 1963.

Hoskin, M. A. "The cosmology of Thomas Wright of Durham." *Journal for the History of Astronomy* 1, 44 (1970).

Hoskin, M. A. "The great debate: What really happened." *Journal for the History of Astronomy* 7, 169 (1976).

Hoskin, M. A. "Newton, providence and the universe of stars." *Journal for the History of Astronomy* 8, 77 (1977).

Hoskin, M. A. "The English background to the cosmology of Wright and Herschel," in *Cosmology, History, and Theology.* Ed. W. Yougrau and A. D. Breck. Plenum Press, New York, 1977.

Huggins, W. "Further observations on the spectra of stars and nebulae, with an attempt to determine therefrom whether these bodies are moving toward or from the Earth." *Philosophical Transactions* 158, 529 (1868).

Huggins, W. "The new astronomy: A personal retrospect." *The Nineteenth Century* 41, 907 (1897).

Huggins, W. and Huggins, M. *Publications of Sir William Huggins's Observations.* Vol. 1, *An Atlas of Representative Stellar Spectra*, 1899. Vol. 2, *The Scientific Papers of Sir William Huggins*, 1909. Wesley, London, 1899 and 1909.

Humboldt, A. von. *Kosmos.* 5 vols. 1845–1862. Translated by E. C. Otte. Bohn, London, 1848–1865.

Huygens, C. *Cosmotheros* and *The Celestial Worlds Discover'd, 1698.* Reprinted: Frank Cass, London, 1968.

Jaki, S. L. *The Milky Way: An Elusive Road for Science.* Science History Publications, New York, 1972.

Kant, I. *Kant's Cosmogony.* Translated by W. Hastie. Greenwood, New York, 1968.

Kant, I. *Universal Natural History and Theory of the Heavens.* Translated by W. Hastie. University of Michigan Press, Ann Arbor, 1969.

Kelvin, see Thomson, W.

Kuhn, T. S. *The Copernican Revolution: Planetary Astronomy in the Development of Western Thought.* Harvard University Press, Cambridge, Massachusetts, 1957.

Kuhn, T. S. *The Structure of Scientific Revolutions.* Chicago University Press, Chicago, 1970.

Lambert, J. H. *Cosmological Letters on the Arrangement of the World Edifice.* Translated by S. L. Jaki. Science History Publications, New York, 1976.

Langley, S. P. *The New Astronomy.* Anchor, Boston, 1888.

Lindberg, D. C. and Numbers, R. L. *God and Nature.* University of California Press, Berkeley, 1986.

Lubbock, C. A. *The Herschel Chronicle.* Cambridge University Press, Cambridge, 1933.

Mandelbrot, B. B. *The Fractal Geometry of Nature.* Freeman, San Francisco, 1982.

Mills, C. E. and Brooke, C. F. Editors. *A Sketch of the Life of Sir William Huggins.* London, 1936.

Newton, I. *Isaac Newton's Mathematical Principles of Natural Philosophy and His System of the World.* 2nd edition. Translated by A. Motte, 1729. University of California, Berkeley, 1934.

Newton, I. *The Chronology of Ancient Kingdoms Amended.* London, 1728.

North, J. D. *The Norton History of Astronomy and Cosmology.* Norton, New York, 1959.

North, J. D. "Chronology and the age of the world," in *Cosmology, History, and Theology.* Editors W. Yourgrau, and A. D. Breck. Plenum Press, New York, 1977.

Numbers, R. L. *Creation by Natural Law: Laplace's Nebula Hypothesis in American Thought.* University of Washington Press, Seattle, 1977.

Pannekoek, A. *A History of Astronomy.* Interscience Publishers, New York, 1961.

Proctor, R. A. *Our Place Among the Infinities: Essays Contrasting Our Little Abode in Space and Time with the Infinities Around Us.* King, London, 1876.

Roberts, F. "Concerning the distance of the fixed stars." *Philosophical Transactions* 18, 101 (1694).

Roemer, O. "A demonstration concerning the motion of light." *Philosophical Transactions* 11, 893 (1677).

Rutherford, E. "Radiation and emanation of radium," 1904, in *The Collected Works of Lord Rutherford.* Volume 1. Allen & Unwin, London, 1962–1965.

Thomson, W. "On ether and gravitational matter through infinite space." *Philosophical Magazine* (Series 6) 2, 161 (1901).

Van Helden, A. "Roemer's speed of light." *Journal for the History of Astronomy* 15, 137 (1983).

Van Helden, A. *Measuring the Universe: Cosmic Dimensions From Aristarchus to Halley.* University of Chicago Press, Chicago, 1985.

Whitrow, G. J. "Kant and the extragalactic nebulae." *Quarterly Journal of the Royal Astronomical Society* 8, 48 (1967).

Wright, T. *An Original Theory or New Hypothesis of the Universe*, 1750. Elsevier, New York, 1971.

5 STARS

"The stars," she whispers, "blindly run:
A web is wov'n across the sky;
From our waste places comes a cry,
And murmurs from the dying sun."
Alfred Tennyson (1809–1892), In Memoriam

THE DISTANT STARS

Light travel time

We look out from Earth and see the Sun, planets, and stars at great distances (see Figure 5.1). The Sun, our nearest star, is at distance 150 million kilometers or 93 million miles. Kilometers and miles, suitable units for measuring distances on the Earth's surface, are much too small for the measurement of astronomical distances (see Table 5.1).

Almost all information from outer space comes to us in the form of light and other kinds of radiation that travel at the speed 300 000 kilometers per second (see Table 5.2). Light from the Sun takes 500 seconds to reach the Earth, and we see the Sun as it was 500 seconds ago. We say the Sun is at distance 500 light seconds. The time taken by light to travel from a distant body is called the light travel time. Light travel time is an attractive way of measuring large distances and has the advantage that we know immediately how far we look back into the past when referring to a distant body. A star 10 light years away (almost 100 trillion kilometers) is seen now as it was 10 years ago. Always, when looking out in space, we look back in time.

Light takes approximately 10 hours to travel across the Solar System (the diameter of Pluto's orbit). Our system of circling planets, so large by ordinary terrestrial standards, is dwarfed by the great distances to the nearest stars, which are several light years away. An amazing variety of tens of billions of stars stretches away tens of thousands of light years. They are the stars of our Galaxy. Beyond our Galaxy lie other galaxies at distances of millions of light years; most of these galaxies are themselves vast systems of billions of stars.

Even astronomers accustomed to thinking of large distances marvel at the lavish use of space in the design of the universe. Time, it seems, has been used rather sparingly, for who can marvel at only a few thousand, or a few million, or even a few billion years of light travel time? But this lavishness in space and economy in time is actually the result of the peculiar units of measurement that we use on Earth. With light travel time as a way of measuring distances we set aside our terrestrial bias of favoring small units of distance and large units of time. One second of light travel time that seems so small is equivalent to 300 000 kilometers that seems so large.

The greatest distances on the Earth's surface are only 1/20 of a light second. We live for 3 score and 10 years and in that time the Sun with its retinue of planets travels a distance of only 20 light days in the Galaxy. From the cosmic viewpoint we are confined to a small region of space but endure for a long period of time. This explains why we are more impressed with the vastness of space than with the vastness of time.

Distances to the stars

There are various methods of determining astronomical distances and at this

Table 5.1. *Distances and sizes*

Distance to Sun	1 astronomical unit
Radius of Sun	6.96×10^{10} centimeters
Radius of Earth	6370 kilometers
Fingerwidth	1 centimeter approximately
Size of flea	1 millimeter approximately
Wavelength of yellow light	6×10^{-5} centimeters
Radius of hydrogen atom	0.5×10^{-8} centimeters

1 parsec $= 3.26$ light years $= 206\,265$ astronomical units
1 light year $= 6.33 \times 10^4$ astronomical units $= 9.46 \times 10^{17}$ centimeters
1 astronomical unit $= 1.50 \times 10^{13}$ centimeters
1 kilometer $= 10^5$ centimeters $= 0.62$ miles
1 meter $= 10^2$ centimeters
1 centimeter $= 0.39$ inches
1 millimeter $= 10^{-1}$ centimeters
1 angstrom $= 10^{-8}$ centimeters

Figure 5.1. Star cluster in the constellation Cancer (Mount Wilson and Las Campanas Observatories; Mount Wilson Observatory photograph).

Table 5.2. *Velocities*

Light (denoted by c)	300 000 kilometers a second $= 3 \times 10^{10}$ centimeters a second
Sun in Galaxy	300 kilometers a second
Earth in Solar System	30 kilometers a second
Escape velocity from Sun	618 kilometers a second
Escape velocity from Earth	11 kilometers a second

Table 5.3. *Time*

Hour	3600 seconds
Day	86 400 seconds
Year	3.2×10^7 seconds
Decade	10 years
Century	100 years
Millennium	1000 years
Age of Earth	4.6 billion years
Age of universe	10–20 billion years

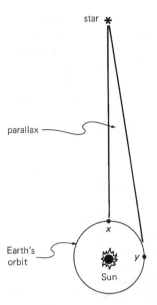

Figure 5.2. From different positions on the Earth's orbit about the Sun an observer sees a star in slightly different directions. A baseline of length one astronomical unit, subtends an angle 1 second of arc when the distance to the star is 1 parsec.

preliminary stage we mention two that are simple but important.

The first method uses the parallax effect for stars within a few hundred light years. The parallax effect is quite obvious when we hold a finger upright at arm's length and see how the finger appears to move to and fro when viewed alternately with the left and right eyes. The nearby stars are also seen in slightly different directions when viewed from different positions on the Earth's orbit about the Sun. A star at a distance of 1 parsec, viewed from two points on the Earth's orbit 1 astronomical unit apart (as shown in Figure 5.2), is seen in different directions separated by an angle of 1 second of arc. A candle flame at 10 miles (16 kilometers) distance, viewed alternately with the left and right eyes, is seen in different directions separated by 1 second of arc. Stellar parallax becomes difficult to detect at distances greater than 300 light years.

The second method may be used for stars too distant to have easily detectable parallax. These distant stars are selected for their close similarity to nearby stars. By comparing apparent brightness it is possible to determine how far away are the distant stars from the known distances of the nearby stars. Every star has an intrinsic or absolute brightness, and an apparent brightness that depends on the intrinsic brightness and the distance from Earth. As an illustration, suppose we find by parallax measurements that a certain star is 100 light years away. A second star is observed of less apparent brightness that is believed to have the same intrinsic brightness as the first. We discover that the apparent brightness of the second star is 1/100 that of the first. The apparent brightness decreases as the inverse square of distance (in the absence of interstellar absorption) and we conclude that the fainter star is at the distance 1000 light years.

A FOREST OF STARS
Multiple stars
Let us leave the Earth and roam through space sight-seeing the different stars. We notice that some stars are extremely bright; the majority, however, are less luminous than the Sun. We also notice that most

stars are grouped into small families of two, three, or more members. Double stars – known as binary systems – are common, and almost half of all stars about us are members of binary systems. Commonly, double stars are separated by many astronomical units and revolve about each other with periods of many years. But some double stars are much closer together and have periods of a only few days; they exchange matter, have eruptive outbursts, and evolve in remarkable and surprising ways.

The average separating distance between neighboring stars (ignoring binary systems) is about 1 parsec. This distance is 200 thousand astronomical units or 20 million times the diameter of the Sun. Thus the distances between stars are immense compared with the size of the stars and their planetary systems.

Color and brightness

What are stars? They are luminous globes of hot gas that pour out radiation into space and are held together by their own gravity. Their observed properties of immediate interest are color and brightness. The color is determined

by the temperature of the star's surface, and the brightness is determined by the amount of light radiated from its surface. The brightness–color diagram for stars is known as the Hertzsprung–Russell diagram (after its originators), or H-R diagram (Figure 5.3). Each point in the H-R diagram corresponds to a particular combination of brightness and color.

The Sun is yellow-white and has a surface temperature 5800 kelvin (see Table 5.4); there are many sunlike stars, yellow-white in color and similar in brightness and size to the Sun. Other stars, red and large, are the red giants with surface temperatures around 3000 kelvin. Yet others, white and small, are the white dwarfs with surface temperatures 10 000 and more kelvin.

Brightness – the rate at which luminous energy is emitted – depends on surface temperature and surface area of the star. A red giant of low surface temperature has high brightness because of its very large surface area. A white dwarf of high surface temperature has low brightness because of its very small surface area. A red giant of low temperature and high brightness is like a charcoal fire; a white dwarf of high

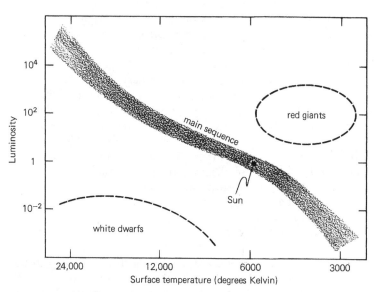

Figure 5.3. The Hertzsprung–Russell diagram in which the brightness (or luminosity) and the color (or surface temperature) of stars are plotted. The luminosity is expressed in units of the Sun's luminosity.

Table 5.4. *Temperatures (kelvin)*

Center of Sun	2×10^7 K
Surface of Sun	5800 K
Filament of electric light bulb	2000 K
Boiling point of water	373 K = 100 °C (degrees celsius)
	= 212 °F (degrees fahrenheit)
Absolute zero	0 K (or −273 °C)

Table 5.5. *The ten nearest stars (Solar luminosity = 4×10^{33} ergs per second = 4×10^{20} megawatts)*

Star	Distance (light years)	Luminosity (solar luminosities)
Sun	1.6×10^{-5}	1
Alpha Centauri	4.3	1.5
Barnard's Star	6.0	5×10^{-4}
Wolf 359	7.6	1.6×10^{-5}
Lalande 21185	8.1	5×10^{-3}
Sirius	8.6	23
Luyten 726	8.9	1×10^{-4}
Ross 154	9.4	4×10^{-4}
Ross 248	10.3	1×10^{-4}
Epsilon Eridani	10.7	0.3

Table 5.6. *The ten brightest stars seen from Earth in order of apparent brightness*

Star	Distance (light years)	Luminosity (solar luminosities)
Sun	1.6×10^{-5}	1
Sirius	8.6	23
Canopus 98	1.5×10^3	760
Alpha Centauri	4.3	1.5
Arcturus	36	114
Vega	26	54
Capella	45	150
Rigel	900	6×10^4
Procyon	11	7.2
Betelgeuse	490	2×10^4

temperature and low brightness is like a flashlight bulb.

Often we are interested in the total amount of radiation emitted by a star, some of which is not visible to the eye. For total radiation (visible and invisible) we use the term *luminosity*. The luminosity of a star is usually stated in solar luminosities (see Tables 5.5 and 5.6). With brightness changed to luminosity and color changed to surface temperature, the brightness–color diagram becomes the astronomer's luminosity–temperature diagram.

Main sequence

When each star is plotted as a point in the H-R diagram, we notice that the points are not scattered randomly, but tend to concen-

Table 5.7. *Masses*

Sun	2×10^{33} grams
	$= 2 \times 10^{30}$ kilograms
Earth	6×10^{27} grams
	$= 2 \times 10^{24}$ kilograms
Water in thimble	1 gram
Flea	1 milligram
Hydrogen atom	1.7×10^{-24} grams

10^3 kilograms $= 1$ ton (metric)
1 kilogram $= 10^3$ grams $= 2.2$ pounds
1 milligram $= 10^{-3}$ grams

trate in certain regions. Most stars, including the Sun, lie in a band called the main sequence that runs diagonally across the diagram. Within the main sequence, red stars (low temperature and cool) have low luminosity, and blue stars (high temperature and hot) have high luminosity. All main-sequence stars derive their energy and maintain their luminosity by transforming hydrogen into helium through nuclear reactions.

Star masses increase steadily up the main sequence. The less massive stars at the lower end of the main sequence transform (or "burn") hydrogen into helium at a slow rate; the more massive at the upper end of the main sequence transform hydrogen at a rapid rate. The majority of stars have masses in the range 1/10 to 10 times the Sun's mass, surface temperatures in the range 2500 to 25000 kelvin, and luminosities in the range 1/1000 to 10000 times the Sun's luminosity. On the main sequence the luminosities of stars vary enormously, but their sizes do not vary greatly (the radius of the most luminous is about 25 times that of the least luminous). An unknown number of stars have masses less than 1/10 the Sun's mass; these dull stars, known as red dwarfs and brown dwarfs, are often the unseen companions of visible stars. A few extreme stars have masses as great as 60 times the Sun's mass and are 10 million times more luminous than the Sun. Such a bright star at the distance of Alpha Centauri would shed as much light on Earth as the full Moon.

Above the main sequence in the H-R diagram lie the red giant stars. The red giants are distended globes of cool gas that may be larger than the Earth's orbit about the Sun. Even though their surface temperatures are low, they are highly luminous because of their extremely large surface areas. They radiate hundreds and often thousands of times more energy each second than the Sun. They have consumed their central supplies of hydrogen and have quit the main sequence, and their central regions are contracting to higher temperature and density in search of further sources of energy.

Below the main sequence in the H-R diagram are the white dwarf stars. The white dwarfs are approximately the size of Earth; they are dense and hot and not very luminous because of their extremely small surface areas. These stars have come to the end of their evolution and are slowly cooling. Large numbers of these dying stars exist (about 10 percent of the nearby stars are white dwarfs), but they are difficult to find because of their low luminosity. Not all stars terminate their careers as dense white dwarfs; many continue to evolve and become neutron stars, and as we shall later see, some become black holes.

Variable stars

Most stars shine with almost constant brightness, but a small proportion – the variable stars – vary periodically in brightness. About a quarter of all variable stars are eclipsing binary systems whose brightness varies because the orbiting stars pass periodically in front of each other. But the majority are pulsating variables, rhythmically expanding and contracting, pulsating in size and brightness.

An important class of pulsating variables consists of luminous yellow giants found above the main sequence and known as cepheids. They shine between 100 and 10000 times as bright as the Sun, and are so named because the first star discovered in this class was Delta Cephei, a faint star seen with the naked eye in the constellation of Cepheus. Over 700 cepheids are known in our Galaxy; most have pulsation periods between 3 and 50 days, and some vary in brightness by a factor as much as 5. Polaris, the Pole Star, is a cepheid that changes in brightness by 10 percent with a period of 4 days. Cepheids are stars more massive than the Sun that have evolved beyond the red giant state. They have discovered that by resorting to oscillation they release more easily the radiation that is dammed up inside.

Cepheids are important because they serve as distance indicators. Their periods of oscillation are related in a known way to their luminosities: the greater the luminosity, the longer the period. This period–luminosity relation was discovered in 1912 by the astronomer Henrietta Leavitt of the Harvard College Observatory. By measuring the period of oscillation of a cepheid, an astronomer discovers the luminosity (intrinsic brightness) from the period–luminosity relation, and is then able to find the distance to the cepheid by comparing the intrinsic brightness with the observed apparent brightness. The determination of distance is always difficult in astronomy, and this neat method explains why the relatively rare cepheids are important stars.

INSIDE THE STARS
Globes of hot gas
Stars are globes of hot gas that radiate energy away into space. This energy, emitted from the surface, originates in the deep interior and diffuses slowly to the surface. Heat flows from hot to cool regions and hence the center of a star is much hotter than its surface. The central temperatures of stars are in fact enormous; in the Sun, for example, the central temperature is around 15 million kelvin. The central temperatures of stars increase up the main sequence going from low-mass to high-mass stars.

Stars are self-gravitating: they are held together by their own gravity. The gravitational force pulling inward is opposed by a force pushing outward. The outward-pushing force is the pressure of the hot gas

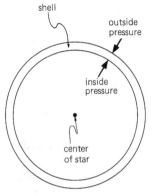

inside pressure − outside pressure
= pressure difference
= weight of shell

Figure 5.4. A thin spherical shell of matter inside a star. The inside pressure pushing outward is greater than the outside pressure pushing inward, and the difference supports the weight of the shell.

in the interior. If there were no pressure inside the Sun, the Sun would collapse in only 1 hour and become a black hole.

The balance of pressure (actually the pressure gradient) and gravity is easy to understand. Consider in a star an imaginary shell consisting of two spherical surfaces, as shown in Figure 5.4. The pressure on the inside surface of the shell pushes outward and the pressure on the outside surface pushes inward. Their difference supports the weight of the shell:

pressure difference = weight of shell. [5.1]

This is the hydrostatic equation for a star. The star consists of a large number of such imaginary concentric shells, and as we proceed inward, the pressure rises each time a shell is crossed. The pressure progressively increases and attains its maximum value at the center. The central pressures of stars are enormous; in the Sun, for example, the central pressure (force per unit area) is equivalent to a weight on Earth of 100 million tons resting on an area equal to that of a dime.

Why the temperature is high
The average density of the Sun is 1.4 grams per cubic centimeter (or 1.4 times the density

of water) and its central density is approximately 150 grams per cubic centimeter. Nothing exists in the normal solid or liquid states that can support the crushing pressure inside the Sun. The only possible form of matter under these conditions is a gas that is both dense and hot.

A gas at a temperature of millions of degrees is unlike any ordinary gas we know. The atoms move at high speeds – hundreds of kilometers per second – and their frequent and energetic encounters with one another strip away their electron clouds. Hence most atoms tend to be fully ionized (all their electrons removed) and the gas consists of negative electrons and positive atomic nuclei moving freely as independent particles. The radiation within this hot and dense gas consists not of the gentle beneficent light emitted from the comparatively cool surface but of intense x-rays. Each ray of this intense radiation in the deep interior travels on the average 1/10 000 of a centimeter before it is captured or deflected by particles of the gas. Pressure in such a gas is proportional to density multiplied by the temperature, and the high central temperatures in stars are the result of the very large pressures needed to support stars against their internal pull of gravity.

Thus we understand why stars are luminous. Their interiors are hot because of the high pressure needed to withstand gravity, and the radiation in the high temperature gas slowly diffuses to the surface and escapes into space. Nuclear reactions replace the lost energy and maintain the stars in a luminous state for long periods of time. Stars are luminous not because of nuclear energy but because their great masses require high internal pressures.

The radiation inside a star is continually scattered by gas particles and its outward flow to the surface is greatly impeded (Figure 5.5). The time taken by radiation to diffuse from the center to the surface of the Sun is about 20 million years. If the generation of nuclear energy suddenly ceased in the center of the Sun, we would not know

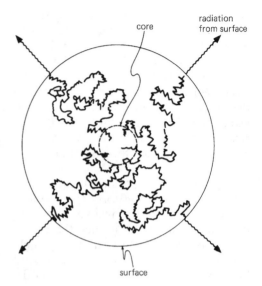

Figure 5.5. Radiation in a star is continually scattered by gas particles and slowly diffuses to the surface. In the Sun the energy generated in the central region takes about 20 million years to reach the surface.

that anything serious had happened until 20 million years later.

Convection and sound waves

Impedance to the flow of radiation is known as opacity. When the opacity becomes high, as often happens, the gas dams up interior radiation and energy is then transported by convection. Gas is stirred into motion, and ascending and descending currents of gas carry energy in the form of heat toward the surface. The outer layers (or envelope) of the Sun have high opacity and radiation cannot easily reach the surface by diffusion. This throws the envelope of the Sun into a state of convection, very much like the water in a boiling kettle, and heat is transported to the surface. Stars of smaller mass than the Sun have deeper convective envelopes; and stars more than twice the Sun's mass lack convective envelopes, but have instead convective cores. In the more massive stars, nuclear energy is released in a small central region, and the core is thrown into a convective state because radiative transport of energy is too slow.

The song of a star outrivals the song of the humpback whale. The star's interior is filled with a symphony of sound ranging from deep reverberating rumblings to quivering high-pitched shrieks. Nobody has told a star what size and shape it must be, or with what brightness and temperature it must shine, or how it must find the energy that is continuously lost from its surface. Sound waves travel through a star, such as the Sun, in about 1 hour, and by making ceaseless adjustments, with each part sending out sound waves to all other parts, the star seeks each moment to rediscover its natural equilibrium state. At the lowest modes of vibration, the star constantly heaves and groans, at the other extreme, at the highest modes 60 octaves higher, it is filled with the hiss of high-speed particles frantically jostling one another. Helioseismology, the study of the vibrations of the Sun's surface, reveals much of the Sun's interior structure.

The Sun acts as an immense loudspeaker. Its density decreases steadily from the center to the surface and each sound wave, as it travels outward, grows in amplitude like a whiplash. An amplified torrent of sound reaches the surface, passes through, and dissipates in the outer thin atmosphere. In the case of the Sun, with its noisy convective envelope, this dumping of acoustic energy raises the corona – the upper solar atmosphere – to a temperature of one million or more degrees kelvin. Because of its low density, the corona cannot radiate away the energy it receives, and does the only thing possible, it expands and carries the energy with it. The corona is like a giant jet engine; it sucks in gas from the Sun, and the gas, heated by acoustic energy, blasts away at high speed. This is the outward-streaming solar wind that carries away each second 100 billion kilograms of gas at a speed of several hundred kilometers a second. Other stars also have stellar winds generated by their internal acoustic tumult. Sometimes these stellar winds are much stronger than the solar wind, and some stars are literally blowing themselves away on a time scale of millions of years.

NUCLEAR ENERGY
Atomic nuclei

Stars are immense nuclear reactors that generate nuclear power. The energy falling on the Earth's surface as sunlight originated as nuclear energy deep inside the Sun. We shall take a detour to try to understand how stars generate nuclear energy.

Atoms combine to form molecules and are held together by electrical forces that result from atoms sharing or exchanging their outermost electrons. The assembly and the rearrangement of atoms in molecules releases chemical energy. Most of the energy used by human beings comes from burning wood, coal, oil, and gas and is therefore chemical.

Each atom consists of a small positively charged nucleus surrounded by a comparatively large cloud of negative electrons. The nucleus itself consists of heavy particles called nucleons that are either protons or neutrons. Protons are positively charged and neutrons have similar mass but no electrical charge. The nucleons in a nucleus are held together by strong nuclear forces. The addition, subtraction, and rearrangement of nucleons in the nuclei of atoms releases or absorbs nuclear energy, and this nuclear energy is generally millions of times greater than the chemical energy released by the addition, subtraction, and rearrangement of atoms in molecules.

Ancient alchemists sought for a way to transmute the elements and we have now realized their dream. The transmutation is done in nuclear reactors, not for the purpose of producing gold from baser metals, but for research and the release of nuclear energy.

Imagine that we have a supply of free nucleons that we combine in various ways to produce the atomic nuclei of the chemical elements. Each time a nucleus is constructed out of free nucleons, no matter what kind of nucleus it is, energy is released because nucleons attract one another with strong short-range nuclear forces. The total energy released in the construction process is the binding energy of the nucleus.

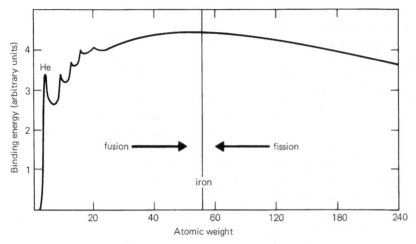

Figure 5.6. The binding energy curve of atomic nuclei. The maximum binding energy per nucleon occurs at iron.

All discrete things have binding energies of one kind or another. A stone is bound to the Earth by gravity and its binding energy is the energy released when the stone falls in from outer space, enters the atmosphere, and strikes the Earth's surface. The attractive force in this case is gravity. The attractive forces between the atoms in a molecule are electrical. The attractive forces between nucleons in a nucleus are known as the strong force or the strong interaction. When a thing is assembled, its binding energy is released; when it is dismantled, the energy expended in breaking it up equals its binding energy. It is more convenient to think of the binding energy per nucleon, that is, the total binding energy of the nucleus divided by the number of nucleons. Figure 5.6 shows the binding energy per nucleon for chemical elements of different atomic weight (the atomic weight is approximately the number of nucleons in the nucleus). Binding energy per nucleon rises rapidly at first for the light nuclei, then slowly increases for nuclei of greater atomic weight, reaches a maximum at iron (which has 56 nucleons), and thereafter steadily decreases. For example, if we start from scratch with 224 free nucleons, more energy is released by making four iron nuclei than from making one radium nucleus of atomic weight of 224.

Fusion and fission

Normally we cannot start from scratch with free nucleons. Protons, the nuclei of hydrogen atoms, are easy to find, but neutrons are scarce because in their free state they decay and have a lifetime of only 10 or so minutes. We must use existing nuclei, and either put them together (this is known as fusion), or break them up (this is known as fission). The aim in the nuclear energy game is to increase binding energy, and the prize is the energy released. To increase nuclear binding energy, we must move toward the iron peak, shown in Figure 5.6. When the move is from the left, energy comes from the fusion of light nuclei into heavier nuclei; and when the move is from the right, energy comes from the fission of heavy nuclei into lighter nuclei. Stars obtain their energy from the fusion of light nuclei. We on Earth at present obtain nuclear energy by fission of the heavy nuclei of uranium and plutonium.

All main-sequence stars obtain their energy by combining hydrogen nuclei (protons) to form the nuclei of helium atoms. Four protons are fused together to produce one helium nucleus that weighs

Table 5.8. *Energies*

Total energy in 1 gram	10^{14} joules
Chemical energy in 1 barrel of oil	10^{10} joules
Energy needed to raise 1 gram of water 1 degree kelvin	4.2 joules
Energy needed to raise flea 1 centimeter	1 erg
1 joule $= 10^7$ ergs	
1 watt $=$ 1 joule per second	
Luminosity of Sun	4×10^{33} ergs per second $= 4 \times 10^{26}$ watts
Power of sunlight incident on Earth	10^{17} watts
Large power station	10^9 watts
Flashlight power	0.2 watts

almost 1 percent less than the four original free protons. The loss in weight is because energy has mass and the released energy carries away a fraction of the total mass. Energy in every form has mass; a kettle of water, for example, when heated to boiling point weighs one billionth of a gram more than when cold because heat is a form of energy and has mass. The law that relates energy and mass is

$$\text{energy} = \text{mass} \times c^2, \qquad [5.2]$$

where c is the speed of light. One thousand kilograms (roughly 1 ton) of matter, if annihilated entirely, could supply the energy needs of the human race for one year. The Sun consumes and radiates away its mass at 4 billion kilograms per second (see Table 5.8).

Barrier penetration

A star on the main sequence generates its energy by slowly converting hydrogen into helium. This energy from nuclear reactions is released slowly in the central region of the star and diffuses toward the surface.

Why is nuclear energy released slowly in stars? Why not suddenly with an immense explosion of energy, as in a nuclear bomb? The explanation is easily understood. Protons are positively charged and their electrical repulsion acts as a barrier that deters them from coming close together. Positively charged protons in the deep interior of a star rush around at speeds of hundreds of kilometers a second; they continually

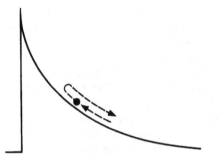

Figure 5.7. The electrical repulsion between two protons is like a hill. Two protons approach, and because of mutual repulsion each, in effect, climbs a hill. But normally they lack sufficient energy to reach the top, so they fall back and go off in new directions.

approach one another, yet, because of their electrical repulsion, rarely come close enough to engage in nuclear reactions.

A proton approaching another proton must in effect climb a hill, as shown in Figure 5.7. It moves up the hill, gets only so far, then comes down again and moves off in a new direction. Protons have different speeds; some move fast, others move slow, and the average speed in the center of the Sun is typically 500 kilometers per second. But this speed is much too small to enable a proton to get anywhere near the top of the hill. To reach the top, a speed of 10 000 kilometers per second is needed, and in the whole of the Sun not a single proton has this high speed.

We have considered protons as if they were bodies just like stones. This is a misleading mechanistic picture. Often atoms are portrayed as miniature solar systems with electrons moving in orbits about the nucleus like planets moving around the Sun. This similarly is a misleading picture and fails to reveal the beauty and intricacy of the subatomic world. All subatomic particles, such as electrons and protons and neutrons, behave like waves spread out in space. These vibrating waves form many patterns. Thus electron waves, waltzing about the nucleus, determine the size and properties of each type of atom.

Protons, like electrons, behave as waves, but their wavelengths are much smaller than the wavelengths of electrons. This explains why the nucleus that contains nucleon waves is so small. All particles have corpuscular and wavelike properties and behave often in strange ways quite unlike the ways of familiar bodies such as stones. A stone thrown against a wall usually rebounds, whereas a wave, such as a radio wave, may penetrate the wall and emerge on the other side with diminished intensity. Because of its wavelike nature, a particle such as a proton may also penetrate a wall. This takes us into the world of quantum mechanics where corpuscles are what we see, and waves explain what we see. Because of its wavelike nature, a particle is spread out in space and the chance of finding it in its corpuscular form at any point is proportional to the square of the wave amplitude at that point. Where the wave amplitude is largest there is the best chance of finding the particle as a discrete entity.

A particle, such as an electron or a proton, spreads itself out in space in the form of waves; when the particle is detected, the waves collapse, and the particle assumes the observed corpuscular form. Of course, we cannot have only a bit of a particle, and in the corpuscular form it must always be either the whole particle or nothing. Consider what happens when a proton encounters a barrier such as the electrical repulsion barrier between it and another proton. As a wave, it is partly reflected by the barrier, and is partly transmitted.

Suppose the amplitude of the wave after penetration is only 1/10 of the original incident amplitude. The chance of finding the particle is proportional to the square of the amplitude, and therefore the chance of finding it on the other side of the barrier is 1/100. It is impossible to have only 1/100 of a particle in corpuscular form on one side of the barrier and 99/100 of a particle on the other side. We therefore say the chance of penetration is 1/100, and of every 100 particles striking the barrier, on the average 99 are reflected and 1 is transmitted.

The hang ups

The first hang up is the repulsion barrier. At each encounter between protons a small chance exists of a wavelike penetration of the electrical repulsion barriers (see Figure 5.8). Even in the center of the Sun, where the temperature is high and protons move fast, the chance of penetration is small. Each proton makes head-on collisions with other protons about one trillion times a second, and about once every second it penetrates a repulsion barrier and comes face to face with another proton.

The second hang up is the weak interaction. When protons meet face to face after penetration they take a long time to react together; before they have made up their minds that they like each other they have separated and gone their different ways. Once in 10 billion years, on the average, each proton in the center of the Sun comes face to face with another proton and together

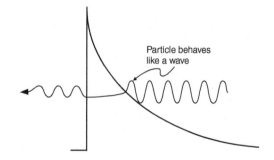

Figure 5.8. A wave penetrating through a barrier illustrates the wavelike nature of an encounter between two particles.

they react violently. Energy is released and they transform into a joint particle known as the deuteron. The deuteron contains one proton and one neutron and forms the nucleus of the heavy hydrogen atom. (Heavy hydrogen is known as deuterium.) The deuteron now quickly picks up another proton – the second hang up no longer applies – and with the release of more energy becomes a helium-3 nucleus that contains two protons and one neutron. Helium-3 nuclei then quickly combine together with the release of yet more energy and become the helium-4 nuclei of ordinary helium atoms (see Figure 5.9).

The nuclear reactions just described are known as the proton chain in which four protons, step by step, become one helium nucleus. Two main hang ups exist: first, positively charged protons have difficulty

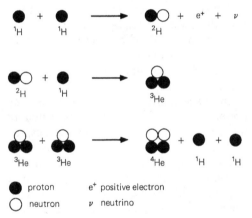

Figure 5.9. The fusion of four protons into one helium nucleus. First, two protons ($^1H + {}^1H$) combine to form a deuteron (^2H) consisting of one proton and one neutron, and a positive electron (e^+) and a neutrino (ν) are created by the reaction. This is a very slow process that takes about 10 billion years in the center of the Sun. Once formed, a deuteron quickly picks up an additional proton and becomes a helium-3 (^3He) nucleus. Two helium-3 nuclei then combine to produce one helium-4 (^4He) nucleus and two free protons. This conversion of four protons into one helium nucleus is the important proton chain of reactions. Competing reactions exist; for example, helium-3 combines with helium-4 to produce a beryllium-7 (^7Be) nucleus that by proton capture becomes two separate helium nuclei.

in penetrating their repulsion barriers; second, protons are slow to engage in a nuclear reaction with each other even after penetration. This second hang up exists because a weak interaction is involved that transforms a proton into a neutron and requires the creation of a positron (a positive electron that is the antiparticle of the common negative electron) and a neutrino that has little or no intrinsic mass and moves at or close to the speed of light.

Elements heavier than helium, such as carbon, oxygen, nitrogen, and so on, add up to 1–2 percent of the matter in stars like the Sun. In main sequence stars, a second method of converting hydrogen into helium – the carbon cycle – works as follows. A carbon-12 nucleus combines with a proton and transforms into nitrogen-13, which decays and becomes carbon-13; the carbon-13 nucleus combines with a second proton and transforms into nitrogen-14; the nitrogen-14 nucleus combines with a third proton and transforms into oxygen-15, which decays and becomes nitrogen-15; the nitrogen-15 nucleus finally combines with a fourth proton and produces nitrogen-16, which is unstable and immediately splits into a carbon-12 nucleus and a helium-4 nucleus. Energy is released at each step of the carbon cycle in which four protons are transformed into one helium nucleus. The carbon-12 nucleus itself acts as a catalyst and is not consumed in the process. The electrical repulsion barriers of carbon-12, carbon-13, nitrogen-14, and nitrogen-15 are higher and more difficult to penetrate than the barriers encountered in the proton–proton chain, and the barrier penetration hang up is therefore greater in the carbon cycle. But the reluctance to engage in a nuclear reaction after penetration is very much less in the carbon cycle because no weak interaction is involved during the fleeting moment of the proton–nucleus encounter. In the carbon cycle there is thus a trade-off in hang ups: the first is increased and the second decreased. In lower main sequence stars, including the Sun, the proton chain dominates, and in upper main sequence stars the carbon cycle dominates. Upper main sequence stars have higher central temperatures and in these stars protons penetrate more easily the strong repulsion barriers of the carbon, oxygen, and nitrogen nuclei.

BIRTH OF STARS
Dark clouds

James Jeans, a physicist, proposed early in the twentieth century a general theory of "fragmentation" (see Figure 5.10). In the beginning, he argued, the universe was filled with chaotic gas, and astronomical systems were formed in succession by a process of fragmentation "of nebulae out of chaos, of stars out of nebulae, of planets out of stars, and of satellites out of planets." We still think the universe fragmented into galaxies, and galaxies fragmented into stars, but we no longer think that stars fragment into planets, and doubt that planets normally fragment into satellites such as the Moon.

The majority of stars in our Galaxy formed long ago. But many are young because new stars are still being born at a rate of one or two each year. They are apparently born in the large and dark clouds of gas in interstellar space. These dark clouds

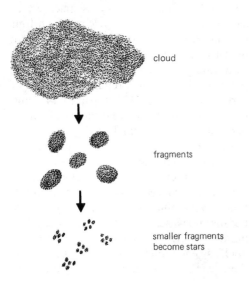

Figure 5.10. The sequential fragmentation of a contracting interstellar gas cloud (or part of a gas cloud) into smaller and smaller fragments.

consist of hydrogen and helium and contain a small amount – 1 or 2 percent by mass – of heavier elements, mostly in the forms of grains of dust. Very young stars are still close to their birthplaces, surrounded by the tattered remnants of clouds from which they were born. Sometimes hundreds of young stars cluster together, as in the Pleiades, and are all born about the same time from the same dark cloud (see Figure 5.11).

Origin of the solar system

The Solar System consists of the Sun, the Earth and other planets, and was born approximately 5 billion years ago when the universe was somewhere between one-third to one-half its present age. We can imagine that the Sun began as a blob, a denser part of a cloud where other blobs were forming into stars. The blob became a globe of gas and dust – a protostar – dark and cool, typically twice as massive as the present Sun. It consisted of hydrogen gas, about 25 percent by mass of helium, and all the heavier elements amounted at most to 2 percent.

In a speculative vein, we shall use the present tense as if we were there watching the scene. The globe of gas rotates, as shown in Figure 5.12, and at first is more or less spherical. But as the globe contracts, it rotates faster, and slowly flattens and becomes oblate. The central region (the core) contracts faster, gets denser, and hence spins faster than the outer regions of the globe. Contraction of the core eventually slows down because of the effect of centrifugal force. But another force now becomes important. This is viscous drag that acts like friction. Because the central regions rotate relatively fast, the globe consists of layers of gas moving at different speeds, rubbing against one another. This rubbing acts as a braking mechanism that slows the spinning core and transfers its rotation to the outer regions of the globe. By means of viscous forces (and the action of magnetic fields) the core continues its slow contraction to higher densities.

Figure 5.11. These majestic stars of the Pleiades were born 60 million years ago and are still festooned with remnants of the gas cloud from which they were born. (Mount Wilson and Las Campanas Observatories. Mount Wilson photograph.)

Outside the central region, or core, the gas remains moderately cool and its chemical elements heavier than hydrogen and helium tend to consolidate into grains of dust. These dust grains, normally about one thousandth of a millimeter in diameter, repeatedly collide with one another and begin to stick together. They aggregate and form small pieces of meteoroidal rock and ice. Perhaps more than 50 percent of this meteoroidal material is in the form of ices of frozen water, ammonia, and methane, and the rest consists of chemical elements such as the metals and their oxides and silicates. When many of the meteoroids have grown and become pebble-sized, they move more or less freely through the gas and begin to behave like tiny planets. But the friction of their motion through the gas causes their orbits to become circular and to settle into a flat disk. Within this thin rotating disk of matter (which lies inside a thick rotating disk of gas) the meteoroids ceaselessly jostle one another and either break into smaller pieces or coalesce to form larger chunks of matter. This is a game of survival of the biggest; the bigger the meteoroid, the more it eats up the smaller meteoroids, and the less likely it is shattered by collisions. The large meteoroids grow into planetesimals, hundreds and even thousands of meters in diameter. All the time there is a downpour into the disk of newly formed meteoroids. The planetesimals themselves occasionally collide and either dissolve into far-flung fragments or

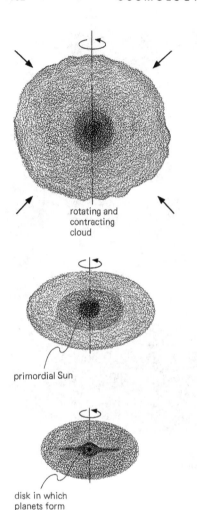

rotating and
contracting
cloud

primordial Sun

disk in which
planets form

Figure 5.12. A possible beginning of the Solar
System. A rotating globe of gas slowly contracts;
as it gets smaller, it spins faster, and as a
consequence the outer regions flatten. The central
region meanwhile becomes dense and hot and
develops into the newborn Sun. The planets form in
an encircling disk of meteoroidal matter.

aggregate into larger bodies. In the planetesimal struggle of survival of the biggest, the
victors sweep up all they encounter and
become the protoplanets. The protoplanets
– massive earthlike bodies – attract and
begin to retain the gaseous elements of the
globe and acquire atmospheres rich in
hydrogen and helium. These atmospheres
grow and the protoplanets become large

planetary spheres of gas, not unlike Jupiter
and Saturn at present.

The core of the globe meanwhile becomes
dense and hot and begins to approach its
final state. The primordial Sun, blanketed
from view by swirling gas and dust, has discovered that by fusing hydrogen into helium
it has access to an immense reservoir of
nuclear energy. It becomes convulsive, a
flaring T Tauri-type star, seeking to find an
internal structure that matches the rate at
which nuclear energy is released in the center
to the rate at which energy is lost from the
surface. (T Tauri is an irregularly varying
star that is surrounded with gas and dust,
and is evidently a newborn star approaching
the main sequence; similar stars are referred
to as T Tauri-type stars.) The primordial
Sun, as it approaches the main sequence, is
in an eruptive state, and from its upheavals
issues an intense wind of fast-moving gas
that rushes outward and carries away large
quantities of matter. The fierce wind and
brilliant radiation from the newborn Sun
thrust the remnants of the gaseous globe
back into interstellar space.

Now commences a final struggle between
the planets and the Sun. The planets try to
hold on to their massive atmospheres while
the bright Sun with its fierce wind tries to
strip away these atmospheres. The inner
planets – Mercury, Venus, Earth, Mars –
lose the struggle and are stripped down to
their rocky cores; the outer planets – Jupiter,
Saturn, Uranus, Neptune – win because of
their greater distances and retain forever
their lighter elements (see Figure 5.13).

The Sun at last settles down into a quies-
cent state. But the Solar System is cluttered
with debris from the birth process. Scattered
planetesimals encircle the Sun and the
planets in great numbers and hordes of
meteoroids drift in interplanetary space.
The mopping up of this debris by the Sun
and planets is at first rapid, and then slows
down after several hundred million years.
This is the bombardment era that lasts for
roughly half a billion years. The results of
this era of intense bombardment are still
visible on the surface of the Moon.

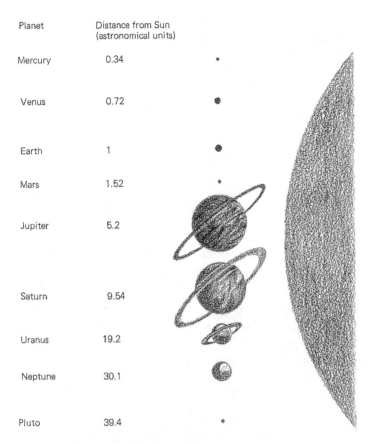

Planet	Distance from Sun (astronomical units)
Mercury	0.34
Venus	0.72
Earth	1
Mars	1.52
Jupiter	5.2
Saturn	9.54
Uranus	19.2
Neptune	30.1
Pluto	39.4

Figure 5.13. The planets of the Solar System, their relative sizes, and their distances from the Sun.

Tens of thousands of surviving planetesimals – the asteroids – still exist between the orbits of Mars and Jupiter; apparently they failed to amalgamate into a planet. Several other planetesimals survive as moons distributed about the various planets; the inner planets have three, of which the Moon is the largest, and the outer planets have at least 29.

THE STAR IS DEAD! LONG LIVE THE STAR!

Hydrogen exhaustion

Stars less massive than the Sun evolve more slowly, and stay on the main sequence longer than 10 billion years. Many low-mass stars have not evolved appreciably in the lifetime of our Galaxy. But more massive stars evolve more quickly, and according to their masses, terminate as either white dwarfs, neutron stars, or black holes. Here we consider briefly the fate of these more massive stars.

After a star consumes its central supply of hydrogen, it leaves the main sequence and moves in the direction of the red giants in the H-R diagram. The core now consists almost entirely of helium and has ceased to generate nuclear energy. But radiant energy is still lost from the hot surface of the star, and this energy drains from the central core. In response, the core does the only thing possible: it contracts to higher density, thus releasing gravitational energy. Because of this contraction, the temperature of the core rises. Two things begin to happen.

First, hydrogen just outside the helium core begins to burn. Surrounding the core a hydrogen-burning shell develops in which the production of helium continues and steadily adds mass to the helium core. Second, energy released by contraction of the core and hydrogen burning in the surrounding shell causes the envelope to expand. As the core contracts, the star swells up, and the distended envelope becomes partly convective. The star – luminous, large, and cool – is now a red giant.

Once a star leaves the main sequence it has entered old age with little time left to live. It realizes too late that life on the main sequence has been dull and sedentary and resolves to have a last glorious fling before death. But not much nuclear energy remains. The burning of hydrogen into helium has consumed 80 percent of the available nuclear energy, and to draw on the remaining reserves requires the prodigious feat of burning helium step by step all the way to nickel and iron. At each step, higher temperatures and densities are needed; hence the star grows in luminosity, and its diminishing nuclear reserves are consumed at an increasing rate.

White dwarfs

We consider an evolved star less massive than about 2 solar masses. Its core continues to contract during the red-giant phase until after tens of millions of years its density and temperature are sufficient to burn helium into carbon. The ignition of helium-burning occurs abruptly at a temperature roughly 100 million kelvin, and is referred to as the helium flash. An intense stellar wind builds up and either at this stage or a little later, depending on the mass of the core, the star throws off its distended envelope, and the bright naked core is all that remains. Contraction ceases, nuclear burning ends, and the core settles down as a white dwarf. The star consists of an ejected gaseous remnant, called a planetary nebula, and a slowly cooling white dwarf that has a size approximately equal to that of the Earth. The Sun in about 5 billion years will become a white dwarf,

and will shine in the sky as a pale light for several billion years while slowly cooling.

Supernovas

We next consider what happens to a more massive star. It squanders its central supply of hydrogen at a more rapid rate and lives on the main sequence for only a few hundred million years. It then becomes a short-lived monstrous red giant. The helium core, surrounded by a hydrogen-burning shell, contracts to higher density and temperature. Soon the helium begins to burn into carbon and oxygen. The star now has a core of carbon and oxygen, surrounded by a helium-burning shell and an outer hydrogen-burning shell. The star evolves beyond the red giant stage, becomes even more luminous, passes through spasms of pulsation, and ejects large quantities of gas into space at high speed.

To meet the ever-growing demand for more energy, the core continues to contract. When the central temperature exceeds 3 billion kelvin, and the density approaches one million grams per cubic centimeter, the carbon and oxygen burn progressively, stage by stage, to neon, magnesium, silicon, phosphorous, sulfur, and so on, to nickel and iron. But the nuclear energy released in this multitude of reactions is soon radiated away.

In these latter stages of advanced evolution an additional loss of energy has emerged and has steadily increased. Hordes of neutrinos, produced in the core by nuclear reactions and the high-temperature gas, stream out through the star in vast numbers. The neutrino luminosity of the core rises and exceeds the radiation luminosity of the surface. Gravitational energy is all that remains, and to meet the growing loss of energy the core contracts faster. The central density and temperature soar and energy spills over and drains away into the production by fusion of elements heavier than iron.

The star is seconds away from death. Neutrinos no longer easily escape from the star; instead, they diffuse outward from the core and transport energy that heats and ignites nuclear reactions in the hydrogen-rich outer layers of the star. The inward-falling core crushes its heavy elements into helium, and the energy previously acquired by fusion of light elements into heavier elements must be paid back by the release of further gravitational energy. The neutrino output from the core intensifies and becomes a blast that lifts the exploding envelope and hurls it into space. During the last brief moments of the imploding core, helium is crushed into free protons and neutrons, and all the energy that was radiated for millions of years while on the main sequence must be paid back immediately. The core obtains this energy by final catastrophic collapse. The electrons are squeezed into the protons and together they become neutrons. The collapsed core, divested of its envelope, emerges as a neutron star.

The titanic blaze of energy unleashed by the imploding core and exploding envelope results in a supernova that for a short time shines as bright as all the stars in a giant galaxy. A supernova at the distance of Alpha Centauri would be as bright as the Sun.

Neutron stars and black holes

A neutron star has a radius of little more than 10 kilometers and a density of nearly 1000 trillion grams per cubic centimeter. A thimbleful of neutron matter would weigh on Earth one billion tons. The neutron star has a magnetic field of 10^{12} gauss – a trillion times stronger than the Earth's magnetic field – and at first rotates rapidly at hundreds of revolutions per second.

The star is dead! Long live the star! From the ashes of the old star a pulsar is born, a star that ululates across space a pulselike message of matter stressed to its uttermost limit. For a million or so years the pulsar electromagnetically radiates away its rotational energy, turning more and more slowly. (See Figure 5.14.)

Neutron stars have masses less than about three times the Sun's mass, and this limit exists because neutron matter cannot withstand the gravitational pull of greater masses. The imploding cores of more massive stars may therefore not terminate as neutron stars, but continue to collapse and become black holes. These intriguing bodies are discussed in Chapter 13. Here it suffices to say that they are enclosed in their own curved space.

REFLECTIONS

1 *A total of roughly 2000 stars can be seen by the naked eye from the Earth's surface. With good binoculars this number increases to 12 000. The Galaxy contains about 100 billion (10^{11}) stars and the visible universe contains about 10^{20} stars. The nearest star, Alpha Centauri, consists of two stars too close together to be resolved by the unaided eye. Actually, it is a triple system; a third star, Proxima Centauri, is faint and slightly closer. Sirius, the brightest star seen from Earth (other than the Sun) has also a faint white dwarf companion.*

2 *Since the time of Hipparchus in the second century BC, astronomers have used a system of units called magnitudes for measuring the brightness of stars. Visible stars were classified into six magnitudes: the first magnitude consists of the brightest observed stars and the sixth magnitude consists of the faintest observed stars. John Herschel in 1836 found that stars in the first magnitude are 100 times brighter than stars in the sixth magnitude, and George Pogson (also an astronomer) showed in 1850 that each successive magnitude corresponds to a decrease in brightness by a factor 2.5. We see this from the relation $(2.5)^5 = 100$, or more precisely $(2.512)^5 = 100$, and an increase in five magnitudes from the first to the sixth corresponds to a hundred-fold decrease in brightness. A star of second magnitude is 2.5 times fainter than a star of first magnitude, and a star of magnitude m is $2.5^{(m-1)}$ times fainter than a star of first magnitude.*

Brightness is proportional to the inverse-square of distance, or $1/L^2$, where L is the distance of the source. Hence L^2 is proportional to $2.5^{(m-1)}$, and because $2.5^5 = 100 = 10^2$, it follows that L is proportional to $10^{(m-1)/5}$. The apparent magnitude m measures the

Figure 5.14. The Crab Nebula in the constellation of Taurus is a strong source of radio waves and has a total luminosity 100 000 times that of the Sun. This immense output of energy from the nebula originates near the center at the pulsar, which rotates 33 times a second. (Mount Wilson and Las Campanas Observatories, Mount Wilson photograph.)

observed brightness of the source – often in a specific spectral range – and the absolute magnitude M is defined as the magnitude the source would appear to have at distance 10 parsecs. Hence

$$\frac{L}{10} = 10^{(m-M)/5},$$

when L is measured in parsecs. Taking the logarithm of both sides, we get

$$m - M = 5\log(L/10).$$ [5.3]

The distance L of a source can be determined when its absolute magnitude M is known. The difference m − M is called the distance modulus.

3   For traditional reasons, astronomers use strange units such as parsecs and magnitudes. To the annoyance of other scientists, particularly physicists, they avoid more fashionable units. Moreover, most scientists use meters and kilograms; astronomers continue to use centimeters and grams, and take the view that astronomical quantities are so immense

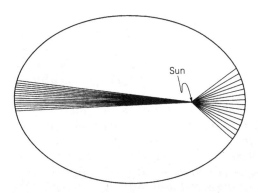

Sun

Figure 5.15. Kepler's first two laws of planetary motion: (1) a planet moves about the Sun in an elliptical orbit with the Sun at one focus of the ellipse; (2) a straight line joining the planet and the Sun sweeps out equal areas within the ellipse in equal intervals of time.

that the difference between centimeters or meters, and between grams and kilograms are trivial. Why trifle with the subject by changing from one system of units to another when each involves numbers that seem incomprehensible? Astronomers stick to the old units or use invented units more compatible with astronomical quantities. Distances are expressed in astronomical units, parsecs (or light years), masses are expressed in $xM_{\odot}$, or x times the Sun's mass $M_{\odot}$, and luminosities in $yL_{\odot}$, or y times the Sun's luminosity $L_{\odot}$.

4 Kepler's three laws of planetary motion (Figure 5.15):

• First law: a planet moves about the Sun in an orbit that is an ellipse with the Sun at one focus of the ellipse.

• Second law: a straight line joining the planet and Sun sweeps out equal areas within the orbit in equal intervals of time.

• Third law (published in 1619 in the Harmony of the Worlds): the square of the period of the planet is proportional to the cube of the average distance of the planet from the Sun. If P is the orbital period in years and R is the average distance of the planet from the Sun in astronomical units, then

$$P^2 = R^3,\qquad\qquad [5.4]$$

for all Solar System planets. For Venus: $P = 0.615$, $R = 0.723$; for Earth: $P = 1$, $R = 1$; and for Mars: $P = 1.881$, $R = 1.524$.

Kepler's laws of planetary motion were later explained by Newton and shown to be the natural consequence of the laws of motion and the inverse-square law of gravity. The laws also apply to double stars. Consider two stars, labeled 1 and 2, in orbit about each other, and let their masses be M_1 and M_2 measured in solar mass units. The period P of revolution of the two stars about each other and their average distance R apart are related by

$$(M_1 + M_2)P^2 = R^3,\qquad\qquad [5.5]$$

where P is measured in years and R is measured in astronomical units. For example, if $M_1 = 20$, $M_2 = 16$, and $R = 1$, then P is 2 months.

5 Matter in bulk exists in four states: solid, liquid, gaseous, and plasma. Solids are rigid, held together by interatomic electrical forces, and consist of atoms vibrating about fixed points. Liquids, gases, and plasmas are fluids in which atoms have mobility and are less controlled by the interatomic forces. When the temperature of a solid rises, the atoms vibrate more strongly and either singly or in clumps begin to slide around one another. The solid has become a liquid. At higher temperature the vibrating atoms break their bonds and evaporate to form a gas of freely moving particles. At even higher temperature, more than a few thousand kelvin, the atoms become ionized and the gas changes into a plasma consisting of free electrons and partially or fully ionized atoms. Most matter in the universe is in the plasma state.

The number of particles (atoms or molecules) in 1 cubic centimeter of gas at room temperature and atmospheric pressure at sea level is 2.7×10^{19}. They rush about, hither and thither, separated from one another by comparatively large regions of empty space, and each collides with the other particles about one billion times a second.

6 The four forces that rule the universe at the present time, in order of increasing

strength, are

gravitational interaction: 1
weak interaction: 10^{25}
electromagnetic interaction: 10^{39}
strong interaction: 10^{40}

Gravitational and electromagnetic forces are long range (decreasing slowly in strength as the inverse square of distance) and account for much of the rich diversity in the universe ranging from atoms to galaxies. Weak and strong forces are short range (operating over very small distances) and account for much of the diversity of the subatomic world.

Electrical forces are much stronger than gravitational forces, as shown by the following illustration. An electron has a negative charge $-e$ and a proton has a positive charge $+e$. The attractive force between a proton and an electron is e^2 divided by the square of their separating distance. The gravitational force between the same two particles, an electron of mass m_e and a proton of mass m_p, is Gm_em_p divided by the square of their separating distance, where G is the universal constant of gravity. The value of G is found by measurement. (No scientist has yet been clever enough to explain its value.) The ratio of these two forces between an electron and a proton is

$$\frac{\text{electrical force}}{\text{gravitational force}} = \frac{e^2}{Gm_em_p}$$

$$= 2 \times 10^{39}. \qquad [5.6]$$

Clearly, in a hydrogen atom the electrical force is vastly stronger than the gravitational force, and gravity is far too weak to hold atoms together. Positive and negative electric charges exist in equal numbers in the world around us, and their repulsions and attractions tend to neutralize each other in systems much larger than atoms and molecules. In our electrically neutral universe, the electromagnetic forces are rarely very strong except over short distances. Gravity, however, cannot be neutralized and gets progressively stronger as the number of particles in a system increases. The gravitational force between

individual particles is the weakest in nature and is negligible in subatomic, atomic, and molecular systems; but in stellar and galactic systems it becomes dominant, and on the cosmic scale is by far the strongest of all forces.

The strong force operates over the very short distance 10^{-13} centimeters (roughly the size of a nucleon) and acts rapidly in a time 10^{-23} seconds (given by the light travel size of a nucleon). It acts between nucleons (protons and neutrons); it is stronger than the electromagnetic force, and is able to hold together the nucleons in a nucleus despite their electric repulsions. If it were not stronger, the universe would consist of hydrogen atoms only and we would not be here discussing the subject. If by mischance it were much stronger, nuclei (diprotons) could form consisting of two protons only, and stars would not exist (nor would we) because most hydrogen would have been consumed in various ways (such as the formation of helium-2 nuclei) in the early universe.

The weak force is of even shorter range and acts very slowly compared with the strong force. Various weak interactions between particles produce neutrinos: ghostly particles with no electric charge and no mass that move at the speed of light. (Neutrinos may have a very small intrinsic mass as yet undetected.) The slowness of the weak interaction is responsible for the main hang up in the proton–proton reaction in stars, and it explains why neutrinos pass easily through ordinary matter without interaction. Nuclear reactions involving weak interactions in the Sun produce 10^{38} neutrinos each second, and "like ghosts from an enchanter fleeing" (in Shelley's words), they stream out freely through the Sun into space. Every second 1000 trillion neutrinos from the Sun pass through each person, even at night, when the Sun is on the other side of the Earth.

7 *The hydrostatic equation of a star can be expressed mathematically as follows. Consider a spherical shell of radius r and infinitesimal thickness dr. The volume of this shell is $4\pi r^2\, dr$, and its infinitesimal mass is*

$dM = 4\pi\rho r^2 \, dr$, where ρ is the mass density at radius r. This relation in the form

$$\frac{dM}{dr} = 4\pi\rho r^2, \qquad [5.7]$$

is the mass equation. The weight of the shell is $GM_r \, dM / r^2$, where M_r is the mass of the star inside radius r, and G is the universal gravity constant. Hydrostatic pressure is force per unit area. Let the pressure at radius r be P and at radius $r + dr$ be $P + dP$. The force $4\pi r^2 \, dP$ pushes outward and the weight $-GM_r \, dM / r^2$, equal to $-4\pi GM_r \rho \, dr$, pulls inward. In hydrostatic equilibrium these two forces balance, and hence,

$$\frac{dP}{dr} = -\frac{GM_r \rho}{r^2}, \qquad [5.8]$$

and this is the hydrostatic equation of a star. So far we have three unknown variables (M_r, ρ, P) but only two relations (Equations 5.7 and 5.8). Dependence of pressure on density ρ and temperature T in a plasma gives an extra relation (the equation of state), but at the cost of an additional variable T. Now we have four unknowns and three equations. The dependence of radiation diffusion on the opacity and temperature gradient (the temperature-gradient equation) and the dependence of energy generation (the luminosity equation) on density and temperature provides two more relations. We arrive finally at the five equations of stellar structure that enable us to solve for the five variables M_r, L_r, P, ρ, and T, where L_r is the interior luminosity at radius r.

8   The value of the universal gravity constant G is 6.7×10^{-8} in centimeter-gram-second units and 6.7×10^{-1} in meter-kilogram-second units. Often this constant is erroneously referred to as Newton's constant, but Newton did not use it, and like later astronomers in the eighteenth century he used a system of units in which $G = 4\pi^2 / M_\odot$, where distance is measured in astronomical units and time is measured in years.

9   The final words in Arthur Eddington's The Internal Constitution of the Stars read: "it is reasonable to hope that in a not too distant future we shall be competent to understand so simple a thing as a star." Is a star a simple thing? "Fred, you would look simple at a distance of ten parsecs!" This remark was made by a member of the audience at a talk given by Fred Hoyle in the library at the Cambridge Observatory in 1954 in response to his statement, "Basically, a star is a pretty simple structure."

True, basically the Sun is a simple structure. In detail, however, much remains to be understood. We do not fully understand the 11-year sunspot cycle: why do sunspots – dark regions on the surface of the Sun – come and go periodically. We do not fully understand the mechanism responsible for the solar wind. And most perplexing of all at present is the neutrino problem.

For decades scientists have tried to solve the riddle of why the Sun appears to generate fewer neutrinos than expected. The Sun derives its energy from the fusion of hydrogen into helium by various reactions. Whatever the reactions, four protons are consumed and two neutrinos emitted in the creation of each helium nucleus. If m_p is the mass of the proton and m_{He} the mass of the helium nucleus, the energy generated in the creation of a helium nucleus is $(4m_p - m_{He})c^2$. The total energy generated per second is the solar luminosity $L_\odot$, and therefore the number of neutrinos emitted per second is

$$N_{neutrinos} = \frac{2L_\odot}{(4m_p - m_{He})c^2}$$
$$= 2 \times 10^{38}. \qquad [5.9]$$

Sensitive experiments have so far succeeded in detecting only a fraction (about half) of the expected neutrino flux reaching the Earth.

10   The technological dream of modern science is to find a way on Earth to produce nuclear energy by fusion as in the Sun. The aim is to use deuterium (plenty of heavy hydrogen exists in the oceans) instead of ordinary hydrogen, and thus avoid the second hang up (the slow weak interaction) in the proton–proton chain. But the first hang up (the electrical repulsion between protons) remains. Although deuterons, when brought face to face, react willingly, releasing nuclear energy, there still exists the difficulty of penetrating their electrical repulsion barriers. It is widely believed that the way to achieve fusion is to imitate the stars. Thus deuterium gas must be heated to a temperature of tens, even hundreds, of millions of kelvin. The thermonuclear release of energy has already been achieved in an uncontrolled manner with the hydrogen bomb using a fission bomb as a detonator. The Sun also would explode if it were made solely of heavy hydrogen. We need a controlled way of releasing fusion energy, but the problem of maintaining a contained gas at the required high temperature for a sufficiently long period of time has not been solved.

11 In 1862, William Thomson (later Lord Kelvin) wrote: "The sun must, therefore, either have been created as an active source of heat at some time of not immeasurable antiquity by an over-ruling decree; or the heat which he has already radiated away and that which he still possesses must have been acquired by a natural process following permanently established laws" (William Thomson, "On the age of the sun's heat"). Following an idea proposed by the German scientist Hermann von Helmholtz, he assumed that the Sun replaces its emitted energy by slow gravitational contraction. He found that the Sun's age, calculated in this way, and now known as the Helmholtz–Kelvin time scale, is approximately 20 million years. Before the discovery of nuclear energy, gravitational contraction seemed the most plausible and natural explanation of the origin of the luminosity of the Sun and stars. A Sun's age of 20 million years, although generous compared with that sanctioned by the Mosaic chronology (Chapters 4 and 25), was too short for geological history and biological evolution, and the debate between physicists, geologists, biologists, and paleontologists became intense in the late nineteenth century. The discovery of radioactivity and the consequent realization that the Solar System is

billions of years old brought the debate to an end.

The Helmholtz–Kelvin time scale plays a useful role in the theory of stellar structure. For example, radiant energy diffuses from the center to the surface of the Sun on this time scale, and the luminosity multiplied by this time equals the heat content of the Sun.

12 Many supernovas have been observed in other galaxies in the twentieth century. Each supernova, for a few weeks, may outshine all the stars in its galaxy. For the last 2000 years we have records of 14 supernovas in our Galaxy, and the last seen was in 1604. Of the famous supernovas, the one seen in 1054 in the constellation Taurus was recorded by the Chinese and its remnant is now the Crab Nebula. Tycho's star of 1572 occurred in the constellation Cassiopeia and Kepler's star of 1604 occurred in the constellation Serpens. In 1572, Tycho Brahe wrote, "One evening, when I was contemplating, as usual, the celestial vault, whose aspect was so familiar to me, I saw with inexpressible astonishment, near the zenith in Cassiopeia, a radiant star of extraordinary magnitude."

PROJECTS

1 Stars have different colors and apparent brightness. Sketch a color–brightness diagram and show where you would put ordinary sources of light, such as a campfire, a 100-watt light bulb, a flash light bulb, a candle flame, a spark, a fire fly, and so on.

2 If two stars of similar mass orbit each other at a distance 4 astronomical units with a period of revolution 8 years, how massive is each of the stars? Suppose later it is discovered that one star has in fact a mass $0.1 \, M_\odot$ (where $M_\odot$ is the mass of the Sun), how massive is its companion?

3 Suppose that in a glass of water we radioactively tag each H_2O molecule. Now suppose that we pour the glass of water in the ocean and wait until it is thoroughly mixed with all the water on the Earth's surface. Show that when the glass is dipped back in the ocean it contains roughly 1000 of the original molecules.

4 A perfect 100-watt light bulb radiates 10^9 ergs of energy each second. How many grams of mass does it radiate in 1 year, and where does this mass come from and go to?

• The Sun's luminosity (energy radiated per second) is $L_\odot = 4 \times 10^{33}$ ergs per second. One watt equals 10^7 ergs per second, and hence the Sun radiates 4×10^{26} watts, or 4×10^{20} megawatts. A large power station generates 1000 megawatts, and the Sun is therefore equivalent to 400 thousand trillion large power stations. How much mass does the Sun radiate each year? What fraction of its mass does it radiate in a lifetime of 10 billion years?

• Show that the Sun contains approximately 2.5×10^{41} joules of heat (1 watt equals 1 joule per second).

5 What would happen to life on Earth if all nuclear reactions in the Sun ceased abruptly at this moment? (Remember, as the Sun contracts, it gets brighter!)

6 Give examples from everyday life of the various states of matter.

• The universe is ruled not by the Furies, nor the Four Horsemen of the Apocalypse, but by the four forces of nature. Discuss these forces.

7 Where did the elements that compose living creatures, such as C, N, O, Na, Mg, P, S, and Ca, come from?

• Write an ode to a coin on when and where its metal was made.

8 If you have difficulty understanding the release of gravitational energy, do the following: tie a heavy stone on the end of a piece of string (Figure 5.16) and let the string slide slowly through the hand. The friction of the sliding string generates heat that can be felt by the hand. The heat is produced by the release of gravitational energy as the stone descends. This is the analogue of the Helmholtz–Kelvin model: the slowly contracting Sun releases gravitational energy in the form of heat, which is radiated from the Sun's surface. From whence comes gravitational energy? Originally from the big bang! Matter in the dense early universe loses energy during expansion to lower den-

Figure 5.16. If you have difficulty in understanding gravitational energy, try the following experiment. Allow a string with a weight attached to the end to slip through the closed hand. The heat generated in the palm of the hand comes from the release of gravitational energy.

sity, and when matter in bounded systems later contracts back from low to higher density some of this energy is recovered. The heat generated in the hand by the sliding string is energy recovered from the big bang.

FURTHER READING

Readers seeking more information on stars (and galaxies) should consult one of the many elementary astronomy texts now available.

Cameron, A. G. W. "Origin and evolution of the Solar System." *Scientific American* (September 1975).

Davies, P. C. W. *The Forces of Nature.* Cambridge University Press, New York, 1979.

Franknoi, A. "The music of the spheres: Astronomical sources of musical inspiration." *Mercury* (May–June 1977).

Graham-Smith, F. "Pulsars today." *Sky and Telescope* 80, 240 (September 1990).

Heel, A. van and Velzel, C. *What Is Light?* McGraw-Hill, New York, 1968.

Henbest, N. and Marten, M. *The New Astronomy.* Cambridge University Press, Cambridge, 1996.

Meadows, A. J. *Stellar Evolution.* Pergamon Press, New York, 1974.

Murdin P. and Murdin, L. *Supernovae.* Cambridge University Press, New York, 1985.

Richardson, R. S. *The Star Lovers*. Macmillan, New York, 1967.

Shapley, H. *Beyond the Observatory*. Charles Scribner's Sons, New York, 1967.

Tayler, R. "The birth of elements." *New Scientist* 16, 25 (1989).

Trimble, V. "White dwarfs: The once and future suns." *Sky and Telescope* 77, 348 (October 1986).

Wheeler, J. A. "After the supernova, what?" *American Scientist* (January–February 1973).

SOURCES

Bahcall, J. N. "The solar neutrino problem." *Scientific American* (May 1990).

Bethe, H. A. "Supernova." *Physics Today* (September 1990).

Burchfield, J. D. *Lord Kelvin and the Age of the Earth*. Science History Publications, New York, 1975.

Burrows, A. "The birth of neutron stars and black holes." *Physics Today* (September 1987).

Eddington, A. S. *The Internal Constitution of the Stars*. Cambridge University Press, Cambridge, 1926.

Harvey, J. "Helioseismology." *Physics Today* (October 1995).

Jeans, J. H. *Astronomy and Cosmogony*. Cambridge University Press, Cambridge, 1929. Page 415.

Kepler, J. "The discovery of the laws of planetary motion." Translated by J. H. Walden from *Harmonices Mundi*. In *A Source Book in Astronomy*. Editors H. Shapley and H. E. Howarth. McGraw-Hill, New York, 1929.

6 GALAXIES

The fires that arch this dusty dot –
Yon myriad-worlded ways –
The vast sun-clusters' gathered blaze,
World-isles in lonely skies,
Whole heavens within themselves amaze
Our brief humanities.
Alfred Tennyson, Epilogue

OUR GALAXY

Milky Way

Our Galaxy, an enormous system of clouds of glowing gas and 100 billion stars, is also known as the Milky Way. Light takes 100 000 years to cross the Galaxy from side to side, and the center of the Galaxy lies in the constellation of Sagittarius, obscured from view by clouds of dusty gas that drift among the stars. Far from the center of the Galaxy is our own star the Sun.

The disk and halo

The Galaxy consists of two basic components: disk and halo (see Figures 6.1 and 6.2). The Milky Way is actually our panoramic view of the disk that has a diameter of about 100 000 light years and a thickness of about one-twentieth, or less, of the diameter. The disk is composed of stars and interstellar gas, and contains over half the visible mass of the Galaxy. The gas amounts to one-tenth of the matter in the disk, and the dust amounts to about 1 percent or more of the mass of the gas. The disk of stars, gas, and dust rotates about the center, or nucleus, of the Galaxy like a giant carousel. Most of the stars seen in the sky are in the disk and are separated from their nearest neighbors by distances of a few light years. The Sun is 30 000 light years from the nucleus; it moves at 300 kilometers per second and takes 200 million years to travel around the Galaxy in a circular orbit. In its lifetime, the Sun has journeyed 25 times around the Galaxy.

The spherical halo, centered on the nucleus of the Galaxy, has a diameter of roughly 200 000 light years. The central region of the halo forms the nuclear bulge of the disk. Outside the nuclear bulge the halo consists of low-density gas, widely separated stars, and about 120 globular clusters. Globular clusters are compact systems of hundreds of thousands of stars, and each globular cluster moves in an elliptical orbit about the nucleus of the Galaxy (see Figure 6.3). The halo does not rotate with the disk.

Two populations of stars

The flat disk, rich in gas and dust, rotates; the spherical halo, poor in gas and dust, rotates very much slower. A further distinction, discovered by Walter Baade in 1942, is that each is populated by a different kind of star. The two kinds of stars are known as populations I and II. The disk contains mostly population I stars and the halo contains mostly population II stars.

Population I stars – the disk stars – are the kind we have considered so far. The Sun is a population I star. These stars are usually much younger than the Galaxy and contain various chemical elements heavier than helium whose total abundance by mass is 1–2 percent. Gas clouds in the disk give birth to new stars. At their death many stars leave a legacy of heavy elements ejected into space. Each newborn population I star inherits heavy elements from stars that have previously died. Over time, the reservoir of gas in the disk decreases,

Figure 6.1. A schematic side view of our Galaxy, showing the disk with its nuclear bulge and the halo containing numerous globular clusters. (With permission, J. S. Plaskett, *Popular Astronomy*.)

and its enrichment with heavy elements increases. In interstellar space, atoms of heavy elements collide, tend to stick together, and form tiny dustlike grains. These grains of dust are usually less than 1 micron (1/10 000 of a centimeter) in size; big enough, however, to absorb visible light. The tiny dust grains ride the atomic winds between the stars and collect in the dark clouds where new stars form.

Population II stars are found in the halo with its numerous globular clusters and in the nuclear bulge. The density of gas in the halo is much too low for the formation of new stars, and most halo stars were born long ago when the Galaxy was young. Their estimated ages lie between 8 and 15 billion years, and they formed from hydrogen and helium gas that contained almost no heavy elements.

Planets cannot form from hydrogen and helium alone, and it seems most unlikely that the halo stars have planetary systems. In the younger population I stars, which formed from hydrogen and helium contaminated with heavy chemical elements, the heavy elements concentrated into dust grains; the dust aggregated into meteoroids that aggregated into planets and the cores of large gaseous planets. Possibly most population I stars, at least those similar to the Sun, possess planets, and these are the places where extraterrestrial forms of life might exist.

In the central region of the Galaxy – the nuclear bulge – exists a mixture of disk (population I) and halo (population II) stars. Although it is convenient to think of two distinctly different populations (the young disk stars and the old halo stars, the first rich and the second poor in heavy elements), naturally many intermediate populations also exist.

Star clusters and their distances
Star clusters are of two types: open and globular. Open clusters, found in the disk,

Figure 6.2. The giant spiral galaxy M 31 in Andromeda at a distance of 2 million light years, which resembles our own Galaxy, and is a member of the Local Group of galaxies. (Las Campanas Observatories, Mount Wilson Observatory photograph.)

are loose aggregations of young population I stars. Usually these clusters consist of hundreds of stars and often, for a few million years, are associated with clouds of gas, as in the Pleiades cluster. Globular clusters, found mostly in the halo and the nuclear bulge, are spherical in shape. They have diameters of about 100 light years and consist of hundreds of thousands of old population II stars. Clusters of both types, open and globular, are important because they enable the astronomer to estimate distances.

The distance of a nearby star is found by parallax measurements. The distance of a star farther away is found by comparing its apparent brightness with the brightness of a similar nearby star at known distance. Better results come from comparing clusters rather than single stars. All stars in a cluster have much the same age, have similar composition of chemical elements, and have approximately the same distance. The cluster consists of stars of various brightness and surface temperatures, and we take its family portrait by constructing an H-R diagram in which the apparent brightness of each star is plotted against its surface temperature. All stars fit beautifully on a well-defined main sequence. By comparing the family portraits, or H-R diagrams, of different clusters we find their relative distances. Thus if the main sequence in one

Figure 6.3. The globular cluster M 3 consists of hundreds of thousands of old population II stars. Held together by its own gravity, it is one of many such clusters in the halo of the Galaxy. (Association of Universities for Research in Astronomy, Inc., The Kitt Peak National Observatory.)

H-R diagram is four times fainter than the main sequence in a second H-R diagram, the first cluster is twice the distance of the second. This method works well, particularly when corrections allow for the difference in element composition of the clusters.

The nearest open cluster is the Hyades (see Figure 19.2). It outlines the face of the Bull in the constellation Taurus and consists of at least 200 stars at 140 light years distance. This important cluster is sufficiently close for its distance to be determined by parallax measurements. Also, its distance is determined by the moving-cluster method. The Hyades cluster is moving away from us and appears to be shrinking slowly in size. By measuring its velocity from the Doppler effect and its apparent rate of shrinking, we are able to determine the distance of the Hyades cluster. From the known distance of the cluster it is then possible, by main-sequence fitting, to determine the distances of other open clusters that are farther away.

A few open clusters at known distances contain cepheid stars. These pulsating stars obey a period–luminosity law; that is, the pulsation period is related to the luminosity. The cepheids are about 10 000 times more luminous than the Sun and can be observed at large distances. With the luminous cepheids as yardsticks we can now take giant strides and measure the distances of nearby galaxies. The Hyades cluster is thus very important, its distance determines the size of our Galaxy and other galaxies, and even the size of the universe.

Clouds of gas and dust

The determination of astronomical distances faces many uncertainties, not the least is the correct allowance for absorption of starlight by dust in interstellar space.

The word nebula – meaning cloud – was once used in astronomy to mean any fuzzy patch of light in the night sky. Charles Messier in the eighteenth century cataloged many conspicuous nebulae, some of which, as we now know, are distant galaxies (such as the Andromeda Nebula). The word

nebulae is now used primarily for clouds of interstellar gas (see Figure 6.4), and we distinguish between reflection nebulae and emission nebulae. Reflection nebulae reflect light from nearby stars and appear bluish in color; emission nebulae, heated by nearby or embedded stars, emit their own light and appear reddish. Many clouds, often large and nonluminous, are widely distributed in the disk, obscuring from view more distant stars and galaxies.

Astronomers early in the twentieth century suspected that starlight is absorbed by dark matter drifting in space between the stars. The existence of this obscuring medium was finally established in 1930 by Robert Trumpler of the Lick Observatory. We now know that the obscuration is caused by small grains of matter, or dust, distributed in the disk, particularly in the clouds of gas. Roughly 1–2 percent of the mass of interstellar matter is in the form of dust. Trumpler found that the intensity of starlight, owing to absorption, is halved every 3000 light years traveled in the disk. Thus a disk star 6000 light years away has an apparent brightness one-quarter the brightness it would have in the absence of absorption. In the disk, stars are seen telescopically to distances of several thousand light years; at greater distances, particularly in the direction of the galactic center, they become obscured from view.

The dust acts like a layer of fog in the disk. Looking in the plane of the Milky Way, we see many nearby stars but no galaxies, looking perpendicularly out of the plane of the Milky Way, we see few stars and many distant galaxies. This holds true in visible light, but not in infrared light and radio signals that can penetrate the fog.

Radiation at wavelengths greater than the size of dust grains is not easily absorbed or scattered (see Figure 6.5 for wavelengths). This explains why red light penetrates better than blue light in foggy weather. The inner region of the Galaxy, obscured from view in visible light, can be observed with long infrared and radio waves. Observations at these longer wavelengths reveal that the

Figure 6.4. This dark interstellar cloud, the Horsehead Nebula in the constellation Orion, is seen silhouetted against a background of stars and luminous clouds. (Mount Wilson and Las Campanas Observatories, Mount Wilson Observatory photograph.)

nuclear bulge, extending 10 000 light years from the nucleus, is a complex system of swirling gas and multitudes of stars. Old and young stars are thickly distributed in this central region and are hundreds of times closer to one another than in our region of the Galaxy. Several strong sources of radiation exist in the center – the nucleus – prompting astronomers to conjecture that the nucleus contains one or more massive black holes. These monsters of the deep cannot by themselves visibly radiate, but infalling gas, compressed and thereby heated, can radiate strongly.

Spiral arms

The striking thing about many external galaxies is their spiral-like appearance (Figure 6.6). Studies of the distribution of gas and stars in the disk shows that our Galaxy also has spiral structure. Spiral arms extend outward from the nuclear bulge and are made luminous by young stars and bright clouds of gas. We see the spiral structure of external galaxies because their spiral arms contain bright stars and gas clouds; other, and less conspicuous, stars in their disks are arranged more uniformly.

At first, astronomers thought spiral arms contained always the same stars, and stars and spiral arms rotated together in the disk. But this idea contradicted appearances. The inner stars of the disk revolve about the galactic center more rapidly than the outer stars in the way that inner planets revolve about the Sun more rapidly than outer planets. Therefore the arms would slowly wind more and more and form a

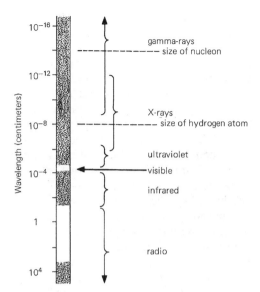

Figure 6.5. The electromagnetic spectrum, showing the gamma-ray, x-ray, ultraviolet, visible, infrared, and radio-wave regions. The Earth's atmosphere is transparent at wavelengths shown in the unshaded regions.

Figure 6.6. The spiral galaxy M 51 of type Sc. (Las Campanas Observatories, Mount Wilson Observatory photograph.)

progressively tighter spiral. Every hundred or so million years each spiral arm would gain an additional turn. Yet many spiral galaxies, apparently billions of years old, have only one or two turns. This "winding problem," as it was known, was solved by the Swedish astronomer Bertil Lindblad who proposed a density-wave theory.

According to the density-wave theory, spiral arms are ripples (or density waves) that travel around the disk like sound waves in air. Each spiral arm is actually a spiral-shaped ripple that moves around in the disk and preserves its spiral shape. Spiral arms are simply regions of higher gas density. The higher gas density triggers star formation (in a way not fully determined) and the spiral arms contain newborn stars. The gas and stars take tens of millions of years to pass through a spiral arm. But the brightest stars have lifetimes of only a few million years, and by the time they move out of a spiral arm, they have died. Between the spiral arms exists a deficiency of bright stars and their associated luminous gas clouds. Hence spiral arms are conspicuous because they contain the brightest stars. The distribution of older stars in the disk is only slightly disturbed by the passing spiral ripple and their aggregate light shows very little spiral appearance.

THE DISTANT GALAXIES
Elliptical and spiral galaxies
The distant galaxies, once known as nebulae, are separated from one another by millions of light years. They stretch away in countless numbers, seemingly endlessly, each a magnificent celestial city of stars moving serenely in the depths of space. Many possess distinguishing features that enable astronomers to classify them as either ellipticals or spirals.

Elliptical galaxies (called ellipticals) have an oval appearance; some are spherical, but most look like oblate spheres (Figure 6.7). Their outer regions fade away and they lack clear boundaries. Many, particularly large ellipticals, have bright centers. They contain almost no gas and dust from which

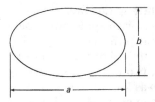

Figure 6.7. The apparent ellipticity of an elliptical galaxy is the quantity ellipticity $= 10 \times (a - b)/a$, where a is the apparent major diameter and b is the apparent minor diameter. An E5 elliptical galaxy, for example, has a major diameter twice its minor diameter.

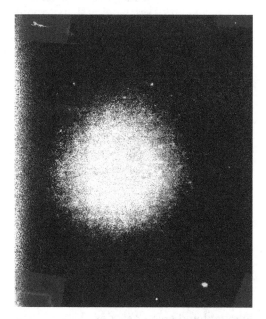

Figure 6.8. The giant elliptical galaxy M 87 in the constellation Virgo is a radio galaxy with a peculiar jet extending from the nucleus. (Association of Universities for Research in Astronomy, Inc., The Kitt Peak National Observatory.)

new stars can form, and consist mainly of old population II stars. The ellipticals cover a wide range of masses and sizes, from dwarf galaxies not much larger than globular clusters to rare giant galaxies, such as M 87 (Figure 6.8), and even rarer supergiant galaxies that are hundreds of times more massive than our Galaxy and have diameters as large as 500 thousand light years. Giant ellipticals, usually type E0 (pronounced E-zero), which are spherical

in appearance, have conspicuously bright nuclei, whereas dwarf ellipticals lack bright nuclei. The majority of galaxies – over 60 percent – are ellipticals, and the majority of the ellipticals are dwarf systems.

Spiral galaxies (called spirals) have, like our Galaxy, disks containing population I stars and halos containing population II stars. The disks have conspicuous spiral arms coiled around the central nuclear bulges. Spirals divide into two distinct sequences: normal spirals, denoted by S, and barred spirals, denoted by SB.

Normal spirals form a sequence of three types: the Sa have large nuclear bulges and tightly wound arms; the Sb have smaller nuclear bulges and less tightly wound arms; and the Sc have the smallest nuclear bulges and the least tightly wound arms. Our Galaxy and M 31 (Andromeda Nebula) are type Sb.

About one third of all spirals are the barred type. They are classified as SBa, SBb, and SBc, according to the size of the nuclear bulge and tightness of the spiral arms, exactly as in normal spirals. They are distinguished by a bright central bar that projects beyond the nuclear bulge and connects with the spiral arms (Figure 6.9). This bar consists of stars and gas; but why it exists is not fully understood.

Figure 6.9 NGC 1300, a barred spiral of type SBb. (Mount Wilson and Las Campanas Observatories, Mount Wilson Observatory photograph.)

Spirals are rich in gas that concentrates in their disks and feeds the birth of new stars. Unlike ellipticals, these galaxies have stars of all ages, with young population I stars in their disks and old population II stars in their halos and nuclear bulges. They do not exhibit the great range of masses and sizes of the ellipticals, and their masses lie usually between 10 and 1000 billion solar masses.

The majority of bright galaxies are spirals. Ellipticals, however, are the most numerous, and the brightest of all galaxies are the rare giant and even rarer supergiant ellipticals.

Tuning fork diagram

Edwin Hubble at Mount Wilson Observatory arranged the galaxies rather neatly in an orderly diagram that looks like a tuning fork (Figure 6.10). Ellipticals form a sequence in one branch, arranged in order of increasing ellipticity, and normal and barred spirals form separate sequences in two parallel branches. The diagram classifies galaxies by their appearance; an E0 galaxy, for example, might be a spherical system or a flattened elliptical system seen face on. At the junction of the three branches, Hubble placed the intriguing S0 (pronounced S-zero) galaxies that combine the properties of ellipticals and spirals. They are disk shaped, like spirals, but lack gas and spiral structure, and therefore resemble flat ellipticals. They in fact look just like spirals swept clean of gas and dust.

The amount of interstellar gas in galaxies increases from left to right in the tuning fork diagram – from ellipticals to spirals. In ellipticals the amount of gas is very small; it is also small in S0 galaxies, and progressively increases in the spirals as we go from Sa and SBa to Sc and SBc. The effect of rotation is more pronounced as we go from ellipticals to spirals. It was originally thought that galaxies evolved along the tuning fork diagram from left to right (that is why galaxies on the left are called early type and those on the right are called late type); then it was thought that they evolve from right to left; astronomers now think that significant evolution along the tuning fork diagram is unlikely, and the basic properties distinguishing ellipticals and spirals were probably determined at the time of their birth.

At least one-tenth of all galaxies have an irregular appearance and are classified as irregulars, denoted by Irr. Our nearest galactic neighbors, the Small and Large Magellanic Clouds, are examples of irregular galaxies. These galaxies have forms similar to those shown in Halton Arp's "Atlas of Peculiar Galaxies," and often their peculiar appearance is the result of tidal interaction with adjacent galaxies.

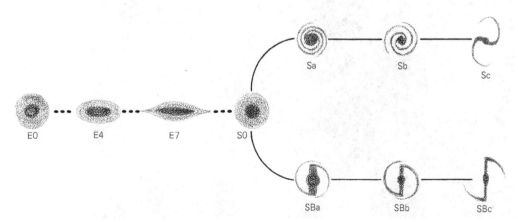

Figure 6.10. The Hubble tuning fork diagram showing the ellipticals arranged on the left, in a sequence of increasing apparent ellipticity, and the spirals arranged on the right in two parallel sequences.

Clusters of galaxies

Galaxies are not uniformly distributed in space, but congregate in clusters of different sizes.

The great regular clusters of galaxies (looking like magnified globular clusters of stars) are spherical in shape. Their galaxies concentrate in the central regions. They are rich – meaning they have many members – and contain thousands of galaxies, mostly of the elliptical and S0 kind. Often in the central regions are found supergiant ellipticals that have conceivably grown to their colossal size by swallowing smaller galaxies. Regular clusters are typically 10 million light years in diameter and lack sharp outer boundaries. They contain intergalactic gas (and intergalactic stars and globular clusters) through which the galaxies rush at speeds often exceeding 1000 kilometers per

Figure 6.11. Coma cluster of galaxies. (Association of Universities for Research in Astronomy, Inc., The Kitt Peak National Observatory.)

second. The wind of intergalactic gas streaming through these fast moving galaxies is strong enough to strip away their interstellar gas. This suggests that S0 galaxies in the great clusters were once ordinary spirals and their gas has been swept out by intergalactic winds.

All other clusters of galaxies are irregular. They have various degrees of richness and are far more numerous than the rare regular clusters. They lack spherical symmetry and strong central condensation and contain a mixture of all types of galaxies. The great irregular clusters are clumpy and actually look like fused aggregations of smaller clusters. Irregular clusters range from rich aggregations of thousands of members, such as the Virgo cluster, to small groups consisting of a few tens of members, such as our Local Group. Small irregular clusters, called groups, have typical sizes of 3 million light years. Our Galaxy is a member of the Local Group, which is a small irregular cluster of approximately 20 galaxies (and many more, if we count midget systems that look little more than escaped globular clusters).

Spread out in space, beyond the Local Group, are multitudes of other groups of galaxies. The nearest great cluster is the irregular Virgo cluster at a distance of approximately 70 million light years. The nearest regular cluster is the Coma cluster (Figure 6.11) at a distance of approximately 450 million light years. Beyond about 100 million light years the galaxies appear to thin out slightly, and we now know that clusters of galaxies are themselves grouped to form superclusters. The Local Supercluster that we occupy has its center somewhere in the vicinity of the Virgo cluster and is known as the Virgo supercluster.

Distances of galaxies

The determination of distances to galaxies is extremely important in cosmology; unfortunately, it is also extremely difficult (Chapter 19).

By measuring the apparent brightness of their most luminous stars, it is possible to determine the approximate distances of nearby galaxies. The bright cepheids aid in charting the immediate extragalactic neighborhood. Distances of galaxies farther away are not so easily determined and uncertainty grows as distance increases. The apparent brightness of the most luminous stars and globular clusters serves as a distance indicator up to about 80 million light years. The apparent size of highly luminous clouds of gas and the brightness of supernovas take us farther out. Beyond 100 million light years all distances are uncertain by a factor 2, and perhaps even more. Galaxies themselves must now be used as distance indicators. What we see far out in space happened far back in time when the galaxies were younger and not the same as now, and allowance for evolutionary changes adds to the many difficulties of determining the distances of very distant galaxies.

BIRTH OF GALAXIES

Most galaxies contain old population II stars and have ages of many billions of years, and most if not all galaxies were born long ago. Matter in intergalactic space is now too low in density to give birth to new galaxies.

An average density of matter in a galaxy such as our own is 1 hydrogen atom per cubic centimeter. This would be the density if all stars were dissolved into gas and spread out in a sphere of diameter 100 000 light years. Roughly, this is 1 million times greater than the average density of ordinary matter in the universe. The universe expands and therefore in the past was denser and the galaxies were crowded closer together. Five billion years ago, when the Solar System was born, the average density of the universe was roughly twice its present value; and 5 billion years earlier still, the galaxies were only half their present average separating distance, and the density of the universe was 8 times its present value. Much earlier, when the average density was greater than 1 million times its present value, galaxies did not exist, at least not in their present form, because they would be crushed together beyond recognition in a universe denser than galaxies. The galaxies originated in rudimentary form when the expanding universe had an average density very much less than 1 hydrogen atom per cubic centimeter. According to cosmological theory, to be discussed in later chapters, this means the galaxies originated in an expanding universe older than 100 million years.

Protogalaxies

Most theories on the origin of galaxies start with small variations of density in the very early universe. The density perturbations grow and eventually develop into protogalaxies. Let ρ represent density and $\delta\rho$ a small perturbation in density. Both ρ and $\delta\rho$ decrease with expansion in the early universe, but $\delta\rho$ decreases slower than ρ. Hence the important ratio $\delta\rho/\rho$, known as the contrast density, steadily increases. Initially, the contrast density $\delta\rho/\rho$ is probably much smaller than 1 trillionth, but eventually, when $\delta\rho/\rho$ has grown and attained a value near unity, we reach the formation stage of protogalaxies.

Protogalaxies begin as large concentrations of hydrogen and helium gas (helium was made from hydrogen in the early universe) with masses ranging from millions to trillions of solar masses. These concentrations (or globes) of cool gas continued to expand, but more slowly than the universe, and their separations widened. At some stage, each globe ceases to expand and begins to collapse. A view of how these collapsing globes of gas might have formed into galaxies is as follows (see Figure 6.12).

Collapse theory

We consider a possible picture of the formation of a giant galaxy. At first, a globe of gas has roughly uniform density. By the time a globe attains its maximum size of about 500 thousand light years diameter, the density is no longer uniform. At about this time, the earliest population II stars begin to form in the central region where the density is highest. The brightest of these first generation stars evolve rapidly and

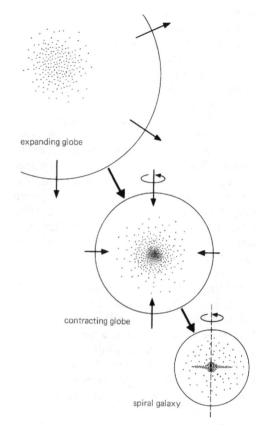

expanding globe

contracting globe

spiral galaxy

Figure 6.12. This figure shows one possible way in which a spiral galaxy might form. The expanding universe, already hundreds of millions of years old, has fragmented into large globes of gas. At first, each globe continues to expand, but slower than the universe, then later it stops expanding and begins to collapse. Population II stars in large numbers form in the central region. Gas from the outer regions of the globe falls inward and forms a rotating disk in which population I stars are slowly born.

erupt as supernovas, ejecting gas enriched with heavy chemical elements. Meanwhile, the whole globe of gas has begun to collapse freely under the influence of its gravity. The collapse lasts hundreds of millions of years, in some cases perhaps billions of years. Stars form continually in the central region and the brightest last only millions of years. The infalling gas of the protogalaxy is therefore steadily enriched with heavy elements.

We must understand that stars, once formed, cease to participate in the general collapse of the protogalaxy. Each newborn star initially moves inward, but after its formation it follows an elliptical orbit and its average distance from the center tends to stay constant. We have a picture of a collapsing protogalaxy, as proposed in 1962 by Olin Eggen, Donald Lynden-Bell, and Allan Sandage, in which the central region becomes a swarm of population II stars. This swarm of stars is roughly spherical, and in the case of our Galaxy has a diameter about 100 thousand light years. Thus the central region of the protogalaxy has become what later will be the galactic halo.

Through the halo of stars falls the rest of the collapsing protogalaxy. This is the "outside-in" theory of galaxy formation. The inside of the protogalaxy forms the halo, then the outside of the protogalaxy falls in and finally forms the nucleus and disk of the newborn galaxy. What happens to the infalling gas determines whether the galaxy is a spiral or an elliptical.

Let us consider a protogalactic globe of gas initially in uniform rotation. The outer parts have higher rotational velocity than the inner parts. The core of the globe, as before, transforms into a swarm of stars, and because of the core's low rotational velocity, the swarm is almost spherical in shape. As the infalling gas from the outer regions falls inward, its rotational velocity increases in the same way that a ballet dancer rotates faster as she lowers her arms. The descending gas sweeps up the previously ejected heavy elements and is partly consumed by the formation of new stars. These new stars are intermediate between population I and population II stars and their distribution in space is less spherical than that of the first-generation stars. The surviving gas cannot fall all the way to the center because of its rotation. Instead it settles into a rotating gaseous disk consisting of hydrogen, helium, and most of the heavy elements ejected from halo stars. Population I stars now begin to form in the gaseous disk. But rotation and the presence of magnetic fields greatly impedes star formation, and to this day only a few stars are born each year. The original gigantic globe of gas has collapsed and produced a spiral galaxy.

Where does the initial rotation come from? We do not know for certain; perhaps the globes of gas pull one another into rotation by their gravitational interactions, or perhaps the initial perturbations in the very early universe have rotation. Whatever the preferred theory, one concludes that some protogalaxies have more rotation than others.

Let us then consider a globe of gas that has very slow rotation. As before, the first generation stars form in the central region, and the infalling gas of the outer regions is partly consumed by the formation of new stars. But the infalling gas now lacks sufficient rotation to create a large disk of swirling gas. Instead, the gas continues to fall and is continually consumed by star formation. The gas that survives converges on the center and settles in the nucleus. The original globe of gas has finally collapsed and produced a large elliptical galaxy. The rotation was never sufficient to form a large gaseous disk, but was sufficient to produce the moderately flattened distribution of stars we see in many ellipticals.

Not only the initial rotation, but also the initial density in protogalaxies is important. The rate of star formation depends on the density of gas; the higher the density, the faster that stars form. If a protogalaxy has higher density (because it separates at an earlier stage in the expanding universe) then stars form more quickly and no infalling gas survives to form a disk or a nucleus. This might explain how small ellipticals form. In small, low-mass galaxies, the ejected gas from firstborn stars never gets incorporated into later stars because all stars form more or less at the same time. Possibly the ejected gas lingers around and is later swept out by galactic winds.

Fragmentation and clustering

We can explain the origin of large-scale astronomical structure in one of three ways. The first is the fragmentation hypothesis, proposed by James Jeans, in which the universe first divides into large pieces, which then fragment successively into smaller and smaller pieces, ultimately terminating with stars. The second is the aggregation hypothesis, advocated by David Layzer of Harvard University, in which the universe first divides into small pieces, which then cluster successively into larger and larger pieces, ultimately terminating with superclusters of galaxies. The third, probably closer to the truth, is a combination of these two pictures. The universe first fragments into globes of gas, some of which may aggregate to form large protogalaxies and clusters of protogalaxies. The collapsing globes subsequently fragment into clusters of stars, which then fragment into isolated stars. The galaxies club together to form clusters of galaxies of different sizes that in turn aggregate to form superclusters. The fragmentation and clustering processes thus appear to be of comparable importance in the structural makeup of the universe.

RADIO GALAXIES AND QUASARS
In 1931, Karl Jansky of the Bell Telephone Laboratories detected radio signals from the Milky Way. This exciting discovery received wide publicity. Signals from Jansky's receiver were relayed and broadcast in a radio program in which the announcer said, "I want you to hear for yourself this radio hiss from the depths of the universe." In 1938, the radio engineer Grote Reber detected radio signals from the Milky Way, and noticed that the signals were strongest from the galactic center and from constellations such as Cygnus and Cassiopeia.

Radio galaxies
Hundreds of radio sources were discovered in the early years of radioastronomy after World War II, and the visible counterparts of many were identified by optical astronomers. Thus Taurus A – the strongest of the observed radio sources in the constellation Taurus – was found to be the Crab Nebula,

a chaotic cloud of gas produced by the supernova of 1054. Cygnus A – or 3C 405 – was identified with a disturbed-looking giant galaxy at a distance of 1 billion light years. The most powerful of the optically identified radio sources were galaxies, and these radio galaxies, such as Cygnus A, emit millions of times more energy in radio waves than ordinary galaxies. Radio galaxies seen on photographic plates have often bright central regions, and sometimes protruding jets (as in M 87) and wispy extensions. The strongest radio galaxies radiate more energy in radio waves than in visible light, and are usually giant ellipticals in rich clusters of galaxies.

Radio waves from these sources are emitted by fast electrons moving in helical orbits in magnetic fields. The waves are known as synchrotron radiation because this radiation is produced in high-energy synchrotron accelerators. Radio galaxies generate, in a way not completely understood, hordes of energetic electrons dispersed throughout their radio emitting regions.

Great strides have been made in radioastronomy and it is now possible to study in detail the structure of radio sources. The majority of sources have double structure; these sources emit radio waves from extended components lying on opposite sides of the source (see Figure 6.13). The two components, or jets, extend to distances of hundreds of thousands, sometimes millions, of light years. The energy radiated by the jets originates in the galaxy, presumably from its nucleus. Often the central region, or

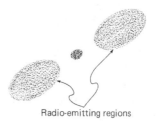

Radio-emitting regions

Figure 6.13. An extended two-component radio source. Radio waves are emitted from the extended components.

nucleus, of a radio galaxy is itself a powerful radio emitter containing two radio-emitting components separated by distances of only hundreds of light years. There are thus two varieties of radio sources: extended sources and compact sources, and some – for example, Cygnus A – are both extended and compact.

Martin Rees of Cambridge University has suggested that the combined output in waves and energetic particles from hundreds of thousands of pulsars in a galactic nucleus is beamed away in two opposite directions. The energy in these intense beams is then distributed over large regions that radiate radio waves. According to another idea, suggested by Soviet astronomers and Philip Morrison of Massachusetts Institute of Technology, the gas in the nucleus of a galaxy contracts and forms one or more supermassive stars that rotate and have strong magnetic fields. These objects, named spinars, are millions of times more massive than the Sun and behave like titanic pulsars radiating intense beams into intergalactic space. Developments in other fields of astronomy soon overtook these early ideas.

Quasars

By 1960, radioastronomers had found and catalogued hundreds of radio galaxies, about 50 of which had been identified by optical astronomers as giant ellipticals. It seemed the unidentified radio galaxies were much too faint to be recorded on photographic plates. Then a sequence of events occurred in 1960 that led to the discovery of some very puzzling objects – the quasi-stellar objects (QSOs) later known as quasars.

It was noticed that the radio source 3C 48 is starlike in appearance (not nebulous like a galaxy) and emits strong ultraviolet radiation. Other objects of similar appearance, such as 3C 273, were soon found. They seemed to be an unusual kind of star in our Galaxy that, unlike ordinary stars, emitted strong radio waves. In 1963 came startling news. Maarten Schmidt of Mount Wilson Observatory had discovered emission lines in the spectrum of 3C 273 that were shifted by 16 percent to longer wavelengths. In other words, the radio source had a redshift 0.16, and was receding from us at about 0.16 the speed of light.

In an expanding universe, all extragalactic objects at great distances are receding from us and as a result the light we receive from these objects is redshifted. This is a subject to be discussed later (Chapter 15), and here we need only remark that a redshift of 0.16 corresponds to a distance of roughly 3 billion light years. The simple and logical explanation of 3C 273 is that it is extragalactic and its redshift is the result of the expansion of the universe. Numerous other radio sources were optically identified as starlike objects with large redshifts, and it became evident that these starlike radio sources were actually remote beacons radiating enormous quantities of energy.

Allan Sandage then discovered that many extragalactic starlike objects are radio quiet. The word quasar, coined by the astronomer Hong Yee Chui at State University of New York at Stony Brook, is a contraction of *quasi*-stell*ar* and is now used to denote all starlike objects of large redshift, whether or not they emit radio waves. The nearest quasar so far discovered is 3C 273; most quasars lie at distances of billions of light years; and the most distant have redshifts greater than 5. (A redshift z means the universe has expanded $1 + z$ fold since emission of the light now seen. Thus $z = 1$, means a twofold expansion.)

It is estimated that tens of millions of quasars are observable with the largest optical telescopes. When we look out in space to such large distances we also look back billions of years into the past. We look back in time and see that quasars were far more numerous in the universe in the past than at present. They were sufficiently numerous at the time when the Solar System formed that at least one was near enough to be seen by the naked eye, gleaming in the sky like a brilliant jewel. Earlier still, shortly after the birth of galaxies, when the universe was one-tenth its present age, the quasars were all around, millions of times more numerous than now.

The quasar puzzle

Quasars radiate energy at approximately 100 times the rate of all the stars in a giant galaxy. They squander energy at this rate for typically a billion or more years. If matter were annihilated and converted entirely into energy (energy equals mass times the square of the speed of light), some quasars would consume in their lifetime the mass of more than 100 million stars.

The light from some quasars fluctuates in brightness, changing on time scales as short as weeks and even days. Normally, a luminous source cannot significantly change its visible appearance in a time less than its size in light travel time. When an object abruptly increases its light output, an observer sees a slower increase because light emitted from the nearest parts of its surface arrives first, and light from its more distant parts arrives later. Thus if light varies in brightness in, say, 1 day, this means a spherical source has a radius smaller than 1 light day and is therefore smaller than the Solar System. Imagine that a large ballroom represents the size of the Galaxy; on this scale a highly luminous quasar is no more than a speck of dust floating in the air. Quasars are powerful sources of radio, infrared, visible, and ultraviolet radiation, and extraordinarily compact by astronomical standards. What are they?

Active galactic nuclei

Radio quasars are similar to radio galaxies; both have either extended or compact radio-emitting regions. Radioastronomers cannot tell the difference. Also optical astronomers cannot tell the difference between quasars and the bright nuclei of radio galaxies; both appear basically the same thing. Possibly the nuclei of most giant ellipticals pass through a quasar state in their early evolution, and most quasars pass through an active radio-emitting phase in their lifetime. Some giant spirals, known as Seyfert galaxies, have bright active nuclei that exhibit violent activity and resemble miniature quasars.

Years have passed since the discovery of quasars and many puzzles remain unsolved concerning how these compact objects generate and emit immense quantities of energy over a wide range of wavelengths. In the excitement that followed their discovery, several theories were proposed, such as annihilation of matter and antimatter; dense stellar systems in which stars continually collide cataclysmically; dense stellar systems in which supernovas occur frequently; and even alterations in the known laws of nature. It is now generally agreed, however, that the radiation is fueled by the release of gravitational energy. Probably the mechanism begins with the formation of numerous stars in the nucleus, many of which evolve and become neutron stars and black holes. These compact bodies then accrete gas and release energy.

There seems little doubt that stars much more massive than the Sun must collapse totally at the end of their evolution and become black holes. The laws of nature as we understand them lead to this conclusion. The most successful explanation of quasars, widely accepted, was proposed by Edwin Salpeter of Cornell University. The idea is that quasars are supermassive black holes that accrete gas. The theory has been broadened to include all highly agitated galactic nuclei – quasars, radio sources, Seyfert galaxies, and other similar strange objects – under the generic name "active galactic nuclei."

Supermassive black holes

Not all infalling gas is consumed in the birth of stars during the formation of a galaxy. The surviving gas finishes in either the disks of spirals or the nuclei of ellipticals. In a giant elliptical galaxy, infalling gas accumulates in the nucleus and has a mass of perhaps one-tenth the total mass of the galaxy. Slow subsequent contraction of this reservoir of matter creates a quasar that shines for the next billion or so years. In a giant spiral, however, the gas accumulating in the nucleus is very much less and nuclear activity is scaled down.

Quite likely the first step is the formation of numerous stars in a gas-rich nucleus, many of which rapidly evolve and become black holes. Imagine then a black hole – initially a few times the mass of the Sun with a diameter of about 10 kilometers – located in a nucleus where gas is dense and millions of stars are huddled close together. It is a situation that conjures up those terrible words "cry Havoc! and let slip the dogs of war." (The cry "Havoc!" was a command to massacre all without quarter. In the reign of Richard II of England the cry was forbidden on pain of death.) Gas swirls in toward the black hole, forming an accretion disk in which matter spirals in on the black hole (Figure 13.10). Incautious stars that wander too close either collide with one other or are torn to shreds by tidal forces, and their gaseous wreckage adds to the headlong rush that spirals in on the black hole. Next time you pull the plug out of a bath and see the water draining away in a dark vortex, think of a voracious black hole. The black hole rapidly grows in mass and occasionally swallows other lesser black holes. Provided the galactic nucleus is sufficiently rich in gas and stars, a black hole will grow in mass to 100 million times that of the Sun in a few hundred million years and attain a size of 1 light hour. During its growth, a torrent of energy is released from the inflowing swirling gas. As the gas swirls in to the black hole, it compresses and heats to high temperature and energy is radiated away. A fraction – a tenth or more – of the mass of the captured gas is converted directly into escaping radiant energy. A monster black hole grows by accretion to a billion solar masses and radiates a total energy equivalent to at least 100 million solar masses. This picture of a quasar accounts for its brilliance and smallness and might also even explain how it becomes a radio source. Possibly the hot and therefore electrically conducting gas that swirls inward on the black hole acts as a vast electrical dynamo that generates oppositely directed beams of high-energy particles. These particles travel out and energize the radio-emitting regions of radio sources.

According to this broad-brush picture, quasars exist in galactic nuclei in highly active states and die only when most matter in the galactic nuclei has been swallowed or ejected from the nucleus. Thereafter, the black hole lies dormant, erupting sporadically whenever fresh supplies of gas come its way.

The concentration of gas in the nuclei of spirals is less, and black holes in spirals are probably smaller than in giant ellipticals. We know from radio studies that the nucleus of our Galaxy is in a disturbed state; black holes might lurk in the nucleus (and in the nuclei of other giant spirals) and have masses hundreds or thousands or even millions of times the mass of the Sun.

REFLECTIONS

1 *James Jeans (1877–1946), a distinguished scientist and author of popular and scholarly books in science and astronomy, wrote in* Astronomy and Cosmogony *in 1929, "We have found that, as Newton first conjectured, a chaotic mass of gas of approximately uniform density and of very great extent would be dynamically unstable: nuclei would tend to form in it, around which the whole of the matter would ultimately condense . . . We may conjecture, although it is improbable that we shall ever be able to prove, that the spiral nebulae were formed in this way. Any currents in the primaeval chaotic medium would persist as rotations of the nebulae, and, as these would be rotating with different speeds, they might be expected to shew all the various types of configurations."*

• *Edwin Hubble (1889–1953), an American astronomer, earned fame for his work on the classification of galaxies and the determination of distances to galaxies, and also for the Hubble law of the expansion of the universe. He studied Roman and English law at Oxford, but decided after a year in the legal profession to "chuck the law for astronomy."*

In his classic book The Realm of the Nebulae
(1936), he wrote: "Research men attempt to
satisfy their curiosity, and are accustomed to
use any reasonable means that may assist
them toward the receding goal. One of the
few universal characteristics is a healthy skep-
ticism toward unverified speculations. These
are regarded as topics for conversation until
tests can be devised. Only then do they attain
the dignity of subjects for investigation." His
book closes with the words: "Thus the
exploration of space ends on a note of uncertainty. And necessarily so. We are, by definition, in the very center of the observable
region. We know our immediate neighborhood rather intimately. With increasing distance, our knowledge fades, and fades
rapidly. Eventually, we reach the dim boundary – the utmost limits of our telescopes.
There, we measure shadows, and we search
among ghostly errors of measurement for
landmarks that are scarcely more substantial."

• Fred Hoyle, in Galaxies, Nuclei, and
Quasars, wrote in 1965: "It is not too much
to say that the understanding of why there
are these different kinds of galaxy, of how
galaxies originate, constitutes the biggest
problem in present-day astronomy. The properties of the individual stars that make up
the galaxies form the classical study of astrophysics, while the phenomenon of galaxy formation touches on cosmology. In fact, the
study of galaxies forms a bridge between conventional astronomy and astrophysics on the
one hand, and cosmology on the other."

• "Astronomical observations now reach far
enough back in time, in enough depth and
detail, to reveal the history of galaxies since
their formation. The early universe contained
a network of gas clouds that filled much of the
space between young galaxies, where stars
were forming at a high rate. Since then, intergalactic space has been swept clean, and
galaxies have continued to convert the
dwindling supply of gas slowly into stars."
M. Fukugita et al., "The history of the
galaxies" (1996).

2 We can estimate the mass of the Galaxy
with Kepler's third law. Let P be the period

of revolution in years and r the radius of the
orbit in astronomical units. Then $P^2 = r^3$ for
planets orbiting around the Sun. For bodies
orbiting around a large mass M, Kepler's
law becomes

$$\frac{M}{M_\odot} P^2 = r^3, \qquad [6.1]$$

where $M_\odot$ is the mass of the Sun. The Sun
moves around the Galaxy in a time $P =
2 \times 10^8$ at distance $r = 2 \times 10^9$ (30 000 light
years) from the center of the Galaxy. Hence

$$M = \frac{(2 \times 10^9)^3}{(2 \times 10^8)^2} M_\odot = 2 \times 10^{11} M_\odot, \qquad [6.2]$$

and the mass of the Galaxy inside the Sun's
orbit is 200 billion times that of the Sun.
Equations [6.1] and [6.2] are only approximate calculations. They assume that all
matter is distributed symmetrically about
the center of the Galaxy as in a sphere. A
sphere exerts a gravitational pull at any
radius as if the mass inside that radius were
concentrated at the center, and the mass outside that radius has no net effect. But the
Galaxy is not a perfect sphere; the error, however, is not very large.

If no mass existed outside the Sun's orbit,
the rotation velocity

$$V_{\text{rot}} = \frac{2\pi r}{P}, \qquad [6.3]$$

would decrease with radius as $r^{-1/2}$, according
to Equation [6.1]. But the rotation velocity
beyond the Sun's orbit, as revealed by observations of 21-centimeter radio emission from
neutral hydrogen gas, does not decrease but
tends to stay almost constant out to a radial
distance of about 150 thousand light years
(Figure 19.3). Other spiral galaxies show
similar constant rotation velocities far beyond
their nuclear bulges. According to Equations
[6.1] and [6.3], a constant rotation velocity
means the mass of the Galaxy increases
linearly with r. Thus most of the mass of the
Galaxy lies outside the Sun's orbit, probably
in the halo, and the total mass of the Galaxy
may be as large as 10^{12} solar masses. If
the mass of the halo increases linearly with

radius r, the density of the halo decreases as $1/r^2$.

3 *The gravitational lensing effect was predicted in 1937 by Fritz Zwicky. He suggested that astronomers might be able to determine the masses of galaxies by their deflection of light emitted by more distant sources. In recent years the study of gravitational lensing has become an important branch of astronomy. The rays of light from a distant source, on passing through an intervening galaxy or cluster of galaxies, are bent very slightly and form a distorted image of the distant source. If the distant source is pointlike (such as a quasar), the image consists of from two to five separate components; if it is nebulous (such as a galaxy), the image consists of arcs. Observed deflections confirm that giant galaxies have masses 1 trillion times that of the Sun.*

4 *Astronomical systems, such as stars, galaxies, and clusters of galaxies, are held together by their own gravity. The virial theorem applies to all self-gravitating systems in equilibrium. This theorem states that the total internal kinetic energy equals half the binding energy of the system. Let M be the mass and R the radius of a spherical system, and let U be a typical internal speed of any component of the system. The virial theorem states*

$$U^2 = \frac{\alpha GM}{R} \qquad [6.4]$$

(see also Equation 19.12), where α is a constant whose value depends on the distribution of matter in the cluster and is usually close to unity. From this expression we find

$$U^2 = \frac{1}{100} \times \frac{\text{mass in solar masses}}{\text{radius in light years}}, \qquad [6.5]$$

where U is in kilometers per second. From U and R, found by observation, we get the mass M. For example, the mass of a rich cluster of radius 10 million light years, in which galaxies have typical speeds 3000 kilometers per second, is 10^{16} solar masses, equivalent to 10 000 large galaxies, each of mass 10^{12} solar masses. The virial mass of galaxies exceeds the visual mass found by counting individual stars, and the virial mass of clusters exceeds the visual mass found by counting individual galaxies, and the larger the astronomical system, the greater the difference between the virial and visual masses. The difference strongly indicates the presence of considerable amounts of unseen matter in astronomical systems. The problem of identifying the nature of this unseen dark matter has not been fully solved.

PROJECTS

1 Imagine the Solar System (100 astronomical units diameter) reduced to the size of a dime (1 centimeter diameter). On this new scale of 100 astronomical units equal to 1 centimeter, give: the average distance (in meters) between stars in the Sun's neighborhood; the diameter of the Galaxy (in kilometers); and the average distance between galaxies (in kilometers). Assume that stars are separated by 5 light years and galaxies by 5 million light years.

2 Try using the moving-cluster method to estimate the distance of a flock of birds flying in the sky. Suppose the flock is flying away at velocity 60 kilometers (40 miles) per hour, and its size is seen to shrink to half in one minute. Show that the distance is 1 kilometer.

3 Suppose that on the average each galaxy contains 100 billion stars and that galaxies are separated from one another by 5 million light years. If we can look out to a distance of 10 billion light years, how many galaxies and how many stars are there in the visible universe? Compare the number of stars with the number of grains of sand on all the beaches and deserts of Earth. (Assume that a grain of sand has a volume of 1 cubic millimeter and sand covers the Earth's surface to a depth of 1 meter.)

4 Why is life more likely to exist in spirals than in ellipticals? Imagine a universe that failed to fragment into galaxies and stars. What might that mean for life?

5 Mass extinctions of life have periodically occurred on Earth, perhaps caused by the infall of massive bodies, such as comets and asteroids. During the lifetime of the

Solar System several supernovas have no doubt also occurred within a distance of several light years from Earth, each filling the sky with a blaze of light and drenching the Earth with high-energy particles and x-rays. Perhaps these celestial explosions have been responsible for at least one or two of the mass extinctions of life. Suppose that a supernova has a luminosity 10^{12} times that of the Sun: (a) How close (in astronomical units and in light years) to the Solar System is the supernova if the Earth receives from it as much energy per second as from the Sun? (b) Assume that 1 percent of all stars in the Galaxy have ended as supernovas; on the average, how many years between each supernova? (c) If the supernovas were distributed uniformly in the disk of the Galaxy, how many, as seen from the Earth, have been as bright or brighter than the Sun? (Assume that the Galaxy is 10^{10} years old, contains 10^{11} stars, and the disk has a diameter 100 000 light years and thickness 1000 light years.)

FURTHER READING

Baade, W. *Evolution of Stars and Galaxies.* Editor C. Payne-Gaposchkin. Harvard University Press, Cambridge, Massachusetts, 1963.

Berendzen, R., Hart, R., and Seeley, D. *Man Discovers the Galaxies.* Science History Publications, New York, 1976.

Bergh, S. van den and Hesser, J. E. "How the Milky Way formed." *Scientific American,* 72 (January 1993).

Binney, J. "The evolution of our Galaxy." *Sky and Telescope* 89, 20 (March 1995).

Bok, B. "The Milky Way Galaxy." *Scientific American* 244, 92 (March 1981).

Henbest, N. and Couper, H. *The Guide to the Galaxy.* Cambridge University Press, Cambridge, 1994.

Herbig, C. H. "Interstellar smog." *American Scientist* (March–April 1974).

Hodge, P. *Galaxies and Cosmology.* McGraw-Hill, New York, 1966.

Scoville, N. and Young, J. S. "Molecular clouds, star formation, and galactic structure." *Scientific American* 250, 42 (April 1984).

Shapley, H. *Flights from Chaos: A Survey of Material Systems from Atoms to Galaxies.* McGraw-Hill, New York, 1930.

Shapley, H. *Through Rugged Ways to the Stars.* Charles Scribner's Sons, New York, 1969.

Shipman, H. *Black Holes, Quasars and the Universe.* Houghton Mifflin, Boston, 1976.

Toomre, A. and Toomre, J. "Violent tides between galaxies." *Scientific American* (December 1973).

Trimble, V. and Woltjer, L. "Quasars at 25." *Science* 234, 10 (1986).

Trimble, V. and Parker, S. "Meet the Galaxy." *Sky and Telescope* 89, 26 (January 1995).

Weaver, H. "Steps toward understanding the large-scale structure of the Milky Way." *Mercury* (September–October 1975, November–December 1975, January–February 1976).

Wright, H. *Explorers of the Universe: A Biography of George Ellery Hale.* Dutton, New York, 1966.

SOURCES

Allen, R. H. *Star Names: Their Lore and Meaning.* 1899. Reprint: Dover Publications, New York, 1963.

Arp, H. "Atlas of Peculiar Galaxies." *Astrophysical Journal Supplement* 14, 1 (1966).

Blitz, L., Binney, J., Lo, K. Y., Bally, J., and Ho, P. T. P. "The center of the Milky Way." *Nature* 361, 417 (February 1993).

Burbidge, M. and Burbidge, G. *Quasi-Stellar Objects.* Freeman, San Francisco, 1967.

Eddington, A. S. *The Internal Constitution of the Stars.* Cambridge University Press, Cambridge, 1926.

Fukugita, M., Hogan C. J., and Peebles, P. J. E. "The history of the galaxies." *Nature* 381, 489 (1996).

Herschel, W. "On the construction of the heavens." *Philosophical Transactions* 75, 213 (1785).

Hey, J. S. *The Evolution of Radio Astronomy.* Neale Watson Academic Publications, New York, 1973.

Hoyle, F. *Galaxies, Nuclei, and Quasars.* Harper and Row, New York, 1965.

Hubble, E. *The Realm of the Nebulae.* Yale University Press, New Haven, Connecticut, 1936.

Jeans, J. *Astronomy and Cosmogony*. Cambridge University Press, Cambridge, 1929.

Pannekoek, A. *A History of Astronomy*. Interscience Publishers, New York, 1961.

Payne-Gaposchkin, C. *Introduction to Astronomy*. Methuen, London, 1954.

Peebles, P. J. E. *The Large-Scale Structure of the Universe*. Princeton University Press, Princeton, 1980.

Rees, M. J. " 'Dead quasars' in nearby galaxies." *Science* 247, 814 (16 February 1990).

Rees, M. J. "Quasars, bursts and relativistic objects." *Quarterly Journal of the Royal Astronomical Society* 35, 391 (1994).

Sandage, A. R. *The Hubble Atlas of Galaxies*. Carnegie Institution, Washington, D.C., 1961.

Struve, O. and Zebergs, V. *Astronomy in the 20th Century*. Macmillan, New York, 1962.

Tayler, R. J. *Galaxies: Structure and Evolution*. Cambridge University Press, Cambridge, New York, 1993.

7 LOCATION AND THE COSMIC CENTER

The Sun is lost, and the earth, and no man's wit
Can well direct him where to look for it.
And freely men confess that this world's spent,
When in the Planets, and the Firmament
They seek so many new; then see that this
Is crumbled out again to his Atomies.
'Tis all in pieces, all coherence gone;
All just supply, and all Relation.
John Donne (1572–1631), The Anatomy of the World

THE LOCATION PRINCIPLE

The Greeks developed the "two-sphere" universe that endured for 2000 years and consisted of a spherical Earth surrounded by a distant spherical surface (the sphere of stars) studded with celestial points of light. This geocentric picture was finally overthrown by the Copernican revolution in the sixteenth century and replaced by the heliocentric picture with the Sun at the center of the cosmos. The sphere of stars remained intact. But revolutions, once begun, do not readily stop, and by the seventeenth century the heliocentric picture had also been overthrown. Out of the turmoil of the revolution emerged an infinite and centerless universe that ever since has had a checkered history. In the eighteenth century the idea arose of a hierarchical universe of many centers, and in the nineteenth came the idea of a one-island universe – the Galaxy – in which the Sun had central location. Once again, in the twentieth century, we have the centerless universe.

As we watch the history of cosmology unfold we see a steady growth in the conviction that the human species does not occupy the center of the universe. The cosmic center was displaced first from the tribe and nation and then from the Earth, the Sun, and finally the Galaxy. Simultaneously, ideas concerning God and the universe became increasingly grand and inflated. Medieval theology developed far-reaching concepts concerning the nature of God that subsequently were transferred to the nature of the universe. Theological ideas of God as unconfined, infinite, and simultaneously everywhere were translated into scientific ideas of the universe as unconfined, infinite, and having its center everywhere. From theology, philosophy, and science has emerged a cosmic outlook expressed by the location principle:

The location principle simply states: *it is unlikely that we have special location in the universe.* How "unlikely" or improbable is "special location," and what special location means, are issues considered in each application of the principle. The principle may, of course, be rejected. In that case evidence should be produced in support of the contrary belief that we enjoy special location, and such evidence nowadays is not easy to find.

The Earth, Sun, Galaxy, and Local Group are all unique, and in this sense we undoubtedly enjoy special location. Yet other creatures living on other planets, encircling other stars, in other galaxies, in other clusters of galaxies have also special location in the same sense. The cosmologist is concerned not with the uniqueness of detail but with the main outline of the cosmic picture, just as hills and valleys are blurred and smoothed out when we think of the overall shape of the

Earth. When all small-scale irregularity of astronomical detail is ignored, special location means location that is special within the universe as a whole.

The location principle does not deny that special location, such as a cosmic center, may exist somewhere. It states that of all the planets, stars, and galaxies of the universe, it is unlikely that the Earth, Sun, and Galaxy are privileged places. It is a revolutionary manifesto proclaiming that mankind is no longer king of the cosmic castle. Renunciation of cosmic privilege was strongly resisted until recent times. Why abdicate the cosmic throne when nothing is gained and all is lost? Advancements in astronomy and biology in the nineteenth and twentieth centuries have forced abdication on us. We can be kings in a medieval universe but not in the modern physical universe.

The Surveyors

As an illustration of the location principle, and of how it works, we imagine that a team of small creatures, called Surveyors, has been placed on a very large surface. Their mission is to determine the large-scale shape of the surface. The Surveyors set up their instruments on hilltops and begin to take measurements. After much observation, calculation, and debate they come to an important conclusion: If the hills and valleys are ignored, or imagined to be smoothed out, the surface appears the same in all directions. On finding that the surface is isotropic – the same in all directions – the amazed Surveyors exclaim, "How fortunate to be on this special central spot!" They have previously agreed that the

surface has no edge for reasons that need not detain us until Chapter 8, and they start theorizing about the shape of the surface. A popular guess is that their camp is on the summit of a large hill that falls away in all directions and extends to infinity, as in Figure 7.1.

After some time one of the more thoughtful Surveyors says, "isn't it odd that we just happen to be on this special spot?" The Surveyors believe that elsewhere on the surface are other Surveyor teams, and reluctantly realize that the chance of occupying a small region, around which the surface is isotropic, is extremely small. They therefore formulate a location principle that states, "It is unlikely that we occupy a special spot on the surface." This means the other Surveyor teams, no less privileged, have perhaps also found that the surface is isotropic about their camps. If isotropy exists at one place, and is an unlikely privilege, then probably it exists everywhere. The Surveyors conclude that if the surface is indeed isotropic at every point, then it is flat.

From the beginning of the mission some Surveyors have disliked the idea of an infinite surface and have said that they much prefer to live on a surface that is finite in extent. They have declared that the measurements, because of uncertainties, are also consistent with the possibility of an egg-shaped surface, as illustrated in Figure 7.2. Their camp, they have pointed out, need not be on the summit of a large hill, but might be situated at the end of a large egg

Surveyors' camp

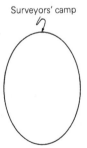

Surveyors' camp

Figure 7.1. A hill-shaped surface that stretches away to infinity. The Surveyors' camp is at the summit of the hill.

Figure 7.2. A finite unbounded surface that is egg-shaped. The Surveyors' camp is at one end of the egg.

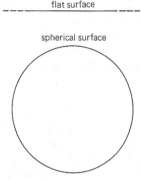

Figure 7.3. After the discovery of the location principle, the "open" group concludes that the surface is actually flat, and the "closed" group concludes that it is spherical.

where the surface is also the same in all directions. These Surveyors, confronted with the location principle and the prospect that the surface is isotropic everywhere, come to the conclusion that they probably occupy the surface of a sphere. The Surveyors thereupon divide into two groups (Figure 7.3): the "open" group that thinks the surface is open, infinite, and flat; and the "closed" group that thinks the surface is closed, finite, and spherical. All agree that more observations are needed to determine which group is correct.

After further surveying it is discovered that the surface is apparently not quite isotropic. An anisotropy of about 1 percent is detected that is possibly due to an overlooked surface irregularity. The alternative explanation, that it is due to a real cosmic center of symmetry at an estimated distance l, is at first a cause of anxiety. Perhaps the surface is hill-shaped after all and their camp is not quite at the summit? Or perhaps the surface is really egg-shaped and their camp is not quite at the end? The Surveyors meet this challenge by arguing as follows. Their camp and also the center of symmetry, if it exists, both occupy a small region of the surface of an approximate area πl^2. Now L is the distance out to which they can survey, and the chance of occupying a small privileged area of πl^2 that contains the center, in a large area of πL^2 is the ratio of these

two areas, equal to l^2/L^2. Their measurements indicate that $L = 100l$ (because l/L is roughly equal to the 1 percent anisotropy), and the chance of special location near a center of symmetry is therefore only 0.0001. The chance of a center existing at all within their field of view, they argue, is only 1 in 10 000, and the probability that the surface is the same everywhere (apart from local irregularities) is 99.99 percent. The Surveyors conclude that the apparent anisotropy is probably the result of an overlooked large-scale irregularity in the surface. The possibility of many centers in the area πL^2 is dismissed as merely another form of irregularity that must be smoothed out.

A few Surveyors are not convinced that the surface everywhere is the same and start to talk about a "hierarchical" surface. Perhaps, they say, we live on a very large hill, and this hill is perhaps a comparatively small irregularity in an even larger hill that extends much farther than distance L, and this larger hill is also perhaps only a comparatively small irregularity in an even larger hill, and so on, without limit. The rest of the Surveyors, finding that such thoughts give them headaches, refuse to listen any further to the hierarchists.

Around the campfire one evening a Surveyor tells of a science-fiction story he has read about a violent universe filled with cosmic radiation. Ferocious two-legged creatures in this universe, the shuddering Surveyors are told, find that this cosmic radiation is 99.2 percent isotropic. They believe that the 0.8 percent anisotropy is because they inhabit a body that whirls through three-dimensional space at very high velocity and is therefore only a sort of local irregularity. The storyteller then says that these warlike creatures worship a terrible god called Big Bang who at the beginning of time spewed forth their universe in fiery radiation. Many of the Surveyors, overcome with horror, have by this time crept away to their tents. Those who remain are told that in this universe $L = 10\,000l$ (after subtracting the anisotropy caused by the whirling motion), and the probability that it has a center is $l^3/L^3 = 0.000\,000\,000\,001$. The listening Surveyors are greatly impressed. But they become appalled and also scatter to their beds when told that among the warlike creatures of the violent universe the "closed" group pray for the return of Big Bang.

THE ISOTROPIC UNIVERSE

In a state of isotropy, things are the same in all directions from a specific point. Thus from a ship at sea we see the surface of the sea stretching away isotropically to the horizon. Although waves ruffle the surface their size is small compared with the distance to the horizon and as far as the sea is concerned they are mere details.

But to the naked eye the universe by comparison is not as isotropic as Lucretius, the Roman poet, declared, and we are surrounded it seems by large waves – a veritable storm of large irregularities. The Galaxy is certainly not isotropic about us. Stars are concentrated in the rotating disk in which we have noncentral location. Beyond the Galaxy, all directions do not look alike because of the dusty interstellar medium, and we look out at the distant universe through "dirty windows" and must allow for the nonisotropic absorption of light.

Around us, outside the Galaxy, are scattered unevenly the galaxies of the Local Group. Beyond the Local Group exist other clusters of galaxies, scattered higgledy-piggledy, that stretch away in multitudes to a distant cosmic horizon, and the irregularities we perceive extend over vast distances. Our Galaxy is like a ship tossed in a stormy ocean. Beyond about a billion light years the optically observed universe looks reasonably isotropic. When allowance is made for obscuration within the Galaxy, what lies at great distances in one direction is not too different from what lies in other directions.

Incoming signals from radio sources are not absorbed by gas and dust in the Galaxy and by the Earth's atmosphere. Radioastronomers find that the very distant and numerous radio sources are apparently distributed about us isotropically. X-rays are generated in galaxies and in the hot and tenuous gas between galaxies in the rich clusters. These x-rays fill the universe and come to us from all directions. They cannot penetrate the Earth's atmosphere and must be detected with instruments mounted on rockets and space vehicles. The x-ray background has been found to be at least 99 percent isotropic.

Our main and most convincing source of evidence for the isotropy of the universe is the low-temperature cosmic background microwave radiation discovered in 1965 by Arno Penzias and Robert Wilson of the Bell Telephone Laboratories. The cosmic background radiation (CBR), which fills the universe, originated in the big bang. This radiation, once extremely hot, has been cooled over billions of years by the expansion of the universe and has a temperature of approximately 3 kelvin (more accurately 2.728 K). This cool radiation has a typical wavelength of 1 millimeter and consists of approximately 400 photons per cubic centimeter. Each of these photons, or waves of energy, has traveled for 10 or more billion years at the speed of light and traversed the observable universe. Careful measurements show that this 3-degree radiation deviates from isotropy by about 1 part in 500.

The discovery of the low-temperature cosmic radiation has opened the way to great developments in cosmology. It provides evidence that the universe was once dense and hot and also enables us to understand what happened in the early universe. We now know, for example, that most of the helium existing at present was produced when the universe was approximately 200 seconds old.

The cosmic background radiation has a small 24-hour anisotropy. Its temperature is very slightly higher in one direction of the sky than in the opposite direction. As the Earth rotates, the temperature of the incoming cosmic radiation rises and falls by about 1 part in 500 in a 24-hour cycle. This amount of anisotropy indicates that the Earth (and the Sun) is probably moving in the universe at a velocity of 600 kilometers per second in the direction of the constellation of Leo.

The cosmic background radiation consists of electromagnetic waves that travel in all directions. Consider an observer who moves at constant velocity. This moving observer sees the waves Doppler shifted. Waves coming from the forward direction have slightly shorter wavelengths, and waves coming from the backward direction have slightly longer wavelengths. Waves from the forward direction are more energetic and are received more frequently, and waves from the backward direction are less energetic and are received less frequently. This means the temperature is slightly greater in the forward direction than in the opposite direction. Let T be the temperature of the cosmic background radiation. The temperature in the forward direction is

$$T_f = T(1 + V/c),$$

and in the backward direction is

$$T_b = T(1 - V/c),$$

where V is the velocity and c the speed of light, and V/c is a small quantity. The difference between the observed temperatures measured in the forward and backward directions is $\Delta T = \frac{1}{2}(T_f - T_b)$, and therefore

$$\frac{\Delta T}{T} = \frac{T_f - T_b}{T_f + T_b} = \frac{V}{c}. \qquad [7.1]$$

Thus, relative to a body moving at $V = 3000$ kilometers a second, $\Delta T/T = 0.01$, and the observed 24-hour anisotropy is 1 percent.

The Earth moves about the Sun at speed 30 kilometers per second and the Sun revolves about the center of the Galaxy at speed of 300 kilometers per second. The Galaxy moves within the Local Group of galaxies at a velocity of about 100 kilometers per second toward the Andromeda Nebula. In addition, the Galaxy moves in the Local Supercluster at velocity 330 kilometers per second toward the Virgo cluster, and the Local Supercluster moves at velocity 310 kilometers per second toward the Hydra–Centaurus supercluster. The resulting approximate velocity of the Galaxy relative to the cosmic background radiation is 600 kilometers per second toward the constellation Leo. Hence, $\Delta T/T = V/c = 0.002$, or 0.2 percent.

The background radiation provides a cosmic framework relative to which we can determine velocities in an absolute sense. This startling discovery in modern cosmology contradicts the Newtonian relativity of velocity. A frame of absolute rest – of zero velocity – actually exists at each place in the universe; it is that frame in which the cosmic background radiation is isotropic. Hence we can find our own absolute velocity.

After subtraction of the 24-hour anisotropy caused by the motion of the Earth, the cosmic background radiation is remarkably isotropic. The small residual anisotropies, which are less than 1 part in 10 000, are the imprint of irregularities of matter in the early universe that subsequently evolved into galaxies and clusters of galaxies.

THE COSMOLOGICAL PRINCIPLE
All places are alike

Most treatments of modern cosmology begin with the cosmological principle. This principle, given its name by the astrophysicist and cosmologist Edward Milne in 1933, is the foundation of modern cosmology. Einstein in 1931 expressed the principle in the words, "all places in the universe are alike." Stated this way, it is reminiscent of Rudyard Kipling's *The Cat That Walked by Himself*: "I am the Cat who walks by himself, and all places are alike to me." The cosmological principle declares that apart from local irregularities the universe is the same everywhere. The principle seems attractively simple; it is, however, a proposition of utmost generality, so sweeping in scope and far-reaching in consequence that it deserves much thought and study.

The observer and the explorer

An ordinary observer in a lifetime travels only short distances as measured on the cosmic scale and is therefore localized to a small region of space. The best that such an observer can do is to look around and see whether the universe is isotropic. This we have done and found that all directions in the universe are alike, seen from our local region of space.

An observer in a cosmic sense is immobile. To escape this constraint we invent, for illustrative purposes, a mobile cosmic "explorer" who is able to move infinitely rapidly from place to place in space. Sex cannot be omitted from cosmology (if only because of the he/she language aspect), and by tossing a coin, I have determined that the stay-at-home observer is he and the gadabout explorer is she.

If the mobile explorer finds in her travels that all places are alike, she will be justified in asserting that the universe is the same everywhere and is hence homogeneous. She will declare that the universe is invariant to translations in space. If the immobile observer finds that all directions are alike, he will be justified in asserting that the universe is isotropic. He will declare that the universe at his place is invariant to rotations in space. Hence the gadabout explorer translates from place to place and discovers homogeneity (according to which all places are alike) in space, and the stay-at-home observer rotates and discovers isotropy (according to which all directions are alike) about a point in space.

When the observer asserts that the universe is homogeneous on the basis of observed isotropy, he in effect postulates the cosmological principle. Lacking essential information, he is not the first to cover his ignorance with an impressive-sounding principle. Our task is to discuss the difference between homogeneity and isotropy, and show how homogeneity can be inferred from an observed state of isotropy.

What homogeneity means

In her travels in a homogeneous universe the explorer sees at every place similar distributions of various objects – living creatures, planets, stars, galaxies, clusters of galaxies

– all evolving and changing in time. Suppose, after waiting one million years, the explorer repeats her instantaneous tour of the universe. Again she sees at every place similar distributions, but now everything is slightly older and more evolved. Repeated tours of the universe separated by long intervals of time reveal a homogeneous universe whose contents are in a continual state of change. The universe remains permanently homogeneous because things everywhere change in the same way. From these observations the explorer draws an important conclusion: to maintain a state of homogeneity, in which things everywhere change in the same way, the laws of nature must everywhere be the same. The laws, whatever they are, might conceivably change in time and be different from tour to tour, but in each tour they are instantaneously everywhere the same. Time-invariant homogeneity means the laws of nature must everywhere be the same.

Homogeneity of the universe also means that all clocks in the universe, apart from local irregularities, agree in their intervals of time. Suppose our imaginary explorer rushes around the universe adjusting clocks everywhere to show a common time. On subsequent tours she finds the clocks all running in synchronism and showing the same time. This universal time is known as cosmic time (Chapter 14). Departures from cosmic time-keeping are caused by local irregularities.

In a homogeneous universe all places are alike and things everywhere evolve according to the same laws. Now suppose that at one place an observer notices that all directions are alike and the universe is therefore isotropic about that place. In a homogeneous universe, if one place is isotropic, then all other places must be isotropic. A homogeneous universe, isotropic at one place, is isotropic at all places. If the observer perceives that all directions are alike, and is told by the explorer that all places are alike, then the observer knows that the universe is isotropic at all places and the universe has no center.

Let us pursue this subject further. A homogeneous universe can also be anisotropic (all directions are not alike). Consider a flat and limitless plain of grass. A prevailing wind causes the grass on the plain to grow tilted in one direction. The plain is still homogeneous – all places are alike – because the tilt is everywhere the same. But the plain is not isotropic. In one direction the grass leans away from, and in the opposite direction it leans toward, a person standing anywhere on the plain. This is an example of a 24-hour anisotropy in which the same state is observed after rotating through 360 degrees. We can imagine a homogeneous but anisotropic universe in which galaxies have their axes of rotation all pointing in the same direction, or in which an intergalactic magnetic field points in one direction only. When the rotation axes of galaxies are randomly oriented, and an intergalactic magnetic field is tangled in all directions, then the universe becomes isotropic, apart from negligible local irregularities.

An inhomogeneous universe (in which all places are not alike) might be isotropic at one place but cannot be isotropic at all places. From the summit of a hill the surrounding countryside can appear isotropic, but when seen from anywhere else in the neighborhood, the countryside is anisotropic because of the hill.

To sum up: a state of homogeneity (all places are alike) proves all places are isotropic when one place is isotropic; but a state of isotropy (all directions are alike) at one place does not prove homogeneity, and a state of anisotropy does not prove inhomogeneity.

How the location principle works

The mobile explorer perceives a state of homogeneity, and the immobile observer perceives a state of isotropy. Can the stay-at-home observer, who represents us, also perceive by careful observations a state of homogeneity? The answer is no, and for a very simple reason. When we look out into the universe we also look back into the past. We see objects at different distances corresponding to different epochs of the past. We see the universe in different states of evolution and expansion. Distant things are different from local things and we observe inhomogeneity.

How then can we prove that the universe is homogeneous? Unfortunately we cannot prove with absolute certainty that the universe is homogeneous. We must either postulate the cosmological principle as an article of faith or become philosophers and use the location argument in the way the Surveyors used it at the beginning of this chapter.

With the location principle the argument becomes as simple as this: From observations we know that the universe is isotropic at our place. If the universe is inhomogeneous, we have special location. But special location is improbable, and hence inhomogeneity is improbable. Observed isotropy and the location principle lead to the conclusion that homogeneity is probable. We can think of the argument as an equation:

observed isotropy + location principle
= homogeneity,   [7.2]

or as a syllogism:

The universe is isotropic,
special location is improbable,
hence the universe is probably homogeneous.

Copernican principle

The Copernican principle, formulated by Hermann Bondi (1960), is similar to the location principle and serves the same purpose. It states, "We are not at the center of the universe." Copernicus believed that the Sun, not the Earth, occupies a central place in the universe. The Copernican principle has therefore the disadvantage of appearing to perpetuate the belief that a center, somewhere or other, exists. In effect, it asserts too much: we may say with certainty only that a central location in the cosmos is improbable. Also, the Copernican principle is perhaps incorrectly named; if historical precedence is important, a more appropriate name is the Aristarchean principle.

PERFECT COSMOLOGICAL PRINCIPLE

Static and expanding steady states

William MacMillan in 1918 assumed and Hermann Bondi and Thomas Gold in 1948 proposed that the universe is homogeneous in both space and time. "All places are alike in space" becomes "all places are alike in space and in time." This sweeping postulate was named the perfect cosmological principle by Bondi and Gold. The name chosen was not inappropriate, for it recalls the Platonic ideal of a universe whose perfection is unmarred by change. The perfect cosmological principle states that the universe is in a steady state (or unchanging state) and over endless periods of time remains the same. The cosmic explorer sees not only that all places are alike in each tour but also that nothing changes from tour to tour. Everything is the same everywhere in space and time, apart from local irregularities. Hence, not only is the state of homogeneity unchanging in time (as in most cosmological models), but also the actual state of the contents of the universe is unchanging in time.

The unevolving Newtonian universe in the eighteenth century and the MacMillan perpetual motion universe in the twentieth century were in a steady state. They were also static, neither expanding nor contracting. (A river can be in a steady state; the water, however, is either nonstatic when flowing, or static when frozen.)

With the advance of science came the understanding that everything evolves and nothing remains eternally the same. The static Newtonian universe became an evolving universe and lost its steady state. The fate of MacMillan's perpetual universe is discussed in Chapter 18.

Astronomers in the twentieth century discovered that the universe is not static but expanding. For almost two decades (1948–1965) the steady-state idea was revived in the context of an expanding universe. In steady-state expansion, the expansion rate is constant. Galaxies grow old and new galaxies replace them. Age distribu-tions never change; as in a long-standing society of zero population growth, births cancel deaths, thus maintaining a steady-state distribution of ages. In a steady-state universe that expands, matter must be created continuously to maintain a constant density: new galaxies form from the created matter, and old galaxies drift apart and are thinned out by the expansion. The expand-ing steady state picture, when looked at in detail, is marvelously self-consistent.

The location principle assures the obser-ver that an isotropic universe is also spatially homogeneous. But the principle is not much help in establishing temporal homogeneity (all epochs are alike). This is because time, unlike space, is peculiarly asymmetric and we cannot perceive the future with the same clarity as the past. If we could see the future as clearly as the past, and see a past–future symmetry in time, then, with the location principle, we could assert that the universe is temporally homogenous. But we cannot and the perfect cosmological principle is by no means as secure as the cosmological principle.

There is now overwhelming evidence to show that the universe is not in a steady state. Radio sources and quasars were more numerous in the past than at present, and the demand by the steady-state cos-mologists to be shown the ashes of the big bang has been met by the discovery of the low-temperature cosmic background radiation.

REFLECTIONS

1 *Edwin Hubble succeeded in resolving cepheid variable stars in M 31 in 1923, and in other nearby galaxies shortly afterward. This was an important step in the history of astronomy and cosmology. Previously it had seemed the Galaxy was immense in size and all other galaxies were midget systems in com-parison. Hubble's observations of cepheids showed that the extragalactic nebulae were at larger distances than previously supposed, and were larger in size, and hence must be ranked as galaxies in their own right. One puzzle still remained: Why was our Galaxy*

still conspicuously larger than other similar galaxies such as M 31? This was not resolved until later when Walter Baade discovered the two populations of stars.

Walter Baade (1893–1960), a German-American astronomer, discovered in 1942 that stars are of two kinds, named population I and population II. During the blackout of Los Angeles in World War II the observing conditions with the 100-inch telescope at Mount Wilson Observatory were exceptional, and with great care and skill, Baade succeeded in resolving many individual stars in the nuclear bulge of the spiral galaxy M 31. He discovered a difference between the bluish population I stars in the disk and the reddish population II stars in the nuclear bulge of M 31. Later, with the 200-inch telescope at Palomar, he found that the cepheids of the two stellar populations obey different period–luminosity laws. With the same period of pulsation, population I cepheids are four times brighter than population II cepheids. This was a very important discovery because Harlow Shapley and Edwin Hubble had assumed that the cepheids in the globular clusters in our Galaxy were the same as the cepheids in the disk of M 31. Baade's discovery implied that M 31 was not 1 million but 2 million light years away, and also all other extragalactic distances needed to be increased twofold. This increased the size of the galaxies, and our Galaxy no longer seemed conspicuously larger than other giant spirals. Baade thus doubled the size of the universe.

2 *Variations on the cosmological principle:*
• *"God is an infinite sphere whose centre is everywhere and circumference nowhere" (Empedocles, fifth century* BC*).*
• *"Whatever spot anyone may occupy, the universe stretches away from him just the same in all directions without limit" (Lucretius,* The Nature of the Universe, *first century* BC*).*
• *"God is that than which nothing greater can be conceived" (Anselm, archbishop of Canterbury, eleventh century). This is the famous ontological proof of the existence of God.*

• *"God is that whose power is not numbered and whose being is not enclosed" (Thomas Bradwardine, archbishop of Canterbury, fourteenth century). The finite Aristotelian system failed to contain an infinite and omnipresent God. See Edward Grant, "Medieval and seventeenth century conceptions of an infinite void space beyond the cosmos."*
• *"The fabric of the world has its center everywhere and its circumference nowhere" (Nicholas of Cusa, fifteenth century,* On Learned Ignorance*). Nicholas of Cusa was influenced by the discovery of the Lucretian poem, and the argument that an infinite all-powerful God could create no less than an infinite universe. See Alexander Koyré,* From the Closed World to the Infinite Universe.
• *"It is evident that all the heavenly bodies, set as if in a destined place, are there formed unto spheres, that they tend to their own centres and that around them there is a confluence of all their parts" (William Gilbert [1544–1603],* New Philosophy*). Gilbert, author of the important book* The Magnet, *was greatly influenced by the work of Thomas Digges and believed the universe is infinite and consists of endless solar systems (Figure 7.4). Bruno was also greatly influenced by Digges's work, and believed in much the same as Gilbert. Bruno was burned at the stake in Rome in 1600, whereas Gilbert was made Queen Elizabeth's physician in London in 1601.*
• *"Wee ... were a consideringe of Kepler's reasons by which he indeavors to overthrow Nolanus [Bruno] and Gilbert's opinions concerning the immensitie of the spheere of the starres and that opinion particularlie of Nolanus by which he affirmed that the eye being placed in anie part of the universe, the appearance would be still all one as unto us here" (William Lower, in a letter, 1610).*
• *"But if the matter was evenly disposed throughout an infinite space, it could never convene into one mass; but some of it would convene into one mass and some into another, so as to make an infinite number of great masses scattered at great distances from one to another throughout all that infinite space"*

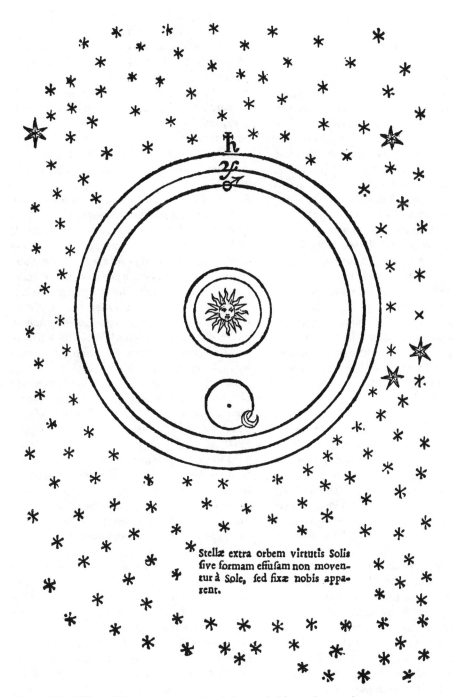

Figure 7.4. William Gilbert's cosmographical diagram in his posthumously published *New Philosophy*. (Reproduced as Figure 9 in Dorothea Singer's *Giordano Bruno: His Life and Thought*.)

(Isaac Newton, in a letter to Richard Bentley, 1692).

• *"It has been said that if the spiral nebulae are islands, our own galaxy is a continent. I suppose that my humility has become a middle-class pride, for I dislike the imputation that we belong to the aristocracy of the universe. The earth is a middle-class planet, not a giant like Jupiter, nor yet one of the smaller vermin like the minor planets. The sun is a middling sort of star, not a giant like Capella but well above the lowest classes. So it seems wrong that we should happen to belong to an altogether exceptional galaxy. Frankly I do not believe it: it would be too much of a coincidence. I think that this relation of the Milky Way to the other galaxies is a subject on which more light will be thrown by further observational research, and that ultimately we shall find that there are many galaxies of a size equal to and surpassing our own. Meanwhile the question does not much affect the present discussion. If we are in a privileged position, we shall not presume upon it"* (Arthur Eddington, The Expanding Universe, *1933).*

• *"Not only the laws of nature, but also the events occurring in nature, the world itself, must appear the same to all observers, wherever they may be"* (Edward Milne, *"World-structure and the expansion of the universe," 1933).*

• *"The nebulae are great beacons, scattered through the depths of space .... Observations give not the slightest hint of a super-system of nebulae. Hence, for purposes of speculation, we may invoke the principle of the Uniformity of Nature, and suppose that any other equal portion of the universe, chosen at random, will exhibit the same general characteristics. As a working hypothesis, serviceable until it leads to contradictions, we may venture the assumption that the realm of the nebulae is the universe – that the Observable Region is a fair sample, and that the nature of the universe may be inferred from the observed characteristics of the sample"* (Edwin Hubble, The Realm of the Nebulae, *1936).*

• *"It is impossible to tell where one is in the universe"* (Edmund Whittaker, From Euclid to Eddington, *1958).*

• *Homogeneity is a cosmic undergarment and "the frills and furbelows required to express individuality can be readily tacked onto this basic undergarment"* (Howard Robertson, "Cosmology," 1957).

• *"The Earth is not in a central, specially favoured position. This principle has become accepted by all men of science, and it is only a small step from this principle to the statement that the Earth is in a typical position"* (Hermann Bondi, Cosmology, *1960).*

3 *"Is it not possible, indeed probable, that our present cosmological ideas on the structure and evolution of the universe as a whole (whatever that may mean) will appear hopelessly premature and primitive to astronomers of the twenty-first century? Less than 50 years after the birth of what we are pleased to call 'modern cosmology,' when so few empirical facts are passably well established, when so many oversimplified models of the universe are still competing for attention, is it, may we ask, really credible to claim, or even reasonable to hope, that we are presently close to a definitive solution of the cosmological problem?"* (Gerard de Vaucouleurs, "The case for a hierarchical cosmology," 1970).

4 *Physical postulates of impotence are rare, perhaps because of the difficulty of making statements of what is truly impossible. Whittaker's postulate, "It is impossible to tell where one is in the universe," is an example that illustrates the cosmological principle. Other postulates of impotence are: "Perpetual motion is impossible;" "it is impossible to measure simultaneously the position and momentum of an electron;" "it is impossible for heat to flow of its own accord from one region to another region of higher temperature." Garrett Hardin, in* Nature and Man's Fate *(1959), has proposed an evolutionary postulate of impotence: "it is impossible" for life "to escape competition." He argues that a species that eliminates all competing species self-destroys by becoming its own competitor. Some postulates of impotence have been shown to be false; for example, "it is impossible to know where one is in* *time" (perfect cosmological principle). Also, "absolute motion is unmeasurable;" we can now determine absolute motion by using the cosmic background radiation as a frame of reference.*

5 Ockham's razor (which states that the preferred theory has the fewest and simplest assumptions) is used in various disguises. One variant is the principle of minimum astonishment (which states that the preferred theory causes the least astonishment). History shows, however, that in retrospect the preferred theory is not always the simplest or the least astonishing.

PROJECTS

1 Why do we have confidence in the truth of the cosmological principle (all places are alike)?

2 (a) Can you invent a postulate of impotence? For instance: it is impossible to verify the truth of Whittaker's postulate of impotence.

(b) Can you invent a principle that might apply in cosmology? Consider David Van Blerkom's variation on the cosmological principle: "the universe is just as bad out there as it is here."

(c) Nicholas of Cusa in the fifteenth century distinguished between unlearned and learned ignorance. In unlearned ignorance, the less we know, the more confident we are in the truth of our knowledge. In learned ignorance, the more we know, the less confident we are in the truth of our knowledge. Discuss: "he will be the more learned the more he comes to know himself for ignorant."

3 Use your imagination and draw one-dimension sketches that illustrate and explain

(a) a homogeneous and isotropic distribution (for example, a row of equally spaced dots),

(b) a homogeneous and anisotropic distribution,

(c) an inhomogeneous and isotropic distribution,

(d) an inhomogeneous and anisotropic distribution.

4 Discuss the difference between (1) a time-invariant spatial homogeneity, and (2) a homogeneity of things in space that is time-invariant.

5 Is the physical universe, as we know it and understand it, something that actually exists apart from us conceiving it? Or is it something that changes as science advances and might never reach a final form that we can call the full truth?

6 Discuss the following: "In every possible way today's standard model of the universe is unlike the standard model of one hundred years ago at the end of the nineteenth century. If cosmology can advance so rapidly in just a single century, and no signs indicate that it is slowing down, what might it be like at the end of the twenty-first century? Dare we hope that we have at last attained a glimmer of secure knowledge of the true nature of the Universe?" (Edward Harrison, "A century of changing perspectives in cosmology," 1992).

FURTHER READING

Berendzen, R., Hart, R., and Seeley, D. *Man Discovers the Galaxies*. Science History Publications, New York, 1976.

Grant, E. "Medieval and seventeenth century conceptions of an infinite void space beyond the cosmos." *Isis* 60, 39 (1969).

Harrison, E. R. "A century of changing perspectives in cosmology." *Quarterly Journal of the Royal Astronomical Society* 33, 335 (1992).

Lucretius. *The Nature of the Universe*. Translated in prose by R. E. Latham. Penguin Books, Harmondsworth, Middlesex, 1951.

SOURCES

Bondi, H. *Cosmology*. Cambridge University Press, Cambridge, 1960.

Bondi, H. and Gold, T. "The steady-state theory of the expanding universe." *Monthly Notices of the Royal Astronomical Society* 108, 252 (1948).

Duhem, P. *Medieval Cosmology: Theory of Infinity, Place, Time, Void, and the Plurality of Worlds*. Edited and translated by R. Ariew. Chicago University Press, Chicago, 1985.

Eddington, A. S. *The Expanding Universe*. Cambridge University Press, Cambridge, 1933.

Grant, E. *Physical Science in the Middle Ages*. Wiley, New York, 1971.

Grant, E. *Much Ado About Nothing: Theories of Space and Vacuum from the Middle Ages to the Scientific Revolution*. Cambridge University Press, New York, 1981.

Guericke, O. von. *Experimenta Nova Magdeburgica de Vacuo Spatio*, Amsterdam 1672. Reprinted: Zellers Verlagsbuchhandlung, Aalen, 1962.

Hardin, G. *Nature and Man's Fate*. Rinehart, New York, 1959.

Herbig, C. H. "Interstellar smog." *American Scientist* (March–April 1974).

Hubble, E. *The Realm of the Nebulae*. Yale University Press, New Haven, Connecticut, 1936.

Johnson, F. R. "Thomas Digges and the infinity of the universe." Reproduced in *Theories of the Universe: From Babylonian Myth to Modern Science*. Editor, M. K. Munitz. Free Press, Glencoe, Illinois, 1957.

Jaspers, K. *Anselm and Nicholas of Cusa*. Editor, H. Arendt. Harcourt Brace Jovanovich, New York, 1966.

Koyré, A. *From the Closed World to the Infinite Universe*. Johns Hopkins Press, Baltimore, 1957.

Lindberg, D. C. Editor. *Science in the Middle Ages*. University of Chicago Press, Chicago, 1978.

Mandelbrot, B. B. *The Fractal Geometry of Nature*. Freeman, San Francisco, 1982.

Miller, P. "Bentley and Newton." In *Isaac Newton's Papers & Letters on Natural Philosophy and Related Documents*. Editor, I. B. Cohen. Harvard University Press, Cambridge, Massachusetts, 1978.

Milne, E. "World-structure and the expansion of the universe." *Zeitschrift für Astrophysik* 6, 1 (1933).

Robertson, H. P. "Cosmology." *Encyclopedia Britannica*, 1957.

Schlegel, R. "Steady-state theory at Chicago." *American Journal of Physics* 26, 601 (1958). A discussion of MacMillan's steady state theory.

Singer, D. *Giordano Bruno: His Life and Thought*, with an Annotated *Translation of his Work on the Infinite Universe and Worlds*. Schumann, New York, 1950.

Toulmin, S. and Goodfield, J. *The Fabric of the Heavens: The Development of Astronomy and Dynamics*. Harper and Row, New York, 1961.

Vaucouleurs, G. de. "The case for a hierarchical cosmology" *Science* 167, 1203 (1970).

Whitehead, A. N. *Science and the Modern World*. Macmillan, London, 1925.

Whittaker, E. *From Euclid to Eddington: A Study of Conceptions of the External World*. Dover Publications, New York, 1958.

8 CONTAINMENT AND THE COSMIC EDGE

To see a World in a grain of sand,
And a Heaven in a wild flower,
Hold Infinity in the palm of your hand,
And Eternity in an hour.
William Blake (1757–1827), Auguries of Innocence

THE CONTAINMENT PRINCIPLE

Much of cosmology in the past has been concerned with the center and edge of the universe (see Figure 8.1), and our attitude nowadays on these matters is expressed by the principles of location and containment. Broadly speaking, the location principle (previous chapter) involves issues concerning the cosmic center, and the containment principle (this chapter) involves issues concerning the cosmic edge. Both principles help us to avoid pitfalls that trapped earlier cosmologists.

The containment principle of the physical universe states: *the physical universe contains everything that is physical and nothing else.* It is the battle cry of the physical sciences (chemistry and physics). To some persons the principle seems so elementary and obvious that it hardly deserves mentioning, to others it is a declaration of an outrageous philosophy. Before condemning the principle as too elementary or too outrageous, we must look more fully at what it means.

Modern scientific cosmology explores a physical universe that includes all that is physical and excludes all that is nonphysical. The definition of physical is sweeping and at first sight exceeds what common sense deems proper. It includes all things that are measurable and are related by concepts that are vulnerable to disproof. Atoms and galaxies, cells and stars, organisms and planets are physical things that belong to the physical world. Particles and their corpuscular-wavelike duality, atoms and their

choreography of electron waves, DNA and its genetic coding, fields and waves that propagate through space, the rich virtual worlds of the vacuum, the special relativity properties of spacetime, the general relativity properties of curved and dynamic spacetime, and the vast astronomical universe are all things of a physical nature.

But there is more. We, as physical creatures, possessing bodies and brains, are imprisoned in the physical universe. Space and time are not just voids into which the universe has been dropped; if they were, we could escape by searching out places in space and time not occupied by the universe. But spacetime, which is the four-dimensional physical combination of space and time, is not a mere receptacle; it is a physically real continuum. A continuum that is real in its own right. Space and time are active participants in the scheme of things, they belong to the physical universe, and do not extend beyond. *The universe contains space and time and does not exist in space and time.* If you believe in a nonphysical realm, such as heaven, you must endow it with its own space and time. You cannot extend our space and time to include heaven, for heaven would then be brought into the physical universe and its existence exposed to the critical methods of scientific inquiry.

The physical nature of space (or rather spacetime) is demonstrated by its dynamic properties. Empty regions of space act on and influence one another! This is the essence of general relativity. Gravitational

Figure 8.1. A nineteenth-century woodcut that supposedly presents the medieval view of the universe. Beyond the sphere of stars lies the celestial machinery and other heavenly wonders.

waves, ripples of space, travel at the speed of light. Gravitation, once a mysterious force that acted instantaneously across empty space, has become the dynamic curvature of space itself that propagates at the speed of light. It is possible, such are the bewildering properties of space, to have a universe containing only gravitational waves, and the dynamic behavior of this universe is governed by the gravitational attraction of the energy in the gravitational waves. A black hole need not contain matter; it may contain only rippling space-waves whose total energy has a mass that accounts for the strong gravity of the black hole.

Space and time in most universes of the past were the stage on which was enacted the cosmic drama. In the modern physical universe space and time are the leading actors. Who can doubt the physical reality of space (or spacetime) when it raises the tides, guides the Moon around the Earth, the Earth around the Sun, and will tear

apart incautious astronauts and their spaceships in the vicinity of neutron stars and black holes?

Space may be finite and yet edgeless. The curved two-dimensional surface of a sphere is an easily visualized analogy. The surface is finite yet has no edge. An ant crawling in a straightforward direction on the surface of a spherical water melon returns to its starting point without encountering an edge. The cosmic explorer traveling in a straight line in a finite, homogeneous, and isotropic universe also returns to the starting point.

Some people will protest that the containment principle leaves out all that is most valuable. What about our souls, our minds, consciousness, and all the richness of the inner mental world, where do they fit in? The response that must be made is quite simple: they do not fit in anywhere. At best only their physical counterparts (such as chemical activities) fit in. All the joys of life are no more than the biochemistry of neurons in the brain. In response to those who protest and want it all put together neatly in a spiritual–psychical–physical universe, we must answer: "You are confusing the Universe with universe. The unknown Universe is everything, including our minds; the known physical universe contains what is physical, including our brains. Mathematicians, physicists, biophysicists, and chemists have made the physical universe, and if you do not like it, despite its extraordinary success, you must make your own universe."

The science of modern cosmology deals only with a physical model of the Universe that is yet another mask on the face of the unknown. But what a fantastic mask it is! All the inventive genius of the greatest thinkers in the history of science has gone into its making. Can one wonder that many people, including scientists, when confronted with the majesty of the physical universe, have mistaken this latest mask for the real face, the physical universe for the Universe?

THE COSMIC EDGE
The cosmic-edge riddle

In the ancient Mediterranean world, the Atomists and Epicureans championed the idea of an infinite, centerless, and edgeless universe; the Aristotelians and Stoics championed the idea of a finite system having a center and an edge. Of primary importance in the long debate was the problem of the cosmic edge.

The cosmic-edge riddle – "what happens to a spear when it is hurled across the outer boundary of the universe?" – was posed in the fifth century BC by Archytas of Tarentum, a Pythagorean soldier-philosopher and friend of Plato. (See Figure 8.2.) "Does the spear rebound or vanish from this world?" he asked. The riddle exposed the logical inconsistency of believing that

Figure 8.2. The cosmic edge riddle: What happens when a spear is thrown across the edge of the universe?

whatever bounds the universe is itself not part of the universe. For more than two thousand years the ablest minds wrestled with the riddle, and it is true to say that Archytas's riddle has shaped much of the history of cosmology.

Epicurus in the fourth century BC stated, "Democritus of Abdera said that there is no end to the universe, since it was not created by an outside power. Moreover, the universe is boundless. For that which is bounded has an extreme point, and the extreme point is seen against something else."

Lucretius the Epicurean, influenced by the riddle of Archytas, wrote in the first century BC in his magnificent poem *De Rerum Natura*: "It is a matter of observation that one thing is limited by another. The hills are demarcated by air, and air by the hills. Land sets bounds to seas, and the seas to every land. But the universe has nothing outside to limit it." Of those who believed in a bounded universe he asked: "Suppose for a moment that the whole of space were bounded and that someone made his way to its uttermost boundary and threw a flying dart. Do you choose to suppose that the missile, hurled with might and main, would speed along the course on which it was aimed? Or do you think something would block the way and stop it? You must assume one alternative or the other.... With this argument I will pursue you. Wherever you may place the ultimate limit of things, I will ask you: 'Well, then, what does happen to the dart?'" Lucretius then gave the Atomist answer: "Learn, therefore, that the universe is not bounded in any direction."

Simplicius in the sixth century AD quoted Archytas in his commentary on Aristotelian physics with the words: "If I am at the extremity of the heaven of fixed stars, can I stretch outwards my hand or staff? It is absurd to suppose that I could not; and if I can, what is outside must be either body or space. We may then in the same way get to the outside of that again, and so on; and if there is always a new place to which the staff may be held out, this clearly means extension without limit."

In the dialogue of *The Infinite Universe*, written by Bruno while he lived in England, he gave Burchio (an imaginary Aristotelian) this argument: "I think that one must reply to this fellow that if a person would stretch out his hand beyond the convex sphere of heaven, the hand would occupy no position in space, nor any space, and in consequence it would not exist." To which Philotheo (Bruno himself) replied that space inside and outside the universe must be continuous and the same; "thus, let the surface be what it will, I must always put the question: what is beyond? And if the reply is: nothing, then call that the void, or emptiness. And such a Void or Emptiness hath no measure nor outer limit, though it hath an inner; and this is harder to imagine than is an infinite or immense universe." The commentary by Simplicius (a Neoplatonist) on Aristotelian physics was little known until translated into Latin in the sixteenth century. The influential poem by Lucretius, after its discovery in 1417 in a monastery, was widely read, in particular by such scholars as Nicholas of Cusa, Giordano Bruno, Thomas Digges, and William Gilbert.

In a finite universe, everything had comprehensible relation to the cosmic center and edge, and this arrangement in which things have absolute location dominated pre-Copernican cosmology. Eventually, astronomical developments abolished the cosmic center and the riddle of Archytas abolished the cosmic edge.

Cosmic edges
Possibly the idea of an infinite universe emerged in response to the cosmic-edge riddle, either in the form posed by Archytas, or in an earlier unrecorded form. Both Atomists and Epicureans certainly had no difficulty with the riddle; they believed space was infinite in extent, endlessly populated with stars and with planetary systems that teemed with life, and the universe had no edge. As we shall later see, solving the cosmic-edge riddle in this way created the

riddle of a dark night-sky (Chapter 24), which also played a prominent role in the history of cosmology.

Aristotelian cosmic edge

Often in ancient cosmology the universe ended abruptly with a wall-like edge. In mythology, the universe was an egg bounded by its shell, or a vast cavern bounded by dark walls. Later, in the Aristotelian universe, the edge became the sphere of fixed stars (Figure 8.3). Even Johannes Kepler believed that the universe was enclosed within a dark cosmic wall, and he was therefore able to explain why the sky at night is dark. Kepler argued

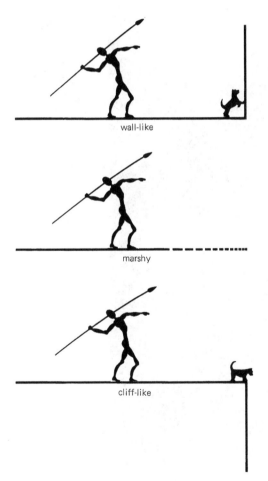

Figure 8.3. Illustrations of the wall-like (Aristotelian), marshy (medieval), and cliff-like (Stoic) cosmic edges.

that in an endless universe of stars the sky at night would not be dark but bright with starlight. We do not know if Kepler had an answer to the cosmic-edge riddle. In the case of a wall-like edge the spear must either rebound or pass through, and according to Epicurean critics, the first is impossible because what bounds space cannot itself be unbounded, and the second is proof that an edge does not exist.

Medieval cosmic edge

In later versions of the Aristotelian universe (Neoplatonic and medieval), space ended not sharply but gradually, beginning in the lunar sphere (Figure 8.3). As one moved outward away from Earth the physical realm slowly transformed into an etheric realm. In the medieval version, the outer etheric-celestial realm was surrounded by the empyrean, a realm occupied by God. To the question, "What happens to a physical body as it moves away from Earth?" two answers were possible. The body's earthly elements either remain unchanged and the body returns to Earth, as when a stone thrown in the air falls back to Earth, or the body transforms into etheric elements and its natural motion is then circular around the Earth and not up and down.

The medieval universe lacked an abrupt boundary and the force of "with this argument I shall pursue you" was lost because the pursuer was led into an etheric outer realm where physical arguments had no force. This kind of cosmic edge was like a gradual fading of firm ground into an infirm marshland. The medieval rebuttal of the riddle is now unacceptable. Observations show that the physical world does not fade into a nonphysical world at great distances.

Stoic cosmic edge

The Stoic universe consisted of a finite cosmos of stars surrounded by an infinite starless void. The Stoic edge was sharp like that of a cliff (Figure 8.3). It divided the universe into two parts: an inner starry cosmos and an outer starless and empty space that extended indefinitely. In this case the answer to the cosmic-edge riddle was quite simple: The act of throwing the spear enlarged the cosmos and extended its outer edge. In early versions of the Stoic system, the infinite void was an addition tacked on to the Aristotelian sphere of stars, presumably in response to the riddle of Archytas.

Inevitably the medieval universe evolved into a Stoic cosmos of stars, and the infinite void became the extramundane realm of God. This cosmic picture subsequently enjoyed considerable popularity. It was Isaac Newton's view of the universe in his early years at Cambridge; it was the Milky Way cosmos of William Herschel in the late eighteenth century; it was the Victorian universe of the nineteenth century; and it survived until the early decades of the twentieth century. One could in principle travel – in imagination at least – outside the Milky Way, look back, and have a magnificent grandstand view of the whole material

Figure 8.4. *The Empyrean* by Gustav Doré. This picture shows Dante and Beatrice standing at the rim of the world and gazing at the angelic spheres on the other side of the universe.

content of the universe. Alas! observations have shown that the material universe extends to vast distances beyond the Milky Way with no sign of an abrupt edge.

Amateur cosmology

Often, on first taking an interest in cosmology, a person has in mind the Stoic cosmic picture. The universe is visualized as a spherical cloud of galaxies that expands in space and has a center and an edge. This simple picture, unfortunately, is quite wrong and violates the containment principle. Space is not a nonphysical receptacle in which the universe expands; space is physical, and expands with the universe. We must think of space as an essential part of the universe and realize that it cannot extend outside the universe.

As an illustration, the big bang did not occur somewhere in space, as seems natural in the Stoic picture, but occupied the whole of space. If space is infinite, the big bang was also infinite. An infinite universe remains always infinite and cannot change and become finite. Wherever we stand, we have only to stay still and travel back in time to find ourselves in the big bang.

A centerless and edgeless infinite three-dimensional space is not too difficult to imagine. Trying to imagine a centerless and edgeless finite three-dimensional space is very difficult. Instead, we think of a two-dimensional surface that is centerless, edgeless, and finite. A spherical surface is finite in extent, and in itself has no center and no edge.

Cosmic edges in space do not exist. We cannot travel to the edge, like Dante and Beatrice in *The Divine Comedy*, and have a grandstand view of the universe (Figure 8.4). In an expanding universe, the galaxies are not rushing away from us through space, but sit in space, and space itself expands in the same way as the surface of an expanding balloon that is slowly inflated. As we later show (Chapter 14), the space between galaxies expands, and the galaxies are carried apart by the expansion of space.

CONTAINMENT OF SPACE AND TIME

Time, like space, is physical and is therefore contained in the universe. It cannot extend outside the universe across a timelike cosmic edge. We must not ask what the universe looks like from outside space and similarly we must not ask what the universe looks like before time begins and after time ends. Such questions violate the containment principle and imply that the physical universe does not contain everything physical.

Cosmogony (the word means the begetting of cosmic progeny) is the subject that deals with the origin of astronomical structures such as planets, stars, and galaxies. It embraces the origin of the elements and even the origin of life. The constraints set on cosmogony by containment are elementary. All things must have sizes smaller than or equal to the size of the universe. The following is a possible cosmogonic space sequence:

> size of atom
>   <size of cell
>   <size of multicellular organism
>   <size of planet
>   <size of planetary system
>   <size of galaxy
>   <size of galaxy cluster
>   <size of supercluster
>   <visible universe

where the symbol < means "less than." Also, all things must have ages shorter than or equal to the age of the universe. The following is a possible cosmogonic time sequence for human beings:

> age of *Homo sapiens*
>   <age of mammals
>   <age of life on Earth
>   <age of Earth
>   ≤age of Sun
>   <age of Galaxy
>   <age of helium produced in early
>   universe
>   <age of universe

where the symbol ≤ means "less than or equal to."

Figure 8.5. "If we don't know how big the whole universe is, then I don't see how we could be sure how big anything in it is either, like the whole thing might not be any bigger than maybe an orange would be if it weren't in the universe, I mean, so I don't think we ought to get too uptight about any of it because it might be really sort of small and unimportant after all, and until we find out that everything isn't just some kind of specks and things, why maybe who needs it?" – John Milligan. (With permission from John Milligan, whose cartoon first appeared in *Saturday Review*, 1971.)

Nucleochronology (the study of the origin and history of the elements) consists of various dating techniques. The light elements, mostly deuterium and helium, were made from protons and neutrons in the big bang while the universe was still young, dense, and very hot. Most other chemical elements were made much later in stars and ejected into space in supernova outbursts. The elements composing the Earth were produced in stars that died before the birth of the Solar System. Most heavy elements when formed are radioactive and decay into daughter elements. Uranium-235, for example, has a half life of 4.5 billion years and decays into lead. By finding how fast radioactive elements decay, and measuring their present relative abundances, it becomes possible to determine the age of the Earth, the Solar System, and the Galaxy. From these studies we find that the Solar System has an age of 4.6 billion years and the Galaxy an age of roughly 15 billion years.

Cosmologists estimate that the universe, from its rate of expansion, has an age of between 10 and 20 billion years. But from the late 1920s until the middle 1950s the estimated age was little more than 1 billion years. A universe younger than the Earth violated containment, and for a quarter of a century this age paradox dominated cosmology. Attempts were made to evade the paradox, as in the hesitation universe (according to which expansion was very slow for a long period in the past) and in the steady-state universe (which had an infinite past).

Cosmologists who favored a big bang universe, such as Georges Lemaître and George Gamow, thought most elements were made in the big bang. This idea proved wrong but had one great virtue: it started Gamow and his colleagues Ralph Alpher and Robert Herman thinking about a hot early universe and led them to the prediction of the cosmic background radiation almost 20 years ahead of its discovery. The steady-state idea proved wrong but also had one great virtue: the steady-state cosmologists could not accept big bang nucleosynthesis and therefore had to show that all heavy elements are made in stars. The pioneers in the successful theory of stellar nucleosynthesis were Alistair Cameron, Margaret and Geoffrey Burbidge, William Fowler, and Fred Hoyle.

Revised estimates of extragalactic distances made by Walter Baade and Allan Sandage in the 1950s, and by others since, have increased the size and age of the universe and it now seems possible to accommodate the ages of the chemical elements, the Earth, Solar System, stars, and galaxies within the lifetime of a big bang type of universe.

DESIGN ARGUMENT

Why is our universe so favorable in numerous ways to the existence of life? Throughout history, mythology and theology have urged the idea of a designed universe. The belief that the universe is intentionally designed to be a fit place for human habitation is as old as the creation myths (Chapter 25).

The design argument claims that we see everywhere evidence of cosmic design, and all the wonders of nature prove the existence of a supreme designer. The design argument emerged in a new and definitive form known as deism in the eighteenth century after the rise of the Cartesian and Newtonian world systems.

Theism is the ancient belief that a supreme being creates and runs the universe. Deism is the new belief that a supreme being creates a universe so perfect in design that supernatural maintenance is unnecessary. Thomas Burnet, a clergyman, wrote in 1687 in *Theory of the Earth*, "We think him a better Artist that makes a clock that strikes regularly every hour from the springs and wheels he puts in the work, than he that hath so made his clock that he must put his finger in it every hour to make it strike." Deists in the eighteenth century often used the clock analogy in support of the design argument. Archdeacon William Paley in 1802, in his book *Natural Theology*, subtitled *Evidence of the Existence and Attributes of the Deity Collected from the Appearances of Nature*, argued that the intricacies of the eye and hand could never have arisen by themselves in response to the blind forces of nature. Suppose, he wrote, that while walking on the heath "I found a watch on the ground;" a natural conclusion would be that "the watch must have a maker; that there must have existed at some time and at some place or other an artificer or artificers who formed it for the common purpose, which we find it actually to answer, who completely comprehended its construction and designed its use." But in the years ahead, the advance of science made it increasingly apparent that the forces of nature are not so blind as Paley believed.

In the *Bridgewater Treatise*, written by eight distinguished authors and dedicated to demonstrating the "Power, Wisdom and Goodness of God, as Manifest in the Creation," the chemist William Prout in 1834 wrote, "the anomalous properties of the expansion of water and its consequences have always struck us as presenting the most remarkable instance of design in the whole order of nature – an instance of something done expressly, and almost (could we indeed conceive such a thing of the Deity) at second thought, to accomplish a particular object." If water did not expand on freezing and ice did not float, as observed, said Prout, the oceans would freeze solid and life on Earth be impossible.

Science progressively revealed a world of astonishing intricacy governed by forces of extraordinary potency. Natural selection (which states those individual differences favoring survival and reproduction are shared increasingly among the members of an interbreeding population) accounts for the excellence of the eye and hand. The miracles of the physical universe are not its structures, such as eyes and hands, but its fundamental properties at the atomic and subatomic levels that miraculously contrive a universe fit for inhabitation by life.

Lawrence Henderson, a distinguished scientist at Harvard University, wrote in 1913 in *The Fitness of the Environment*: "The fitness of the environment results from characteristics that constitute a series of maxima – unique or nearly unique properties of water, carbonic acid, the compounds of carbon, hydrogen, and oxygen, and the ocean – so numerous, so varied, so nearly complete among all things which are concerned in the problem that together they form certainly the greatest possible fitness." Developments in physics show that the fitness of the molecular world discussed by Henderson is the consequence of basic design at the atomic and subatomic levels.

MANY PHYSICAL UNIVERSES

Hitherto we have mentioned universes in the sense that each is a model of the Universe. Let us suppose that the Universe is better represented by not one but many physical universes, and each is self-contained and isolated from all the rest.

Suppose that in each universe the fundamental constants of nature have different values. By constants of nature we mean those basic and unexplained things such as the speed of light c, the universal gravitational constant G, the electric charge e of the proton and the electron, Planck's constant h, and the masses of the subatomic particles (see Chapter 23). We have thus a collection, or an ensemble, of universes containing all values of the constants of nature in all combinations. Each universe serves as a workshop in which we study what happens when the constants are assigned values other than those in our own universe. In some universes G is large and gravity strong, in others G is small and gravity weak, and in others G is negative and gravity repulsive. In this cosmic menagerie, bizarre universes exist in which h is so large that quantum mechanics operates macroscopically, and so small that quantum mechanical effects are negligible.

Even tiny changes in the known values of the fundamental constants cause huge changes in the cosmic scenery. Minute changes in the known values of c, h, and e cause dramatic changes in the structure of atoms and atomic nuclei. Stable atomic nuclei exist in only a few universes, and most universes of the ensemble contain only hydrogen gas and lack elements such as carbon, nitrogen, and oxygen essential for organic life.

Slight changes in the values of c, G, h, e, and the masses of subatomic particles cause large alterations in the structure and evolution of stars. Almost all universes contain no stars, and in those that do, most have stars that are nonluminous, or so luminous that their lifetimes are too short for biological evolution.

Lifeforms – those organisms that we recognize as living – are constructed from various chemical elements and require environments, such as hospitable planets warmed by long-lived stars, where they can originate and evolve. Study of the cosmic ensemble shows that these requirements are fulfilled only in universes very closely similar to our own. We would not exist if the constants of nature had values other than those observed. The "accidentals of nature" (Aristotle's expression) are finely tuned at the fundamental level. Most other universes, differently designed, are starless and lifeless. In the entire ensemble, our universe and those very closely similar are perhaps the only members containing life. The conclusion of this exercise in thought is that life could not exist if the constants of physics had values other than those observed.

Why are the constants of physics so finely tuned? Is the fine-tuning purposive or fortuitous? Or is there some undiscovered theory that connects and explains why the constants of physics have necessarily their observed values? Thoughts along these lines are discussed in Chapter 25.

The basic constants of physics are either intentionally adjusted to be compatible with the existence of life (the theistic principle), or fortuitously compatible with the existence of life (the anthropic principle).

In the first case we may discard the ensemble of universes, in the second case we must retain it.

THEISTIC AND ANTHROPIC PRINCIPLES

Implications of a finely tuned universe are considered in the theistic and anthropic principles.

Theistic principle

A common belief in the Judaic–Christian–Islamic world is that God designed and created our finely tuned universe – the best of all possible worlds – specifically for the containment of life. Why is the universe the way it is? Because God made it that way for the benefit of life. (Deism more than theism stresses the importance of design.) This is the answer that naturally springs to mind. All other universes of the cosmic ensemble can be discarded. They served merely as convenient fictions enabling us to realize that our universe is finely tuned for habitation by life.

Anthropic principle

The universe is the way it is because we exist. This is the anthropic principle expressed in scientific form by Robert Dicke in 1961 (Chapter 23), developed in a definitive form by Brandon Carter, and explored and discussed by many authors. In *Twelfth Night* the clown uttered only half the truth when he declared, "That that is is." The full truth is, "I that am am, hence that that is is."

Brandon Carter introduced the term *anthropic principle* in 1974 with the definition: *what we can expect to observe must be restricted by the conditions necessary for our presence as observers*. He called this non-controversial statement the weak anthropic principle. The strong anthropic principle was defined in the controversial form: *the universe necessarily has the properties requisite for life – life that exists at some time in its history*. The strong form of the principle opens up a maze of philosophical issues out of which we have separated the theological issues under the heading *theistic principle*.

Why is the universe the way it is?

If one is asked why the universe is the way it is, or asked almost any other probing existential question that begins with "why," such as why do atoms exist?, three responses are possible. First response. The question is replaced by a functional question that begins with "how," such as how do atoms work?, and this is the usual response in science: existential questions are replaced by functional questions.

Second response. The theistic principle: the cosmic design is determined by God who ensures that the design is compatible with the existence of life. This theological response terminates scientific inquiry. Further questions, such as why did God make it that way?, and who made and designed God?, are countered by arguments that the nature of God is beyond rational inquiry.

Third response. The anthropic principle: things are the way they are because we exist. The cosmic design is determined by our existence; if the design were different, we would not be here. The implication is that the accidentals of nature are distributed with random values in different universes. Why is the universe the way it is? Because we exist. But why do we exist? Because the universe is the way it is. The circularity of the argument is broken in the theistic principle at the cost of introducing a supreme being, and in the anthropic principle at the cost of introducing an ensemble of universes.

Neither principle is limited to the existence of human beings. All other creatures on Earth are manifestations of a finely tuned universe. Wild flowers do not exist in other universes. When given a wild flower, one is also given our universe. On the face of it, intelligent creatures impose no more constraint on the design of the universe than do wild flowers.

The fundamental constants have values that are either intentional (theistic principle)

Figure 8.6. "Man is one world and hath/Another to attend him" (George Herbert, 1593–1633). *Hand with a Reflecting Globe*, by M. C. Escher. (Courtesy of the Collection Haags Gemeentemuseum, The Hague.)

or fortuitous (anthropic principle), and otherwise are inexplicable. But science forever advances by explaining what previously was thought to be inexplicable. Perhaps in the future the constants of nature will have rationally explained values that are neither intentional nor fortuitous.

At present we may prefer either anthropic or theistic ideas or ignore the whole subject and concentrate on functional issues. We

Figure 8.7. The self-aware universe.

have on one side the idea that our universe is just one of a multitude of universes and is necessarily the way it is because we exist, or we have on the other side the idea that our universe is supernaturally designed, an idea that terminates further scientific inquiry. Given a choice between a supreme being beyond rational inquiry and a wasteland of mostly dark and barren universes, most persons would probably choose the former.

WHITHER THE LAWS OF NATURE?
The realms of physics
The laws of nature are those symmetries and regularities whereby we perceive harmony and order in the universe. Often the laws represent the basic symmetries of spacetime and those ensuing from its relativistic decomposition into three-dimensional space and one-dimensional time. Often the laws are mathematical codifications that mimic the structure of the world as observed by the human mind. We believe in them until they fall into discord with observation, and are then either modified or discarded. The known laws were different in the past and no doubt in the future will be different from those of today.

Most branches of science are occupied with the study of objects ranging in size from atoms to galaxies. Atoms have sizes typically 10^{-8} centimeters and galaxies have sizes typically 10^{23} centimeters (100 000 light years). Galaxies are 10^{31} times larger than atoms and this entire range of sizes covers 31 orders of magnitude.

Most phenomena of interest in science lie in this middle realm of physics. At the atomic end of the scale almost everything is explained by electromagnetic forces, and at the galactic end of the scale almost everything is explained by gravitational forces. By moving up 5 orders of magnitude to a typical cosmic size 10^{28} centimeters (10 billion light years), and down 5 orders of magnitude to a typical subatomic size 10^{-13} centimeters (size of a nucleon), we quit the middle realm and enter the less familiar outer realms of physics.

At the upper end, in the cosmic realm, gravity in the guise of curved dynamic space-time continues to be the ruling force. Yet at the lower end, in the subatomic realm, electromagnetism is joined by other forces, weak and strong. The universe thus seems oddly asymmetric: simplicity in the cosmic realm and complexity in the subatomic realm. But the electromagnetic and weak forces are the twin aspects of an electroweak force, and both were once united at high energy in the early universe (Chapter 20). The strong and electroweak forces are also twin aspects of a hyperweak force, and were also once united at even higher energy at an earlier time in the very early universe. Hence some of the asymmetry between the atomic and cosmic realms may be apparent and not real. After all, we must not forget that the universe contains the subatomic particles and is therefore at least as complex as the particles themselves.

Doctrines of internal and external relations
It is probable that the universe contains hitherto undiscovered physical properties that unify particles and the cosmos. This idea is implicit in the bootstrap principle that has been around for a long time.

The bootstrap principle states that all things are immanent within one another. Anaxagoras had the idea when he said, "In everything there is a portion of everything." The physicist Geoffrey Chew said, "Nature is as it is because this is the only possible nature consistent with itself." The way things are on the largest scales determines the way things are on the smallest scales, and vice versa. In Gottfried Leibniz's theory of "this is the best of all possible worlds" the microcosm mirrors the macrocosm. The answer to Victor Hugo's question – "Where the telescope ends, the microscope begins. Which of the two has the grander view?" – might be that in an ultimate sense both views are equivalent. Bishop Berkeley in the eighteenth century proposed a bootstrap principle when he argued that the inertia of a body is determined by the distribution and masses of all other bodies in the universe. Ernst Mach proposed a similar idea in the nineteenth century, and this particular bootstrap argument, which inspired Albert Einstein, is now called Mach's principle (Chapter 12).

It seems that we have organized the physical universe in accordance with the old Atomist principle:

$$\text{universe} = \text{sum of all particles.} \qquad [8.1]$$

The philosopher Alfred Whitehead referred to this kind of arrangement as the doctrine of external relations. Everything works in a push–pull manner, and this way of explaining the universe derives from the human experience of living in a competitive environment. Existential questions such as why do all electrons have the same electric charge are not easily answered in the Atomist world picture and tend to be suppressed. Functional questions beginning with how, which relate to the way things are pushed about, are answered much more readily.

The bootstrap principle, on the contrary, states:

$$\text{universe} \equiv \text{particle,} \qquad [8.2]$$

(where the symbol $\equiv$ means "is equivalent to"), and each particle in some way represents a facet of the universe and is not just a small part of it. Whitehead referred to this kind of arrangement as the doctrine of internal relations. David Bohm refers to it as the implicate universe. Why do all electrons have the same electric charge? Because all electrons represent a single aspect of the universe.

We see signs of internal relations emerging in physics. Newtonian gravity was an external relation of the push–pull kind. General relativity (Chapter 12) is a much subtler scheme of things; gravity, as we know it, vanishes and is replaced by the dynamic curvature of spacetime exhibiting properties of internal relationship. But for our push–pull minds it is much too abstract and we continue to talk of gravity as a force even though we know it has vanished. Quantum mechanics uncovers a strange world of internal relations. As Richard Feyman said, nobody understands quantum mechanics. An excited atom in the depths of space decays and while decaying emits a photon wave in all directions. A million years later, a million light years away, the wave enters a telescope on Earth and creates an electronic signal in a detection instrument. Instantly, the entire spherical wave throughout space collapses into the photon emitted by the decaying atom and received at the detector. The photon can be observed only once and therefore the entire wave, spread out over millions of light years, collapses at the moment of observation of the photon. In the push–pull world of external relations, a world invented long ago by primitive human beings, such behavior is totally mysterious. The old Atomist principle tells us how, but not always why; the implicate principle tells us why, but not always how. Possibly, in a hypothetical world of intelligent and nonaggresive creatures, things are explained entirely by means of the doctrine of internal relations.

Heaven in an electron

The basic parameters of the physical world, such as the constants of nature, seem accidental. But in the bootstrap picture the universe is a self-consistent whole and therefore should contain nothing accidental of a fundamental nature. In this picture we are not free to distribute accidental properties in random combinations among universes as in the anthropic principle.

Until better and more general theories are found, we should note that the anthropic principle serves as a makeshift or poor-man's bootstrap. It relates organisms and the universe in an existential manner, and although it is only a partial bootstrap, it nonetheless is useful until better ideas are found. At present, when given a wild flower, we are also given our universe. Perhaps, in the future, when given an electron, we shall also be given our universe. We shall then have not only heaven in a wild flower, but heaven in an electron, or in any other particle.

CONTAINMENT RIDDLE

Where in a universe is the cosmologist studying that universe?

The physical universe is remarkably success-ful in explaining how various things work and in enabling us to control our environ-ment. But it fails to explain psychic things, such as consciousness and self-awareness. A person who persistently asks where life and mind are in the physical world will be passed from science to science like a person with a mysterious illness who is passed from specialist to specialist. The biologists will say that life and mind are the manifesta-tions of elaborate physical systems of billions of organized cells, and each cell is an organization of billions of atoms. If this answer fails to satisfy, the biologists will probably recommend a visit to the psychol-ogists. No two psychologists will give the same answer; some will talk of mental processes of the mind, others will talk of neurological processes of the brain, and others, admitting that they know very little about the physical universe, will recommend a visit to the physicists who seem to be doing some pretty deep things these days. But the physicists will either deny that there is a problem (what is not physical does not exist) or will recommend a visit to the biolo-gists. In desperation, the inquirer might then think of consulting the cosmologists.

The cosmologists (at least some) confess that they have not the foggiest notion of an answer. They know instead that universes come and go, springing from the imaginative fertility of the human mind, and they

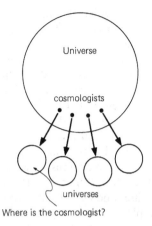

Where is the cosmologist?

Figure 8.8 The containment riddle. Where in the many universes can be found the cosmologists conceiving and studying those universes?

wonder whether a product of the mind is capable of adequately explaining the mind. The buck stops with the cosmologists because of the containment riddle: *Where in a universe is the cosmologist studying that universe?* (see Figure 8.8). Other scientists can avoid the riddle by claiming that it does not belong to their province of special knowledge. But cosmologists study the whole universe, and the riddle stares them in the face.

Consider a painter who paints a picture of his studio. The painter stands within the studio and yet does not include himself within the picture in the act of painting the picture. An attempt to portray the act of painting leads to the absurdity of an infinite regression of pictures: the picture contains the painter painting a picture, which con-tains the painter painting a picture, and so on, indefinitely. A universe is a world picture and the cosmologist is in the same situation as the painter. The cosmologist constructs a world picture that contains his physical body and physical brain but not his mind that constructs the picture. If his mind is not excluded, or he thinks of himself as only a physical brain, he also encounters the absurdity of an infinite regression: the universe contains the cosmologist studying a universe, which in turn contains the cosmologist studying a universe, and so on, indefinitely. Where then is the cosmologist studying the universe?

Can an image contain the image maker? The physical universe, consisting of multitudes of facts woven together in a web of ideas, apparently does not contain the thing that shapes the facts and spins the ideas. Those persons who cannot agree and claim that life and mind are no more than a collective dance of atoms must answer the containment riddle. Generally, those who think that life and mind are fully contained in the physical universe confuse the physical universe with the unknown Universe and mistake the mask for the face.

The containment riddle applies to universes of all kinds – the theologian's theocosmos and the psychologist's psychocosmos – and not just the physicist's physical universe.

The answer to the containment riddle is that we, who create the universes, occupy the unknown Universe. We, the image makers, occupy the Universe and our images are the universes that often, mistakenly, we believe we live in. Each universe contains not life and mind but the physical representations of life and mind, such as wild flowers and brains. The world of matter contains body and brain but not the agent that conceives the world of matter containing body and brain.

REFLECTIONS

1 Know then thyself, presume not God to scan,
The proper study of mankind is man.
Placed on this isthmus of a middle state,
A being darkly wise, and rudely great:
With too much pride for the sceptic side,
With too much weakness for the stoic's pride,
He hangs between; in doubt to act or rest;
In doubt to deem himself a god, or beast;
In doubt his mind or body to prefer;
Born but to die and reasoning but to err;
Alike in ignorance, his reason such,
Whether he thinks too little or too much;
Chaos of thought and passion, all confused:
Still by himself abused, or disabused;
Created half to rise, and half to fall;
Great lord of all things, yet a prey to all;

Sole judge of truth, in endless error hurled;
The glory, jest, and riddle of the world!
Alexander Pope (1688–1744), An Essay on Man

2 *"Behold a universe so immense that I am lost in it. I no longer know where I am. I am just nothing at all. Our little world is terrifying in its insignificance!" (Bernard de Fontenelle [1657–1757],* Conversations with a Lady on the Plurality of Worlds*).*

• *"It is impossible to contemplate the spectacle of the starry universe without wondering how it was formed: perhaps we ought to wait, and not look for a solution until we have patiently assembled the elements ... but if we were so reasonable, if we were curious without impatience, it is probable we would never have created Science and we would always have been content with a trivial existence. Thus the mind has imperiously laid claim to this solution long before it was ripe, even while perceived in only faint glimmers – allowing us to guess a solution rather than wait for it" (Henri Poincaré, French mathematician, 1913).*

3 *The physical universe contains only physical things controlled by natural laws. Is the law of natural selection a physical law? Since the time of Charles Darwin, many persons have claimed we cannot prove that fossils of various forms are linked by evolution and natural selection is unfounded speculation. But the law of natural selection is as physical as the law of gravity. The theory of natural selection requires three components:*

• *an interbreeding population,*
• *genetic variations among the members of the population,*
• *a selective environment that causes differential survival.*

Those differences that enhance survival become dominant in the population, and the species therefore changes (evolves) as it continually adapts to a changing environment. We have everyday proof of the effectiveness of natural selection. Widespread use of antibiotics, such as penicillin, has decreased the potency of antibiotics by the natural selection of more resistant populations of pathogenic bacteria.

4 *Timelike edges appear to pose problems that are more philosophical than physical. A big bang universe has a beginning and has therefore a timelike edge. If this universe collapses back to a big bang, it has two timelike edges: a beginning and an ending. These cosmic time edges are presumably not cliff-like as in the Stoic universe, or wall-like as in the Aristotelian universe, but more like the gradual fading of the medieval edge. Possibly, the physical world with its orderly historical sequence of events in cosmic time dissolves into disorderly metrical chaos at the beginning and the end. In truth, we do not understand timelike cosmic edges, perhaps because we do not fully understand time itself. Insofar as time, like space, is physical and part of the physical universe, we can say that physical time is contained within the physical universe and cannot extend "outside." Hence, as we understand and experience it, time cannot exist either before the beginning or after the end of the universe.*

• *Despite these remarks there is a way in which we might talk sensibly about what happened before the beginning of our universe. Suppose that our universe is the offspring of another physical universe – reproduced either spontaneously or deliberately by intelligent life in the parent universe – then, in that sense, we can talk of what happened in the parent universe before reproduction. Joe Rosen at the University of Central Arkansas has shown ("Self-generating universes and many worlds") that the offspring universe can also reproduce the parent universe. Thus parent and offspring universes form a cosmic unity that is self-creating: universe X creates universe Y in X's time, and universe Y creates universe X in Y's time.*

• *Contained creation is different from uncontained creation. Contained creation refers to coming into existence at a moment in time and at a place in space of something that previously did not exist. Thus the creation and annihilation of particles, the creation of a flower, the birth of a child are examples. Uncontained creation refers to the timeless and placeless creation of the universe. When people talk of the creation of*
the universe they often have in mind contained creation, which is a wrong way of looking at the subject, and are misled into asking: what came before the creation of the universe? (See Chapter 25.)

5 *Thomas Huxley (1825–1895), famous biologist and champion of Charles Darwin's theory of natural selection, wrote in 1869 for the first issue of the science journal* Nature: *"It seemed to me that no more fitting preface could be put before a Journal, which aims to mirror the progress of that fashioning by Nature of a picture of herself, in the mind of man, which we call the progress of Science."*

• *"An adequate cosmology will only begin to be written when an adequate philosophy of mind has appeared, and such a philosophy of mind must provide full satisfaction both for the motives of the behaviorists who wish to make mind material for experimental manipulation and exact measurement, and for the motives of idealists who wish to see the startling difference between a universe without mind and a universe organized into a living and sensitive unity through mind properly accounted for" (Edwin Burtt,* The Metaphysical Foundations of Modern Physical Science*).*

• *"Human sciences face an impasse because their central concept of the self is transcendental" (G. Stent, "Limits to the scientific understanding of man").*

6 *At the roots of all knowledge lies the containment riddle (where in the image is the image-maker?). In various guises the containment riddle has haunted philosophy down the ages, and now it haunts physics. Leon Rosenfeld, in "Niels Bohr's contribution to epistemology," explains how in the early days of formulating quantum theory, Bohr was very much impressed by Paul Møller's "The tale of a Danish student." The student is perplexed by problems in epistemology (the study of the validity and limits of knowledge):*

"... man divides himself into two persons, one of whom tries to fool the other, while a third one, who in fact is the same as the other two, is filled with wonder at this

confrontation. In short, thinking becomes dramatic and quietly acts the most complicated plots with itself and for itself; and the spectator again and again becomes actor ... and then I come to think of my thinking about it; again I think that I think of my thinking about it, and divide myself into an infinitely retreating succession of egos observing each other. I don't know which ego is the real one to stop at, for as soon as I stop at any one of them, it is another ego again that stops at it. My head gets all in a whirl with dizziness, as if I were peering down a bottomless chasm, and the end of my thinking is a horrible headache."

• The burning issue of what is observing what lies at the heart of quantum mechanics. In Mind and Matter, Erwin Schrödinger wrote: "Without being aware of it, we exclude the Subject of Cognizance from the domain of nature that we endeavour to understand. We step with our own person back into the part of an onlooker who does not belong to the world, which by this very process becomes an objective world." Schrödinger continued: "a satisfying picture of this world has only been reached at the high price of taking ourselves out of the picture, stepping back into the role of a non-concerned observer." Schrödinger then turned to religion: "Can science vouchsafe information on matters of religion? Can the results of scientific research be of any help in gaining a reasonable and satisfactory attitude towards these burning questions which assail everyone at times? Some of us ... succeed in shoving them aside for long periods; others, in advanced age, have satisfied themselves that there is no answer and have resigned themselves to giving up looking for one; while others again are haunted throughout their lives by this incongruity of our intellect, haunted also by serious fears raised by time-honoured superstition. I mean mainly the questions concerned with the 'other world,' with 'life after death' and all that is connected with them."

7   In Men Like Gods, H. G. Wells introduced the idea of parallel universes occupying what might be called "superspace." Normally they are isolated from one another, but occasionally, according to Wells and a host of more recent science-fiction writers, they make contact. Recent theoretical developments have also enlarged the notion of containment to embrace many universes as self-contained members of a superuniverse. Andre Linde, a Russian physicist, treats the universe as an eternal, self-reproducing multi-universe in which universes spontaneously generate new universes.

• Consider an ensemble of self-contained universes covering all possible values of the fundamental constants of physics. Suppose that all these universes are "virtual." That is, they all exist in potential form, and each only becomes "real" by observation by conscious life. Because each is self-contained, the only possible observations are by internal forms of life. Thus only those universes are real in which the fundamental constants are compatible with the existence of life. What happens in a universe such as our own that passes through an early lifeless stage? In the early stage there exists no conscious life to observe it; does it exist? Presumably, it evolves in a virtual state and becomes real by observation when conscious life comes into existence. This seemingly weird picture of virtual (potential) versus real (actual) is common in the theory of quantum mechanics. A wavefunction consists of many evolving, parallel, virtual states, all of which are potential candidates for a final observed state in the real world.

8   The theistic principle holds that the universe is the way it is because it is created by a supreme being for inhabitation by living creatures; the anthropic principle holds that the universe is necessarily the way it is because we exist. According to the first principle, an extracosmic agent determines the cosmic design; according to the second principle, our existence determines the cosmic design. The two principles are not in direct conflict, and it may be argued that the first encompasses aspects of the second.

9   Consider an excited atom in which an electron makes a transition from an initial state to any one of a number of permissible states. The

actual final state is unpredictable, only its probability. Hugh Everett in 1957 proposed that at such a transition the universe splits so that all permissible states are reached, each in a separate universe. Thus from each atom there exfoliates many new universes each time a transition occurs, and these many universes are identical except that in each the atom is in a different final state. A human being, composed of many quantum transitions, continually splits into numerous human beings occupying different universes, in each of which the human being is in a slightly different state and follows a different history.

• *"Could the solution to the dilemma of indeterminism be a universe in which all possible outcomes of an experiment actually occur?" (Bryce DeWitt, "Quantum mechanics and reality"). Why not apply this idea to free will? When a person must make a choice either to do this or to do that, the universe at that moment branches into two or more universes such that all choices are made? In this way we solve the conflict of free will and determinism. Rigid determinism operates in each separate universe, but a person enjoys complete free will by moving from universe to universe. All universes are virtual except that which the person occupies, which is real.*

• *We demand that our universes are rational and deterministic even at the cost of creating a bizarre many-world picture. My own preferred view is that at each decision the branching universes are virtual except for the particular one that contains consciousness. Thus free will reigns supreme and determinism dwells in the shadows. Alternatively, free will belongs to the Universe and determinism, as conceived by the human mind, belongs to the universes. We do not know what is consciousness, except that it is not part of the physical universe.*

10 *Kurt Gödel in 1931 showed that mathematical systems are not fully self-contained. In a self-consistent logical system (free of internal contradictions), statements can be formulated whose truth is undecidable. When the system is enlarged with additional axioms, the previous statements of uncertain truth can be proved to be true. But the enlarged system contains new undecidable statements that can only be proved to be true by making the system still larger. One conclusion is that the mathematician is inseparable from mathematics, just as the cosmologist is inseparable from cosmology. For a discussion of Gödel's ideas see* Gödel's Proof *by Ernst Nagel and James Newman, and* What is the Name of This Book? *by Raymond Smullyan.*

11 *"Scientists have answered some questions beyond reasonable doubt, and have stitched these findings into a compelling, if not terribly detailed, map and history of existence" (John Horgan,* The End of Science*).*

• *Is a final theory of the physical world possible? It has been suggested that string theory offers the prospect of a final theory. String theory attempts to unify gravity and quantum mechanics in such a way that subatomic particles consist of tiny strings having various modes of vibration. In* Dreams of a Final Theory, *Steven Weinberg writes: "Our present theories are of only limited validity, still tentative and incomplete. But behind them now and then we catch glimpses of a final theory, one that would be of unlimited validity and entirely satisfying in its completeness and consistency."*

• *Against the modern conceit of a "final theory" we may contrast the medieval principle of learned ignorance. In his work* On Learned Ignorance, *written in 1440, Cardinal Nicholas of Cusa argued that the more we know the more aware we become of our ignorance. This form of enlightened ignorance he called learned ignorance. The darkness of unlearned ignorance disperses and the brightness of learned ignorance grows with knowledge and wisdom. He wrote: "No man, even the most learned in his discipline, can progress farther along the road to perfection than the point where he is found most knowing in the very ignorance that characterizes him; and he will be the more learned, the more he comes to know himself for ignorant."*

One might dare to hazard that glimpses of a final theory have frequently occurred in history, particularly when the current universe was ripe for change.

• *"Consider the unlearned, unaware of their ignorance, who think they know everything! As knowledge increases, ignorance decreases, but this kind of ignorance – unlearned ignorance – is no more than the lack of knowledge. With knowledge comes awareness of ignorance – learned ignorance – and the more we know, the more aware we become of what we do not know"* (Edward Harrison, *Masks of the Universe*).

PROJECTS

1 "When you can measure what you are speaking about, and express it in numbers, you know something about it; but when you cannot express it in numbers, your knowledge is of a meagre and unsatisfactory kind . . . you have scarcely, in your thoughts, advanced to the stage of Science, whatever the matter may be" (William Thomson [Lord Kelvin], in a lecture to the Institute of Civil Engineers, 1883). Does this express the containment principle of the physical universe?

2 Many persons – such as materialists and reductionists – overwhelmed by the grandeur of the physical universe, believe that what is not contained does not exist. What do you think?

3 "Mr. Podsnap settled that whatever he put behind him he put out of existence . . . . Mr. Podsnap had even acquired a peculiar flourish of his right arm in often clearing the world of its most difficult problems, by sweeping them behind him" (Charles Dickens, *Our Mutual Friend*). Have you ever noticed the Podsnap flourish when a scientist is asked, "What is life?" and a humanist, "What is physics?" The Podsnap flourish is a useful mannerism in many walks of life; try practicing it.

4 Delve into the containment riddle. Comment on the statement that the universes are the Universe seeking to understand itself. René Descartes said, "I think, therefore I am." The Universe contains us, who think,

hence we may say, "the Universe thinks, therefore it is." Can we say the same thing about the physical universe?

5 Things to think and talk about:

• The cosmic-edge riddle.

• The beginning and the end of the universe are cosmic edges of a timelike nature. What comes before the beginning of the universe? And what comes after the end? Should we say nothing? What is "nothing"?

• Contrast the anthropic and theistic principles.

• Does a final theory, if it can be found, have to explain the origin of life on earth, the origin of language, consciousness, the nature of mind, and the nature of time with its complimentary aspects of being and becoming?

6 A particularly significant postulate of impotence is expressed in the words: "It is impossible by any physical experiment to determine whether an object possesses consciousness." The object may be a star, a stone, a flower, a bird, or a human being. If I am the object, I know I have consciousness, but you, the experimenter, with all your instruments, cannot prove it. If you are the object and I am the experimenter, I know you have consciousness, but there is no physical way that I can prove it. Consciousness is not a property of the physical world and cannot be explained as a physical phenomenon. Some scientists think consciousness does not exist, or at best is a metaphysical illusion. What do you think?

7 Norman Campbell in *What is Science?* (1955), writes, "Science deals with judgments concerning which it is possible to obtain universal agreement." If Campbell is correct, and if cosmology is indeed a science, then it cannot claim exemption from this rule. Yet "universal agreement" is impossible concerning the reality of an ensemble of physical universes that can never be verified by direct observation. Cosmology is a workshop in which experimental universes are invented and investigated as potential representations of the Universe. In this sense cosmology is a science. But when cosmology invents a plurality of physical universes, and claims that each is self-contained and real in its own right, it ceases to be a science according to Campbell's dictum. Cosmology takes on metaphysical aspects and the notion of containment becomes vague. Once again cosmology is at the mercy of metaphysics, as in earlier ages. What do you think?

8 Nicholas of Cusa distinguished between unlearned and learned ignorance. In unlearned ignorance, the less we know the more confident we are in the truth of our knowledge. In learned ignorance, the more we know the less confident we are in the truth of our knowledge. Discuss: "he will be the more learned, the more he comes to know himself for ignorant."

9 Consider the following: "The more the universe seems incomprehensible, the more it also seems pointless.... The effort to understand the universe is one of the very few things that lifts human life a little above the level of farce, and gives it some of the grace of tragedy" (Steven Weinberg, *The First Three Minutes*). Do you agree that the universe is only a model of the Universe and cannot contain everything; in particular, it cannot contain the model-maker, and in that sense must always seem pointless?

FURTHER READING

Cornell, J. Editor. *Bubbles, Voids, and Bumps in Time: The New Cosmology.* Cambridge University Press, Cambridge, 1909.

Crick, F. H. C. "Thinking about the brain." *Scientific American* (September 1979).

Davies, P. C. W. *God and the New Physics.* Pelican, London, 1984.

Davies, P. C. W. *The Cosmic Blueprint.* Heinemann, London, 1987.

Dawkins, R. *The Blind Watchmaker: Why the Evidence of Evolution Reveals a Universe Without Design.* Norton, New York, 1987.

Flew, A. *Body, Mind, and Death.* Macmillan, New York, 1979.

Gribbin, J. and Rees, M. J. *Cosmic Coincidences.* Black Swan, London, 1991.

Harrison, E. R. *Masks of the Universe.* Macmillan, New York, 1985.

Leslie, J. *Universes.* Routledge, New York, 1989.

Polkinghorne, J. *Science and Creation.* New Science Library, Boston, 1988.

Robson, J. *Origin and Evolution of the Universe: Evidence for Design.* McGill-Queens University Press, Montreal, 1988.

Rosenfeld, L. "Niels Bohr's contribution to epistemology." *Physics Today* (October 1963).

Wald, G. "Fitness in the universe: the choices and necessities," in *Cosmochemical Evolution and the Origin of Life.* Editors, J. Oró, S. L. Miller, and C. Ponnamperuma. Reidel Publishing Co., Dordrecht, The Netherlands, 1974.

Wheeler, J. A. "The universe as home for man." *American Scientist* (November–December 1974).

SOURCES

Barrow, J. D. and Tipler, F. J. *The Anthropic Principle.* Oxford University Press, London, 1982.

Bohm, D. *Wholeness and the Implicate Order.* Routledge and Kegan Paul, New York, 1980.

Burchfield, J. D. *Lord Kelvin and the Age of the Earth.* Science History Publications, New York, 1975.

Burtt, E. A. *The Metaphysical Foundations of Modern Physical Science.* 1924. Revised edition: Humanities Press, New York, 1932.

Campbell, N. *What Is Science?* Dover Publications, New York, 1955.

Carr, B. J. and Rees, M. J. "The anthropic principle and the structure of the physical world." *Nature* 278, 605 (April 12, 1979).

Carter, B. "Large number coincidences and the anthropic principle in cosmology," in *Confrontation of Cosmological Theories with Observational Data.* Editor, M. S. Longair. Reidel, Dordrecht, Netherlands, 1974.

Chew, G. F "'Bootstrap': a scientific ideal." *Science* 161, 762 (1968).

DeWitt, B. S. "Quantum mechanics and reality." *Physics Today* (September 1970).

Eddington, A. S. *New Pathways in Science.* Cambridge University Press, Cambridge, 1935.

Ellis, G. F. R. "The theology of the anthropic principle," in *Quantum Cosmology and the Laws of Nature.* Editors, R. J. Russell et al. Vatican Observatory, Rome, 1993.

Frank, P. *Modern Science and Its Philosophy.* Harvard University Press, Cambridge, Massachusetts, 1949.

Gardner, M. *Mathematical Magic Show.* Alfred A. Knopf, New York, 1977. See chapter 1, "Nothing"; chapter 2, "More about Nothing"; chapter 19, "Everything."

Haber, F. C. *The Age of the World, Moses to Darwin.* Johns Hopkins University Press, Baltimore, 1959.

Hawking, S. "Origin of the universe," in *Black Holes and Baby Universes.* Bantam, New York, 1993.

Henderson, L. J. *The Fitness of the Environment: An Inquiry into the Biological Significance of the Properties of Matter.* Macmillan, New York, 1913.

Horgan, J. *The End of Science: Facing the Limits of Knowledge in the Twilight of the Scientific Age.* Addison-Wesley, New York, 1996.

Huxley, T. H. "Nature: Aphorisms by Goethe." *Nature* 1, 9 (1869).

James, W. "Does consciousness exist?" in *Essays on Radical Empiricism.* Holt, New York, 1912.

Laird, J. *Theism and Cosmology.* Philosophical Library, New York, 1942.

Linde, A. "Particle physics and inflationary cosmology." *Physics Today* (September 1987).

Mason, F. and Thomson, A. *The Great Design: Order and Progress in Nature.* Duckworth, London, 1934.

Murphy, N. "Evidence of design in the fine-tuning of the universe," in *Quantum Cosmology and the Laws of Nature.* Editors, R. J. Russell et al. Vatican Observatory, Rome, 1993.

Nagel, F. and Newman, J. E. *Gödel's Proof.* New York University Press, New York, 1958.

Paley, W. *Natural Theology.* Editor, F. Ferre. Bobbs-Merrill, New York, 1963.

Planck, M. *Survey of Physical Theory.* Dover Publications, New York, 1960.

Popper, K. *The Logic of Scientific Discovery.* Harper and Row, New York, 1965.

Prout, W. In *The Bridgewater Treatise on the Power, Wisdom and Goodness of God, as Manifest in the Creation, on the Adaptation of External Nature to the Physical Condition of Man.* Editor, J. Kidd. Bohn, London, 1852.

Rosen, J. "Self-generating universes and many worlds." *Foundations of Physics* 21, 977 (1991).

Russell, R. J., Murphy, N., and Isham, C. J. Editors. *Quantum Cosmology and the Laws of Nature: Scientific Aspects on Divine Action.* Vatican Observatory, Rome, 1993.

Sandage, A. R. "The time scale for creation," in *Galaxies and the Universe.* Editor, L. Woltjer. Columbia University Press, New York, 1968.

Schrödinger, E. *Mind and Matter.* Cambridge University Press, Cambridge, 1958.

Smullyan, R. *What is the Name of this Book?* Prentice Hall, Englewood Cliffs, New Jersey, 1978.

Stapledon, O. *The Star Maker.* 1937. Reprinted in *Last and First Men and Star Maker.* Dover Publications, New York, 1968.

Stent, G. S. "Limits to the scientific understanding of man." *Science* 187, 1052 (March 21, 1975).

Weinberg, S. *The First Three Minutes: A Modern View of the Origin of the Universe.* Basic Books, New York, 1977.

Weinberg, S. *Dreams of a Final Theory.* Pantheon Books, New York, 1992.

Whitehead, A. N. *Adventure of Ideas.* Macmillan, New York, 1933.

Wigner, E. F. *Symmetries and Reflections.* Indiana University Press, Bloomington, 1967.

Yourgrau, W. and Breck, A. D. Editors. *Cosmology, History, and Theology.* Plenum Press, New York, 1977.

9 SPACE AND TIME

I do not define time, space, place and motion, as being well known to all.
Isaac Newton (1642–1726), Principia

Our knowledge of time as of space owes more to the labours of mathematicians and physicists than to those of professional philosophers.
C. D. Broad (Philosophy, 1938)

SPACE
Dressed and undressed space

From the Heroic Age of Greece until modern times we see the development, side by side, of two views on the nature of space: "dressed space" and "undressed space."

Space as a void – undressed, existing in its own right, independent of the things it contains – was at first a lofty abstraction that many persons could not take seriously. It seemed more natural to think of space as dressed and made real with a continuous covering of material and ethereal substances. Aristotle, who believed in dressed space, regarded the notion of a vacuum as nonsense and said that a vacuum is nothing and what is nothing does not exist. This enabled him to argue in favor of a finite universe. The ether – the fifth element – ended at the sphere of fixed stars. Beyond the sphere of stars, because there was no ether, there could be no space. At first this was the view of scholars in the Middle Ages who later succeeded in extending space beyond the sphere of fixed stars by inhabiting it with God. It was also the view of René Descartes (and the Cartesians holding similar views) who extended space indefinitely by suffusing it with tenuous and continuous matter. Numerous Continental scientists and philosophers in the eighteenth and even nineteenth centuries clung to the Cartesian view.

But experiments with air pumps in the seventeenth century (Figure 9.1) convinced many other scientists, particularly the Newtonians, that space exists without material covering. The Cartesians resisted such evidence and argued that atoms, which are isolated specks of matter surrounded by voids, cannot exist because empty space is illogical and matter must always be continuous, however low in density. Matter exists everywhere, for if there were no matter, there could be no space.

Aristotelian beliefs, in modified forms, persisted until the early twentieth century. Even if empty space exists by itself, many thought, space is still a sort of nothing, lacking substance, and possessing only extension. Elegant equations of the electromagnetic field, with which James Clerk Maxwell unified electricity and magnetism in the nineteenth century, showed that waves of light travel at finite speed. All electromagnetic waves – radio waves to x-rays – travel in empty space at the universal speed of 300 000 kilometers per second. Sound waves, as we know, traveling in the atmosphere, consist of undulations of air. How can waves of light, traveling in empty space, consist of undulations of emptiness? In response, physicists filled the universe with a subtle luminiferous ether able to support the propagation of light and all other electromagnetic waves (Figure 9.2). The advent of special relativity helped to shake off the Aristotelian heritage, and nowadays we think in terms of the propagation of electromagnetic fields and see no need for an ether.

Figure 9.1. Otto von Guericke, mayor of Magdeburg, constructed in 1650 the first air pump and in subsequent years made spectacular demonstrations of the properties of the vacuum. He showed that sound does not propagate and candles do not burn in the absence of air. In the experiment shown in this illustration he measured the atmospheric pressure holding together the two halves of an evacuated sphere.

An undressed space – a continuum existing in its own right, absolute, and independent of all contained things – is an abstract idea that originated with the Atomists of the ancient world. Accompanied by the idea that forces can act at a distance, it later became the master concept of the Newtonian universe.

In the twentieth century, undressed space developed into the spacetime of special relativity. Then, with the rise of general relativity, spacetime became once more dressed, dressed not only with dynamic geometric curvature, but also with the virtual states of the vacuum.

The algebra connection

Descartes combined geometry and algebra into analytical geometry. He used coordinate systems in which the position of a point in three-dimensional space is specified by the values of the three coordinates x, y, and z. Two-dimensional grids or graphs were used by the ancient Egyptians in land surveys, and by others in later ages in map making, and were used for calculations by scholars at Merton College in Oxford in the Late Middle Ages. Descartes, however, was the first to use coordinates in the study of geometry. When the coordinate axes are orthogonal (at right angles to each other),

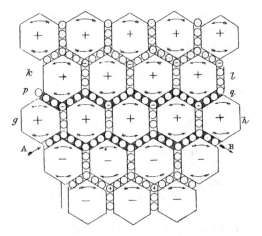

Figure 9.2. James Clerk Maxwell's mechanical ether from his paper "On physical lines of force" (1861). This model, with its vortices and intermediate vortices, "serves to bring out the actual mechanical connections between the known electromagnetic phenomena; so that I venture to say that any one who understands the provisional and temporary character of this hypothesis, will find himself rather helped than hindered by it in his search after the true interpretation of the phenomena."

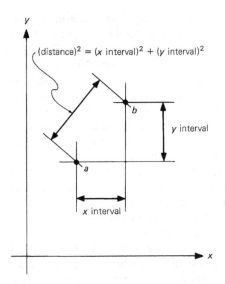

Figure 9.3. The distance between points a and b is given in terms of the x interval and y interval by the Pythagorean rule.

the distance between two points in the x and y plane is given by the Pythagorean rule

$$(\text{distance})^2 = (x \text{ interval})^2 + (y \text{ interval})^2, \quad [9.1]$$

as in Figure 9.3; and the distance between two points in the three-dimensional x, y, and z space is

$$(\text{distance})^2 = (x \text{ interval})^2 + (y \text{ interval})^2 + (z \text{ interval})^2. \quad [9.2]$$

The equation of circle A of radius r in Figure 9.4 is

$$r^2 = x^2 + y^2, \quad [9.3]$$

and that of circle B, also of radius r, displaced distance a in the x direction and distance b in the y direction, is

$$r^2 = (x - a)^2 + (y - b)^2. \quad [9.4]$$

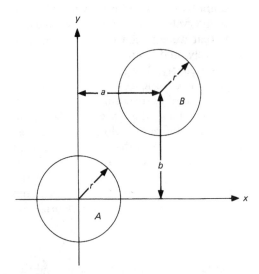

Figure 9.4. Diagram illustrating the equations of circles (Equations 9.3 and 9.4) in terms of x and y coordinates.

SPACE AND TIME

Nothing puzzles me more than time and space: and yet nothing troubles me less.
Charles Lamb, in a letter (1810)

Scholars at Oxford and Paris in the fourteenth century made great strides in clarifying our notions of space, time, and motion.

They invented diagrams combining space and time in their studies of motion. Later, in the seventeenth century, the Cartesians and Newtonians established scientific pre-relativity concepts of space, time, and motion. Mathematicians in the nineteenth century studied the theoretical properties of multi-dimensional spaces, and Charles Hinton in 1887 advanced the idea of time as a fourth dimension. Then came the relativity theories in the early twentieth century that indissolubly combine space and time into a spacetime continuum.

We think we know what space is like: It is that thing all around us, stretching away, in which objects are visibly distributed. It is spanned by intervals of distance and measured in units such as meters with meter sticks that can be directly observed. Time is not quite so simple because we cannot objectively observe bodies distributed in it, and we cannot directly observe with the five senses intervals of time such as seconds. It seems that we experience intervals of time subjectively and cannot directly observe them objectively. The impression gained is that all our experiences are of two kinds, consisting of objective things (trees, clouds, and mountains) that are diversified in space, and subjective things (sensations, emotions, and ideas) that are diversified in time. Somehow these opposite kinds of experience come together to make up the phenomenal world in which we live and the physical world about which we theorize.

TIME
Physical time
The nature of time is a perplexing subject that provokes endless philosophical discussion, and the words of Saint Augustine of Hippo in the fifth century still strike a responsive chord: "What, then, is time? If no one asks me, I know what it is. If I wish to explain what it is to him who asks me, I do not know." The time that we experience as human beings, the "time that devours all things," according to Ovid's metaphor, is not quite the same as that used in science. Science simplifies the time that we experi-

ence into a continuous one-dimensional space that conforms to the Hausdorff axioms (see Reflections and Figure 9.15). We must not be too surprised if physical time lacks some of the characteristics of the time that we experience. The neglected characteristics are usually regarded as psychological or metaphysical.

The fourth dimension
Physics seizes time, strips away many of the characteristics ascribed to it in everyday life, and makes it akin to space. Our physical world has become a four-dimensional continuum that decomposes into a three-dimensional space and a one-dimensional time stretching from the past to the future.

The restless world is in a state of continual change. At any instant in time there is one distribution of things in space, and then at a later instant there is another distribution. Hinton initiated a revolution in our understanding of the world around us when he showed that its continual change can be tamed and made orderly by displaying it in a joint space and time diagram. Time has the properties of a one-dimensional space, said Hinton, and events in the past, present, and future form a continuous sequence in much the same way as the points of a straight line in space.

Drawing three-dimensional figures on a two-dimensional sheet of paper is not always easy. Drawing on a sheet of paper four-dimensional figures, with time included as the fourth dimension, is even less easy (some might say impossible). Frequently, in pre-relativity space-and-time (and relativity spacetime) diagrams, only one of the three dimensions of space is shown, as in Figures 9.5 and 9.6. Hinton in his diagrams used two dimensions of space with time as an extra dimension perpendicular to space, as in Figure 9.7.

In space-and-time diagrams, a point represents an event, such as the flash of a firefly or the blink of an eyelid, and is something that happens at a point in space at an instant in time. A line – called a world line – represents something that endures, such

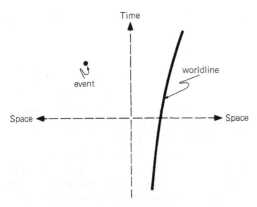

Figure 9.5. A space-and-time diagram. A point represents an event that occurs at an instant in time at a position in space. A line represents a body that endures and shows its position in space at each instant in time; such a line is called a world line.

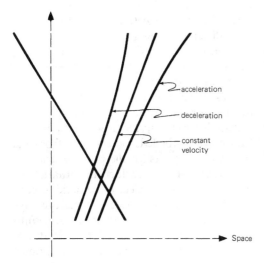

Figure 9.6. Inclined world lines represent bodies in relative motion. If the velocity changes, because of acceleration or deceleration, the world line is curved.

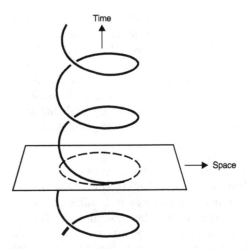

Figure 9.7. Charles Hinton's illustration of a world line (which he called a filamentary atom) of a particle in circular motion. The world line is a helix, and as time advances the particle describes a circle in space.

as a stone or an atom, and occupies a position in space at each successive instant in time. Figure 9.7 reproduces Hinton's 1887 illustration of the world line of a particle moving in a circle in space.

Of paramount importance is the fact that nothing changes in a space-and-time (and spacetime) representation. Nothing ever happens! We have tamed the restless world by displaying it in a frozen state throughout space and time. Everything is displayed as it was, as it is, and as it will be in a tenseless world. Nothing changes or moves in a space-and-time diagram because time, used once, cannot be used twice. The frozen four-dimensional world in which time is one of the dimensions was dubbed the "block universe" by the philosopher William James, who emphasized the complete absence of change in a space-and-time world.

The four-dimensional world of space and time is a fixed and unchanging world. We must refer to all its regions in a common tense. In a tenseless world it is wrong to say that an event has happened, another is happening, and yet another will happen, for all are present together. It is also wrong to say that a body moves along a world line from the past to the future, for the body exists at all segments along its world line. We must guard our tongues and continually remember that time is already embodied in the four-dimensional world and cannot be used again.

If we make a mistake and say that a particle moves along its world line, we are confronted by the question of the speed at which it moves through time. This compels us to invoke a second time, and having admitted the possibility of movement through time, we must decide at what speed the particle moves through this second time. Motion through time opens up the prospect of serial time consisting of an infinite regression of timelike dimensions. We avoid this absurdity by realizing that nothing physical changes in spacetime.

The arrow of time

Worldlines do not have arrows attached to them pointing in the direction of the future. In spacetime pictures it is difficult to see what determines the past and the future. When a spacetime picture is turned upside down, there is little or nothing in the picture that prevents us from relabeling the new top as the future and the bottom as the past (see Figure 9.8). Past and future are labels that often have no intrinsic spacetime meaning. Yet obviously a time-reversed universe is completely different: Life begins in the grave and ends in the cradle, cool objects grow hot, candles and stars absorb radiation, and everything is in a topsy-turvy

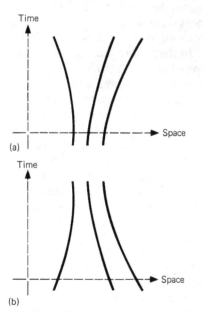

Figure 9.8. (a) A spacetime picture, and (b) the same picture turned upside down with the past and the future interchanged. In spacetime pictures there is often little that distinguishes between the real past and future.

state. Presumably such time-reversed worlds lack life and are hence ruled out by the anthropic principle.

Most of the fundamental laws of physics fail to distinguish between the past and the future and are said to be invariant to time-reversal. The equations of motion of bodies, and many laws of microscopic physics, are reversible in time. Their effect remains the same when the spacetime picture is turned upside down and past and future labels are interchanged. Thus, two particles come together, interact, and move away from each other; in the time-reversed picture, the same thing happens: two particles again come together, interact, and move away from each other. Their motions and interactions, if they can be displayed graphically, are time-reversible. Our equations of physics, like our spacetime diagrams, use a spatialized form of time with no intrinsic direction. Even the equations that govern the propagation of light are time-reversible. According to experience, absorbed light arrives from the past and emitted light departs for the future. When the spacetime picture is turned upside down and past and future are relabelled, light arrives from the future and departs for the past, contrary to experience. This bewildering situation is not forbidden by electromagnetic theory and nothing in the spacetime picture says that light rays must propagate only into the future. Time has properties that not only in life but also in the laboratory cannot be entirely spatialized.

The flow of heat

Thermodynamics is a branch of macroscopic physics devoted to the study of all aspects of heat. The first law of thermodynamics deals with the conservation of thermal and other forms of energy in dynamic systems. The second law deals with the time-reversibility of thermal systems. Strictly, reversible systems are idealizations that never exist but enable us to perform simple calculations. Nonreversible systems are the real world in which heat is always lost at every stage.

Reversible systems have constant entropy. In the real world entropy always increases, while energy, although always conserved, assumes forms that are less and less effective. The flow of heat from hot regions to cooler regions is not time-reversible because entropy would decrease. No scientist would dream of contradicting the cherished second law of thermodynamics by saying that heat flows from cool regions to hot regions. A mug of coffee on my desk gets cooler in the direction of the future and hotter in the direction of the past, and in the process, entropy increases. Total energy remains the same, but is less accessible, and this another way of saying that the entropy has increased. (See Chapter 19 for a discussion on what entropy in the universe means.) If this did not happen then you and I would not exist. The laws of thermodynamics that govern the mug of coffee and its environment are not time-reversible; and yet, strange to say, the laws that govern the behavior of the individual particles that constitute the mug of coffee and its environment are time-reversible. Heat continues to flow from hotter to cooler regions when the direction of time is reversed in the equations that control the microscopic motions and behavior of individual particles. It seems that the collective behavior of many particles acts as an arrow of time, whereas the behavior of individual particles lacks any arrow of time (Figure 9.9).

Let us try to understand in an elementary way what happens in the spacetime picture when heat flows from hot regions to cooler regions. We can regard heat in this case as an agitation of particles: the hotter the particles, the more agitated their motions. Agitated particles move about and have wavy or crinkled world lines, and crinkled world lines communicate their agitation by collisions and other interactions to neighboring and less agitated world lines. Hot (or crinkled) world lines, surrounded by cooler (less crinkled) world lines, as in Figure 9.10, are cooler in the direction of the future. This is another way of stating the second law. If you are shown a space-and-time diagram with many crinkled

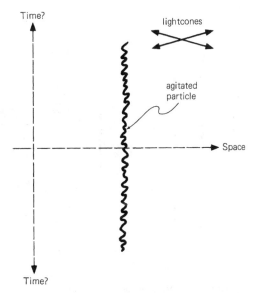

Figure 9.9. Heat is particle agitation, and an agitated particle is represented by a crinkled world line. Notice that the spacetime picture looks similar when turned upside down, thus indicating that time reversal does not affect the behavior of isolated individual particles.

world lines, you immediately know which is the past and which is the future. The agitation of isolated individual particles is unaffected by time reversal, but their collective behavior – the spread of agitation from hot world lines to cooler world lines – is governed by the second law of thermodynamics and is not time-reversible. Single particles are time-reversible but systems of particles are not. It is like hanging a picture on a wall. If you look closely at individual dabs of paint you will never know which way up to hang the picture; you have to stand back and look at large areas of many dabs to determine the right way to hang the picture.

The universe consists of numerous many-particle systems (for example, living creatures, planets, stars, galaxies), and most scientists believe that the arrow of time is determined either partly or fully by the collective (statistical) behavior of these many-particle systems. In the course of time, the organized energies of these systems become disorganized and dispersed, and their states of order tend toward disorder.

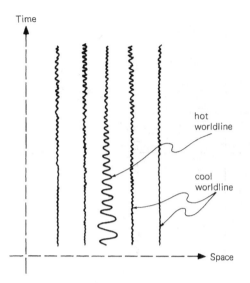

Figure 9.10. A hot (agitated) world line surrounded by cooler (less agitated) world lines. Agitation is communicated to neighboring world lines by particle collisions, and the hot world line loses agitation and the cooler world lines gain agitation in the direction of the future. The spacetime picture now contains an arrow of time because of things contained in spacetime but not because of any property of spacetime itself.

In trillions of years, when all stars have died and all systems have attained their lowest accessible energy levels, it will be difficult to determine the direction of time.

In our universe, where matter is distributed irregularly in the form of stars and galaxies, we see energy continually cascades into dispersed states of less accessible energy and are provided with a cosmic arrow of time. This raises the question: In a universe that is perfectly uniform and without irregularities of any kind, is there no arrow of time? Such a universe might be expanding and the change in its density will identify which is the past and which is the future. But how would we know that it is expanding? It might be contracting, and the higher density lies in the future and not the past (see Figure 9.11). Thomas Gold of Cornell University has argued that the universe gets its sense of time direction from its expansion, and the future always lies in the direction of diverging world lines. It is

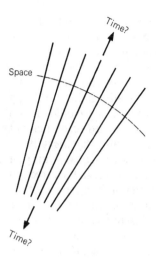

Figure 9.11. Spacetime picture of diverging (converging) world lines in an expanding (contracting) universe. Does this picture fix the direction of time?

not clear in this case what happens if the universe ceases to expand and begins to collapse.

THE "NOW"

The shadow by my finger cast
Divides the future from the past:
Before it, sleeps the unborn hour,
In darkness, and beyond thy power:
Behind its unreturning line,
The vanished hour, no longer thine:
One hour alone is in thy hands, –
The NOW on which the shadow stands.
Henry van Dyke, "The Sun-Dial at Wells College" (1904)

The river of time

In our everyday experiences we are acutely aware of the "now" as a moment or segment of time. We (but apparently not Aborigines, Mayans, Hopi, and many other societies) imagine time as a river. Our boat, the "now," drifts on the river of time, gliding past a changing landscape, with always the future approaching and the past receding.

Alternatively, we can think of the now as a wave of vividity that divides the known past from the unknown future. We have an awareness of transience: of things becoming, bursting forth into actuality, and then

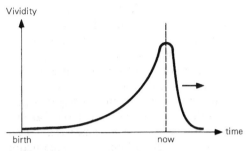

Figure 9.12. A wave of vividity of consciousness advances into the future. This picture illustrates our usual way of thinking and speaking about experiences in time. And yet it is nonsensical! It implies the existence of a second time that determines the speed of movement through time. Also, why does the vividity wave of awareness move in only one direction?

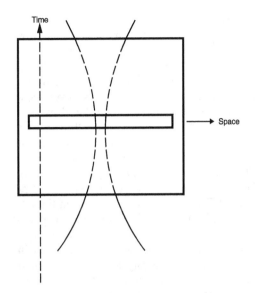

Figure 9.13. A cardboard sheet with a narrow horizontal slit is placed over a spacetime diagram. The slit represents the "now" and what we see through the slit is the physical world of space. As the slit moves upward, we see the particles (small segments of world lines) move around in space.

fading into a limbo of memory. Our awareness seems like an asymmetric wave of vividity advancing into the future, having an intensity that is almost zero ahead, rises sharply to a maximum at the present moment, and then falls off slowly into the past, as illustrated in Figure 9.12.

Our languages embody a mixture of primitive and sophisticated metaphors concerning the nature of time, and their conflict leads to confusion and paradox. Amidst this confusion we discuss "being" and "becoming," free will versus determinism, and the arrow of time. We have so far failed to harmonize the discordant concepts of temporal continuity (states of being) and transience (states of becoming). We only need to ask at what speed the "now" moves through time? – at what speed does our boat drift on the river of time? – to realize that we still do not understand time.

Each of us in spacetime is a world line, or rather a bundle of interwoven world lines, like the fibers in a rope. This representation, however, does not explain the "now" that moves up the world line from the past to the future. Hermann Weyl (1949) wrote, "The objective world simply is, it does not *happen*. Only to the gaze of my consciousness, crawling upward along the lifeline [world line] of my body, does a section of the world come to life as a fleeting image in

space which continually changes." A simple model of this process consists of a spacetime diagram of world lines (as shown in Figure 9.13), over which lies a cardboard sheet having a horizontal slit. The slit, which exposes a thin slice of spacetime, reveals a three-dimensional world of space. The world lines passing through this space are particles occupying points in space. As the slit moves upward, we see the particles moving around in space.

By fully spatializing time we are left with no way of physically explaining our experience of transience: of a restless world of things forever changing. If indeed the "now" moves upward along a world line, at what speed does it move? Must we invent a second time, to account for this motion in time? If this second time is spatialized like the first, then we need a third time, and so on, and we have the frightful prospect of serial time. If time is fully spatialized, then no matter how many dimensions we assign to it, we encounter the paradox of motion in time.

Transience

Scientists, confronted with the question, how do we move in time, tend to dismiss it as a metaphysical and not a physical problem. Many incline to the view that it is a psychological illusion. But such views never seem convincing, and indeed can be dismissed with Gerald Whitrow's words, "how do we get the illusion of time's transience without presupposing transient time as its origin?" (*The Natural Philosophy of Time*, 1980). Whitrow also makes the perceptive comment, "Time is not itself a process in time."

The physical time incorporated in spacetime is a refinement of Newtonian time; and Newtonian time, which is still used in sciences other than physics, is a refinement of colloquial time used in cultures of European origin. Many persons take the view that spacetime incorporates real time, and that an awareness of transience is a nonphysical peculiarity of living creatures. This view revives the Parmenidean doctrine according to which the real world is changeless and our transitory experiences are illusions. Which is the real thing – the idea or the experience? – opens endless discussion. If ideas derive from experiences, and experiences are illusory, how can ideas be more real than experiences? The real problem is: If living creatures are nothing but bundles of world line fibers in spacetime, how can world lines have an illusion of transience when nothing in spacetime changes? How can we hope to explain the illusion of transience without presupposing the existence of transience? We still do not know the answer to Whitrow's problem.

TIME TRAVEL

The Time Machine

We travel backward and forward in space. Can we travel backward and forward in time? Metaphysical travel in time as a story-telling device is as old as human beings. Often the journey is told in the form of a dream (usually into the past) as in *An Ancient Captivity* by Neville Shute, or in the form of suspended animation (usually into the future) as in *The Sleeper*

Awakes by H. G. Wells. The notion of a time machine that actually transports a person into the past or the future began with H. G. Wells's *The Chronic Argonauts* (1888) that was later updated in *The Time Machine* (1895). Wells's thoughts were inspired by Charles Hinton's writings proposing that time is spacelike and exists as a fourth dimension. Ever since *The Time Machine*, a large volume of science fiction literature has been devoted to physical transport in time.

Travel back in time raises problems concerning causality: what happens in the past determines what exists and happens now. If we travel back in time and alter what once happened, on our return the world would be different. Perhaps a world that has not yet acquired the skill to make a time machine. Or even more paradoxical, by arranging that the time traveler's parents never meet, a world that does not contain the time traveler.

We need not worry. Actual physical travel in time as presented by Wells and hosts of science fiction writers is totally and utterly impossible. Not for technical reasons, but because it is based on a misunderstanding of spacetime. In *The Time Machine*, the Time Traveler shows a circle of friends a small model of his much larger time machine in which he later travels. With his friends watching, the Time Traveler activates the model and sends it into the future. "There was a breath of wind and the lamp flame jumped. One of the candles on the mantel was blown out, and the little machine suddenly swung round, became indistinct, was seen as a ghost for a second perhaps, as an eddy of faintly glittering brass and ivory, and it was gone – vanished! Save for the lamp the table was bare."

Where did the time machine go? If it went into the future, it would be on the table one minute later, two minutes later, ... one hour later, two hours later, ... and so on, and as we ourselves move into the future, we should still see it on the table one minute later, two minutes later, and so on. Wells has overlooked the fact that the time machine has a world line. He and all other time-traveling science fiction writers ignore two unbreakable rules:

(i) Nothing changes in spacetime and all world lines are fixed;
(ii) movement along a world line, as in the case of the "now," is metaphysical and cannot be physical.

Wells's story of time travel and all similar stories break both laws. The basic idea of physical transport in time violates (i); the idea that a physical object can disappear into the past or the future violates (ii). How can a physical thing, as it travels in time, roll up its own world line and take it with it in a spacetime world in which nothing changes? We need not worry about traveling back into the past and creating paradoxes; at most we can metaphysically travel along our world lines, but nothing can be altered.

Kurt Gödel, a mathematician, discovered in 1949 a solution in general relativity in which world lines in specially contrived cases can form closed loops. Metaphysical travel along a world line into the future brings a person back, through the past, to the present. This also has been used as a popular literary device in numerous tales in which a person, trapped in a time loop, repeatedly experiences the same sequence of events, as in the movie comedy *Groundhog Day*. In all these tales, the trapped person remains fully aware of the cyclic repetition. How is this possible? A person's thoughts each time around the loop are repetitious. If the trapped person is unaware of the time loop in the first circuit, then, because of the repetition of identical events, the person remains unaware in the second, third, ... circuits. Trapped persons therefore never know that they are trapped. In many tales, the trapped person alters the sequence of events and escapes from the loop. How is this possible when world lines never change?

ATOMIC TIME
Being and becoming
Greek philosopher-scientists in the sixth and fifth centuries BC identified two basic aspects

of time – being (the Parmenidean continuity aspect in which time stretches from the past to the future) and becoming (the Heraclitean transience aspect in which everything forever changes) – that to this day remain unreconciled in language and in science. Time extends continuously from the past to the future (the being aspect used in the spacetime of physics) and things change in time (the becoming aspect that is omitted from spacetime because it cannot be spatialized). By failing to separate these two aspects of time we constantly encounter the Augustinian riddle: how can things in time change in time?

Zeno's paradoxes
Zeno of Elea in the fifth century BC denied that truth can be attained through perceptions by the senses. He is best known for his paradoxes of motion that attempt to prove that all apparent change in the sensible world is illusory. Only the states of being exist, the states of becoming are an illusion. His various paradoxes are similar and the one best known is that of the race between Achilles and the tortoise. Suppose the tortoise has a 100-meter head start and Achilles runs 100 times faster than the tortoise. While Achilles runs the 100 meters, the tortoise moves 1 meter, while Achilles runs 1 meter, the tortoise moves 1 centimeter, and so on, in an infinite number of steps, and although Achilles always gets closer, according to Zeno he never reaches and overtakes the tortoise. Zeno's paradox in this case is usually solved by showing with infinitesimal calculus that the infinite number of steps occupies only a finite interval of time and Achilles is hence able to reach and overtake the tortoise, as observed by the senses. Many philosophers still debate Zeno's paradoxes; what is at issue is the assumption that time is continuous and infinitely divisible.

Moses Maimonides, a Jewish scholar of the twelfth century, discussed atomic time in *The Guide for the Perplexed* and referred to a resolution of Zeno's paradox that had originally been proposed by the Atomists.

According to this resolution, time is itself divided into finite indivisible atomic intervals. "An hour is divided into sixty minutes, the minute into sixty seconds, the second into sixty parts, and so on; at last, after ten or more successive divisions by sixty, time-elements are obtained, which are not subjected to division, and in fact are indivisible, just as is the case with space." Motion is therefore not continuous but consists of a series of small jerks (like motion in a movie), and Achilles is able to overtake the tortoise with jerky movements. Maimonides, like most Arab philosophers, was greatly influenced by Aristotle and therefore hostile to the philosophy of the Atomists. The Atomist theory of time, he said, meant that the universe is created not once, but repeatedly in each interval of atomic time (see Chapter 25).

The kalam universe
In the tenth and eleventh centuries, the ilm al-kalam, a rebel school of Arab theologians, rejected the Aristotelian philosophy of orthodox Muslim theology. With the atomic theory of the Epicureans of the Greco-Roman world, the mutakallimun (the scholars of the kalam) tried to show that the world is totally dependent on the will of the supreme being – the sole agent. Atoms, they said, are isolated from one another by voids and their configurations are determined not by natural forces but by the will of the sole agent.

Bakillani of Basra, who lived in Baghdad where he died in 1013, suggested that time also is atomic, and in each atom of time the sole agent recreates the world in slightly different form. The world is created not once but repeatedly. The kalam theory attributes all explanation to the sole agent and nothing to the natural world. No forces activate the dead material world. Despite this anti-scientific attitude, the kalam theory succeeds in reconciling the dual aspects of time. In each "now," or atom of time, the material world stretches away in space, and memories of the past and anticipations of the future stretch away in time. Everything exists in a

state of static being. This state of being dissolves, and in a new atom of time a new state of static being exists. Transient acts of becoming transform whole states of static being. Shorn of its extreme theology, the kalam theory accounts moderately well for our complex experience of time.

Conjugate time
The kalam atomic theory of time succeeds in harmonizing the extensive and nonextensive properties of time. This suggests that perhaps we need a new theory of time in which the conjugate aspects of being and becoming play equally important roles. At this point we should remember that physics advances by modifying and discarding old ideas that were once mistaken for reality. Physicists more than most scientists are aware of the impermanence of theories, and many express uneasiness when they hear philosophers, biologists, and psychologists discuss the nature of time in the physical world. If, in the future, we succeed in changing the nature of physical time in a way that incorporates its dual or conjugate aspects, then undoubtedly the physical universe and cosmology will greatly change.

REFLECTIONS
1 *"I don't say that matter and space are the same thing. I only say there is no space where there is no matter; and that space in itself is not an absolute reality." Gottfried Leibniz (1646–1716) wrote these words in a letter to Samuel Clarke, who vigorously argued in defense of Newton's ideas on the reality of an absolute space that is independent of the existence of matter. Leibniz shared the views of many Continental philosophers (the Cartesians) who believed that undressed space was meaningless, and that no force could act at a distance unless conveyed by a material medium.*

2 *In the Cartesian coordinate system, shown in Figure 9.14, the point A is at the position denoted by the coordinates x, y, and z measured from the origin O. Note that from the three perpendicular axes X, Y,*

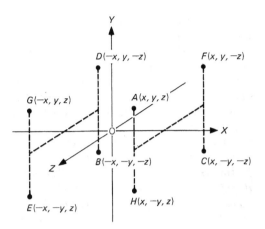

Figure 9.14. This shows how the point A at x, y, z is reflected through the origin O; the axes X, Y, and Z; and the planes XY, YZ, and ZX.

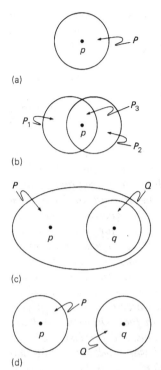

Figure 9.15. Diagram illustrating the Hausdorff axioms of continuous space.

and Z we construct the three perpendicular planes XY, YZ, and ZX. Now consider the following reflection symmetries:

• *Reflection through the origin:* Point A at position x, y, z becomes the point B at $-x$, $-y$, $-z$.

• *Reflection through the axes:* Point A at position x, y, z becomes the point C at x, $-y$, $-z$, the point D at $-x$, y, $-z$, and the point E at $-x$, $-y$, z.

• *Reflection through the planes:* Point A at position x, y, z becomes the point F at x, y, $-z$, the point G at $-x$, y, z, and the point H at x, $-y$, z.

Starting from A, notice that the points C, D, and E can be obtained by reflecting twice through the planes. Notice that the point B can be obtained by reflecting thrice through the planes but cannot be obtained by reflections through the axes. Notice that points B, C, D, E, F, G, and H can be obtained from A by multiple reflections through planes, or by a combination of reflections through the axes and the origin.

The above geometric reflections are known as discrete isotropies. If the point A, for all values of x, y, and z, is reflection-symmetric through one of the XY, YZ, and ZX planes, we have plane symmetry about that plane; if it is reflection-symmetric about one of the X, Y, and Z axes, we have cylindrical symmetry about that axis; and if it is

reflection-symmetric through the origin, we have spherical symmetry about the origin. These three cases are examples of continuous isotropies, and spherical symmetry is the isotropy in which all directions in space from the origin O are alike. A uniform (homogeneous and isotropic) space is reflection-symmetric through all points (the origin O may be anywhere).

3   What does continuous space mean? The Hausdorff axioms (Felix Hausdorff, Basic Features of Set Theory) express many of our deepest intuitions concerning space (see Figure 9.15). Hausdorff considered dots, and imagined them surrounded by circles enclosing neighborhoods whose size can expand or contract. The axioms are:

(a) To a point p there is at least one neighborhood P that contains p.

(b) If P_1 and P_2 are two neighborhoods of the same point p, there exists a neighborhood P_3 that is contained within P_1 and P_2.

(c) *If point p has a neighborhood P, and point q is contained within P, there exists a neighborhood Q of q that is contained within P.*

(d) *The points p and q have neighborhoods P and Q with no points in common.*

From these axioms, Hausdorff developed a formal treatment of the concept of spatial continuity.

4 *In the* Confessions, *Book XI, St. Augustine of Hippo (354–430) wrote, "What did God do before He made heaven and earth? I answer not as one is said to have done merrily (eluding the question), 'He was preparing hell (saith he) for pryers into mysteries.' It is one thing to answer inquiries, another to make sport of inquiries. So I answer not...." This statement is sometimes misquoted and made to seem that Augustine answered "He was preparing hell for pryers into mysteries." In fact, elsewhere in Book XI, Augustine makes the prescient remark, "verily the world was made with time and not in time, for that which is made in time is before some time and after some time." Thus time is created with the world in a timeless manner.*

5 *We move backward and forward with equal ease in space. When time is made akin to space we are puzzled that we cannot move backward and forward with equal ease in time.*

- Time goes, you say? Ah no!
    Alas, Time stays, we go.
  Austin Dobson (1840–1921), "The Paradox of Time"

- *The philosopher Alfred Whitehead wrote, "In every act of becoming there is the becoming of something with temporal extension, but ... the act itself is not extensive" (*Process and Reality*). Temporal acts of becoming cannot be spatialized in the same way as temporal states of being. The being aspect of time can be spatialized, as in spacetime, but the becoming aspect defies spatializing.*

- *There are undoubtedly deep biological and psychological aspects to our experience of time. "When a man is racked with pain, or with expectation, he can hardly think of anything but his distress; and the more his*

mind is occupied by that sole object, the longer the time appears. On the other hand, when he is entertained with cheerful music, and lively conversation and brief sallies of wit, there seems to be the quickest succession of ideas but the time appears shortest." Thomas Reid, Essays on the Intellectual Powers of Man *(1785).*

6 *What in physics distinguishes the past from the future?* Arthur Eddington *in* The Nature of the Physical World *(1928) introduced the term "arrow of time." He wrote: "The great thing about time is that it goes on. But this is an aspect of it which the physicist seems inclined to neglect.... I shall use the phrase 'time's arrow' to express this one-way property of time which has no analogue in space. It is a singularly interesting property from a philosophical standpoint. We must note that:*

(1) It is vividly recognized by consciousness.

(2) It is equally insisted on by our reasoning faculty, which tells us that a reversal of the arrow would render the external world nonsensical.

(3) It makes no appearance in physical science except in the study of the organization of a number of individuals. Here the arrow indicates the direction of progressive increase of the random element."

7 *"As all our means of sense-perception extend only to a space of three dimensions, and a fourth is not merely a modification of what we have, but something perfectly new, we find ourselves, by reason of our bodily organization, quite unable to represent a fourth dimension" (*Hermann Helmholtz, *"Origin and meaning of geometrical axioms," 1876).*

The idea of time as the fourth dimension first emerged in a series of essays by Charles Hinton, starting with "What is the fourth dimension?" (1887). After discussing the possibility of four dimensions, he wrote, "Why, then, should not the four-dimensional beings be ourselves, and our successive states ... the three-dimensional space to which our consciousness is confined?" His essays contained imaginative thoughts about the fourth dimension, and how ghosts might be explained as visitors from other three-dimensional regions of a four-dimensional world, and how we must adjust our thinking of space and time if time is the fourth dimension. Hinton's essays inspired H. G. Wells to write the Chronic Argonauts in 1888 (developed into The Time Machine in 1895), in which the fourth dimension is time, and The Invisible Man in 1897, in which the fourth dimension is an additional space dimension.

8 *The following are some extracts from the first chapter in Wells's* The Time Machine *in which the Time Traveler explains to his circle of friends how time travel is possible.* " 'There are really four dimensions, three of which we call the three planes of Space, and a fourth, Time. There is, however, a tendency to draw an unreal distinction between the former three dimensions and the latter, because it happens that our consciousness moves intermittently in one direction along the latter from the beginning to the end of our lives.... Well, I don't mind telling you I have been at work on this geometry of Four Dimensions for some time. Some of my results are curious. For instance, here is a portrait of a man at eight years old, another at fifteen, another at seventeen, another at twenty-three, and so on. All these are evidently sections, as it were, Three-Dimensional representations of his Four-Dimensional being, which is a fixed and unalterable thing.... Our mental existences, which are immaterial and have no dimensions, are passing along the Time-Dimension with a uniform velocity from the cradle to the grave.' " In these opening remarks Wells shows clearly that only mental consciousness travels along a physically fixed world line. Very soon, almost imperceptibly, Wells's account of time travel moves from the metaphysical to the physical, from the plausible to the implausible. The Time Traveler offers to prove to his friends that time travel is possible: "the Time Traveler held in his hand ... a glittering metallic framework, scarcely larger than a small clock, and very delicately made.... 'This little affair,' said the Time Traveler, resting his elbows on the table and pressing his hands together above

the apparatus, 'is only a model. It is my plan for a machine to travel through time.... 'Now I want you clearly to understand that this lever, being pushed over, sends the machine gliding into the future, and this other reverses the motion. This saddle represents the seat of a time traveler. Presently I am going to press the lever, and off the machine will go. It will vanish, pass into the future Time, and disappear.' " And that is what happened, the lever was pushed, and the machine disappeared from sight into the future. Thus in only a page or two Wells passes from metaphysical time travel, which contradicts no physical principle, to physical time travel, which contradicts the physical principles of causality and timeless spacetime. In metaphysical time travel, mental consciousness travels along a world line; in physical time travel, an object travels along its world line and takes its world line with it.

9 There once was a man who said "Damn!
  It is borne in upon me I am
  An engine that moves
  In predestinate grooves,
  I'm not even a bus, I'm a tram."
  Maurice Hare, "Limerick"

• *Determinism is the doctrine that all events have their causes. Fatalism is the doctrine that all events are preordained, that whatever happens is inevitable, and we might as well submit for nothing we do can change the future.*

• *The age-old cosmological debate concerning free will deserves thought. Human beings instinctively believe that they have some degree of control over their lives, and are free to influence at will events of the external world. But if they are part of the world, they are subject to deterministic laws, and their belief in free will is an illusion. Either we have no free will – we are puppets – and are not responsible for the things we do, or we truly have free will according to principles not yet accessible to rational inquiry. This argument applies not only to the physical universe but also to all rational universes. Even if we believe in a rational spiritual realm, we cannot escape the dilemma of either free will in an indeterministic world or no free will in a deterministic world. Augustine of Hippo,* architect of Christian theology, promoted the latter and believed in fate.

• *"Only two possibilities exist: either one must believe in determinism and regard free will as a subjective illusion, or one must become a mystic and regard the discovery of natural laws as a meaningless intellectual game"* (Max Born, "Man and the atom," 1957).

• *"If it is necessary, then it is not a sin; if it is optional, then it can be avoided"* (Pelagius, 360–420). Pelagius was a British theologian who lived at the time of Saint Augustine. He was opposed to Augustinian predestination because it exonerated sinful behavior. His argument in support of free will and individual responsibility, which diminished the omnipotence of God, was condemned by Christian authorities and became known as the Pelagian heresy. Pelagius was forced to flee from the hostility of Rome.

PROJECTS

1 Discuss: "nature abhors a vacuum" (René Descartes). The Latin proverb states, "a vacuum is repugnant to reason."

2 Use the Hausdorff axioms to discuss one-dimensional continuous time. Do the axioms assign an arrow to time, and do they satisfactorily account for our experience of time?

3 At a point in time we see things at many points in space, but at a point in space we cannot see things at many points in time. Why?

4 Discuss: "Past and future must be acknowledged to be as real as the present, and a certain emancipation from slavery to time is essential to philosophical thought" Bertrand Russell, *Mysticism and Logic* (1918).

5 Does growth in entropy explain the direction of time, as often claimed, or is the growth of entropy a consequence of time having a direction?

6 Discuss the following. The Time Traveler tells his circle of friends of what it is like traveling into the far future. Some of the time while traveling he was underground

and it was dark because of changes in the Earth's surface. "The peculiar risk lay in the possibility of my finding some substance in the space which I, or the machine, occupied. So long as I traveled at a high velocity through time, this scarcely mattered; I was, so to speak, attenuated – was slipping like a vapor through the interstices of intervening substance! But to come to a stop involved the jamming of myself, molecule by molecule, into whatever lay in my way... "

7 Discuss the saying: The future is not there awaiting us, but is something we make as we go along.

• Concerning fatalism, Michael Dummett in "Causal loops" writes: the fatalist argument "was frequently applied, during the bombing of London in the Second World War, to being killed by a bomb. So applied, it ran like this. Either your name is written on any given bomb, or it is not. If it is, then you will be killed by that bomb whatever precautions you take. If it is not, you will survive the explosion, whatever precautions you neglect. In the first case, any precautions you take will be fruitless; in the second, they will be redundant: hence it is pointless to take precautions." According to this common form of fatalism, if an event is going to occur, any action taken to bring it about is redundant and any action taken to prevent it will be fruitless. If it is not going to occur, any action taken to bring it about will be fruitless and any action taken to prevent it is redundant. Can fatalism sustain us in time of adversity? (The bombing of London in World War II occurred while I was a student at London University, and my impression is that the answer is yes.) Does it diminish motivation in normal times?

• Each society, each generation, writes its own history. The past is infinitely malleable and always the victim of personal and national vanities. Distortions and equivocations form the mythologies that we call history. Can history ever be more than what we and the historians want it to be?

• Discuss: Free will belongs to the Universe, determinism belongs to our universes.

FURTHER READING

Butler, S. T. and Messel, H. Editors. *Time: Selected Lectures on Time and Relativity.* Pergamon Press, Oxford, 1965.

Fraser, J. T. *Of Time, Passion and Knowledge.* Brazilier, New York, 1975.

Fraser, J. T. *Time, The Familiar Stranger.* University of Massachusetts Press, Amherst, Massachusetts, 1987.

Gale, R. M. Editor. *The Philosophy of Time. An Anthology.* Macmillan, London, 1968.

Gardner, M. *The Ambidextrous Universe: Mirror Asymmetry and Time-Reversed Worlds.* Basic Books, New York, 1964.

Gold, T. "The arrow of time." *American Journal of Physics* 30, 403 (June 1962).

Landsberg, P. T. Editor. *The Enigma of Time.* An anthology. Adam Hilger, Bristol, 1982.

Nahin, P. J. *Time Machines: Time Travel in Physics, Metaphysics, and Science Fiction.* American Institute of Physics, New York, 1993.

Priestley, J. B. *Man and Time.* Aldus Books, London, 1964.

Wells, H. G. *The Time Machine.* Random House, New York, 1931. This edition has an informative preface by Wells, written 36 years after the book was first published.

Whitrow, G. J. *What Is Time?* Thames and Hudson, London, 1972.

Whitrow, G. J. *Time in History: Views of Time from Prehistory to the Present Day.* Oxford University Press, Oxford, 1988.

SOURCES

Augustine. *Confessions.* Book Eleven. Translator, A. C. Outler. In *Problems of Space and Time. From Augustine to Albert Einstein.* Editor, J. J. C. Smart. Macmillan, New York, 1964.

Bergonzi, B. *The Early H. G. Wells: A Study of the Scientific Romances.* Manchester University Press, Manchester, 1961.

Bork, A. M. "The fourth dimension in nineteenth-century physics." *Isis* 55, 326 (1964).

Born, M. "Man and the atom." *Bulletin of the Atomic Scientists* (June 1957).

Boyer, C. B. *A History of Mathematics.* John Wiley, New York, 1968. Gives an elementary discussion of the Hausdorff axioms, p. 668.

Broad, C. D. *Philosophy* 10, 168 (1938).

Burnet, J. *Early Greek Philosophy.* Black, London, 1920.

Dummett, M. "Causal loops," in *The Nature of Time.* Editors, R. Flood and M. Lockwood. Blackwell, Cambridge, Massachusetts, 1986.

Eddington, A. S. *The Nature of the Physical World.* Cambridge University Press, Cambridge, 1928.

Ellis, G. F. R. and Harrison, E. R. "Cosmological principles I: symmetry principles." *Comments on Astrophysics and Space Physics* 6, 23 (1974).

Grünbaum, A. *Philosophical Problems of Space and Time.* Routledge and Kegan Paul, London, 1964.

Grünbaum, A. *Zeno's Paradoxes in Modern Science.* Wesleyan University Press, Middletown, 1967.

Guericke, O. von. *Experimenta Nova Magdeburgica de Vacuo Spatio.* Amsterdam 1672. Reprinted: Zellers Verlagsbuchhandlung, Aalen, 1962.

Harrison, E. R. "Atomicity of time." *Encyclopedia of Time.* Editor, S. L. Macey. Garland, New York, 1994.

Helmholtz, H. L. F. von. "Origin and meaning of geometrical axioms." *Mind* (July 1876).

Hinckfuss, I. *The Existence of Space and Time.* Oxford University Press, Clarendon Press, Oxford, 1975.

Hinton, C. H. *Speculations on the Fourth Dimension: Selected Writings.* Editor, R. B. Rucker. Dover Publications, New York, 1980.

Hook, S. Editor. *Determinism and Freedom in the Age of Modern Science.* New York University Press, New York, 1958.

James, W. *A Pluralistic Universe.* Holt, New York, 1909.

Jammer, M. *Concepts of Space: The History of the Theories of Space in Physics.* Harvard University Press, Cambridge, Massachusetts, 1954.

Jammer, M. *Concepts of Force: A Study in the Foundations of Dynamics.* Harvard University Press, Cambridge, Massachusetts, 1957.

Kline, M. *Mathematics and the Physical World.* Thomas Y. Crowell, New York, 1969.

Lanczos, C. *Space Through the Ages.* Academic Press, New York, 1970.

Lucas, J. *Space, Time and Causality*. Oxford University Press, Oxford, 1984.

Lucretius. *The Nature of the Universe*. Translated in prose by R. E. Latham. Penguin Books, Harmondsworth, Middlesex, 1951.

MacDonald, D. B. "Continuous recreation and atomic time in Muslim scholastic theology." *Isis* 9, 326 (1927).

Maimonides, M. *The Guide for the Perplexed*. Dover, New York, 1972. Part I, Chapters 72–76.

Maxwell, J. C. "On physical lines of force. Part II. The theory of molecular vortices applied to electric currents." *Philosophical Magazine* 281, 21 (1861).

Reichenbach, H. *The Philosophy of Space and Time*. Dover Publications, New York, 1957.

Reichenbach, H. *The Direction of Time*. University of California Press, Berkeley, 1971.

Rucker, R. B. Editor. *Speculations on the Fourth Dimension: Selected Writings of Charles Hinton*. Dover Publications, New York, 1980.

Smart, J. J. C. Editor. *Problems of Space and Time: From Augustine to Albert Einstein*. Macmillan, New York, 1964.

Whitehead, A. N. *Process and Reality. An Essay in Cosmology*. Editors D. R. Griffin and D. W. Sherburne. Macmillan, New York, 1978.

Whitrow, G. J. *The Natural Philosophy of Time*. 2nd edition. Thomas Nelson, London, 1980.

Wigner, E. P. "Violations of symmetry in physics." *Scientific American* (December 1965).

Wolfson, H. A. *The Philosophy of the Kalam*. Harvard University Press, Cambridge, Massachusetts, 1976.

Part II

10 CURVED SPACE

Man has weav'd out a net, and this net throwne
Upon the Heavens, and now they are his owne.
John Donne (1571–1631), Ignatius His Conclave

EUCLIDEAN GEOMETRY

In the ancient Delta civilizations, geometry was the art of land measurement, and indispensable in the construction of such mammoth works as the Great Pyramid of Giza and Stonehenge. Geometry at first consisted of trial-by-error and rule-of-thumb methods. According to the sacred Rhind Papyrus, the Egyptians of 1800 BC used for π, the ratio of the circumference and the diameter of a circle, the value $(16/9)^2 = 3.1605$, as compared with its more exact value 3.1416. The Babylonians of 2000 BC and the Chinese of 300 BC used the rule that the circumference of a circle is three times its diameter, and this value for π is found in Hebraic scripture. The Greeks, in their thorough fashion, developed geometry into a science that climaxed in the axiomatic and definitive treatment presented by Euclid at the Museum in Alexandria in the third century BC.

Axioms

The axiomatic method starts with a set of self-consistent propositions (called postulates or axioms), which are often the simplest and most obvious truths, and examines their logical consequences. Suppose that we wish to persuade someone that statement S is true. We might try to show that this statement follows logically from another statement R that the person already accepts. But if the person is unconvinced of the truth of R, then we must try to show that R follows logically from yet another statement Q.

This process might have to be repeated several times until eventually a statement A is reached that is accepted as obviously true and has no further need of logical justification. The basic statement is called an axiom. Sometimes a residue of doubt remains, then the axiom, more suitably called a postulate, stands on the merits of its logical consequences. Euclid, in his textbook *The Elements* (of which more copies have been published than any other textbook), recorded in a systematic manner the advances in geometry made since the time of Thales. Euclid used five basic axioms, and from them, with accompanying definitions, deduced all that was known of geometry in 465 theorems.

Parallel postulate

The axiom of most interest to us, which remained controversial until the nineteenth century, is the Euclidean parallel postulate. This postulate asserts that through any point there is one and only one parallel to a given straight line. The definition of *parallel* states that two straight lines drawn in the same plane are parallel if they do not intersect (see Figure 10.1).

For more than 2000 years most persons acquainted with geometry accepted the parallel postulate as intuitively obvious. A few, however, including even Euclid himself, confessed uneasiness because the parallel postulate cannot be verified by direct appeal to experience. We always encounter only segments of straight lines, never straight

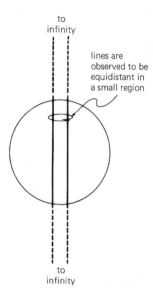

to
infinity

lines are
observed to be
equidistant in
a small region

to
infinity

Figure 10.1. Two parallel lines, extended to unlimited distances, remain equidistant. Do they? How do we know if this is true when we only have experience with straight lines of relatively short length?

lines of indefinite length, and hence we cannot assert with utmost confidence that two straight lines will always remain equidistant when extended to unlimited distances. Through the ages the few geometricians who felt uneasy sought for a more basic axiom from which the parallel postulate could be derived. But all attempts failed. John Wallis in the seventeenth century thought that our ordinary senses inform us directly that the geometric relations of a figure remain unchanged when the figure is scaled in size. He postulated: "If the sides of a triangle are changed in the same ratio, the angles of the triangle remain unchanged." From this axiom it can be shown that the parallel postulate follows immediately. But like all other attempts to find a more acceptable statement, it fails because it is merely the Euclidean postulate restated in alternative form. Clearly, on the basis of experience, it is impossible to declare that the geometric relations of a triangle remain unchanged when the triangle is made indefinitely large. Gerolamo Saccheri was also convinced that the parallel postulate could be proved

by appeal to more obvious truths. In his work *Euclid Vindicated*, published in 1733, he wrongly thought that he had at last established the parallel postulate as a transparent truth. Others did not find his argument at all transparent. He showed that two straight lines, intersecting at an infinite distance, have an angle of intersection that is zero; but if the angle of intersection is zero, he argued, the lines are indistinguishable; therefore distinguishable straight lines that do not intersect are necessarily parallel. In this work, Saccheri derived and discussed many non-Euclidean theorems but failed to realize that non-Euclidean geometry can have a theoretical validity equal to that of Euclidean geometry.

Immanuel Kant shared the prevailing belief that Euclidean geometry is transparently true and no alternative system of geometry is conceivable by the human mind. In the *Critique of Pure Reason* he attempted to place Euclidean geometry on a firm foundation by arguing that its axioms are "a priori" (prior to experience) and "an inevitable necessity of thought." Kant believed that what is unimaginable is automatically impossible. But in mathematics and physics what is possible today was unimaginable yesterday.

We now know that the parallel postulate is a fundamental statement and cannot be reduced to a more basic axiom. It is basic to Euclidean geometry and singles out Euclidean space from other possible spaces. In Euclidean space the circumference of a circle is π times its diameter, and the sum of the interior angles of a triangle is equal to two right angles (or π radians). In other spaces these relations are not necessarily true.

NON-EUCLIDEAN GEOMETRY
Uniform spaces

Countless theoretical spaces exist, all having their own peculiar geometry. Euclidean space is one of the simplest spaces: it is uniform, which means it is homogeneous (all places are alike) and isotropic (all directions are alike), and has therefore a congruence geometry. In a congruence geometry all spatial forms are invariant under translations and rotations. Thus if the ratio of the circumference and diameter of a circle is $x\pi$, this value is everywhere the same for that circle; and if the sum of the interior angles of a triangle is $y\pi$ radians, this sum is everywhere the same for all orientations of that triangle.

Of all possible non-Euclidean spaces, there are only two that are uniform in the same way as Euclidean space. Both were discovered in the nineteenth century. The first – *hyperbolic geometry* – was discovered by Johann Gauss, Nikolai Lobachevski, and Janos Bolyai; the second – *spherical geometry* – was discovered by Georg Riemann. Hyperbolic and spherical spaces, like Euclidean space, are uniform and therefore have congruence geometries. But unlike Euclidean space, they have an intrinsic scale length that is denoted by the symbol R. In all regions of these two spaces, small in size compared with the length R, the local geometry closely resembles Euclidean geometry. When R is very large compared with regions of familiar experience, it is very difficult to distinguish between the three uniform spaces. They all appear to be the same. We now know why the axioms of uniform geometries must contain a postulate, such as the parallel postulate, that refers to what happens on large scales and at large distances. It is the only way that the geometries of uniform spaces can be distinguished.

Our knowledge of the world derives directly from our experience with small-scale phenomena (that is, small on the cosmic scale), and this local experience contains no apparent information concerning a large intrinsic scale length R. This explains our preference for the Euclidean geometry that has no intrinsic scale length. All people who live in hyperbolic and spherical spaces, and have experience of only small-scale local phenomena, think in terms of Euclidean geometry. We still do not know whether the uniform space of our universe is Euclidean, hyperbolic, or spherical. Cosmological surveys over very large distances are

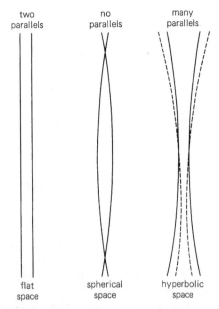

Figure 10.2. Parallel lines in uniform spaces. Through a given point: (a) in flat space there is only one parallel to a straight line; (b) in spherical space there are no parallels to a straight line; and (c) in hyperbolic space there are many parallels to a straight line.

not yet precise enough to make the distinction.

The three uniform spaces are distinguished by the following postulates (see Figure 10.2):

(i)   In hyperbolic space there are many parallels to a straight line through a given point.

(ii) In Euclidean space there is one parallel to a straight line through a given point.

(iii) In spherical space there is no parallel to a straight line through a given point.

In hyperbolic space the circumference of a circle is greater than π times its diameter (x greater than 1), and the sum of the interior angles of a triangle is less than two right angles (y less than 1); in spherical space the circumference of a circle is less than π times its diameter (x less than 1), and the sum of the interior angles of a triangle is greater than two right angles (y greater than 1). See Figure 10.3. One right angle

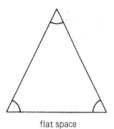
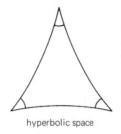

flat space   hyperbolic space   spherical space

Figure 10.3. Triangles in flat, hyperbolic, and spherical spaces.

equals 90 degrees, and 1 radian is $180/\pi = 57.30$ degrees.

Curvature of space

We are familiar with two-dimensional surfaces in our three-dimensional space and we have little or no difficulty imagining uniform (homogeneous and isotropic) surfaces. A uniformly flat and infinite surface illustrates the nature of Euclidean space, and the uniformly curved surface of a sphere illustrates the nature of spherical non-Euclidean space. Straight lines drawn on the surface of a sphere are called great circles. Circles of constant longitude on the Earth's surface are great circles, but circles of constant latitude are not straight lines. A plane flying from London to Los Angeles takes the shortest route and follows a great circle; first it flies north, and later it flies south. Great circles always intersect one another at finite distances, and it is therefore apparent that parallel straight lines do not exist on the surface of a sphere (see Figure 10.4). Furthermore, the sum of the interior angles of a triangle on the surface of a sphere always exceeds two right angles (see Figure 10.5). A spherical surface neatly illustrates the geometry of a finite spherical space.

But what kind of surface illustrates the geometry of hyperbolic space? The mathematician David Hilbert showed that we cannot construct in Euclidean space a two-dimensional surface that accurately represents the geometry of a uniform hyperbolic space. The surface of a pseudosphere (Figure 10.6) has hyperbolic geometry, but is not homogeneous (all places are not alike) and is not isotropic (all directions are not

alike). A saddle-shaped surface, as in Figure 10.7, has hyperbolic geometry, but is homogeneous and isotropic in only a central small region, and all places are not alike. The surface is therefore not uniform. On these surfaces, which illustrate the properties of hyperbolic geometry and have at each point their radii of curvature in opposite directions, the sum of the interior angles of a triangle is less than two right angles.

A saddle-shaped surface shows that hyperbolic space, like Euclidean space, is "open" and of infinite extent, whereas a spherical surface shows that spherical space is "closed" and of finite extent. Note that all three uniform spaces, finite and infinite, are unbounded: they have no edge.

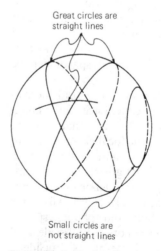

Great circles are
straight lines

Small circles are
not straight lines

Figure 10.4. Straight lines on the surface of a sphere are great circles that always intersect one another and cannot therefore be parallel. Lines of constant longitude are great circles, lines of constant latitude are not.

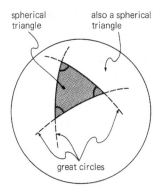

spherical triangle

also a spherical triangle

great circles

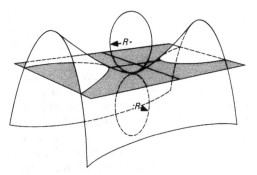

Figure 10.7. A saddle-shaped surface has negative curvature. Only the central region of the saddle is representative of uniform hyperbolic space; far from the central region, however, the surface is neither homogeneous nor isotropic. The radii R of curvature are on opposite sides of the surface, thus making the curvature ($K = -1/R^2$) negative.

Figure 10.5. Three intersecting great circles form a triangle on the surface of a sphere. The sum of the interior angles of the triangle is greater than two right angles. Notice that the surface of the sphere exterior to the triangle is also a triangle. Thus three intersecting straight lines divide the surface into two triangles. This can be shown by allowing the small triangle to expand and become large; the large exterior triangle to contract and become small.

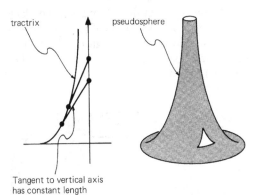

tractrix

pseudosphere

Tangent to vertical axis has constant length

Figure 10.6. The surface of a pseudosphere has hyperbolic geometry of negative curvature. The sum of the interior angles of a triangle is less than two right angles. The curvature of the surface, however, is not homogeneous (the same at all places).

The three uniform spaces are defined as follows:

spherical space (closed): K is positive;
Euclidean space (open): K is zero;
hyperbolic space (open): K is negative;

where the quantity K is the symbol for curvature. The positive curvature of a uniform surface is equal to $1/R^2$, where R is the radius of curvature and is the intrinsic scale length mentioned earlier. A surface

has two radii of curvature measured in directions perpendicular to each other. In a uniform surface they are both equal and everywhere have the same value. When both radii of curvature are on the same side of the surface (as in spherical space) the curvature K is positive, and when on the opposite sides of the surface (as in hyperbolic space) the curvature K is negative.

A flat surface has an infinite radius ($R = \infty$) and the curvature is therefore zero. This explains why Euclidean space, which has zero curvature, is often said to be flat. A spherical surface has both its radii of curvature on the same concave side, and the curvature ($K = 1/R^2$) of spherical space is positive. A saddle-shaped surface has its radii of curvature on opposite sides (see Figure 10.7), and the curvature ($K = -1/R^2$) of hyperbolic space is negative. A three-dimensional space has three perpendicular surfaces and hence at every point has three curvatures and six radii of curvature. When the space is uniform, all radii are equal in magnitude, and the curvature K is the same everywhere and in all directions.

Two-dimensional curved surfaces, embedded in three-dimensional flat space, illustrate the properties of non-Euclidean geometries. But we need not adopt a three-dimensional picture to study the geometry of two-dimensional surfaces. Flatlanders living in flat surfaces, spherelanders in spherical surfaces, and hyperbolanders in hyperbolic surfaces are two-dimensional creatures who are unaware of the existence of a third dimension. By means of measurements, however, they can determine the curvature of their two-dimensional worlds. Also, we need not think of our three-dimensional world embedded in a higher-dimensional space to study its curvature.

The curvature of space is a loose expression meaning that space has non-Euclidean geometry. As a descriptive term it originated in the nineteenth century when the non-Euclidean geometries of curved surfaces were first studied. It is a figurative and harmless term provided we do not take it too literally. Our universe contains three-dimensional space, or rather four-dimensional spacetime, which is not necessarily embedded and curved in a higher-dimensional flat space. Curvature should be understood as an intrinsic geometric property and does not require the existence of a higher-dimensional space.

MEASURING THE CURVATURE OF SPACE

In a space of two or more dimensions, the sum of the three interior angles of a triangle, minus two right angles, is equal to the curvature K multiplied by the area of the triangle. The curvature measured in this way is the curvature of the two-dimensional surface containing the triangle. The sum of two right angles equals π radians, and hence the general rule is:

sum of interior angles of a triangle $- \pi$

$$= K \times \text{area of triangle}. \qquad [10.1]$$

This rule enables us to determine the curvature anywhere in any space of two or more dimensions. In a nonuniform three-dimensional space, K has at every point three different values that vary from place to place, corresponding to the three orthogonal

two-dimensional planes in which triangles can be drawn.

Dense triangulation

On a surface we draw numerous small triangles (see Figure 10.8). If the surface is flat, the sum of the interior angles of each triangle is equal to two right angles. If the surface is curved the interior angles do not add up precisely to two right angles. Let the sum of the interior angles of a triangle, minus two right angles, be θ:

$$\theta = \text{sum of interior angles} - \pi, \qquad [10.2]$$

where π radians (equal to 180 degrees) is the sum of two right angles, and θ is measured in radians. The angle difference θ, which we shall call the curvature angle, is therefore, according to Equation [10.1],

$$\theta = K \times \text{area of triangle}, \qquad [10.3]$$

and the area is that of a triangle at the place of curvature K. In regions of positive curvature θ is positive and in regions of negative curvature (as when the surface is saddle-shaped) θ is negative. By measuring θ and

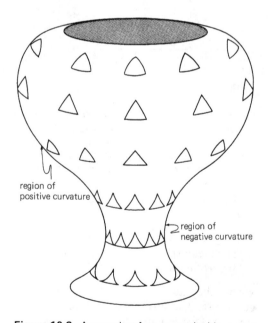

Figure 10.8. A curved surface covered with numerous triangles. Each triangle tells us the value of the local curvature.

the area of each triangle we can determine the curvature everywhere without making measurements outside the surface.

Actually, we can use any figure constructed from triangles. If a figure contains N triangles, the curvature angle of the figure is

$\theta = $ sum of interior angles
of figure of N triangles $- N\pi$,

and therefore, by adding the triangles,

$$\theta = K \times \text{area of figure.} \qquad [10.4]$$

Thus a four-sided figure (quadrilateral) can be constructed from $N = 2$ triangles, and the sum of its interior angles minus 2π is the curvature angle θ, which equals the curvature K times the area of the quadrilateral.

Measuring the interior angles and areas of small triangles is arduous and not very practical. To determine the shape of a surface we need its curvature at every point. We must shrink the triangles to very small sizes and cover the surface with dense triangulation (Figure 10.8). Ideally, the triangles should be infinitesimally small. But triangles of infinitesimal size have infinitesimal areas, and interior angles whose sum is infinitesimally different from two right angles, and cannot easily be used for measuring curvature.

A neat sort of way

The curvature of a fixed surface can be determined with a simple instrument known as a spherometer. This instrument rests on the surface at three points forming a triangle, and an adjustable screw determines the height of the central point of the triangle. This is a three-dimensional method not available to the two-dimensional creatures who live in the surface.

Another way available to three-dimensional creatures examining a two-dimensional surface is to compare circles on a curved surface. A flat disk of aluminum foil, with radial cuts, is placed on a curved surface and shaped to fit snugly on the surface. The disk, now curved, has more area than the surface if K is positive and

less area if K is negative. By adding together the small angles of the overlaps (K positive) or the gaps (K negative), we obtain the curvature angle. This result is proved by supposing that the disk consists of many triangles. Hence, we have

$$\theta = K \times \text{area of disk,} \qquad [10.5]$$

and the curvature K is found by dividing the measured angle θ by the area of the circle. Notice that θ is positive (overlaps) if K is positive, and negative (gaps) if K is negative. Unfortunately, this "neat sort of way" measures mechanically the curvature of a rigid surface, such as that of a large vase or amphora, and is impractical for measuring the curvature of empty space.

Parallel transport and vector deviation

Mathematicians take a more serious approach to the problem of determining curvature. They want to compare not only scalars but also vectors (and higher-order tensors) at adjacent points. A scalar has at each point in space a single value: examples are density, temperature, and hydrostatic pressure. Finding the scalar difference for adjacent points (in order to determine density, temperature, and pressure gradients) is not very difficult, even in curved spaces.

A vector has direction and magnitude, and finding the vector difference for adjacent points is not so easy, particularly in curved spaces. A fluid in motion has a velocity vector that is different at different places in space. To see how its velocity varies in space we must find the velocity difference for adjacent points. To do this, we take the vector at one place and move it without change (parallel transport it) to an adjacent point of the fluid where the velocity is different, and find in this way the velocity difference. Parallel transport is not too difficult in flat space; the direction of a parallel transported vector remains unchanged, as shown in Figure 10.9. When, however, a vector is parallel transported in curved space, it changes direction because of spatial curvature. Thus some of the velocity difference at adjacent points is due to the curvature

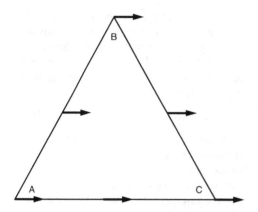

Figure 10.9. In parallel transport, the direction of a vector relative to a straight line (a geodesic) remains constant as the vector is moved along the straight line. A vector in flat space does not rotate when transported around a closed circuit such as a triangle.

of space. It can be shown that the change in direction of a vector (its deviation) as it is transported around a closed circuit is the curvature angle θ:

$$\theta = K \times \text{area enclosed by circuit.} \quad [10.6]$$

A vector, parallel transported along a straight line (called a geodesic), remains unchanged relative to the straight line; this is true in flat and curved spaces. The angle of inclination of the vector to the geodesic is constant in parallel transport. Consider the surface of a sphere on which straight lines are great circles. Let us construct a triangle covering a fraction f of the surface. The surface area of the sphere is $4\pi R^2$ and the area of the triangle is $4\pi f R^2$. The curvature of the sphere is $K = 1/R^2$, and therefore, according to Equation [10.6], the deviation in direction of a vector transported around the triangle is $\theta = 4\pi f$. Thus when $f = \frac{1}{8}$, as in Figure 10.10, the deviation is $\frac{1}{2}\pi$, or 90 degrees.

THE "OUTSTANDING THEOREM"

The mathematician Johann Gauss used differential calculus in the study of curved spaces. When he participated in a large land survey at the invitation of the Hanoverian government he was confronted

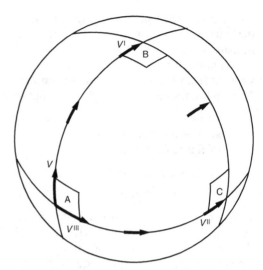

Figure 10.10. Parallel transport of a vector in curved space causes the vector to rotate by an amount equal to the curvature angle. This figure shows a triangle on a spherical surface in which each of the three interior angles is a right angle. Parallel transport of vector V from A to B gives V', from B to C gives V'', and from C back to A gives V'''. Clearly, in this case, the rotation equals a right angle (90 degrees).

with a surface deformed into hills and valleys. Ordinary geometry is not of great help in the theoretical study of an inhomogeneous surface, and Gauss had a novel idea.

To explain Gauss's discovery we start with a flat surface and lay out a network of imaginary lines that form a coordinate system. If the coordinates are perpendicular to each other, as in the Cartesian coordinate system (see Figure 10.11a), and labeled x and y, the distance between any two points is given by the Pythagorean rule:

$$(\text{space interval})^2 = (x \text{ interval})^2$$
$$+ (y \text{ interval})^2. \quad [10.7]$$

Another person might choose a different set of perpendicular coordinates, labeled x' and y', and the distance between the same two points would be

$$(\text{space interval})^2 = (x' \text{ interval})^2$$
$$+ (y' \text{ interval})^2, \quad [10.8]$$

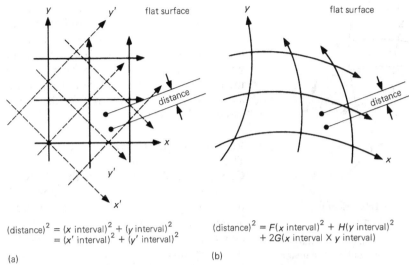

$$(distance)^2 = (x \text{ interval})^2 + (y \text{ interval})^2$$
$$= (x' \text{ interval})^2 + (y' \text{ interval})^2$$

(a)

$$(distance)^2 = F(x \text{ interval})^2 + H(y \text{ interval})^2$$
$$+ 2G(x \text{ interval} \times y \text{ interval})$$

(b)

Figure 10.11. Coordinate systems in flat surfaces. When coordinate lines are perpendicular to each other, as are x and y or x' and y' shown in (a), the distance between any two points is given by the ordinary Pythagorean rule. When coordinates are curvilinear, as shown in (b), and not necessarily perpendicular, the distance between two adjacent points is given by a more general rule involving metric coefficients that vary from point to point and depend on the arbitrary coordinates chosen.

and, although the coordinates are different, this distance is the same as that given by Equation [10.7].

Coordinate systems are obviously just a convenient way of marking out a surface and do not affect the actual distances between points in the surface. We may use any coordinate system consisting of a network of arbitrarily curved lines (not necessarily intersecting perpendicularly), as shown in Figure 10.11b, and the Pythagorean rule becomes

$$(\text{space interval})^2$$
$$= F(x \text{ interval})^2$$
$$+ 2G(x \text{ interval} \times y \text{ interval})$$
$$+ H(y \text{ interval})^2. \qquad [10.9]$$

Because F, G, and H are functions of x and y (i.e., they vary from place to place), the intervals must be kept small. An equation of this kind, which gives the distance between two adjacent points with arbitrary coordinates, is known as a metric equation, and the

functions F, G, and H are metric coefficients. In a flat surface, the metric coefficients depend on the kind of coordinates selected, and we are free to choose Cartesian coordinates ($F = 1$, $G = 0$, and $H = 1$), as in the ordinary Pythagorean rule of Equation [10.7].

We now deform the surface into any desired smooth shape, avoiding tears and wrinkles, as in Figure 10.12. The coordinate lines on the surface also deform with the surface. The distance between two points close together continues to be given by the metric Equation [10.9], but the metric coefficients F, G, and H have changed. Thus we see that altering the coordinate system changes the metric coefficients, and altering the shape of the surface also changes the metric coefficients.

If, after deforming the surface, the angles between the coordinate lines remain unchanged, for example the angles remain right angles for Cartesian coordinates, the deformation from a flat to a curved space is called a conformally flat transformation.

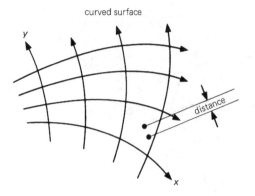

$$(\text{distance})^2 = f(x \text{ interval})^2 + 2g(x \text{ interval}$$
$$\times y \text{ interval}) + h(y \text{ interval})^2$$

Figure 10.12. A curved surface showing the distance between two adjacent points. The metric coefficients F, G, and H have values that vary with position and depend not only on the arbitrarily chosen coordinates but also on the curvature of the surface. The curvature can be determined from the way the metric coefficients vary from point to point in the surface, as shown by Gauss.

A surveyor confronted with an undulating landscape does not start with a flat surface. As it happens, however, this is not a great disadvantage. The surveyor lays out on the curved surface an imaginary network of intersecting lines and from the beginning, with this coordinate system, has a metric equation in which the metric coefficients F, G, and H depend on the coordinates chosen and also on the shape of the surface. Different surveyors will choose different coordinates having different metric coefficients and their metric equations will always agree on the actual space intervals between adjacent points. Gauss discovered with differential geometry how to determine the shape of the surface from the way in which the metric coefficients vary from place to place in the surface, despite the arbitrariness of the coordinates chosen and the dependence of the metric coefficients on the coordinates. The values of the metric coefficients are not important – they can always be changed by altering the coordinates – but the way in which they vary from place to place in the surface contains all the

information needed to determine the geometry of the surface. Gauss, excited by his discovery, called it the *theorema egregium* – the outstanding theorem. In a letter he wrote, "These investigations deeply affect many other things; and I would go so far as to say they are involved in the metaphysics of the geometry of space."

RIEMANNIAN SPACES
The genius of Riemann

Georg Bernard Riemann saw that the work by Gauss opened an entirely new approach to the study of geometry. Riemann's great achievement was the discovery of how this new approach of differential geometry can be generalized and applied to three-dimensional and higher-dimensional spaces. The distance between adjacent points is expressed by a metric equation, as before, which is really nothing more than a glorified Pythagorean rule. The metric coefficients in this metric equation depend on the arbitrary coordinates chosen, and the geometric properties of the space are expressed in the form of differential equations showing how the metric coefficients vary from place to place. In a two-dimensional space there are in general three metric coefficients (F, G, and H); in a three-dimensional space there are in general six metric coefficients; and in a four-dimensional space there are in general ten metric coefficients.

Riemann developed differential equations for the variation of the metric coefficients. One of these equations gives us the Riemann curvature, which is a more general expression for the curvature K that we introduced earlier. In a two-dimensional curved space the curvature has a single value K, which varies from place to place when the space is inhomogeneous. In a three-dimensional space the curvature is a more complicated expression containing six components that in general have different values; when the space is uniform (homogeneous and isotropic), three of the components are zero and the other three are everywhere constant and equal to K; when the space is flat, and hence Euclidean, all components are zero.

In a four-dimensional space the curvature has twenty components, and in a five-dimensional space it has fifty components, and all components are zero if these spaces are flat.

If we think of curvature as the nonflatness of one space embedded in a higher-dimensional flat space, then a nonflat two-dimensional space requires a flat space of three dimensions (this is the case with which we are familiar); a nonflat three-dimensional space requires a flat space of six dimensions; and a nonflat four-dimensional space requires a flat space of ten dimensions. Space-time is four-dimensional; yet when thinking of its geometric properties we do not try to visualize it embedded in a ten-dimensional flat space. This superhuman feat of the imagination is quite unnecessary because flat space is no more fundamental than curved space.

Riemann foresaw the possibility of science evolving beyond Newtonian mechanics and its three-dimensional flat space, and that science might one day draw on the more general theory of space that mathematicians had developed. At the end of his inaugural doctoral lecture "On the hypotheses forming the foundation of geometry," delivered in 1854, he said, "This leads us into the domain of another science, that of physics, into which the object of this work does not allow us to go today." Einstein paid to Riemann the tribute: "Only the genius of Riemann, solitary and uncomprehended, had already won its way by the middle of the last century to a new conception of space, in which space was deprived of its rigidity, and in which its power to take part in physical events was recognized as possible."

Clifford's dream

William Clifford, a young mathematician who translated Riemann's work into English, enthusiastically championed the idea of a fusion of geometry and mechanics. In *The Common Sense of the Exact Sciences*, published posthumously in 1885, he made the prophetic remark: "Our space may be really the same (of equal curvature), but its degree of curvature may change as a whole with time. In this way our geometry based on the sameness of space would still hold good for all parts of space, but the change of curvature might produce in space a succession of apparent physical changes." This was a remarkable anticipation of the expanding homogeneous space of modern cosmology. He continued: "We may conceive our space to have everywhere a nearly uniform curvature, but slight variations of the curvature may occur from point to point, and themselves vary with time. These variations of the curvature with time may produce effects which we not unnaturally attribute to physical causes independent of the geometry of our space. We might even go so far as to assign to this variation of the curvature of space 'what really happens in that phenomena which we term the motion of matter.'" In 1876, Clifford wrote: "I wish here to indicate a manner in which these speculations may be applied to the investigation of physical phenomena. I hold in fact

(i)   That small portions of space are of a nature analogous to little hills on a surface which is on the average flat; namely, that the ordinary laws of geometry are not valid in them.

(ii) That this property of being curved or distorted is continually being passed on from one portion of space to another after the manner of a wave.

(iii) That this variation of the curvature of space is what really happens in that phenomena which we call the motion of matter, whether ponderable or ethereal.

(iv) That in the physical world nothing else takes place but this variation, subject (possibly) to the laws of continuity."

Clifford was an outstanding mathematician whose speculative ideas, outrageous in their day, anticipated general relativity by forty years. He died at age 34 in the year that Einstein was born.

REFLECTIONS

1 *Nikolai Lobachevski (1793–1856), an outstanding Russian mathematician at the*

University of Kazan, developed hyperbolic geometry, which he referred to as "imaginary geometry," and published his results in 1829. Lobachevski said, "There is no branch of mathematics, however abstract, that may not someday be applied to phenomena of the real world." Although an outstanding teacher and university administrator, he was dismissed from his academic position in 1846 without explanation, possibly because of unfavorable reviews of his unorthodox work.

Janos Bolyai (1802–1860) of Hungary, unaware of Lobachevski's recently published work, also developed the theory of hyperbolic geometry. "Out of nothing," he said, "I have created a strange new universe." His enthusiasm was quenched, however, when Gauss replied in a letter that he himself had discovered similar results many years previously.

Johann Karl Friedrich Gauss (1777–1855), of humble origin, was an infant prodigy in mathematics who rose to become the prince of nineteenth-century mathematicians and the leading professor at the University of Göttingen. He also made important contributions in physics and astronomy. Gauss hesitated to publish his researches in non-Euclidean geometry. He wrote in a letter that he had "a great antipathy against being drawn into any sort of polemic," and in 1817 he said, "Perhaps in another world we may gain other insights into the nature of space which at present are unattainable to us. Until then we must consider geometry of equal rank not with arithmetic, which is purely logical, but with mechanics, which is empirical."

Georg Friedrich Bernhard Riemann (1826–1866), the son of a Lutheran minister, made many advances in several fields, mainly in mathematics, and initiated the subject of tensor calculus. A tensor is a mathematical term having at every point n^m components. Here n is the number of dimensions of the space and m is the order of the tensor. For a scalar (single component), $m = 0$; and a vector (n components), $m = 1$. Riemann's investigations into the structure of space were ignored by his contemporaries who regarded this aspect of his work as excessively abstract and speculative. Riemann suffered from poor health and died at the age of 39. Theories of space curvature and tensor calculus were further developed by the mathematicians Elwin Christoffel (1829–1900) of Germany and Giuseppi Ricci (1811–1881) and Tullio Levi-Civita (1873–1941) of Italy.

• William Clifford (1845–1879) died of tuberculosis while still relatively young. In the introduction to Clifford's book The Common Sense of the Exact Sciences, James Newman wrote, "All his life he had burdened his physical powers. The abundant but self-consuming nervous energy, the warfare against false beliefs, the self-goading search for new riddles and new challenges, the full submission to the demands of his intellect, were altogether out of proportion to what the physical machine could endure."

• After combining space and time in the theory of special relativity, Albert Einstein (1879–1955) developed the theory of general relativity drawing on the mathematical developments of Riemann, Christoffel, Ricci, and Levi-Civita, and succeeded in endowing spacetime with a varying curvature that explains gravity, thus realizing Clifford's dream.

2 In non-Euclidean space the sum of the interior angles of a triangle will not equal 180 degrees (or π radians). Gauss performed an experiment, using three mountain peaks as the vertices of a large triangle, and found that the interior angles of the triangle added up to 180 degrees within the uncertainties of the measurements. We now know that his triangle was much too small and his survey instruments much too inaccurate.

• The curvature of space was a hot subject in mathematical and scientific journals of the late nineteenth century, and Karl Schwarzschild, a German astronomer and mathematician, attempted in 1900 to measure the curvature of space. Light rays from a distant star, intersecting the Earth's orbit at two widely separated points, form a triangle. By measuring angles and distance to the star,

Schwarzschild hoped to determine the curvature of space. He concluded that space, if curved, has an extremely large radius of curvature, and wrote, "One finds oneself here, if one will, in a geometrical fairyland, and the beauty of this fairy tale is that one does not know but that it may be true. We accordingly address the question of how far we must push back the frontiers of this fairyland; of how small we must choose the curvature of space, how great its radius of curvature." In 1916, shortly before he died as a soldier in World War I, he found the first exact solution of Einstein's equation of general relativity. The solution is for the exterior geometry of a spherical body, such as a star. This solution (Chapter 13), of fundamental importance, is the equivalent of the Newtonian inverse-square law of gravity.

3   The Euclidean parallel postulate, as stated in the text, is known as Playfair's axiom. John Playfair (1748–1819) expressed the parallel postulate in this improved form.

• David Hilbert (1862–1943) of Germany, the world's leading mathematician in the early decades of the twentieth century, advocated the axiomatic method and showed that the Euclidean axioms are complete and self-consistent. The axiomatic method no longer requires that axioms be transparent truths. It is necessary only that they are free of contradiction (i.e., be self-consistent) and are sufficient to construct a theoretical system. The theory so devised stands or falls by its correspondence with the real world. Many axioms may at first seem strange, but common sense is no longer an infallible guide, and only the theoretical consequence of the axioms can now determine their validity.

4   On the word "curvature," introduced by Riemann, Arthur Eddington wrote, "Space-curvature is something found in nature with which we are beginning to be familiar, recognizable by certain tests, for which ordinarily we need not a picture but a name" (The Expanding Universe 1933). Howard Robertson wrote in "Geometry as a branch of physics" (1949), "This name and this representation are for our purpose at least psychologically unfortunate, for we propose ultimately to deal exclusively with properties intrinsic to the space under consideration – properties that in the later physical applications can be measured within the space itself – and are not dependent upon some extrinsic construction, such as its relation to an hypothesized higher-dimensional embedding space. We must accordingly seek some determination of K – which we nevertheless continue to call curvature – in terms of such inner properties."

5   Consider circles and spheres of radius r (distances determined by a stretched tape measure) in a uniform three-dimensional space of curvature K. When r is small compared with R, the circumference of a circle is given by

$$\text{circumference of circle} = 2\pi r\left(1 - \frac{Kr^2}{6}\right),$$

[10.10]

and the area of the circle is

$$\text{area of circle} = \pi r^2\left(1 - \frac{Kr^2}{12}\right). \qquad [10.11]$$

The surface area of a sphere of radius r is

$$\text{surface area of sphere} = 4\pi r^2\left(1 - \frac{Kr^2}{3}\right),$$

[10.12]

and the volume of the sphere is

$$\text{volume of sphere} = \tfrac{4}{3}\pi r^3\left(1 - \frac{Kr^2}{5}\right).$$

[10.13]

The departure from Euclidean geometry in all cases depends on the value $(r/R)^2$. In our universe, R has a value of possibly 10^{10} light years. Even if r is as large as 100 million light years, the departure is only about 1 part in 10^4, and when we remember that observational estimates of cosmologically large distances can be in error by a factor of two or more, we realize the difficulty of deciding whether our universe has positive, zero, or negative curvature.

When r/R is not small, the exact form of the above expressions is a little more complicated. It is interesting to note that in spherical three-dimensional space the distance to the antipode (the opposite side of the universe) is πR, and the circumnavigation distance (the distance traveled round the universe) is $2\pi R$. The total volume of three-dimensional spherical space is $2\pi^2 R^3$.

6 It was the custom once to refer to Euclidean geometry as parabolic to distinguish it from the elliptical and hyperbolic geometries. Elliptical and spherical spaces possess similar geometries that are different in the following sense: The surface of a sphere is analogous to spherical space, whereas the surface of a hemisphere is analogous to elliptical space, as shown in Figure 10.13. In elliptical space the antipodal hemisphere is regarded as the same as one's own hemisphere, and when a receding body reaches halfway to the

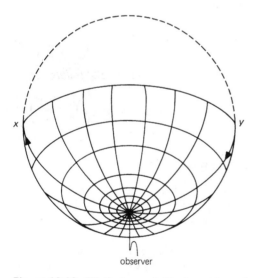

Figure 10.13. Elliptical space is like the surface of a hemisphere and, unlike spherical space, has no antipode. Points x and y, halfway to the antipode in opposite directions in spherical space, are equivalent in elliptical space. A receding particle reaching x vanishes and reappears at y as an approaching particle. Elliptical space is nowadays an historical curiosity.

antipode, it disappears and reappears on the other side of the universe, and is seen approaching.

7 A zero-order tensor, known as a scalar, has a single component whose value varies from place to place in space. Temperature is one example of a zero-order tensor, and the Newtonian gravity potential is another. A first-order tensor is a vector with n components at each point, where n is the number of dimensions of space. The motion of a simple fluid has at each point three components of velocity and is an example of a vector in three-dimensional space. A second-order tensor has n^2 components. Many fluids have complex dynamic properties that require the descriptive power of second-order tensors. The metric tensor is also second order, but owing to certain symmetries in our conception of space, it contains not n^2 components (or metric coefficients), but $n(n+1)/2$ independent components. There are three components (as used by Gauss) when $n = 2$, six components when $n = 3$, and 10 components when $n = 4$ (as in relativity spacetime). The Riemann curvature is a fourth-order tensor of n^4 components, or 256 components at each point in four-dimensional spacetime. But various symmetries reduce the total number of independent components to $n^2(n^2 - 1)/12$ in n-dimensional space. Thus when $n = 2$, there is one component, when $n = 3$, there are six components, and when $n = 4$, there are 20 components.

8 Let the separating distance between two points close together be dL. On the surface of a sphere of radius R the distance dL between any two adjacent points at colatitude α (angle from the pole) and longitude θ, separated by small angles $d\alpha$ and $d\theta$, is given by the equation

$$dL^2 = R^2(d\alpha^2 + \sin^2\alpha\, d\theta^2). \qquad [10.14]$$

Note that $R\, d\alpha$ and $R\sin\alpha\, d\theta$ are tape-measure distances made in the surface. This result can be generalized for a spherically curved three-dimensional space:

$$dL^2 = R^2[d\alpha^2 + \sin^2\alpha(d\theta^2 + \sin^2\theta\, d\phi^2)], \qquad [10.15]$$

in which the additional dimension is measured in ϕ coordinates. In hyperbolic space, the metric equation is

$$dL^2 = R^2[d\alpha^2 + \sinh^2\alpha(d\theta^2 + \sin^2\theta d\phi^2)],$$
[10.16]

in which sin α has been changed to hyperbolic sinh α.

PROJECTS

1   Find out what are the Euclidean axioms.
2   Discuss the statement that Euclidean geometry is the only geometry without an intrinsic length scale. In Euclidean space we have no way of knowing how big we are!
3   Why do people who live in spherical and hyperbolic spaces of small curvature (large radius of curvature) think always in terms of Euclidean geometry?
4   Note that on the surface of a sphere the outside of a triangle is also a triangle (see Figure 10.5). What is the sum of the interior angles in (a) the smallest triangle, and (b) the largest triangle, on the surface of a sphere? The surface of a sphere of radius R has a curvature $K = 1/R^2$ and a total surface area $4\pi R^2$. Show that a triangle becomes a hemisphere when the sum of its interior angles is 3π.
5   The surface of a cylinder is homogeneous (all places are alike) but not isotropic. Show that the surface has zero curvature (see Figure 10.14). (Hint: Draw a triangle and a circle on a flat sheet of paper and then wrap the sheet around a cylinder.)
6   The following experiment demonstrates the nature of a congruence geometry and requires only a wall and a flashlight (see Figure 10.15). On the glass front of the flashlight is attached a transparent cover on which a figure, such as a triangle or a circle, is heavily inked. The flashlight is held at a constant distance from the wall, and is moved around always with the beam perpendicular to the wall. The figure seen on

Figure 10.15. A flashlight beam is projected on a wall or large screen. The flashlight projects a figure, in this case a triangle. If the flashlight is moved about at constant distance from the wall, with the beam perpendicular to the wall but free to rotate, the projected figure demonstrates the nature of a congruence geometry.

Figure 10.16. Altering the distance of the flashlight from the wall and varying the angle of incidence as the beam scans the wall demonstrates the nature of an affine geometry, which is more general than a congruence geometry.

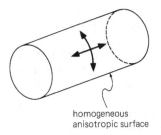

homogeneous
anisotropic surface

Figure 10.14. The surface of a cylinder is homogeneous but anisotropic. It has zero curvature and is topologically different from Euclidean space.

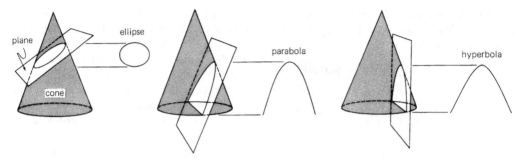

Figure 10.17. Ellipses, parabolas, and hyperbolas are conic sections produced by the intersection of a plane and a circular cone.

the wall preserves its shape and size and is "invariant to translations and rotations." This shows the properties of a uniform or congruent space.

With this simple apparatus we can also demonstrate some properties of affine geometry (see Figure 10.16). Riemannian spaces have affine geometry. In these geometries a point remains always a point and a geodesic (straight line) is always the shortest distance between two points. As the flashlight beam incident on the wall moves around, we are free to do two things: vary the distance of the flashlight from the wall, and vary the angle of incidence of the flashlight beam. The first causes the projected figure to alter in size, and the second changes its shape. When circles can be projected not only as ellipses but also as parabolas and hyperbolas, we have an example of a projective geometry (see Figure 10.17). An affine geometry is more general than a congruence geometry, and a projective geometry is more general than an affine geometry.

FURTHER READING

Abbott, E. A. *Flatland*. Dover Publications, New York, 1952.

Bergmann, P. G. *The Riddle of Gravitation*. Scribner's Sons, New York, 1968.

Burger, D. *Sphereland: A Fantasy about Curved Spaces and an Expanding Universe*. Thomas Crowell, New York, 1969.

Ellis, G. F. R. and Williams, R. M. *Flat and Curved Space-Times*. Clarendon Press, Oxford, 1988.

Greenberg, M. J. *Euclidean and Non-Euclidean Geometries: Development and History*. W. H. Freeman, San Francisco, 1973.

Hilbert, D. and Cohn-Vossen, S. *Geometry and the Imagination*. Chelsea Publishing Co., New York, 1952.

Jaggi, M. P. "The visionary ideas of Bernhard Riemann." *Physics Today* (December 1967).

Kline, M. *Mathematics and the Physical World*. Thomas Crowell, New York, 1969.

Manning, H. P. *Geometry of Four Dimensions*. Macmillan, London, 1914. Dover Publications, New York, 1956.

Resnikoff, H. L. and Wells, R. O. *Mathematics in Civilization*. Holt, Rinehart and Winston, New York, 1973.

Rucker, R. B. *Geometry, Relativity and the Fourth Dimension*. Dover Publications, New York, 1977.

Stewart, I. "Gauss." *Scientific American* (July 1977).

Waerden, B. L. van der. *Science Awakening*. Nordhoff, Groningen, Holland, 1954.

SOURCES

Beckmann, P. *A History of π (Pi)*. Golem Press, Boulder, Colorado, 1970.

Bell, E. T. *Men of Mathematics: The Lives and Achievements of the Great Mathematicians from Zeno to Poincaré*. Simon and Schuster, New York, 1937.

Clifford, W. *The Common Sense of the Exact Sciences*. 1885. Dover Publications, New York, 1955.

Eddington, A. S. *The Expanding Universe*. Cambridge University Press, Cambridge, 1933.

Hall, R. S. *About Mathematics*. Prentice-Hall, Englewood Cliffs, N. J., 1973.

Le Lionnais, F. *Great Currents of Mathematical Thought*; Vol. I, *Mathematics: Concepts and Developments*. Vol. II, *Mathematics in the Arts and Sciences*. Dover Publications, New York, 1871.

Misner, C. W., Thorne, K. S., and Wheeler, J. A. *Gravitation*. Freeman, San Francisco, 1973.

Rindler, W. *Essential Relativity: Special, General, and Cosmological*. Van Nostrand Reinhold, New York, 1969.

Robertson, H. P. "Geometry as a branch of physics," in *Albert Einstein: Philosopher-Scientist*. Editor, P. A. Schilpp. Library of Living Philosophers, Evanston, Illinois, 1949.

Sommerville, D. M. Y. *The Elements of Non-Euclidean Geometry*. Dover, New York, 1958.

SPECIAL RELATIVITY

Where I am not understood, it shall be concluded that something useful and profound is couched underneath.
Jonathan Swift (1617–1745), Tale of a Tub

NEW IDEAS FOR OLD

Old ideas

Newtonian space and time were public property, which all observers shared in common. Its intervals of space and intervals of time separating events were absolute. They were the same for everybody. One person in an apple orchard would see an apple fall from a tree and take 1 second to drop 5 meters. Another person in motion relative to the tree also would see it drop 5 meters in 1 second, no matter how fast that person moved. Now things have changed. The old Newtonian universe, with its ideas on the fixity of intervals of space and time, is no longer the universe in which we live.

Space-and-time diagrams, displaying events and world lines, were used in the Middle Ages, and there is nothing particularly frightening or difficult about them. Until the beginning of this century they were a convenient graphical way of representing things in motion. Then came the theory of special relativity, and diagrams of this kind acquired a new physical meaning.

New ideas

The theory of special relativity emerged toward the end of the nineteenth century and was brought into final form in 1905 by the genius of Albert Einstein. It has withstood countless tests and is now in everyday use by physicists. Yet even nowadays, when we pause to reflect, the theory is as astonishing as when it first emerged. Relativity requires that we abandon the belief that

intervals of time and space are the same for everybody. We trade in these two old invariants for two new invariants.

The first of the new invariants is the speed of light in empty space denoted by c. The speed (300 000 kilometers per second) is the same everywhere and for everybody and is independent of the state of motion of the observer. In 1887, Albert Michelson and Edward Morley found that the speed of light is the same in all directions on the Earth's surface. This was an unexpected result. The Earth moves at orbital speed $V = 30$ kilometers per second around the Sun, and Michelson and Morley expected to find that the measured speed of light would be $c + V$ and $c - V$ in opposite directions parallel to the Earth's orbital motion. Instead, they found that the speed of light is the same in both directions. The Sun, as we now know, moves at 300 kilometers per second around the center of the Galaxy, and the Galaxy itself moves in the Local Group of galaxies, and the Local Group moves in the Local Supercluster. Despite these motions the speed of light remains the same in all directions. Even at the high particle speeds in high-energy experiments, the speed of light relative to the particles is always c.

The fact that light moves at speed c, and c is a speed limit for all particles moving at high energy, is not in itself greatly astonishing. It is the constancy of c for all observers, even for observers moving in opposite directions at high speed, that astonishes. A body moves past us at, say, half the speed of light, and relative to that body light has the same speed that it has relative to us. Even when the body moves at $0.9c$, or $0.99c$, or $0.999c$, or $0.9999c$, the speed of light relative to it has the invariant value $c = 300\,000$ kilometers per second.

Physicists of the nineteenth century thought that space was just an empty nothing, and electromagnetic radiation such as light could not travel through space unless its emptiness was filled with a medium called the luminiferous ether. Waves of light traveling through the ether were much like waves of sound traveling through the air. The observed speed of sound depends on how fast the observer moves through the air; it is even possible to fly faster than sound waves. It was very puzzling that the luminiferous ether refused to behave like air or any other medium. The discovery of the invariance of the speed of light killed the ether theory and led to the revolutionary realization that space is more than mere emptiness and waves of light are more than mere undulations of an etheric medium. Waves of light became an intricate combination of propagating electromagnetic fields. Space – wedded to time – became an intricate physical structure. We no longer dress space with an ether to give it physical reality, for it has its own reality that makes the speed of light independent of relative motion.

The second invariant is the spacetime interval. Intervals of space and intervals of time by themselves are no longer invariant for all observers. Instead, together they form an invariant spacetime interval:

(spacetime interval)^2

$$= \text{(time interval)}^2 - \text{(space interval)}^2.$$

$$[11.1]$$

Notice that if time is measured in seconds (or years), then distance is measured in light seconds (or light years). We can use units of time in the measurement of space because, according to the first invariant, the speed of light is universal.

All observers, independent of their relative motion, are in complete agreement on the value of the spacetime intervals between events when determined in the way shown in Equation [11.1]. Space and time are fused together to form a unified four-dimensional spacetime continuum. Observers in relative motion to one another have their own private spaces and times, and only spacetime is the public domain that all share in common. This new way of looking at the physical world was stressed by Hermann Minkowski in 1908 when he said, "Henceforth, space by itself and time by itself are doomed to fade away into mere shadows, and only a kind of union of the two will preserve an independent reality."

Two old ideas, the invariance of distances in space and the invariance of periods in time, have gone, and in their place we have two new ideas, the invariance of the speed of light and the invariance of the spacetime interval. Once, space and time were each separately absolute; now only spacetime is absolute, and space and time are its decompositions relative to each observer.

THE STRANGENESS OF SPACETIME

Albert Einstein (1879–1955), once a student of Minkowski, crafted the theory of special relativity and formulated its algebra. In 1908, Minkowski showed that diagrams of space and time are not just convenient mathematical fictions, but actual representations of a four-dimensional physical reality. The universe consists of spacetime and not just space and time.

Lightcones

Spacetime consists of three basic elements: points (events), lines (world lines), and null-geodesics (light rays), as shown in Figure 11.1. Brief happenings are events denoted by points in spacetime diagrams. Things that endure, such as observers, are world lines extending from the past to the future. An observer receives light signals that come in from the past and transmits light signals that go out into the future.

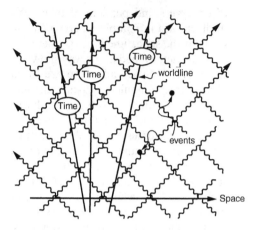

Figure 11.1. The spacetime of special relativity contains points (transitory events), world lines (things that endure in time), and light rays. World lines are strings of events and each event has lightcones.

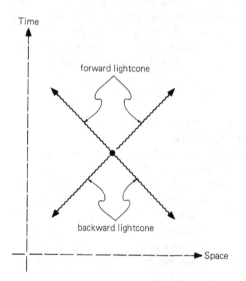

Figure 11.2. A lightcone at a point in spacetime. Light comes in from the past on the backward lightcone and goes out into the future on the forward lightcone.

These light signals come in and go out on the observer's lightcone, as shown in Figures 11.2 and 11.3, and travel in each second a distance of one light second. (1 light second equals 300 000 kilometers.) Each point in spacetime has its backward and forward lightcones, and all are similar.

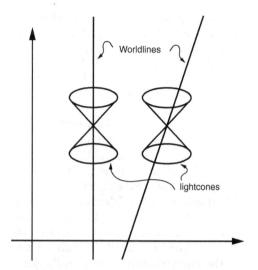

Figure 11.3. All observers, no matter how tilted their world lines, have similar lightcones.

The backward lightcone extends out in space and back in time, the forward lightcone extends out in space and forward in time. Each observer at each world line point has lightcones, and all lightcones are similar, no matter what the inclination of the world line.

All information comes to the observer at speeds less than or equal to the speed of light. Information conveyed by sound waves, for example, travels at much less than the speed of light. The events observed by the observer must therefore lie either inside or on the backward lightcone. The events influenced by the observer must lie either inside or on the forward lightcone. Observers in relative motion at the same point in spacetime always agree that the events inside or on their backward lightcone belong to the past and the events inside or on their forward lightcone belong to the future, but they do not agree on the past–future ordering of all events outside their lightcone.

Many spaces, many times, one spacetime
Observers in relative motion have world lines inclined to one another (Figure 11.4). Thus observers A for Albert and B for Bertha have relative motion, as shown by their inclined world lines. They do not

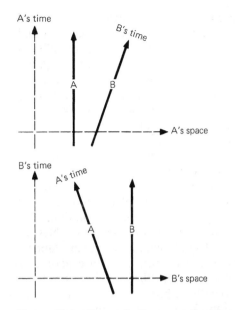

Figure 11.4. We can display a spacetime diagram so that either A's space is shown perpendicular to A's world line or B's space is shown perpendicular to B's world line, but we cannot show simultaneously in a diagram the fact that both spaces are perpendicular to their world lines. This is because the geometry of spacetime is unlike the geometry of a blackboard or a sheet of paper.

share a common space and a common time but have different spaces and different times within a common spacetime. Observer A measures time along his world line (spacetime intervals along his world line are his intervals of time), and his space is perpendicular to his world line. Similarly, observer B measures time along her world line, and her space is perpendicular to her world line.

A's space is perpendicular to his world line; B's space is perpendicular to her world line. Unfortunately, in a single diagram we can only show that space is perpendicular to one world line. This is because the sheets of paper on which we draw our diagrams do not have the geometry of spacetime. Much of the confusion about special relativity is the result of not understanding that the geometry of spacetime is not our familiar Euclidean geometry of space.

A's time is composed of contributions from B's time and space, and B's time is composed of contributions from A's time and space. Similarly, their spaces are composed of contributions from each other's spaces and times. Two events at different places occurring simultaneously (at the same time) in A's space are not simultaneous in B's space, and two events occurring at the same place in A's space, but at different instants of time, do not occur at the same place in B's space.

What then, one might ask, is the true decomposition of spacetime into space and time? There is none. All decompositions into space and time, when made according to the rules, are valid and equally real. Hence the name relativity. The rules are very simple: (i) the speed of light is the same for all observers, and (ii) all spacetime intervals are also the same for all observers. The four-dimensional world of spacetime contains events, world lines, and light rays; each event has its lightcones, and each world line has its space and time. This is the theory of special relativity, and the algebraic details (often without Minkowski's insight) are found in many books on elementary physics.

Invariance and covariance

Quantities that are the same for all observers (like the speed of light and spacetime intervals) are said to be invariant. Laws and equations that are the same for all observers, independent of their relative motion, are said to be covariant.

The Newtonian universe provides us with simple examples of invariance and covariance. The intervals of space and time are the same for everybody and are invariant. If we are on a ship moving at sea, or in a plane flying in the air, and toss a coin in the air, it falls back into the hand exactly as when we are standing on the ground. The laws governing the motion of the coin are the same in the plane and on the ground, they are covariant and independent of relative uniform motion.

Motion at constant velocity is said to be inertial, meaning that no inertial force exists. A passenger with closed eyes in an automobile that travels at constant velocity

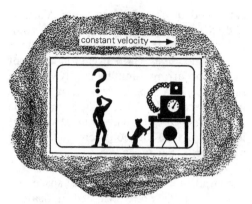

Figure 11.5. In a windowless laboratory moving at constant velocity the laws are covariant, which means they are independent of the velocity. No experiment can be performed in the laboratory that will determine the velocity. The laboratory possesses inertial motion, which means no inertial forces are present and the motion is unaccelerated.

cannot tell how fast the automobile travels because the motion is inertial. Only when the speed increases or decreases, or the driver steers to the left or right, can the passenger experience inertial forces and know that the velocity is changing. In the Newtonian universe all velocities are relative and all accelerations are absolute.

In special relativity the laws of nature are covariant in a wide sense. Consider a windowless laboratory that moves freely and at constant velocity. The motion is inertial. If, in this laboratory, we measure the speed of light, or perform any experiment whatever, we get always the same result, independent of the velocity of the laboratory (see Figure 11.5). No experiment can be devised that will tell us the velocity of the laboratory. The laws of nature are covariant and the same for observers in all laboratories moving at constant velocity.

In special relativity the laws of nature are covariant if the observer's velocity is constant (the motion is inertial). Despite the physical structure of spacetime, with its speed limit that is the same for everybody, we are still not too far removed from Newtonian ideas. Velocity is relative but acceleration is absolute. An accelerated observer

experiences a force known as an inertial force, which is peculiar to the observer and is not shared by other observers of different motion. The laws of nature in special relativity, as in the Newtonian universe, are not covariant for accelerated observers, only for a special class of observers whose motion is inertial (of constant velocity). We shall see in the next chapter that the laws in general relativity are covariant for the more general class of observers who are in free fall.

Despite its constant-velocity constraint, physicists use special relativity all the time for accelerated (noninertial) motion. The trick is quite simple and comes from Newtonian mechanics. Suppose we are in a laboratory that is accelerated. In our experiments we obtain results that are not strictly in accord with the predictions of our covariant equations. We do not throw away the results and give up in despair. (Remember, a laboratory on the Earth's surface is in noninertial motion because of the Earth's rotation and revolution about the Sun.) Instead, we imagine ourselves moving inertially (at constant velocity), while studying the events occurring in the accelerated laboratory. The covariant laws, which now hold in our inertial state, tell us what are the true results and what are the corrections to be made to the observations obtained in the accelerated laboratory. Physicists constantly use imaginary inertial motion to understand and correct the results obtained in accelerated motion.

TRAVELS IN SPACE AND TIME
The golden rule
The geometry of spacetime is not the same as that of a sheet of paper, and generally the shortest distance between two events is not a straight line.

Speeds close to the speed of light
Spacetime reveals its most interesting properties when bodies have relative speeds close to the speed of light. Most people find the results surprising and incomprehensible. How is it possible for a space traveler to journey in an ordinary lifetime to a galaxy

millions of light years away? How is it possible that a space traveler can return to Earth and find that his twin sister has grown old while he is still young? The answer is actually quite simple and springs from the fact that the geometry of spacetime is not the same as the geometry of the blackboard or the sheets of paper on which we draw spacetime diagrams. All the confusion attending special relativity comes from the simple fact that distances between points in spacetime are not measured in the same way as distances between points on a sheet of paper.

The distance between two points on a sheet of paper ruled with x and y Cartesian coordinates is given by the Pythagorean formula:

$$(\text{space interval})^2 = (x \text{ interval})^2$$
$$+ (y \text{ interval})^2,$$

and this can be extended to the three-dimensional space of x, y, and z, as shown by Descartes:

$$(\text{space interval})^2$$
$$= (x \text{ interval})^2 + (y \text{ interval})^2$$
$$+ (z \text{ interval})^2. \qquad [11.2]$$

If we go to four-dimensional space, in which the extra dimension is time, we might expect distances in this space-and-time to be measured in the same way, according to the Pythagorean formula:

$$(\text{space-and-time interval})^2$$
$$= (\text{time interval})^2 + (\text{space interval})^2, \qquad [11.3]$$

in which the space interval is given by Equation [11.2], and distances in space are measured in light-travel time. This would give us the space-and-time interval in terms of Newtonian space and time, but it most certainly does not give the spacetime interval in the world of relativity. The spacetime interval that is the same for all observers,

independent of their velocity, is

$$(\text{spacetime interval})^2$$
$$= (\text{time interval})^2 - (\text{space interval})^2. \qquad [11.4]$$

The minus sign (instead of a plus sign), which contradicts our Euclidean intuition, accounts for all the bewildering results of special relativity.

We are accustomed to the idea that the shortest distance between two points is a straight line. This is true in ordinary space but not in spacetime. To understand why this is so let us consider the lightcones that are the paths on which light rays travel. In 1 second of time light travels 1 light second in space. For a light ray this means

$$\text{interval of time} = \text{interval of space}, \quad [11.5]$$

and therefore the spacetime interval between any two events on a lightcone, given in Equation [11.4], is always zero. Consider a star at a distance of 1000 light years. Light from this star hurries to us at great speed across a wide gulf of space, and after 1000 years it enters the eye. No one can disagree that the light has traversed immense intervals of space and time. Yet the spacetime interval between the eye and the star is zero! Spacetime distances along the backward lightcone to all events that send us signals, and along the forward lightcones to all events that receive our signals, are always zero. This may seem bizarre, but it explains why the speed of light is invariant and forms an upper limit to all material motion.

Consider a straight world line connecting two events labeled a and b. Also consider the alternative path acb (i.e., a to c and then c to b) forming the lightcones that intersect a and b, as shown in Figure 11.6. The length of acb, as drawn on paper, is longer than the straight line ab. But in spacetime the distance acb is zero. The longest spacetime distance between a and b is measured along the straight line that passes through a and b; all other paths between a and b that bend out toward the lightcones are of

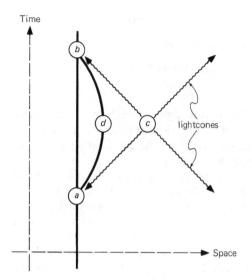

Figure 11.6. The straight world line between *a* and *b* is the longest distance. The lightcone distance *acb* (i.e., *a* to *c* and then *c* to *b*) is of zero length. The closer the bent world line *adb* approaches the lightcones *acb*, the shorter its spacetime length. The observer's time is measured along the observer's world line, and the time taken to go from *a* to *b* is the length of the world line. The longest time is along the straight segment, and the time taken gets shorter as the segment *adb* approaches the lightcones. This explains the twin paradox. Different routes in spacetime between two points result in different ages.

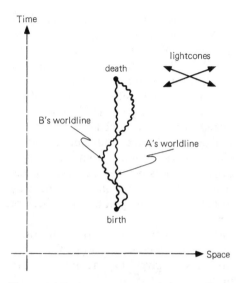

Figure 11.7. The crinkled world lines of twins A and B. As they travel around, their world lines follow different paths in spacetime and are slightly unequal in length. The twins therefore have slightly different ages whenever they meet. The difference in age becomes large if one twin takes a trip at high speed in a spaceship.

shorter length. Spacetime contradicts common sense and this leads to the so-called twin paradox.

Twin paradox

The time that is peculiar to a body is measured along its world line and its age is the length of the world line from its beginning. But bent (or curved) world lines, connecting any two points in spacetime, are shorter than straight world lines. Thus a person going from *a* to *b* via a bent world line, such as *adb*, arrives sooner than another person who goes via a straight world line. Both persons measure time with a wristwatch, or other physical means, such as the number of heartbeats, and the lapsed time in each case is equal to the length of the world line measured in spacetime.

Twins A and B are born together and their world lines begin at the same event. But for the rest of their lives their world lines will follow different paths and whenever they meet they will have slightly different ages as measured by the clocks they carry (see Figure 11.7). Let us suppose that A stays at home all his life, and that B travels a great deal by plane during her life and has a more crinkled world line. At a reunion to celebrate their eightieth birthdays, her world line is slightly shorter than his, and she has therefore lived a slightly shorter life. The difference in ages in this example is very small, perhaps as small as 100 microseconds.

Now suppose that B travels in a spaceship at extremely high speed and eventually returns to Earth (see Figure 11.8). Before the journey, the twins A and B were the same age, but when B returns, it is immediately apparent that A is older than B. "You have cheated!" says A; "you have gained and I have lost time!" "No I haven't," says B; "I am younger than you,

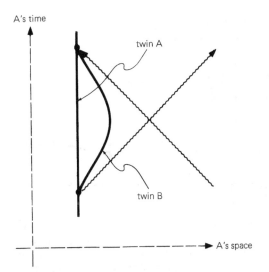

Figure 11.8. Twin B takes a long trip in a fast spaceship while twin A stays at home.

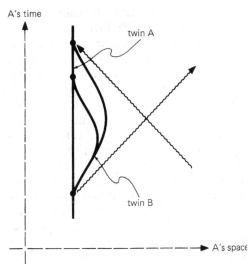

Figure 11.9. Twin B takes two journeys of unequal length. In both journeys the accelerations are the same. But the age difference experienced by the twins increases with the length of the journey, thus showing that it has nothing to do with the acceleration.

my heart has beat fewer times than yours, you have brushed your teeth more times and read more books than I have."

The space-traveling twin B experiences periods of acceleration (including deceleration) in her travels, but acceleration by itself cannot be the cause of the asymmetric aging of the twins. If B takes a second, longer journey at the same speed as the first journey (see Figure 11.9), with similar periods of acceleration as previously, the difference in the ages of A and B is greater. Twin B ages less because of travel near the lightcones; and the greater the distance traveled, the greater the age difference on her return.

Let us view A's world line in B's outward-bound space and time, as shown in Figure 11.10. Twin A is now seen traveling away close to the lightcone. Surely this proves that A should be younger than B? But at half-time (in B's time) B must start traveling toward A, and in order to catch up with A, B must travel even faster and closer to the lightcone. This makes B younger than A when they finally meet. The calculated age difference in either A's space and time or B's space and time is the same.

The technical problems of space travel at speeds close to that of light are formidable,

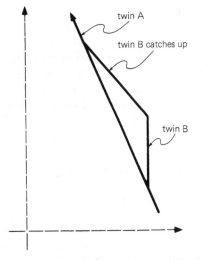

Figure 11.10. The outward-bound trip of B as viewed in B's space. B sees A moving away. To catch up with A, B must later travel even faster than A, and hence at their reunion B is again younger than A.

and will not be solved in the immediate future. Probably, human beings will not experience in any significant way the bizarre phenomenon of asymmetric aging for many centuries to come. But asymmetric aging occurs repeatedly with high-energy particles. A subatomic particle known as the muon decays in a millionth of a second into other particles, and this intrinsic lifetime is measured along its world line. When a muon moves relativistically (close to the speed of light) it travels a distance much greater than 300 meters before decaying. In its own time, measured along its world line, it decays in a millionth of a second; but to us, who are not traveling with it, it is seen to decay much more slowly. The decay time is multiplied by $1/(1 - V^2/c^2)^{1/2}$, where V is the speed of the muon in the laboratory. Thus, if $V = 0.9998c$, then the observed decay time is increased by 50 and the muon travels 15 kilometers. Similarly, if space travelers journey away and then return at this high speed, on their arrival on Earth 1 year later in their time, they would find that everybody on Earth had aged by 50 years.

REFLECTIONS

1 *The algebra of special relativity is based on the two invariants that we have discussed: (i) the invariance, for all observers, of the speed of light that allows us to express distances linearly in light-travel time; (ii) the invariance, for all observers, of the spacetime interval (Equation 11.4):*

(spacetime interval)^2

= (time interval)^2 − (space interval)^2.

Consider observers A and B in relative motion. Although they have different spaces and times they always agree on the value of the spacetime interval separating any two events. Thus, between any two events in spacetime,

(A's time interval)^2 − (A's space interval)^2

= (B's time interval)^2

− (B's space interval)^2.   [11.6]

In Figure 11.11, the intervals of space and time between events a and b (i.e., events at

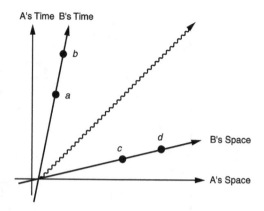

Figure 11.11. This diagram shows the space and time frames of A and B at the instant B passes A at relative velocity V. Notice that in spacetime B's time and space axes appear tilted toward the lightcone. B's space interval Δx_B between events a and b is zero, and B's time interval Δt_B between events c and d is also zero.

the same place and different time in frame B) are:

A's time interval $= \Delta t_A$

A's space interval $= \Delta x_A = \dfrac{V}{c}\Delta t_A$

B's time interval $= \Delta t_B$

B's space interval $= \Delta x_B = 0$

where Δ denotes a small interval, space intervals are expressed in light-travel time (on dividing by c), and the space interval Δx_B in B's frame is zero because a and b occur at the same place in B's frame. From these expressions and Equation [11.6], we find

$$(\Delta t_A)^2 - \left(\frac{V}{c}\Delta t_A\right)^2 = (\Delta t_B)^2, \qquad [11.7]$$

and the relation between intervals of A's time Δt_A and intervals of B's time Δt_B between events a and b is

$$\Delta t_B = \Delta t_A (1 - V^2/c^2)^{1/2}. \qquad [11.8]$$

In Figure 11.11, the intervals between events c and d (i.e., events at the same time and different place in frame B) are

A's time interval $= \Delta t_A = \dfrac{V}{c} \Delta x_A$

A's space interval $= \Delta x_A$

B's time interval $= \Delta t_B = 0$

B's space interval $= \Delta x_B$

where time interval Δt_B is zero because c and d are synchronous (exist at the same time) in B's frame. From these expressions and Equation [11.6] we find

$$\left(\frac{V}{c}\Delta x_A\right)^2 - (\Delta x_A)^2 = -(\Delta x_B)^2, \qquad [11.9]$$

and the relation between A's space and B's space for simultaneous events in B's space is

$$\Delta x_B = \Delta x_A (1 - V^2/c^2)^{1/2}. \qquad [11.10]$$

Equations such as [11.8] and [11.10] are often expressed in the form

$$\Delta t_A = \gamma \Delta t_B, \qquad [11.11]$$

$$\Delta x_A = \gamma \Delta x_B, \qquad [11.12]$$

where

$$\gamma = (1 - V^2/c^2)^{-1/2}. \qquad [11.13]$$

These are the abbreviated Lorentz transformations applying to events that are either at the same place (Equation 11.11) or at the same time (Equation 11.12) in a moving system. Equation [11.11] tells us what B's unit of time is in A's frame, which we use in the twin paradox, and Equation [11.12], sometimes referred to as the FitzGerald contraction, tells us what B's unit of length is in A's frame.

• *"A sphere moving with great speed appears as a flattened ellipsoid which collapses into a disk if light velocity is reached.... An observer attached to the sphere perceives the sphere as a sphere, while an observer relative to whom the sphere is in motion perceives the same sphere as an ellipsoid because his space and time measurements differ from those of the other observer who is attached to the sphere" (Cornelius Lanczos,* Albert Einstein and the Cosmic World Order*).*

• *When $V/c = 0.6$ and therefore $\gamma = 1.25$, then $\Delta t_B = 0.8\Delta t_A$, $\Delta x_B = 0.8\Delta x_A$; and when $V/c = 0.8$ and therefore $\gamma = 1.66$, then $\Delta t_B = 0.6\Delta t_A$ and $\Delta x_B = 0.6\Delta x_A$. If V/c is small, then approximately*

$$\gamma = 1 + \frac{V^2}{2c^2}. \qquad [11.14]$$

We can write $V/c = 1 - \varepsilon$, and when V/c is close to unity, ε is small, and we find approximately

$$\gamma = (2\varepsilon)^{-1/2}. \qquad [11.15]$$

Thus when $V/c = 0.98$, $\varepsilon = 0.02$, then $\gamma = 5$.

• *Transformations of space and time between systems in relative motion are known as Lorentz transformations and usually involve the term γ. We should note that a stationary particle of mass m, has a mass γm when moving at a relative velocity V. Energy has mass given by $E = mc^2$, and the energy of a particle of mass m moving at speed V is $m\gamma c^2$. When V is small compared with c, the energy of a nonrelativistic particle, according to Equation [11.14], becomes $mc^2 + \frac{1}{2}mV^2$, in which the first term is the rest-mass energy and the second term is the kinetic energy.*

2 *Figure 11.12 illustrates the twin paradox. Twin A stays at home; twin B travels to a distant star at velocity V, and immediately returns at velocity V. The intervals between events a and d are*

A's time interval $= \frac{1}{2}\Delta t_A$

A's space interval $= \frac{1}{2}\Delta x_A = \frac{1}{2}\dfrac{V}{c}\Delta t_A$

B's time interval $= \frac{1}{2}\Delta t_B$

B's space interval $= \frac{1}{2}\Delta x_B = 0$

and similar intervals apply to the return journey between d and b. Hence, from Equation [11.6], for the outward journey

$$\left(\frac{1}{2}\Delta t_A\right)^2 - \left(\frac{1}{2}\frac{V}{c}\Delta t_A\right)^2 = \left(\frac{1}{2}\Delta t_B\right)^2,$$

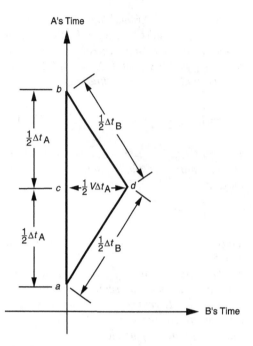

A's Time

b

$\frac{1}{2}\Delta t_A$

$\frac{1}{2}\Delta t_B$

c $-\frac{1}{2}V\Delta t_A$ d

$\frac{1}{2}\Delta t_A$

$\frac{1}{2}\Delta t_B$

a

B's Time

Figure 11.12. Twin A stays at home and has a straight world line from *a* to *b* (*acb*) and experiences a total interval of time Δt_A. Twin B travels at speed *V* to a star at distance $L = \frac{1}{2}V\Delta t_A$ and immediately returns at speed *V*. B's bent world line is *adb* and B experiences a total interval of $\Delta t_B = (1 - V^2/c^2)^{1/2}$.

and similarly for the return journey, thus giving

$$\Delta t_A = \gamma \Delta t_B. \qquad [11.16]$$

Thus if $V/c = 0.1$, then $\gamma = 1.005$, and if A's interval of time is 50 years, then B's interval of time is 49.75 years, and B is 3 months younger than A on her return from a star at distance 2.5 light years. If $V/c = 0.9998$, and $\varepsilon = 0.0002$, then $\gamma = 50$, and if A's interval of time is again 50 years, then B's interval is 1 year, and B is 49 years younger than A on returning from a star at a distance of almost 25 light years.

3   *Twin A stays at home on Earth, twin B travels away from Earth at a constant acceleration of 1g. This acceleration means the inertial force on a body in the spaceship is equal to its weight on the Earth's surface. Halfway to her destination, B decelerates at*

1g, and arrives at a distant star. The distance traveled in B's time is shown in Figure 11.13, curve a. Because B travels most of the time close to the speed of light, the distance traveled in light-travel time is approximately A's time.

According to Figure 11.13, a traveler can journey 10 billion (10^{10}) light years in slightly less than 50 years. But out in the vastness of the expanding universe the theory of special relativity is inadequate and we must use the theory of general relativity. If we adopt a simple relativistic model of the expanding universe known as the Einstein–de Sitter model, the more relevant theory shows that the distance traveled close to the speed of light is

$$L = 3\alpha c t_0 (1 - \alpha^{-1/3}), \qquad [11.17]$$

where $\alpha = t/t_0$, t_0 is the age of the universe when the journey begins, and t is the age of the universe when the journey ends. If we assume that the universe is 10 billion years old when the journey begins, and the space traveler journeys for 10 billion more years (i.e., $\alpha = 2$), we find that the actual distance traveled is 12.4 billion light years. Figure 11.13 can still be used, but for the abscissa coordinates "distance traveled" we must use L from Equation [11.17]. Hence in B's time the journey to the "edge of the universe" takes slightly less than 50 years, and B travels 12.4 billion light years (more than 10 billion light years because the universe is expanding).

4   There was a young lady named Bright,
   Who could travel much faster than light.
   She set out one day
   In the relative way
   And came back the previous night.
   Anonymous

If faster-than-light travel were possible, this limerick could be true. We could take an interstellar journey and return to Earth at any desired time in the past or future. But according to current theory nothing physical can travel faster than the speed of light. If faster-than-light communication were possible, we could perform "impossible" things, such as send information back into the past that

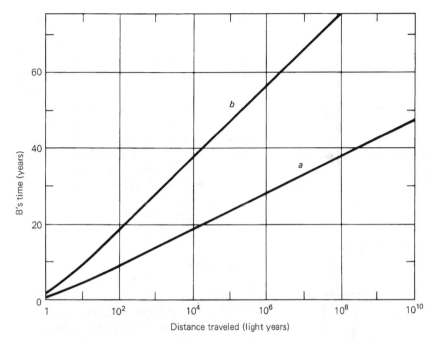

Figure 11.13. B accelerates away from Earth at 1g, and halfway, decelerates at 1g. When B is once again stationary relative to A, the total distance traveled in B's time is shown by curve a. Most of the time A travels very close to the speed of light and the distance traveled is approximately A's time multiplied by the speed of light. If the distance traveled is 1 million light years, then A's time is 1 million years. If B returns to Earth by accelerating and decelerating as before, the trip takes twice as long, as shown by curve b. In 60 years, a space traveler can journey to the Andromeda galaxy (distance 2 million light years) and back, and the Earth will be 4 million years older on return.

would change present conditions; or if faster-than-light travel were possible, we could journey into the past and do "impossible" things, such as prevent our parents from meeting. Fortunately for us, but unfortunately for science fiction writers, Wellsian time travel is itself impossible, as explained in Chapter 9.

• To keep the argument simple, let us go to the extreme and suppose that M-rays (M for mystic) travel at infinite speed. In Figure 11.14, people on Earth and on a distant planet in another solar system communicate with each other by M-rays. The distant planet is at distance L measured in Earth's space.

In diagram (a), the distant planet is stationary ($V = 0$) relative to Earth and two-way communication occurs instantaneously without delay.

In diagram (c), the distant planet is shown moving toward the Earth at velocity V. Signals sent from Earth at time $t = 0$ travel instantaneously in Earth space and time. The return signals travel instantaneously in the planet's time and arrive at the Earth in the future at time LV/c^2. If $L/c = 10^4$ light years and $V/c = 10^{-3}$, we could send messages 10 years into the future. By traveling at infinite speed to a distant planet, spacetime travelers could return to Earth in the future. A more realistic way of traveling into Earth's future is by means of the time dilation effect discussed in the twin paradox. By traveling close to the speed of light, the traveler arrives back on Earth at time $t = L/c = 10^4$ years.

In diagram (b), the distant planet is shown moving away from Earth at velocity V.

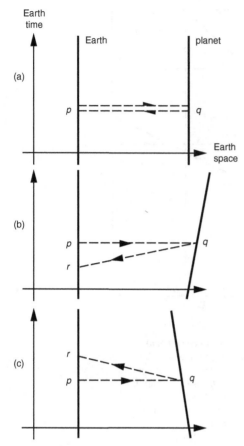

Figure 11.14. People on Earth and on a distant planet communicate with each other by M-rays (M for mystic) that travel at infinite speed. In (a) the distant planet is stationary relative to Earth. Signals sent from Earth at instant p arrive instantaneously in Earth time at the planet at instant q and their return is also instantaneous. In (b) the planet is shown moving away from Earth. Signals sent from Earth at instant p arrive at the planet at instant q, and the return signals travel instantaneously in the planet's time and arrive on Earth at instant r, which lies in Earth's past. In diagram (c) the distant planet is shown moving toward the Earth. Signals sent from Earth at instant p travel infinitely rapidly and arrive at the planet at instant q, and the return signals travel instantaneously in the planet's time and arrive on Earth at instant r, which lies in Earth's future. If M-rays existed, we could communicate directly with our descendants and ancestors by using distant mirrors moving away and toward the Earth.

Signals sent from Earth at time $t = 0$ travel instantaneously in Earth's space and time. The return signals travel instantaneously in the planet's time and arrive at the Earth in the past at time $-LV/c^2$. With the previous numbers, $V/c = 10^{-3}$, $L/c = 10^4$, we could send signals back 10 years into the past. By traveling at infinite speed, spacetime travelers could return to Earth in the past. Travel faster than light is impossible according to the laws of physics, and therefore causality is preserved and the "impossible" avoided. Notice that by traveling close to the speed of light, one cannot travel into the past by means of the twin paradox.

PROJECTS

1   A stays at home; B travels for 1 year in B's time to the planet of a distant star at 0.8 times the speed of light, stays for 1 year on the planet, and then returns home at 0.8 times the speed of light. B arrives back 3 years older. How much has A aged while B was away? How far is the planet?

2   Twins A and B both live for 100 years in their own time. At birth B is taken on a journey to a planet of a distant star at 0.9998 times the speed of light. The planet is unfit for habitation and the spaceship returns immediately at the same speed as on the outward journey. A remains all the time on Earth. How far is the planet in light years if B returns in the year that A dies? How far is the planet if B returns in the year that B dies?

3   Traveling with an acceleration and deceleration of 1g, how long in your time will it take to reach the center of the Galaxy, and how much older will the Earth be when you return and step out of your spaceship?

FURTHER READING

Einstein, A. *Relativity: The Special and the General Theory*. Methuen, London, 1920.

Ellis, G. F. R. and Williams, R. M. *Flat and Curved Space-Times*. Clarendon Press, Oxford, 1988.

French, A. P. *Special Relativity*. Norton, New York, 1968. A clear mathematical treatment of special relativity.

Gardner, M. *The Relativity Explosion*. Random House, New York, 1976.

Holton, G. "On the origin of the special theory of relativity." *American Journal of Physics* 28, 627 (October 1960).

Kaufmann, W. J. "Traveling near the speed of light." *Mercury* (January–February 1976).

SOURCES

Born, M. *Einstein's Theory of Relativity*. 1924. Reprint: Dover Publications, New York, 1962.

Lanczos, C. *Albert Einstein and the Cosmic World Order*. Wiley, New York, 1965.

Rindler, W. *Introduction to Special Relativity*. Clarendon Press, Oxford, 1995.

Taylor, E. F. and Wheeler, J. A. *Spacetime Physics*. W. H. Freeman, San Francisco, 1966.

12 GENERAL RELATIVITY

It is as if a wall which separated us from the truth has collapsed. Wider expanses and greater depths are now exposed to the searching eye of knowledge, regions of which we had not even a presentiment. It has brought us much nearer to grasping the plan that underlies all physical happening.
Herman Weyl (1885–1955), Space, Time, and Matter

PRINCIPLE OF EQUIVALENCE

Gravitational and inertial forces produce effects that are indistinguishable – this is the principle of equivalence. It serves as an essential stepping-stone to the theory of general relativity, and makes a basic connection between motion and gravity. It leads to a second stepping-stone: the realization that geometry and gravity have much in common. Then, in an inspired leap across the gulf of non-Euclidean geometry, we enter a country into which comparatively few explorers have ventured. No person entering the third millenium may claim to have a liberal education who has not glimpsed, however briefly, the universe of general relativity.

An inertial force, such as centrifugal force, exists when a body is accelerated. We recall from Newtonian theory that when a body is in free fall, and hence moves freely in space under the influence of gravity, it follows a path of such a kind that the sum of the inertial and gravitational forces is zero. With items of knowledge such as these, sufficient to land men on the Moon, we have made our first step toward the theory of general relativity.

We begin by considering an imaginary laboratory that is out in space. It is equipped with scientific apparatus, and the experimenters conduct various investigations. The laboratory has no windows and the experimenters cannot look outside to see the external world.

The laboratory is out in space and far from the nearest star where gravity is virtually zero (see Figure 12.2). It moves freely, and because there is nothing to accelerate it, the inertial force experienced in the laboratory is zero. This kind of motion, free and unaccelerated, is known as inertial motion, and the laboratory is said to be in an inertial state. The experimenters perform tests designed to detect acceleration and announce that the laboratory is unaccelerated and in an inertial state. After a period of time the laboratory approaches a star, swings around the star in a curved orbit, and then moves away (see Figure 12.3).

Figure 12.1. "Let us then take a leap over a precipice so that we may contemplate Nature undisturbed" (Arthur Eddington, *The Nature of the Physical World*, 1928).

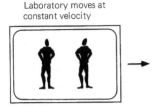

Laboratory moves at constant velocity

Figure 12.2. A windowless laboratory moves freely and at constant velocity out in space far from any star. This is an inertial system and the experimenters inside cannot determine its velocity.

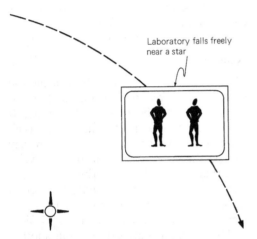

Laboratory falls freely near a star

Figure 12.3. The laboratory now falls freely in the vicinity of a star. The inertial force caused by the acceleration cancels the force of gravity, and the experimenters think they still have inertial motion and are moving at constant velocity.

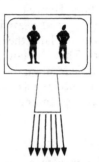

Figure 12.4. Objects in a laboratory at rest on the Earth's surface experience a gravitational force called weight. Objects in an accelerated laboratory out in space experience an inertial force. Experimenters in the laboratories cannot distinguish between the two forces.

While this happens, the experimenters, unaware of the nearby star, continue to perform tests and continue to announce that the laboratory is unaccelerated and in an inertial state. This is because the laboratory follows a free-fall orbit and the inertial force resulting from its acceleration exactly cancels the gravitational force of the nearby star. The principle of equivalence in a windowless laboratory is in effect a postulate of impotence: it is impossible in a windowless laboratory to distinguish between gravitational and inertial forces.

Our inability to distinguish between inertial and gravitational forces is also illustrated in the following way (see Figure 12.4). We suppose the windowless laboratory in space is accelerated by an applied force, such as the thrust of a rocket engine. Everything inside the laboratory now experiences an inertial force owing to the acceleration. With a piece of apparatus, such as an ordinary pendulum, the experimenters are able to measure the acceleration. Let the thrust of the rocket engine be adjusted so that the acceleration equals the value of g at the Earth's surface; hence the velocity increases 9.8 meters per second every second. The force now acting on the experimenters and their apparatus is the same in magnitude as when the laboratory is at rest on the surface of the Earth. The principle of equivalence states that the experimenters cannot determine, with experimental apparatus of any kind, if the laboratory is accelerating in space or at rest on the Earth's surface. In one case the force is inertial and in the other gravitational, and the experimenters are unable to distinguish between the two.

According to special relativity, the laws of nature and the equations of physics are the same in all inertial systems (that is, in systems that are not accelerated). In all laboratories moving at arbitrary but constant velocities, experimenters obtain identical results when performing identical tests. But we have just seen that these experimenters are unable to tell if their laboratories are inertial or free falling. Hence the experimenters are free to use special relativity in free-falling systems to explain the results of their experiments. That the laws of nature are the same in inertial and free-falling systems, and that special relativity can be used in both, indicates an amazing state of affairs in nature.

A CLOSER LOOK

The principle of equivalence can be broken down into what Robert Dicke at Princeton University has called the weak and strong principles. The weak principle of equivalence can be traced back to the Middle Ages; it lies at the heart of Newtonian theory, and we shall refer to it as the Newtonian principle of equivalence. The strong principle was introduced by Einstein in 1911, and we shall refer to it as the Einstein principle of equivalence.

Newtonian principle of equivalence

The Newtonian form of the principle of equivalence states that the trajectory followed by a small body in free fall is independent of the mass of the body. We see this idea emerging in the late Middle Ages in the work of the scholars of Merton College and the University of Paris, and in the demonstration by Simon Stevinus that bodies of unequal weight, when dropped, reach the ground in equal time. Galileo rolled balls of different weights down an inclined plane and observed that they accelerate in the same way. Newton experimented with pendulums of equal length but different weights and confirmed that they swing with the same period. For a body in free fall, Newton's equation of motion is

$$\text{mass} \times \text{acceleration} = \text{gravitational force}, \quad [12.1]$$

where the mass is that of the accelerated body. The gravitational force acting on the body is proportional to its mass, and we have

$$\text{mass} \times \text{gravity} = \text{gravitational force}, \quad [12.2]$$

where gravity (or gravitational field) is the force acting on a unit of mass. The mass of the moving body is on both sides of the equation of motion (Equation 12.1) and can therefore be canceled:

$$\text{acceleration} = \text{gravity}. \quad [12.3]$$

This is the Newtonian equation of motion for a body in free fall and we see that it is independent of the mass of the body. It shows why bodies of different masses accelerate in the same way and why, when they start at the same place with the same velocity, they follow identical orbits. The acceleration of 9.8 meters a second every second (equal to g, the gravity at the Earth's surface) is an illustration of Equation [12.3].

These thoughts are expressed in another way. The mass of a body acts in two ways: it has an inertial property, and is affected by gravity. We may say a body has both an inertial mass and a gravitational mass. For a body in free fall,

$$\text{inertial mass} \times \text{acceleration}$$
$$= \text{gravitational force},$$

and

$$\text{gravitational mass} \times \text{gravity}$$
$$= \text{gravitational force},$$

and hence

$$\text{inertial mass} = \text{gravitational mass}, \quad [12.4]$$

for all bodies in free fall. The Newtonian principle of equivalence, enshrined in Equation [12.3], can be interpreted to mean that the inertial and gravitational masses are equal. Equation [12.1] thereby becomes Equation [12.3] because of Equation [12.4].

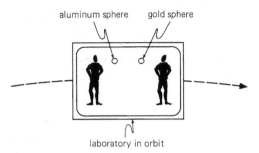

aluminum sphere   gold sphere

laboratory in orbit

Figure 12.5. Two spheres float side by side in a space vehicle orbiting the Earth. One sphere is made of gold and the other of aluminum, yet they fall freely about the Earth in identical orbits.

Equality of inertial and gravitational masses is an expression of the fact that inertial and gravitational forces cancel in a free-falling laboratory, and that also the inertial force experienced in an accelerated laboratory is indistinguishable from the gravitational force experienced in a laboratory resting on the surface of a planet.

Astronauts in a spaceship orbiting the Earth are in free fall and follow the same orbit as their spaceship. All objects inside their spaceship are also in free fall and follow the same orbit. Consider, for example, two spheres, one of gold and the one of aluminum (see Figure 12.5). When placed side by side in the spaceship they float stationary relative to each other because they follow the same free-fall orbit. If their diameters are equal, the gold sphere has a mass slightly more than seven times that of the aluminum sphere, and yet both fall freely in exactly the same way.

The Newtonian principle of equivalence causes no surprise to all those accustomed to using the equation of motion in a gravitational field. Yet despite our indifference bred of long familiarity, on deep reflection, the principle still surprises us. The nucleus of an atom has a positive charge that produces an intense electric field inside the atom. An electric field has energy, and energy has mass, and hence atomic electric fields have mass. In the case of gold, the atomic electric fields have a mass approximately 0.5 percent of the total mass. Thus 1 part in 200 of the weight of gold is due solely to the weight of atomic

electric fields. The nucleus of an atom of aluminum has a smaller positive charge, and the weight of the atomic electric fields of aluminum is a much smaller fraction of the total weight. Yet we find that gold and aluminum bodies behave in exactly the same way in free fall. Electric fields are therefore subject to the same inertial and gravitational forces as all other forms of mass, and the gravitational and inertial masses of electric fields are exactly equal. Consider also two bodies of the same material floating side by side in free fall. When one of the bodies is heated, while the other remains cold, we observe that both continue to float side by side. Heat energy has mass, and because the hot and cold bodies follow similar free-fall orbits, we conclude that the mass associated with heat energy is subject to the same gravitational and inertial forces as all other forms of mass. Considerations of this kind lead us to the conclusion that all forms of energy – electromagnetic, thermal, chemical, nuclear, and so forth – have a mass that obeys the Newtonian principle of equivalence.

Newton experimented with pendulums and found that their gravitational and inertial masses are equal to less than 1 part in 1000 (see Figure 12.6). Friedrich Bessel,

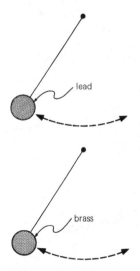

lead

brass

Figure 12.6. Two pendulums of the same length and made of different materials have equal periods of oscillation.

who was the first to succeed in measuring the parallax of a star, showed in 1827 that the gravitational and inertial masses of pendulums are equal to less than 2 parts in 100 000. The Hungarian nobleman Roland von Eötvös greatly improved on this result in 1890 by using a torsion balance to measure forces, and found for different substances – such as snakewood and platinum – that their gravitational and inertial masses are equal to a few parts in 1 billion. In more recent years the Eötvös experiment has been greatly refined, and results show that the gravitational and inertial masses of gold and aluminum are equal to a few parts in 1 trillion.

Einstein principle of equivalence

The Einstein form of the principle of equivalence states that inertial and free-falling systems are entirely equivalent. In inertial and free-falling laboratories there are no experiments of any kind capable of distinguishing between inertial and free-falling motion. The Einstein principle of equivalence declares that the acceleration of a free-falling laboratory cancels completely the effect of gravity, not only dynamically, as in the weaker form of the principle, but also in all conceivable physical experiments in every branch of science. Hence, special relativity, and not just Newtonian mechanics, may be used in free-falling systems as well as in inertial systems, and this is the essence of the principle of equivalence in its modern or strong form.

GEOMETRY AND GRAVITY

Curved surfaces are analogous to gravity

If the principle of equivalence is the first stepping-stone to general relativity, the second is the realization that geometry and gravity have much in common. To illustrate the similarity of geometric curvature and gravity, let us consider a large rubber sheet that is stretched and initially flat. The curvature of the sheet is everywhere zero and is like the flat spacetime that exists far from a star where gravity is practically zero and a laboratory moves inertially at constant velocity. If we roll a small ball, such as a ball bearing, on the surface of the sheet, it will also move inertially at constant velocity (we ignore friction). The ball bearing follows a straight line at constant speed, just like a freely moving laboratory far from a star. In the center of the sheet we now place a heavy ball that produces a large depression, as shown in Figure 12.7. Far from the central body the surface of the sheet is almost flat and a ball bearing follows a path that is almost straight. This situation resembles the almost-flat spacetime that exists far from a star. Close to the central body the curvature of the surface is large and a ball bearing in motion on the surface accelerates in a way analogous to the acceleration of a body in the vicinity of a star. By altering

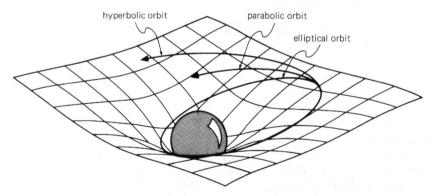

Figure 12.7. A horizontal, stretched rubber sheet is depressed by a heavy spherical body. The curvature of the sheet mimics the effect of gravity, and a ball bearing follows an orbit that is either elliptical, parabolic, or hyperbolic.

the initial speed of the ball bearing we can make it describe elliptical-like, parabolic-like, and hyperbolic-like orbits about the central body. The ball bearing moves in a way that is similar to a free-falling laboratory, and the curvature of the surface mimics the properties of gravity. Where curvature of the surface is zero, and where gravity far from a star is practically zero, the ball bearing and the laboratory move inertially; where curvature is not zero, and where gravity exists, the ball bearing and the laboratory follow similar curved orbits.

Bodies moving inertially have constant velocities and straight world lines in the flat Minkowski spacetime of special relativity. Bodies in free fall in the same spacetime have curved world lines. We have seen that experimenters in a free-falling laboratory (having no windows) do not know that their laboratory is accelerating and think it has a straight world line. By peeping through a hole cut in the wall of the laboratory, the experimenters suddenly become aware of the presence of a nearby star. This discovery reveals that the laboratory is actually accelerating, and has therefore a curved world line in the flat spacetime of special relativity – the spacetime that the experimenters use inside the laboratory. The experimenters might wonder if perhaps there exists some grand theory of gravity that reduces locally to special relativity for all systems in a state of free fall.

What might this grand theory be? If we think about it long enough, we might stumble on the idea that all free-falling bodies have in fact straight world lines, just as inertial bodies have in the absence of gravity. This would mean abandoning flat spacetime and finding a theory in which gravity alters the geometry of spacetime in a way that makes all free-falling bodies have straight world lines, a theory that would take us from the old Newtonian picture of curved world lines in flat spacetime to a new picture of straight world lines (known as geodesics) in curved spacetime. If we were like Einstein, we might eventually succeed in discovering the grand theory that governs the geometry

of spacetime, and would then have accomplished the most imaginative feat in the history of science.

TIDAL FORCES
Tidal forces are complications that help us understand gravity

Before proceeding with our main theme – the theory of general relativity – we must turn aside and consider tidal forces. To illustrate the principle of equivalence we have performed thought experiments with imaginary laboratories. Thought experiments often consist of idealizations that enable us to isolate and study basic principles. In both Newtonian and special relativity theory, for example, we perform thought experiments with systems having inertial motion; yet this kind of motion is an idealization because it rarely if ever exists in nature. However far we escape into intergalactic space, the gravitational pull of nearby galaxies will produce acceleration and destroy the assumed state of inertial motion. There are a few places, hardly larger than points, where the pull and counterpull of galaxies exactly balance and gravity is zero, and at these places we can say that motion is truly inertial. Bodies in motion, however, occupy these places only momentarily, and their motion, like that of all bodies, is then noninertial. Inertial motion, although it rarely exists, is nevertheless a useful idealization.

The principle of equivalence, as presented earlier, is also an idealization. When we perform experiments in imaginary laboratories and other regions of finite volume, the principle is strictly true only when the gravitational field is uniform and the same everywhere in the laboratory. This actually was the way that Einstein first defined the equivalence principle: in terms of a gravitational field that does not vary from place to place. But a uniform gravitational field is an idealization, because gravity always varies in space and is never uniform. (Point masses, for example, have gravitational fields that decrease as the inverse square of distance.) Finite regions of space in which the gravitational field is uniform, as we assumed in our treatment of the equivalence principle, simply do not exist.

In a laboratory resting on the Earth's surface, gravity is not everywhere the same. When a body is raised a height $h = 3$ meters above the floor its weight decreases by a part in 1 million. (The change in gravity is $\Delta g/g = 2h/R$, and the Earth's radius is $R = 6371$ kilometers.) Such a variation of weight does not exist in a laboratory in space that is accelerated by a rocket engine, because all parts of the interior are equally accelerated, and the inertial force acting on an object is the same wherever that object is placed in the laboratory. Contrary to what has been previously said, there are actually many experiments that can distinguish between stationary laboratories in nonuniform gravitational fields and accelerated laboratories out in space. All we need do is search for a small variation in the gravitational force acting on an object as its position is changed in the laboratory. The larger the laboratory, the easier it is to detect any such variation.

Consider a solid body moving freely under the influence of gravity (see Figure 12.8). Its atoms occupy different positions, and because the gravitational field is not exactly the same at every position, the atoms are acted upon by slightly different gravitational forces. The atoms, however, are all stuck together and compelled to follow a common orbit – the orbit of its center of mass. If the atoms were not stuck together, but able to move freely and independently, they would all follow slightly different orbits. The force that tends to distort and even tear bodies apart because each part tries to follow its own free-fall orbit is known as the tidal force. The center of mass is the only point actually in a state of free fall, and at this point the tidal force is zero. All other points of the body are not in perfect free fall; at these points there exists a detectable tidal force. What happens is obvious. Atoms are constrained to follow the center of mass, and the inertial forces created by these motions fail to cancel

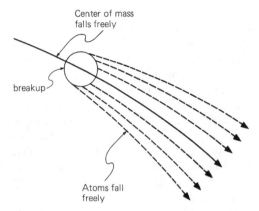

Figure 12.8. A body falls freely under the influence of gravity, but only its center of mass follows a perfect free-fall trajectory. All its atoms are stuck together and compelled to follow the center-of-mass orbit. If the atoms were free and not stuck together they would all follow slightly different trajectories because gravity is not exactly the same everywhere in the body.

exactly the gravitational force at positions other than the center of mass. The tidal force is mainly the result of gravity not being uniform, and at any point the tidal force is approximately the gravitational force at that point minus the gravitational force at the center of mass.

Let us return to our imaginary laboratory in free fall out in space. By searching for and detecting the presence of a tidal force, the experimenters can tell whether their laboratory is in an inertial state or is free falling under the influence of gravity. Our imaginary laboratory could be an elevator that falls freely in a vertical shaft penetrating deep into the Earth. We can place in this elevator two tennis balls that float side by side at the same height above the floor (see Figure 12.9). As the elevator plunges toward the center of the Earth, we see the tennis balls accelerate slowly toward each other. Both fall freely toward the center of the Earth and their separating distance slowly decreases. The elevator accelerates toward the center of the Earth, and the tennis balls, driven by the tidal force, accelerate toward each other. By observing the way in which floating bodies move relative to one

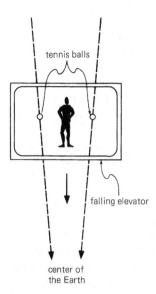

Figure 12.9. Two tennis balls float, as shown, in a free-falling elevator. As the elevator plunges toward the center of the Earth at an accelerating rate, the tennis balls move toward each other at an accelerating rate. To an observer in the elevator it seems that a force exists that acts on the two balls. This force, which is the result of the nonuniformity of gravity, is the tidal force.

another, as in this case, we can determine if a laboratory is inertial or free-falling.

Tidal forces are normally negligible. Usually, scientists pay little attention to the tidal forces in their laboratories on the Earth's surface. A principle, however, is a principle, and if equivalence is fundamental there should be some way of stating the principle without the bother of tidal forces. The time has come to let the reader into a secret. All our talk of experiments in imaginary laboratories was for the sake of illustrating equivalence in a simple and understandable way. But equivalence of results obtained by experimenters in inertial and free-falling laboratories of finite volume, though interesting, is not of great theoretical importance. What is important is to have equivalence in the equations of physics as a property of infinitesimal volumes.

The principle of equivalence cannot apply everywhere in a free-falling laboratory, simply because all parts are not in

perfect free fall. Gravitational and inertial forces are actually indistinguishable only in free-falling regions of extremely small volume. This means we are unable to perform experiments with apparatus of finite size to establish with utmost precision the truth of the equivalence principle. There are, however, other ways. The equation of motion and other equations of physics are statements about what happens in regions of extremely small size, and the predictions of these equations have been amply verified. Let dx, dy, dz indicate the size of a very small region, measured as intervals in the x, y, and z directions, as shown in Figure 12.10. (Here dx means difference or differential in x, and dy and dz are defined similarly.) The equations of physics are cast into a form that tells us what happens when dx, dy, dz, and other differences, such as dt in

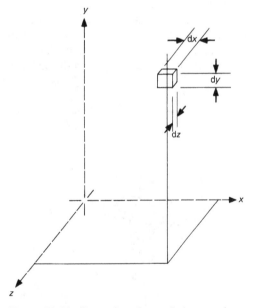

Figure 12.10. Illustration of a small element of volume of size dx, dy, dz. The equations of physics refer to what happens in such small elements when dx, dy, and dz shrink to zero. The principle of equivalence, free of the bother of tidal forces, states that the equations of physics do not distinguish between inertial and free-falling elements of infinitesimal volume, and the equations are the same in either case.

time, all shrink to zero. This is why they are called differential equations. The laws of physics are expressed in terms of differential equations that determine what happens in infinitesimally small regions. The principle of equivalence now means that the inertial and gravitational forces cancel each other in infinitesimally small free-falling regions, and the basic equations of physics do not distinguish between inertial and free-falling states.

Tidal forces and curvature variations

We start by considering a surface, such as that of a vase, having a curvature K that varies from place to place. On the surface we draw a figure, such as a circle, and notice that inside this figure the curvature K varies by small amounts. This closed region is analogous to a laboratory in free fall, and the variation of K inside the closed region resembles the tidal force that varies inside the laboratory.

We next take a small piece of thin malleable material cut in the shape of a rectangle. This piece of material, labeled "lab" in Figure 12.11, is fitted snugly on the surface.

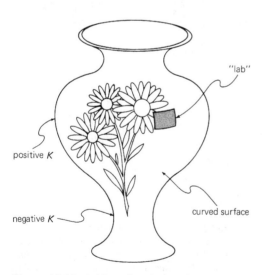

Figure 12.11. A thin and malleable piece of material, labeled "lab," fits snugly on a curved surface. The variation of curvature within the lab is analogous to the variation of gravity (the tidal force) within a free-falling laboratory.

We notice that the curvature K is not everywhere the same in the lab, and varies also in time as we move the lab around on the surface. This again is analogous to what happens in a free-falling laboratory. The tidal force in the laboratory varies from place to place and varies also with time.

The principle of equivalence applies only to infinitesimally small regions (i.e., laboratories of very small volume); similarly, Euclidean geometry (which is flat and has zero curvature) can be used only in extremely small regions on curved surfaces (i.e., in labs of extremely small size). Thus tidal forces, which at first seemed to be a bothersome complication, do not detract from the beauty of the principle of equivalence; they in fact help us to understand the relationship between gravity and geometry. We have on the one hand variations in the tidal force and on the other variations in curvature, both occurring in regions of finite size. The tidal forces vanish in an extremely small laboratory and the geometry becomes flat and Euclidean in an extremely small lab. Euclidean geometry in a plane tangent to a curved surface is like special relativity in a free-falling system; both apply in small regions.

Although experiments performed in laboratories of finite size fail to obey the equivalence principle under all circumstances, we shall continue to use this convenient pedagogic way of performing imaginary experiments.

THEORY OF GENERAL RELATIVITY
Einstein's equation

Einstein's theory of general relativity was developed in the early years of the twentieth century and reached its final form in 1916. The Newtonian universe with its Euclidean geometry and gravitational forces was at last overthrown and replaced with a relativistic universe of spacetime of varying curvature. The curved orbits of free-falling bodies in the Newtonian universe became the straight orbits in the curved spacetime of the Einstein universe.

A straight-line orbit is known as a geodesic; it is the shortest distance in space between two points; it is a straight line in the local geometry, but to an observer elsewhere, whose local geometry is different, it appears curved. A geodesic in flat space (as in special relativity) is the familiar Euclidean straight line, and a particle following such an orbit is unaccelerated and has a straight world line. A geodesic on the surface of a sphere is a great circle. In general relativity, free-fall motion follows geodesic paths.

The Einstein equation of general relativity states that the curvature of spacetime is influenced by matter. Stated differently, the strain (deformation) of spacetime is related to the stress induced by matter. Expressed in the simplest manner possible,

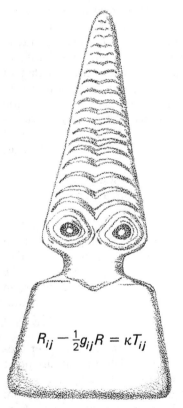

$$R_{ij} - \tfrac{1}{2}g_{ij}R = \kappa T_{ij}$$

Figure 12.12. The Eye Goddess (Syria, 2800 BC) displaying the Einstein equation as a modern addition. The Einstein equation is explained qualitatively in the text.

the equation states

curvature of spacetime

$$= \text{constant} \times \text{matter.} \qquad [12.5]$$

This is a mathematical equation that breaks down into 10 separate equations, and not a great many exact solutions have been discovered. We interpret the Einstein equation to mean that curvature is equivalent to gravity.

The "matter" on the right side includes all forms of energy (including pressure) that have mass. When the curvature is only slight (hence gravity is weak as in the Solar System) the Einstein equation reduces to special relativity, and when also velocities are small the Einstein equation simplifies further and becomes the Newtonian laws of gravity and motion. General relativity by itself does not tell us the value of the "constant" in Equation [12.5] that couples together curvature and matter; comparison with Newtonian theory shows that it contains the universal gravitational constant G. What determines G, we do not know, and its value must be found by observations.

The Einstein equation does not say that curvature and matter are the same. They are, of course, distinctly different. Instead, it shows how curvature and matter influence each other. The Riemann curvature of four-dimensional spacetime, discussed in the previous chapter, has 20 components. Hence the Einstein equation, which has built into it 10 equations, is unable to determine the values of all components of the curvature at every point of spacetime. This is just as well, because in the empty space around the Sun there is no matter, and yet a gravitational field exists that controls the motions of the planets. In the absence of matter, spacetime is therefore not necessarily flat, as in special relativity; it may have curvature determined by the presence of distant bodies such as the Sun. All components of the Riemann curvature are determined when we take into account distant as well as local matter. A rubber sheet, stretched and initially flat, illustrates clearly what happens

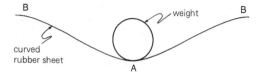

Figure 12.13. Spacetime is curved by local matter (as at A) and also by distant matter (as at B), as illustrated by a curved rubber sheet.

in spacetime (see Figure 12.13). A heavy ball placed on the sheet produces a depression and the curvature of the sheet diminishes with distance from the ball. This demonstrates how curvature is produced by distant as well as local matter.

Gravity, acting in a mysterious way across a vacuum, has vanished and been replaced by the geometrical curvature of physical spacetime. Yet we still use the old Newtonian language of gravity. One reason is that the language of gravity is more vivid and familiar than the language of differential geometry; another is that the old language helps us to distinguish gravity from other forces, such as the electromagnetic, strong, and weak forces that are not fully geometrized in general relativity.

Curvature produces curvature

In the Newtonian universe, gravity obeys the principle of superposition. This means the gravitational force at a point is the sum of the forces produced by bodies everywhere, and the force produced by each body is unaffected by the presence of other bodies. If one body by itself produces a force F_1, and another body by itself produces a force F_2, the two bodies existing at the same time in their respective places produce a combined force $F_1 + F_2$. The force that each body produces is independent of the presence of the other. But in the Einstein equation the geometric curvature produced jointly by two or more bodies is not exactly the arithmetical sum of the curvatures produced by the bodies when taken separately. This is also true of a stretched rubber sheet depressed at several places by different weights: the total depression at any point is not exactly the sum of the depressions

taken separately. The Einstein equation does not obey the principle of superposition, and we cannot add together the curvatures of simple arrangements to find the curvature of a complex arrangement. In general relativity, bodies curve spacetime, and the curvatures they produce act on one another, and this self-interaction of spacetime is what is so distinctive and important about general relativity. Furthermore, the curvature in one region affects the curvature in another region. This self-interaction exists because the curvature of spacetime is itself a form of energy, which produces its own gravitational field, and is thus the source of further curvature. This explains why gravity in the vicinity of a body such as the Sun does not obey exactly the inverse square law. The energy that resides in the spacetime curvature outside the Sun makes its own contribution to the distant gravitational field. Curvature generates curvature, whereas in the Newtonian universe gravity does not generate gravity. In the case of weak gravitational fields, such as that of the Sun, the Einstein equation simplifies to the Newtonian equations with the addition of small corrections. Tests of general relativity are usually difficult because regions of strong gravity are not readily available to us.

Gravity travels at the speed of light

Gravity in the Newtonian universe propagates at infinite speed. A star, or any other body, produces a gravitational field that exists instantaneously everywhere. But in the Einstein universe, gravity – or rather spacetime curvature – propagates at the speed of light. The Einstein equation is in fact a dynamic wave-equation that generates and propagates the curved deformations of spacetime.

Two stars in orbit about each other produce a gravitational field that at any point varies periodically with time (see Figure 12.14). But gravity is spacetime curvature that contains energy, and if gravity varies periodically, then the spacetime curvature varies periodically, and energy is constantly redistributed in the surrounding region of

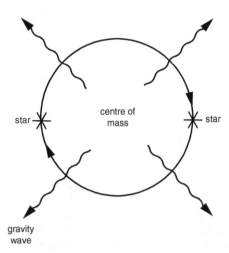

gravity
wave

Figure 12.14. Two stars in orbit about each other radiate gravitational waves. The waves – ripples in spacetime – are caused by periodic variation of the curvature of spacetime. Curvature contains energy, and outward-traveling ripples or gravitational waves carry away energy from the binary system at the speed of light. Because of a steady loss of energy and angular momentum the two stars slowly spiral toward each other.

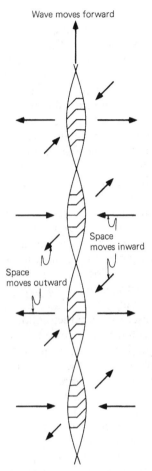

Figure 12.15. A gravitational wave is a ripple of spacetime curvature. This picture illustrates the way that space is periodically varied by the passage of a gravitational wave.

spacetime. This cyclic bending and warping of spacetime streams away as ripples in all directions at the speed of light, and energy and angular momentum are lost from the orbiting stars in the form of gravitational waves. This loss of energy means that the two stars slowly spiral toward each other on typical time scales of billions and even trillions of years. Gravitational waves are generally very weak and extremely difficult to detect.

For many years physicists endeavored to detect in the laboratory with sensitive instruments the gravitational waves generated by energetic events in the Galaxy. But such waves have so far not been observed with laboratory instruments, at least not in a manner that is beyond doubt. The problem is not just to observe the time-variation of gravity – the oceanic tides on Earth reveal that – but to detect the energy in spacetime variations that travel at the speed of light (see Figure 12.15). The discovery by Joseph Taylor and Russell Hulse at the University

of Massachusetts of a pulsar in a close binary system now provides astronomers with an opportunity to observe the effects of gravitational fields much stronger than previously available. The loss of energy from this binary system, carried away by gravitational waves, causes the orbits of the pulsar and its companion to shrink and the orbital frequency to increase at a measurable rate. Taylor and Hulse have shown that the increase in orbital frequency is in agreement with general relativity, and this is the first observational proof of the existence of gravitational radiation.

Laboratories with windows

We have seen that in a free-falling laboratory gravity is abolished locally in the laboratory, and all experiments performed in the laboratory give results identically the same as obtained in the flat spacetime of special relativity. (We ignore tidal forces or, equivalently, curvature variations in the laboratory.) If we use the Einstein equation in a state of free fall, we find that locally it reduces to the same equations that are used in special relativity: gravity vanishes and spacetime is flat.

Let our experimenters now cut holes in the walls of their laboratory and observe all that is happening outside. What they see is in accordance, not with special relativity, which applies only locally inside the laboratory, but with general relativity, which applies globally. Everything of a dynamic nature observed from their laboratory is explained with the Einstein equation. The Einstein equation is covariant (the same) for all free-falling observers, and because a state of free fall is more general than the inertial state of special relativity, the equation is covariant for a class of observers more general than the inertial class of observers in special relativity.

The Sun moves freely in the Galaxy and the Earth moves freely about the Sun, yet we ourselves on the surface of the Earth are not in free fall in the gravitational field of the Earth. This does not mean that we on Earth cannot use the Einstein equation. Some mathematicians and physicists use it all the time. What they do is quite simple. When they want to explain our observations from the Earth's surface, they imagine themselves in a free-falling system, and are able to calculate with the Einstein equation the corrections that must be used in a state that is not free falling, as on the Earth's surface.

Einstein's quest for a unified theory

Einstein once said that the Riemannian geometry on the left side of the Einstein equation is like an elegant marble hall. On the right side is what might be called an outside yard into which is put almost anything we choose, subject to certain elementary constraints. It has become the custom, since the time when Einstein first formulated the theory, to place on the right side "matter" and other things that have an energy (therefore mass) content. But Einstein was cautious on this aspect of the theory and at one time allowed for a more general interpretation of the meaning of "matter." General relativity by itself does not tell us what is the fundamental nature of matter, and Einstein was not entirely satisfied with his choice. He regarded the right side as incomplete, and in the later years of his life he said, "The right side is a formal condensation of all things whose comprehension in the sense of a field theory is still problematic. Not for a moment, of course, did I doubt that this formulation was merely a makeshift in order to give the general principle of relativity a preliminary expression. For it was essentially not anything more than a theory of the gravitational field, which was somewhat artificially isolated from a total field of as yet unknown structure."

Einstein viewed his great theory as incomplete and for years sought for a way in which geometry could be equated more directly with the fundamental properties of matter. Maxwell had previously unified electricity and magnetism into electromagnetism, and Einstein sought to unify electromagnetism and gravity, but his quest for a unified theory never succeeded. In recent decades, eminent physicists have combined the electromagnetic and weak forces into an electroweak force. Grand unified theories have also emerged that unify the electroweak and strong forces into a single hyperweak force. The hyperweak force was important at the beginning of the early universe when the energy density was extremely high, but in the universe at present it operates in different ways that appear to us as three distinct forces.

Clifford's idea that everything can be explained in terms of spacetime may yet come true. Multidimensional theories of space have been proposed and explored.

One particularly interesting example is a ten-dimensional space in which six dimensions collapse to form the subatomic structures of particles distributed in the remaining four dimensions.

TESTS OF GENERAL RELATIVITY

What is needed is a homely experiment which could be carried out in the basement with parts from an old sewing machine and an Ingersoll watch, with an old file of *Popular Mechanics* standing by for reference.
Howard Robertson, in Albert Einstein: Philosopher-Scientist *(1949)*

Strength of gravity

Various tests of general relativity have been successfully conducted and have yielded results in close agreement with the predictions of the theory. Such tests are usually not easy to perform, mainly because, with the weak gravitational fields generally available to us, the difference between the Einstein and Newtonian theories is quite small.

Gravity is weak, generally speaking, when the escape speed from a body is small compared with the speed of light. We can say

$$\text{strength of gravity} = \left(\frac{V_{esc}}{c}\right)^2, \qquad [12.6]$$

where V_{esc} is the escape speed and c is the speed of light. At the surface of the Sun, where the escape speed is 618 kilometers a second, the strength of gravity is 4.24×10^{-6}, or roughly 4 parts in a million. At the Earth's orbit the escape speed from the Sun is 42 kilometers a second and the strength of gravity has fallen to 2×10^{-8}. Gravity is weak not only at the surface of the Sun but also throughout the Solar System, and the effects peculiar to general relativity are exceedingly small. When the strength of gravity approaches unity, as in the vicinity of black holes, gravity is strong and the effects of general relativity become pronounced.

Einstein proposed three famous tests of general relativity, which are now discussed briefly.

Deflection of starlight

The first of Einstein's tests is the deflection of starlight in the Sun's gravitational field (see Figure 12.16). Light from a distant star, on passing close to the Sun, should be deflected through a small angle that is twice the amount predicted by Newtonian arguments. General relativity theory predicts a deflection of 1.75 seconds of arc for starlight grazing the edge of the Sun's disk – an angle about equal to that subtended by a person's small finger at a distance of 1 kilometer. The deflection in radians is equal to twice the strength of gravity,

$$\text{deflection of light} = 2\left(\frac{V_{esc}}{c}\right)^2 \text{ radians,}$$

$$[12.7]$$

where 1 radian is 57.3 degrees of arc. There are 206 265 seconds of arc in 1 radian, and if this is multiplied by twice the Sun's strength of gravity, we obtain 1.75 seconds of arc.

The light-deflection test was proposed by Einstein in 1916 but not performed until after World War I in 1919. Stars are seen

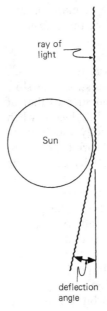

ray of light

Sun

deflection angle

Figure 12.16. The deflection of a ray of light, grazing the Sun's disk, is 1.75 seconds of arc.

close to the Sun only for a short time during a solar eclipse, and two eclipse expeditions were organized at the suggestion of Frank Dyson, the Astronomer Royal of Britain, and of Arthur Eddington, who was the first person outside Germany to champion the theory of general relativity. The expeditions observed the eclipse from Sobral in Brazil and from the island of Principe off the coast of West Africa. The expedition to Sobral measured a displacement of 1.98 seconds of arc and the expedition to Principe measured a displacement of 1.61 seconds of arc. Observations of this kind are difficult to make in the short time during an eclipse, and the results were considered to be in reasonable agreement with the Einstein prediction. Both measurements were significantly greater than the Newtonian prediction of 0.87 seconds of arc.

Precise measurements have since been made in other eclipse expeditions. The most precise results, which fully confirm the Einstein prediction, have been made by radioastronomers. Radio waves from distant radio sources are deflected in exactly the same way as starlight, and radioastronomy has the advantage that observations close to the Sun can be made on any day without waiting for an occasional eclipse.

Precession of planetary orbits

The second test is the precession – or slow drift – of a planetary orbit such as that of Mercury (see Figure 12.17). Always, when gravity fails to obey exactly the inverse-square law, an elliptical orbit precesses. According to general relativity theory, the gravitational field of the Sun is not exactly of the inverse-square form except at very large distances from the Sun. The precession, because of general relativity, is therefore most obvious in the case of Mercury, the planet closest to the Sun. For planetary orbits that are almost circular, as in the Solar System, the precession is independent of the eccentricity of the orbit. The amount of the precession, measured in radians per revolution, is equal to the strength of gravity

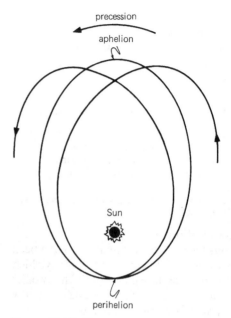

Figure 12.17. The precession of an elliptical orbit of a planet about a star. Mercury's precession about the Sun is 43 seconds of arc per century.

of the Sun:

precession per revolution

$$= \left(\frac{V_{esc}}{c}\right)^2 \text{ radians per revolution. [12.8]}$$

In this case the strength of gravity is determined at the planetary orbit and not at the surface of the Sun. If V_{orb} is the orbital speed of the planet, then $V_{esc} = \sqrt{2}V_{orb}$ at the radius of the planetary orbit, and

precession per revolution

$$= 2\left(\frac{V_{orb}}{c}\right)^2 \text{ radians per revolution.}$$

$$[12.9]$$

The precession of planetary orbits in the Solar System resulting from this effect is quite small. For Mercury, which has an orbital speed of 48 kilometers per second and an orbital period of 88 days, the precession equals 43 seconds of arc per century. This amounts to a complete revolution once every 3 million years. The gravitational fields of other planets, which also cause deviations from the inverse-square law of the Sun's gravity, produce a much larger precession, and this must first be subtracted from the observed precession of Mercury's orbit. What remains after this subtraction is in close agreement with the predicted precession of 43 seconds of arc per century.

Gravitational redshift

The third test is the gravitational redshift effect. Radiation escaping from the surface of a body like a star or a planet loses energy because of the pull of gravity. The energy of a particle of light (a photon) is proportional to its frequency, and as the energy of the escaping photon decreases, the frequency also decreases. Hence the wavelength of a light ray increases as it travels away from a gravitating body. A light ray falling on a gravitating body, like a star or a planet, gains energy because of the pull of gravity, and its wavelength decreases.

Let λ be the wavelength of the radiation emitted from the surface of a body, and λ_0 be the observed wavelength when the radiation has escaped to a great distance. The redshift, denoted by z, is the fractional amount by which the wavelength has increased:

$$z = \frac{\lambda_0 - \lambda}{\lambda}. \qquad [12.10]$$

When gravity is weak, the redshift is equal to half the strength of gravity:

$$z = \frac{1}{2}\left(\frac{V_{esc}}{c}\right)^2. \qquad [12.11]$$

This equals 2.12×10^{-6} for radiation escaping from the Sun, and 7×10^{-10} for radiation escaping from the Earth (where the escape speed from Earth is 11.2 kilometers a second).

The redshift that occurs when radiation travels vertical distances of tens of meters at the Earth's surface has been observed using methods developed in nuclear physics. Slow-decaying nuclei of certain atoms embedded in crystals emit short-wavelength radiation of sharply defined frequency (this is known as the Mössbauer effect), and the shift in frequency caused by a change in

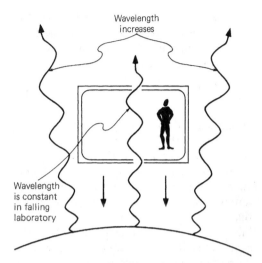

Figure 12.18. A ray of light loses energy as it moves away from a star or a planet and its wavelength steadily increases. Inside a free-falling laboratory the wavelength is constant. This enables us to calculate the redshift from the known acceleration of the laboratory.

height at the Earth's surface has been measured and found to be in agreement with prediction.

The gravitational redshift is a direct consequence of the equivalence principle. This is shown by the following thought experiment. We consider a light ray of increasing wavelength that is escaping vertically from a gravitating body, as shown in Figure 12.18. We suppose that an experimenter is in a free-falling laboratory and the light ray enters through a hole in the floor and passes upward and out through a hole in the ceiling. Within this laboratory the effect of gravity is abolished and the experimenter therefore observes the light ray enter and leave with its wavelength unchanged. We must take into account the acceleration of the downward falling laboratory. While the light ray passes upward from the floor to the ceiling, the velocity of the falling laboratory increases. The wavelength, in effect, is squeezed by the increasing velocity of the laboratory and stays constant in value, as seen by the experimenter inside. Because the wavelength is constant in the free-falling laboratory, it follows that to an outside observer, standing on the surface of the body, the wavelength increases as the light ray travels upward. Calculation shows that when gravity is weak, the redshift is

$$z = \frac{GM}{Rc^2},$$ [12.12]

at the surface of a body of mass M and radius R. With $V_{esc}^2 = 2GM/R$, this result is the same as Equation [12.11].

Imagine that we are out in space observing what happens on the surface of a gravitating body. All wavelengths of radiation that we receive from the surface are increased by the same factor and have the same redshift. All frequencies are also decreased. Out in space we see everything happening slower on the surface of the body. Let us ignore the practical aspects of the situation and take an extreme case in which the redshift is as large as unity. The atoms on the surface now appear to vibrate twice as slowly as the same atoms around us and in our bodies. Everything on the surface appears to happen twice as slowly as out in space. A person on the surface, communicating to us by radio, has a very deep voice and talks painfully slowly. That person, instead of living for 80 years, lives for 160 years in our time. The redshift not only increases wavelengths and decreases frequencies, but also it slows up the apparent rate at which everything happens. All intervals of time are increased in the same way as wavelengths, and two events separated by 1 second on the surface are seen out in space to be separated by 2 seconds.

Let us reverse the situation and imagine that we are living on the surface of a body of redshift equal to unity. Everything around us happens at what seems to be a normal rate as measured by our pulse rates and wrist watches, and if we are not unlucky we shall live for the usual 80 years. (We are of course ignoring the practical aspects of living on the surface of a body of very strong gravity.) Incoming radiation from distant space is now blueshifted (redshift $z = 0.5$); all wavelengths are decreased by a common factor and shifted toward the blue (short

wavelength) end of the spectrum. A person in space, communicating with us by radio, has a high pitched voice and talks rapidly in a very squeaky voice. In our time, that person out in space lives for only 40 years.

Summing up: To an observer out in space, all things in the vicinity of gravitating bodies are redshifted and appear to happen more slowly; to an observer in the vicinity of a gravitating body, all things out in space are blueshifted and appear to happen more rapidly.

Other tests of general relativity

Many tests of general relativity have been suggested in addition to those proposed by Einstein. Several of these tests, such as the measurement of time delays of radar signals in the Solar System, have been successfully accomplished. With the advance of technology and observational precision, the results in all cases have progressively been more reassuring concerning the validity of general relativity.

In the Hulse–Taylor binary system, the precession of the pulsar's orbit about the companion star is approximately 4 arc degrees a year, which is about 35 thousand times faster than the precession of Mercury's orbit. The loss of energy from this binary system, carried away by gravitational waves, causes the orbits of the pulsar and its companion to shrink and the orbital frequency to increase at a measurable rate. Taylor and Hulse have shown that the increase in orbital frequency is in agreement with general relativity.

MACH'S PRINCIPLE
Absolute space

Mach's principle, which nowadays is only of historical interest, states that all inertial forces are due to the distribution of matter in the universe. This intriguing principle inspired Einstein while he was developing the theory of general relativity.

Newton said, "Absolute space in its own nature, without relation to anything external, remains similar and immovable." These thoughts of an absolute space, of a space existing in its own right without need of material support, at that time seemed contrary to common sense and was challenged by philosophers and scientists. Gottfried Leibniz regarded Newton's ideas of space as outrageous, and responded by asserting, "There is no space where there is no matter." But Newton believed that he had proof of absolute space, and in the controversy that followed he was the only person who submitted his ideas to experimental tests.

Everybody agreed that uniform motion (motion at constant velocity) is relative, and to many persons it seemed natural therefore that nonuniform motion (accelerated motion) must also be relative. Newton, however, declared that accelerated motion is absolute in absolute space. We can understand what Newton had in mind by the following argument. We imagine that only a single body exists in the whole of space. If the body has uniform motion, we cannot determine how fast it moves or the direction in which it moves. But when the body has nonuniform motion, we can determine how fast it accelerates and the direction in which it accelerates because of the existence of inertial forces. Consider rotation. The acceleration is toward the axis of rotation and the inertial force – centrifugal force – is away from the axis. This inertial force distorts a rotating body. A rotating planet or star bulges at the equator and is flattened at the poles. The measurement of effects of this kind enables us to determine how fast the body rotates, and this rotation, said Newton, is relative not to other bodies but to absolute space.

Absolute space, existing in its own right, was the overarching concept of the Newtonian universe. It was implicit in the Newtonian equation of motion, and no rival philosopher or scientist was able to suggest an alternative idea that could match it. Of the experiments dealing with the absolute nature of rotation, which were discussed by Newton, the most famous is the rotating-bucket-of-water experiment, discussed in the Reflections at the end of this chapter.

Bishop Berkeley

The idea of absolute space was vigorously attacked by the Irish philosopher George Berkeley (1685–1753) in a work entitled *Motion*, published in 1721. Berkeley's dislike of absolute space stemmed from the old Aristotelian belief that space exists by virtue of its association with matter, and undressed space had no physical properties of its own. Space, as argued by Descartes, was a sideless box that vanishes when nothing is contained; space by itself was emptiness, was nothing; its only property was extension, and this property, without designation by material content, was by itself meaningless.

Berkeley's principle, which lies at the heart of his argument, can be stated as follows: A single body in an otherwise empty universe has no measurable motion of any kind (see Figure 12.19). The principle, unfortunately, cannot be verified by observation. If we were to suppose, said Berkeley, that "the other bodies were annihilated and, for example, a globe were to exist alone, no motion could be conceived in it; so necessary is it that another body should be given by whose situation the motion should be understood to be determined. The truth of this opinion will be very clearly seen if we shall have carried out thoroughly the supposed annihilation of all bodies, our own and that of others, except that solitary globe."

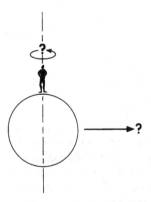

Figure 12.19. Berkeley's "solitary globe" in an otherwise empty universe. All forms of motion, uniform and accelerated, are unobservable and meaningless, according to Berkeley.

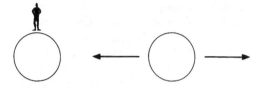

Figure 12.20. Berkeley's "two globes" in an empty universe. The only detectable motion, according to Berkeley, is their one-dimensional movement toward and away from each other; their revolution about a common axis, for example, is unobservable and even meaningless.

A single body, all alone, is thus denied conceivable motion of any kind, relative and absolute. If two bodies alone exist, said Berkeley, only their relative motion toward and away from each other can be observed. "Let two globes be conceived to exist and nothing corporeal besides them. Let forces then be conceived to be applied in some way; whatever we may understand by the application of forces, a circular motion of the two globes around a common center cannot be conceived by the imagination" (see Figure 12.20). If there are three bodies alone, according to this argument, only their relative motion in a common plane is observable. And if there are four bodies alone, then their relative motion in three dimensions is observable, but we cannot determine if they have revolution about a common axis. "Then let us suppose that the sky of the fixed stars is created; suddenly from the conception of the approach of the globes to the different parts of the sky the motion will be conceived."

Instead of an absolute space of independent reality, Berkeley invoked a "sky of fixed stars." The fixed stars were the reference points relative to which all motion, uniform and nonuniform, was defined. To Berkeley and many other philosophers of science it seemed natural to suppose that space in all respects was subordinate to matter and that the properties attributed by Newton to absolute space were in fact the result of the material content of the universe.

Ernst Mach

Ernst Mach (1838–1916), an Austrian physicist, apparently unaware of previous arguments, expressed ideas essentially similar to George Berkeley's. Berkeley had stressed the relativity of all motion, uniform and accelerated, and Mach developed this theme and stressed the relativity of inertial forces.

Once again let us suppose there is only a single body in an empty universe. Motion of any kind, according to Mach, is inconceivable and therefore inertial forces cannot exist. Because rotation is inconceivable, centrifugal force does not exist. The addition of one, two, three, or more bodies now allows relative motion to exist. This raises a problem. The additional bodies serve as reference points and can be made as small as specks of dust. How then can they account for the sudden creation of inertial forces? Having denied inertial forces to the first body, it becomes absurd to suppose that they are suddenly acquired in full strength by the addition of a few small bodies. How then, at the same time that rotation becomes measurable, can centrifugal force become real? Mach's answer was that the inertial forces increase only slightly, and that all inertial forces are determined by, and are proportional to, the total amount of matter in the universe. Hence the universe of stars is responsible for the inertial forces of nonuniform motion. A single body by itself has no measurable rotation and no detectable centrifugal force, but when the sky of fixed stars is created, the body has measurable rotation relative to the stars and acquires a centrifugal force because of the stars.

Mach's influence on Einstein

Mach's work helped Einstein to realize the importance of the equivalence principle. Einstein found Mach's cosmological argument highly suggestive and helpful during the years he was developing the theory of general relativity, and referred to it as Mach's principle. The principle asserts that all local inertial forces are fully determined by the distribution of distant matter in the universe.

The notion that inertial forces are determined by distant matter was for Einstein a signpost pointing to a possible connection between geometry and matter, in the sense that distant matter affects geometry, and geometry affects local motion. If the geodesics (straight lines) of Riemannian spacetime are like the orbits of free-falling bodies in Newtonian space, and these orbits are affected by inertial forces, then, according to Mach's principle, the distant matter that affects geometry should also account for inertial forces. Einstein initially took the tentative view that spacetime was fully determined by matter and all inertial forces were the consequence of matter interacting with matter. "In a consistent theory of relativity," said Einstein as late as 1917, "there can be no inertia relative to 'space,' but only an inertia of masses relative to one another. If, therefore, I have a mass at a sufficient distance from all other masses in the universe, its inertia must fall to zero." At that time it was thought that the equation of motion of a body had to be postulated in addition to the Einstein equation, and only later was it realized that the equation of motion, as a geodesic equation, was already implicit in the theory of general relativity.

Soon it became clear to Einstein and other scientists that general relativity had outgrown the suggestions of Mach's principle. Spacetime had achieved a physical reality of its own. Although its geometry is influenced by matter, and motion is controlled by geometry, the nature and existence of spacetime are not dependent on the existence of matter. Spacetime can exist without matter and inertial forces can exist in a universe containing only a single body. Einstein abandoned Mach's principle. It had served its purpose. Instead of continuing to seek for a way to materialize spacetime, Einstein took a new departure and sought for a way to geometrize matter. Some physicists have advanced the view that perhaps matter exists only by virtue of the geometrical properties of spacetime.

Charles Misner and John Wheeler wrote in 1957: Either we must think the spacetime "continuum serves only as an arena for the struggles of fields and particles," or there "is nothing in the world except empty curved space. Matter, charge, electromagnetism, and other fields are manifestations of the bending of space. Physics is geometry." At least, as a consequence of general relativity, we can now say with confidence that spacetime is real with its own physical properties.

REFLECTIONS

1 Albert Einstein (1879–1955) was awarded the Nobel Prize in 1922, not for his work on relativity theory, but for work published in 1905 on the photoelectric effect. (In the photoelectric effect incident light ejects electrons from metal surfaces, and the energy of the ejected electrons depends on the frequency of the radiation, whereas the number of electrons depends on the intensity of the radiation.) Einstein, son of an unsuccessful businessman, was an inattentive pupil and a dropout from high school. He was mainly self-taught. After getting his doctorate degree in 1905 he sought in vain for an academic appointment, and had to accept a clerical position in a Swiss patent office. In that year, 1905, he published three pioneering scientific papers, in one of which he advanced the theory of special relativity. He secured an academic position four years later and thereafter his professional progress was rapid. Einstein had a great power for sustained concentration, a clear insight into the fundamentals of physics, and an immensely creative and disciplined imagination. One of his famous sayings concerning quantum mechanics is that "God does not play with dice." Several aspects of the theory of quantum mechanics were developed by Einstein, but in his later years he viewed quantum theory as insufficiently profound. He believed that beyond its uncertainties lie undiscovered rational laws. In this sense, he was a modern Plato.

• In 1930, in an entertaining after-dinner toast to Albert Einstein, who was present as principal guest, the eminent Irish playwright

George Bernard Shaw said, "Religion is always right. Religion solves every problem and thereby abolishes problems from the universe. Religion gives us certainty, stability, peace and the absolute. It protects us against progress which we all dread. Science is the very opposite. Science is always wrong. It never solves a problem without raising ten more problems." Shaw continued, "Copernicus proved that Ptolemy was wrong. Kepler proved that Copernicus was wrong. Galileo proved that Aristotle was wrong. But at that point the sequence broke down, because science then came up for the first time against that incalculable phenomenon, an Englishman. As an Englishman, Newton was able to combine a prodigious mental faculty with the credulities and delusions that would disgrace a rabbit. As an Englishman, he postulated a rectilinear universe because the English always use the word 'square' to denote honesty, truthfulness, in short: rectitude. Newton knew that the universe consisted of bodies in motion, and that none of them moved in straight lines, nor ever could. But an Englishman was not daunted by the facts. To explain why all the lines in his rectilinear universe were bent, he invented a force called gravitation and then erected a complex British universe and established it as a religion which was devoutly believed in for 300 years. The book of this Newtonian religion was not that oriental magic thing, the Bible. It was that British and matter-of-fact-thing, a Bradshaw [a railway timetable]. It gives the stations of all the heavenly bodies, their distances, the rates at which they are traveling, and the hour at which they reach eclipsing points or crash into the earth. Every item is precise, ascertained, absolute and English.

"Three hundred years after its establishment a young professor rises calmly in the middle of Europe and says to our astronomers: 'Gentlemen: if you will observe the next eclipse of the sun carefully, you will be able to explain what is wrong with the perihelion of Mercury.' The civilized Newtonian world replies that, if the dreadful thing is true, if the eclipse makes good the blasphemy, the next thing the young professor will do is to question the existence of gravity. The young professor smiles and says that gravitation is a very useful hypothesis and gives fairly close results in most cases, but that personally he can do without it. He is asked to explain how, if there is no gravitation, the heavenly bodies do not move in straight lines and run clear out of the universe. He replies that no explanation is needed because the universe is not rectilinear and exclusively British; it is curvilinear. The Newtonian universe thereupon drops dead and is supplanted by the Einstein universe. Einstein has not challenged the facts of science but the axioms of science, and science has surrendered to the challenge" (Blanche Patch [Shaw's secretary], Thirty Years With G.B.S.).

2   Tensors were discussed briefly in Chapter 10. Many equations in physics are expressed in terms of tensors. Scalars are zero-order tensors. A scalar field is a continuous variable in space and time and has only a single value at each point. Air or water or any other fluid has at each point in space three components of velocity corresponding to the three dimensions of space. Velocity is a vector – it has magnitude and direction at each point – and a vector field has at each point in space three values. Maxwell's equations of the electromagnetic field are vector equations. More generally, in relativity, a vector field has four components at each point of spacetime, corresponding to the four dimensions of spacetime. Vectors are first-order tensors. Fluids may have complex motions that involve expansion, rotation, and shear and require a second-order tensor field. A second-order tensor in relativity has at each point $4^2 = 16$ components, and many basic equations in physics, such as the Einstein equation, use second-order tensors. The metric equation (Chapter 10), which contains the metric coefficients and determines the geometry of spacetime, is a second-order tensor equation. Owing to the symmetries of spacetime – such as the shortest distance from A to B is the same as from B to A, and the distance around a circle is the same clockwise as counterclockwise – the metric tensor has at most 10 different components at each point.

The number of components in a tensor at each point of spacetime is 4^m, where 4 is the number of dimensions of spacetime and m indicates the order of the tensor. A scalar is zero order with $m = 0$ and has 1 component; a vector is first order with $m = 1$ and has four components; and so on. Tensors higher than second order are not uncommon. The gravitational field of general relativity – the Riemann curvature tensor – is fourth order and has 256 components at each point of spacetime. Owing to various spacetime symmetries, some of which are ingrained in us as common sense, many Riemann curvature components are either zero or have similar values, and only 20 are distinctly different at each point in spacetime.

3   The deflection of starlight by the Sun was predicted in 1801 by the German mathematician Johann von Soldner. He used the idea that light consists of particles that obey the Newtonian equations and estimated a deflection half that given by the theory of general relativity. The principle of equivalence, when considered by itself, is unable to explain why general relativity gives a deflection twice that obtained from Newtonian theory. We can understand why this is so in a qualitative way. Consider a laboratory on the Earth's surface in which a beam of light travels horizontally from one side to the other and is deflected downward by gravity, as in Figure 12.21. This suggests, according to the principle of equivalence, that in a free-falling laboratory the same beam of light will travel in a straight line and not be deflected. Such would be true if the gravitational field were uniform and everywhere the same. It is not, and all parts of the laboratory and all parts of the beam of light are not simultaneously in the same state of free fall. Here is an instance where we cannot ignore tidal forces, or equivalently, the variation of curvature of space within the laboratory. A beam of light in a free-falling laboratory is slightly curved because the laboratory has finite size. When we take into account the variation of curvature we obtain a deflection within the laboratory twice that obtained with the principle of equivalence in a uniform gravitational field.

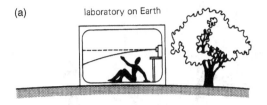

(a) laboratory on Earth

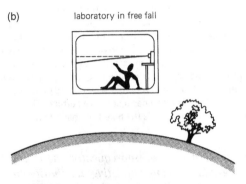

(b) laboratory in free fall

Figure 12.21. (a) A horizontal beam of light in a laboratory on the Earth's surface is deflected downward by gravity. (b) In a free-falling laboratory the beam of light is also deflected, but only half the original amount. This is because in a free-falling laboratory of finite size not all parts of the laboratory are in free fall.

4   Newton wrote: "The effects which distinguish absolute from relative motion are centrifugal forces, or those forces in circular motion which produce a tendency of recession from the axis. For in a circular motion which is purely relative no such forces exist, but in a true and absolute circular motion they do exist, and are greater or less according to the quantity of the absolute motion."

Newton then described the rotating-bucket-of-water experiment (see Figure 12.22). "If a bucket, suspended by a long cord, is so often turned about that finally the cord is strongly twisted, then is filled with water, and held at rest together with the water; and afterwards, by the action of a second force, it is suddenly set whirling about the contrary way, and continues, while the cord is untwisting itself, for some time in this motion; the surface of the water will at first be level, just as it was before the vessel

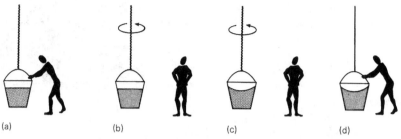

(a)   (b)   (c)   (d)

Figure 12.22. Newton's water-bucket experiment. A bucket of water is suspended by a rope and rotated until the rope is tightly twisted. In (a) the bucket is stationary and the surface of the water is level. In (b) the bucket is released and it begins to rotate. The surface of the water is still level, showing that centrifugal force is not created by the rotation of the bucket. In (c) the water is now rotating with the bucket and its surface is concave, thus showing the existence of a centrifugal force acting on the water. In (d) the bucket is brought to rest, but the water still rotates and has a concave surface. This experiment, said Newton, shows that centrifugal force is the result of absolute motion, and not relative motion.

began to move; but, subsequently, the vessel, by gradually communicating its motion to the water, will make it begin sensibly to rotate, and the water will recede little by little from the middle and rise up at the sides of the vessel, its surface assuming a concave form (as I have experienced) and the swifter the motion becomes, the higher will the water rise, till at last, performing its revolutions in the same times as the vessel, it becomes relatively at rest to it." In this experiment, the surface of the rotating water is depressed in the center, and the depth of the depression depends on how fast the water rotates in an absolute sense, not on the rotation of the water relative to the bucket. Newton concluded that rotation can be determined absolutely, without reference to other bodies.

5 Ernst Mach (1838–1916), a physicist and leading philosopher of science, believed that observations are of primary importance in science, and that all things that cannot be directly perceived exist subjectively only in our minds and have no objective reality. This positivist philosophy led him to a rejection of the atomic theory because atoms are not directly perceived. He also rejected the notion of space and time existing independently of observed material things, and he did not accept the special theory of relativity. Mach

is best known for his study of supersonic flight. When a plane flies at the speed of sound in air we say that it moves at Mach 1. A Mach number M means that an object moves through a medium at M times the speed of sound in that medium.

In 1872, Mach wrote, "For me only relative motion exists.... When a body rotates relative to the fixed stars, centrifugal forces are produced; when it rotates relative to some different body and not relative to the fixed stars, no centrifugal forces are produced. I have no objection to just calling the first rotation so long as it be remembered that nothing is meant except relative rotation with respect to the fixed stars."

6 In The Science of Mechanics, Mach wrote, "We ourselves, when we jump or fall from an elevation, experience a peculiar state, which must be due to the discontinuance of the gravitational pressure of the parts of the body on one another." Michael Heller, in "The happiest thought of Einstein's life," suggests that this led Einstein to realize the importance of the equivalence principle. "Then there occurred to me the happiest thought of my life, in the following form. The gravitational field has only a relative existence.... Because for an observer falling freely from the roof of a house there exists – at least in his immediate surroundings –

no gravitational field" *(Einstein's italics, 1907, quoted by A. Pais,* Subtle is the Lord*). In a speech delivered in Kyoto in 1922, Einstein said, "The breakthrough came suddenly one day. I was sitting on a chair in my patent office in Bern.... Suddenly a thought struck me: if a man falls freely, he would not feel his weight. I was taken aback. This simple thought experiment made a deep impression on me. This led to the theory of gravity."*

7 *In the neighborhood of a rotating body a particle tends to be dragged around with the body. This tendency to share rotation, known as inertial dragging, was first discovered in 1918 by Thirring and Lense and is an effect explained by the theory of general relativity. Consider a hollow spherical shell of matter of mass M, as shown in Figure 12.23, enclosing a suspended rod. The suspended rod acts as an inertial compass. Normally, when the shell rotates, we are able to detect its rotation relative to the suspended rod. When, however, the mass M of the shell is large, the rod tends to be dragged into corotation. In the limit, when M becomes very large (dense enough to form a black hole), there is no apparent rotation because the compass of inertia rotates with the shell. The Thirring–Lense dragging of the inertial frame shows that general relativity possesses a sort of Machian property. But many situations can be found that do not conform to Mach's principle. In a rotating universe, for example, the compass of inertia should corotate, and cosmic rotation should be undetectable according to Mach. But in all rotating models so far constructed with general relativity, the compass of inertia does not corotate in a Machian manner, and rotation of the universe is detectable relative to a body in inertial motion. This theoretical discovery, first made by Kurt Gödel in 1949, delivered a fatal blow to Mach's principle and showed that it is not an indispensable aspect of general relativity.*

• *Philosophical, scientific, and even political interpretations of Mach's principle exist. Lenin (1870–1924), the Russian revolutionary leader and communist dictator, wrote an*

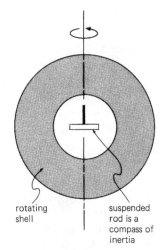

Figure 12.23. A hollow sphere of mass *M* rotates. Inside is suspended a rod free to rotate independently. The rod acts as a compass of inertia. It tends to be dragged around by the rotating sphere; this is the Thirring–Lense effect. When *M* is large and also the density of the body is large, the compass of inertia rotates with the sphere and rotation by relative motion cannot then be detected from inside.

*article in 1909 criticizing Mach's ideas and pointing out that they were politically incorrect according to communistic doctrine. Cosmology has escaped the constraint of religiously correct views, and hopefully it will never fall prey to politically correct constraints. Lenin's article was posthumously translated into English in 1927 (*Materialism and Empiricism: Critical Comment on a Reactionary Philosophy*).*

PROJECTS

1 Discuss the principle of equivalence.

2 Suggest an experiment that can be performed in a free-falling laboratory to check the equivalence of inertial and gravitational mass. How would you verify that the energy in a magnetic field has an inertial mass equal to its gravitational mass?

3 Stretch a thin rubber sheet in a large frame. Now place a heavy spherical body in the center. In the saucer-shaped depression of the rubber sheet roll small ball bearings and notice that they follow trajectories

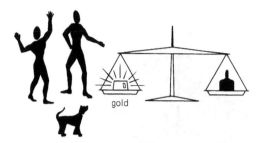

gold

Figure 12.24. "Take out the electric fields, they cost too much!"

similar to those of bodies in motion in the gravitational field of the Sun.

4 What are tidal forces, and why are there tides twice daily on the Earth's surface?

5 Far from the Solar System, what is the redshift of radiation emitted from the Sun's surface? What is the redshift of radiation emitted from the Earth's surface? What is the redshift of the Sun's radiation at the Earth's surface?

6 What is the precession in arc-seconds per century of Venus, Earth, and Mars?

7 In the Newtonian picture, gravity acts mysteriously and instantaneously across a vacuum. In the Einstein picture, gravity is the curvature of physical spacetime and propagates at the speed of light. Which picture is the more attractive?

8 Albert Einstein, a man of simple tastes, was indifferent to prizes and other honors. By contrast, many if not most scientists are consumed by a desire to win prizes and academic distinctions, even though they teach students that scientists are motivated primarily by a spirit of inquiry and a desire to probe the mysteries of nature. Motivation in science makes an interesting psychological study. What do you think? Are members of science departments different from members of art departments?

FURTHER READING

Bergmann, P. G. *The Riddle of Gravitation: From Newton to Einstein to Today's Exciting Theories.* Charles Scribner's Sons, New York, 1968.

Bernstein, J. *Einstein.* Viking Press, New York, 1973.

Davies, P. C. W. *Space and Time in the Modern Universe.* Cambridge University Press, Cambridge, 1977.

Davies, P. C. W. *The Search for Gravity Waves.* Cambridge University Press, Cambridge, 1980.

Eddington, A. S. *Space, Time, and Gravitation.* Cambridge University Press, Cambridge, 1920.

Einstein, A. *Out of My Later Years.* Philosophical Library, New York, 1950.

Einstein, A. "How I created the theory of relativity." *Physics Today* 35, 45 (1982).

Einstein, A. and Infeld, L. *The Evolution of Physics: From Early Concepts to Relativity and Quanta.* Simon and Schuster, New York, 1960.

Frank, P. *Einstein: His Life and Times.* Alfred Knopf, New York, 1947.

Gardner, M. *The Relativity Explosion.* Vintage Books, New York, 1976.

Hoffman, B. *Albert Einstein: Creator and Rebel.* Viking Press, New York, 1972.

Infeld, L. *Albert Einstein: His Work and Influence on Our World Order.* Charles Scribner's Sons, New York, 1950.

Kaufman, W. J. *Relativity and Cosmology.* Harper and Row, New York, 1973.

Lanczos, C. *Albert Einstein and the Cosmic World Order.* John Wiley, New York, 1965.

Mach, E. *The Science of Mechanics: A Critical and Historical Account of its Development.* La Salle, Open Court, New York, 1974.

Pais, A. *"Subtle is the Lord ..." The Science and Life of Albert Einstein.* Oxford University Press, New York, 1980.

Sciama, D. W. *The Unity of the Universe.* Faber and Faber, London, 1959.

Will, M. C. *Was Einstein Right? Putting General Relativity to the Test.* Basic Books, New York, 1986.

SOURCES

Born, M. *Einstein's Theory of Relativity.* 1924. Reprint: Dover Publications, New York, 1962.

Eddington. A. S. *The Nature of the Physical World.* Cambridge University Press, Cambridge, 1928.

Einstein, A. *Relativity: The Special and the General Theory: A Popular Exposition.* Methuen, London, 1954.

Einstein, A. *The Meaning of Relativity*. Princeton University Press, Princeton, N.J., 1955.

Einstein, A., Lorentz, H. A., Weyl, H., and Minkowski, H. *The Principle of Relativity*. Dover Publications, New York, 1952. Some of Einstein's important papers on relativity are reprinted in this book.

Heller, M. "The happiest thought of Einstein's life." *The Astronomy Quarterly* 8, 177 (1991).

Holton, G. "Einstein and the crucial experiment." *American Journal of Physics* 37, 968 (October 1969).

Howard, D. and Stachel, J. Editors. *Einstein and the History of General Relativity*. Birkhäuser, Boston, 1989.

Jaki, S. L. "A forgotten bicentenary: Johann Georg Soldner." *Sky and Telescope* (June 1978).

Klein, H. A. *The New Gravitation*. Lippincott, Philadelphia, 1971.

Lenin, V. I. *Materialism and Empiricism: Critical Comment on a Reactionary Philosophy*. International Publications, New York, 1927.

Mach, E. *The Science of Mechanics*. Open Court, LaSalle, Illinois, 1942.

Misner, C. W. and Wheeler, J. A. "Classical physics as geometry." *Annals of Physics* 2, 525 (1957).

Patch, B. *Thirty Years With G.B.S.*, Gollancz, London, 1951.

Pearce, W. L. Editor. *Relativity Theory: Its Origin and Impact on Modern Thought*. John Wiley, New York, 1968.

Rindler, W. *Essential Relativity: Special, General, and Cosmological*. Van Nostrand Reinhold, New York, 1969.

Schilpp, P. A. Editor. *Albert Einstein: Philosopher-Scientist*. Library of Living Philosophers, Evanston, Illinois, 1949.

Thorne, K. S. "Gravitational radiation," in *300 Years of Gravitation*. Editors S. W. Hawking and W. Israel. Cambridge University Press, Cambridge, 1987.

Weisberg, J. M., Taylor, J. H., and Fowler, L. A. "Gravitational waves from orbiting pulsars." *Scientific American* 245, 74 (October 1981).

13 BLACK HOLES

Confinement to the Black Hole ... to be reserved for cases of Drunkenness, Riot, Violence, or Insolence to Superiors.
British Army regulation (1844)

GRAVITATIONAL COLLAPSE

Stars are luminous globes of gas in which the inward pull of gravity matches the outward push of pressure. The nuclear energy released in the interior at high temperature is radiated from the surface at low temperature and this low-temperature radiation sustains the chemistry of planetary life.

But to each star comes a day of reckoning. Its central reservoir of hydrogen approaches exhaustion and the star begins to die. The tireless pull of gravity causes the central regions to contract to higher densities and temperatures, and as a consequence the outer regions swell up and the star becomes a red giant. A star like the Sun then evolves into a white dwarf in which most of its matter is compressed into a sphere roughly the size of Earth. Many stars end as white dwarfs, slowly cooling, supported internally against gravity by the pressure of electron waves (as in ordinary metals).

More massive stars do not give up the game so easily. Gravity is stronger in these stars and their central regions continue to contract to even higher densities and temperatures, thus enabling them to draw on the last reserves of nuclear energy. These stars become luminous giants squandering energy at a prodigious rate. Soon their reserves of nuclear energy are exhausted. Only gravitational energy remains with its fatal price of continual contraction. In their final throes, the cores rush inward, the mantles explode outward, and for a brief ecstatic moment these stars become blazing supernovas. Out of these cataclysms are born neutron stars in the form of rapidly rotating pulsars.

The most massive stars have imploding cores that cannot be arrested by any known state of matter. Gravity, normally the weakest of forces, overwhelms all opposition and the end is the birth of a black hole.

If we fall with the imploding core of a massive star, following its final moments, in tens of microseconds the core hits virtually infinite density. It becomes a singularity, in some ways similar to the extreme early universe, about which we still understand very little. According to some arguments the density of the singular state is 10^{94} grams per cubic centimeter, or

10 000 000 000 000 000 000 000 000 000
  000 000 000 000 000 000 000 000 000
  000 000 000 000 000 000 000 000 000
  000 000 000 000

times the density of water. The only surviving particles under these extreme conditions are perhaps the basic quanta of spacetime. Spacetime itself is perhaps a dense foam in which space and time are inextricably scrambled together and an orderly timelike sequence of events vanishes (see Chapter 20).

But to a distant observer in the outside world the falling star never reaches the singular state. The gravitational redshift steadily increases and the star appears to fall more and more slowly. As the star approaches a critical size, known as the

Schwarzschild radius, the redshift approaches infinity. The star reddens, darkens quickly into blackness, and stays forever at the critical size, frozen in a permanent state of free-fall collapse. Nothing, not even light, can now escape to the outside world (according to classical theory). Within the collapsing star itself doom lies only a fraction of a second away. To a distant observer, however, the star is a black hole where time stands still and the fate of gravitational collapse is forever concealed from view.

Let observer A fall into a black hole and observer B be far away in the outside world. B thinks, "Why should I worry about the fate of A? In my space and time A is suspended forever in a frozen state and will never reach the singularity." But is this true? Theory shows that in a closed universe of normal pressure and density there is only one future singularity. The observer in the collapsing star sees the outside world blueshifted, and the blueshift attains its extreme value when the star reaches the critical size (the Schwarzschild radius) and becomes a black hole. Everything in the outside world is seen speeded up by observer A, getting faster and faster as the collapse progresses, and, at the critical size, he sees that everything outside happens with extreme rapidity. The future history of the universe passes in a flash. Suppose that in the far future, in tens or hundreds of billions of years in our time (B's time), the universe ceases to expand and collapses back into a second big bang. The inside observer A sees all this happen in little or no time; the galaxies streak away and then streak back again, and the outside world soars in density until it matches the density of the black hole. Time inside the black hole and time in the outside world now tick away together at the same rate, and A and B, holding hands, descend together into a cosmic singularity. Observer B outside, who congratulates herself on not falling into the black hole, now finds herself on equal footing with the inside observer A, and together they meet their doom.

Binary systems

Most massive stars are members of binary systems, and in Chapter 5 we saw that stellar evolution in close binary systems is complicated by the exchange of matter from star to star. When star X in a close binary system swells into a red giant it spills matter onto its companion star Y, and then, with reduced mass, X evolves into a neutron star or perhaps a black hole. The companion star Y in its turn swells up and returns matter to the collapsed star X, which grows in mass, and if it is a neutron star might implode and become a black hole. The companion star Y continues to evolve and finally collapses into either a neutron star or a black hole. The most massive stars are 50 or more times the mass of the Sun, and it is therefore difficult to escape the conclusion that many stars, either isolated or in binary systems, have evolved into black holes (see Figure 13.1).

Collapsed stars in binary systems are of great interest to astronomers. The gas from a companion star that spills over and falls on to a neutron star or a black hole is heated and radiates considerable energy as x-rays.

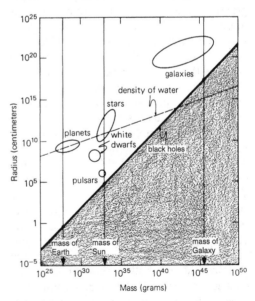

Figure 13.1. Mass and radius of black holes. When ordinary bodies contract they move downward in this diagram and approach the black hole line.

Numerous x-ray sources are known to be binary systems and in each case it seems that one of the companions is either a neutron star or a black hole. Neutron stars have masses usually less than three solar masses and collapsed stars of greater mass are probably black holes. A good candidate for a black hole is the collapsed star in the x-ray binary source known as Cygnus X-1. This was the first x-ray source discovered in the constellation Cygnus, and the collapsed star has a mass of about eight times that of the Sun.

CURVED SPACETIME OF BLACK HOLES
Curvature of spacetime
John Mitchell and Pierre de Laplace showed in the late eighteenth century that Newtonian theory warns us that something serious happens when the escape speed from the surface of a body equals the speed of light.

According to general relativity, spacetime is curved by matter and the curvature is a measure of the strength of gravity. When a body of fixed mass contracts, its surface gravity increases and spacetime becomes more curved. At a critical size, known as the Schwarzschild radius, spacetime becomes so curved that in effect it entirely encloses the body. The body has become a black hole, wrapped in curved spacetime, and nothing, not even light, can leave it and escape to the outside world.

A stretched horizontal rubber sheet helps us to visualize what happens (see Figure 13.2). We imagine a ball of fixed mass resting on the sheet and suppose that the ball slowly shrinks. While we watch the ball on the rubber sheet shrink, we notice how the sheet becomes more depressed and curved in the central region. Eventually, when the imaginary ball is very small and very dense, the sheet enfolds the ball, except for a narrow connecting neck. The ball is not completely isolated because of the neck connecting it with the rest of the sheet; a black hole is also not entirely isolated because it also remains connected with the outside world.

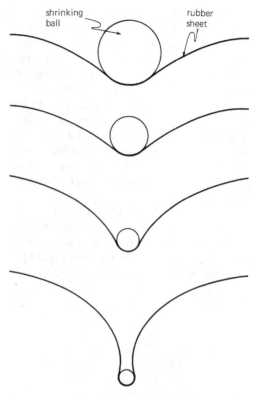

Figure 13.2. A rubber sheet is depressed in the center by a heavy ball. The ball is of constant mass and slowly shrinks in size. When the ball has shrunk to a very small size, as shown, the sheet encapsulates the ball and forms a connecting neck. This is analogous to the deforming of space in general relativity theory.

The amount of depression in the sheet at a fixed distance from the ball is determined by the ball's mass, and because this is fixed, the curvature of the sheet remains unchanged while the ball shrinks. This effect is similar in general relativity. At a fixed distance the gravitational field, or curvature of space, remains unchanged as a spherical body of fixed mass contracts. The curvature of distant space remains unchanged, and hence no gravitational disturbances travel away from a collapsing spherical body.

The Schwarzschild radius of a collapsed body of mass M is

$$R_S = \frac{2GM}{c^2},$$ [13.1]

where G is the gravitational constant. When M is measured in solar masses and R_S in kilometers,

$$R_S = 3\frac{M}{M_\odot} \text{ kilometers.} \qquad [13.2]$$

Thus the radius of a black hole of the Sun's mass is 3 kilometers, and of the Earth's mass is about 1 centimeter. If the Sun collapses to a sphere of 3 kilometers radius, or the Earth collapses to a sphere slightly smaller than a table-tennis ball, both would become black holes and have infinite redshifts.

Redshift

Rays of light leaving a luminous gravitating body are affected in two ways: they are redshifted and deflected. As a spherical body of fixed mass contracts, its surface gravity increases in strength, and light rays emitted from the surface are increasingly redshifted and deflected. Let R be the radius of a body whose Schwarzschild radius is R_S; the redshift of light rays escaping to large distance (infinity) is given by the equation

gravitational redshift

$$= \frac{1}{\sqrt{(1 - R_S/R)}} - 1. \qquad [13.3]$$

Thus, when the radius is one-third greater than the Schwarzschild radius (i.e., $R = 4R_S/3$), the redshift is equal to 1. We saw in Chapter 12 that if λ is the emitted wavelength and λ_0 the received wavelength, then

$$\text{redshift} = \frac{\lambda_0 - \lambda}{\lambda}, \qquad [13.4]$$

and for a redshift of 1 an emitted ray's wavelength is increased twofold. When the radius R differs from R_S by only one-millionth, the redshift is 1000, and when the difference is one-trillionth, the redshift is one million, and so on to infinity.

Light deflection

The light rays leaving the surface perpendicularly are undeflected, as shown in Figure 13.3. Light rays moving tangential to the surface are deflected the most. Again we imagine a contracting spherical body of

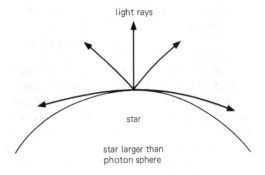

light rays

star

star larger than photon sphere

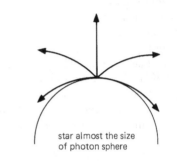

star almost the size of photon sphere

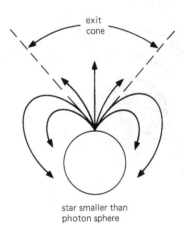

exit cone

star smaller than photon sphere

Figure 13.3. Rays of light leaving a gravitating body are curved, as shown. As the body shrinks in size the rays become more curved. When the radius of the body is less than the radius of the photon sphere, the exit cone begins to close. Rays within the exit cone escape; those outside are trapped and fall back.

fixed mass. When the contracting body reaches a radius 1.5 times the Schwarzschild radius, all rays emitted tangential to the surface are curved into circular orbits. This is the radius of the *photon sphere*; an atom at this distance from a black hole "sees the back of its head" when looking in a tangential direction. On further contraction, the emitted rays become more strongly deflected and many now fall back to the surface. Only the rays emitted within an *exit cone* can escape, and this exit cone narrows as contraction continues. When the body reaches the Schwarzschild radius, the exit cone closes completely and no light rays escape. Redshift and deflection conspire to ensure that no radiation escapes from a black hole.

The photon sphere – of radius 1.5 times the Schwarzschild radius – is the surface where light rays can travel in circular orbits around a black hole. These orbits are unstable: if a circulating ray is disturbed slightly, it either spirals in and is captured, or spirals out and then escapes at radius $\sqrt{3} = 1.732$ times that of the photon sphere

(see Figure 13.4). The redshift of light leaking outward from the photon sphere is $\sqrt{3} - 1 = 0.732$. All light rays approaching a black hole closer than $\sqrt{3}$ times the radius of the photon sphere spiral inward and are captured (see Figure 13.5).

Again we consider what happens at the surface of a contracting spherical body, and this time we imagine that particles having rest mass are shot out horizontally from a point on the surface. Usually we can find a velocity such that a particle will follow a circular orbit. Particles in circular orbits still obey Kepler's law (P^2 is proportional to R^3) in general relativity, provided that the period P and the orbit radius R are measured in the space and time of a distant stationary observer. Circular orbits, however, are not possible when the radius is less than three times the Schwarzschild value, and particles within this region spiral inward and are captured by the black hole.

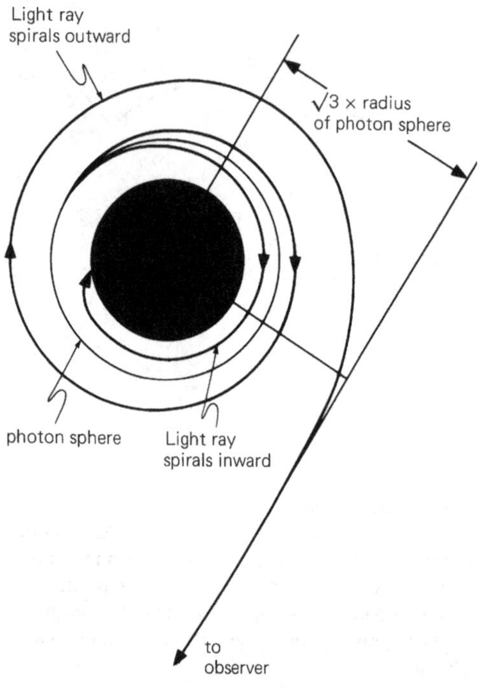

Light ray
spirals outward

$\sqrt{3} \times$ radius
of photon sphere

photon sphere   Light ray
spirals inward

to
observer

Figure 13.4. The photon sphere has a radius 1.5 times the Schwarzschild radius. Circular rays at the photon sphere either spiral in and are captured or spiral out and escape. Rays escaping from the photon sphere have redshift $\sqrt{3} - 1 = 0.732$.

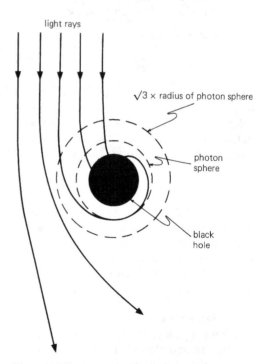

light rays

$\sqrt{3} \times$ radius of photon sphere

photon
sphere

black
hole

Figure 13.5. Deflection of rays by a nonrotating black hole. Rays approaching closer than $\sqrt{3}$ times the radius of the photon sphere are captured.

Lightcone tilting

Gravity is weak at large distances from a black hole and spacetime is approximately flat and the same as the spacetime used in special relativity. But close to a black hole spacetime is greatly deformed. The intervals of space and time of a stationary observer are not the same as those of the distant observer. It is obviously more convenient to discuss a black hole always in terms of the normal space and time of the distant observer. When we say that a black hole has a certain radius, such as 3 or 30 kilometers, we are using not the space of a nearby observer but that of the distant observer. Let us consider a particle falling toward a black hole in terms of the space and time of a distant stationary observer.

At large distances from the black hole the particle has lightcones similar to those of the distant observer. As the particle approaches the black hole its lightcones become tilted and its future lightcone tips forward toward the black hole, as shown in Figure 13.6. This tilting of the lightcone is caused by the curvature of spacetime. When the particle reaches the Schwarzschild surface, its future lightcone is tipped so far forward that all light emitted by the particle falls into the black hole and none escapes to the outer world. The past lightcone is tipped so far backward that light is received only from the outside world, and the particle therefore sees only the world it leaves and not the fate that awaits it. Inside the Schwarzschild surface the lightcones are tilted even more; all light emitted by the particle moves ahead into the singularity and the particle cannot see the singularity into which it is plummeting.

The free-falling particle's own spacetime, in its immediate vicinity, is flat and the same as that of ordinary special relativity. Thus the particle passes smoothly into the black hole without knowing that something terrible has happened. In its own time, it reaches the singularity in a time equal to the Schwarzschild radius divided by the speed of light. To the distant observer, however, it takes an infinite time for the particle to reach and enter the black hole.

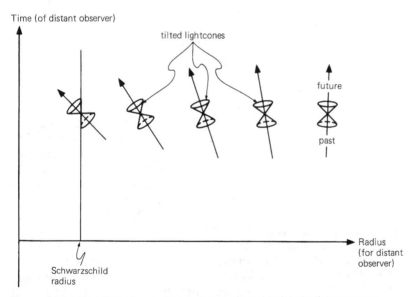

Figure 13.6. The effect of spacetime curvature near a black hole: lightcones are tilted so that the future lightcone tips toward the black hole. At the surface (the event horizon) of the black hole, all rays emitted fall into the black hole, and no rays from the past are received from the black hole. An observer passing into a black hole receives no information of what lies ahead.

The point–circle survey

Emitting points and wavefront circles help us to understand the effect of spacetime curvature (see Figure 13.7). We imagine that a point in space, stationary relative to the black hole, emits a brief pulse of radiation. The wavefront circle shows us where the rays of radiation have reached a moment later. Far from the black hole, space is almost flat and the emitting point is in the center of the wavefront circle. Near the black hole the emitting point is off-center and displaced away from the black hole. Curved space shifts the wavefront circle; in

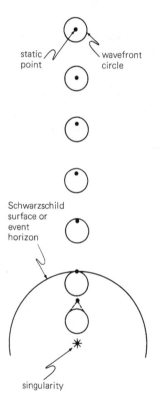

Figure 13.7. Static emitting points and wavefront circles. A static point (i.e., static in the spacetime frame of a distant observer) emits a pulse of radiation in all directions. A moment later the rays reach a surface that is shown as the wavefront circle. Near the black hole the rays are dragged inward and the wavefront circle is displaced toward the black hole. At the surface of the black hole (the event horizon), the emitting point lies on the wavefront circle and no rays escape outward.

effect, gravity drags the light rays toward regions of stronger gravity.

Dynamic space

This dragging effect is the same as if space itself flows into the black hole and carries the light rays with it. That space is dynamic forms an important part of general relativity, and is of utmost significance in cosmology; it will be encountered later in the "expanding space paradigm" in Chapter 14. Special relativity tells us that nothing moves through space faster than light; general relativity agrees, but tells us that space, being dynamic, may also move and its motion is not subject to the rules of special relativity. In some instances, as inside a black hole and outside the Hubble sphere, space flows faster than the speed of light.

At the Schwarzschild surface the emitting point is located on the wavefront circle itself, just what one would expect with space itself flowing inward at the speed of light. Light rays moving outward at the Schwarzschild surface remain in the same place; they move at the speed of light and travel through space that is itself falling in at the speed of light. The surface of the black hole is the country of the Red Queen where one must move as fast as possible in order to remain on the same spot.

The event horizon

The surface of a black hole, where space in effect falls inward at the speed of light, is known as the *event horizon*. Events inside the horizon are unobservable and can never communicate with observers outside because light signals cannot travel faster than the speed of light (see Figure 13.8). A signal traveling outward at the speed of light remains static at the event horizon. Inside the horizon space flows inward faster than the speed of light, and outward-moving signals traveling through space at the speed of light are dragged inward and cannot reach the horizon. Special relativity is still valid locally in the frame of the free-falling particle, and in this free-falling frame the local speed of light is the same as that used

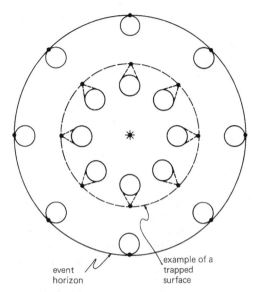

event
horizon

example of a
trapped
surface

Figure 13.8. A surface on which emitting points are in contact with their wavefront circles forms an event horizon of a nonrotating black hole. The event horizon in general is a one-way membrane. Light and everything else can move inward but not outward. Inside the event horizon a static emitting point lies outside its wavefront circle. A closed surface on which emitting points are outside their wavefront circles is called a trapped surface. The event horizon is the outermost trapped surface. Roger Penrose and Stephen Hawking showed that inside a trapped surface collapse to a singularity is inevitable (provided $rc^2 + 3P$ is positive, where r is density and P is pressure).

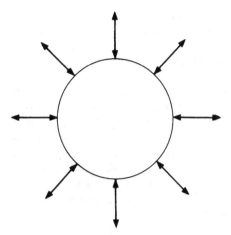

Figure 13.9. The "no hair theorem." Only "hair" that sticks out straight survives collapse to a black hole, and all else is drawn in or radiated away. Thus radial electric lines of force survive and a black hole may have an electric charge.

by the distant observer in flat space. Free-falling observers who use special relativity locally are carried along in the flow of space. A spaceship that falls into a black hole can never escape its doom; no matter how strong the thrust of its rocket engines, the spaceship cannot exceed the speed of light in its own local space, which is itself falling faster than the speed of light.

Perfect symmetry

Nonrotating black holes are perfectly spherical. This raises an interesting question. If the collapsing body is not exactly spherical, does this mean the final black hole is also not exactly spherical? No. During the final moments of collapse all nonspherical irregularities are radiated

away into the outside world as gravitational and electromagnetic waves, leaving behind a perfectly spherical black hole. This is quaintly referred to by the phrase "black holes have no hair." Long-range fields of force, such as gravitational and electric fields, survive the collapse, but nonradial short-range fields of force, such as magnetic fields, are either pulled in or radiated away (see Figure 13.9).

If nothing can escape, it has been asked, how is it possible for gravity to reach out beyond a black hole? A way of looking at this question is to think of gravity as radial lines of force; the collapsing body slides down the lines of force, leaving them to survive in the outside world. Another and better way is to understand that gravity consists of curved spacetime, which is continuous and cannot terminate at an edge. A black hole is not an isolated universe of its own, but remains connected with and part of our universe.

ROTATING BLACK HOLES
Rotation and electric charge

Up till now we have considered black holes as having mass and nothing else. More generally, there are three basic quantities that determine a black hole: mass, rotation (or rather angular momentum), and electric charge. These are the only properties that survive when a collapsing body becomes a black hole.

Strong concentrations of electric charge are rare in astronomy because they are easily neutralized by electrically charged particles in the interstellar medium, and most massive bodies tend to be electrically neutral. We shall therefore not explore the effect of electric charge in black holes. Rotation, however, is common in astronomy and cannot be ignored.

Rotation

Almost all matter in the universe has large-scale bulk rotation. Stars generally rotate, and as they contract they spin faster. Most stars would fly apart because of centrifugal force long before they had contracted to the size of a black hole. But stars in their death throes have various tricks for getting rid of the rapid rotation of their contracting cores. The neutrinos that stream out of the core carry away some rotation (angular momentum), and the spinning core is also braked by viscous forces and magnetic fields that link it with the slower rotating mantle. By various means the core sheds its rapid rotation and continues to collapse and become a neutron star and even a black hole. There is little doubt that neutron stars and black holes are generally born with rapid rotation, and some initially may spin as fast as 10 000 revolutions a second.

The gas dispersed between stars is generally in a swirling state of motion and cannot fall from large distances directly into a black hole. Gas instead will accumulate around a black hole and rotate in the form of a disk-shaped cloud. Such a rotating cloud, called an accretion disk (see Figure 13.10), is supported against gravity by centrifugal force, just like the rings of Saturn. The inner regions of the disk, in accordance with Kepler's third law, revolve about the black hole more rapidly than the outer regions. Friction within the disk brakes the inner regions and speeds up the outer regions. As a result, matter in the inner regions of the disk loses rotation and slowly spirals inward until it is accreted by the black hole, and matter in the outer regions gains rotation and slowly spirals outward carrying away into space excess angular momentum. The accretion disk is continually replenished with gas falling in from large distances. Even if a black hole initially had no rotation, it would soon acquire rotation from the accretion of swirling gas. Accretion disks develop in regions where an abundance of

Figure 13.10. This shows an accretion disk encircling a black hole. Gas and dust spiral inward, angular momentum spirals outward, and the black hole grows in mass.

gas exists, such as close binary systems and the nuclei of galaxies.

The Kerr metric

Karl Schwarzschild in 1916 solved the Einstein equation for the exterior spacetime of a nonrotating black hole. Roy Kerr in 1963, the year that quasars were discovered, solved the Einstein equation for the exterior spacetime of a rotating black hole. The Kerr spacetime, or Kerr metric, is not as simple as the Schwarzschild spacetime. In the discussion of nonrotating black holes we saw that spacetime curvature causes the future lightcone to be tilted in the direction of the black hole. Now, in the neighborhood of a rotating black hole, we have the additional effect that spacetime curvature causes the future lightcone to tilt also in the direction of rotation.

We have previously imagined space flowing into a nonrotating black hole, creating an event horizon at the Schwarzschild surface where space flows inward at the speed of light. We must now imagine that space also rotates like a whirlpool as it flows inward into a rotating black hole. Particles caught in the whirlpool are carried around as they fall inward.

We can visualize more easily what happens by using emitting points and wavefront circles (see Figure 13.11). At large distances from a rotating black hole, space is flat and the static emitting point is in the center of the wavefront circle. Near the black hole the rays emitted by the point are dragged inward and around, and consequently the wavefront circle is displaced partly in the inward direction and partly in the direction of rotation. This situation forms two distinctly different surfaces about the black hole.

The first surface, known as the static-limit surface (or just the *static surface*), is where space flows at the speed of light. At this surface the static emitting points are at the edges of their wavefront circles. The second surface is the event horizon where space flows in the inward radial direction at the speed of light. At this surface all emitting

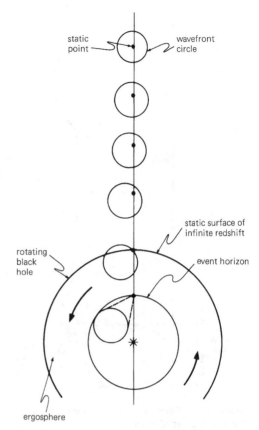

Figure 13.11. In the vicinity of a rotating black hole all wavefront circles are displaced inward and also in the direction of rotation. We may imagine space as a whirlpool in which it rotates as it flows into the rotating black hole. There are now two surfaces enclosing a black hole: the outer static surface, which has the shape of an oblate ellipsoid, and the inner event horizon, which is spherical. At the static surface all static emitting points lie on their wavefront circles. At the event horizon all static emitting points have their wavefront circles touching the horizon.

points and their wavefront circles touch the horizon.

The outer static surface has the shape of an oblate (flattened) spheroid, as in Figure 13.12, and an equatorial radius equal to the Schwarzschild radius. It is the surface where space flows at the speed of light, and a particle, stationary in this flow of space, is dragged along at the speed of light relative to the distant observer. If the particle moves against the flow of space at the speed of light

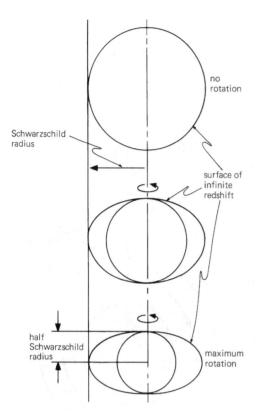

Figure 13.12. Black holes of the same mass but different amounts of rotation.

event horizon shrinks to a smaller radius, and at maximum rotation has a radius half the Schwarzschild value. At maximum rotation the equator of the event horizon rotates at the speed of light relative to the distant observer.

The ergosphere

Between the event horizon and the static surface lies a region known as the ergosphere; it consists of swirling space spiraling inward to the horizon (see Figure 13.13). Space in the ergosphere flows faster than the speed of light, but the inward radial component of velocity of the flow is less than the speed of light. Static points lie outside their

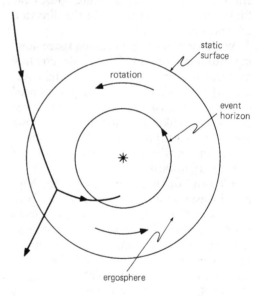

Figure 13.13. The ergosphere is the region between the outer static surface and the inner event horizon. A body entering the ergosphere can divide into two components in such a way that one component escapes with more energy than the initial infalling body, and the other component falls into the black hole. The energy gained by the escaping component is at the expense of the rotational energy of the black hole. Thus the escaping component subtracts more rotational energy than the incident body contributes. Similarly, an incident ray of light may split into two components, one component is scattered with increased energy, and the other is absorbed. This is known as superradiant scattering.

it remains static on this surface, hence the name static surface. The wavefront circle protrudes outside the static circle, and some rays of light can therefore still escape to infinity. Inside the static surface space flows faster than light, and a particle, no matter how fast it moves against the flow, is always dragged along and can never be stationary relative to the distant observer.

The inner surface, the event horizon, is spherical in shape. Across this surface space moves inward with a radial component of velocity that equals the speed of light. Light rays traveling in any direction cannot escape to the outside world. Both surfaces, the static surface and the event horizon, are in contact at the poles. In a nonrotating black hole both surfaces merge together and have the Schwarzschild spherical configuration. As rotation increases, the wavefront circles, but some rays can still move outward and escape to the outside world. Roger Penrose of Birkbeck College, London, showed that in principle it is possible to gain energy in the ergosphere from the black hole. A body is shot into the ergosphere where it divides into two components; one component is shot backwards against the rotation and the other component moves forward with increased energy. The first component plunges into the event horizon, and the second component escapes across the static surface carrying energy that comes from the rotational energy of the black hole.

Some of the mass of the black hole is due solely to its rotational energy. A black hole of maximum rotation has $(1 - 1/\sqrt{2}) = 0.29$ of its mass contributed by the rotational energy, and this is the maximum amount of mass that can be extracted as energy by the Penrose method. Robert Wald in *Space, Time, and Gravity* cautions us with these words: "the energy extraction idea discussed here – as well as all the other ideas in this book – are (at least at the present time) of interest for understanding nature rather than for possible practical, technological advances."

Cosmic censorship

At maximum rotation the event horizon has a radius half the Schwarzschild value; its surface at the equator rotates at the speed of light, and space falls inward at the speed of light. Suppose that it were possible to increase the rotation even further. The horizon would vanish! In effect, the rotational velocity would increase, but the inward velocity of flowing space would decrease and be less than the velocity of light. Radiation would then escape and the singularity become exposed to the outside world. The idea of naked singularities is so alarming that it has been banned by what Penrose calls the "cosmic censorship hypothesis." Cosmic censorship states that all singularities are cloaked from view by event horizons and the natural laws conspire to avoid naked singularities. Perhaps universes exist in

which naked singularities are common, but the eruption of energy from such singularities might be so enormous that life could not exist in these scorched universes. We exist because naked singularities do not exist, and the cosmic censorship hypothesis conforms with the anthropic principle.

SUPERHOLES

All-devouring, all-destroying,
Never finding full repast,
Till I eat the world at last.
Jonathan Swift, On Time

A voyage to Brobdingnag

Once born, a black hole grows by accretion to Brobdingnagian size, and its growth halts only when the ambient supply of matter becomes exhausted. In the nuclei of giant galaxies, where stars and gas clouds abound, the possibilities are awesome. Black holes draw in the gaseous wreckage of tidally disrupted stars, and black holes even swallow one another. They swell into superholes of thousands, millions, and perhaps even billions of solar masses. The compressed and heated gas drawn in toward the black hole radiates energy. The radiated energy may be as high as 40 percent of the mass of the accreted gas. This conversion of mass into energy is much more efficient than the 1 percent conversion in nuclear reactions. The released energy covers a wide spectrum of optical, ultraviolet, and x-ray radiation, and may include infrared radiation from surrounding clouds of dust.

Superholes of hundreds of millions of solar masses can grow in a time of hundreds of millions of years in the matter-rich nuclei of giant galaxies, and while growing release energy continually at a prodigious rate. An accreting superhole can radiate more energy than an entire galaxy of luminous stars. The "best-buy" theory of quasars is that they are accreting superholes, and possibly the powerful radio sources are driven by energy also released from superholes.

The tidal forces of a black hole will tear apart a body, such as a planet or a star, when they are stronger than the gravity that

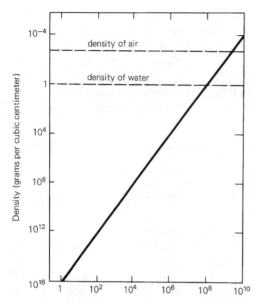

Figure 13.14. The average density of a black hole decreases as mass increases. A black hole 100 million solar masses has an average density equal to that of water, and a black hole of 3000 million solar masses has an average density equal to that of air at sea level.

holds the body together. As a rough guide we may say that a body held together by its own gravity will be tidally disrupted when its average density is less than the density of the black hole. The size of a black hole is proportional to its mass, and this means that its density is inversely proportional to the square of its mass (Figure 13.14):

density

$$= 10^{16} \left(\frac{M_\odot}{M} \right)^2 \text{ grams per cubic centimeter.} \qquad [13.5]$$

Thus a black hole of 10^8 solar masses has a density of only 1 gram per cubic centimeter. Black holes more than 3 billion solar masses have densities less than the Earth's atmosphere at sea level. A star similar to the Sun, having a density of 1 gram per cubic centimeter, is torn apart by tidal forces when it approaches closely a black hole of mass less than 10^8 solar masses. The gaseous wreckage of disrupted stars accumulates in the accretion disk and eventually drains

into the black hole. The gas, compressed as it drains inward, radiates energy efficiently. But when the mass has grown to the point where the tidal disruption of stars ceases, and the black hole swallows stars whole, the energy output wanes. Stars approaching black holes of mass greater than 10^8 solar masses survive tidal disruption and dive straight in without releasing very much energy.

A spaceship could plunge into a superhole of millions of solar masses without its human occupants experiencing great discomfort from tidal forces. In the case of a superhole of much greater mass the occupants at first might not even know that they had passed the event horizon.

MINIHOLES
A voyage to Lilliput
All black holes so far discussed have masses greater than that of the Sun, and it is unlikely that Lilliputian black holes of lesser mass are ever produced by gravitational collapse. A planet, such as the Earth, for example, can never collapse and become a black hole because gravity in these smaller bodies is not strong enough to overcome pressure-gradient forces exerted by matter.

Yet, conceivably, black holes of a wide range of masses were once created in the early universe. In the early stages of expansion the density of the universe was extremely high, and cosmologists have conjectured that black holes might have formed when their densities were comparable with the density of the universe. Thus the formation of primordial black holes does not depend on catastrophic collapse from low densities. A black hole of mass M forms at an approximate time

$$t = 1 \times 10^{-5} \frac{M}{M_\odot} \text{ seconds.} \qquad [13.6]$$

At time t of 1 microsecond, M is 0.1 solar mass, and at a time of 1 picosecond (10^{-12} seconds), M is 10^{-7} solar mass (roughly 0.1 of the Earth's mass). The formation of primordial black holes depends on the nature of the irregularities and other conditions prevailing in the early universe, and the extent to which they actually form is still uncertain.

A primordial black hole of radius R centimeters has a mass M in grams of approximately

$$M = 1 \times 10^{28} R \text{ grams.} \qquad [13.7]$$

Miniholes the size of an atom (10^{-8} centimeters) have a mass of 10^{20} grams and weigh on Earth 100 trillion tons (the weight of a small mountain); miniholes the size of a nucleon (10^{-13} centimeters) have a mass of 10^{15} grams and weigh on Earth 1 billion tons. Note that if we divide the radius of a minihole by the speed of light we obtain roughly the age of the universe at the time it was formed, as given by Equation [13.6]. The smallest of all miniholes formed during the very earliest moments of the universe are the quantum black holes having a Planck mass 10^{-5} grams and a size 10^{-33} centimeters. These primordial quantum miniholes have a mass very roughly 10^{20} times that of a nucleon mass and a size very roughly 10^{-20} that of a nucleon, and it is extremely unlikely that they now exist in the universe, as we shall see below. Miniholes of atomic size or less do not easily accrete matter; thus an atom-sized minihole could pass easily through the Earth, and its mass, equal to a weight of 100 trillion tons, would increase by only 1 microgram.

Possibly the universe contains primordial black holes, not only miniholes of atomic size or less, but also black holes of greater mass.

BLACK-HOLE MAGIC
Various speculative ideas have emerged that involve black holes. One idea is the time-reversed black hole. It has been suggested that a collapsing black hole can be reversed to make a white hole, in which everything rushes out in an immense release of energy. This theory was popular in astronomical circles in the 1970s. What happens when matter falls in to a singularity? One answer, some said, is that it emerges elsewhere in the

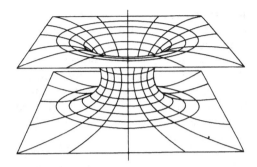

Figure 13.15. Figurative representation of two black holes bridging two universes, or separate spacetime regions of the same universe. Science fiction writers and even some scientists have proposed that bridges may be used for travel in space and time by technologically advanced civilizations.

universe as a white hole, at a different place and time, or perhaps even in another universe. Black holes in other universes, possessing equal rights, pop out unexpectedly in our universe as white-hole eruptions. One bold suggestion was that matter is siphoned from black holes back into the big bang, which is hot because the matter reappears as white holes. There is no evidence that matter can be transported in this manner and the concept of white holes no longer seems tenable.

There is also the argument that black holes form bridges connecting widely separated regions of spacetime (Figure 13.15). Such bridges, sometimes called wormholes, are supposedly used by technologically advanced civilizations as a transportation system for faster-than-light travel in space and backward and forward travel in time. The idea of using bridges as time machines has drawn attention in recent years, particularly in the entertainment industry. A time traveler journeys almost instantaneously to a space station at distance L that recedes at velocity V; the time traveler then returns almost instantaneously via a second bridge and arrives back on Earth in the past at LV/c^2 units of time preceding departure; if the space station is approaching at velocity V, the time traveler arrives at LV/c^2 units of time in the future. (The impossibility of time travel as originally conceived by H. G. Wells in *The Time Machine* is mentioned briefly in Chapter 9, and time travel in general is considered in Chapter 11.) The construction of bridges that stay open and are suitable for transport requires a source of exotic matter of positive density and negative pressure (a pressure less than $-\frac{1}{3}\rho c^2$, where ρ is density).

That black holes are the birthplaces of offspring universes is yet another suggestion (Chapter 25). Black-hole singularities are probably of the same nature as the cosmogenic foam of the extreme early universe. The suggestion is that each black hole collapses into a singularity that then inflates to form a new or offspring universe. Black holes in the offspring universe give birth to further universes, which in turn give birth to more universes, and so on. Lee Smolin of Syracuse University proposed that this reproduction of universes forms the basis of a natural selection theory that explains why the constants of nature (the gravitational constant G, the speed of light c, the Planck constant h, the electric charge of the electron, and the subatomic masses) have their finely tuned values. The anthropic principle says the constants are compatible with our existence and are necessarily finely tuned. Smolin's natural selection theory says the constants are finely tuned because black holes exist, and black holes imply stars, stars imply planets, and planets imply the existence of life and human beings.

The most amazing piece of black-hole magic is the discovery that black holes have a temperature and radiate energy.

HAWKING RADIATION

Sir, I have found you an argument, I am not obliged to find you an understanding.
Samuel Johnson (1709–1784), in Boswell's
Life of Johnson

The richness of the vacuum
In 1974, Stephen Hawking at Cambridge University showed that black holes emit thermal radiation. This remarkable discovery revealed that black holes are less black than previously supposed.

To understand why black holes radiate we must look at the nature of empty space. The vacuum is not so empty as we might think. It is actually a dense sea of virtual particles of every kind, each existing for only a fleeting moment of time. According to the uncertainty principle, energy may always be borrowed, but must be repaid within a limited time. The greater the amount borrowed, the quicker it must be repaid. For example, the energy needed to make two electrons – one positive and the other negative – can be borrowed for 10^{-21} seconds, and for two nucleons the time limit for repayment is 10^{-24} seconds. The whole of space is filled with virtual particles that incessantly appear on borrowed energy and then disappear when payment is due. Collectively they account for many observed effects in the structure of atoms, yet each does not exist long enough to produce its own gravitational field.

Subatomic particles are more than just packets of energy. They possess conserved quantities such as spin and electric charge that cannot be borrowed. Virtual particles are therefore always accompanied by their antiparticles. A virtual electron has a negative electric charge and is accompanied by a virtual positron that has a positive charge and is the antielectron. A particle spins in one direction and its antiparticle spins in the opposite direction and their combined spin is zero. The vacuum is densely populated with virtual electron pairs (electrons and positrons) whose combined spin and electric charge is zero. Particles and antiparticles have the same mass but their conserved quantities are always the opposite of each other. Thus, when a particle and its antiparticle appear briefly in a virtual state, only their joint energy has been borrowed.

Photons – particles of quantized radiation – are their own antiparticles. The antiparticle of a photon is another photon of opposite spin. Both kinds are equally abundant and are emitted equally by nonrotating luminous bodies. Luminous bodies of either

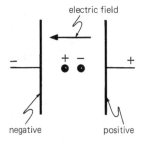

Figure 13.16. A virtual pair of electrons is pulled apart by an electric field. But in the brief time they exist, their chance of gaining energy from the electric field sufficient to become real is usually extremely small. The probability of becoming real increases as the strength of the electric field increases. A similar situation exists with tidal fields.

matter or antimatter emit similar photons. The vacuum seethes with virtual photons in addition to all other virtual particles.

Whenever a virtual particle and its antiparticle are able to gain sufficient energy to become real, they do not return to the limbo of the vacuum state (see Figure 13.16). The required energy can be gained by various methods. We discuss two methods.

Electric fields

Consider an electric field between two parallel conducting plates. Increasing the potential difference between the two plates increases the strength of the electric field. In the space between the plates exist hordes of virtual particle pairs, including electron pairs, each existing for only a brief moment of time. During the short existence of a virtual electron pair, the electron is attracted to the positive plate and the positron is attracted to the negative plate. Usually, in their short lifetime, they move apart hardly at all because the electric field is too weak. But when the electric field is strong, approaching 10^{16} volts per centimeter, the oppositely charged electron and positron move apart a sufficient distance to gain from the electric field the requisite energy to become real. The vacuum now springs to life and a deluge of electrons, positrons, and photons is created. Stronger fields create heavier particles.

Gravitational fields

Consider now virtual particles in a gravitational field. Particle pairs are accelerated in the same direction, but gravity tends to tear the pairs apart because of the tidal force. Black holes produce strong tidal forces, and the smaller the mass (and therefore size) of the black hole, the stronger is the tidal force. When the tidal force close to the surface is sufficiently strong, a virtual particle and its antiparticle are pulled apart a sufficient distance to gain the necessary energy to become real. Most of the real particles created in this way fall inward and no energy is lost from the black hole. But some virtual pairs separate sufficiently for one member to move outward with enough energy to escape. Photons are also created in the same way from the vacuum state. Black holes therefore quantum mechanically emit particles, including photons and neutrinos, and continually lose energy.

Particles in their wavelike manner penetrate seemingly impenetrable barriers. This behavior offers another way of looking at particle emission by black holes. A particle can tunnel through the event horizon and emerge as a wave (see Figure 13.17). The

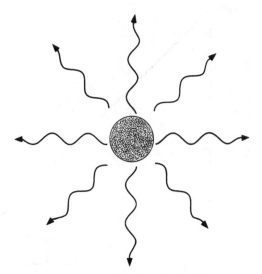

Figure 13.17. Black holes emit radiation by a quantum mechanical process similar to that shown in Figure 13.16. The principal wavelength is roughly the size of the black hole.

most probable wavelength of the emerging particle is roughly the radius of the black hole. Hawking discovered that the emission of particles and photons is thermal – as from a hot body – and black holes have a temperature of roughly

$$\text{temperature} = 1 \times 10^{-7} \frac{M_\odot}{M} \text{ kelvin.} \quad [13.8]$$

Black holes of solar mass have extremely low temperature. The temperature increases as mass decreases, but even a black hole of radius 1 centimeter, of mass equal to that of the Earth, has a temperature of only 0.01 of a kelvin. A black hole the size of a speck of dust, of radius 10^{-3} centimeter and 0.1 lunar mass, has a temperature 10 kelvin. Miniholes of atomic size have temperatures of 1 million degrees and emit intense radiation. Hawking's discovery has shown us that black holes are actually not black but bright when small. They radiate energy and slowly lose mass.

As their mass decreases, black holes get hotter and radiate faster (see Figure 13.18).

With sufficient time it seems possible that all black holes will radiate away their entire mass and leave behind nothing but the radiation they have emitted. The evaporation time is approximately

evaporation time

$$= 1 \times 10^{62} \left(\frac{M}{M_\odot} \right)^3 \text{ years.} \quad [13.9]$$

In most cases the time is extremely long. Even an intensely bright minihole of atomic size lasts for 10^{25} years. A minihole of nucleon size (10^{-13} centimeters) and mass 10^{16} grams has a lifetime of 10^{10} years, which is roughly the age of the universe. This indicates that all primordial black holes of mass less than 10^{16} grams have vanished, and only miniholes of greater mass still survive.

As a minihole diminishes in mass its temperature and luminosity rise and in its final moments it erupts in an outburst of high-energy particles. Maybe in the future astronomers will discover these minihole outbursts.

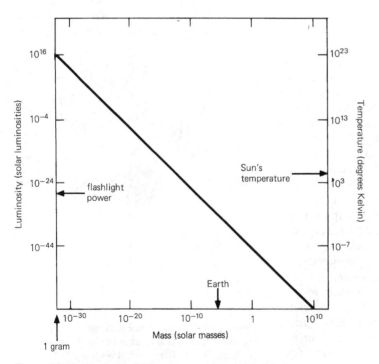

Figure 13.18. The luminosity and temperature of black holes.

BLACK HOLES ARE HEAT ENGINES

A black hole has mass, angular momentum, and electric charge, and nothing else. When a body falls into a black hole, it affects these three properties, and whatever else belongs to the falling body is forever lost. Each time a body falls into a black hole a vast amount of information vanishes, never to be recovered.

A black hole might form from matter or antimatter, or from smaller black holes, or conceivably from radiation, or just gravitational waves. In the course of time it loses mass, which is radiated away mainly as photons. The black hole also emits neutrinos and other particles, such as electrons and positrons, and always equal numbers of particles and antiparticles are emitted. Its emission does not tell from what the black hole was originally made. It slowly evaporates and even in its final outburst it still emits equal numbers of particles and antiparticles. The emitted particles and antiparticles may then annihilate each other and produce further photons. It comes to this: A black hole is a machine for converting all forms of mass into radiation and has no memory of whether the mass was originally that of matter or antimatter. Except for mass, angular momentum, and electric charge, it ignores the conservation of things such as baryon number and lepton number.

Black holes are heat engines that obey the laws of thermodynamics. They have temperature and therefore have also entropy. Entropy is a measure of information lost; it is also a measure of thermal disorder; and in a closed and isolated system, entropy either remains constant or increases and never decreases. We can try to understand entropy by thinking of energy in all its many forms forever cascading into less useful and accessible states; in the process, entropy increases. For example, the energy of a burning candle is converted into radiation, heat, and molecules in lower energy states, and entropy increases.

A black hole is a sink that drains away order and information from the external world. All the detailed structure of the things

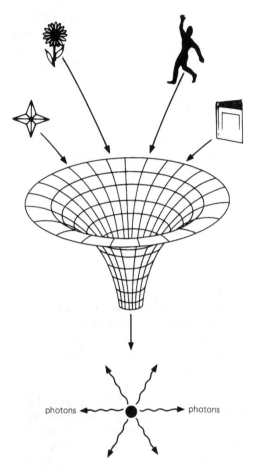

Figure 13.19. A black hole is a sink of lost information. It is also a machine that generates entropy.

from which it was made, such as flowers, crystals, books, and planets, has vanished and in return it gives us photons of high entropy that lack order and information (see Figure 13.19).

The information lost to a black hole is proportional to the surface area of its event horizon, and for a nonrotating black hole this is

$$\text{area} = 4\pi R_S^2 = \frac{16\pi G^2 M^2}{c^4}. \qquad [13.10]$$

Hence entropy is proportional to M^2. When two black holes encounter each other and coalesce, they form a single black hole whose surface area is greater than the sum of the horizon areas of the two initial black holes. A black hole can never split into two black holes because the total final surface area would decrease and entropy would also decrease.

Yet we have seen that a black hole is hot and radiates away its mass. As its mass decreases, its surface area decreases, and its entropy therefore gets less. This does not violate the laws of black-hole thermodynamics. The entropy of the black hole decreases because the emitted radiation carries away entropy into space. The closed system in which entropy is either constant or increases is not just the black hole by itself, but the whole universe, and in the universe the entropy of the black hole plus the entropy of the emitted radiation does not decrease. Black-hole thermodynamics, securely founded on the theoretical discovery that black holes have temperature, has opened up a new realm of physics.

REFLECTIONS

1 *Newtonian laws forewarn us that something odd happens when the escape speed from a star equals the speed of light. This was realized in 1784 by John Mitchell, rector of Thornhill in Yorkshire, who was an innovative astronomer. To escape from the Sun's surface a particle must have a speed of 1/500 of the speed of light. Mitchell argued that if a star has the same average density as the Sun, but a radius 500 times greater, then light would be unable to escape from the surface of the star. "All light emitted from such a body would be made to return to it by its own power of gravity," he wrote. William Herschel was impressed with Mitchell's argument and used it to interpret some of his own observations (erroneously as we now know). The search for evidence of black holes is more than 200 years old!*

Mitchell used Newton's idea that light rays consist of particles, and he assumed that they obey the Newtonian equations of motion. The escape speed V_{esc} from the surface of a body of mass M and radius R is given by

$$V_{esc}^2 = \frac{2GM}{R}, \qquad [13.11]$$

and when V_{esc} is equal to the speed of light c, the radius is

$$R_S = \frac{2GM}{c^2}, \qquad [13.12]$$

and light is extinguished because it cannot escape. It is interesting that Newtonian theory in this instance gives exactly the same result as general relativity theory. Mitchell's argument was as follows: When the density is kept constant, the mass M is proportional to R^3, and therefore, from Equation [13.11], V_{esc} is proportional to R. The escape speed from the Sun is $V_{esc} = 600$ kilometers a second, or 1/500 of the speed 300 000 kilometers per second of light. When the radius increases by a factor 500, the escape speed equals the speed of light. Pierre Simon, Marquis de Laplace, in his classic Exposition of the System of the World *published at the end of the eighteenth century, aware perhaps of Mitchell's work, made a similar prediction: "the attractive force of a heavenly body could be so large that light could not flow out of it." Laplace's discussion is translated into English in* The Large Scale Structure of Space-Time *by Stephen Hawking and George Ellis.*

2 *When gravity in a star overwhelms all possible forms of pressure, the star collapses, and nothing can stop it. The time taken to collapse in free fall is roughly $t_{col} = R/V_{esc}$, where V_{esc} is the escape speed of Equation [13.11]. If we write $M = 4\pi\rho R^3/3$, where ρ is the average density, we find*

$$t_{col} = \left(\frac{3}{8\pi G\rho}\right)^{1/2}.$$

More precisely, as shown by Kelvin in 1902, the gravitational collapse time is

$$t_{col} = \left(\frac{3\pi}{32G\rho}\right)^{1/2}$$

$$= 2 \times 10^3 \rho^{-1/2} \text{ seconds}, \qquad [13.13]$$

where ρ is the density in grams per cubic centimeter. A star similar to the Sun having an average density 1 gram per cubic centimeter collapses to zero radius in 35 minutes, according to Newtonian theory. According to general relativity the star cannot collapse to zero radius in a finite time of the distant observer. The result in Equation [13.13] gives, however, the approximate time for a body of average density ρ to collapse to a black hole, and the approximate time to reach the singularity for an observer who falls with the body.

3 Karl Schwarzschild in 1916, shortly after Einstein had published the final version of general relativity, solved the Einstein equation for the exterior spacetime of a spherical nonrotating body. This solution yields the redshift result quoted in Equation [13.3] and shows that redshift is infinite when a body has a radius $R_S = 2GM/c^2$, now known as the Schwarzschild radius. It is interesting that general relativity gives the same result as Newtonian theory, although for very different reasons.

In 1930, Subrahmanyan Chandrasekhar showed that white dwarfs of mass greater than $1.4M_\odot$ cannot be supported against gravity by electron pressure gradients and must therefore collapse.

In 1934, Walter Baade and Fritz Zwicky advanced the concept of neutron stars. They proposed that these very dense bodies, having the density of the nucleus of an atom, are born in catastrophic stellar events called supernovas. This was only two years after the discovery of the neutron by James Chadwick.

Robert Oppenheimer and George Volkoff in 1939 used general relativity to investigate the structure of neutron stars. In the same year, Oppenheimer and Hartland Snyder studied the collapse of a spherical body from the viewpoints of the external stationary and the internal free-falling observers.

In 1963, Roy Kerr discovered the general relativity solution for rotating black holes, equivalent to the Schwarzschild solution for nonrotating black holes. The discovery of quasars in 1963 stimulated renewed interest in gravitational theory, and in the same year Fred Hoyle and William Fowler argued that the energy released is gravitational in origin and comes from supermassive bodies in the nuclei of giant galaxies. Edwin Salpeter of Cornell University and Y. Zel'dovich in Moscow proposed in 1964 that the supermassive objects are black holes. Pulsars were discovered in 1967 and Thomas Gold in 1968 proposed that pulsars are rotating neutron stars.

The term black hole was first published by John Wheeler in 1968 in an article entitled "Our universe: the known and the unknown." Roger Penrose in 1969 showed that energy can be extracted from rotating black holes and also proposed the cosmic censorship hypothesis that the laws of physics prevent naked singularities from forming. Hawking suggested in 1971 that primordial black holes were formed in the early universe. Many other discoveries have been made on the nature of black holes, of which the most outstanding is Hawking's demonstration in 1974 that black holes emit thermal radiation.

4 Some astronomers have suggested that the Galaxy contains billions of black holes. A black hole might therefore one day come our way and the particles of our bodies in company with the wreckage of the Solar System will be engulfed in a black hole. Instead of continuing to exist until the end of the universe, the particles, once engulfed, will exist for a fraction of a second in their own time before ending in the black hole singularity. But if the universe comes to an end by eventually collapsing, all the particles in the universe, in company with those that previously have fallen into black holes, will together end in the universal singularity. A black hole is a quick way to reach the end of the universe.

5 The word antimatter was first coined in 1898 by the scientist Arthur Schuster in a letter to Nature entitled "Potential matter: a holiday dream." He discussed the possible properties of antimatter and speculated on the existence of antistars. He wrote:

"Astronomy, the oldest and yet most juvenile of the sciences, may still have some surprises in store. May antimatter be commended to its care!"

• Paul Dirac in 1930 developed a relativistic equation in quantum theory that revealed the possibility of antielectrons and antiprotons. Two years later the positron was discovered and the antiproton was found in 1955. Numerous subatomic particles and their antiparticles have since been discovered with high-energy accelerators. Particles and antiparticles are distinguished by their conserved intrinsic properties such as electric charge and baryon and lepton numbers. Baryons, for example nucleons (proton and neutrons), are heavy particles (their name derives from the Greek bary, meaning heavy). Each has a baryon number +1. Antibaryons, such as antiprotons and antineutrons, have each a baryon number −1. A nucleon and its antinucleon have a combined baryon number equal to zero, and can therefore annihilate and produce photons that have no baryon number. Leptons, such as electrons and neutrinos, are light particles (their name derives from the Greek lepto, meaning small or light). Each has a lepton number +1; their antiparticles, the positrons and antineutrinos, each have a lepton number −1. An electron and a positron have a combined lepton number equal to zero and can therefore annihilate, and their energy goes into photons that have no lepton number. The baryon and lepton numbers are conserved in all particle interactions. But black holes apparently ignore these powerful laws of conservation. Once a body has passed beyond the event horizon we have forever lost all information on whether originally it was made of matter or antimatter.

• All matter consists of particles and all antimatter consists of antiparticles. Everything in our part of the universe – the Earth, Solar System, and Galaxy – is apparently made solely of matter. A substantial amount of antimatter in the Galaxy, if it exists. would betray itself by violent and recognizable interactions with ordinary matter. We do not observe significant emission of high-energy photons – gamma rays – of the expected energy. What about other places in the universe? Antistars in antigalaxies emit the same kind of photons as stars in galaxies, and we cannot distinguish them by their radiation. Most galaxies are members of clusters, however, and if antigalaxies are mixed with galaxies we should be able to see unmistakable signs of their violent interactions. There are indications of violence, it is true, but not of the kind that emits gamma rays at distinctive energies. Gas dispersed between galaxies in the rich clusters appears not to contain an admixture of antigas because we have not observed the distinctive radiation that results from annihilation processes. The evidence so far indicates that antimatter is rare in the universe. According to grand unified theories the universe is made only of matter because in the very early expanding universe the dominant hyperweak force distinguished very slightly between matter and antimatter in their particle decay schemes.

6 The Planck mass referred to in the text is easily calculated. We suppose that a particle exists of mass m_P whose gravitational length $\lambda_P = Gm_P/c^2$ is equal to its Compton wavelength $\lambda_P = \hbar/m_P c$. Thus,

$$m_P = \left(\frac{\hbar c}{G} \right)^{1/2} = 2 \times 10^{-5} \text{ grams}, \quad [13.14]$$

$$\lambda_P = \left(\frac{G\hbar}{c^3} \right)^{1/2}$$

$$= 2 \times 10^{-33} \text{ centimeters}. \quad [13.15]$$

A Planck unit of time is

$$t_P = \frac{\lambda_P}{c} = 10^{-43} \text{ seconds}. \quad [13.16]$$

Our familiar units of mass (grams and kilograms), length (centimeters and meters), and time (seconds and years) are determined by human values and are not universal or "natural." Extraterrestrial intelligent beings have other values and different units. Max Planck showed that the physical constants G, c, and $\hbar$ provide a set of universal and natural units.

7 *Thermal radiation at temperature T has a characteristic wavelength $\lambda = 0.3/T$ centimeters, where T is measured in kelvin. This is the part of the spectrum of maximum intensity. A black hole has a radius R_S given by $3 \times 10^5 M/M_\odot$ centimeters. Because a black hole and the thermal radiation it emits have characteristic length scales of R_S and λ, it would not be surprising to find that they are comparable in value. By equating the two we obtain the result $T = 1 \times 10^{-6} M_\odot/M$, which is not greatly different from Hawking's more exact result in Equation [13.8]. The agreement is better if we equate the wavelength to the circumference of the black hole.*

In more detail, we may argue as follows. A photon of wavelength λ has an energy $\hbar c/\lambda$. Let λ be the characteristic wavelength of thermal radiation at temperature T; the characteristic energy of a photon is then $kT = \hbar c/\lambda$, where k is the Boltzmann constant. If we equate the wavelength to the circumference $4\pi GM/c^2$ of a black hole of mass M, we find

$$T = \frac{\hbar c^3}{4\pi GMk}, \qquad [13.17]$$

which is almost exactly the same as Hawking's result. With $k = 1.4 \times 10^{-16}$ ergs per kelvin, we obtain a temperature similar to that quoted in the text (Equation 13.8). Note that this result can be expressed in the form

$$kT = m_P c^2 (m_P/4\pi M), \qquad [13.18]$$

thus the energy of emitted photons (and of other particles) is the Planck energy $m_P c^2$ multiplied by $m_P/4\pi M$.

The luminosity can be estimated in a similar rough-hewn manner. The black hole emits one photon of energy $\hbar c/\lambda$ in time λ/c. Therefore the luminosity is $L = \hbar c^2/\lambda^2$, and with λ equal to $4\pi GM/c^2$, we obtain

$$L = \frac{\hbar c^6}{16\pi^2 G^2 M^2} = \frac{c^5}{G}\left(\frac{m_P}{4\pi M}\right)^2, \qquad [13.19]$$

where $c^5/G = 9 \times 10^{25}$ solar luminosities. The lifetime t of the black hole can be estimated in a similar approximate manner, and from $t = Mc^2/L$, we find

$$t = \frac{16\pi^2 G^2 M^3}{\hbar c^4}, \qquad [13.20]$$

from which we obtain the result shown in Equation [13.9].

8 *We may estimate the entropy of a black hole in a similar very approximate manner. Entropy in the universe is discussed in Chapter 19, where it is shown that the entropy of the universe is related to the number of photons it contains multiplied by the Boltzmann constant k. The number N of photons (and other particles) emitted by the black hole is the total energy Mc^2 divided by the characteristic energy $\hbar c/\lambda$ of individual photons. With the wavelength simply equal to the circumference $4\pi GM/c^2$, we find*

$$N = 4\pi(M/m_P)^2, \qquad [13.21]$$

where, as before, m_P is the Planck mass. Thus a black hole of solar mass emits a total of 10^{78} photons. This is a measure of its entropy, and when the black hole has evaporated, it has given all this entropy to the universe. Notice that the number of photons emitted is proportional to M^2 and therefore proportional to the surface area $4\pi R^2$, and hence the entropy of a black hole is proportional to its surface area.

PROJECTS

1 Talking points:
   (a) Does an object falling into a black hole vanish from the universe?
   (b) How can the presence of a black hole be detected?
   (c) What is seen as one enters a black hole?
   (d) How large would the Galaxy be if it were a black hole?
   (e) Can a black hole be composed of smaller black holes?

(f) When is a minihole not black but white?

(g) Consider two particles, one inside a black hole and the other outside. Which reaches the singularity first in a collapsing universe?

(h) "The very hairs of your head are all numbered" (Matthew 10:30). Is this true of a friend who has fallen into a black hole?

(i) Suppose that the mind is more than just the physiological brain. Imagine that a person falls into a black hole. The black hole slowly evaporates and is converted mainly into photons. Where has the mind gone?

2 Discuss matter and antimatter.

3 Why are black holes like heat engines? What is so strange about them?

4 Is the universe a black hole? (Remember, an essential feature of a black hole is that its interior spacetime is continuous with the spacetime of the outside world.)

5 The totalitarian doctrine was first enunciated by the journalist John T. Whittaker in "Italy's Seven Secrets" (*Saturday Evening Post*, December 23, 1939, p. 53). He wrote: "Coffee is forbidden, the use of motorcars banned and meat proscribed twice a week, until one says of Fascism, 'Everything which is not compulsory is forbidden.'" This is the principle of prohibition that attends authoritarian rule. The inverse is the principle of plenitude: everything that is not forbidden is compulsory. Whatever is possible must exist. In science we see that nature is plenitudinous rather than parsimonious. What do you think?

FURTHER READING

Carr, B. "A brief history of time warps." *Nature* 371, 212 (1994).

Davies, P. C. W. *Space and Time in the Modern Universe.* Cambridge University Press, Cambridge, 1977.

Greenstein, G. *Frozen Star.* Freundlich Books, New York, 1984.

Harrison, E. R. "'Black holes' in history." *Quarterly Journal of the Royal Astronomical Society* 29, 87 (1988).

Hawking, S. "The quantum mechanics of black holes." *Scientific American* (January 1977).

Penrose, R. "Black holes." *Scientific American* (May 1972).

Ruffini, R. and Wheeler, J. A. "Introducing the black hole." *Physics Today* (January 1971).

Sciama, D. W. "The ether transmogrified." *New Scientist* (February 2, 1978).

Smarr, L. L. and Press, W. H. "Our elastic spacetime: black holes and gravitational waves." *American Scientist* (January–February 1978).

Thorne, K. S. *Black Holes and Time Warps: Einstein's Outrageous Legacy.* Norton, New York, 1994.

Wheeler, J. A. "Our universe: the known and the unknown." *American Scholar* 37, 248 (1968) and *American Scientist* 56, 1 (1968). First mention of the term "black hole."

SOURCES

Blandford, R. D. "Astrophysical black holes," in *300 Years of Gravitation.* Editors, S. W. Hawking and W. Israel. Cambridge University Press, Cambridge, 1987.

Hawking, S. W. and Ellis, G. F. R. *The Large Scale Structure of Space-Time.* Cambridge University Press, Cambridge, 1973.

Israel, W. "Dark stars and the evolution of an idea," in *300 Years of Gravitation,* Editors, S. W. Hawking and W. Israel. Cambridge University Press, Cambridge, 1987.

Mackeown, P. K. and Trevor, C. W. "Cosmic rays from Cygnus X-3." *Scientific American* 253, 60 (November 1985).

Misner, C. W., Thorne, K. S., and Wheeler, J. A. *Gravitation.* Freeman, San Francisco, 1973.

Oppenheimer, J. R. and Snyder, H. "On continued gravitational contraction." *Physical Review* 56, 455 (1939).

Penrose, R. "Gravitational collapse: the role of general relativity." *Revista Nuovo Cimento,* numero special 1, 252 (1969).

Schaffer, S. "John Mitchell and black holes." *Journal for the History of Astronomy* 10, 42 (1979).

Schuster, A. "Potential matter: a holiday dream." *Nature* 58, 367 (August 18, 1898).

Sciama, D. W. "Black holes and their thermo-dynamics." *Vistas in Astronomy* 19, 385 (1976).

Thomson, W. (Lord Kelvin). "On the clustering of gravitational matter in any part of the universe." *Philosophical Magazine* 3, 1 (1902).

Trimble, V. "Neutron stars and black holes in binary systems." *Contemporary Physics* 32, 103 (1991).

Wald, R. M. *Space, Time, and Gravity*. University of Chicago Press, Chicago, 1977.

14 EXPANSION OF THE UNIVERSE

The theory of relativity brought the insight that space and time are not merely the stage on which the piece is produced, but are themselves actors playing an essential part in the plot.
Willem de Sitter, "The expanding universe" (1931)

THE GREAT DISCOVERY

Doppler effect

From a historical viewpoint the Doppler effect paved the way to the discovery of the expanding universe. Nowadays we do not use the Doppler effect in cosmology, except in its classical Fizeau–Doppler form as a rough and ready guide. We examine the Doppler effect briefly and defer to Chapter 15 a more searching inquiry.

The spectrum of light from a luminous source contains bright and dark narrow regions, as shown in Figure 14.1, that are the emission (bright) and absorption (dark) lines produced by atoms. When a luminous source such as a candle or a star moves away from an observer, all wavelengths of its emitted radiation, as seen by the observer, are increased. Its spectral lines are moved toward the longer wavelength (redder) end of the spectrum and it is said to have a redshift. This redshift is detected by comparing the spectrum of the luminous source with the spectrum of a similar source that is stationary relative to the observer. The source may move away from the observer, or the observer may move away from the source, and in either case the separating distance increases and there is an observed redshift.

When the luminous source moves toward the observer, all wavelengths of its emitted radiation, as seen by the observer, are decreased. Its spectral lines are moved toward the shorter wavelength (bluer) end of the spectrum and it has a blueshift. The source may move toward the observer, or the observer may move toward the source, and in either case the separating distance decreases and there is an observed blueshift.

The shift in the observed wavelengths can be expressed in terms of the relative velocity by means of the classic Fizeau–Doppler relation (Figure 14.2). Let V be the relative velocity of a luminous source moving away; also let λ be the wavelength of an emitted ray of light and λ_0 the wavelength of the same ray received by the observer. According to the Fizeau–Doppler formula,

$$\frac{\text{observed wavelength}}{\text{emitted wavelength}} = \frac{\lambda_0}{\lambda} = 1 + \frac{V}{c}, \quad [14.1]$$

where c is the speed of light. The wavelength of a line in the spectrum of the light from the source, as seen by the observer, is measured and compared with the wavelength of the same line emitted by atoms in the observer's laboratory. The relative velocity V of the source is determined from the fractional difference in the emitted and received wavelengths:

$$\frac{V}{c} = \frac{\lambda_0 - \lambda}{\lambda}, \quad [14.2]$$

given by Equation [14.1]. But this classic formula can be used only when V is very much less than c (V less than $0.01c$). For higher velocities we must use a formula (referred to as the special relativity Doppler formula or just the Doppler formula) derived from special relativity theory and given in the next chapter (Chapter 15). Notice that when the source moves toward

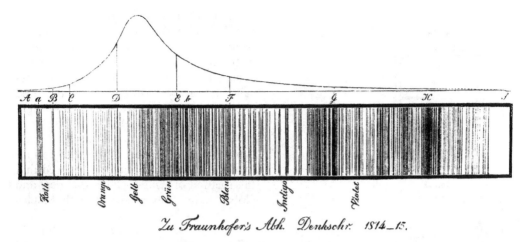

Figure 14.1. A spectrum of light from the Sun in which wavelengths increase from violet at the right to red at the left. This spectrum shows the dark absorption lines, as observed in 1814 by the German physicist Joseph von Fraunhofer. These lines are now known as Fraunhofer lines. The curve on the top shows that the intensity of sunlight peaks between yellow and green.

the observer the velocity V changes sign and becomes negative, and we can still use Equations [14.1] and [14.2]. The redshift of a source is defined as the fractional increase in wavelength:

$$z = \frac{\lambda_0 - \lambda}{\lambda} \qquad [14.3]$$

and is always denoted by z. From Equations [14.2] and [14.3] we find

$$V = cz. \qquad [14.4]$$

Thus a redshift of 0.01 means that the source is receding at 1 percent of the velocity of light, or 3000 kilometers per second.

Slipher's celestial speed champions
Vesto Slipher, an astronomer at the Lowell Observatory at Flagstaff, Arizona, in 1912 began to measure the shift in the spectral lines of light received from spiral nebulae. By 1923, as a result of his painstaking measurements, it was known that of the 41 galaxies studied, 36 had redshifts, and the other five, which included the Andromeda Nebula, had blueshifts. His measured redshifts multiplied by the velocity of light indicated recession velocities of many thousands of kilometers a second. The discovery of galaxies moving with these

amazingly high velocities received wide publicity in the press. Slipher's results were also surprising because, if galaxies had random motions, one would expect that those moving away with redshifts would be approximately equal in number to those approaching with blueshifts. His observations revealed that most galaxies were moving away from us.

Discoverers of the expanding universe
The discovery of the expanding universe did not occur abruptly. To unfold the historical record we must anticipate certain developments that will be clearer in later chapters.

The first intimation of an expanding universe came in 1917 in the work of the Dutch astronomer-cosmologist Willem de Sitter. On the basis of theoretical studies, he predicted the existence of a "systematic displacement of spectral lines toward the red" in the light received from distant nebulae. There is little doubt that Slipher's redshift measurements and de Sitter's studies gave birth to the idea of an expanding universe.

In 1917, Einstein and de Sitter proposed two different kinds of universe, both based on Einstein's theory of general relativity that had been published in its final form

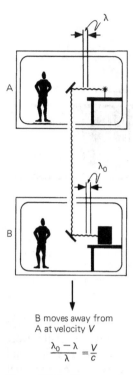

B moves away from
A at velocity V

$$\frac{\lambda_0 - \lambda}{\lambda} = \frac{V}{c}$$

Figure 14.2. Observers A and B are in separate laboratories that move apart at velocity V. Both study the light emitted by atoms in their laboratories and find their emission and absorption lines have identical wavelengths in the two laboratories. Let λ be the wavelength of one of these spectral lines. A sends light to B, who finds that the wavelength λ of the transmitted line is received at λ_0. The classical Fizeau–Doppler formula states that the fractional increase $(\lambda_0 - \lambda)/\lambda$ is equal to V/c, where c is the velocity of light.

the previous year. Einstein's universe was uniform: it contained uniformly distributed matter and had uniformly curved spherical space. The main feature of Einstein's model of the universe was its static nature; it was unchanging, neither expanding nor collapsing. We must remember that at that time the astronomical universe was believed to be static on the cosmic scale. To conform with this belief, and to enable his universe to maintain a static state, Einstein introduced a cosmological constant into the theory of general relativity. The cosmological constant is equivalent to a repulsive force that opposes the force of gravity. By adjusting

the value of the cosmological constant it is possible to make the repulsion counterbalance the gravity due to the uniform distribution of matter. The cosmological constant, denoted by Λ and sometimes called the lambda constant, was introduced in a rather *ad hoc* way, and Einstein sought to justify its use by appeal to Mach's principle. In his static universe, the local mass density was related directly to the cosmological constant, and Einstein believed that this method of tying together the local and global realms was in accord with Mach's philosophy.

But in the same year came the completely different de Sitter universe. It incorporated the cosmological constant, it was assumed to be static, and unlike the Einstein universe it contained no matter. This alternative universe, derived from general relativity theory, showed clearly that the cosmological constant did not guarantee a unique universe, as Einstein had hoped. The empty de Sitter universe might have been ignored as a curious freak were it not for one arresting property: when particles are sprinkled in it, they accelerate away from one another. It was thought that this "de Sitter effect," as it became known, has perhaps some bearing on the recessions and redshifts observed by Slipher.

The astronomer Carl Wirtz, inspired by Slipher's redshift measurements and the de Sitter effect, proposed in 1922 a velocity–distance relation. He used the apparent diameters of galaxies as distance indicators (the larger the distance the smaller the average apparent diameter) and the Fizeau–Doppler formula of Equation [14.4] and found that the recession velocity V increased with distance.

The apparent static nature of the de Sitter universe was a mathematical fiction. This universe appeared to be static because it contained nothing that could exhibit its actual dynamic state. Howard Robertson, a mathematician, later showed that a simple readjustment in the distinction between space and time made the de Sitter universe spatially homogeneous and flat, and in this form it was found to be expanding. An apt distinction then emerged: the Einstein universe was "matter without motion," and the de Sitter universe was "motion without matter."

Fundamental theoretical developments by the Russian scientist Alexander Friedmann in 1922 and 1924, and the Belgian cosmologist Georges Lemaître in 1927 and 1931, opened wide the door to a large class of candidate universes, all homogeneous, isotropic, expanding, and all containing matter.

Hubble: measurer of the universe

Edwin Hubble was an exacting observer who made many discoveries at the 100-inch Mount Wilson telescope. We have recounted how Hubble classified the galaxies by their appearance and how he showed that the Andromeda Nebula is a galaxy beyond the Milky Way. He developed into a fine art the distance-measuring techniques pioneered by Harlow Shapley and in 1924 began to determine the distances of galaxies. In 1929 he announced the momentous discovery that the redshifts of galaxies tend to increase with their distances. From this redshift–distance relation, and from the Fizeau–Doppler formula of Equation [14.4], he arrived at a velocity–distance law stating that the recession velocity of galaxies increases with distance. The result seemed incredible: the observable universe is expanding! Figures 14.3 and 14.4 show Hubble's results.

In the following years, Hubble extended his distance measurements. With redshifts determined by Milton Humason, Hubble made secure the concept of an expanding universe. By 1955, using the 200-inch telescope of Mount Palomar, Humason, Nicholas Mayall, and Allan Sandage had observed and analyzed the spectra of more than 800 galaxies and detected redshifts up to 0.2.

But Hubble was not the first to discover the expansion of the universe, nor did he ever claim that he had made such a discovery, and in *The Realm of the Nebulae*

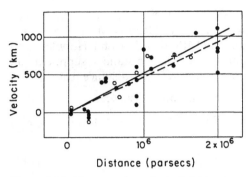

Figure 14.3. The recession velocity in kilometers per second of extragalactic nebulae plotted against their distances in parsecs, taken from Hubble's 1929 paper. The velocity used by Hubble is *cz* from the Fizeau–Doppler formula.

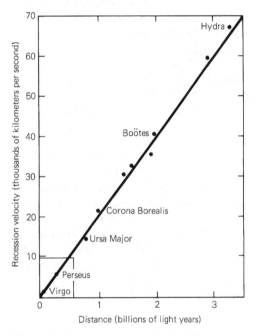

Figure 14.4. A more recent velocity–distance diagram showing the expansion of the universe. The lower left corner is the region surveyed by Hubble shown in Figure 14.3.

published in 1936 he expressed reservations concerning the velocity interpretation of extragalactic redshifts. Many astronomers since the time of Hubble have viewed with reservation the velocity–redshift relation of Equation [14.4]. This relation is valid for only very small values of redshift and, as we shall see, their reservations have been justified. Howard Robertson in 1928 showed that Slipher's redshifts and Hubble's previously published distances supported an approximate redshift–distance relation,

$$zc = \text{constant} \times \text{distance},$$

and this relation in the form $zc = H \times$ distance, or

$$zc = HL, \qquad [14.5]$$

where L is the distance of the galaxy, is now known as the Hubble law, where the constant H is called the Hubble term. The Hubble term changes in time and its value at the present cosmic epoch is denoted by H_0. Using a Doppler-like formula, such as Equation [14.4], Robertson derived the velocity–distance law,

$$V = HL. \qquad [14.6]$$

According to this law, the more distant a galaxy, the faster it moves away. When distance is doubled, the recession velocity V is also doubled. Robertson's "rough verification," as he called it, was tucked away in a theoretical paper in a physics journal not widely read by astronomers.

At this stage in the development of modern cosmology it had yet to be realized that the velocity–distance law of Equation [14.6] is rigorously true at all distances in all expanding uniform universes. This realization emerged in the context of the expanding space paradigm that was established in the mid-1930s.

Decline of the Hubble term

Both Robertson in 1928 and Hubble in 1929 found for H a value of 150 kilometers a second per million light years (or nearly 500 kilometers a second per megaparsec). In other words, the recession velocity increased with distance by 150 kilometers per second for every million light years (or 500 kilometers per second for every megaparsec).

In 1952, Walter Baade discovered the two stellar populations. We have seen how his discovery showed that the distances of galaxies had previously been greatly underestimated. A second revision in distances came in 1958 when Allan Sandage discovered that what had previously been supposed to be bright stars in more distant galaxies were in fact very luminous regions of hot gas, and these more remote galaxies were therefore at even greater distances than previously supposed. These revisions in distance estimates have enlarged the scale of the universe and reduced the original value of H by a factor between 5 and 10. Most estimates nowadays place the value of the Hubble term between 15 and 30 kilometers a second per million light years (or 50 and 100 kilometers per megaparsec). A main theme in the history of twentieth-century cosmology has been the progressive decline in the value of the Hubble term determined by astronomers. The uncertainty in H stems from the extraordinary difficulty of measuring the distances of remote extragalactic systems. The Hubble term is customarily expressed in the form:

$$H_0 = 100h \text{ kilometers a second per} \\ \text{megaparsec}, \qquad [14.7]$$

or roughly

$$H_0 = 30h \text{ kilometers a second per} \\ \text{million light years}, \qquad [14.8]$$

and the zero subscript denotes the present cosmic epoch. The parameter h lies probably somewhere between 0.5 and 1. The Hubble term is everywhere the same in uniform space but varies in time. For this reason, in this book, we avoid using the confusing term "Hubble constant." We are not sure of the exact value of H_0 and shall sometimes assume that $h = 0.5$ and hence $H_0 = 50$ kilometers a second per megaparsec, or roughly 15 kilometers a second per million light years. Thus, at a distance of 1 billion light years, the recession velocity is 15 000 kilometers a second, or one-twentieth the velocity of light.

Two laws with the same name

The *redshift–distance law* $zc = HL$, Equation [14.5], is the observers' linear law first established by Slipher's redshift measurements and Hubble's distance determinations. Its proper name is the *Hubble law*. From the time of its discovery most cosmologists have realized that in its linear form it is only approximately true.

On the other hand, the *velocity–distance law* $V = HL$, Equation [14.6], is the theorists' linear law that follows automatically from the assumption that expanding space is uniform (isotropic and homogeneous). This law, often improperly referred to as the Hubble law, is of central importance in modern cosmology and is rigorously true in all uniform universes. At any moment in cosmic time the velocity of recession of the galaxies increases linearly with their distance. Considerable confusion exists because the approximate redshift–distance law and the exact velocity–distance law are indiscriminately referred to as the Hubble law.

The connecting link between the two laws is the linear velocity–redshift relation $V = cz$ from the Fizeau–Doppler formula of Equation [14.4]. This formula is an approximation only, thus explaining why Hubble's redshift–distance law is also only an approximation, valid for small redshifts of z much less than unity. This will become clearer in the next chapter. No linear velocity–redshift relation exists that is true for all redshifts in all universes, and the correct velocity–redshift relation must be derived from basic principles for each universe.

The technical definitions of recession velocity and distance were ambiguous, and the relation between recession velocity and redshift was obscure, until the mid-1930s. To this day they remain ambiguous and obscure to those not actively engaged in cosmological research. In the following, we introduce the *expanding space paradigm*, and define recession velocity, distance, and various other terms by means of imaginary experiments with an expanding rubber sheet.

THE EXPANDING SPACE PARADIGM

The expanding space paradigm emerged in the 1930s amid much controversy concerning the meaning of extragalactic redshifts. Arthur Eddington in "The expansion of the universe" in 1931 enunciated the paradigm when he wrote of the galaxies, "it is as though they were embedded in the surface of a rubber balloon which is being steadily inflated." Slowly emerged the idea that the universe consists of expanding space! The lesson we must learn from general relativity is that space can be dynamic as well as curved.

According to the expanding space paradigm, the universe does not expand in space, instead it consists of expanding space. From this statement flows all the simplicity and complexity of modern cosmology. The galaxies do not move through space, but instead float stationary in space. Their separating distances increase because the space between the galaxies expands. The paradigm helps us to understand the velocity–distance law, and also, in the next chapter, the nature of cosmological redshifts.

THE EXPANDING RUBBER SHEET UNIVERSE

The time has come to introduce ERSU – short for "Expanding Rubber Sheet Universe" – a make-believe two-dimensional flatland with which we shall perform imaginary experiments to illustrate the properties of an expanding universe.

A two-dimensional model has one drawback. It consists of a surface that expands in three-dimensional space. Our universe of three-dimensional space does not expand in a universe of higher-dimensional space, instead it consists of expanding space. This can be deduced from the containment principle: the universe is not in space but contains space.

In our imaginary experiments we stand back and with a godlike view survey the surface as it is everywhere at an instant in time. We see the universe in much the same way as seen by the cosmic explorer (Chapter 9) who moves from place to place instantaneously. But occasionally we must quit our godlike view and take the wormlike view of two-dimensional observers living in the surface.

Because there is no cosmic edge we must imagine an expanding surface that is indefinitely large. Instead of an expanding sheet we may imagine, if we wish, a spherical balloon that is steadily inflating, as suggested by Eddington. A flat surface serves just as well and in some ways is simpler.

Experiment 1: Dilation, shear, and rotation

We start by supposing that the flat surface exhibits every kind of two-dimensional motion consistent with flatness. On the surface we draw a large number of triangles and notice how in the course of time the triangles change in size, shape, and orientation. Three basic kinds of motion exist: dilation, shear, and rotation.

Those triangles that only dilate (or contract) exhibit shape-preserving motions, and the regions they occupy expand (or contract) isotropically (see Figure 14.5).

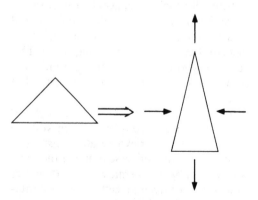

Figure 14.5. Dilation only: triangles change their area but not their shape and orientation.

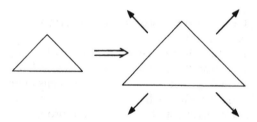

Figure 14.6. Shear only: triangles change their shape but not their area and orientation.

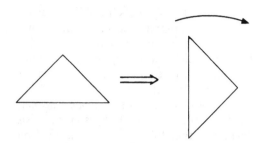

Figure 14.7. Rotation only: triangles change their orientation but not their area and shape.

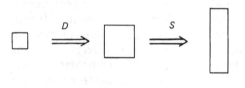

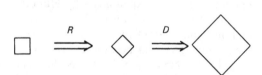

Figure 14.8. Combinations of dilation (D), shear (S), and rotation (R).

Those triangles that only shear exhibit area-preserving motions, and in the regions they occupy there is no dilation or rotation (see Figure 14.6).

Those triangles that only rotate exhibit both shape-preserving and area-preserving motions, and in the regions they occupy there is no dilation or shear (see Figure 14.7).

In general, complex motions combine dilation, shear, and rotation, and when all three vary from place to place, we have inhomogeneous motion (see Figure 14.8).

Experiment 2: Homogeneity

We draw identical triangles everywhere on the surface at a particular moment in time.

At this particular moment, ERSU represents a universe of homogeneous constitution. The preservation of this homogeneity requires that all triangles change in the same way and by the same amount in each interval of time. This means the motion of the surface is also homogeneous and has at every place in space at a common instant in time the same amount of dilation, shear, and rotation. We have now extended the meaning of homogeneity to include all forms of motion. A homogeneous universe in which all places are alike remains homogeneous when the motion is also homogeneous. We are of course ignoring the small-scale irregularities of astronomical systems.

A homogeneous universe having dilation only is shape-preserving; it expands (contracts) equally in all directions, and its motion is isotropic as well as homogeneous. This means, in ERSU, that all triangles increase (decrease) in size and retain their shapes and orientations. A homogeneous (all places are alike) and isotropic (all directions are alike) universe displays the simplest kinematics: shearfree and irrotational. We have previously referred to such a universe as uniform.

A homogeneous universe having only shear is area-preserving; it expands unequally in different directions, and its motion is anisotropic. This means, in ERSU, that all triangles change their shapes but preserve their areas: circles become ellipses and squares become oblongs. The observed highly isotropic cosmic background radiation shows that our dilating universe has little or no shear.

A homogeneous universe having only rotation is shape-preserving and area-preserving. Dilation and shear are zero, and in ERSU triangles preserve their shapes and areas but change their orientations. Because the motion is homogeneous, the rotation is not about a single central point but about all points in the surface. Remember, as observers we are in the surface and must not think of three-dimensional space. Rotation of our three-dimensional universe of space is not easy to imagine. It is the

same everywhere and consists of anisotropic motion about one of three perpendicular axes. Such rotation can be detected by shooting a particle at a distant target and noticing that its trajectory curves away from the target. The compass of inertia does not rotate with the universe (as Mach claimed); if it did, we would not know if the universe were rotating or not. The observed highly isotropic cosmic background radiation shows that the universe has little or no rotation.

Experiment 3: Uniformity and cosmic time

When the expansion is homogeneous and isotropic, triangles in ERSU dilate everywhere in the same way, and do not change their shape or rotate. In such uniform expansion, figures change everywhere in space in exactly the same way and preserve their form (shape) in time. In subsequent experiments we shall assume that ERSU expands uniformly.

Homogeneity of the universe also means that all clocks in the universe – apart from timekeeping variations owing to local irregularities – agree in their intervals of time. With our godlike view we see clocks everywhere ticking away in constant agreement; but as denizens of flatland, with only a wormlike view, we must summon the explorer to our aid, who goes around the universe at infinite speed adjusting all clocks to show a common time. On subsequent tours the explorer finds the clocks running in synchronism, showing the same time. This universal time is known as *cosmic time*. All local departures from cosmic time are due to the motions and gravitational fields of individual astronomical systems. Because we are disregarding local irregularities, we shall disregard also local irregularities in time.

Experiment 4: Drawn circles are not "galaxies"

With chalk we draw a circle on the surface of the uniformly expanding ERSU and declare that it represents a galaxy (see Figure 14.9). As the sheet expands we observe that the "galaxy" gets bigger. This result is

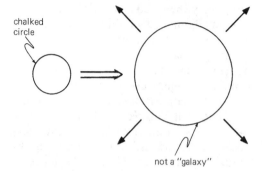

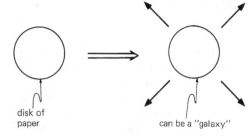

Figure 14.10. A disk of paper retains its size when placed on the surface of the expanding sheet and therefore can represent a galaxy of fixed size.

Figure 14.9. A circle drawn on the surface of the expanding rubber sheet also expands and cannot therefore represent a galaxy.

misleading. A real galaxy is held together by its own gravity and is not free to expand with the universe. Similarly, if the chalked circle is labeled "Solar System," "Earth," "atom," or almost anything, the result would be misleading because most systems are held together by various forces in some sort of equilibrium and cannot partake in cosmic expansion. If we label the chalked circle "cluster of galaxies" the result could also be wrong because most clusters are bound together and cannot expand. Superclusters are vast sprawling systems of numerous clusters that are weakly bound and can expand almost freely with the universe.

This experiment teaches us a useful lesson. We detect expansion because our measuring instruments do not expand but have fixed sizes. If everything were like the chalked circle, free to expand, then clearly there would be no way of detecting expansion. It is an amusing thought that perhaps the universe is not expanding but is static, and we fail to notice this because all atoms – and this means ourselves, our laboratories, and observatories – are all shrinking. With tongue in cheek, Eddington in 1933, in *The Expanding Universe*, said the theory of the "expanding universe" might also be called the "theory of the shrinking atom." He said: "We walk the stage of life, performers of a drama for the benefit of the cosmic spectator. As the scenes proceed he notices that the actors are growing smaller and the action quicker. When the last act opens the

curtain rises on midget actors rushing through their parts at frantic speed. Smaller and smaller. Faster and faster. One last microscopic blur of intense agitation. And then nothing."

Experiment 5: Disks of paper are "galaxies"

We place on the surface of the rubber sheet a disk of paper to represent a galaxy or any other bound system (see Figure 14.10). As the sheet expands the disk stays constant in size. This result is not misleading, and we have found a way in ERSU of correctly representing a galaxy or a cluster of galaxies of fixed size. We sprinkle uniformly over the surface disks of various sizes to represent the galaxies (see Figure 14.11). Strictly speaking, the disks should represent the largest bound systems, the clusters of galaxies, but for convenience we shall continue to refer to them as "galaxies."

Experiment 6: World map and world picture

A selected galaxy is surrounded by receding galaxies and its wormlike inhabitants might therefore think that they occupy the cosmic center from which everything is flying away. But because ERSU, like our universe, is uniform, with no cosmic center and no edge, this impression of being at the center is shared with the inhabitants of all galaxies.

We, the ERSU experimenters, look down on the surface and observe that it is uniform; we see that the sprinkled disks, on average, are everywhere the same, and the surface

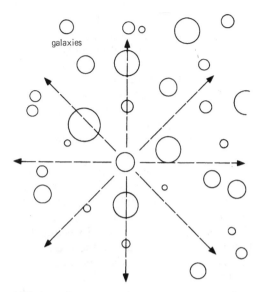

Figure 14.11. A large number of disks are placed on the expanding surface. About any one disk the other disks recede isotropically. This gives a godlike view of space at an instant in time and is what Milne called the world map.

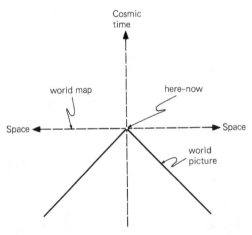

Figure 14.12. A spacetime diagram that shows the backward lightcone on which the observer sees the universe. This wormlike view of the universe is what Milne called the world picture.

everywhere expands in the same way. We are godlike spectators seeing everything everywhere in space at an instant of cosmic time. This is what Edward Milne called the *world map* (see Figure 14.12). It is also what is seen by the cosmic explorer who rushes

around and instantaneously sees that all places are alike.

Wormlike denizens in a particular galaxy look out in space and back in time and cannot see the way things are everywhere in space at the moment of observation. They cannot observe the world map. Instead, they observe things distributed on their backward lightcone. They observe what Milne called the *world picture*. The world map is godlike, the world picture is wormlike. Both would be similar if the speed of light were infinitely great.

Regrettably, we in our universe, like the inhabitants in ERSU, are limited to the wormlike view. From the Galaxy we see other galaxies scattered about us isotropically and moving away isotropically. All directions are alike. Only by invoking the location principle can we conclude that probably all places are alike. The location principle bridges the world picture and the world map.

Experiment 7: Velocity–distance law

Our next experiment shows that ERSU obeys the velocity–distance law (see Figure 14.13). We choose any disk and label it A. A second disk, labeled B, at a certain distance, moves away from A at a certain velocity. A third disk, labeled C, in the same direction as B and at twice the distance, moves away from A at twice the velocity. It must. The expansion is homogeneous, and therefore C moves away from B at the same velocity that B moves away from A. This argument, extended to disks E, F, G,..., all equally spaced in the chosen direction at a moment in time, shows that the recession velocity relative to A is always strictly proportional to distance. The equal spaces between disks all increase in the same way and the disks remain equally spaced. This result is independent of the location of disk A. ERSU thus shows us that homogeneity is preserved when the expansion is homogeneous:

recession velocity = constant × distance.

The linear expansion law of Equation [14.6]

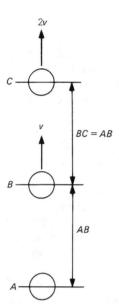

Figure 14.13. Three disks A, B, and C are arranged in a straight line with the distance between A and B equal to the distance between B and C. In uniform expansion the velocity that B recedes from A is the same as the velocity that C recedes from B. Hence C recedes from A at twice the velocity that B recedes from A, showing that the recession velocity is proportional to distance.

is a direct consequence of time-invariant homogeneity. The recession velocity relative to any disk at rest on the surface obeys the same law. If the expansion is anisotropic (faster in one direction than in another), the "constant" has different values in different directions. But in isotropic expansion, which is our primary interest, the constant has the same value for recession in all directions. The law we have derived by this experiment is the velocity–distance law:

$$V = HL, \qquad [14.9]$$

and at the time of observation H is denoted H_0.

We must pause and take note that recession velocity V, Hubble term H_0, and distance L all require thought and careful interpretation, for each is open to misunderstanding. The velocity–distance law is obviously true in the world map for us looking down on ERSU, but not so for the poor observers inhabiting their disks who see only the world picture.

The expression recession velocity needs careful handling. On the surface of ERSU the disks (our imaginary galaxies) are stationary; they move apart because the surface is expanding. The disks do not move on the surface, but are at rest and are carried apart by the expansion of the surface. Similarly, the galaxies in the universe are stationary, yet recede from one another because intergalactic space expands. The galaxies are not hurling through space; they are at rest in space and are carried apart by the expansion of space. Recession velocity is therefore not an ordinary velocity in the usual sense and is unlike the velocities encountered on Earth, in the Solar System, or in the Galaxy. They are not Newtonian velocities or the velocities used in special relativity. For this reason we must be careful in cosmology when using the word "velocity," and to avoid confusion we shall most of the time use recession velocity or just recession to indicate relative motion owing to the expansion of space.

The Hubble term H (present value denoted by H_0) is the same everywhere in space at a common instant in cosmic time, but is usually not constant in time. The expansion may have been faster in the past, in which case the value of H was greater than H_0; and if the expansion was slower, the value of H was smaller than H_0. Observers look out from their galaxy to great distances and look back great periods of time, and see H having different values at different distances. To them the velocity–distance law is true only for short distances; at larger distances the law breaks down because H appears to change with distance. The velocity–distance law is true in the world map visualized by the theorist but not in the world picture seen by the observer.

The measurement of distances is no great problem to the theorist who can always use a tape measure (see Chapter 10). But the observer who looks out in space and back in time uses rough-hewn scales of distance and has a dreadful problem trying to

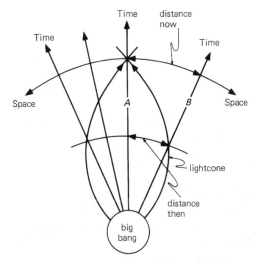

Figure 14.14. Galaxies shown as world lines diverging from the big bang. An observer in galaxy A looks out into space and back in time and sees galaxy B as it was in the past. There are thus two distinctly different distances between galaxies A and B: the "distance now" of B from A, and the "distance then" at the time B emitted the light that A now sees. The distance now is the tape-measure distance in the world map that is used in the velocity–distance law.

determine how far away are the faint galaxies. The difficulty is twofold. First, everything seen is distributed on the sky and distances are not apparent but must be inferred. Second, distances in an expanding universe change with time (see Figure 14.14). We may speak of the distance of a galaxy now at the time of observation or at the time when the galaxy emitted the light now seen. In an expanding universe the distance a galaxy now has is greater than the distance it had when it emitted the light now seen. The distance now is the theorists' tape-measure distance in the world map, the distance at the time of emission is the observers' distance in the world picture. In the velocity–distance law we must use the theorists' distances that galaxies have now at a common instant in time in order to determine H_0. All world-picture distances must be adjusted before they can be used in the velocity–distance law of the world map. The determination of the Hubble

term H_0 requires the mapping of the world picture into the world map.

Experiment 8: The Hubble sphere

The recession velocity increases with distance and equals at a certain distance the velocity of light (see Figure 14.15). This distance is c/H_0 and is the Hubble length L_H:

$$L_H = \frac{c}{H_0}.$$ [14.10]

This expression is obtained by writing $V = c$ and $L = L_H$ in Equation [14.6]. With a Hubble term of H_0 equal to $30h$ kilometers a second per million light years, we find

$$L = 10h^{-1} \text{ billion light years.}$$ [14.11]

If $h = 0.5$, the recession equals the velocity of light at distance 20 billion light years. Notice that the velocity–distance law can be written in the form:

$$\frac{V}{c} = \frac{L}{L_H},$$ [14.12]

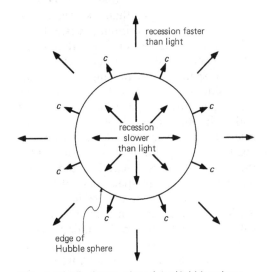

Figure 14.15. At the edge of the Hubble sphere the recession velocity of the galaxies is transluminal (equal to the velocity of light); inside the Hubble sphere all galaxies recede subluminally (slower than the velocity of light); and outside all galaxies recede superluminally (faster than the velocity of light). The observable universe is approximately the size of the Hubble sphere. A more exact definition is given in Chapter 21.

where V is the recession velocity of a galaxy at distance L in the world map. A Hubble length $L_H = c/H_0$ serves as a cosmic yardstick, and when we speak of distances of cosmic magnitude we have in mind distances comparable with the Hubble length.

In ERSU we draw about any disk a large circle whose radius we call the Hubble length. Inside this Hubble circle, which we shall call a sphere, the recession is subluminal (less than the velocity of light), outside the recession is superluminal (greater than the velocity of light), and at the edge of the Hubble sphere the recession is transluminal (equal to the velocity of light). Each disk (galaxy) is the center of its Hubble sphere.

The universe, like ERSU, has no edge and cannot terminate abruptly at the boundary of the Hubble sphere. A cosmic edge at which the recession from our Galaxy equals the velocity of light, even if it existed (as often implied in popular literature), could not be at the same distance from all other galaxies. There is no cosmic edge and galaxies farther away than the Hubble distance recede faster than the velocity of light. How, the beginning cosmologist asks, can galaxies move faster that light? The answer is that galaxies are not moving through space but are moving apart by the expansion of intergalactic space. No galaxy can move through space faster than light and in its local space it obeys always the rules of special relativity. But recession is a result of the expansion of space that obeys the rules of general relativity, and is not like motion through space that obeys the rules of special relativity. Recession velocity is without limit, and in an infinite universe a galaxy at infinite distance has infinite recession. Those persons who find it difficult to understand that recession is without limit usually make the mistake of thinking that the receding galaxies are like projectiles shooting away through space. This is an incorrect view. The correct view is of galaxies more or less at rest in expanding space.

This important experiment demonstrates that the expansion of space does not obey the rules of special relativity, and the recession velocity is not limited by the speed of light.

Experiment 9: The steady-state expanding universe

The steady-state expanding universe is easily simulated with ERSU. As the surface expands and the disks move apart, we continually sprinkle new disks on the surface so that the average separating distance between disks remains always the same. The surface presents an unchanging appearance because of the "continuous creation" of disks. Expansion also never changes and the Hubble term H_0 therefore stays constant and the Hubble sphere has constant radius.

Experiment 10: Comoving galaxies and peculiar motion

Galaxies stationary in expanding space are said to be comoving. They comove with the expansion. Clocks on comoving galaxies all measure cosmic time. Our gadabout cosmic explorer has set all clocks on comoving bodies to read the same time. In subsequent tours of the homogeneous universe the explorer finds that these clocks are in constant agreement. All comoving bodies have their world lines perpendicular to a cosmic space of uniform curvature and uniform expansion (Figure 14.14).

In all previous experiments with ERSU we have supposed that the disks comove with the expanding sheet. But few galaxies are exactly comoving. They dither around relative to their neighbors and have independent motion in their local space. This independent motion superposed on the expansion is known as peculiar motion. Because of peculiar motion the world lines of galaxies are not straight but slightly crinkled. We can easily imagine that all disks in ERSU have independent motion and jitter around slowly on the expanding surface.

Observers see the peculiar velocities of other galaxies superposed on the flow of recession. Normally the observers in one galaxy cannot distinguish between the recession and peculiar motion of other nearby galaxies. Peculiar motion tends to be random and the velocities of many galaxies in a particular region can be averaged to find the recession velocity of that region. Observers must take into account the peculiar motion of their own galaxy, and this can in principle be done by averaging the peculiar motion of neighboring galaxies, or better still, by determining the anisotropy of the cosmic background radiation.

It sounds easy, particularly for us looking down on ERSU, but in the real universe the determination of peculiar velocities is very difficult. Fortunately, recession dominates at large distances and peculiar motions can then often be neglected. Some nearby galaxies are approaching us and others are moving away: farther away, a few are approaching and most are receding; and even farther away, none are approaching and all are receding. Typical peculiar velocities of galaxies in small clusters are 300 kilometers a second. Beyond $10h^{-1}$ million light years, at redshifts greater than 0.001, recession dominates. In rich clusters many galaxies have peculiar velocities as great as 3000 kilometers a second, and in their case the recession dominates beyond $100h^{-1}$ million light years, at redshifts greater than 0.01.

Our own peculiar motion in the universe consists of the Earth moving about the Sun (30 kilometers a second with annual variation in direction), the Sun moving around in the Galaxy (200 kilometers a second), the motion of the Galaxy in the Local Group (approximately 100 kilometers a second), the peculiar motion of the Local Group (350 kilometers a second) toward the Virgo cluster, and the motion of the Local Supercluster (300 kilometers a second) toward the Hydra Centaurus supercluster, giving a net peculiar velocity of 600 kilometers a second in the direction of the constellation Leo. Determining the peculiar velocity of the Local Group in the universe by optical means is not easy, and the most reliable information comes from the anisotropy of the cosmic background radiation.

The implications of the experiments performed so far are startling. First, in Newtonian theory we are taught there is no such thing as absolute rest and all motions are purely relative. Yet in cosmology a comoving body is in a state of absolute rest that can in principle be verified. All peculiar velocities have absolute values that can be determined relative to the state of rest in the local comoving frame. Second, in special relativity we are taught that there is no preferred way of decomposing spacetime into space and time and that all decompositions are relative. Yet in cosmology spacetime separates naturally into uniformly curved expanding space and orthogonal cosmic time. Third, we are taught that a body cannot move faster than light, but in cosmology we find that comoving bodies obey a velocity–distance law in which recession velocities can exceed the velocity of light and are without limit.

What we have been taught in Newtonian mechanics and special relativity theory applies to local peculiar motions in the laboratory, the Solar System, and the Galaxy, and not at all to cosmic motion. All local velocities are peculiar within the cosmic frame of reference and cannot exceed the velocity of light. Recession velocity, however, is not a local phenomenon; it is the result of the expansion of space and does not conform to the rules of special relativity. In summary, we may say that motion in an expanding universe is compounded from recession and peculiar velocities; recession velocities are due to the expansion of space and are without limit, and peculiar velocities are due to motion through space and conform to special relativity.

Experiment 11: Sub-Hubble sphere

The sub-Hubble sphere contains the nearby universe in which astronomical peculiar motions dominate over cosmic recession. Thus

$$L_{\text{sub-}H} = \frac{V_{\text{pec}}}{H_0} = \frac{V_{\text{pec}}}{c} L_H, \qquad [14.13]$$

where $L_{\text{sub-}H}$ is the sub-Hubble radius and V_{pec} is a typical large peculiar velocity. A typically large peculiar velocity, let us say, is 1000 kilometers a second, and the present radius $L_{\text{sub-}H}$ of the sub-Hubble sphere is hence $10h^{-1}$ megaparsecs, or approximately $30h^{-1}$ million light years. Meaningful distant cosmological observations are made beyond the sub-Hubble sphere where the Hubble flow becomes fully developed and dominates over peculiar motions. Galactic redshifts less than 0.003 are sub-Hubble and not cosmologically significant. If we take into account peculiar velocities as large as 3000 kilometers a second, the sub-Hubble sphere approaches $100h^{-1}$ million light years and sub-Hubble redshifts are as large as 0.01. This raises troubling questions: Does the linear Hubble law of Equation [14.5] break down before leaving the sub-Hubble sphere? Is it possible that the linear Hubble law applies nowhere, neither in the sub-Hubble sphere (where measurements are of no cosmological significance) nor in the Hubble sphere itself? Must we interpret observations with a more exact redshift–distance formula appropriate to a specific model of the universe? We glimpse here one of the reasons why the determination of H_0 is so difficult.

The sub-Hubble sphere is dominated by the irregularities of astronomical systems. Beyond the sub-Hubble sphere, astronomical irregularities are much less pronounced and tend to be less important. The length $L_{\text{sub-}H}$ is a measure of the scale of the largest irregularities. In response to the question, over how large a region should we average matter to discover the cosmic density, the answer is the size of the sub-Hubble sphere.

Experiment 12: Comoving coordinates

With chalk we draw on the expanding surface of ERSU a network of lines. These intersecting lines form what is called a comoving coordinate system (see Figure 14.16). The coordinate lines are fixed on the surface. Whether the lines are straight or curved is not very important; what

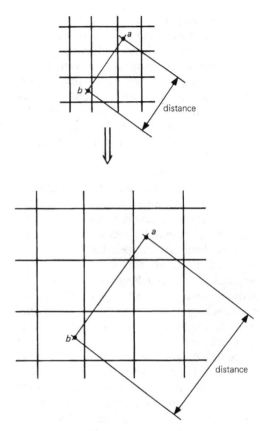

Figure 14.16. This shows a comoving coordinate system consisting of a network of intersecting lines drawn on the expanding surface. All coordinate distances, such as that between points a and b, remain constant in value. The real distance, of course, increases.

matters is that comoving disks are fixed in their relation to these coordinate lines, and their positions are specified by appropriate coordinate values. All distances between comoving disks remain constant when measured in comoving coordinates, and these constant distances are called comoving coordinate distances. A comoving coordinate system enables us to distinguish between recession and peculiar motion. Recession velocities apply to bodies stationary in the comoving coordinate system, and peculiar velocities apply to bodies moving relative to the comoving coordinates.

Experiment 13: An empty universe

We remove all disks from the surface of ERSU except the one occupied by our observers. These observers are left with no way of determining their peculiar motion; moreover, they cannot tell if the surface is expanding, static, or contracting. Confronted with this situation of undressed space of indeterminate kinematic behavior, they might ruefully recall Mach's words: "When, accordingly, we say that a body preserves unchanged its direction and velocity in space, our assertion is nothing more or less than an abbreviated reference to the entire universe." By the entire universe, Mach had in mind the "remote heavenly bodies" that we have, figuratively, removed from ERSU. Even the ubiquitous explorer is puzzled and tempted to believe that empty space is meaningless and the physical properties of space are dependent on the presence of matter. Then we learn the trick of scattering around a few tiny comoving particles, or (what amounts to the same thing) of chalking on the sheet a network of comoving coordinates. This illustrates what happened originally with the empty universe proposed by Willem de Sitter: nobody knew that it was expanding until it was sprinkled with test particles and equipped with a network of comoving coordinates. It was then found to have kinematic properties even though it contained no matter.

Experiment 14: Idealized universe

The distribution of disks in ERSU is clumpy on small scales and not obviously uniform except on large scales. These random irregularities are distracting when we wish only to study the large-scale behavior of the universe. In the last experiment we removed all the disks. Let us now take these disks, grind them into powder, and then smoothly and uniformly distribute the powder over the expanding surface of ERSU. This represents an idealized universe in which all galaxies are smoothed out into a continuous fluid of uniform density.

Idealized universes have their uses. It is of interest, for instance, to know what happens to light when it propagates in a universe free of all irregularity. The large-scale effects of the universe are first determined and corrections for irregularities can be added later. Cosmologists take the view that an idealized universe is a convenient fiction, useful for easy calculations, and in Howard Robertson's words a sort of cosmic undergarment onto which the ostentatious detail of the real world is tacked.

An idealized universe is also useful for studying the origin of galaxies. These vast celestial systems have not always existed, and prior to their formation the universe was much less irregular than at present. The idealized universe may therefore resemble the way things were once upon a time. How the original unstructured universe evolved into its present highly structured state is a major research area in cosmology.

Idealized universes – perfectly homogeneous and isotropic – are known as Robertson–Walker models after Howard Robertson and Arthur Walker showed rigorously that universes obeying the cosmological principle have a spacetime that uniquely separates into a curved expanding space and a cosmic time that is common to all comoving observers.

By studying the expansion of a small region of an idealized universe we automatically learn how all other small regions expand, and by piecing these regions together we learn how the whole universe expands. The behavior of large regions, even the universe itself, is mirrored in the behavior of small regions. We shall exploit this intriguing aspect of cosmology in Chapter 17.

MEASURING THE EXPANSION OF THE UNIVERSE

The universal scaling factor

Distances between galaxies (or clusters of galaxies) increase in an expanding universe, whereas distances inside galaxies and even clusters of galaxies do not increase. This difference, of vital importance in astronomy and other sciences, is a distraction in cosmology. The solution is simple. Because we are interested only in cosmic phenomena, we abolish all astronomical systems and use instead an idealized universe. In this way we become free to consider how all distances vary in time without the bother of distinguishing between large expanding and small nonexpanding regions. In a smoothed universe, distances can be as small or as large as we please.

We consider only uniform (homogeneous and isotropic) expansion. Over an interval of time all distances between comoving points increase by the same factor. If one distance increases by 1 percent, then all distances increase by 1 percent. A comoving triangle, for example, has its three sides scaled by the same factor and the dilated triangle retains its original shape. There exists a universal scaling factor, often (for historical reasons) denoted by R, that increases in time in a uniformly expanding universe, and at any instant in cosmic time has the same value everywhere in space (see Figure 14.17). All distances of the tape-measure kind between

comoving points increase in proportion to R, all areas of two-dimensional figures increase in proportion to R^2, and all volumes of three-dimensional figures increase in proportion to R^3.

Comoving coordinates and the scaling factor

The scaling factor has many important uses. We can best show this by beginning with comoving coordinates. A network of intersecting lines drawn on the surface of ERSU is an example of a comoving coordinate system. All comoving points are separated by coordinate distances that stay constant during expansion. Such a coordinate system is fixed in expanding space. With this coordinate system we can say that the actual distance L is the coordinate distance multiplied by the scaling factor:

$$L = R \times \text{coordinate distance}. \qquad [14.14]$$

The "actual distance" is the tape-measure distance: the distance that would be measured by stretching a tape measure in a uniformly curved surface; it is the straight-line (shortest or geodesic) distance between two points. The coordinate distance stays constant, whereas the distance itself increases in proportion to R. Observers use other distances, such as luminosity distance and distance by apparent size (see Chapter 19), but these distances unfortunately offer no help in understanding the fundamentals of cosmology.

Instead of a flat surface with a network of lines drawn on it, as in ERSU, we could use a rubber balloon with latitude and longitude coordinates drawn on its surface. Latitude and longitude are comoving coordinates, and a point on the surface of an expanding balloon retains its position relative to these coordinates. In this application, the scaling factor becomes the radius of the balloon's surface. The distance between two points on the surface is the constant coordinate distance (expressed in terms of latitude and longitude) multiplied by the radius R. Originally the scaling factor was referred to as the radius of the universe, and this is

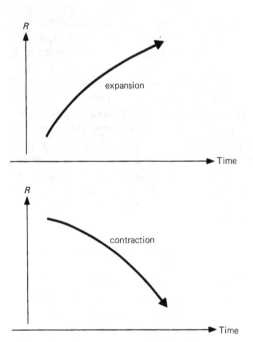

Figure 14.17. The scaling factor R – everywhere the same in space – changes in time; its value increases in an expanding universe and decreases in a contracting universe.

why it is denoted by the symbol R. The expression "radius of the universe" may be misleading, however, because some universes are flat, and a more neutral expression, such as "scaling factor," is less ambiguous.

Distances, areas, volumes, and densities
Usually we are not interested in the actual value of the scaling factor R, only in how it changes, and how its value compares at different stages in cosmic time.

Let R_0 be the value of the scaling factor at the present time, and let L_0 be the present distance between two comoving points:

$$L_0 = R_0 \times \text{coordinate distance.}$$

At any other time the scaling factor is R, and the distance between the same two comoving points is

$$L = R \times \text{coordinate distance,}$$

and only the scaling factor has changed. Therefore

$$\frac{L}{L_0} = \frac{R}{R_0}, \qquad [14.15]$$

which is obvious, for if the universe doubles its size, then the scaling factor is increased twofold and all distances between comoving points are also increased twofold. Pursuing similar arguments, we can say that if an area comoves, then its value A in terms of its present value A_0 is

$$\frac{A}{A_0} = \left(\frac{R}{R_0} \right)^2, \qquad [14.16]$$

Similarly with volumes:

$$\frac{V}{V_0} = \left(\frac{R}{R_0} \right)^3, \qquad [14.17]$$

where V (not to be confused with recession velocity) is a comoving volume whose present value is V_0. When distances double in size, comoving areas increase fourfold, and comoving volumes increase eightfold.

Suppose an expanding volume V contains N particles and that no particles are created or destroyed. The density of particles, call it n, is the number in a unit of volume, such as a cubic centimeter. Hence $n = N/V$. The present density n_0 is the fixed number N divided by the present volume: $n_0 = N/V_0$. We know how volumes vary, and hence densities vary as:

$$n = n_0 \left(\frac{R_0}{R} \right)^3. \qquad [14.18]$$

With this important result it is possible to find the density in the past or the future from the present density. The average density of matter in the universe is about 1 hydrogen atom per cubic meter. Back in the past when the scaling factor was 1 percent of its present value, the density was a million times greater and equal to 1 hydrogen atom per cubic centimeter. This is a typical value for the density of galaxies, and we infer that galaxies as we know them had not formed when the universe was smaller than 1 percent of its present size. Incidentally, when we say that the universe "changes in size" we imply not that it is finite but only that the scaling factor R changes.

THE VELOCITY–DISTANCE LAW
The scaling factor increases with cosmic time in an expanding universe. But how fast does it increase? Thought on this matter soon makes clear that the rate of increase of R must have something to do with the Hubble term.

Consider a comoving body at a fixed coordinate distance and at an actual distance $L = R \times$ coordinate distance. As R increases, the distance L increases and the body recedes. The faster R increases, the faster the body recedes. The recession velocity V of a comoving body is the rate at which its distance L increases. This equals the rate of increase of R multiplied by the constant coordinate distance; that is

$$V = \text{rate of increase of } R$$
$$\times \text{ coordinate distance.} \qquad [14.19]$$

For convenience we use Newton's notation and let $\dot{R}$ stand for the rate of increase of

R; therefore

$$V = \dot{R} \times \text{coordinate distance.} \qquad [14.20]$$

On inserting the expression $L = R \times$ coordinate distance into Equation [14.20], we obtain

$$V = L\frac{\dot{R}}{R}. \qquad [14.21]$$

This is the velocity–distance law in which the Hubble term is given by the expression

$$H = \frac{\dot{R}}{R}. \qquad [14.22]$$

The Hubble term is everywhere the same in space and in most universes varies in time. This important derivation of the velocity–distance law:

$$V = HL, \qquad [14.23]$$

reveals that this fundamental law is nothing more than the consequence of uniform expansion. A scaling factor that is everywhere the same in space, and varies in time, automatically yields the linear velocity-distance law.

Because H changes in time, the velocity–distance law at the present cosmic epoch is

$$V = H_0 L,$$

as in Equation [14.6]. As stressed previously, this is the theorists' law and is not the observers' Hubble law:

$$zc = H_0 L,$$

of Equation [14.5].

Hubble time and the age of the universe

The Hubble time (or period) would be the age of the universe if expansion were constant (see Figure 14.18). The Hubble time is

$$t_H = \frac{1}{H_0} = \frac{L_H}{c}$$

$$= 10h^{-1} \text{ billion years,} \qquad [14.24]$$

and is the Hubble distance divided by the speed of light. Often the Hubble time is referred to as the expansion time. If h has a

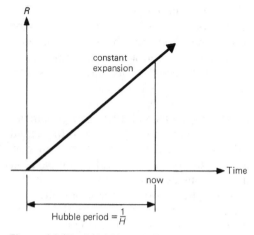

Figure 14.18. A Hubble period would be the age of the universe if the universe expanded at a constant rate.

value somewhere between 0.5 and 1, t_H lies between 10 and 20 billion years.

In almost all universes studied by cosmologists the scaling factor does not increase at a constant rate in time but either accelerates or decelerates. In an accelerating universe R increases more rapidly in time and the actual age is greater than a Hubble time. In a decelerating universe R increases more slowly in time and the actual age is shorter than a Hubble time. At present we cannot tell precisely the age of the universe, and a Hubble time serves as a rough measure of age. We must be on our guard, however, because in some universes a Hubble time is a grossly misleading indicator of age. The de Sitter and steady-state universes of infinite age are examples. In the steady-state universe nothing ever changes in a cosmic sense, and therefore the Hubble term stays constant (see Figure 14.19). Because $\dot{R} = HR$, and H is constant, $\dot{R}$ increases as R, and this is an accelerating universe.

The Hubble sphere and the observable universe

Broadly speaking, the observable universe spans the Hubble sphere. If the age of the universe is roughly a Hubble period, the

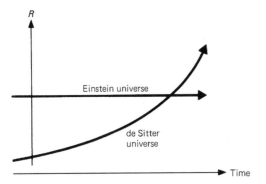

Figure 14.19. The static Einstein universe and the exponentially expanding de Sitter universe. In the Einstein universe the Hubble term is zero and the scaling factor is constant in time. In the de Sitter universe, as in the steady-state universe, the Hubble term is constant; and the scaling factor R increases exponentially with time.

distance light travels in this time is approximately a Hubble length. We cannot see things outside the observable universe because their light is still traveling and has yet to reach us. The older the universe, the more we can see of it (Chapter 22).

ACCELERATING AND DECELERATING UNIVERSES

We have seen that a tape-measure distance increases according to the rule

$$L = R \times \text{coordinate distance},$$

and the recession velocity of a comoving body of constant coordinate distance is the rate of increase of distance ($V = \mathrm{d}L/\mathrm{d}t = \dot{L}$) according to

$$V = \dot{R} \times \text{coordinate distance}.$$

Acceleration is just the rate of increase of velocity ($\mathrm{d}V/\mathrm{d}t$), and if we use the symbol $\ddot{R}$ to denote the rate of increase of $\dot{R}$, we have

$$\text{acceleration} = \frac{\mathrm{d}V}{\mathrm{d}t}$$

$$= \ddot{R} \times \text{coordinate distance} \qquad [14.25]$$

because, as before, the coordinate distance of the comoving body is constant. We now use our first relation, $L = R \times \text{coordinate}$

distance, and find

$$\text{acceleration} = L\frac{\ddot{R}}{R}. \qquad [14.26]$$

The term $\ddot{R}/R$ was once referred to as the acceleration. More popular is the deceleration term, indicated by the symbol q and defined by

$$q = -\frac{\ddot{R}}{RH^2}. \qquad [14.27]$$

The deceleration term, like the Hubble term, is constant everywhere in space at a common instant in time, and generally changes in time.

When the rate of expansion never changes, and $\dot{R}$ is constant, the scaling factor is proportional to time t ($R = \text{constant} \times t$), and the deceleration term is zero, as in Figure 14.18. When the Hubble term is constant, the deceleration term q is also constant and equal to -1, as in the de Sitter and steady-state universes, shown in Figure 14.19. In most universes the deceleration term changes in time, as illustrated in Figure 14.20.

When the deceleration term is positive, there is deceleration (slowing down of expansion), and when it is negative, there is acceleration (speeding up of expansion). From the curves in Figure 14.21 we see that in a decelerating universe, where q is positive, the age of the universe is shorter than a Hubble period; and in an accelerating universe, where q is negative, the age of the universe is longer than a Hubble period.

The present values of the Hubble term H_0 and deceleration term q_0 show how the scaling factor is now changing. The precise value of the Hubble term is unknown and it is believed that the h parameter lies somewhere between 0.5 and 1. The rate of increase in the scaling factor is possibly slowing down and the expansion is decelerating. From observations it is very difficult to determine precisely the value of the deceleration term. The available evidence at present suggests that q_0 is perhaps as small as 0.05, but many cosmologists believe that it may be as large as 0.5 because of "missing

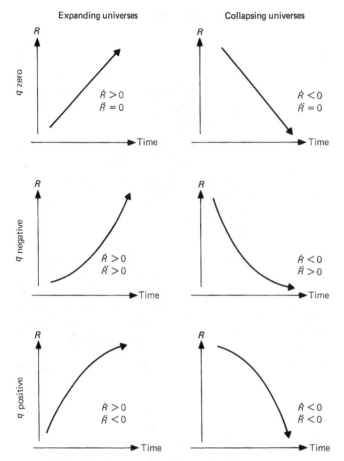

Figure 14.20. This array of diagrams shows universes classified by their values of the Hubble term H and the deceleration term q.

mass." The estimated values of H_0 and q_0 have changed frequently over the decades, and even current estimates must be viewed with reservations. Fortunately, much in cosmology can be discussed in general kinematic and geometric terms without having an exact knowledge of the present values of the Hubble and deceleration terms.

CLASSIFYING UNIVERSES

Geometrical classification

Universes are classified in various ways. Here we mention three quite simple methods of classifying uniform universes. The first is the geometrical method based on curvature. In this method there are three classes:

(a) $k = 0$:   flat space (open)

(b) $k = 1$:   spherical space (closed)
(c) $k = -1$:   hyperbolic space (open)

defined by the curvature constant k and discussed in Chapter 10.

Kinematic classification

The way in which the scaling factor varies, based on the values of H and q, gives us a second method of classification, as shown in Figure 14.20. All models can be characterized by whether they expand or contract, and accelerate or decelerate. To the four classes

(a) $(H > 0, q > 0)$:   expanding and
                        decelerating
(b) $(H > 0, q < 0)$:   expanding and
                        accelerating

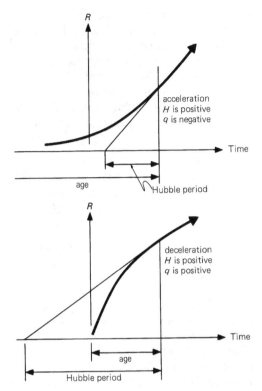

Figure 14.21. Expansion in accelerating and decelerating universes. Notice that in accelerating universes the age is greater than the Hubble period, and in decelerating universes the age is less than the Hubble period.

(c) $(H < 0, q > 0)$:   contracting and decelerating
(d) $(H < 0, q < 0)$:   contracting and accelerating

we can add three classes

(e) $(H > 0, q = 0)$:   expanding, zero deceleration
(f) $(H < 0, q = 0)$:   contracting, zero deceleration
(g) $(H = 0, q = 0)$:   static.

There is little doubt that we live in an expanding universe, and hence only (a), (b), and (e) are possible candidates.

Bang–whimper classification
Another simple method is the bang–whimper classification. Our universe was more dense in the past than now, and also much hotter. The cosmic background radiation is generally accepted as evidence of a dense and hot early universe. Universes that start or end at high density, or pass through a high-density phase, are of the big bang type, and the descriptive name "big bang" was coined by Fred Hoyle. The universes that begin or die "not with a bang but a whimper" (in T. S. Eliot's words) we shall call whimper universes.

A big bang occurs whenever the scaling factor R is either zero or extremely small, and a whimper is a long-drawn-out state that occurs when R is large and without limit. Thus a big bang means: at some time R is close to 0 and density is close to ∞; a whimper means R approaches ∞ and the density approaches 0; and the symbol ∞ denotes infinity. Because a universe can begin either as a bang or a whimper, and can end either as a bang or a whimper, there are altogether four classes (see Figure 14.22):

(a) bang–bang: has finite lifetime
(b) bang–whimper: has infinite lifetime
(c) whimper–bang: has infinite lifetime
(d) whimper–whimper: has infinite lifetime.

Of these four classes, we note that only (a) has a finite lifetime.

Because we live in an expanding universe we can rule out the whimper–bang class (c)

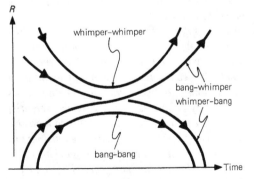

Figure 14.22. The four classes of universes in the bang–whimper classification scheme.

that continually contracts. The whimper–whimper class (d), although possible, is unlikely because the turnaround, or bounce, must presumably occur at a density great enough to create the cosmic background radiation, and a dense turnaround qualifies as a big bang. We are left with only two possible classes of universe: the bang–bang class (a) that expands from a big bang and then collapses back to a big bang, with us at present in the expanding phase; and the bang–whimper class (b) that expands continually for an infinite period of time.

We shall see later in the case of the Friedmann universes that the possible classes are: geometrically, all three curvatures are possible; kinematically, only $(H > 0, q > 0)$ and $(H < 0, q > 0)$ are possible; and physically, only the bang–bang and bang–whimper classes are possible.

REFLECTIONS

1 *In 1886, in his observatory at Tulse Hill just outside London, William Huggins was the first to observe the displacement of stellar spectral lines predicted in 1848 by the French scientist Armand Fizeau. The "classical Doppler" formula (Equation 14.1) was in fact formulated by Fizeau. Previously, in 1843, Christian Doppler had argued that the pitch of sound waves should be affected by the velocity of the source. Doppler argued that this effect should occur not only with sound waves but also with light waves. Thus he correctly predicted the "Doppler effect" for both sound and light. But Doppler erred when he proposed that the color difference between stars in binary systems was due to this Doppler effect. He argued that the approaching star would be blue and the receding star would be red. Although on the right track, he was a long way off in estimating the amount of spectral displacement, as shown by Fizeau. The French scientist Armand Fizeau, who was the first to measure the speed of light in a terrestrial environment, formulated the expression (Equation 14.1) often attributed to Doppler. Through the 19th century the velocity displacement of spectral lines was the Doppler effect and the*

actual amount of displacement was given by the Fizeau–Doppler formula. I shall continue this forgotten custom and in the case of light use the term "Fizeau–Doppler formula" for nonrelativistic velocities and "relativistic Doppler formula" for relativistic velocities.

2 *The following article appeared in the* New York Times, *page 61, on January 19, 1921:*

"DREYER NEBULA NO. 584 INCONCEIVABLY DISTANT Dr. Slipher Says the Celestial Speed Champion Is 'Many Millions of Light Years' Away.

By Dr. Vesto Melvin Slipher, Assistant Director of the Lowell Observatory, Flagstaff, Ariz.

FLAGSTAFF, Ariz., Jan. 17. – The Lowell Observatory some years ago undertook to determine the velocity of the spiral nebulae – a thing that had not been previously attempted or thought possible. The undertaking soon revealed the quite unexpected fact that spiral nebulae are far the most swiftly moving objects known in the heavens. A recent observation has shown that the nebula in the constellation Cetus, number 584 in Dreyer's catalogues, is one of very exceptional interest.

"Like most spiral nebulae, this one is extremely faint, and to observe its velocity requires an exceedingly long photographic exposure with the most powerful instrumental equipment. This photograph was exposed from the end of December to the middle of January in order to give the weak light of the nebula's spectrum time to impress itself upon the plate. It is necessary to disperse the nebular light into a spectrum in order to observe the spectral lines, and to measure the amount that they are shifted out of their normal positions, for it is this displacement of the nebula's lines that discloses and determines the velocity with which the nebula is itself moving. The lines in its spectrum are greatly shifted showing that the nebula is flying away from our region of space with a marvelous velocity of 1100 miles per second. This nebula belongs to the spiral family,

which includes the great majority of the nebulae. They are the most distant of all celestial bodies, and must be enormously large.

"If the above swiftly moving nebula be assumed to have left the region of the sun at the beginning of the earth, it is easily computed, assuming the geologists' recent estimate of the earth's age, that the nebula now must be many millions of light years distant.

"The velocity of this nebula thus suggests a further increase to the estimated size of the spiral nebulae themselves as well as to their distances, and also further swells the dimensions of the known universe."

3 Gerald Whitrow wrote in "Hubble, Edwin Powell" that Hubble's work "made as great a change in man's conception of the universe as the Copernican revolution 400 years before. For, instead of an overall static picture of the cosmos, it seemed that the universe must be regarded as expanding, the rate of the mutual recession of its parts increasing with their relative distance."

• In The Expanding Universe (1933), Eddington wrote: "The unanimity with which the galaxies are running away looks almost as though they had a pointed aversion to us. We wonder why we should be shunned as though our system were a plague spot in the universe.... But the theory of the expanding universe is in some respects so preposterous that we naturally hesitate to commit ourselves to it. It contains elements apparently so incredible that I feel almost an indignation that anyone should believe in it – except myself."

4 We have previously seen that matter affects the geometry of spacetime. We now see that matter also affects the dynamics of spacetime. Indeed, in some expanding universes containing uniformly distributed matter, space is flat and the matter affects only the dynamics and not the geometry of space. In general, in the presence of a uniform distribution of matter: a curved and static space is possible; a flat and expanding space is possible; but a flat and static space is impossible.

• Consider two comoving particles separated by distance L. From general relativity, their relative acceleration in uniform space is given by the equation

$$\ddot{L} = \frac{\Lambda}{3}L - \frac{4\pi G}{3}\left(\rho + \frac{3P}{c^2}\right)L, \qquad [14.28]$$

where Λ is the cosmological constant, ρ the density, and P the pressure. In the empty de Sitter universe, ρ and P are both zero, and therefore $\ddot{L} = (\Lambda/3)L$, and this acceleration is the "de Sitter effect." In the Einstein static universe, $\ddot{L} = 0$, and if $P = 0$ (as Einstein assumed), then $\Lambda = 4\pi G\rho$.

5 In 1930, Arthur Eddington introduced the idea of an expanding rubber surface in an article entitled "On the instability of Einstein's spherical world." He wrote, "Observationally, galaxies 'at rest' will appear to be receding from one another since the scale of the whole distribution is increasing. It is as though they were embedded in the surface of a rubber balloon which is being steadily inflated." Let us inflate a spherical balloon, and with a soft pen draw on its surface coordinate circles of latitude and longitude, and a few scattered "galaxies" (see Figure 14.23). Notice, as the balloon inflates and deflates, that the lines of latitude and longitude behave as a comoving system of coordinates, and the mutually receding galaxies are stationary relative to the coordinate lines. The balloon analogy was also mentioned by Eddington in 1933 in The

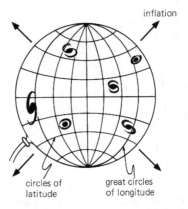

Figure 14.23. These circles of latitude and great circles of longitude on the surface of an expanding balloon illustrate the nature of comoving coordinates.

Expanding Universe: *"For a model of the universe let us represent spherical space by a rubber balloon. Our three dimensions of length, breadth, and thickness ought all to lie in the skin of the balloon; but there is room for only two, so the model will have to sacrifice one of them. That does not matter very seriously. Imagine the galaxies to be embedded in the rubber. Now let the balloon be steadily inflated. That's the expanding universe."*

6   *Hermann Weyl, a mathematician and philosopher, who was a pioneer in general relativity theory and quantum mechanics, wrote in 1922 (at the end of World War I) in the preface of his book* Space, Time, Matter, *"To gaze up from the ruins of the oppressive towards the stars is to recognize the indestructible world of laws, to strengthen faith in reason, to realize the 'harmonia mundi' [harmony of the worlds] that transcends all phenomena, and that never has been, nor will be, disturbed." Weyl in 1923 supposed that the galaxies have diverging world lines, as shown in Figure 14.24, and the galaxies in effect are stationary in a uniform space that is perpendicular to the world lines. In this space the galaxies share a common time (cosmic time). The velocity–distance law, implicit in Weyl's principle, emerged a decade later*

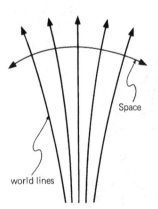

Figure 14.24. The Weyl principle: nebulae have diverging world lines, and are stationary in a space that is perpendicular to the world lines. Hermann Weyl argued in 1923 that the nebulae would recede from one another with apparent velocities that increase with their separation.

amidst considerable controversy. The debate began at a British Association science meeting in 1931 and was published as a collection of contributions in Nature *under the title "The evolution of the universe." From this symposium Edward Milne emerged as a principal contributor, and in a series of publications he formulated the cosmological principle and stressed the fact that the velocity–distance relation is the consequence of time-invariant homogeneity. It seems to have been forgotten by many cosmologists that Hubble's approximate redshift–distance law derives from observation, whereas the exact velocity–distance law derives from theory and the cosmological principle. The two laws are not the same and should not be referred to indiscriminately as Hubble's law. Why? Because one might then think either that the redshift–distance law is rigorously true for all distances (which it is not), or the velocity–distance law is only approximately true (which it is not).*

The cosmological principle formed the foundation of the work by Howard Robertson, Arthur Walker, Richard Tolman and others on homogeneous and isotropic spacetimes (now enshrined in the Robertson–Walker metric) that consist of uniformly curved or flat expanding three-dimensional space with cosmic time as a fourth dimension.

But Milne rejected general relativity and strenuously opposed the expanding space paradigm. He refused to attribute to space (which "by itself has no existence") the properties of curvature and expansion. In protest he developed his own theory, which he called kinematic relativity. Of the expanding space paradigm, he said in 1934, "This concept, though mathematically significant, has by itself no physical content; it is merely the choice of a particular mathematical apparatus for describing and analyzing phenomena. An alternative procedure is to choose a static space, as in ordinary physics, and analyze the expansion phenomena as actual motions in this space" ("A Newtonian expanding universe"). But a bounded finite cloud of galaxies expanding at the boundary at the speed of light in an infinite static space restores the cosmic center and the cosmic edge, and is contrary to modern cosmological beliefs.

7 *The determination of the distances of very distant galaxies is extremely difficult. If we rely only on less distant galaxies, which have redshifts small enough to justify the use of the Fizeau–Doppler formula, the determination of H_0 becomes uncertain, as history has shown. But the greater the redshifts, the greater the problem of determining the distances; furthermore, the nonlinear corrections to the Fizeau–Doppler formula, necessary at larger redshifts, depend on the characteristics of the cosmological model that observations have yet to determine. It seems like a no-win situation; corrections to the observations depend on knowing the model, and the choice of a model depends on knowing the correct observations.*

• *Observers use various kinds of distance indicators, such as "luminosity distance," "apparent-size distance," "number-count distance," and "redshift distance." But the Hubble term H_0 of the linear velocity–distance law is defined in terms of geometric tape-measure distances, and therefore the observers' distance indicators must be translated into tape-measure distances to determine H_0. Translating distance indicators into geometric distances is tricky. In this book, unless otherwise stated, we use only the geometric distances (as in Chapter 19) of the kind one would obtain with a tape measure stretched in the world map. These are the unambiguous, clearly defined, and easily understood distances used in the velocity–distance law, and in terms of these distances the velocity–distance law is linear and recession velocities are without limit.*

8 *Can we prove that all places are alike in space? Can we, in other words, prove that the universe is homogeneous in the world map? Remember, we cannot observe the world map; we see only a world picture that slices back through space and time. From this world picture, which is isotropic but not homogeneous, we try to construct a world map showing what the universe is like everywhere at the present moment in cosmic history. We must project the world picture onto the world map of the present cosmic epoch by transforming the observers' distance indicators into tape-measure distances and by allowing for evolution and expansion. Hence we must assume that the laws of physics are everywhere the same and that things evolve everywhere as they do in our neighborhood. This presupposes the existence of an underlying homogeneity. We prove geometric homogeneity with arguments that presuppose physical homogeneity. At best, we can show that the observed world picture is consistent with a homogeneous world map; we can never prove by direct observation that all places are alike. It is possible for an infinite number of isotropic universes to mimic at some instant our world picture and yet have world maps that are inhomogeneous. Yet each of these deceptive universes requires that we have special location at a cosmic center. On philosophical grounds and by appeal to the location principle we dismiss these possibilities as unlikely. We favor homogeneity because special location is improbable.*

• *It is a curious consequence of homogeneity that on the average everybody in the universe thinks alike. Our gross averaging is done over a cosmologically significant element of volume of hundreds of millions of light years in size and hundreds of millions of years in duration. All variety in thought and outlook is to be found within a cluster of galaxies and not by exploring the uttermost depths of the universe.*

• *An object – an organism or a planet – consists of a finite number of particles. A finite number of particles can be arranged into only a finite number of distinctly different configurations of finite size. If the particles are rearranged an infinite number of ways, then each configuration of finite probability occurs an infinite number of times. Consider now a uniform universe containing space (flat or hyperbolic) of infinite extent. The observable part extends out roughly a Hubble distance of between 10 and 20 billion light years. The unobservable universe that lies beyond extends endlessly. Trillions of trillions of Hubble distances are nothing compared with* infinity. And if all places are alike, there exists out there an infinite number of identical copies of all things that exist here: an infinite number of Solar Systems having identical Earths, having identical human populations living identical lives. All things of finite probability are repeated an infinite number of times in an infinite universe. The principle of plenitude returns with a vengeance! See "Life in the infinite universe" by G. F. R. Ellis and G. B. Brundrit (1974).

This form of infinite plenitude was discussed by the German philosopher and poet Friedrich Nietzsche in "the great game of chance that constitutes the universe." He wrote: "In infinity, every possible combination must have been realized, and must also have been realized an infinite number of times" (The Will to Strength, 1886).

While on this theme we should consider also the steady-state universe in which everything goes on forever in the same way. It is a universe of infinite and uniform space in which everything is eternally the same. Out there in space at this instant are an infinite number of identical Harrisons writing this identical book. Moreover, in every cosmic element of proper volume every configuration of finite probability has been repeated in the past and will be repeated in the future an infinite number of times. Uniqueness is a forbidden word. At this moment in time an infinite number of Harrisons exists in space and at this place in space an infinite number of Harrisons have existed in the past and will exist in the future. An infinite space of homogeneous content has never appealed to me, and I have felt repelled by the steady-state theory of the universe from its inception because of its eternal sameness in time. What's the point of infinite plenitude when once is usually more than enough? It's easy to understand why some cosmologists, if only for philosophical considerations, favor homogeneous universes that are finite in space and time.

9 The term big bang, used to denote a dense beginning, was first used by Fred Hoyle in 1949 in his series of BBC radio talks on astronomy, first published in The Listener and later in The Nature of the Universe, 1950. The word whimper, used to denote a universe that does not begin or end with a big bang, was used by T. S. Eliot in "The hollow men":

This is the way the world ends
Not with a bang but a whimper.

George Ellis has used whimper differently to indicate universes that collapse nonuniformly, with different regions inside their own trapped surfaces (Chapter 20), and do not terminate in a single big bang. Our alternative use of the word is not likely to confuse the reader, and is perhaps more in accord with Eliot's poem. Instead of bang and whimper we could use fire and ice from Robert Frost's "Fire and ice":

Some say the world will end in fire,
Some say in ice.
From what I've tasted of desire
I hold with those who favor fire.
But if it had to perish twice,
I think I know enough of hate
To say that for destruction ice
Is also great
And would suffice.

• In classical theory, a cosmic singularity occurs when density is infinitely great. A bang type of universe has a singularity when the scaling factor R is zero. What happens at infinite density is not known, and for physical reasons (see Chapters 13 and 20) it is likely that a singular state of this nature is unattainable. The extreme density attained in gravitational collapse, however, is still referred to as a singularity.

• Many persons have disliked the notion of a big bang. In The Expanding Universe (1933), Arthur Eddington wrote, "Since I cannot avoid introducing this question of a beginning, it has seemed to me that the most satisfactory theory would be one which made the beginning not too unaesthetically abrupt." In 1931, in "The expansion of the universe," he wrote, "Philosophically, the notion of a beginning of the present order of Nature is repugnant to me." His aversion to a big bang was shared by others, including advocates of the steady-state universe.

Cosmic birth and death (Chapter 25) were common notions in mythology, and fears of such ideas may (I am guessing) stem from the way that people now live. Most of us are no longer members of extended family communities, surrounded by relatives, young and old, among whom birth and death are common events. We live instead singly or in small families, isolated from one another, and birth and death are unfamiliar events occurring out of sight in hospitals. Eddington, who was outspoken in his dislike of cosmic birth and rejected catastrophic cosmic death, was a bachelor who lived with his sister. When a cosmologist presents an argument for a particular type of universe, perhaps we should not read too much into the science but wonder about that person's religion, philosophy, and even psychology.

10 *The Hubble sphere contains all galaxies receding subluminally (less than the velocity of light); galaxies at the edge of the Hubble sphere recede transluminally (at the velocity of light); and galaxies outside the Hubble sphere recede superluminally (greater than the velocity of light). The edge of the Hubble sphere at radial distance $L_H = c/H_0$ recedes at velocity $U_H = dL_H/dt$, or*

$$U_H = c(1 + q). \qquad [14.29]$$

Galaxies at the edge recede at the velocity of light c, and the edge overtakes the galaxies at relative velocity

$$U_H - c = cq.$$

In all decelerating universes $(q > 0)$, the Hubble sphere expands faster than the universe and contains an increasing number of galaxies. In all accelerating universes $(q < 0)$ the Hubble sphere expands slower than the universe and contains a decreasing number of galaxies. If N_H is the number of galaxies in the Hubble sphere, it can be shown that

$$\frac{dN_H}{dt} = 3qHN_H. \qquad [14.30]$$

This expression cannot be used in the steady-state universes of $q = -1$ because of the

continual creation of new galaxies; in this universe U_H and N_H are constant.

11 *The metric equation for the distance dL between two adjacent points in a spherical space of curvature $1/R^2$ and comoving coordinates a, θ, and ϕ is*

$$dL^2 = R^2[da^2 + \sin^2 a(d\theta^2 + \sin^2 \theta \, d\phi^2)],$$

from Equation [10.15]. The coordinates a, θ, and ϕ are physically dimensionless (as for colatitude and longitude on the surface of a sphere), and with distances measured in light-travel time we see that R has the dimensions of time. By changing the symbol a into r, we have

$$dL^2 = R^2[dr^2 + \sin^2 r(d\theta^2 + \sin^2 \theta \, d\phi^2)].$$

$$[14.31]$$

To any point the distance from an arbitrary origin $r = 0$, is $L = R \int dr = Rr$ and the comoving coordinate distance is simply r. In Equation [14.31], R becomes the scaling factor that varies in time, and r, θ, and ϕ are the fixed coordinates of comoving bodies. More generally,

$$dL^2 = R^2[dr^2 + S^2(d\theta^2 + \sin^2 \theta \, d\phi^2)],$$

$$[14.32]$$

in which

$S = r$:	flat space $(k = 0)$
$S = \sin r$:	spherical space $(k = 1)$
$S = \sinh r$:	hyperbolic space $(k = -1)$

and Equation [14.32] applies to the three homogeneous and isotropic spaces of curvature constant k and curvature k/R^2.

The relativity line-element expressing ds in terms of intervals of time dt and space dL is

$$ds^2 = dt^2 - dL^2, \qquad [14.33]$$

as in Equation [11.4]. With dL from Equation [14.32] and dt understood as an interval in cosmic time, we obtain the famous Robertson–Walker line element

$$ds^2 = dt^2 - R^2[dr^2 + S^2(d\theta^2 + \sin^2 \theta \, d\phi^2)],$$

$$[14.34]$$

anticipated by many cosmologists, notably by Georges Lemaître, Howard Robertson, Richard Tolman, and Arthur Walker. Because of the study of its significance by Robertson and Walker it is referred to as the Robertson–Walker geometry, or metric, or line element. Various forms of Equation [14.34] are obtained by transforming the radial coordinate r. The form shown is most convenient because the radial coordinate distance r corresponds to a linear tape-measure distance.

• Much of modern cosmology flows from the Robertson–Walker (or R–W) line element. At this stage we mention three issues:

(i) The R–W equation defines cosmic time t as orthogonal to uniformly curved, expanding space.

(ii) The R–W equation automatically yields the velocity–distance law. Let us arrange the coordinate system such that we are at the origin $r = 0$. A comoving galaxy at distance $L = Rr$ recedes at velocity $V = dL/dt$. Because r is constant for the galaxy, the change in its distance in time dt is

$$dL = r\,dR = L\,dR/R = LH\,dt,$$

where $H = \dot{R}/R$, and hence the recession velocity of the galaxy is $V = LH$. At a fixed value of H (and therefore at a fixed instant in cosmic time) the recession velocities of all galaxies are proportional to their tape-measure distances L. Thus we see that the velocity–distance law is implicit in the R–W line-element because it applies to spaces of time-invariant homogeneity.

(iii) The R–W line element enables us to relate the world picture (on the backward lightcone) and the world map (in which the velocity–distance law applies). Light and anything moving at the velocity of light propagates on null-geodesics defined by $ds = 0$. Again we assume for convenience that we are at the origin $r = 0$ and consider radial rays of light for which $d\theta = 0$, $d\phi = 0$. The R–W line element, with $ds = 0$, reduces to $dt = \pm R\,dr$ for increments in radial distance, and the plus sign applies to outgoing rays on the forward lightcone and the negative sign to incoming rays on the backward lightcone.

Hence, a coordinate distance on the backward lightcone (the world picture) is given by

$$r = \int_t^{t_0} dt/R, \qquad [14.35]$$

where t is the time of emission of the ray observed at the present time t_0. Thus the actual distance to the source at the present time is

$$L = R_0 r = R_0 \int_t^{t_0} dt/R, \qquad [14.36]$$

and its distance at the time of emission is

$$L_{emit} = Rr$$
$$= R \int_t^{t_0} dt/R = (R/R_0)L. \qquad [14.37]$$

These equations help us to relate observed distances in the world picture to tape-measure distances in the world map and provide the bridge that must be crossed to determine H_0 and q_0 from observations.

12   The way in which the scaling factor R varies in time must be found by means of a dynamic model of the universe. This will be discussed in later chapters. In the power-law models, R varies as t^n, where n is a constant. Of these bang–whimper universes, the most important is the Einstein–de Sitter model of $n = 2/3$. If we assume that

$$R = R_0(t/t_0)^n, \qquad [14.38]$$

we find

$$H = n/t, \qquad [14.39]$$
$$q = (1 - n)/n, \qquad [14.40]$$

hence $n = (1 + q)^{-1}$. The subscript zero is added when we wish to denote present values. The deceleration q is constant in all power-law models. The universe expands when $H > 0$, or $n > 0$, and decelerates when $q > 0$, or $n < 1$. The Hubble length in the power-law models is

$$L_H = ct/n = ct(1 + q), \qquad [14.41]$$

and the Hubble time is

$$t_H = t/n = t(1 + q). \qquad [14.42]$$

Notice that $t/t_H = 1/(q+1)$, illustrating the general rule that the age of the universe t is less than the Hubble time t_H in a decelerating universe of $q > 0$. Finally, the Hubble sphere expands at velocity

$$U_H = c/n = c(1+q). \qquad [14.43]$$

In the Einstein–de Sitter model of $n = 2/3$, we find $q = 1/2$, $L_H = 3ct/2$, $t_H = 3t/2$, $U_H = 3c/2$, and the edge of the Hubble sphere overtakes the galaxies at $c/2$.

Using Equations [14.35] and [14.38], we find that the present distance of an observed body that emitted light at time t is

$$L = \frac{n}{1-n} L_H(1 - x^{1-n}), \qquad [14.44]$$

and the distance of this body at the time it emitted the light that we now see is

$$L_{emit} = \frac{n}{1-n} L_H x^n (1 - x^{1-n}), \qquad [14.45]$$

where $x = t/t_0$. These equations apply in a big bang universe of $0 < n < 1$. We see from Equation [14.44] that the maximum present distance (the maximum value of L) of a body is $nL_H/(1-n) = t_0/(1-n)$, and this value, because of expansion, is more than the maximum distance t_0 in a static universe of $n = 0$. And from Equation [14.45] we see that the emission distance is zero when $t = t_0$ and also when $t = 0$.

PROJECTS

1 Take an elastic cord, fix markers to it, such as clothes pegs, and slowly stretch it, as in Figure 14.25. The comoving markers represent the galaxies in an expanding

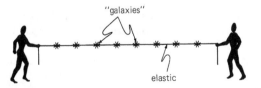

Figure 14.25. A one-dimensional expanding universe consisting of an elastic string having attached markers representing galaxies. The markers have fixed comoving coordinates and their relative motions illustrate the velocity–distance law.

universe. Notice that distances are the tape-measure kind and that the markers move apart according to the velocity–distance law. We see how separating distances between the markers are related to comoving coordinate distances by means of a scaling factor. We see also that when stretched slowly or rapidly the elastic string demonstrates the change in time of the scaling factor.

2 The term big bang was first used by Fred Hoyle in his series of BBC radio talks on astronomy published in *The Nature of the Universe*, 1950. "This big bang idea seemed to me to be unsatisfactory even before examination showed that it leads to serious difficulties. For when we look at our own Galaxy there is not the smallest sign that such an explosion ever occurred. This might not be such a cogent argument against the explosion school of thought if our Galaxy had turned out to be much younger than the whole Universe. But this is not so. On the contrary, in some of these theories the Universe comes out to be younger than our astrophysical estimates of the age of our own Galaxy.... On philosophical grounds too I cannot see any good reason for preferring the big bang idea. Indeed it seems to me in the philosophical sense to be a distinctly unsatisfactory notion, since it puts the basic assumption out of sight where it can never be challenged by direct appeal to observation." Discuss Hoyle's remarks.

3 Let the smoothed-out density of the universe be equivalent to 1 hydrogen atom per cubic meter. Now gather this matter together into uniformly distributed marbles and find their separating distance. Gather the matter together into uniformly distributed stars similar to the Sun and find their separating distance. Now do the same for galaxies of 10^{11} solar masses.

4 Explain why, in a universe 1 year old, we cannot see farther than a distance of approximately 1 light year.

5 What is wrong with the idea of a universe beginning at a point in space?

6 Explain the difference between recession velocity and ordinary or peculiar velocity, and comment on the following remark taken from a textbook in astronomy: "There are mathematical models of the universe that have galaxies ... going even faster than the velocity of light. Of course, the laws of relativity forbid this, and such models are only of academic interest."

7 Draw diagrams showing how the scaling factor R may vary in time. Show in a single diagram the variation of R with time in an imaginary universe that passes through successive periods of expansion, contraction, deceleration, and acceleration, and label these periods with $H > 0$, $H < 0$, $q > 0$, and $q < 0$.

8 In what universe is (a) R constant? (b) H constant? (c) q constant?

9 Derive Hubble's redshift–distance law from the mathematicians' velocity–distance law. Explain the approximations used. What is the distance of galaxies of redshift $z = 0.01$ and 0.1? (Give the answer in terms of h and in light years.)

10 Show that when $R = t$, and therefore $H = 1/t$, $q = 0$, the Hubble sphere contains a constant number of galaxies.

FURTHER READING

Davies, P. C. W. *The Runaway Universe*. Dent, London, 1978.

Eddington, A. S. *The Expanding Universe*. Cambridge University Press, Cambridge, 1933.

Ellis, G. F. R. "Innovation, resistance and change: the transition to the expanding universe," in *Modern Cosmology in Retrospect*. Edited by B. Bertotti et al. Cambridge University Press, Cambridge, 1990.

Gamow, G. *The Creation of the Universe*. Viking Press, New York, 1952.

Hetherington, N. S. "Hubble's cosmology." *American Scientist* 78, 142 (1990).

Huchra, J. P. "The Hubble constant." *Science* 256, 321 (1992).

North, J. D. *The Measure of the Universe: A History of Modern Cosmology*. Oxford University Press, Clarendon Press, Oxford, 1965.

Sandage, A. R. "Cosmology: a search for two numbers." *Physics Today* (February 1970).

SOURCES

Andrade, E. N. da C. "Doppler and the Doppler effect." *Endeavour* p. 14 (January 1959).

Brush, S. G. *The Temperature of History: Phases of Science and Culture in the Nineteenth Century*. Franklin and Company, New York, 1978.

Burbidge, G. "Modern cosmology: the harmonious and discordant facts." *International Journal of Theoretical Physics* 28, 983 (1989).

Eddington, A. S. "On the instability of Einstein's spherical world." *Monthly Notices of the Royal Astronomical Society* 90, 668 (1930).

Eddington, A. S. "The evolution of the universe." Symposium. *Nature*, Supplement, p. 699 (October 24, 1931).

Eddington, A. S. "The expansion of the universe." *Monthly Notices of the Royal Astronomical Society* 91, 412 (1931).

Eddington, A. S. *The Expanding Universe*. Cambridge University Press, Cambridge, 1933.

Ellis, G. F. R. "The expanding universe: A history of cosmology from 1917 to 1960," in *Einstein and the History of General Relativity*. Editors, D. Howard and J. Stachel. (Einstein Study Series.) Birkhauser, Boston, 1988.

Ellis, G. F. R. and Brundrit, G. B. "Life in the infinite universe." *Quarterly Journal of the Royal Astronomical Society* 20, 37 (1974).

Friedmann, A. "On the curvature of space." *Zeitschrift für Physik* 10, 377 (1922). "On the possibility of a world with constant negative curvature." *Zeitschrift für Physik* 21, 326 (1924). Both translated in *Cosmological Constants*. Editors, J. Bernstein and G. Feinberg. Columbia University Press, New York, 1986.

Harrison, E. R. "A century of changing perspectives in cosmology." *Quarterly Journal of the Royal Astronomical Society* 33, 335 (1992).

Harrison, E. R. "Hubble spheres and particle horizons." *Astrophysical Journal* 383, 60 (1991).

Harrison, E. R. "The redshift–distance and velocity–distance laws." *Astrophysical Journal* 403, 28 (1993).

Hoyle, F. *The Nature of the Universe*. Blackwell, Oxford, 1950.

Hubble, E. *The Realm of the Nebulae*. Yale University Press, New Haven, 1936.

Hubble, E. *The Observational Approach to Cosmology*. Oxford University Press, Clarendon Press, Oxford, 1937.

Huggins, W. "Further observations on the spectra of some stars and nebulae, with an attempt to determine therefrom whether these bodies are moving towards or away from the Earth." *Philosophical Transactions* 158, 529 (1868).

Hujer, K. "Sesquicentennial of Christian Doppler." *American Journal of Physics* 23, 51 (1955).

Kragh, H. *Cosmology and Controversy: The Historical Development of Two Theories of the Universe*. Princeton University Press, Princeton, 1996.

McCrea, W. H. "Willem de Sitter, 1872–1934." *Journal of the British Astronomical Association* 82, 178 (1972).

Metz, W. D. "The decline of the Hubble constant: a new age for the universe." *Science* 178, 600 (1972).

Milne, E. A. "A Newtonian expanding universe." *Quarterly Journal of Mathematics*, 5, 64 (1934).

Milne, E. A. *Relativity, Gravitation and World Structure*. Oxford University Press, Oxford, 1935.

Murdoch, H. S. "Recession velocities greater than light." *Quarterly Journal of the Royal Astronomical Society* 18, 242 (1977).

Nietzsche, F. See Brush, S. G.

Robertson, H. P. "On relativistic cosmology." *Philosophical Magazine* 5, 835 (1928).

Robertson, H. P. "The expanding universe." *Science* 76, 221 (1932).

Sandage, A. R. "Observational cosmology." *Observatory* 88, 91 (1968).

Sandage, A. R. "Distances to galaxies: the Hubble constant, Friedmann time, and the edge of the world." *Quarterly Journal of the Royal Astronomical Society* 13, 282 (1972).

Sitter, W. de. "The expanding universe." *Scientia* 49, 1 (1931).

Sitter, W. de. *Kosmos*. Harvard University Press, Cambridge, Massachusetts, 1932.

Smith, R. W. "The origin of the velocity–distance law." *Journal for the History of Astronomy* 10, 133 (1979).

Tolman, R. C. "The age of the universe." *Reviews of Modern Physics* 54, 374 (1949).

Weyl, H. *Space, Time, Matter*. Methuen, London, 1922.

Whitrow, G. J. "Hubble, Edwin Powell." *Dictionary of Scientific Biography* 6, 528 (1967).

O ruddier than the cherry,
O sweeter than the berry.
O nymph more bright
Than moonshine night,
Like kidlings blithe and merry.
John Gay (1685–1732), Acis and Galatea

COSMIC REDSHIFTS
Wavelength stretching
In the previous chapter we saw how the Fizeau–Doppler (known more briefly as the Doppler) formula played a vital role in the discovery of the expansion of the universe. Distant galaxies have redshifted spectra, and their redshifts were interpreted to mean the galaxies are rushing away from us. Then in the late 1920s and early 1930s Georges Lemaître, Howard Robertson, and other cosmologists discovered a totally new interpretation of extragalactic redshifts based on the expanding space paradigm.

The new expanding space redshift is simple and very easy to understand. We suppose that all galaxies are comoving and their emitted light is received by observers who are also comoving. Light leaves a galaxy, which is stationary in its local region of space, and is eventually received by observers who are stationary in their own local region of space. Between the galaxy and the observer, light travels through vast regions of expanding space. As a result, all wavelengths of the light are stretched by the expansion of space (see Figure 15.1). It is as simple as that.

A light ray, emitted by a distant galaxy, travels across expanding space and is received by the observer. If, while the light ray travels, all comoving distances are doubled, it follows that all wavelengths of the light ray are also doubled. Waves are stretched by the expansion of space and their increase is proportional to the increase

in the scaling factor. Let λ represent a wavelength of an emitted wave of light and λ_0 represent the wavelength of the same wave when it is received by an observer. If R is the value of the scaling factor at the time of emission and R_0 is the value at the time of reception, then

$$\frac{\lambda_0}{\lambda} = \frac{R_0}{R},$$ [15.1]

and wavelengths increase in just the same way as comoving distances in an expanding universe. This consequence of the expanding space paradigm applies to radiation of all wavelengths and is independent of the velocity of recession. Equation [15.1] is of great importance in cosmology.

Expansion redshifts
A spectrum shows how the intensity of radiation varies with wavelength, and contains recognizable features in the form of spectral lines (see Figure 15.2). These lines are the result of electrons in atoms jumping between different energy levels and thereby absorbing and emitting radiation of specific wavelengths. From measurements made in the laboratory we know the standard wavelengths of the spectral lines of atoms of different elements in various states of excitation. Because the universe is homogeneous, atoms everywhere in the universe are alike and their wavelengths of emitted radiation are identical with those observed in our laboratories on Earth. When a recognizable spectral line from a distant galaxy is

Figure 15.1. A wave of radiation is continually stretched as it travels in expanding space.

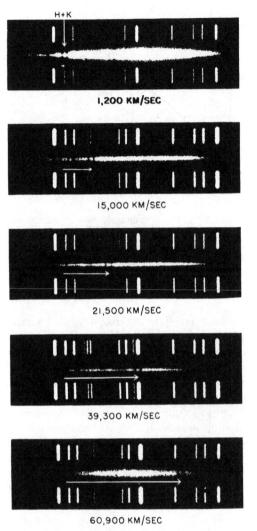

Figure 15.2. The spectra of five galaxies showing the displacement of the H and K absorption lines of calcium. The recession velocities shown are found by multiplying redshift by the velocity of light.

observed at wavelength λ_0, we know that far away and long ago it originated at wavelength λ. From the ratio λ_0/λ we obtain the ratio R_0/R, and we know immediately how much the universe has expanded since the time of emission (Figure 15.3). This is the marvel of cosmic redshifts: they directly measure the expansion of the universe.

The redshift z, defined earlier, is the increase of wavelength $\lambda_0 - \lambda$ divided by the emitted standard wavelength λ:

$$z = \frac{\lambda_0 - \lambda}{\lambda}, \qquad [15.2]$$

and hence

$$1 + z = \frac{\lambda_0}{\lambda}. \qquad [15.3]$$

All Doppler, gravitational, and expansion redshifts are defined this way. The expansion redshift is given by the expression

$$1 + z = \frac{R_0}{R}, \qquad [15.4]$$

from Equations [15.1] and [15.2], see Figure 15.3. The velocity–distance law ($V = HL$) of Equation [14.6] in the preceding chapter and the expansion–redshift law ($1 + z = R_0/R$) of Equation [15.4] in this chapter are the two most important laws in cosmology. Note that according to

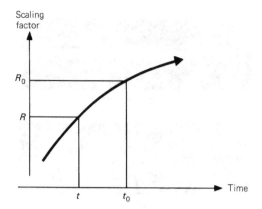

Figure 15.3. In an expanding universe the scaling factor increases with time. Radiation is emitted by a source at time t when the scaling factor has a value R, and is received at time t_0 when the scaling factor has a value R_0. The redshift z of the received radiation is given by $1 + z = R_0/R$.

Equation [15.4] the expansion redshift is independent of the actual velocity of recession.

The wavelengths of all kinds of radiation received from a source have identical expansion redshifts, not only the wavelengths of optical radiation but also the shorter wavelengths of x-rays and the longer wavelength of radio waves. If the universe doubles in size, which means R_0/R equals 2, then all wavelengths, short and long, are increased twofold and the redshift is $z = 1$. If the universe trebles in size (i.e., $R_0/R = 3$), all wavelengths are increased threefold and $z = 2$.

In an expanding universe the cosmological redshift ranges from zero to infinity. This is because in Equation [14.4] the scaling factor R ranges from zero to R_0. Small redshifts indicate that light has traveled for relatively short periods of time and large redshifts indicate that light has traveled for long periods of time. Redshifts of optically observed galaxies have measured values as large as 0.5, or greater, and the observed redshifts of luminous quasars extend to values as large as 5.

In a collapsing universe the cosmological redshift is negative and ranges from 0 to −1. This is because R_0 is now less than R, and at the time of reception the universe is smaller

in size than at the time of emission. Negative redshifts are blueshifts. If a universe contracts to half its size in the time between the emission and reception of radiation, then $R_0/R = 0.5$ and the redshift is −0.5.

On measuring the redshift of a luminous extragalactic source we know immediately the ratio of the scaling factors for the epochs of emission and reception of the radiation from the source. We know, in other words, how much the universe has expanded since the time the radiation was emitted. The quasar 3C 273 has redshift 0.16, which means the universe is now 1.16 times larger than when this quasar emitted the light we now see. Hence the universe has increased in size by 16 percent since 3C 273 emitted its light. Quasar 3C 48 has a redshift 0.37 and the universe has expanded by 37 percent since its light was emitted. The rule is quite simple: multiply the expansion redshift by 100 to get the percentage increase in size of the universe since the radiation was emitted.

Red means slow

A wavelength λ of radiation, multiplied by frequency, equals the speed of light:

$$\lambda f = c, \qquad [15.5]$$

where f stands for frequency at wavelength λ. (The light speed c is constant and always the same when measured locally in the region of space through which the radiation passes.) When wavelengths are increased, frequencies are decreased. From Equations [15.4] and [15.5] we find that

$$1 + z = \frac{f}{f_0}, \qquad [15.6]$$

where f is the emitted frequency and f_0 the received frequency.

It must be understood that the slowing down of vibrations applies to all frequencies, not only the ultraviolet, optical, infrared, and radio, but also the variations in luminosity having periods of hours, days, or even years. Some quasars fluctuate in brightness with periods ranging from days to years; these are observed periods that have been increased by the cosmological redshift.

If a quasar of redshift $z = 1$ is seen to vary in brightness with a period of 1 month, the original period of variation at the quasar was 2 weeks.

Things appear to happen slower the farther we probe into the depths of outer space. This strange result can be understood more easily in the following way. Suppose that a distant galaxy emits short pulses of radiation at intervals of 1 second. In a static universe we would receive these pulses at 1 second intervals. This is because the separating distance between successive pulses as they travel through space remains constant and equal to 1 light second. But in an expanding universe, the distance between successive pulses, which initially is 1 light second, continually increases and finally is $1 + z$ light seconds on arrival. The pulses are emitted at one-second intervals and received at intervals of $1 + z$ seconds, as shown in Figure 15.4. If something happens in time Δt at redshift z, we observe it happen in time Δt_0, given by

$$\frac{\Delta t_0}{\Delta t} = 1 + z. \qquad [15.7]$$

Clocks at great distances appear to us to run slow, and their intervals of time are

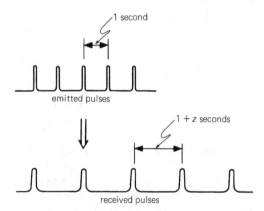

Figure 15.4. Pulses of radiation are emitted by an extragalactic source at 1 second intervals. After traveling in expanding space the pulses become more widely separated and arrive in our Galaxy at intervals of $1 + z$ seconds. This time-stretching illustration explains how redshift makes everything at great distance appear to happen more slowly than here.

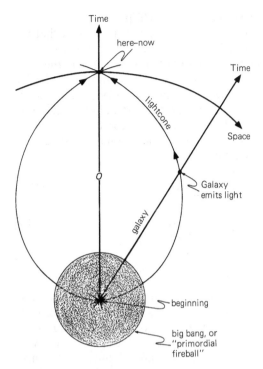

Figure 15.5. This diagram shows the world map and the world picture. We (world line O) look out in space and back in time and all information comes to us on our backward lightcone. The farthest we look back in time is the fireball of the big bang from which comes the cosmic background radiation of redshift $z = 1000$. If we could look back even farther, to the beginning of the universe, the scaling factor R at the time of emission would be zero, making the redshift infinite.

increased by a factor $1 + z$. If something happens in 1 second at redshift $z = 1$, it is seen by us to happen in 2 seconds.

At the frontier of the observable universe, where the expansion redshift rises to infinity, nothing seems to change; everything is apparently in a frozen static state (see Figure 15.5). Infinite redshifts of course are not observed, and the maximum encountered so far is roughly $z = 1000$ of the cosmic background. A second at the time of decoupling is seen by us to last 17 minutes. If we succeed in detecting neutrinos from the big bang then we shall look back in time to a redshift about $z = 10$ billion, and a second in time will be seen by us to last 300 years.

Usefulness of redshifts

In the previous chapter it was said that the actual value of the scaling factor R is rarely important and often we need only the ratio of the values it has at different epochs. Whenever the ratio R_0/R occurs, for example, we can substitute the observable quantity $1 + z$. Thus for a comoving length L, comoving area A, and comoving volume V we may write:

$$\frac{L}{L_0} = \frac{1}{1+z}, \qquad [15.8]$$

$$\frac{A}{A_0} = \frac{1}{(1+z)^2}, \qquad [15.9]$$

$$\frac{V}{V_0} = \frac{1}{(1+z)^3}. \qquad [15.10]$$

A comoving length L at the time of emission has now a length $L_0 = L(1 + z)$; a comoving area A at the time of emission has now an area $A_0 = A(1 + z)^2$; and a comoving volume V at the time of emission has now a volume $V_0 = V(1 + z)^3$. In the case of $z = 1$, at the time of emission comoving lengths were one-half their value now, comoving areas were one-quarter their value now, and comoving volumes were one-eighth their value now.

The change in density of the universe provides an important example of the relation between expansion redshift and the scaling factor. The number n of things (atoms or galaxies) in a unit of volume equals the present density n_0 multiplied by $V_0/V = (R_0/R)^3$, and therefore the density at redshift z is given by

$$\frac{n}{n_0} = (1 + z)^3. \qquad [15.11]$$

For $z = 1$, the density at the epoch of emission was eight times greater than the present density. For a redshift $z = 100$ the density was 1 million times greater than now, or approximately 1 hydrogen atom per cubic centimeter, which is typical of the average density of matter in galaxies. At redshifts greater than 100, galaxies in their present form did not exist. We cannot normally look out in space and back in time to such large redshifts because luminous galaxies and quasars did not exist. The cosmic background radiation that fills the universe is a remarkable exception; it has been redshifted by approximately 1000 since it last interacted with matter, at which time the universe was hot and had a density 1 billion times greater than at present.

THE THREE REDSHIFTS
Gravitational redshift
The gravitational redshift is observed when the emission of light occurs in regions where the strength of gravity is greater than in the observer's region. Light emitted from the surface of a spherical body of radius R and mass M and observed at large distances has redshift

$$z = \frac{1}{\sqrt{1 - R_S/R}} - 1$$

$$\text{(gravitational redshift)} \qquad [15.12]$$

where R_S is the Schwarzschild radius equal to $2GM/c^2$. We have discussed this redshift in Chapters 12 and 13.

Doppler redshift
The Doppler redshift occurs because of relative motion in space. A body moving away from the observer in the laboratory, the Solar System, or the Galaxy moves through space and the radiation it emits is seen redshifted. If V is the radial velocity of the luminous body moving away, then

$$z = \frac{V}{c} \quad \text{(Fizeau–Doppler formula)}.$$

$$[15.13]$$

This is the classical Fizeau–Doppler formula that is true only when V is very small compared with the velocity of light c. The exact formula – the special relativity Doppler formula – which must be used when V/c is not small, is

$$z = \frac{1 + V/c}{\sqrt{1 - V^2/c^2}} - 1$$

$$\text{(special relativity Doppler formula)}.$$

$$[15.14]$$

This equation reduces to Equation [15.13] when V/c is small.

Most of the peculiar velocities encountered in astronomy of stars in galaxies and galaxies in clusters are small in comparison with the velocity of light, thus allowing us to use the original Fizeau–Doppler formula. When z is not small, we find from the special relativity Doppler formula of Equation [15.14] that

$$\frac{V}{c} = \frac{z^2 + 2z}{z^2 + 2z + 2}. \qquad [15.15]$$

With the classical formula (Equation 15.13) we find $V = 0.1c$ when $z = 0.1$; and with the special relativity formula (Equation 15.15) we find $V = 0.6c$ when $z = 1$, $V = 0.8c$ when $z = 2$, $V = 0.88c$ when $z = 3$, and $V = c$ when z equals infinity.

Expansion redshift

The expansion redshift, as we saw earlier, is the result of the expansion of space in an expanding universe. Comoving bodies, stationary in expanding space, receive radiation from one another that is redshifted. The radiation propagates through expanding space and all wavelengths are stretched. This redshift is determined by the amount of expansion according to the law

$$z = \frac{R_0}{R} - 1 \quad \text{(expansion redshift)} \qquad [15.16]$$

where R is the value of the scaling factor at the time of emission and R_0 the value at the time of reception. Once the expansion redshift of a distant galaxy has been determined, the ratio R_0/R tells us immediately how much the universe has expanded during the time the light has been traveling.

Despite widespread confusion between expansion and Doppler redshifts, the difference is quite marked and easily understood. Doppler redshifts are the result of relative motion of bodies moving through space; they depend on the velocity of the emitter at the instant of emission relative to the velocity of the receiver, and on the velocity of the receiver at the time of reception relative to the velocity of the emitter; they are produced by peculiar and not by recession velocities, and are governed by the rules of special relativity. Expansion redshifts are caused by the expansion of space between bodies that are stationary in space; they depend on the increase of distance between the emitter and the receiver during the time of propagation; they are produced by recession and not peculiar velocities, and they are governed by the rules of general relativity.

It is interesting to note that the expansion redshift is independent of the way the universe expands. This is perhaps not very surprising; after all, when a length of elastic is stretched by a certain amount, it is unimportant how the stretching is done, whether slowly, quickly, or in a series of jerks. In the end it is always stretched by a stated amount. The time taken to expand from a given value R of the scaling factor to the present value R_0 and the way the expansion occurs do not affect the expansion redshift.

TWO BASIC LAWS

Velocity–distance and expansion–redshift laws

We have two basic laws: the velocity–distance law (Equation 14.6)

$$V = H_0 L, \qquad [15.17]$$

of the last chapter, and the expansion–redshift law (Equation 15.4)

$$1 + z = \frac{R_0}{R}, \qquad [15.18]$$

of this chapter. Both laws are the consequence of uniformly expanding space, and are sanctioned by general relativity, not by special relativity. These simple laws, although easy to understand, mean that modern cosmology does not entirely conform to pre-twentieth century modes of thought.

The two basic laws cannot be combined into a redshift–distance relation without a theory that relates redshifts in the world picture to distances in the world map. The velocity–distance law ($V = H_0 L$) applies to behavior in the world map and the expansion–redshift law ($z + 1 = R_0/R$) applies to observations in the world picture. We need to know how the universe evolves (how R changes in time) to project observations made in the world picture on to the world map in which H_0 must be determined. Only when redshifts are small can the two basic laws be combined to give the linear redshift–distance ($zc = H_0L$) and the linear redshift–velocity ($V = cz$) relations that were so useful in early years of modern cosmology.

Small expansion redshifts

An extragalactic source of light that has a small expansion redshift is not, cosmologically speaking, very far away; its distance is small compared with the Hubble length L_H, and the light-travel time (or lookback time) is small compared with the Hubble period t_H. The increase $\Delta R = R_0 - R$ in the scaling factor between emission and reception is small compared with R itself, and the redshift

$$z = \frac{R_0 - R}{R} = \frac{\Delta R}{R},$$

is therefore approximately

$$z = H_0 \Delta t,$$

where $H_0 = \Delta R / R \Delta t$. The distance L to a nearby source is approximately the light-travel time Δt multiplied by the speed of light: $L = c\Delta t$, and hence

$$zc = H_0 L. \qquad [15.19]$$

This is Hubble's linear redshift–distance relation and we see that it applies only to sources of small redshift. Equivalent expressions are

$$z = \frac{\Delta t}{t_H} = \frac{L}{L_H}. \qquad [15.20]$$

If we use the velocity–distance law $V = H_0L$ in Equation [15.19], we have

$$z = \frac{V}{c}, \qquad [15.21]$$

which is the Fizeau–Doppler formula. This shows that when comoving bodies have small separating distances (small compared

with the Hubble length), the basic expansion–redshift and the velocity–distance laws can be combined to give Hubble's redshift–distance relation (Equation 15.19) and the Fizeau–Doppler formula (Equation 15.21).

Alternatively, the Fizeau–Doppler formula can be used to combine the two basic laws of Equations [15.17] and [15.18] when expansion redshifts are less than about 0.1. At $z = 0.1$ the recession velocity (equal to cz) is 30 000 kilometers per second; and with $h = 0.5$, the distance (equal to zL_H) is 2 billion light years and the lookback time (equal to zt_H) is 2 billion years.

The redshift ballet

We have a picture of light emitted by extragalactic comoving sources and received by comoving observers. The observed redshift in this idealized situation is purely a cosmological effect caused by expansion of the universe. In reality, the picture is more complex. Light is emitted by stars in a galaxy and has therefore an initial gravitational redshift caused by the stars and the galaxy. It also has an initial Doppler shift caused by the peculiar motion of luminous stars and the galaxy. After traversing expanding space and acquiring an expansion redshift, the radiation is again Doppler shifted by the peculiar motion of the observer, and receives a final gravitational redshift contribution from the system in which the observer is located. The total redshift is the expansion redshift with gravitational and Doppler additions at both ends, and one of the astronomer's many skills consists of knowing how to average out or otherwise discount the noncosmological contributions. In most cases the gravitational effects are small and unimportant. Light entering our Galaxy and striking the Earth's surface receives the negligible gravitational redshift of -0.001. Doppler shifts in distant sources are often small in comparison with the expansion redshift and either are negligible or can be averaged out over many sources in the same region. Our own peculiar velocity of the Earth moving in the Solar System, the

Solar System moving in the Galaxy, the Galaxy moving in the Local Group, and the Local Group moving in the complex of clusters known as the Local Supercluster must all be taken into account.

In cosmology, the three Redshift Graces – Gravity, Doppler, and Expansion – are usually together and rarely is one alone without the others.

DISTANCES AND RECESSION VELOCITIES

Once the cosmological redshift of a distant galaxy has been found, there are several things we would like to know, such as the distance of the galaxy, its recession velocity at the present epoch, and the age of the universe at the epoch of emission (see Figure 15.6). This information, regrettably, cannot in general be found immediately from the observed redshift unless, as we have seen, the redshift is small, thus justifying the use of Hubble's redshift–distance law. In general, such information requires that we

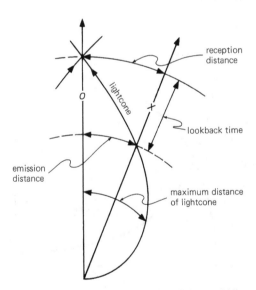

Figure 15.6. Our world line is O and the world line of a distant galaxy is X. Our backward lightcone intersects X at the time of emission of the light we now see from X. The reception (or present) distance and the emission (or past) distance of the galaxy are shown. The lookback time, from the present to the instant of emission, is also shown.

know the sort of universe in which we live; in other words, that we know the global geometry of space and the way the scaling factor evolves in time. Because we do not know for certain these things, we must make calculations for various tentative universes. Each universe yields different answers.

A popular universe, because of its simplicity, is that jointly proposed by Einstein and de Sitter in 1931. This Einstein–de Sitter universe belongs to the bang–whimper class, it possesses flat space (Euclidean geometry) and a fixed deceleration term q of value 0.5. Nothing could be simpler. The scaling factor R varies in time as $t^{2/3}$ and the Hubble term H at any time is $2/3t$. Redshifts corresponding to distances, recession velocities, and ages at epochs of emission and reception are shown in Figures 15.7, 15.8, and 15.9. Our universe is possibly not of the Einstein–de Sitter type, but the graphical results shown are far more meaningful than the erroneous results based on the Doppler effect.

COSMOLOGICAL PITFALLS

O Thou who didst with pitfall and with gin
Beset the road I was to wander in.
Edward FitzGerald (1809–1883), The Rubáiyát of Omar Khayyám

Two pitfalls beset the road to an understanding of modern cosmology.

First pitfall

The first, discussed in the previous chapter, is the common failure to distinguish between the Hubble redshift–distance law ($zc = H_0L$) and the velocity–distance law ($V = H_0L$). On the one hand, the Hubble law, which is valid only for small redshifts, is employed by observers to determine the Hubble term H_0. On the other hand, the velocity–distance law, which is employed by theorists, is a direct consequence of homogeneity. Both laws become the same when the classic Fizeau–Doppler formula is invoked for small redshifts. This is a source of confusion: the known limitations (small redshifts) of the Hubble law get transferred to the velocity–distance law, and even

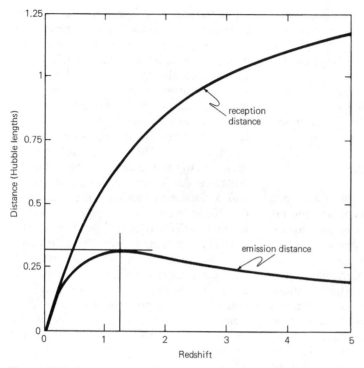

Figure 15.7. The reception and emission distances of a source of redshift z in the Einstein–de Sitter universe. The Hubble length L_H equals $10h^{-1}$ billion light years, where h, still unknown, is probably between 0.5 and 1.

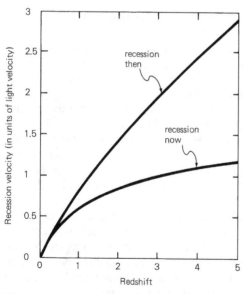

Figure 15.8. The present recession velocity at the time of reception and the past recession velocity at the time of emission of a source of redshift z in an Einstein–de Sitter universe.

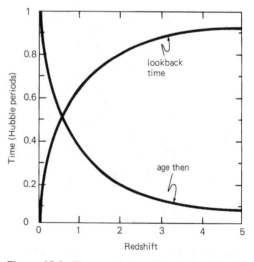

Figure 15.9. The age of the universe ("age then") at the time a source of redshift z emitted the radiation we now see, and also the lookback time, in an Einstein–de Sitter universe. The present age t_0 equals $\frac{2}{3}t_H$, and a Hubble period t_H equals $10h^{-1}$ billion years.

eminent cosmologists have been known to say that the velocity–distance law breaks down at large redshifts. By keeping the two laws separate and realizing that one is approximate (valid only for small redshifts) and the other exact (for all tape-measure distances) we avoid this pitfall.

Second pitfall

The second, discussed in this chapter, concerns the nature of cosmological redshifts. When a distant galaxy or quasar is mentioned in the popular literature, it is the custom to state its recession velocity and distance. Usually, the velocity and distance are not stated in the context of a particular cosmological model, nor hedged with reservations concerning the validity of the treatment, for fear of puzzling the audience. Instead, the Fizeau–Doppler formula of Equation [15.13] is used. The redshift is multiplied by the light velocity to give the recession velocity. Thus if $z = 0.2$, the recession velocity is said to be one-fifth the velocity of light, or 60 000 kilometers per second. The velocity–distance law then states that the distance is one-fifth the Hubble length, or 4 billion light years (assuming $h = 0.5$), and light has traveled for 4 billion years. Clearly, the classical Fizeau–Doppler formula is very convenient for making quick estimates, but the results obtained in this way are incorrect, unless z is small, less than, say, 0.1. All that may be said correctly is that the universe has expanded 20 percent since the emission of light from the source. More accurate information cannot be obtained by mistakenly using the special relativity Doppler formula of Equation [15.14]. It must be found in the framework of a particular cosmological model, such as the Einstein–de Sitter universe.

Suppose that the redshift of a quasar is larger than unity. The Fizeau–Doppler formula fails at large velocities and the special relativity Doppler formula of Equation [15.14] or [15.15] is sometimes used instead. For $z = 1$, the recession velocity becomes $0.6c$, or 180 000 kilometers per second, and the distance is said to be 0.6 times the Hubble length, or 12 billion light years ($h = 0.5$). For $z = 2$, the velocity becomes $0.8c$, or 240 000 kilometers per second, and the distance is said to be 0.8 times the Hubble length, or 16 billion light years. These statements regrettably are grossly in error. All that can be stated correctly is the universe has expanded 100 percent for $z = 1$, and 200 percent for $z = 2$, since the time of emission.

Although the classical Fizeau–Doppler formula enables us easily and quickly to make statements about recession velocities and distances, the statements made for redshifts greater than about 0.1 tend to be erroneous; their only virtue is that they capture the interest of the audience. The damage, however, is that students are misled into thinking that this is the correct method. But the Doppler effect applies only to bodies moving in space and not to stationary bodies comoving in expanding space. By ignoring the expansion of space and using the Doppler effect we depict a universe in which galaxies are shooting away from us through static space with redshifts that are the result of motions limited by the velocity of light. The recession velocity is thus reduced to the status of an ordinary velocity – such as that of an automobile or a rocket – and is made subject to the rules of special relativity. This picture violates the principles of containment and location. For if the recession velocity cannot exceed the velocity of light, then the velocity–distance law terminates at the edge of the Hubble sphere, creating a cosmic edge with us at a cosmic center.

By failing to distinguish between recession velocity and ordinary velocity, and between expansion redshift and Doppler redshift, a confused student is faced with irrefutable evidence that the edge of the universe is at a distance of about 20 billion light years and we occupy the center of a bounded universe.

REDSHIFT CURIOSITIES
Misinterpretation
Many investigators since the time of Hubble have seriously questioned the reality of an

expanding universe. Often, their doubts about the expansion of the universe are based on a dislike of the Doppler interpretation of extragalactic redshifts. The Fizeau–Doppler formula, they declare, when extrapolated, leads to cosmological absurdity. Generally, these investigators, unfamiliar with the expanding space paradigm, fail to realize that the appropriate interpretation is not the Doppler effect but the expansion–redshift effect that leads to a rational cosmology.

Tired light

The expansion interpretation of galactic redshifts, though delightfully simple, has been challenged many times. Fritz Zwicky, a famed astronomer who, among many other things, pioneered the study of supernovas, advanced in 1929 the theory that light steadily loses energy while traveling across large regions of extragalactic space. This "tired-light" idea has been resurrected repeatedly since Zwicky first proposed it. Various reasons have been given to explain why light might suffer fatigue while traveling in the universe, but none so far has been very successful. Authors have either failed to look fully into the consequences (if interaction with intergalactic gas is the cause, scattering must occur, and quasars would not be seen as starlike sources), or they have failed to appreciate that a hypothetical unknown law is rarely an attractive substitute of a known law that works.

According to the tired-light theory, extragalactic redshifts are the result of light fatigue and not expansion, and hence the universe is static and does not expand. It is a quaint idea that creates problems and solves none. If true, it must explain why the universe is static, and why the redshift is the same for a wide range of wavelengths. Furthermore, it must explain the origin of the thermal cosmic background radiation produced in the big bang. In the past, the cosmic background radiation had a much higher temperature, and the tired-light theory confronts us with the startling prospect of a big bang in a static universe. One

is left wondering where all the energy has gone. A more subtle question is where all the entropy has gone. In an expanding universe, the entropy of the cosmic background radiation remains constant; but in a static universe, in which radiation suffers from growing fatigue and is reddened by old age, the entropy declines and no tired-light advocate has yet been able to say where it all goes.

Discordant redshifts

In more recent years further controversy has arisen and the expansion theory of redshifts has been further challenged. Halton Arp has attacked the hypothesis that all extragalactic redshifts are primarily due to expansion. Undoubtedly many galaxies and quasars have redshifts caused by expansion of the universe, but others, argues Arp, appear to have redshifts produced by unknown causes. Galaxies within a group are all at practically the same distance from us and should therefore have almost equal redshifts. But Arp claims that chains and groups of galaxies exist whose members have widely different redshifts (see Figure 15.10). Some galaxies appear to have companion galaxies connected by bridges and filaments of luminous material. The members of these systems must therefore be at the same distance; sometimes, however, the smaller companions have redshifts larger than their associated galaxies. Arp suggests that these companions have been ejected from their parent galaxies and possibly all such young ejected systems have large intrinsic redshifts of unknown origin. Arp's claims to have discovered discordant redshifts are contested by other astronomers who argue that the apparent connections between galaxies of different redshifts are the result of seeing galaxies superposed on more distant galaxies.

Arp's arguments have one important virtue. The possibility of discordant redshifts prompts us to scrutinize carefully the nature of the three redshifts and the roles they play in cosmology. We can never prove beyond doubt that a thing is true,

Figure 15.10. This group of galaxies is known as Stephen's Quintet. The largest galaxy, lower left of center, has a recession velocity 5000 kilometers a second less than that of the companion galaxies. Possibly it is a foreground galaxy and not a true member of the group. (Association of Universities for Research in Astronomy, Inc., The Kitt Peak National Observatory.)

only that it is false, and science is lost without those bold enough to interrogate its scientifically correct beliefs.

Are quasars near or far?

Soon after the discovery of quasars an attack on the cosmological interpretation of redshifts came from a different quarter. We recall that Maarten Schmidt in 1963 discovered a redshift of 0.16 for the bright starlike object 3C 273, previously known to be a radio source. Quasars are among the most puzzling of all known celestial bodies. Nowadays most astronomers believe that quasar redshifts are cosmological and the result of the expansion of the universe.

This means quasars are at large extragalactic distances and are therefore extremely powerful sources of radiation. The light from some quasars varies rapidly on a time scale of only days, indicating that quasars are compact bodies about the size of the Solar System. This made it difficult to understand their source of immense energy. Some astronomers argued that quasar redshifts are not entirely due to expansion of the universe.

Jesse Greenstein and Maarten Schmidt showed that it is unlikely that quasars have large gravitational redshifts. Many of the emission lines come from the gas clouds surrounding quasars and are formed at different depths in the gas clouds. If the redshifts were mainly gravitational the emission lines would be spread over a continuous range of redshifts.

An entirely different interpretation, advanced by James Terrell and advocated by Geoffrey Burbidge and Fred Hoyle, is that quasar redshifts are due almost entirely to the Doppler effect. According to this idea, quasars are nearby extragalactic bodies that have been expelled at high velocity by violent explosions in the Galaxy and neighboring galaxies. This local hypothesis, as opposed to the cosmological hypothesis, reduces the distances of quasars from billions to millions of light years and alleviates the problem of explaining their immense output of energy. The local hypothesis, however, does not explain the powerful radio sources that undoubtedly are at cosmologically large distances.

If quasars are flying out of galaxies, as suggested by Terrell, then some will move away and some will move toward us. But no quasars have blueshifts. They are all moving away, which is possible only if they have originated in our Galaxy. The local hypothesis faces two severe difficulties. First, estimates show that millions of quasars are observable with the largest telescopes, and this number is obviously far too great for a local origin. Second, large numbers of quasars must also originate in other galaxies similar to our own. In that case the night sky would be very much brighter than is observed. The local hypothesis has been weakened further by the discovery of quasars with redshifts similar to those of galaxies in the same region of the sky, which suggests that these quasars are members of distant clusters of galaxies and are not merely nearby bodies. Also, many quasars are radio sources. There now seems little doubt that quasars are at cosmologically large distances.

Multiple redshifts

In addition to bright emission lines, the spectra of quasars often contain a forest of dark absorption lines. The bright lines are produced by atoms emitting radiation in hot gaseous regions in the vicinity of the quasar, and the dark lines are produced by atoms absorbing radiation in cooler regions between us and the quasar. In many cases the emission and absorption lines have similar redshifts. But the spectra of some quasars are complex and have absorption lines of different redshifts. These absorption redshifts are usually much less than the emission redshift of the quasar. A typical case is the quasar PHL 957: it has an emission redshift 2.69, and its spectrum contains numerous dark lines all grouped into eight absorption redshifts having values between 2.0 and 2.7. This suggests that light from these quasars has passed through absorbing gas at different distances and therefore at different redshifts, and the accepted explanation is that the absorption occurred in extended intergalactic clouds and in halos of galaxies between the quasar and us.

Very curious

In cosmology we have two basic and beautiful laws: the velocity–distance law (Equation 15.17) and the expansion redshift law (Equation 15.18). These basic laws cannot be combined to give recession velocities and distances in terms of redshifts except when redshifts are small. When the redshifts are not small we must use a particular model, such as the Einstein–de Sitter universe, in which we know how the scaling factor changes with time. The custom of referring to the velocity–distance law as the Hubble law and the expansion–redshift as a Doppler redshift rank among the most curious aspects of modern cosmology.

REFLECTIONS

1 *We must distinguish between redshift and reddening. The redshift effect consists of moving the entire spectrum toward longer wavelengths. The reddening effect occurs when radiation of shorter wavelengths is removed from the spectrum by absorption or scattering; the values of the wavelengths are not altered, as in the redshift effect. A fog,*

which transmits red light better than blue light, is one example. The red appearance of the setting Sun is another example. Interstellar dust grains absorb radiation of short optical wavelengths and starlight often appears reddened because of this effect.

2 *"Nevertheless, the possibility that the redshift may be due to some other cause, connected with the long time or distance involved in the passage of light from nebula to observer, should not be prematurely neglected. ... Until further evidence is available, both the present writers wish to express an open mind with respect to the ultimately most satisfactory explanation of the nebular red-shift ... [and] if the red-shift is not due to recessional motion, its explanation will probably involve some quite new physical principle"* (Edwin Hubble and Richard Tolman, "The nature of the nebular redshifts," 1935).

• *In 1937, Edwin Hubble wrote in* The Observational Approach to Cosmology, *"The study of many nebulae has shown that, on the average, the red-shifts increase with the apparent faintness of the nebulae in which they are measured. Therefore, we conclude that, on the average, red-shifts increase with distance. Extensive investigations have demonstrated that the relation is approximately linear,*

red-shifts = constant × distance,

or $d\lambda/\lambda = k \times r$. This relation is called the law of red-shifts .... Most of the theoretical investigations ... accept without question the interpretation of red-shifts as velocity-shifts. They are fully justified in their position until evidence to the contrary is forthcoming. But these lectures will present a remarkable situation. The familiar interpretation of red-shifts seems to imply a strange and dubious universe, very young and very small. On the other hand, the plausible and, in a sense, familiar conception of a universe extending indefinitely in space and time, a universe vastly greater than the observable region, seems to imply that red-shifts are not primarily velocity-shifts." By *"velocity-shifts"* Hubble meant Doppler shifts.

• *"The red shift is something that happens to the light on its journey, along with the expansion of space, not really a Doppler effect"* (Erwin Schrödinger, Expanding Universes, *1957*).

• *"Note that the cosmological red shift is really an expansion effect rather than a velocity effect"* (Wolfgang Rindler, Essential Relativity, *1969*).

3 *In* The Origin and Evolution of the Universe, *the French astrophysicist Evry Schatzman remarks: "For anyone who does not accept the expansion of the universe, the red shift of spectral lines remains an important but totally unexplained physical phenomenon."* We note that tired-light redshifts and expansion redshifts have much in common. Unlike the Doppler effect, both consist of a steady displacement toward the red end of the spectrum as light traverses vast regions of space. One might even with justice call the expansion redshift a theory of tired light. Light rays are progressively robbed of energy and become fatigued by the expansion of the universe. It is ironic that all the difficulties of the tired-light theory are banished by changing its name and attributing fatigue to expansion: we have fatigue but no scattering; entropy is conserved but not energy (see Chapter 17).

4 *Slipher, Humason, and other early observers used the Doppler effect to interpret the significance of extragalactic redshifts. The galaxies they observed had small redshifts and their use of the Fizeau–Doppler formula was justified. The de Sitter effect indicated an expanding universe. But unfortunately the de Sitter universe was misleading. The Hubble redshift–distance law in the de Sitter universe is true for all redshifts, large and small, thus supporting the belief that recession redshifts in all universes are a Doppler effect. The custom of referring to expansion redshifts as Doppler redshifts has survived and is now widespread. Astronomers add to the confusion by multiplying redshifts by the velocity of light and cataloguing their results as velocities. Thus, if the observed redshift of a galaxy is 0.05, the galaxy is catalogued with a recession velocity of 15 000 kilometers per second.*

Professional cosmologists know what they are doing and avoid the pitfalls that by a misuse of words they inadvertently dig for others. The truth is that expansion redshifts are different from Doppler redshifts, and the velocities catalogued by astronomers, except for small redshifts, are not true recession velocities used in the velocity–distance law.

5 *Consider the following symmetrical arrangement (see Figure 15.11). Two widely separated comoving galaxies X and Y emit identical signals of the same wavelength and at the same instant in cosmic time. X receives the signal from Y and Y receives the signal front X at the same instant and same wavelength. This state of symmetry exists because the homogeneity of the universe applies not only to expansion but also to the laws and fundamental constants of nature that determine the wavelengths of radiation emitted and absorbed by atoms. If by mischance the laws and constants of nature were not the same everywhere, then symmetry between X and Y would be lost. We are moderately confident that symmetry does exist for the reasons that led us to believe in homogeneity: the universe is isotropic, hence the laws and constants of nature are the same in all directions, and if we are not privileged occupants of a cosmic center, it follows that the laws and constants must be the same everywhere in space.*

●   Perhaps the laws and constants are the same everywhere in space but change in time? If so, we retain homogeneity, and symmetry between X and Y is preserved, but the interpretation of cosmological redshifts as an expansion effect is then possibly in error. But any theory postulating a change in time in the structure of atoms must be well contrived: all wavelengths, short and long, produced by various physical processes must exhibit the same redshift. For example, we know that the 21-centimeter wavelength radiation from hydrogen atoms in distant galaxies exhibits the same redshift as optical wavelengths almost a million times shorter from the same galaxies. Yet the emission mechanisms are entirely different and involve the basic constants of nature in different combinations. The expansion interpretation of cosmological redshifts seems therefore secure and will not easily be discarded in favor of a "shrinking atom" interpretation.

6 *Distant galaxies in an expanding universe are seen redshifted and in a collapsing universe are seen blueshifted. Imagine that our universe ceases to expand and begins to collapse. What do we see? Galaxies near us are seen approaching with blueshifts, but distant galaxies, because we look back to a time when the universe was still expanding, are seen receding with redshifts. The boundary between blueshifts and redshifts recedes with time, and more and more galaxies are seen with blueshifts. In the last moments of a bang bang universe, just before we plunge into the second bang, everything is seen blueshifted.*

7 *We do not know exactly the kind of physical universe in which we live, but probably it is a member of the big bang class. There are numerous big bang universes, and we do not necessarily live in the simple Einstein–de Sitter kind that has flat space. But the differences between the redshift–distance curves of the various big bang universes are generally not very great, and the Einstein–de Sitter curves shown in Figures 15.7–15.9 are reasonable   approximations   for   most*

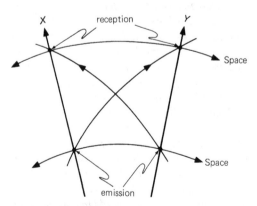

Figure 15.11. Galaxies X and Y have world lines that diverge in expanding space. Each sends a signal to the other at the same instant in cosmic time. Because of symmetry, the signals arrive at the same instants in cosmic time and also have identical redshifts.

universes. *These curves should be used when quoting distances and recession velocities of galaxies of large redshift. At least they are more reliable than the misleading Doppler curves.*

• *Examine the Einstein–de Sitter curves. They reveal several interesting facts. Notice that a receding body has two distances and two recession velocities, corresponding to the world map and the world picture, and we should always bear in mind which of the two is being used. Notice that in the world map the present distances (reception distances) of the galaxies increase steadily with redshift, and at infinite redshift the reception distance attains a maximum value equal to twice the Hubble distance and is therefore 40 billion light years (h = 0.5). This maximum reception distance is the particle horizon (discussed in Chapter 21). Notice that the world-picture distances (emission distances) of the galaxies do not continually increase with redshift (see Figure 15.12). They attain a maximum value at a redshift of 1.25, and at greater redshifts the emission distances get smaller. When the* redshift *is very small, you are close to it now; when the redshift is very large, you were close to it long ago.*

8 *Perhaps you are not convinced that a difference exists between expansion and Doppler redshifts? Let us then demonstrate the difference in the following imaginary experiments.*

Consider first the Doppler redshift. Two bodies, call them X and Y, are separated by a fixed distance in the Galaxy or the laboratory (see Figure 15.13). Let X emit a pulse of radiation toward Y. After the pulse has left X, and while it travels toward Y, let the distance between X and Y increase. Before the pulse arrives at Y, let the separating distance again become fixed. The two bodies X and Y have zero relative velocity at the instants when the pulse is emitted and received. Hence Y receives the pulse of radiation at the wavelength it was emitted by X and the Doppler redshift is zero.

Consider now the expansion redshift (see Figure 15.14). Two comoving bodies X and

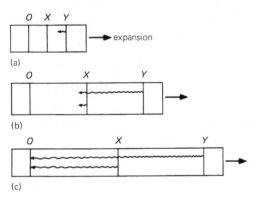

(a)

(b)

(c)

Figure 15.12. On an elastic strip let O represent our position, and X and Y the positions of two galaxies. If signals from X and Y are to reach us at the same instant, then Y, which is farther away, must emit before X. In (a), Y emits a signal. In (b), X emits a signal at a later instant when it is farther away than Y was when it emitted its signal. In (c), both signals arrive simultaneously at O. Y's signal has the greater redshift (it has been stretched more) although Y was closer than X at the time of emission. This odd situation occurs at large redshifts in all big bang universes.

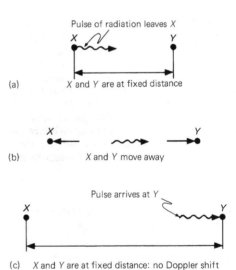

(a) X and Y are at fixed distance

(b) X and Y move away

(c) X and Y are at fixed distance: no Doppler shift

Figure 15.13. (a) A fixed distance separates X and Y in the laboratory. X emits a pulse of radiation toward Y. (b) While the pulse is traveling, X and Y move apart and again come to rest at a wider separating distance, (c) The pulse arrives at Y when the separating is again fixed. No Doppler effect occurs because X and Y have zero relative velocity at the instants of emission and absorption.

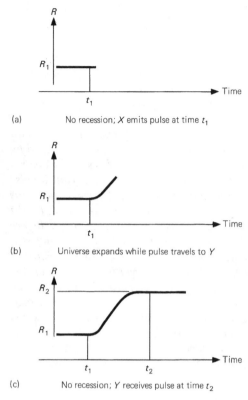

Figure 15.14. (a) A static universe has a scaling factor R_1. (b) The universe now expands. (c) The universe becomes static again and has a scaling factor R_2. Consider a pulse of radiation emitted by a comoving body X while the universe is static the first time, and received by a comoving body Y while the universe is static the second time. At the instants of emission and reception the recession velocity is zero. Yet the pulse of radiation is received by Y with a redshift z given by $1 + z = R_2/R_1$. This proves that the Doppler and expansion redshifts are not identical.

Y are stationary in expanding space. In this experiment we suppose that the universe is initially static and that the distance between X and Y is therefore constant and the recession velocity is zero. The body X now emits a pulse of radiation toward Y, and while this pulse travels, we suppose that the universe ceases to be static and expands. Before the pulse arrives at Y the universe stops expanding and again becomes static, and the distance between X and Y has again a fixed value and the recession velocity is zero. When the pulse

eventually arrives at Y its wavelength has been increased by the expansion of space through which it has traveled. If the scaling factor is R_1 when the universe is static the first time, and R_2 when the universe is static the second time, the wavelength has increased by the amount R_2/R_1, and the redshift is given by

$$1 + z = \frac{R_2}{R_1}.$$

Although the distance between X and Y is constant at the instants of emission and reception, and the relative recession velocity is zero, an expansion redshift nonetheless has occurred. Yet in the previous experiment, the Doppler effect vanished under similar conditions. These two imaginary experiments show that the Doppler redshift depends on motion through space at the instants of emission and reception, whereas the expansion redshift depends on the expansion of space between the instants of emission and reception.

9 *The wave-stretching relation $\lambda_0/\lambda = R_0/R$ can be derived in various ways. We have used the expanding space paradigm to derive this result. Max Planck early in the 20th century obtained a similar result by considering radiation in a slowly expanding cavity. The accumulation of small Fizeau–Doppler redshifts caused by repeated reflections off the walls of the cavity produce a steady increase in wavelength. A formal method uses the null-geodesic equation of motion based on the Robertson–Walker metric. A simpler procedure uses Equation [14.35] in the previous chapter:*

$$r = \int_t^{t_0} dt/R, \qquad [14.35]$$

which gives the comoving coordinate distance r to a luminous source whose radiation is emitted at time t and received at time t_0. Imagine that a short pulse of radiation is emitted at time t_1 and received at time t_2, and a second pulse is emitted a moment later at $t_1 + dt_1$ and received at $t_2 + dt_2$. Because r is constant,

$$r = \int_{t_1}^{t_2} dt/R = \int_{t_1+dt_1}^{t_2+dt_2} dt/R,$$

and therefore:

$$\int_{t_2}^{t_2 + dt_2} dt/R - \int_{t_1}^{t_1 + dt_1} dt/R = 0;$$

and hence

$$\frac{dt_1}{R_1} = \frac{dt_2}{R_2} \qquad [15.22]$$

for small dt_1 and dt_2, and dt/R is a constant along the lightcone. Hence all wavelengths, wave trains, and time intervals are stretched while propagating in an expanding universe.

• Imagine a large number of equally spaced comoving observers strung out in a straight line between us at origin $r = 0$ and a receding extragalactic source at comoving coordinate distance r. Let dr be the small separating coordinate distance between any two adjacent observers. The radiation from the source of wavelength λ arrives at the first and nearest observer, at distance $dL = R\,dr$, at increased wavelength $\lambda + d\lambda$. Because dL is small, we may use the Fizeau–Doppler formula to obtain $d\lambda/\lambda = H\,dL/c$, or $d\lambda/\lambda = dR/R$ because $H = dR/R\,dt$ and $dL = c\,dt$. On integration we get the wave-stretching result of Equation [15.1]. Thus the expansion redshift is equivalent to a large number of small Fizeau–Doppler shifts. But we may not regard the expansion redshift as equivalent to a large single Fizeau–Doppler redshift.

10 Various relations (Equations 14.38–14.45) were derived in the previous chapter for the power-law models in which $R = R_0(t/t_0)^n$, and n is a constant in the range $0 \leq n < 1$. In these models, the velocity–redshift relation is

$$\frac{V}{c} = \frac{n}{1-n}\left(1 - \frac{1}{(1+z)^{(1-n)/n}}\right), \qquad [15.23]$$

and the recession velocity at the time of emission is

$$\frac{V_{emit}}{c} = \frac{n}{1-n}[(1+z)^{(1-n)/n} - 1], \qquad [15.24]$$

and we see that when z is small both recession velocities reduce to $V = cz$; and when $z = \infty$, $V = nc/(1-n)$ and $V_{emit} = \infty$.

The redshift–distance relation is

$$L = \frac{n}{1-n}L_H\left(1 - \frac{1}{(1+z)^{(1-n)/n}}\right) \qquad [15.25]$$

and the distance at the time of emission is

$$L_{emit} = \frac{n}{1-n}\frac{1}{1+z}$$
$$\times L_H\left(1 - \frac{1}{(1+z)^{(1-n)/n}}\right), \qquad [15.26]$$

and we see that when z is small, both distances reduce to the linear Hubble law $L = zL_H$; and when $z = \infty$, $L = nL_H/(1-n)$ and $L_{emit} = 0$.

The age of the universe at the time of emission is

$$\frac{t_{emit}}{t_0} = \frac{1}{(1+z)^{1/n}} \qquad [15.27]$$

or

$$t_{emit} = nt_H\frac{1}{(1+z)^{1/n}}, \qquad [15.28]$$

where t_0 is the present age and $t_H = 1/H_0$, and the lookback time

$$t_0 - t_{emit} = nt_H\left(1 - \frac{1}{(1+z)^{1/n}}\right), \qquad [15.29]$$

equals zt_H for small redshifts.

In the Einstein–de Sitter universe of $n = 2/3$, the recession velocity is

$$V = 2c\left(1 - \frac{1}{\sqrt{1+z}}\right), \qquad [15.30]$$

and at the time of emission in the world picture is

$$V_{emit} = 2c(\sqrt{1+z} - 1). \qquad [15.31]$$

The distance in the world map (the present distance) of an extragalactic body of redshift z is

$$L = 2L_H\left(1 - \frac{1}{\sqrt{1+z}}\right), \qquad [15.32]$$

and its distance in the world picture (emission distance) is

$$L_{emit} = \frac{2L_H}{1+z}\left(1 - \frac{1}{\sqrt{1+z}}\right). \qquad [15.33]$$

As the redshift increases, the emission distance L_{emit} at first increases, then it attains a maximum value $8L_H/27$ at redshift $z = 5/4$, and thereafter decreases as the redshift increases. The age of the universe now is $t_0 = 2t_H/3$, and at the time of emission it was

$$t_{emit} = \frac{2t_H}{3(1+z)^{3/2}}, \qquad [15.34]$$

and the lookback time is

$$t_0 - t_{emit} = \frac{2}{3}t_H\left(1 - \frac{1}{(1+z)^{3/2}}\right). \qquad [15.35]$$

PROJECTS

1 In a city at night one hears the wail of approaching and receding sirens, and the Doppler effect is often pronounced (see Figure 15.15). Have you noticed how at first sirens are always high-pitched and then become low-pitched as they fade away into the distance? The opposite effect – first a low pitch that later fades away into a high pitch – never occurs. Why is this? Bend a length of stiff wire into a wave shape figure. Now slowly stretch the figure and notice how the wavelength increases. This experiment illustrates the stretching of waves of radiation as they travel through expanding space.

2 Use Figures 15.7, 15.8, and 15.9 to derive as much information as possible about the galaxies and quasars listed in Table 15.1.

3 Draw redshift–velocity curves using the Fizeau–Doppler and relativity Doppler formulas. Using these curves and Hubble's redshift–distance law, draw the corresponding velocity–distance curves. Notice that in this comedy of errors we do not know if we are in the world map or the world picture.

4 What, in the Einstein–de Sitter universe, is the recession velocity in the world map at the particle horizon? (Redshift is infinity at the particle horizon.) The maximum emission distance occurs at $z = 1.25$. What is this distance, and what is the recession velocity at this redshift in the world picture?

5 Can you understand how it is possible for bodies at large redshift to be nearer to us at the time of their emission than bodies of lesser redshift? Two quasars have redshifts 1 and 3, respectively; which of the two is closer at the time of emission?

6 For the mathematically inclined: Show that $dt/R = -dz/R_0H$, and the distance in the world map as a function of z, from Equation [14.36], becomes

$$L = \int_0^z \frac{dz}{H}, \qquad [15.36]$$

(in light-travel time) and we must regard H as a function of z. For example, in the power-law models, $H = H_0(1 + z)^{1/n}$; in the Einstein–de Sitter model, $H = H_0(1 + z)^{3/2}$; and in the de Sitter model, H is constant and equal to H_0. From Equation [15.36] and the velocity–distance law, we see that the velocity–redshift relation is

$$\frac{V}{c} = H_0\int_0^z \frac{dz}{H}. \qquad [15.37]$$

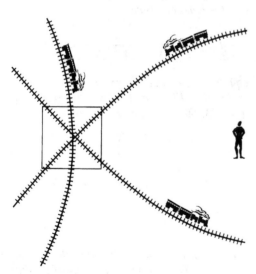

Figure 15.15. The "train whistle" effect. Trains run through a railway station from all directions blowing their whistles. People who live nearby always hear the pitch of the whistles steadily decreasing from high to low.

Table 15.1. *Sample redshifts*

Object	Redshift	Remarks
Virgo	0.016	rich cluster
BL Lacertae	0.07	radio galaxy, AGN
Boötes	0.13	rich cluster
Ton 256	0.13	quasar
3C 273	0.16	first radio source identified as a quasar
PKS 2251 + 11	0.32	quasar
3C 48	0.37	quasar
3C 295	0.46	radio galaxy
3C 9	2.01	quasar
PHL 957	2.69	quasar, absorption redshifts: 2.67, 2.55, 2.54, 2.31, 2.23
PC 1247 + 3406	4.89	quasar

AGN: Active galactic nucleus
3C: Third Cambridge survey (England)
PKS: Parkes radio survey (Australia)
PHL: Palomar, Haro, Luyten (USA)

Equations [15.36] and [15.37] are the general redshift–distance and velocity–redshift relations for all uniform models of the universe. Notice that Equation [15.37] bears no resemblance to the improper Fizeau–Doppler formula. Show that in the de Sitter and steady-state universes $L = zL_H$, $L_{emit} = zL_H/(1 + z)$, and $V = cz$ for all redshifts.

7 The principle of prohibitional correctness characterizes much of human behavior. History is full of examples of the prohibitions of the religiously correct, the artistically correct, the politically correct, and, of course, the scientifically correct. Relativity theory, quantum theory, and the theory of the expanding universe all had to struggle against existing scientifically correct beliefs. Give examples from history of the principle of prohibitional correctness.

SOURCES

Arp, H. C. *Quasars, Redshifts, and Controversies*. Interstellar Media, Berkeley, California, 1987.

Arp, H. C., Burbidge, G., Hoyle, F., Narlikar, J. V., and Wickramsinghe, N. C. "The extragalactic universe: An alternative view." *Nature* 807, 346 (1990).

Burbidge, G. "The harmonious and discordant facts." *International Journal of Theoretical Physics* 28, 983 (1989).

Eddington, A. S. *The Expanding Universe*. Cambridge University Press, Cambridge, 1933.

Einstein, A. and Sitter, W. de. "On the relation between the expansion and the mean density of the universe." *Proceedings of the National Academy of Sciences* 18, 213 (1932).

Ellis, G. F. R. "The expanding universe: A history of cosmology from 1917 to 1960," in *Einstein and the History of General Relativity*. Editors, D. Howard and J. Stachel. (Einstein Study Series.) Birkhauser, Boston, 1988.

Field, G. B., Arp, H., and Bahcall, J. *The Redshift Controversy*. Benjamin, Reading, Massachusetts, 1973.

Friedmann, A. "On the curvature of space." *Zeitschrift für Physik* 10, 377 (1922). "On the possibility of a world with constant negative curvature." *Zeitschrift für Physik* 21, 326 (1924). Both translated in *Cosmological Constants*, editors J. Bernstein and G. Feinberg. Columbia University Press, New York, 1986.

Harrison, E. R. "Hubble spheres and particle horizons." *Astrophysical Journal* 383, 60 (1991).

Harrison, E. R. "A century of changing perspectives in cosmology." *Quarterly Journal of the Royal Astronomical Society* 33, 335 (1992).

Harrison, E. R. "The redshift–distance and velocity–distance laws." *Astrophysical Journal* 403, 28 (1993).

Hubble, E. *The Observational Approach to Cosmology*. Oxford University Press, Clarendon Press, Oxford, 1937.

Hubble E. and Tolman, R. C. "The nature of the nebular red-shifts." *Astrophysical Journal* 82, 302 (1935).

Lemaître, G. "A homogeneous universe of constant mass and increasing radius accounting for the radial velocity of extra-galactic nebulae." *Monthly Notices of the Royal Astronomical Society* 91 483 (1931).

McVittie, G. C. "Distance and large redshifts." *Quarterly Journal of the Royal Astronomical Society* 15, 246 (1974).

North, J. D. *The Measure of the Universe: A History of Modern Cosmology*. Oxford University Press, Clarendon Press, Oxford, 1965.

Planck, M. *The Theory of Heat Radiation*. First published in 1913. Dover Publications, New York, 1959.

Rees, M. J. "Understanding the high-redshift universe: progress, hype, and prospects."

Quarterly Journal of the Royal Astronomical Society 34, 279 (1993).

Rindler, W. *Essential Relativity: Special, General, and Cosmological*. Van Nostrand Reinhold, New York, 1969. See p. 239.

Sandage, A. R. "The red-shift." *Scientific American* (September 1956).

Sandage, A. R. "Travel time for light from distant galaxies related to the Riemannian curvature of the universe." *Science* 134, 1434 (1961).

Schatzman, E. *The Origin and Evolution of the Universe*. Basic Books, New York, 1957.

Schrödinger, E. *Expanding Universes*. Cambridge University Press, Cambridge, 1957. See p. 62.

Terrell, J. "Quasi-stellar objects: possible local origin." *Science* 154, 1281 (1966).

Zwicky, F. "On the redshift of spectral lines through interstellar space." *Proceedings of the National Academy of Science* 15, 773 (1929). *Physical Review* 33, 1077 (1929).

16 NEWTONIAN COSMOLOGY

All are but parts of one stupendous whole.
Alexander Pope (1688–1744), An Essay on Man

STATIC NEWTONIAN UNIVERSE

Until the 20th century everybody believed that the universe is naturally static: not expanding and not contracting. Even Albert Einstein, after the discovery of general relativity, continued to hold this belief for several years.

In the late 17th century, belief in a static order remained unshaken when Newton advanced the theory of universal gravity. In response to a question in a letter from the young clergyman Richard Bentley (Chapter 3), Newton wrote in reply that in an infinite universe it would be impossible for all matter to fall together and form a single large mass, but "some of it would convene into one mass and some into another, so as to make an infinite number of great masses, scattered at great distances from one to another throughout all that infinite space."

The Newtonian theory of universal gravity, in which all bodies attract one another, reinforced the growing belief that the universe must be edgeless and therefore infinite. For if the universe were finite and bounded by a cosmic edge, it would have a center of gravity, and the attraction between its parts would cause it, said Newton, to "fall down into the middle of the whole space, and there compose one great spherical mass." This argument led him finally to abandon the finite Stoic cosmos in favor of the infinite Atomist universe. In an infinite, uniform universe, no preferred direction exists in which gravity can pull and make matter fall into a single "middle." Newton wrote later in the second edition of the *Principia*, "The fixed stars, being equally spread out in all points of the heavens, cancel out their mutual pulls by opposite attractions." Thus each particle of matter is pulled in all directions by gravitational forces and stays undisturbed in equilibrium. Thus the theory of universal gravity reinforced the belief that the universe is static on the cosmic scale and fostered the idea that on smaller scales gravity could cause matter to condense and form astronomical systems.

Warring cosmic forces

The Newtonian universe was beset by two riddles that Edmund Halley in 1721–2 recognized as being similar. The first is the dark night-sky riddle that began with the astronomer Thomas Digges in 1576 and is known nowadays as Olbers' paradox; this riddle has played a conspicuous part in the history of cosmology and is discussed in Chapter 24. The second is the gravity riddle that has played a less conspicuous part, but is nonetheless interesting and deserves our attention.

In a universe uniformly populated with stars we construct about a point, anywhere in space, a set of concentric shells in the manner introduced by Halley. All shells have equal thickness, as shown in Figure 16.1, and the thickness dr of each shell is small compared with its radius r. The number of stars in any shell is proportional to the square of the radius of the shell (that is,

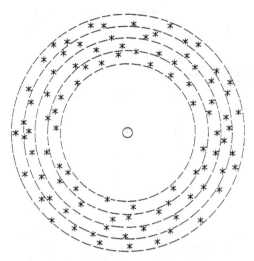

Figure 16.1. Halley's shells of equal thickness. The thickness is assumed to be small compared with the radius of a shell. In a universe uniformly populated with stars, the number of stars contained in any shell is proportional to the square of the shell's radius. But a star exerts a gravitational force on a central body that is proportional to the inverse square of its distance, and hence all shells have equal effects on the central body.

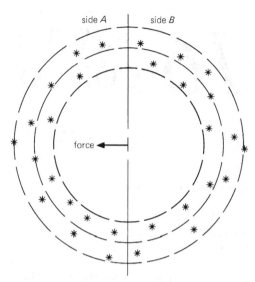

Figure 16.2. Imagine that each star on side A has on the average one more atom than on side B. Each shell therefore exerts a slight residual gravitational force on a body at the center, as shown. This residual force is the same for each shell. In an infinite universe of stars there is an infinite number of shells, and the total force exerted by all shells is infinitely large, even though the force of each shell is extremely small. For the cosmic forces to be in equilibrium, exact isotropy must exist, as was argued by the young theologian Richard Bentley in 1692.

r^2), and each star exerts on a body at the center a gravitational pull inversely proportional to the square of the shell's radius (that is, r^{-2}). The gravitational effect of the stars in the shell is therefore independent of the radius of the shell. Because the universe is assumed to be isotropic about the chosen point, the stars in a shell on one side of the sky exert an attraction that equals and opposes the attraction of stars on the opposite side of the sky. Hence the shell exerts no net force. On adding up all shells we find as their number increases that the opposing forces get stronger, and when the shells are added up to an infinite distance, these forces become infinitely great. The exact isotropy that must therefore exist in the Newtonian universe is puzzling. Suppose that the universe is not precisely isotropic, and on one side of the sky each star contains on the average 1 atom more than stars on the other side. A single shell now exerts on a body at the center a tiny unbalanced force in one

direction (Figure 16.2). All shells exert similar finite forces in the same direction, and therefore an infinite number of shells together must exert an infinitely large unbalanced force. Clearly, according to this argument, isotropy of the universe is necessarily perfect about every point. Cosmic forces, in effect, do not exist, and peculiar motions are caused by local forces.

Resolutions of the gravity riddle
In 1872 the German physicist-astronomer Johann Zöllner, inspired by Riemann's work on curved space, suggested that space was spherically curved and finite in extent. Hence the total amount of matter in the universe was finite and all gravitational forces were also finite. This was a remarkable anticipation of Einstein's finite but unbounded static universe of 1917. Zöllner

argued that at each point in such a space the gravitational force in any direction was finite. He naturally did not know that curved space is already gravity, and under whatever name, gravity cannot be used twice.

Toward the end of the 19th century the German physicist Carl Neumann and the astronomer Hugo von Seeliger sought in separate ways to break the deadlock of infinite cosmic forces by simply abolishing them. They proposed that gravity at large distances decreased faster than the inverse square law, and Newton's inverse square law r^{-2} was altered to $r^{-(2+\varepsilon)}$, where ε is a small positive constant. Using again Halley's shells, in which the number of stars in each shell increases as r^2, we now find that the force exerted by each shell decreases as $r^{-\varepsilon}$. According to this idea, gravitational attraction diminishes faster than the inverse square law, and is negligible at cosmic distances. Thus in Figure 16.1 the shells at large distances exert forces much smaller than the shells at small distances, and the unbalanced force in Figure 16.2 remains negligible.

The Swedish astronomer Charles Charlier tried in 1908 and 1922 to resolve the deadlock of cosmic forces with hierarchical structure (see Chapter 7). In an infinite hierarchical universe, the density of matter progressively decreases when averaged over larger and larger regions. The hierarchy can be arranged in such a way that in the limit, on infinitely large scales, the average density goes to zero. Gravity operates normally on small scales, gets weaker on larger scales, and vanishes on the largest scales. Charlier was also interested in the dark night-sky riddle, and showed that in a hierarchical universe in which cosmic forces fade away at large distances, the night sky is dark because stars fail to cover the entire sky. The hierarchical universe solved, so it seemed at the time, two riddles: it ended the war of cosmic forces and gave us a dark night sky.

In Newtonian theory, gravity acts simultaneously at all distances, and when any one particle changes its position, all other particles in the universe respond instantaneously. The theory of general relativity has taught us that gravitational forces cannot act instantaneously at a distance. Gravity is the dynamic curvature of spacetime that propagates at the same speed as light. When the darkness of the night sky first became a problem, nobody knew that light travels at finite speed; also, while the gravity riddle held sway for more than two centuries, nobody knew that gravity also travels at finite speed.

Let us alter the Newtonian universe to the extent of admitting that gravity propagates at the speed of light. This elementary alteration, sanctioned by the theory of general relativity, immediately solves most of the perplexing aspects of the gravity riddle. The large Newtonian forces pulling in every direction at every point, produced by matter at vast distances, have traveled for vast periods of time. But in a modified Newtonian universe of finite age, the forces have traveled only finite distances and are of finite magnitude. For example, in a universe 10 billion years old, we experience no Newtonian forces beyond 10 billion light years, and the war of cosmic forces abates and becomes a tussle between forces a trillion times weaker than the Earth's gravitation pull at its surface. A hierarchical arrangement of matter beyond the distance that gravity travels in the lifetime of the universe contributes nothing to solving the gravity riddle. Also, as we shall see in Chapter 24, hierarchically arranged matter beyond the distance that light travels in the lifetime of the universe, contributes nothing to resolving the dark night sky riddle.

The notion of a gravity riddle is questionable on mathematical grounds. In a uniform, finite, and bounded distribution of matter, the gravitational force has a direction at each point determined by the boundary conditions. If the uniform distribution is of infinite extent and has no boundary, as in the edgeless and centerless Newtonian universe, the gravitational force is indeterminate: it has no direction and cannot be calculated. This is known as the Dirichilet problem in the theory of potential functions. A mathematician, confronted with the gravity problem, knows immediately that the riddle is ill conceived. The Dirichilet problem is one reason why we cannot rest content with Newtonian cosmology.

In this chapter, using Newtonian theory, we obtain equations remarkably similar to those obtained with general relativity. We must admit, however, that the method used for deriving these dynamical equations of the universe does violence to Newtonian theory. Solely on the basis of Newtonian theory, no pre-20th century mathematician would have predicted that the universe was in a synchronized state of expansion (or contraction). General relativity predicted a nonstatic universe and endorsed a Newtonian treatment of limited validity.

EXPANDING COSMIC SPHERE
General relativity leads the way

An unbounded universe must be cosmically static according to Newtonian theory. But according to general relativity an unbounded universe, whether finite or infinite, is made static only by contrived conditions, and even then can be unstable. General relativity, not Newtonian theory, first opened the door to the realization that the universe is nonstatic. In 1917, Willem de Sitter of Holland approached the door; in 1922, Alexander Friedmann of Russia opened the door; in 1927, Georges Lemaître of Belgium passed through the door, and was almost immediately followed by Arthur Eddington in England and Howard Robertson and Richard Tolman in the United States. In the meantime, in 1929, Edwin Hubble made familiar the idea of an expanding universe.

General relativity in the 1920s and early 1930s was a confusing and difficult theory to apply to the whole universe. Only the ablest theoreticians worked on the subject. In 1934 a surprising development occurred. The cosmologists Edward Milne and William McCrea in Britain showed that the equations controlling the dynamics of the universe, which previously had been derived

from the theory of general relativity by mathematical labor and skill, could be derived directly from simple Newtonian theory. This startling development created a cosmological puzzle that is still not fully solved: Why should Newtonian theory and general relativity theory, when applied to a uniform universe, yield identical results? At best, Newtonian theory is only approximately true, and yet in this most unlikely of all applications it gives the correct result. We return to this problem at the end of the chapter.

An expanding sphere

We use Newtonian ideas in the manner shown by Milne and McCrea. The picture obtained is referred to as Newtonian cosmology.

Our first step would have dismayed Newton, who, as Master of the Mint, would have regarded us as little better than coin clippers (see Reflections). We consider the behavior in Euclidean space of a sphere of matter of uniform density. We forget all about the rest of the universe (we in effect abolish it) and let the behavior of the sample sphere, free-falling in its own gravitational field, represent the behavior of the universe. This nonstatic sphere, or cosmic ball, exists in an otherwise empty Euclidean space. The sphere has an edge and a center of symmetry and therefore the gravitational force everywhere is easily calculated.

To help us understand what happens we first suppose that the sphere is static, of fixed radius a, and imagine that particles leave the surface and travel away in a radial direction (see Figure 16.3). The escape speed V_{esc} from the surface of the sphere of mass M and radius a is given by

$$(V_{esc})^2 = \frac{2GM}{a}, \qquad [16.1]$$

where G is the gravitational constant. The escape speed is the speed needed by the particles to escape and reach infinity with zero speed. Particles moving with the escape speed follow parabolic orbits (see Figure 16.4) and ultimately reach infinity with

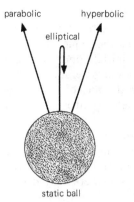

Figure 16.3. Particles move away radially from the surface of a gravitating sphere. The initial velocity determines whether the orbit is parabolic, elliptical, or hyperbolic.

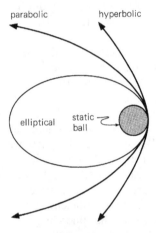

Figure 16.4. Particles move away with initial tangential velocities from the surface of a gravitating sphere. This illustrates more clearly the distinction between parabolic, elliptical, and hyperbolic orbits.

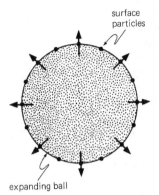

Figure 16.5. Here the gravitating sphere of uniform density is shown expanding, with its surface following the particles shown in Figure 16.3. All particles in the sphere now move radially in free fall in the gravitational field of the sphere. Motion inside the sphere obeys the velocity–distance law.

In this first step we have considered the motion only of the surface particles, and have assumed and will continue to assume that particles follow straight radial trajectories.

The strength of the gravitational field at a point outside the sphere depends on the mass M of the sphere and on the distance r of the point from the center of the sphere, and does not depend on the radius a of the sphere. Hence we may imagine that the sphere itself expands, keeping up with its surface particles whose motion remains unaffected by the expansion of the sphere (see Figure 16.5). The surface particles stay at the surface of the expanding sphere and follow the orbits we considered previously. Let the density of matter in the sphere be uniform (the same everywhere at any instant in time). In that case, all particles in the sphere follow orbits of a similar nature; for example, if the surface particles follow elliptical orbits, all interior particles follow elliptical orbits, and the density of the sphere stays uniform.

When the radial velocity of the surface is less than the escape velocity the sphere expands and then collapses; when the velocity equals or exceeds the escape velocity the sphere expands continually and never

zero kinetic energy. When the particles have a speed less than escape speed, they cannot reach infinity and eventually turn around and fall back, following elliptical orbits. When the speed exceeds the escape speed, the particles ultimately reach infinity with finite speed, following hyperbolic orbits. If V is the actual speed of the particles when they leave the surface, we have

V less than V_{esc}: elliptical orbit
V equal to V_{esc}: parabolic orbit
V greater than V_{esc}: hyperbolic orbit.

collapses. The mass M of the sphere equals its density ρ multiplied by its volume (equal to $\frac{4}{3}\pi a^3$). As the sphere expands and its radius a increases, its mass stays constant and its density decreases as a^{-3}. We found in Chapter 14 that the velocity–distance law, which states that velocity is proportional to distance, preserves uniformity in density. Therefore the farther a particle is from the center of an expanding sphere, the faster it moves. In the expanding sphere of radius a, let r denote the distance of a particle from the center. The radial velocity of the particle is $V(r/a)$, where V is the velocity of a surface particle. This linear form of expansion ensures that the density remains everywhere the same in the sphere during expansion; it also ensures that all particles follow orbits of a similar nature.

The total energy (the sum of the kinetic and gravitational energies) is constant for each particle:

total energy $=$ kinetic energy

$+$ gravitational energy.

The kinetic energy of a unit of mass at the surface is just $\frac{1}{2}V^2$; the gravitational energy is $-GM/a$. When the expanding sphere reaches infinite radius ($a = \infty$) its gravitational energy is zero. Hence the total energy of a unit of mass equals its kinetic energy at infinity.

First, we consider the motion of the surface particles. Each unit of mass has a kinetic energy $\frac{1}{2}V^2$ and a gravitational energy $-GM/a$. Hence,

$$V^2 = \frac{2GM}{a} + C,\qquad\qquad [16.2]$$

where the constant C is twice the total energy of a unit of mass. When the velocity V equals the escape velocity, the total energy is zero, and hence C is also zero, in agreement with Equation [16.1]. When V is less than the escape velocity, the total energy is negative, hence C is negative, and the particles lack sufficient energy to reach infinity. And when V is greater than the escape velocity, the total energy is positive, hence C is

positive, and the particles have more than sufficient energy to reach infinity. Our theory will not tell us the value of C, for that depends on the initial conditions; the same limitation applies in general relativity.

We now consider the motion of an interior particle. The velocity at r is $V(r/a)$, where V is the surface velocity, and a unit of mass has kinetic energy $\frac{1}{2}V^2(r/a)^2$. Because the density is uniform, the mass enclosed in a sphere of radius r is $M(r/a)^3$, and hence the gravitational energy per unit mass is $-GM(r/a)^3/r$. The constant C, which is equal to twice the total energy per unit mass, must also be reduced by the factor $(r/a)^2$. On equating the total energy to the sum of the kinetic and gravitational energies for an interior particle, and canceling out the common factor $(a/r)^2$, we again arrive at Equation [16.2].

Because of homogeneity, and the velocity–distance law that preserves homogeneity, Equation [16.2] represents the motion of all particles in the sphere. When C is zero, all particles follow parabolic orbits and ultimately have zero kinetic energy at infinity; when C is negative, all particles follow elliptical orbits and lack sufficient kinetic energy to reach infinity; and when C is positive, all particles follow hyperbolic orbits and have more than sufficient kinetic energy to reach infinity. The sphere is gravitationally bound (particles cannot escape) when C is negative, and gravitationally unbound when C is either zero or positive.

The expanding sphere represents the universe!

We have gained some idea of how an expanding sphere of matter of uniform density behaves under the influence of its own gravity when all particles fall freely and obey at any instant the velocity–distance law. We now take the bold step of assuming that the sphere represents a part of the universe.

Suppose at some instant t_0 in time the radius of the sphere is a_0 and the scale factor R (discussed in Chapter 14) has the value R_0. The radius at any other instant is

Table 16.1. *Comparison of Newtonian and relativistic cosmologies*

Curvature constant k	Newtonian	Relativistic
1	elliptical orbits	spherical space
0	parabolic orbits	flat space
−1	hyperbolic orbits	hyperbolic space

$a = a_0(R/R_0)$, and the radial velocity $(V = da/dt)$ is $V = a_0(\dot{R}/R_0)$, where $\dot{R}$ is the rate at which R increases with time. Hence $V = a\dot{R}/R$. The mass of the sphere is its volume multiplied by its density: $M = \frac{4}{3}\pi\rho a^3$. With these two expressions, Equation [16.2] becomes

$$\dot{R}^2 = \frac{8\pi G\rho R^2}{3} + C\left(\frac{R}{a}\right)^2. \qquad [16.3]$$

The constant $C(R/a)^2$ is the same for all particles in a uniform sphere obeying the velocity–distance law, and we can write $C(R/a)^2 = -k$, where k is a new constant, and the negative sign is conventional.

The equation for the expanding sphere now takes the universal form

$$\dot{R}^2 = \frac{8\pi G\rho R^2}{3} - k, \qquad [16.4]$$

and because the mass M does not change, the density obeys the equation

$$\rho R^3 = \text{constant.} \qquad [16.5]$$

The scale factor R as a function of time tells us how distances vary with time, and usually we are interested only in comparing the different values of R at different times. The actual value of R itself is often not of primary importance. We make use of this freedom to adjust R_0 and "normalize" k to its simplest values. When the sphere is gravitationally bound and particle orbits are elliptical, k is positive, and adjustment of R_0 makes k equal to $+1$. Obviously, when the particle orbits are parabolic, the constant k is zero. When the sphere is not gravitationally bound and particle orbits are hyperbolic, k is negative, and adjustment of R_0 makes k equal to -1. Hence for the values

$$k = 1, 0, -1, \qquad [16.6]$$

the corresponding orbits are elliptical, parabolic, and hyperbolic, as in Table 16.1.

Our final Equations [16.4], [16.5], and [16.6] show how the sphere expands as a dynamical system. They are obtained by very simple arguments. Only the scale factor R, the density ρ, and the constants k and G appear in the equations. The radius a of the sphere has dropped out. Thus all spheres of large and small mass, large and small radius, are governed by identical equations and behave in similar ways.

Even more remarkable, the set of Equations [16.4]–[16.6] is identical with those obtained by general relativity theory. In honor of Alexander Friedmann, who first obtained such cosmological equations in 1922 and 1924 from general relativity, they are known as the Friedmann equations.

We have used Newtonian theory to derive the Friedmann equations. In our treatment of a free-falling sphere of matter – a treatment known as Newtonian cosmology – space is flat and static, particles have motion in this space, and k determines the nature of their orbits. Two views are possible: either the sphere represents a small sample of the universe, or the sphere is indefinitely large and is the entire universe. Milne and McCrea in their original work took the latter view. We shall return to this controversial matter shortly.

Friedmann universes

In modern cosmology – or relativistic cosmology – the Gaussian curvature of space is k/R^2, and $k = 1$, 0, or -1 is the curvature constant introduced in Chapter 10. In both the Newtonian and relativistic treatments, k always has one of the three values 1, 0, −1. The meaning of these values in the two treatments is summarized in Table 16.1.

When $k = 1$, orbits are elliptical, expanding space is spherical; when $k = 0$, orbits are parabolic, expanding space is flat; and when $k = -1$, orbits are hyperbolic, expanding space is hyperbolic. In the Newtonian treatment the constant k distinguishes between orbits, and in the relativistic treatment it distinguishes between geometries. Orbits and geometries have similar names and similar topologies: elliptical orbits and spherical space (which was sometimes known as elliptical space) are closed; parabolic orbits and flat space (which was once known as parabolic space) are open; and hyperbolic orbits and hyperbolic space are open. For convenience, in the Newtonian treatment we shall refer to k as the curvature constant.

The way the scale factor R changes in time shows how the universe expands. Because there are many possible ways in which R can change in time, there are potentially many universes (not just the Friedmann universes), and one of the main aims in cosmology is to determine the universe (or model of the Universe) that best fits the observations.

In the present treatment there are basically three classes of universes, known as the Friedmann universes, corresponding to the three values of the curvature constant k. These geometrically distinct universes are as follows:

- $k = 1$. This universe, shown in Figure 16.6, was discovered by Friedmann in 1922 and rediscovered by Lemaître in 1927. In the relativistic treatment, this kind of universe has spherical expanding space that is finite and unbounded. It starts as a big bang ($R = 0$, $t = 0$), expands to a maximum size R_{max}, then collapses back to a second big bang, and has a finite lifetime. In the Newtonian treatment, a finite sphere expands and then collapses in Euclidean space, and its free-falling particles have velocities less than their escape velocities and follow elliptical orbits.

- $k = 0$. This universe, shown in Figure 16.7, was proposed jointly by Einstein and de Sitter in 1932, and though the simplest of all universes, was considered neither by Friedmann nor Lemaître. It is known as the Friedmann universe of zero curvature or the Einstein–de Sitter universe. In the relativistic treatment, this kind of universe has flat expanding space that is infinite and unbounded. It starts as a big bang, expands continually, has an infinite lifetime, and is of the bang–whimper kind. In the Newtonian treatment, a finite sphere expands continually in Euclidean space, and its free-falling particles have velocities equal to their escape velocities and follow parabolic orbits.

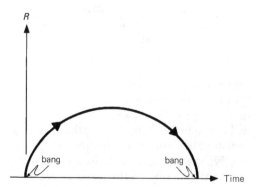

Figure 16.6. A Friedmann universe of spherical space ($k = 1$).

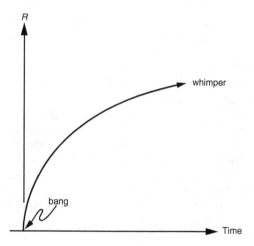

Figure 16.7. A Friedmann universe of zero curvature ($k = 0$), also known as the Einstein–de Sitter universe.

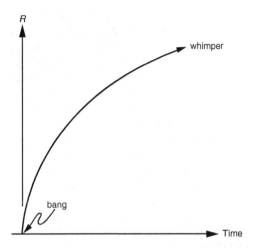

Figure 16.8. A Friedmann universe of hyperbolic space ($k = -1$).

• $k = -1$. This universe, shown in Figure 16.8, was discovered by Friedmann in 1924 and investigated in 1932 by the German cosmologist Otto Heckmann. In the relativistic treatment, this kind of universe has hyperbolic expanding space that is infinite and unbounded. It starts as a big bang, expands continually, has an infinite lifetime, and is of the bang–whimper kind. In the Newtonian treatment, a finite sphere expands continually in Euclidean space, and its free-falling particles have velocities greater than their escape velocities and follow hyperbolic orbits.

A word of warning: although the relativistic and Newtonian pictures bear similarities, their underlying concepts should not be confused. We may not in the relativistic treatment talk of the kinetic and total energies of comoving particles, nor may we talk of the gravitational energy of the universe. Once we have changed from Newtonian to relativistic theory, force, velocity, and energy in the cosmological setting lose much of their usual meaning.

COSMOLOGICAL CONSTANT

In the Newtonian treatment, at any instant in time, the gravitational force inside our sphere of uniform density is proportional to the distance r from the center of the sphere. The gravitational force per unit mass in a sphere of mass M and uniform density ρ is:

$$\text{gravitational force} = -\frac{GM}{r^2}$$

$$= -\frac{4\pi G\rho r}{3}. \qquad [16.7]$$

Any additional force that is proportional to r will produce a mock gravitational effect. Einstein introduced in 1917 a cosmological term Λ (the lambda force) that when translated from the relativistic picture to the Newtonian picture acts like a repulsive gravitational force:

$$\Lambda \text{ force} = \frac{\Lambda r}{3}. \qquad [16.8]$$

When Λ is positive, the new force opposes gravity and is equivalent to a cosmic repulsion; when Λ is negative, the force reinforces gravity and is equivalent to a cosmic attraction. The total force becomes

$$\text{total force} = -\frac{4\pi G\rho r}{3} + \frac{\Lambda r}{3}. \qquad [16.9]$$

In the particular case where Λ has the value

$$\Lambda = 4\pi G\rho, \qquad [16.10]$$

the net force is zero. This was just what Einstein wanted to achieve a static universe.

Let us include the cosmological term in our previous discussion. Arguing as before in the Newtonian treatment, we have a sphere of mass M, radius a, uniform density ρ, expanding with a radial surface velocity V given by

$$V^2 = \frac{2GM}{a} + \frac{\Lambda a^2}{3} + C, \qquad [16.11]$$

and on using the scaling factor this relation transforms to

$$\dot{R}^2 = \frac{8\pi G\rho R^2}{3} + \frac{\Lambda R^2}{3} - k. \qquad [16.12]$$

Our general equations are now Equations [16.12], with (Equation 16.5)

$$\rho R^3 = \text{constant}, \qquad [16.13]$$

and (Equation 16.6)

$$k = 1, 0, -1, \qquad\qquad [16.14]$$

where k is the curvature constant. Although Friedmann derived cosmological equations in this form (Equation 16.12), we refer to them as the Friedmann–Lemaître equations (as distinct from the Friedmann equations 16.4) in honor of Georges Lemaître who derived them independently in 1927.

WHY DOES NEWTONIAN COSMOLOGY GIVE THE SAME ANSWER?

In the Newtonian picture our sphere expands in flat Euclidean space under the influence of Newtonian gravity, whereas in the relativistic picture the whole universe consists of expanding curved space. Yet these two pictures, despite their basic difference, yield identical equations.

We have noticed that the radius a of the sphere seems irrelevant. A small sphere the size of a tennis ball and a large sphere millions of light years in radius behave similarly if they have similar uniform densities and curvature constants. It is tempting therefore to go the whole way and suppose that the sphere is of unlimited size and represents an unbounded universe having no center and no edge. This was the viewpoint adopted by Milne and McCrea when they first introduced Newtonian cosmology.

Later, in 1954, David Layzer of Harvard University pointed out that gravitational forces are indeterminate in a uniform unbounded universe. At each point in a uniform universe without center and edge the cosmic force has no determinate direction. How then can matter on a global scale move in any particular direction? Two views are possible.

Bounded yet vast in size

One way of evading this problem in the Newtonian treatment is to regard the universe as vast yet finite in size. The gravitational force remains determinate. Although the universe has a center of gravity, the edge is far enough away that most observers can ignore it. But there are objections to this way of explaining why Newtonian cosmology gives the same answer as relativistic cosmology. First is the unappealing fact that a large but finite universe is surrounded by an edge and is embedded in infinite space devoid of matter beyond the edge. With such a "clipped" universe we are back again to the Stoic cosmos. Another serious objection is that in the Newtonian picture, particles move through space (not with expanding space) faster than the speed of light at very large distances. Particles moving through space faster than light contradicts the principles of special and general relativity. A vast expanding sphere with its edge moving faster than the speed of light leads to contradiction and not reconciliation between Newtonian and relativistic theory. One can argue that a third objection is the magnitude of the gravity potential (equal to $-V_{esc}^2$, where V_{esc} is given by Equation 16.1). Newtonian theory can be used only when $-V_{esc}^2$ is small compared with c^2, otherwise it must yield to relativity theory. These three fatal objections suggest that we are looking at Newtonian cosmology in the wrong way.

Bounded yet small in size

The other way of trying to understand Newtonian cosmology is to regard the sphere as representing only a very small part of the expanding universe. The cosmological equations (Equations 16.4, 16.5, and 16.6) apply to a perfect universe entirely free of irregularity. In such an ideal universe we are free to study a representative sample as small as we please. This sample, taken from anywhere, may have a radius of 1 light year, or 1 kilometer, or 1 centimeter, or even less. The expansion velocity is small when the sphere is small. Also V_{esc}^2 is small when the sphere is small. In such a small sphere we have no need to distinguish between motion through space and comoving motion in expanding space. Moreover, we can ignore spatial curvature and justly assume that space is Euclidean. We see that smallness reconciles the Newtonian and relativistic theories. Thus when Newtonian theory is used in cosmology, we must regard the sample region as small in comparison with the Hubble sphere.

All regions of a uniform universe expand in the same way – "all are but parts of one stupendous whole" – and by knowing how a small sample region expands, we know how the whole universe expands. In a small sample region, the spatial curvature and expansion velocities are extremely small. Newtonian dynamics is then in agreement with general relativity. In this way, by piecing the small regions together, we know how large regions behave. This way of viewing Newtonian cosmology – of regarding the sphere as a very small sample of an idealized universe – is much more enlightening than regarding the sphere as a vast expanding Stoic cosmos that violates every tenet of relativity theory.

Caution, however, must be exercised. The Newtonian treatment breaks down when the pressure is not small compared with the energy density ρc^2. Pressure (stress) is a form of energy density and is therefore a source of gravity. Newtonian cosmology is not in agreement with relativistic cosmology when the universe is inhomogeneous, and also not in agreement when the universe is homogeneous but anisotropic. Motional anisotropy (strain) is also a source of gravity not included in Newtonian theory.

Newtonian clarification

The Newtonian treatment helps us to understand the relativistic picture of the universe with the aid of familiar concepts. Elementary accounts of cosmology before the discovery of Newtonian cosmology were obscure and confusing. Readers gained the impression that the universe was expanding because spacetime was endowed with mysterious power. The Newtonian treatment has made clear that the universe does not expand because spacetime insists that it must; it expands for the same reason that the Newtonian sphere expands; both the sphere and the universe are in dynamic states determined by their initial conditions, and both are initially launched into a state of expansion. Both could just as easily be in states of contraction, and the equations governing these dynamic states show no more preference for expansion than contraction.

REFLECTIONS

1 *The Newton–Bentley correspondence (Chapter 3) led Newton to abandon the Stoic cosmos of a finite distribution of matter in infinite space and to adopt the Atomist universe in which matter is distributed throughout infinite space. If the distribution of matter were finite, wrote Newton to Bentley, "the matter on the outside of this space would by its gravity tend toward the matter on the inside, and by consequence, fall down into the middle of the whole space, and there compose one great spherical mass." From this correspondence emerged a picture in which the discovery of universal gravity implied that the universe was infinite and unbounded.*

2 *Now that Edward Milne and William McCrea have shown that Newtonian theory can be used in cosmology, the subject looks deceptively simple, so simple that some persons have wondered why Newton himself did not predict the expansion of the universe. (Heinrich Olbers has been faulted for not making this prediction by those believing that the darkness of the night sky proves the universe is expanding.) But the Newtonian laws are incapable of resolving the deadlock of cosmic forces in an infinite uniform universe. The same laws spring to life, and motion is no longer indeterminate, in the finite-sphere treatment, and we stumble on equations that general relativity endorses as correct.*

3 *Dennis Sciama in* Modern Cosmology *writes, "There is one important difference between a large cloud and an infinite one. The large cloud has a unique centre while the infinite one does not. We do not really want a special point to be picked out in this way, and we can minimize the effect of having one by making the cloud uniform out to its edge, isotropic about its centre and much larger than any distance that has yet been measured. Under these circumstances* *any galaxy or QSO [quasar] that we can detect would see around itself with arbitrarily high precision a uniform isotropic universe. In the last analysis we can never distinguish observationally between an infinite universe and a finite one that is suitably larger than any distance yet surveyed.*" But John North in The Measure of the Universe *writes:* "*However large the system, there must be observers within sight of the boundary. In this respect the theory seems to have more loose ends than are justified, even in a theory whose principal function is merely one of suggestion.*"

4 *Albert Einstein in his first cosmological paper in 1917 wrote of the cosmic forces that their differences in various directions "must really be of so low an order of magnitude that the stellar velocities generated by them do not exceed the velocities actually observed." He proposed a static finite universe made possible by the cosmological term, and at the end of the paper wrote that the cosmological term was "necessary only for the purpose of making possible a quasi-static distribution of matter, as required by the fact of the small velocities of the stars"* (The Principle of Relativity).

5 "*It is therefore clear that from the direct data of observation we can derive neither the sign nor the value of the curvature, and the question arises whether it is possible to represent the observed facts without introducing a curvature at all*" (Albert Einstein and Willem de Sitter, "On the relation between the expansion and the mean density of the universe," *1932). In this paper the Einstein–de Sitter universe of zero curvature and zero cosmological term was first proposed.*

6 *We need to explain why in Newtonian cosmology a small sample of an unbounded universe must be transplanted to an empty Euclidean space in order to study its dynamic behavior. This is not done in relativity cosmology, why must it be done in Newtonian cosmology? The Newtonian and relativity theories of gravity are expressed in the form of differential equations; in the former, a single gravity potential varies in space and time, and its value is determined by the*

distribution of matter; in the latter, the 10 metric coefficients vary in spacetime, and their values depend on the distribution of matter and energy. The Newtonian gravity potential is the negative value of the binding energy per unit mass, and in a system such as a star, it has everywhere a determinate finite value. But in a uniform and unbounded system, it becomes indeterminate. This is the gravity riddle: how does a system behave when the cosmic forces are indeterminate? No such riddle exists in general relativity. In general relativity theory, gravity consists of dynamic curved space. The 10 metric coefficients, which reduce to 4 in an isotropic and homogeneous distribution, determine the curvature at each point in space and unlike the gravity potential are finite and determinate everywhere. Within a small region of a uniform universe, the metric coefficients can be further simplified and made equivalent to the single Newtonian gravity potential, provided the small region is isolated and embedded in empty Euclidean space. The Einstein equation, when applied to the small region, is the same as the Newtonian equation applied to the same region treated as an isolated small cosmic sphere.

7 *Anything that travels faster than light, such as Newtonian gravity, can be made to do impossible things and is therefore ruled out by modern physics. Let us imagine that in the fictional world of Star Wars every person has a faster-than-light gun. To make the point clear, we suppose that the guns shoot rays at infinite speed. Albert (A) and Bertha (B), in separate spaceships, are fleeing side-by-side at very high speed from enemy X. The enemy fires and destroys A. In X's space the firing and the destroying of A are simultaneous events. But not simultaneous in the space of A and B where A's destruction occurs before X fires. Bertha sees Albert destroyed, and fires back, and is thus able to destroy the enemy X before X has fired. In this way she saves Albert. We have here a situation where an effect, after occurring, is canceled by the elimination of its cause. This violates what is called causality. The situation we have described becomes* farcical if a second enemy Y is in the neighbor-hood of X. Y sees X destroyed, and fires back and eliminates B, thus saving X; A sees B destroyed, and fires back and eliminates Y, thus saving B; X sees Y destroyed, and fires back and eliminates A, thus saving Y; and the struggle goes on, with each side creating effects and eliminating their causes.

With faster-than-light travel one can easily journey forth into the future (in Earth's time). Suppose that we travel instantaneously (in Earth's space) to a star at distance L. On arrival, we turn our spaceship around and accelerate toward Earth at a speed V close to that of light c. Our space is now different from Earth's space, and by entering into instantaneous travel, we arrive back on Earth LV/c^2 in the future. If the star is in the Andromeda galaxy, and V is close to c, we would arrive back two million years in the future. We can also travel back into the past a time LV/c^2; in this case, while at the distant star, we accelerate away from Earth to a velocity V, and then in this new space turn our spaceship around and return instantaneously to Earth. Can we then do things that would change Earth's past history? The answer is no, as discussed in Chapter 9: we each have an immutable worldline in spacetime on which we can travel only meta-physically but never physically. When we devise paradoxes by means of Wellsian time travel, we contradict the basic tenets of spacetime.

8 In a uniform universe the density and pres-sure are the same everywhere in space and no pressure gradients influence the motion. The equation of motion of a radially moving parti-cle of unit mass at the surface of a sphere of radius a and density ρ is hence

$$\frac{dV}{dt} = -\frac{4\pi G\rho a}{3},$$

where the velocity is $V = da/dt$ and the density obeys $\rho a^3 = $ constant. With $a = a_0(R/R_0)$, this equation becomes

$$\ddot{R} = -\frac{4\pi G\rho R}{3}, \qquad [16.15]$$

and the behavior of the sphere is seen to be independent of the radius a. Using

$$\rho R^3 = \text{constant}, \qquad [16.16]$$

we can integrate Equation [16.15] to give

$$\dot{R}^2 = \frac{8\pi G\rho R^2}{3} - k, \qquad [16.17]$$

as in Equation [16.4]. Notice that the scaling factor in our treatment has the dimension of time. The three solutions of Equations [16.16] and [16.17] are the Friedmann models.

The Einstein–de Sitter model of zero cur-vature ($k = 0$) is

$$R = R_0 \left(\frac{t}{t_0}\right)^{2/3}, \qquad [16.18]$$

shown in Figure 16.7, where t is the age of the universe and the zero subscript denotes the present epoch.

The Friedmann model of positive curvature ($k = 1$) is the cycloid solution shown in Figure 16.6:

$$R = A \sin^2 \phi, \qquad [16.19]$$

$$t = A(\phi - \sin\phi\cos\phi), \qquad [16.20]$$

where $A = 8\pi G\rho R_0^3/3 = R_{max}$.

The Friedmann model of negative curva-ture ($k = -1$), shown in Figure 16.8, is

$$R = A \sinh^2 \phi, \qquad [16.21]$$

$$t = A(\sinh\phi\cosh\phi - \phi), \qquad [16.22]$$

where again $A = 8\pi G\rho R_0^3/3$.

9 When the Λ force is included, Equations [16.15] and [16.17] become respectively,

$$\ddot{R} = -\frac{4\pi G\rho R}{3} + \frac{\Lambda R}{3}, \qquad [16.23]$$

$$\dot{R}^2 = \frac{8\pi G\rho R^2}{3} + \frac{\Lambda R^2}{3} - k. \qquad [16.24]$$

The Friedmann–Lemaître set of equations (Equations 16.16, 16.23, and 16.24) are important and we shall return to them in later chapters.

10 In general relativity the pressure cannot always be neglected, and more generally,

Equation [16.15] becomes

$$\ddot{R} = -\frac{4\pi G}{3}\left(\rho + \frac{3P}{c^2}\right)R \qquad [16.25]$$

and the source of gravity includes the pressure P. The pressure P (or rather 3P) is an energy density that has mass $3P/c^2$. Normally the mass contributed by the pressure is negligible. But in the early universe, and any situation in which the speed of sound approaches the speed of light, the pressure contribution to mass is important. In thermal radiation of density ρ, for example, the pressure is $P = \frac{1}{3}\rho c^2$, and the source of gravity is not just $4\pi G\rho$, but $8\pi G\rho$. From general relativity we find that the density ρ and the pressure P are related by the expression

$$\frac{d(\rho R^3)}{dt} + P\frac{dR^3}{dt} = 0. \qquad [16.26]$$

When P is zero, we obtain the conservation of matter, Equation [16.15]. We need not go to general relativity for this relation, however, because it is no more than the first law of thermodynamics in the form:

$$\frac{dE}{dt} + P\frac{dV}{dt} = 0,$$

where $E = \rho c^2 V$ is the energy in an expanding volume V that is proportional to R^3. This subject is discussed more fully in Chapter 17. With Equation [16.26] we can eliminate the pressure from Equation [16.25]:

$$\ddot{R} = \frac{4\pi G}{3}\frac{d(\rho R^2)}{dR},$$

and with $\ddot{R} = \frac{1}{2}d\dot{R}^2/dt$, we obtain

$$\dot{R}^2 = \frac{8\pi G\rho R^2}{3} - k, \qquad [16.27]$$

which is the same as Equation [16.17]. The set of Equations [16.25]–[16.27] are the relativistic versions of Equations [16.15]–[16.17]. We see that Equations [16.27] and [16.22] are identical. When the cosmological term is added, the Friedmann–Lemaître

Equations [16.23] and [16.24] become

$$\ddot{R} = -\frac{4\pi G}{3}\left(\rho + \frac{3P}{c^2}\right)R + \frac{\Lambda R}{3}, \qquad [16.28]$$

$$\dot{R}^2 = \frac{8\pi G\rho R^2}{3} + \frac{\Lambda R^2}{3} - k, \qquad [16.29]$$

and these equations, when combined, give Equation [16.26].

11   *Consider a box of volume V, having perfectly reflecting walls and containing radiation of mass density ρ. The mass of the radiation in the box is $M = \rho V$. We now weigh the box and find that its mass, because of the enclosed radiation, has increased not by M but by an amount 2M. For example, 1 gram of radiation in a box increases the mass of the box by 2 grams. This unexpected increase in mass occurs because the radiation exerts pressure on the walls of the box and the walls contain stresses. These stresses in the walls are a form of energy that equals $3PV$, where P is the pressure of the radiation. The pressure equals $\rho c^2/3$, and the energy in the walls is therefore $\rho c^2 V$ and has a mass equivalent of $M = \rho V$. The mass of the box is therefore increased by the mass M of the radiation and the mass M of the stresses in the walls, giving a total increase of $2M$. In the universe there are no walls: nonetheless, the radiation still behaves as if it had a gravitational mass twice what is normally expected. Instead of using ρ, we must use $\rho + 3P/c^2$, as in the first of the relativistic Friedmann–Lemaître equations (Equation 16.28). This feature of general relativity explains why in a collapsing star, where all particles are squeezed to high energy, increasing the pressure, contrary to expectation, hastens the collapse of the star.*

PROJECTS

1   What is the gravity riddle?

2   Explain why Newton felt it necessary to abandon the Stoic universe in favor of the infinite Atomist universe. To answer this question you may have to refer to Chapter 3.

3   Many commentators on cosmology have wondered why pre-20th century mathematicians failed to predict the expansion of the universe on the basis of Newtonian theory. What do you think?

4 It has been suggested that in Newtonian cosmology we may regard the universe as a Stoic cosmos of galaxies surrounded by an infinite and empty space. In the Stoic cosmos, however, a large number of inhabitants live near the edge, and they observe that all places are not alike. The larger the system, however, the smaller the fraction of inhabitants living near the edge, and the larger the fraction who will believe the universe is edgeless and everywhere the same. Is this a satisfactory reason for retaining the cosmic edge?

5 Derive the Friedmann and the Friedmann–Lemaître equations by Newtonian arguments and discuss the effect of the cosmological term.

6 Suppose that the theory of general relativity had never been discovered. What might have happened in response to the discovery that the universe is expanding?

FURTHER READING

Bondi, H. *Cosmology*. Cambridge University Press, Cambridge, 1960.

Callan, C., Dicke, R. H., and Peebles, P. J. E. "Cosmology and Newtonian mechanics." *American Journal of Physics* 33, 105 (February 1965).

Sciama, D. W. *Modern Cosmology*. Cambridge University Press, Cambridge, 1971.

Tinsley, B. M. "The cosmological constant and cosmological change." *Physics Today* (June 1977).

SOURCES

Charlier, C. V. L. "How an infinite world may be built up." *Arkiv för Matamatik, Astronomi och Fysik* 16, no. 22 (1922).

Dicke, R. H. "Gravitational theory and observation." *Physics Today* (January 1967).

Einstein, A. *The Principle of Relativity*. Methuen, London, 1923.

Einstein, A. and Sitter, W. de. "On the relation between the expansion and the mean density of the universe." *Proceedings of the National Academy of Sciences* 18, 213 (1932).

Ellis, G. F. R. "The expanding universe: A history of cosmology from 1917 to 1960," in *Einstein and the History of General Relativity*. Editors D. Howard and J. Stachel. (Einstein Study Series.) Birkhauser, Boston, 1988.

Friedmann, A. "On the curvature of space." *Zeitschrift für Physik* 10, 377 (1922). "On the possibility of a world with constant negative curvature." *Zeitschrift für Physik* 21, 326 (1924). Both translated in *Cosmological Constants*. Editors J. Bernstein and G. Feinberg. Columbia University Press, New York, 1986.

Guerlac, H. and Jacob, M. C. "Bentley, Newton, and Providence." *Journal of the History of Ideas* 30, 307 (1969).

Halley, E. "Of the infinity of the sphere of fix'd stars." *Philosophical Transactions* 31, 22 (1720–1721). "Of the number, order, and light of the fix'd stars." *Philosophical Transactions* 31, 24 (1720–1721). Both papers reproduced in Harrison, *Darkness at Night*.

Harrison, E. R. "Newton and the infinite universe." *Physics Today*, 24, 39 (February 1986).

Harrison, E. R. *Darkness at Night: A Riddle of the Universe*. Harvard University Press, Cambridge, Massachusetts, 1987.

Hoskin, M. A. "Newton, providence and the universe of stars." *Journal for the History of Astronomy* 8, 77 (1977).

Kellogg, O. D. *Foundations of Potential Theory*. Dover Publications, New York, 1929.

Layzer, D. "The significance of Newtonian cosmology." *Astronomical Journal* 59, 168 (1954).

Lemaître, G. "A homogeneous universe of constant mass and increasing radius accounting for the radial velocity of extragalactic nebulae." First published in 1927 and translated in *Monthly Notices of the Royal Astronomical Society* 91, 483 (1931).

McCrea, W. H. "On the significance of Newtonian cosmology." *Astronomical Journal* 60, 271 (1955).

McCrea, W. H. and Milne, E. "Newtonian universes and the curvature of space." *Quarterly Journal of Mathematics* 5, 73 (1934).

Milne, E. "A Newtonian expanding universe." *Quarterly Journal of Mathematics* 5, 64 (1934).

Newton, I. *Sir Isaac Newton's Mathematical Principles of Natural Philosophy and his System of the World* (1729). Translated by E. Motte. Revised by F. Cajori. University of California Press, Berkeley, 1960.

Newton, I. *Opticks: A Treatise of the Reflections, Refractions, Inflections and Colours of Light.* 4th edition 1730. Dover Publications, New York, 1952.

Newton, I. *A Treatise of the System of the World.* Translated by I. B. Cohen. Dawsons, London, 1969.

North, J. D. *The Measure of the Universe: A History of Modern Cosmology.* Oxford University Press, Oxford, 1965. See Chapter 15, "Newtonian cosmology."

Ramsey, A. S. *Newtonian Attraction.* Cambridge University Press, Cambridge, 1964.

Richardson, R. S. "Edmund Halley: To fix the frame of the world," in *The Star Lovers.* Macmillan, New York, 1967.

Robertson, H. P. "The expanding universe." *Science* 76, 221 (1932).

Tolman, R. C. *Relativity, Thermodynamics and Cosmology.* Oxford University Press, New York, 1934.

Tropp, E. A., Frenckel, V. Ya., Chernin, A. D. *Alexander A. Friedmann: The Man Who Made the Universe Expand.* Translated by A. Dron and M. Burov. Cambridge University Press, Cambridge, 1993.

Zöllner, J. C. F. *Über die Natur der Cometen.* Staackmann, Leipzig, 1883.

17 THE COSMIC BOX

I could be bounded in a nutshell and count myself king of infinite space, were it not that I have bad dreams.
William Shakespeare (1564–1616), Hamlet

THE UNIVERSE IN A NUTSHELL

Reflecting walls

We look out in space and back in time and the things seen at large distances are similar to things that existed in this part of the universe long ago. The scenery billions of light years away, as we see it, is the same as the scenery here billions of years ago. With a time machine that could travel back into the past we would have less need of large telescopes that strain to reach the limits of the observable universe.

This argument prompts the following thought. Things are very much the same everywhere at the same time, why not then confine our attention to a single region, concentrate on its history and ignore the rest of the universe? The history of what happens in this single region is the same as the history of what happens everywhere.

But this argument has an apparent drawback. Any chosen sample region is influenced by other regions near and far, how then can we afford to ignore the affect of these other regions? Light, for instance, travels great distances and influences what happens in the sample region. If we are to pay undivided attention to a single region, ignoring all other regions, we must in some way allow for their influence.

The things at great distances that influence the sample region are on the average the same as the things already in the sample region that existed long ago. Can we therefore contrive a way in which things in the sample region of long ago are substituted for the things at great distances?

Cosmologists cannot potter around botanizing and experimenting like most other scientists and must compensate by becoming adept at performing imaginary experiments. Isolating a region of the universe and making it self-influencing offers no great difficulty. The trick is as follows.

Our own region of the universe, or any other sample region, is isolated by surrounding it with imaginary reflecting walls (see Figure 17.1). The chosen region becomes enclosed within a cosmic box. Light rays emitted inside the box are mirrored to and fro by the reflecting walls and are not allowed to escape. Things inside the box are now influenced by the light emitted long ago by local things, and on the average this light is the same as that normally received from distant things. We have thus succeeded in isolating a sample region in such a way that its present condition is influenced by its own past conditions, and these conditions are identical with those existing elsewhere. The imaginary reflecting walls of the cosmic box must, of course, be perfect in every sense: They must transmit nothing, absorb nothing, and reflect everything, such as light, gravitational waves, particles, neutrinos, and whatever else moves from place to place in the universe.

Partitions do not affect the universe

A partitioned universe helps to clarify our ideas on this subject. Imaginary partitions,

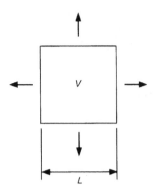

Figure 17.1. The expanding cosmic box of volume V. The box is shown as a cube of side length L.

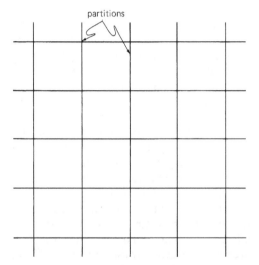

Figure 17.2. The universe is divided into cells with imaginary, massless, and perfectly reflecting partitions. The contents of all cells are in identical states and we need study only what happens in any single cell.

comoving and perfectly reflecting, are used to divide the universe into numerous separate cells (see Figure 17.2). Each cell encloses a representative sample and is sufficiently large to contain galaxies and clusters of galaxies. Each cell is larger than the largest scale of irregularity in the universe, and the contents of all cells are in identical states.

A partitioned universe behaves exactly as a universe without partitions. We assume that the partitions have no mass and hence their insertion cannot alter the dynamical behavior of the universe. The contents of all cells are in similar states, and in the same state as when there were no partitions. Light rays that normally come from very distant galaxies come instead from local galaxies of long ago and travel similar distances by multiple reflections. What normally passes out of a region is reflected back and copies what normally enters a region. By such "detailed balancing" arguments we see that partitions have no effect whatever on the behavior of the universe.

We have shown that comoving and reflecting partitions do not change the nature of the universe. Let us now remove all partitions and leave only those walls that enclose a single cell. Clearly, what happens inside this cell – or cosmic box – is similar in every way to what happens everywhere outside. Observers inside and observers outside the cosmic box perceive essentially the same scenery. The cosmic box is the universe in a nutshell and an observer inside can truly say with Hamlet "I am king of infinite space." In his "bad dreams" he lives in a looking-glass universe created by a cosmic jester!

Cosmic box in an idealized universe

The comoving walls of the cosmic box move apart at a velocity given by the velocity–distance law. If the box is a cube with sides of length L, then L is proportional to R, where R is the scaling factor that is everywhere the same in space. Opposite walls move apart at relative velocity HL, where $H = \mathrm{d}R/R\,\mathrm{d}t$ is the Hubble term. If, say, L is 100 megaparsecs, this relative velocity is $10\,000h$ kilometers per second for a Hubble term H equal to $100h$ kilometers per second per megaparsec.

Irregularities, we must admit, are a distraction and serve little purpose in the present discussion. For simplicity we shall ignore all irregularities and assume that the universe is ideally smooth. Effects caused by irregularities can always be considered at some other time if they ever become important. In an idealized universe the cosmic box can be small, as small as we please, and its contents will still remain in a representative state.

When L is as little as 1 kiloparsec, the walls move apart with a relative velocity of $100h$ meters per second; when L is 1 parsec, the relative velocity is $10h$ centimeters per second; when L is 1 meter, the velocity is approximately $1h$ centimeter per million years; and so on. The velocity of the receding walls may be made as small as we please. Naturally, the box must always be larger than the things under investigation. If we study the behavior of radiation, for example, the box must be sufficiently large to contain the longest wavelengths of interest. A star, of course, cannot be put into a box having a volume of only 1 cubic centimeter or 1 cubic kilometer. But matter in small boxes can be made luminous in the same average way as matter in the unsmoothed universe. When the smoothing process becomes more than we are willing to tolerate, L can be made sufficiently large to encompass things of interest. The important point is that L must be much smaller than the Hubble length L_H of $10h$ billion light years.

Advantages of the cosmic box

Some of the numerous advantages of the cosmic box are mentioned now and others will become apparent later.

In a small box we are free to use Euclidean geometry. Space may be curved with either spherical or hyperbolic geometry and yet remain virtually flat and Euclidean in any small region. It is a considerable advantage to study cosmic phenomena in a box without the bother of taking into account the large-scale curvature of space. Whatever is adequately explicable in the confines of a cosmic box is essentially independent of the large-scale geometry of space. This is an important rule. We shall see in Chapter 24 that the solution of the night-sky riddle of darkness (Olbers' paradox) does not depend on whether the universe is open or closed, and this explains why we are able to use the cosmic box to solve the riddle.

Because the size of the box L is small compared with the Hubble distance L_H, the walls have a recession velocity that is small compared with the velocity of light. Inside the box there is no need to distinguish between Doppler and expansion redshifts. Light bounces to and fro repeatedly, and because the walls are moving apart slowly, the sum of many small Fizeau–Doppler shifts is equivalent to the cosmological expansion redshift. Also, there is no need to distinguish between peculiar and recession velocities. All things in motion within the box can be regarded as moving through space as in the laboratory.

Inside a relatively small cosmic box we use ordinary everyday physics and are thus able to determine easily the consequences of expansion. We can even use Newtonian mechanics to determine the expansion if we embed a spherical cosmic box in Euclidean space. The physics of what happens in a slowly expanding box with nonabsorbing walls are well understood, and this knowledge becomes immediately available for use in cosmology. A uniformly distributed gas in the universe serves as an example. It behaves in exactly the same way as a sample of gas in a slowly expanding cell, and because we know how gases behave in expanding cells, we also know how gases behave when uniformly distributed in an expanding universe.

The cosmic box is the universe in a nutshell. We need study only what happens in a relatively small region surrounded by perfectly reflecting walls to know what happens everywhere.

PARTICLES AND WAVES
Freely moving particles

Let us consider a particle moving freely in an expanding universe. It moves freely in the sense that it is not trapped in a gravitationally bound system such as a galaxy. It has peculiar motion as seen by a local comoving observer who is stationary in expanding space. Strange to say, this freely moving particle slowly loses its peculiar motion; it slows down, and eventually becomes stationary in

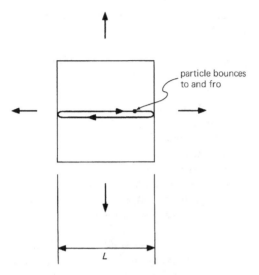

particle bounces
to and fro

Figure 17.3. A particle bounces to and fro in an expanding box and slowly loses energy. It loses energy in exactly the same way as it would when moving freely in an expanding universe. All peculiar motion relative to the comoving state decelerates and ultimately comes to rest.

expanding space. All freely moving particles, including galaxies (when not bound in clusters), slowly lose their peculiar motion and ultimately become stationary in expanding space.

We shall try to understand what happens by considering a moving particle inside an expanding box (see Figure 17.3). The cosmic box must be small enough that the walls move apart slower than the particle and the particle repeatedly rebounds from the expanding walls.

A particle moving freely in the universe travels in a straight line. The expanding regions through which it passes are identical with the regions inside the cosmic box at each instant of time. Its speed in the box is the same as its speed when moving freely in a straight line outside. The particle inside the box continually changes its direction, but otherwise behaves the same as when outside. In effect, the straight-line trajectory is folded up inside the box and the rebounding particle has the same speed and energy at each instant, as if the reflecting walls did not exist.

When a tennis player runs forward while striking a tennis ball, the ball crosses the court faster and with more energy than when the player runs backward. Similarly, a particle bouncing backwards and forwards in a contracting (expanding) box gains (loses) energy each time it strikes an approaching (receding) wall.

For simplicity we suppose the particle moves in a direction perpendicular to two opposite walls of an expanding box. Normally, of course, the particle rebounds in different directions, but the final result is just the same. The walls are perfect reflectors and therefore, relative to the wall, the particle rebounds with the same speed as when it strikes the wall. During the collision, the direction of motion is reversed, but the speed relative to the wall remains unchanged. Because the wall is receding, the particle returns to the center of the box with slightly reduced speed. Each time the particle strikes a receding wall it returns with reduced speed.

It can be shown that a particle of mass m and speed U, moving within an expanding box, obeys the law that mU is proportional to $1/L$. The product mU is the momentum. As the box gets larger the momentum gets smaller. The length L expands in the same way as the scaling factor R, and the momentum therefore obeys the important law:

$$mUR = \text{constant.} \qquad [17.1]$$

This law holds not only for particles in an expanding box but also for particles moving freely in an expanding universe. Remarkably, the general relativity equation of motion of a freely moving particle in the uniformly curved space of an expanding universe gives exactly the same result. This illustrates how the cosmic box not only helps us to understand what happens but also allows us to employ very simple methods to derive important results.

When the speed is much less than the speed of light the mass m remains constant and

$$UR = \text{constant} \quad \text{(nonrelativistic)} \qquad [17.2]$$

or, in terms of the redshift z,

$$U = U_0(1 + z) \quad \text{(nonrelativistic)} \quad [17.3]$$

where U_0 is the present speed at $z = 0$. The kinetic energy of the particle varies as the square of the speed and therefore

$$\text{kinetic energy} \times R^2 = \text{constant.} \quad [17.4]$$

In terms of the redshift this gives

$$E = E_0(1 + z)^2 \quad \text{(nonrelativistic)} \quad [17.5]$$

where E represents the kinetic energy of the particle. These results apply to all freely moving bodies. Even a field galaxy (a galaxy not in a cluster of galaxies) loses its kinetic energy in this fashion.

Consider a particle with a speed close to the speed of light, a relativistic particle of high energy, such as a cosmic ray particle. In this case, the speed U is almost constant and the particle mass is proportional to energy, and we get from Equation [17.1]

$$\text{energy} \times R = \text{constant.} \quad [17.6]$$

In terms of the redshift this gives

$$E = E_0(1 + z) \quad \text{(relativistic).} \quad [17.7]$$

At redshift $z = 1$, when the universe is half its present size, the energy of a freely moving nonrelativistic particle is four times its present value, and the energy of a relativistic particle is twice its present value.

Light waves
The cosmic box also helps us to understand why light loses energy in an expanding universe. Each time a ray of light is reflected from a receding wall it is slightly redshifted because of the Doppler effect. In 1913, Max Planck showed that the cumulative effect of repeated small Doppler redshifts in a uniformly expanding box obeys the law that wavelengths grow in proportion to L. All wavelengths are stretched as L slowly increases. Because L is proportional to the scaling factor R, this yields the relation

$$\text{wavelength} = \text{constant} \times R,$$

and therefore, in terms of redshift,

$$\lambda = \lambda_0 \frac{1}{1 + z}, \quad [17.8]$$

where λ is the emitted wavelength and λ_0 the received wavelength. Light rays bouncing to and fro in an expanding box and light rays traveling in straight lines in expanding space have their wavelengths stretched according to the same law. The cumulative Doppler redshift is independent of how the box expands, just as the cosmological redshift is independent of how the universe expands, and in both cases the redshift depends on the amount of expansion.

Another way of understanding the behavior of radiation in an expanding box is to consider a resonant cavity containing radio waves. These standing waves have zero amplitude at the walls where they are reflected, as shown in Figure 17.4, and as

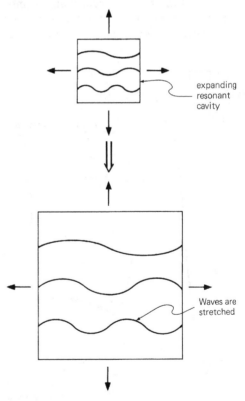

expanding resonant cavity

Waves are stretched

Figure 17.4. Waves of radiation in an expanding box (such as a resonant cavity). As the box slowly expands, the waves are stretched.

the cavity slowly expands all wavelengths increase in step with the size of the cavity.

Yet another way is to consider light – and all electromagnetic radiation – as composed of photons. These are particles that travel at light speed and have energy proportional to their frequency. A photon may be viewed as a relativistic particle, and its energy must therefore change as $1/R$. Because frequency is inversely proportional to wavelength, it follows that the wavelength grows with the scaling factor R. This argument applies to all particles that move at the speed of light and therefore applies to neutrinos. In an expanding universe, neutrinos lose energy in the same way as photons.

THERMODYNAMICS AND COSMOLOGY
Temperature
The full power of the cosmic box is realized when we turn to the science of thermodynamics. In thermodynamics we study the properties of heat in isolated systems under various conditions. How does a uniformly distributed gas behave in an expanding universe? By isolating some of it in an expanding cosmic box we can harness thermodynamics in the service of cosmology.

A system consisting of particles in thermal equilibrium has everywhere the same temperature. The particles move in all directions and have various speeds, as shown in Figure 17.5, and their average energy is everywhere the same. Temperature is a measure of thermal energy. Radiation in thermal equilibrium, referred to as blackbody radiation, consists of photons of various energies, as shown in Figure 17.6, and their average energy is also the same everywhere and proportional to temperature.

We have already seen that individual particles, moving freely, lose their energy when enclosed in an expanding box. Exactly the same thing happens to a gas consisting of many particles. Particles composing a gas continually collide with one another;

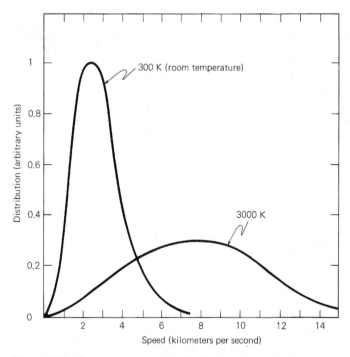

Figure 17.5. These curves show the distribution of speeds of hydrogen atoms in two gases of the same density but different temperatures.

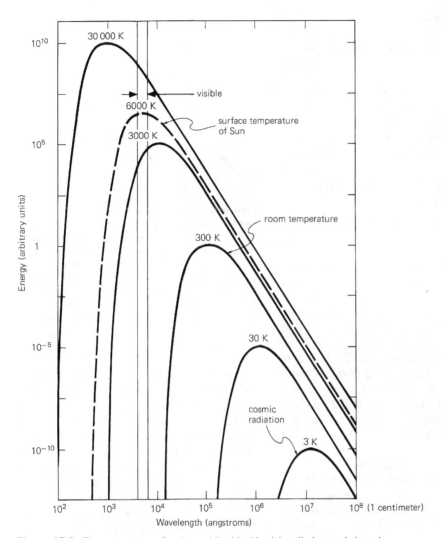

Figure 17.6. These curves are for thermal (or blackbody) radiation and show how energy is distributed at different wavelengths for different temperatures. A typical average separating distance between photons is approximately the wavelength at peak energy. For example, at 3 kelvin, the separating distance is roughly 1 millimeter and 1000 photons occupy 1 cubic centimeter; at 300 kelvin (room temperature), the separating distance is 10^{-3} centimeters and 10^9 photons occupy 1 cubic centimeter; and at 3000 kelvin, the separating distance is 10^{-4} centimeters and 10^{12} photons occupy 1 cubic centimeter. Note that as radiation gets squeezed into a smaller and smaller volume, the temperature rises, but the total number of photons (which is proportional to entropy) stays constant.

between collisions they move freely and lose energy in the way described for free particles; during their encounters they exchange energy, but collisions do not change the total energy. The temperature of a gas therefore varies with expansion in the same manner as the energy of a single (nonrelativistic) particle:

gas temperature is proportional to $\dfrac{1}{R^2}$.

If T denotes temperature, and T_0 the present

temperature, then

$$T = T_0(1 + z)^2 \quad \text{(nonrelativistic gas)}.$$
[17.9]

The temperature decreases with expansion and the gas stays in thermal equilibrium.

Thermal radiation consists of photons in thermal equilibrium. We may think of it as a gas of photons. The temperature of thermal radiation is a measure of the average photon energy; we know how photon energy varies with expansion, and therefore

$$T = T_0(1 + z) \quad \text{(thermal radiation)}$$
[17.10]

and the temperature is in kelvin, the absolute zero of temperature, 0 kelvin, corresponding to −273 celsius. The cosmic background radiation is in thermal equilibrium at temperature 2.728 kelvin, and its temperature in the past was much higher. At redshift $z = 100$ its temperature was 272.8 kelvin, which is almost the freezing point of water at 273 kelvin. At redshift $z = 1000$ it was incandescent at temperature 2728 kelvin, and this was roughly when it last interacted intimately with matter.

First law of thermodynamics

The nonabsorbing walls of the cosmic box are perfectly reflecting; nothing passes through the walls, including heat and all other forms of energy (see Figure 17.7). No energy passes into or out of the box, and such an isolated system, which neither gains nor loses energy across its boundaries, is said to be adiabatic. Clearly, the universe as a whole is adiabatic. Also each representative region is adiabatic because energy that leaves is balanced by incoming energy of the same kind and same amount.

The first law of thermodynamics can be broken down into two parts. The first part states that energy entering a closed volume increases the energy already inside the closed volume. Because a cosmic box is adiabatic, and no energy enters or leaves, we can ignore this first part and consider only an adiabatic system. The second part states that the

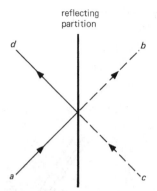

reflecting partition

Figure 17.7. Perfectly reflecting partitions change nothing in a uniform universe. A particle that would normally have traveled from *a* to *b* is deflected by a partition and travels instead to *d*. In a uniform universe there exists on the average a similar particle of the same energy that would normally have traveled from *c* to *d*, and the partition deflects this second particle in direction *b*. By "detailed balancing" arguments of this kind we see why reflecting partitions do not affect the contents of a uniform universe.

energy in an expanding volume is steadily reduced because of the work performed by pressure. A classic example is the steam engine. Steam in a cylinder exerts pressure on the piston, and as the steam expands and pushes out the piston, the steam performs work and loses energy by getting cooler (see Figure 17.8). The cosmic box has pressure exerted on the inside walls, and hence the contained energy decreases with expansion. This tells us that energy in an expanding universe is decreasing.

Let V be the volume of the cosmic box, E the total energy inside, and P the pressure. For adiabatic changes, the first law of thermodynamics states that a small increase in energy (call it dE) plus the pressure times a small increase in volume (call it dV), is equal to zero:

$$dE + PdV = 0.$$
[17.11]

As V increases (dV is positive) the energy E decreases (dE is negative). The greater the pressure, the more the energy is decreased by the expansion. The first law, expressed by Equation [17.11], is of fundamental importance in cosmology (see Equation

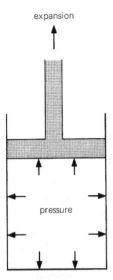

expansion

pressure

Figure 17.8. Steam (or rather hot water vapor) in a cylinder pushes against a piston. As the piston withdraws and the volume expands, the heat energy in the water vapor is converted into mechanical energy.

[16.26] and subsequent discussion in the previous chapter).

Second law of thermodynamics

The second law of thermodynamics states that entropy either remains constant or increases in an isolated system. Perfect systems can be devised in which entropy stays constant, but they are idealizations. The most perfect system realized in nature is the universe, but even the universe is not totally perfect because stars generate entropy and the entropy of the universe is increasing.

Energy forever cascades into less useful and effective forms, which is just another way of saying that entropy forever increases. In practice, one cannot use the heat discarded by the house next door to heat one's own house to the same temperature because, in the process, energy is lost to the external world and the neighbor's discarded heat has lower temperature.

The cosmic box is adiabatic – no energy enters or leaves – and we must ask how it is possible for entropy to increase in such an isolated system. The isolated system, which is a representative part of the universe, contains stars, and stars are the main generators of entropy. Their hydrogen burns to helium and nuclear energy is transformed into heat. Energy is released in the center at a temperature of millions of kelvin and is radiated away from the surface at a temperature of thousands of kelvin. Stars mine energy at high temperature and discard it at low temperature; they are entropy-generating machines. They pour out radiation into space and the amount of starlight in space is a measure of the entropy they generate.

Most of the entropy of the universe resides in the radiation that fills space. Everything emits radiation, even black holes, and entropy steadily increases. Crudely, to measure entropy in cosmology, all we need do is count photons. The total number N of photons in the cosmic box, multiplied by the Boltzmann constant k, gives Nk, which is an approximate measure of its total entropy.

I cannot use my next-door neighbor's discarded heat to warm my house effectively because some heat escapes into the environment. The environment radiates this heat into space in the form of infrared photons. My inability to exploit fully my neighbor's discarded heat is registered in the universe by an increase in the number of photons (Figure 17.9).

Now comes the surprise. The number of photons in the 2.7 kelvin cosmic background radiation is at least 10 000 times greater than the number of all other photons that have been emitted by stars. Most of the entropy of the universe is therefore already in the cosmic background radiation. Whatever stars may do cannot affect the total entropy very much. A popular theme in science in the first half of the 20th century was the eventual heat death of the universe, how all things must fade and die, and how entropy must inexorably rise to its maximum value. We now know that the heat death actually occurred long ago and we live in a universe that has very nearly attained maximum entropy.

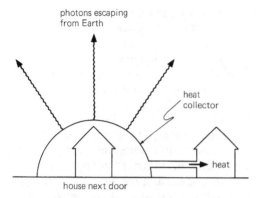

photons escaping
from Earth

heat
collector

heat

house next door

Figure 17.9. My inability to utilize completely my neighbor's discarded heat is registered in the universe by an increase in the number of photons in space. The escaping photons increase the entropy of the universe.

The number of photons per cubic centimeter of cosmic thermal radiation at absolute temperature T is given by

$$n_\lambda = 20 \times T^3 \text{ per cubic centimeter,} \quad [17.12]$$

where T is in kelvin. The 2.7-degree cosmic background radiation has about 400 photons per cubic centimeter. If you go out of doors in the daytime or at nighttime and hold up the palm of the hand to the sky, a thousand trillion (or 10^{15}) photons of the cosmic background radiation will strike it in just 1 second. These photons continually bombard everything in the universe, and yet their number is so enormous that this loss is completely negligible.

The number of photons in the universe is a measure of its entropy. The total number of photons N_λ in the cosmic box is n_λ multiplied by the volume V:

$$N_\lambda = n_\lambda V$$

$$= \text{constant} \times \text{entropy in cosmic box.}$$

$$[17.13]$$

We assume no sources and N_λ is constant. This means the entropy in V is also constant. According to Equation [17.12], the density n_λ varies as T^3, and therefore varies as $1/R^3$. But V varies as R^3, hence also VT^3 is constant. Thus the entropy of the thermal radiation in the cosmic box is constant during expansion. This is just another way of saying that the total number of photons N_λ in the box is constant. Actually, their number is slowly increased by the light emitted by stars and other sources, but this contribution is so small that for most purposes it can be ignored.

A convenient way of measuring cosmic entropy is to count photons and nucleons and take the ratio of the two numbers. The universe contains on the average roughly 1 atomic nucleon per cubic meter, and with 400 photons per cubic centimeter, this ratio is between 10^8 and 10^9. The number of photons per nucleon, almost 1 billion, is the specific entropy of the universe. The specific entropy does not alter much during expansion, and either the universe was created with a large initial reservoir of entropy (a high-temperature big bang) or the entropy was generated during the earliest moments of the big bang. An advantage of the big bang type of universe is that we are able to see how the specific entropy determines the nature of the universe. If the specific entropy were much smaller, almost all hydrogen would have been converted into helium in the big bang, and if it were much larger, the universe would have been too hot for the formation of galaxies; in the first case, stars would not be luminous over long periods of time, and in the second case, stars would not exist; in either case, the universe would not contain life.

WHERE HAS ALL THE ENERGY GONE?

Gases, light rays, freely moving particles (including galaxies), and thermal radiation (including neutrinos) lose energy in an expanding universe. Where does this energy go? We take for granted that light is redshifted and usually do not bother ourselves about where its energy has gone.

It is easy to see why energy is lost in an expanding box. Light rays and particles pushing on the walls exert pressure, and as the box expands, this pressure performs work and the energy of the rays and particles decreases. Many engines make use of this principle. Steam engines and internal combustion engines are familiar examples of how pressure on a piston produces mechanical energy at the expense of heat energy in the cylinder.

The cosmic box contains a representative sample of the universe, and what is inside is always in the same state as what is outside. Rays of light and particles push on the inside of the walls, and similar rays and particles push back with equal force on the outside of the walls. The rays and particles inside lose energy because they push against the receding walls and they therefore also push against the rays and particles outside. But those outside do not gain the energy lost by those inside; and, vice versa, those inside do not gain the energy lost by those outside. To try to understand what is happening we return to the partitioned universe of many cells. In any expanding cell there is a progressive loss of energy, which cannot be gained by adjacent cells because they also are expanding and losing energy. The pressures inside the cells work against one another and energy is lost everywhere in all cells and reappears nowhere as useful work.

The universe is not in the least like a steam engine and we must never jump to the conclusion that pressure is the cause of expansion. Pressure has nothing to do with why the universe expands. The universe could just as easily contract, and in the future may pass from its present state of expansion into a state of contraction. If the universe possessed a cosmic edge, the situation would be different; the pressure at the edge could then do work, and we would have a universe similar to a steam engine. But the universe has no edge, and the pressure everywhere is therefore impotent to produce mechanical energy. We conclude that energy in the universe is not conserved.

The familiar world contains all sorts of things surrounded by all sorts of boundaries and gradients that separate them from one another. Energy flows across these boundaries or down these gradients, and wherever energy is lost in one place in the familiar world it reappears elsewhere, either in the same form or in some other form, such as mechanical energy. A fraction of the energy we use escapes into the atmosphere and the oceans and is written off as an inevitable loss that nonetheless balances the energy budget. In bounded systems energy is conserved. But the energy that escapes from these systems into outer space is not conserved and will slowly vanish because the expanding universe is not a bounded system.

Science clings tenaciously to principles and concepts of conservation, the foremost of which is the conservation-of-energy principle. Whenever scientists find that energy has apparently vanished, they search for its reappearance in some other form, and these searches have led in the past to the discovery of new and hitherto unrecognized forms of energy. The discovery of the neutrino is the classic example. Radioactive nuclei decay into lower-energy states, and in the 1920s it was known that the electrons emitted by the decaying nuclei fail to carry away all the energy released. Wolfgang Pauli, discoverer of the exclusion principle (not more than two electrons have the same waveform in an atom), suggested in 1931 that an undetected particle is also emitted in radioactive decay that carries away the missing energy. Enrico Fermi shortly afterward named the hypothetical particle the neutrino. The neutrino was experimentally found in 1956.

The conservation-of-energy principle serves us well in all sciences except cosmology. In bound regions that do not expand with the universe (because they are dense compared with the average density of the universe), we can trace the cascade and interplay of energy in its multitudinous forms and claim that it is conserved. But in the universe as a whole it is not conserved. The total energy decreases in an expanding universe and increases in a contracting universe. Where does energy go in an expanding universe? And where does it come from in a contracting universe? The answer is nowhere, because in the cosmos, energy is not conserved.

REFLECTIONS

1 *The cosmic box helps us to understand in a simple way how things change in an expanding universe. We do not worry about whether space is curved because in a small enough sample region space can always be considered flat. We do not have to distinguish between peculiar motion through space and comoving motion in expanding space, because the recession velocity of the walls is very small compared with the velocity of light, and the two forms of motion are indistinguishable. Nor do we have to distinguish between Doppler and expansion redshifts. Furthermore, we can use all the familiar laws of science that govern the behavior of things in isolated expanding systems. If a cosmological problem cannot be solved with the cosmic box, then the solution will probably involve curvature, distinction between peculiar and recession velocities, and distinction between the Doppler and expansion redshifts.*

• *The convenience of the cosmic box is made clear in the following way. Let Albert be an observer outside the cosmic box and Bertha an observer inside. Albert outside tries to estimate the radiation he receives from all sources in the universe. He adds together the contributions received from individual sources everywhere, taking into account their distances in expanding curved space, their redshifts, and the absorption of their rays of light while traveling through curved space, and he has to be alert to the possibility of rays circumnavigating in a closed universe. Imagine his surprise when, having finished this tedious calculation, he finds that the result is independent of the spatial curvature of the universe! Bertha inside the cosmic box finds out how much light has been emitted by local sources, makes an allowance for the loss owing to absorption, and by a much simpler method obtains exactly the same result as Albert. She is not in the least surprised that her result is independent of the curvature of space. Albert outside has the difficult problem of evaluating double integral equations, Bertha inside has the easy problem*

of solving standard differential equations that are common in physics.

2 *Consider a particle moving to and fro at speed U in an expanding box of side length L. We assume that U is small compared with the speed of light, and for simplicity we consider only normal reflections, as shown in Figure 17.3. Oblique-angle reflections yield identical results. Each wall recedes at velocity $W = \frac{1}{2}HL$. The particle approaches a receding wall at relative velocity $U - W$, and because the wall perfectly reflects, the particle velocity reverses during the bounce and the speed remains unchanged. The particle rebounds at speed $U - W$ relative to the wall, or at speed $U - 2W$ relative to the center of the box. The change in speed per bounce is therefore $dU = -2W$, which is equal to $-HL$. The time between bounces is $L/U = dt$, and the rate of change in speed is hence*

$$\frac{dU}{dt} = -UH, \qquad [17.14]$$

and because $H = dR/R\,dt$, we find

$$\frac{d(UR)}{dt} = 0. \qquad [17.15]$$

Thus UR is constant during expansion. The more general relativistic treatment gives the result: $mUR = $ constant, where m is the particle mass.

Suppose the particle is a photon. The Fizeau–Doppler shift at each reflection is

$$\frac{d\lambda}{\lambda} = \frac{2W}{c} = \frac{HL}{c},$$

where λ is the wavelength and $d\lambda$ a small increase in wavelength. The Fizeau–Doppler formula may be used, as in this equation, provided the wall velocity W is small compared with the velocity of light c, and W meets this condition when the size L of the expanding box is small compared with the Hubble distance. The interval of time between successive reflections is $dt = L/c$, and because $H\,dt = dR/R$, we obtain

$$\frac{d\lambda}{\lambda} = \frac{dR}{R}. \qquad [17.16]$$

Hence the wavelength λ is proportional to the scaling factor R, and the redshift becomes

$$z = \frac{\lambda_0}{\lambda} - 1 = \frac{R_0}{R} - 1, \qquad [17.17]$$

as in an expanding universe, where the zero subscript denotes the present epoch. As discussed in Chapter 15, the cosmological redshift may be regarded as a series of small incremental Fizeau–Doppler shifts.

3 *"Once Nernst, the great physiochemist, pondering on the great quandary of the physical sciences that all energy forms in nature seem to be converted eventually into heat and the universe is aging more and more without visible signs of rejuvenation, concocted an ingenious scheme which allowed a reconversion of heat into matter and thus made a periodic universe possible in which the beginning and end of time were eliminated. In his enthusiasm he called up Einstein (with whom he had the best of scientific and the worst of personal relations: the personalities of these two great men of science were utterly clashing) and explained to him how he envisaged the evolution of the world over billions of years, asking his opinion about the theory. Einstein's comment was: 'I was not present'"* (Cornelius Lanczos, Albert Einstein and the Cosmic World Order, *1965*).

4 *"Physics tells the same story as astronomy. For, independently of all astronomical considerations, the general physical principle known as the second law of thermodynamics predicts that there can be but one end to the universe – a 'heat death' in which the total energy of the universe is uniformly distributed, and all the substance of the universe is at the same temperature. This temperature will be so low as to make life impossible. It matters little by what particular road this final state is reached; all roads lead to Rome, and the end of the journey cannot be other than universal death.... Thus, unless this whole branch of science is wrong, nature permits herself, quite literally, only two alternatives, progress and death; the only standing still she permits is in the stillness of the grave.*

Some scientists, although not, I think, very many, would dissent from this last view. While they do not dispute that the present stars are melting away into radiation, they maintain that somewhere in the remote depths of space this radiation may be reconsolidating itself again into matter. A new heaven and a new earth may, they suggest, be in process of being built, not out of the ashes of the old, but out of the radiation set free by the combustion of the old. In this way they advocate what may be described as a cyclic universe: while it dies in one place the products of its death are busy producing new life in others. This concept of a cyclic universe is entirely at variance with the well-established principle of the second law of thermodynamics, which teaches that entropy must forever increase, and that cyclic universes are impossible in the same way, and for much the same reason, as perpetual motion machines" (James Jeans, The Mysterious Universe, *1930*). See also the last static steady state universe in the next chapter (Chapter 17).

5 *The first law of thermodynamics for the cosmic box is*

$$dE + P\,dV = 0$$

(Equation 17.11), as discussed in the text. This is the adiabatic form in which no heat is gained from or lost to the outside world. Often we are interested only in energy density, such as the energy in a cubic centimeter. Let $\varepsilon = E/V$ be the energy density; the first law then becomes

$$d\varepsilon + (\varepsilon + P)\frac{dV}{V} = 0. \qquad [17.18]$$

The total energy density ε is equal to the mass density ρ times the square of the speed of light, or $\varepsilon = \rho c^2$.

• *The pressure P in an ordinary gas is negligibly small compared with the energy density ε of matter, or P is small in comparison with $\varepsilon = \rho c^2$ (or the speed of sound is small compared with the speed of light). The total energy E therefore remains unchanged during expansion, and the total mass M of the gas in the cosmic box, equal to ρV, also stays constant. This is what happens in our* *ordinary everyday experience: matter has constant mass because the pressures exerted are normally negligible compared with energy densities.*

Thermal radiation, however, has a pressure P that equals one-third its energy density ε, and is by no means negligible. Equation [17.18] in this case gives $\varepsilon \, dV + \frac{3}{4} V \, d\varepsilon = 0$, and by integration we find that $\varepsilon V^{4/3}$ is constant. Because the volume V varies as R^3, we find that εR^4 is constant during expansion, and in the case of radiation

$$\varepsilon = \varepsilon_0 (R_0/R)^4 = \varepsilon_0 (1+z)^4, \qquad [17.19]$$

where ε_0 is the present radiation density. Notice that this result is in agreement with $TR = constant$, because ε is proportional to T^4.

- *Let ρ_m be the mass density of matter and ρ_r the mass density of radiation. Let us suppose that matter and radiation are non-interacting; in other words, the emissions and absorptions by matter are of negligible effect. Then*

$$\rho_m = \rho_{m0}(1+z)^3, \qquad [17.20]$$

$$\rho_r = \rho_{r0}(1+z)^4, \qquad [17.21]$$

from previous arguments, where the zero subscript denotes present values. The ratio of the radiation and matter densities is

$$\frac{\rho_r}{\rho_m} = \frac{\rho_{r0}}{\rho_{m0}}(1+z), \qquad [17.22]$$

and as redshift increases (and we look father back in time), the radiation density rises faster than the density of matter. The present average density of matter is $\rho_{m0} = 2 \times 10^{-29} \Omega h^2$ grams per cubic centimeter (where Ω is the density parameter discussed in Chapter 18, and h is the Hubble coefficient), and the present density of the cosmic background radiation (of temperature 2.73 kelvin) is $\rho_{r0} = 4.6 \times 10^{-34}$ grams per cubic centimeter. The ratio ρ_{m0}/ρ_{r0} is $4 \times 10^4 \Omega h^2$, and therefore when $1 + z$ equals this value the two densities are equal. Roughly, at the epoch of equal densities, the redshift z_{eq} is 4000 (when $\Omega h^2 = 0.1$) and the temperature T_{eq} is 10 000 kelvin. Earlier

still, at higher temperatures, the radiation was more dense than matter and the universe was radiation dominated. Dense radiation that behaves like a thick fluid is not easy to imagine. The radiation in the early universe has fluid-like properties because photons are constantly scattered by the relatively scarce electrons (one electron to every billion photons).

6 *"Uniform radiation, such as the cosmic background radiation, is subject to the cosmological redshift effect, and in this instance the adiabatic form of the first law applies, but leaves unresolved the problem of the lost internal energy in an expanding, homogeneous and unbounded universe. Does the energy totally vanish, or does it reappear, perhaps in some global dynamic form? The tentative answer based on standard relativistic equations is that the vanished energy does not reappear in any other form, and therefore it seems that on the cosmic scale energy is not conserved"* *(Harrison, "Mining energy in an expanding universe").*

Consider a universe containing matter of density ρ at pressure P in which the cosmological constant Λ is zero. From Equation [16.25] of the previous chapter we have:

$$\ddot{R} = -\frac{4\pi G}{3}\left(\rho + \frac{3P}{c^2}\right)R. \qquad [17.23]$$

We notice that as the pressure increases, the acceleration $\ddot{R}$ decreases. This is contrary to normal experience. Ordinarily, we expect the universe to expand faster as the pressure increases, not slower. But the universe has no edge and the pressure can do no work. Moreover, increasing the pressure increases the energy density – the source of gravity – and hence slows the expansion. Thus the universe is not in the least like a steam engine.

7 *Consider a box of volume V, having perfectly reflecting walls and containing radiation of mass density ρ. The mass of the radiation in the box is $M = \rho V$. We now weigh the box and find that its mass, because of the enclosed radiation, has increased not by M but by an amount 2M. For example, 1 gram of radiation in a box increases the* *mass of the box by 2 grams. This unexpected increase in mass occurs because the radiation exerts pressure on the walls of the box and the walls contain stresses. These stresses in the walls are a form of energy that equals $3PV$, where P is the pressure of the radiation. The pressure equals $\rho c^2/3$, and the energy in the walls is therefore $\rho c^2 V$ and has a mass equivalent of $M = \rho V$. The mass of the box is therefore increased by the mass M of the radiation and the mass M of the stresses in the walls, giving a total increase of $2M$. In the universe there are no walls: nonetheless, the radiation still behaves as if it had a gravitational mass twice what is normally expected. In general, instead of using ρ, we must use $\rho + 3P/c^2$, as in the first of the Friedmann–Lemaître equations (Equation 16.28). This feature of general relativity explains why in a collapsing star, where all particles are squeezed to high energy, increasing the pressure, contrary to expectation, hastens the collapse of the star.*

PROJECTS

1 Consider the following puzzle. In cosmology, we look for those things that will unify the universe. But the partitioned universe shows that when regions are completely isolated from one another in cells they continue to behave as before as if nothing had changed. What then unifies the universe?

2 David Wilkinson of Princeton University points out that about 1 percent of the noise on a television screen, when not tuned to a station, is the cosmic background radiation. Here is a cosmological observation that everyone can make: Tune in to the big bang!

3 Another cosmological observation consists of nothing more than looking at the sky at night. We look out vast distances, far beyond the Galaxy, the Local Group, and neighboring clusters, and we look far back in time, and all around we see a wall of darkness. What is this wall? Can it be nothing? Can we ever see nothing? Once, long ago, the wall glowed with bright light and was what Edgar Allen Poe called "the golden wall of the universe" (Chapter 24).

In this simplest of cosmological observations we look in all directions and realize that we are looking out to the limit of the visible universe. We are actually looking at the big bang. We see it spread over the whole sky as it was in its last moments at the decoupling epoch. The radiation that comes to us from the big bang has been mercifully redshifted by expansion into the far infrared and is invisible to the unaided eye. The "wall" though invisible is not nothing!

4 Most photons in the universe belong to the cosmic background radiation that survives from the big bang. Their total number remains almost constant and the entropy of the universe is also almost constant. The entropy per unit volume (call it s) equals $4/3T$ times the thermal energy aT^4 per unit volume, where T is temperature and a the radiation energy density constant, and s is therefore equal to $4aT^3/3$. Hence s/T^3 is constant, and because T varies as $V^{1/3}$, the total entropy $S = sV$ in volume V stays constant during expansion. The thermal radiation energy in V, equal to aT^4V, however, decreases at the same rate as T and is proportional to $1/V^{1/3}$. The energy in the cosmic background radiation, once very large, is now quite small. Where has this energy gone? Can you think of an answer that conserves total energy? (The author has tried and failed.) Do you think that the second law of thermodynamics is a better conservation principle in cosmology than the familiar conservation of energy principle?

SOURCES

Davidson, W. "Local thermodynamics and the universe." *Nature* 206, 249 (1965).

Dutta, M. "A hundred years of entropy." *Physics Today* (January 1968).

Harrison, E. R. "Olbers' paradox." *Nature* 204, 271 (1964). Introduction of the cosmic box method of treating radiation in the universe. More fully explained in "Why the sky is dark at night," *Physics Today* (February 1974).

Harrison, E. R. *Darkness at Night: A Riddle of the Universe.* Harvard University Press, Cambridge, Massachusetts, 1987.

Harrison, E. R. "Mining energy in an expanding universe." *Astrophysical Journal* 446, 63 (1995).

Jeans, J. *The Mysterious Universe*. Cambridge University Press, Cambridge, 1930.

Klein, M. J. "Maxwell, his demon, and the second law of thermodynamics." *American Scientist* (January–February 1970).

Lanczos, C. *Albert Einstein and the Cosmic World Order*. John Wiley, New York, 1965.

McVittie, G. C. "Distance and large redshifts." *Quarterly Journal of the Royal Astronomical Society* 15, 246 (1974).

Planck, M. *The Theory of Heat Radiation*. First published in 1913. Dover Publications, New York, 1959.

Tolman, R. C. *Relativity Thermodynamics and Cosmology*. Oxford University Press, Oxford, 1934. See Chapter 10, part 2.

18 THE MANY UNIVERSES

Hereafter, when they come to model Heav'n
And calculate the stars: how they will wield
The mighty frame: how build, unbuild, contrive
To save appearances . . .
John Milton (1608–1674), Paradise Lost

STATIC UNIVERSES

Before the twentieth century most Europeans and people of European descent believed the universe was created only a few thousand years ago, or at most a few hundred thousand years. Some people in the 18th and 19th centuries, more radical in outlook, thought the static Newtonian universe was in a steady state – everything remaining eternally unchanged – and the stars would shine endlessly. The realization in the late 19th century that stars have finite energy resources brought to an end the idea of a perpetually unchanging cosmos.

Static Einstein universe

In 1917, Einstein contrived an ingenious static universe using his recently developed theory of general relativity. In this universe, as in all universes we discuss, all places are alike and matter is distributed with uniform density.

Space and time in the new theory of general relativity had at last been awakened from the dead and become active participants in the world at large. Einstein, believing the universe to be static, tranquilized spacetime with a counteracting agent. In his 1917 paper, "Cosmological considerations on the general theory of relativity," he wrote, "I shall conduct the reader over the road I have myself traveled, rather a rough and winding road, because otherwise I cannot hope that he will take much interest in the result at the end of the journey. The conclusion that I shall arrive at is that the field equations of gravitation that I have championed hitherto still need a slight modification." The modification that he referred to was the introduction of the cosmological constant Λ. When this new constant is positive, it acts as antigravitational force that leaves unaffected the curvature of space. A universe made static at one moment, however, is not necessarily static at earlier and later moments. To ensure that the universe remained static, in a state of permanent equilibrium, the curvature of space must be positive. The static Einstein universe therefore has spherical space: it is closed and finite, and contains a mysterious Λ force that opposes the attraction of gravity.

If distances are measured in light-travel time, the radius of curvature of space in the Einstein universe equals the scaling factor R (see Figure 18.1). The distance around the universe, or the circumnavigation time of a light ray, is $2\pi R$. The antipode of an observer, or point on the opposite side of a spherical universe, is at distance πR.

A static spherical universe, when ideally smooth, acts as a giant lens. A body moving away appears at first to get smaller in the usual way; when halfway to the antipode, however, it ceases to get smaller, and thereafter, as it recedes, it appears to get bigger. Objects in the antipodal region are seen imaged as if they were close by in the local region (see Figure 18.2). People at the antipode see us as if we were close to them and we see them as if they were close to us. Because light rays circumnavigate the

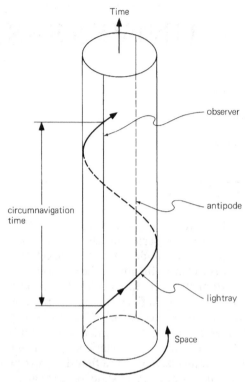

Figure 18.1. A spacetime diagram of the Einstein static universe, showing only one of the three dimensions of space. A light ray, moving in space and advancing in time, describes a helical path on the surface of a cylinder of radius R. (Note: the scaling factor is measured in units of time, and distances are measured in light-travel time and have also the same units as time.)

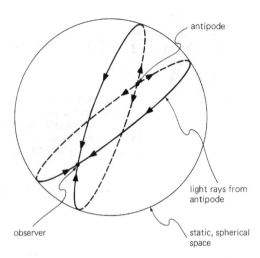

Figure 18.2. The Einstein static universe of spherical space. Light rays from the antipode first diverge and then converge, and we see objects at the antipode as if they were close to us.

cosmic globe we also see ourselves from behind.

The equations of the Einstein universe are obtained from the Friedmann–Lemaître equations (Chapter 16) by setting $H = 0$, $q = 0$; hence

$$K = 4\pi G\rho, \qquad [18.1]$$

$$\Lambda = 4\pi G\rho, \qquad [18.2]$$

where ρ is the smoothed density, curvature is $K = k/R^2$, and the curvature constant k is 1. The time that light rays take to travel once around the Einstein universe is given by the circumnavigation time

$$t_{\text{circ}} = 2\pi R = \left(\frac{\pi}{G\rho}\right)^{1/2}, \qquad [18.3]$$

where G is the universal constant of gravity. When density is measured in grams per cubic centimeter the circumnavigation time is approximately $\sqrt{(4/\rho)}$ hours. Water has a density of 1 gram per cubic centimeter, and in a universe filled with transparent water, the time light takes to circulate once is 2 hours. In this hydrocosmos, which is smaller than the Solar System and has a radius of curvature R of about 20 light minutes, antipodal objects are seen as they were 1 hour ago and observers see themselves as they were 2 hours ago. Light repeatedly circulates and the bemused inhabitants are reminded of what they were doing 2 hours ago, 4 hours ago, 6 hours ago, and so on. These Einsteinian creatures see their past in graphic detail and are deprived of the convenience of short memories. The circumnavigation time in an Einstein universe filled with a gas of density equal to that of our atmosphere is slightly more than 60 hours; and when the density is considerably less, the inhabitants observe the ghostly antics of their ancestors.

The distance around spherical space of radius R is $2\pi R$, or $2\pi c R$ in space units, but the volume of spherical space is $2\pi^2 c^3 R^3$, and not the familiar $4\pi c^3 R^3/3$ in flat space of Euclidean geometry. The total mass M of the universe is its volume multiplied by the density; thus $M = 2\pi^2 \rho c^3 R^3$, and with Equation [18.1], we find

$$R = \frac{2GM}{\pi c^3}. \tag{18.4}$$

In solar masses, this gives

$$R = 1 \times 10^{-13} \frac{M}{M_\odot} \text{ light years} \tag{18.5}$$

and in a universe of 10^{10} light years radius, the mass is 10^{23} times the Sun's mass, or a trillion times the mass of the Galaxy. The radius of a black hole is $2GM/c^3$ in light-travel time, and the radius of an Einstein universe of the same mass is therefore $1/\pi$ times smaller. Although it is sometimes said that closed universes are black holes, this is misleading because there is no external space and there are no external observers to say "a universe is over there" in the way they can say "a black hole is over there."

Arthur Eddington showed in 1930 that the Einstein universe is unstable, and its inhabitants must tread on tiptoe and speak in hushed voices (Figure 18.3). It stands on the edge of a razor: when nudged slightly one way, gravity begins to dominate and the universe collapses into a big bang in a time equal to a circumnavigation period; if nudged slightly the other way, the repulsive Λ force begins to dominate and the universe

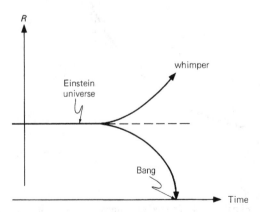

Figure 18.3. The Einstein universe is unstable, and when disturbed, it either collapses or expands.

expands forever into a whimper. If a person lights a match it will start collapsing, or if some radiation is absorbed it will start expanding.

Last static steady-state universe

Static universes neither expand nor contract. Steady-state universes, whether static or not, never change their cosmic appearance, and what happens now always has happened and will happen. Steady-state universes, in which the endless vistas of the future perpetuate those of the past, obey what many of their inhabitants out of sheer boredom might call the principle of cosmological monotony.

Static universes in a steady state neither expand nor collapse and their contents, on the average, never appear to change. Universes of this kind are not uncommon in the history of cosmology. They seem to be most popular in great empires that have gone into decline, in which men and women desire the perpetuity of past glory, and are keenly favored by aristocracies whose privileged members are of the umpteenth generation. The last of the static steady-state universes, strange to say, was conceived at the University of Chicago in 1918 and elaborated in the 1920s by the astronomer William MacMillan (1861–1948).

The foremost problem that besets any static universe in a steady state is the eternal brightness of the stars. The radiation streaming away from stars accumulates in space and the night sky cannot therefore remain dark. It was this celebrated problem – the puzzling darkness of the night sky (see Chapter 24) – that prompted MacMillan to think of the idea that atoms are "generated in the depth of space through the agency of radiant energy." He proposed the theory that stars are formed in the usual way out of interstellar gas; they then evolve over a long period of time and slowly radiate away their entire mass. Out in the depths of space, by an unknown mechanism, starlight is reconstituted into atoms of matter. The interstellar gas, thus continually replenished with newborn atoms, condenses to form new stars, which in turn melt away into radiation, thus maintaining a perpetual steady state. MacMillan in one great swoop was able to conserve energy and explain the darkness of the night sky.

MacMillan's perpetual motion universe was popular in certain sections of the public, and was enthusiastically adopted by the physicist Robert Millikan, famous for measuring in 1905 the electric charge of an electron. Millikan believed that cosmic rays, discovered in 1911 by Victor Hess, were the "birth cry" of newly created matter in the depths of space and were proof that "the Creator is still on the job."

We know now that a star radiates away at most only one percent of its mass in its luminous lifetime on the main sequence and cannot dissolve away into nothing in the manner proposed by MacMillan. A star has a fixed number of baryons (in this case nucleons), and however much it radiates at the expense of its nuclear and gravitational energy, the number of baryons remains constant (apart from particles escaping from the surface) and the star cannot dissolve entirely into radiation. We also know that matter cannot be recycled in the ingenious way proposed by MacMillan. Highly energetic radiation is capable of creating particles and antiparticles – particle pairs and not just particles by themselves – and these particle pairs then annihilate and convert back into radiation. Most starlight in space is far too weak to create particle pairs. Although total energy in MacMillan's static universe is conserved, total entropy is certainly not. A static, steady state of conserved entropy exists only in a universe in perfect thermal equilibrium in which there are no galaxies, stars, or planets, and also no arrow of time, and hence no way of distinguishing the past from the future. And, of course, there are no inhabitants. The last static steady-state universe enjoyed fame until the 1930s when, confronted with growing evidence of an expanding universe, it quietly faded away.

DE SITTER UNIVERSE
An expanding, empty, steady-state universe

The de Sitter universe, proposed in 1917 – the same year that Einstein proposed his static universe – was at first thought to have diminished the status of Einstein's cosmological theory. It consisted of flat space and was slightly absurd in the sense that it contained no matter. A space of Euclidean geometry without matter should not exhibit unusual properties. Yet, surprisingly, it springs to life when the cosmological term Λ is included.

The equations of the de Sitter universe are

$$q = -1, \qquad\qquad [18.6]$$

$$\Lambda = 3H^2, \qquad\qquad [18.7]$$

obtained by setting both the curvature K and the density ρ equal to zero in the Friedmann–Lemaître equations. Space expands exponentially rapidly and the Hubble term $H = (\Lambda/3)^{1/2}$ is constant. The repulsive effect of the Λ force causes space to expand with constant acceleration of $q = -1$, and the scaling factor R increases as shown in Figure 18.4. The de Sitter universe is in a steady state: the Hubble and deceleration terms are constant, and without matter, nothing appears to change.

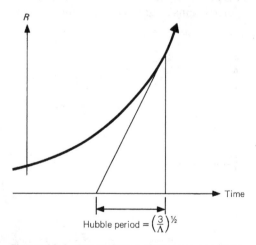

Figure 18.4. The de Sitter universe has an infinite past and an infinite future, and accelerates at $q = -1$.

The "steady-state universe" label is commonly applied only to the model proposed by Hermann Bondi, Thomas Gold, and Fred Hoyle that expands in the same way as the de Sitter universe, but contains matter, and instead of the matter being diluted by expansion, the matter is replenished by the continuous creation of new matter. But we should not forget that other universes, such as the de Sitter and MacMillan universes, are also in steady states.

The Einstein universe with its matter without motion and the de Sitter universe with its motion without matter were at first the leading cosmological models. Then, in 1927 and 1931, the Belgium cosmologist Georges Lemaître made known the general cosmological equations (first discovered and published by the Russian scientist Alexander Friedmann in 1922 and 1924 and forgotten) and cosmology entered a new era.

FRIEDMANN UNIVERSES

Alexander Friedmann

In the Friedmann family of universes, or models of the Universe, the Λ force is ignored and assumed to be zero. There are three basic types of Friedmann universe, corresponding to the values 0, 1, and −1 of the curvature constant k.

Alexander Friedmann (1888–1925), a Russian scientist of many interests, was professor of mathematics at the University of Leningrad. According to George Gamow, a onetime Friedmann student, Friedmann spotted an error in Einstein's 1917 cosmology paper, an error that had led Einstein to the conclusion that the universe is necessarily static when the Λ force is introduced. Friedmann wrote to Einstein about his own more general treatment, but received no reply. Through a colleague visiting Berlin, Friedmann succeeded in obtaining from Einstein a "grumpy letter" agreeing with his conclusions. Friedmann then published an article "On the curvature of space" in 1922 in the German journal *Zeitschrift für Physik*. A second article "On the possibility of a world with constant negative curvature" appeared in the same journal in 1924. Both articles were timely because of the discovery of extragalactic redshifts, but sadly, Friedmann's work was ignored and made no impact on contemporary cosmology. In honor of Friedmann's pioneering work we refer to uniform expanding universes of zero cosmological constant ($\Lambda = 0$) as Friedmann universes.

Friedmann equations

As shown in the Reflections, the Friedmann equations are

$$4\pi G\rho = 3qH^2, \qquad [18.8]$$

$$K = H^2(2q - 1), \qquad [18.9]$$

where $q = -\ddot{R}/RH^2$ is the deceleration term, $K = k/R^2$ is the curvature and the density varies with expansion as

$$\rho = \rho_0(R_0/R)^3, \qquad [18.10]$$

and the zero subscript denotes the present epoch. The three Friedmann classes of universe correspond to $k = 1$, 0, and −1. We see from Equation [18.8] that in the Friedmann universes the deceleration term q is always positive.

Einstein–de Sitter universe

The simplest of the Friedmann universes is the Einstein–de Sitter model (see Figure 18.5) having the flat space $k = 0$. This

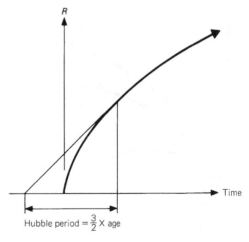

Figure 18.5. The Einstein–de Sitter universe in which $q = 0.5$.

means, according to Equation [18.9], that the deceleration term q is 0.5, and hence, according to Equations [18.8] and [18.10], $H^2 R^3 =$ constant, and by integration we find the scaling factor varies as

$$R = R_0 \left(\frac{t}{t_0} \right)^{2/3}, \qquad [18.11]$$

where t is the age of the universe. Hence the Hubble term is $H = 2/3t$. The age of the universe at any time is $t = 2t_H/3$ (two-thirds of a Hubble period $t_H = 1/H$). With a Hubble term $H = 100h$ kilometers a second per megaparsec, the Hubble period equals $9.8h^{-1}$ billion years, and the present age is therefore $t_0 = 6.6h^{-1}$ billion years.

Friedmann universes

The Einstein–de Sitter universe of $q = 0.5$ divides closed universes of positive curvature K from open universes of negative curvature K (see Figures 18.6 and 18.7). Thus we have

$q > \frac{1}{2}$: closed, spherical space $k = 1$

$q = \frac{1}{2}$: open, flat space $k = 0$

$q < \frac{1}{2}$: open, hyperbolic space $k = -1$.

From Figure 18.7 we see that the greater the deceleration, the shorter the age of the

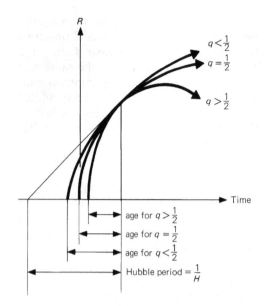

Figure 18.7. Comparison of Friedmann universes of the same Hubble term H but different values of the deceleration term q. The universes of higher q have shorter ages.

universe, and therefore

$q > \frac{1}{2}$: age $< t_{\text{crit}}$

$q = \frac{1}{2}$: age $= t_{\text{crit}}$

$q < \frac{1}{2}$: age $> t_{\text{crit}}$

where the age of the Einstein–de Sitter universe is

$$t_{\text{crit}} = \tfrac{2}{3}t_H = 6.6h^{-1} \text{ billion years.} \qquad [18.12]$$

With an expansion parameter $h = 0.5$, a closed universe has an age less than 13 billion years and an open universe has an age equal to or greater than 13 billion years.

We can define a critical density corresponding to $q = 0.5$ in Equation [18.8]:

$$\rho_{\text{crit}} = \frac{3H^2}{8\pi G}. \qquad [18.13]$$

The present critical density is $2 \times 10^{-29} h^2$ grams per cubic centimeter. We see

$$\rho = 2q\rho_{\text{crit}}, \qquad [18.14]$$

and Equation [18.9] becomes

$$K = H^2 \left(\frac{\rho}{\rho_{\text{crit}}} - 1 \right). \qquad [18.15]$$

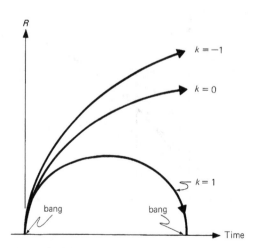

Figure 18.6. All Friedmann universes begin with big bangs.

Thus we can say

$q > \frac{1}{2}$: $\rho > \rho_{crit}$

$q = \frac{1}{2}$: $\rho = \rho_{crit}$

$q < \frac{1}{2}$: $\rho < \rho_{crit}$.

When $h = 0.5$, the critical density is 5×10^{-30} grams per cubic centimeter, and is about equal to 3 hydrogen atoms per cubic meter.

Closed Friedmann universes

All Friedmann universes decelerate while expanding, and when the deceleration term is greater than 0.5 they eventually cease to expand and commence collapsing. These closed expanding and collapsing universes are of great interest. The cycloid curve in Figure 18.8 shows how the scaling factor R changes in time. A cycloid is a curve traced by a point on the rim of a wheel rolling on a flat surface. The scaling factor R in these closed universes is the radius of curvature and the circumnavigation distance in light-travel time is $2\pi R$. Equations [16.19]–[16.20] show

$$R = A \sin^2 \phi,$$ [16.19]

$$t = A(\phi - \sin \phi \cos \phi),$$ [16.20]

where the constant $A = 8\pi G \rho_0 R_0^3 / 3$ is the maximum radius R_{max}. At maximum

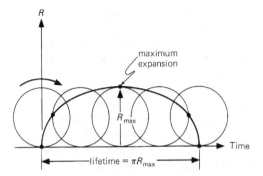

Figure 18.8. The closed Friedmann universe ($k = 1$) that begins and ends with big bangs. The curve shown is a cycloid and similar to that generated by a point on the rim of a rolling wheel.

expansion

$$R_{max} = R_0 \left(\frac{2q_0}{2q_0 - 1} \right),$$ [18.16]

where R_0 is the present radius and q_0 is the present value of the deceleration term. If $q_0 = 1$, then R_{max} is twice R_0 and the universe is halfway to maximum size. At maximum radius the density has a minimum value

$$\rho_{min} = \rho_0 \left(\frac{2q_0 - 1}{2q_0} \right)^3,$$ [18.17]

thus if $q_0 = 1$, the minimum density is one-eighth the present value. Although the universe at maximum expansion is momentarily stationary ($H = 0$), it is not in the least like the Einstein static universe. For universes of the same mass we easily find that R_{max} is only two-thirds the radius of the Einstein universe. We find also that the radius is $2/3\pi$ the radius of a black hole of the same mass.

The total lifetime of a closed Friedmann universe, from bang to bang, is πR_{max}, or half the circumnavigation time at maximum expansion. We find, by juggling with the equations that

$$\text{lifetime} = t_H \frac{2\pi q_0}{(2q_0 - 1)^{3/2}},$$ [18.18]

and when $q_0 = 1$, the lifetime is 2π Hubble periods, or about 120 billion years for $h = 0.5$. The half-age, or time to reach maximum size, is 60 billion years and will be reached in about 47 billion years. If this is our universe, its present age is still only about 10 percent of its total lifetime; its constituents, however, are already middle-aged, and by the time the universe begins to collapse, galaxies will have grown old and dark.

Open Friedmann universes

Friedmann universes of q less than 0.5 have hyperbolic geometry; they extend in space to infinite distances and have an infinite lifetime of expansion. Their density is less than the critical density and their age is greater than the critical age. According to Equation [18.15], when the density is much less than the critical density, then approximately

$$K = -H^2,$$

hence, with $K = -1/R^2$, we find that $HR =$ constant, and R increases linearly with time t. All Friedmann universes of hyperbolic geometry eventually expand linearly with time when the density gets low. More exactly, from Equations [16.21]–[16.22],

$$R = A \sin h^2 \phi, \qquad [16.21]$$

$$t = A(\sinh \phi \cosh \phi - \phi), \qquad [16.22]$$

as shown in Figure 16.8, where $A = 8\pi G\rho_0 R_0^3/3$.

OSCILLATING UNIVERSES

Universes that begin and end in big bangs were once referred to as oscillating universes (see Figure 18.9). Because the universe had expanded phoenix-like out of one big bang, it was thought that it should also expand out of the next big bang, and so on, repeatedly, bouncing from bang to bang. According to this picture, the universe has already oscillated an indefinite, perhaps infinite, number of times and will continue to oscillate endlessly in the future. A closed universe, which we thought existed for only a finite lifetime, has acquired an infinite lifetime by perpetual reincarnation.

Yet each period of oscillation cannot be exactly the same as the previous period, as was shown by Richard Tolman in 1932 and 1934. Stars and other luminous sources pour out radiation and the number of

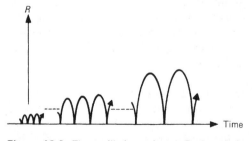

Figure 18.9. The oscillating universe. Each cycle is slightly larger than the preceding cycle because of the growth of entropy.

photons in space increases during any period. The amount of radiation, or entropy, in one cycle is thus slightly greater than in the preceding cycle. Despite the devastating nature of a big bang and its obliteration of all structural detail, we can at least claim that each bang is a little hotter than the preceding bang because of the inexorable increase in total entropy. It is like inflating a tire with a bicycle pump; the pump steadily gets warmer. In the next cycle of our universe the specific entropy (photons per nucleon) will have risen by about 0.001. Tolman showed that because of this slow and steady growth in background radiation, the universe expands to a slightly greater size in each succeeding cycle and the lifetime of each cycle increases.

The cosmic background radiation, now at a low temperature of almost 3 kelvin, will get steadily hotter in future cycles until eventually galaxies cease to form and stars cease to exist. Thereafter, in these future bright-sky starless cycles, the growth in radiation will be much slower and the amplitude and period of the oscillations will change very slowly. Life as we know it will have ceased to exist. Back in the past the bangs were cooler and most hydrogen burned into helium during each bang. Thus luminous stars, starved of hydrogen fuel, were short-lived, and there was not sufficient time for biological evolution. If you were to ask: why are we so fortunate to live in the present hospitable cycle? – the answer would be that in an oscillating universe of an infinite number of cycles there is only a finite number of cycles that can be occupied by life, and we naturally live in one of those habitable cycles.

Although oscillating universes form a fascinating subject, we must treat with considerable reservation the whole idea of a universe preserving its basic identity through a series of big bangs. We have no knowledge of what happens ultimately in a big bang when the universe reverts to indescribable primordial conditions. The only thing we can be sure about is that the universe preserves its topology; if it is spatially closed, it remains closed, and if it is open, it remains open; a finite universe cannot become infinite, and an infinite universe cannot become finite.

Perhaps a cosmogenic genie lurks in the primordial big bang and conjures up multitudes of universes, each equipped with unique laws and fundamental constants. What comes out of a big bang may have no relation to what goes in.

FRIEDMANN–LEMAÎTRE UNIVERSES
Georges Lemaître

Georges Lemaître (1894–1966) was ordained as a priest in 1922, and in 1927, the year he received his doctorate degree at the Massachusetts Institute of Technology, he published in Belgium his major work on the expansion of the universe. In the midst of discussions in the late 1920s on the merits of the Einstein and de Sitter universes, Lemaître's work at first went unnoticed, until Eddington drew attention to it by arranging for its translation into English.

Lemaître was the first to advocate an initial high-density state that he called the *primeval atom*, and he is therefore said to be "the father of the big bang." He rediscovered the cosmological equations that had been developed earlier by Friedmann, and we shall refer to these more general equations that include the cosmological constant as the Friedmann–Lemaître equations.

Friedmann–Lemaître equations

The equations that apply to most zero-pressure universes based on general relativity and the cosmological principle are

$$K = 4\pi G\rho - H^2(q + 1), \qquad [18.19]$$

$$\Lambda = 4\pi G\rho - 3qH^2, \qquad [18.20]$$

and we must add Equation [18.10] showing how the density changes as $1/R^3$. When the cosmological term Λ is made zero, these equations reduce to the Friedmann equations (Equations 18.8 and 18.9). The motionless Einstein universe and the matterless de Sitter universe automatically come from these equations; thus in the static Einstein universe (Equations 18.1 and 18.2), we have $H = 0$, and in the empty de Sitter universe (Equations 18.6 and 18.7), we have $\rho = 0$, $K = 0$.

Lemaître universe

Lemaître singled out from the many solutions of the Friedmann–Lemaître equations a closed universe (K positive) containing a repulsive force (Λ positive). This universe has the same basic ingredients as the Einstein universe, with the important difference that Λ is slightly greater than the value chosen by Einstein, and the Lemaître universe therefore cannot attain a static state. It starts as a big bang and evolves through two stages of expansion, as shown in Figure 18.10. Expansion in the first stage decelerates because gravity is stronger than the repulsion of the Λ force and the radius of the universe slowly approaches the Einstein radius. Expansion in the second stage accelerates because the repulsion of the Λ force is now stronger than the attraction of gravity. The Lemaître universe begins as a big bang, finally becomes a whimper, and in between hesitates and loiters as a sort of Einstein universe. It neatly combines the properties of the rival Einstein and de Sitter universes: it is closed and has cosmic repulsion like the Einstein universe; under the urge of the

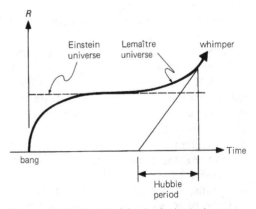

Figure 18.10. The Lemaître hesitation universe that begins as a big bang, passes through a hesitation stage, and then becomes a de Sitter universe.

repulsion force, it expands into a whimper like the de Sitter universe. Lemaître thought that galaxies formed during the hesitation era.

Lemaître's "hesitation universe" was popular and had something to offer everyone. Its great virtue was that in the second stage its age is greater than a Hubble period. By adjusting the cosmological constant so that it exceeds only slightly the Einstein value, we can greatly extend the hesitation (or loitering) period. Until the 1950s cosmologists believed that the Hubble period was 2 billion years, and consequently a universe of prolonged age owing to delayed expansion was very attractive. Subsequent downward revisions in the value of H reduced the attraction of the Lemaître universe.

Late in the 1960s the Lemaître universe was revived to explain why quasars appeared to have redshifts concentrated near the value of 2. The idea was that quasars formed during a long hesitation period and had a maximum redshift of 2 because the radius of the universe during hesitation was one-third its present value. Note that

$$z = \frac{R_0}{R} - 1,$$

and with $R_0 = 3R$, the redshift z has a value 2. Possibly, because light circles the universe more than once during a long hesitation period, quasars and radio sources would produce multiple images.

Numerous quasars have since been found and the clustering of redshifts at $z = 2$ is now much less pronounced. Many quasars have redshifts much greater than 2 and the evidence for a hesitation era is less impressive. The trouble with a long hesitation period is that it suffers from the fatal instability that besets the Einstein universe. The hesitation period has been described as an age ruled by Titans, in which galaxies emerge and quasars reign supreme, an age in which tumultuous events shake the universe. Hence hesitation cannot greatly extend the age of the universe.

Eddington universe

Lemaître was attracted by the creation implication of a big bang, presumably for religious reasons, whereas Eddington was repelled by the creation implication and thought a big bang was esthetically unattractive. Both men worshipped in different temples of cosmology and to this day there exist two main cosmological cults: the "bangers" (now in the majority) and the "antibangers" (now in the minority). Instead of believing in an abrupt beginning, Eddington in 1930 professed faith in a universe where "evolution is allowed an infinite time to get started," which is "necessary if the universe is to have a natural beginning." The Eddington universe (see Figure 18.11) is closed and has a cosmological constant equal to the Einstein value. It exists initially for an infinite or at least indefinite period of time as a static Einstein universe; then, because of random disturbances, it awakes from a long sleep and begins to expand. (It might instead collapse and then would not be our universe.) It exists embryonically in the Einstein state in which gravity and repulsion are exactly balanced, and later in the active de Sitter state in which repulsion dominates over gravity. Like the Lemaître universe, the Eddington universe conveniently combines the Einstein and de Sitter universes that had earlier preoccupied cosmologists. Eddington discovered the instability of the Einstein universe and it

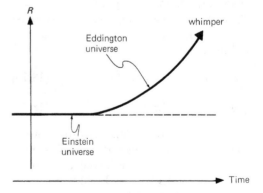

Figure 18.11. The Eddington antibang universe that begins as an Einstein universe, and then, on being disturbed, becomes a de Sitter universe.

seems odd that he favored an unstable universe existing initially for an indefinitely long period of time. No galaxy formed and no life evolved in that precariously balanced dormant world that suddenly awoke only 10 or so billion years ago.

Eddington felt forced to postulate an infinite or an indefinitely long past to exorcise the specter of a primordial beginning. He was the first but not the last of modern cosmologists to be repelled by cosmic birth and death. His preferred world exists in a slumber state; it awakes; ages gracefully; and ends in a whimper. But however one twists and turns there is no escape from the implacable law of cosmogenesis, discovered by St Augustine of Hippo in the fourth century (Chapter 25). Creation cannot be set aside as an event within time that occurred in the infinite past, for the universe contains time, and time, like space, whether finite or infinite, is created with the universe. Nowadays, the Augustinian law of cosmogenesis derives from the containment principle.

The Lemaître and Eddington worlds have a common virtue. They contain the cosmological constant that serves as a yardstick by which everything is measured. How do particles know what size to be? How do the fundamental constants know what values to have? According to Eddington, the cosmological constant determines in a natural but unknown way the scale of the universe.

CLASSIFICATION OF UNIVERSES
Various classifying schemes exist, some of which have been discussed in a preliminary way in Chapter 14. Here we consider the geometric, kinematic, and dynamic classifications.

Geometric
The geometric system is based on curvature: spherical and closed ($k = 1$), flat and open ($k = 0$), and hyperbolic and open ($k = -1$).

From a popular viewpoint the main geometric distinction lies between the spatially closed and open universes. Closed (finite) universes have a curvature constant $k = 1$,

and open (infinite) universes have either $k = 0$ or $k = -1$. This primary topological difference between finite and infinite space as a method of classification was stressed in 1931 by Ernest Barnes, bishop of Birmingham, in a cosmological discussion entitled "Evolution of the universe" at a British Association meeting. He expressed the view, "It is fairly certain that our space is finite, though unbounded. Infinite space is simply a scandal to human thought," for only in a closed universe could we hope to understand the "range of God's activity."

Kinematic
The kinematic system of classification ignores geometric and dynamic considerations and asks: Is a particular universe expanding, static, or collapsing? We have previously used the bang and whimper terminology, and must now extend this system to include static and oscillating states. By combining the bang, static, and whimper states in all realistic combinations we obtain fourteen classes:

1. bang–bang   Friedmann ($k = 1$)
2. bang–static
3. bang–whimper   Friedmann ($k = 0, -1$)
4. bang–static–bang
5. bang–static–   Lemaître whimper
6. static   Newton ($k = 0$), Einstein ($k = 1$)
7. static–bang
8. static–whimper   Eddington ($k = 1$)
9. whimper   de Sitter ($k = 0$)
10. whimper–bang
11. whimper–static
12. whimper–whimper
13. whimper–static–bang
14. whimper–static–whimper.

The manner in which the scaling factor varies with time in these various classes is shown in Figure 18.12. Our universe is now expanding, and the acceptable nine classes are hence 1, 2, 3, 4, 5, 8, 9, 12, and 14; the remaining

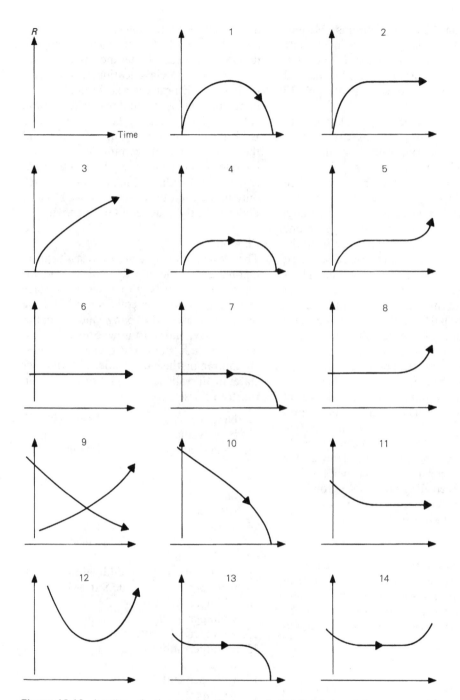

Figure 18.12. A gallery of universes classified according to their bang–static–
whimper states.

five classes 6, 7, 10, 11, and 13 are ruled out because they have no expansion periods. Some classes, such as 1, 6, and 12, may exhibit oscillatory properties.

Dynamic

The third system is dynamic and based on the parameters of the Friedmann–Lemaître equations. To each of the three values of the curvature constant k, the cosmological constant Λ can have two specific values:

A. Λ equals the Einstein value ($\Lambda = \Lambda_E$)
B. Λ equals zero ($\Lambda = 0$)

and three significant ranges of value:

C. Λ greater than the Einstein value ($\Lambda > \Lambda_E$)
D. Λ greater than zero but less than the Einstein value ($0 < \Lambda < \Lambda_E$)
E. Λ less than zero ($\Lambda < 0$).

Fifteen classes are thus possible. The classes A, C, and D, for $k = 0$ and $k = -1$, are similar and we are left with eleven distinct classes, as shown in Figure 18.13. In the classes A, C, and D, the Λ force is repulsive and opposes gravity, and in E the Λ force

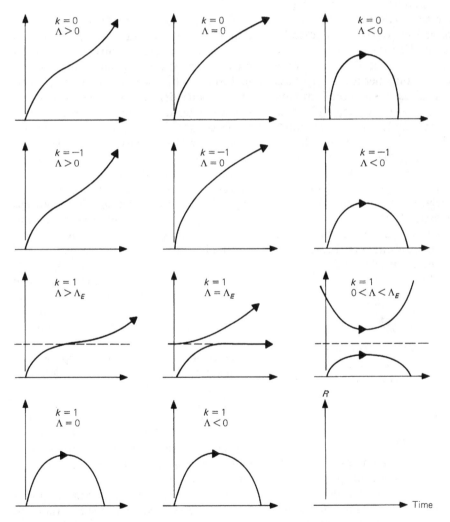

Figure 18.13. A gallery of universes classified according to their curvature and cosmological constants.

is attractive and augments gravity. We notice that not all kinematic classes are possible in the dynamic classification scheme.

UNIVERSES IN COMPRESSION
Pressure increases deceleration
A pressure gradient in a star opposes the pull of gravity. An outward force exists because pressure increases toward the star's center. A pressure difference, or pressure gradient, causes water to move in a pipe and drives a piston in a steam engine. But in a uniform universe, where all places are alike and pressure is everywhere the same, no pressure difference exists and no forces are exerted.

Because pressure in the universe produces no net force, at least on the cosmic scale, how then does it affect the expansion of the universe? We must realize that pressure is a form of energy, and energy – or its mass equivalent – acts as a source of gravity. When a sealed vessel containing gas is heated, the gas molecules move more rapidly, and pressure increases. This is another way of saying that energy in the container has increased. A hotter gas has a greater mass and weighs more than when colder. Similarly, a universe containing pressure has increased gravity.

In cosmology, we gauge the importance of pressure by its equivalent energy density. The nonrelativistic formula for the speed of sound in a fluid is

$$\text{sound speed} = \left(\frac{\gamma P}{\rho} \right)^{1/2},$$

where P is the pressure, ρ is density, and γ the ratio of specific heats (which in a perfect nonrelativistic gas has a value 5/3). From this we see

$$\frac{\text{sound speed}}{\text{light speed}} = \left(\frac{\gamma P}{\rho c^2} \right)^{1/2}, \qquad [18.21]$$

where ρc^2 is the total energy in a unit volume. The importance of pressure, as compared with energy density, depends on the ratio of the speeds of sound and light. When the sound speed is relatively small, as in the world around us, the pressure is

also relatively small. The sound speed in a gas is about equal to the average speed of the constituent particles, which normally move much slower than light. Most of the time the pressure in cosmology is negligible. But in the early universe the temperature is high, particles move at relativistic speeds, and pressure is not negligible.

Universes that contain only radiation have the maximum pressure for a given energy density. When radiation is uniform – and radiant flux is isotropic – the pressure is equal to one-third the energy density:

$$P = \tfrac{1}{3} \rho c^2.$$

Richard Tolman in the early 1930s found that a universe containing only radiation behaves much like a universe containing only matter of low pressure, but with one important difference: a radiation universe of the same density as a matter universe has a greater deceleration. This is because the large radiation pressure acts as an additional source of gravity and the expansion slows down more quickly than in the matter universe. We saw that a closed Friedmann universe containing zero-pressure matter expands to a maximum size R_{max} and has a lifetime πR_{max}. A closed Friedmann universe containing radiation only expands and collapses in the same way and has a finite lifetime (see Figure 18.14). If we suppose that R_{max} is the same for the two universes, the radiation universe has a lifetime $2R_{max}$, which is shorter than the lifetime of the matter universe because of the greater deceleration.

Thus pressure has an effect opposite to what we might expect. Common sense suggests that pressure in an expanding universe should hasten the expansion. And in a collapsing universe pressure should slow the collapse, and even arrest the collapse when sufficiently great. But instead, pressure does the opposite; it causes slower expansion and faster collapse.

This unexpected result is because a uniform universe has no pressure gradients; furthermore, unlike a boiler, the universe has no walls against which pressure can

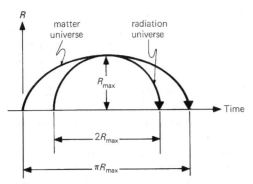

Figure 18.14. Two closed universes, one containing radiation only and the other containing matter only. When they both expand to the same maximum radius R_{max}, and therefore have the same density at maximum expansion, the radiation-dominated universe has the shorter lifetime.

push, and the only remaining dynamic effect of pressure is its contribution to the gravitational forces that control the universe. Uniform radiation, for example, produces a gravitational force that is twice as strong as that produced by zero-pressure matter of the same mass density.

The Einstein–de Sitter universe, having flat space and no cosmological constant, is convenient for studying the effect of pressure. When it contains matter of zero pressure the scaling factor R is proportional to $t^{2/3}$, where t is the age measured from the big bang. We have previously seen that

$$t = \tfrac{2}{3} t_H, \qquad [18.22]$$

where $t_H = 1/H$ is the Hubble period. The deceleration term q is constant and equal to 0.5. In a similar universe containing radiation only, the scaling factor R is proportional to $t^{1/2}$, and the age is reduced to

$$t = \tfrac{1}{2} t_H, \qquad [18.23]$$

and the deceleration term has increased to $q = 1$.

Now consider a more realistic universe, such as our own, which contains both matter and radiation. The principal contribution to the radiation in our universe is the cosmic background radiation. Starlight and radio waves are minor additions. Matter at the present epoch is more dense than radiation

by a factor of about 1000, and our universe is matter-dominated. In Chapter 17 we saw how the density of zero-pressure matter varies as $1/R^3$, whereas the density of radiation varies as $1/R^4$. In the past, when R was more that 1000 times smaller, the density of radiation was greater than the density of matter. Far back in time, at redshifts greater than 1000, the universe was radiation-dominated. Universes containing cosmic background radiation, however little, are radiation-dominated at some time in their early history.

UNIVERSES IN TENSION
The strange worlds of negative pressure

We are all familiar with positive pressure, as in stars and steam engines, and the notion of negative pressure is at first startling. The cosmologist William McCrea argued in 1951 that negative pressure in the universe, equivalent to a state of cosmic tension, cannot be ruled out on the basis of our normal experience. A cosmic tension, everywhere the same and therefore without gradients to pull things around, does not participate directly in determining the behavior of galaxies, stars, and steam engines. We are aware of pressures when they vary from place to place, as in the Earth's atmosphere and oceans, and when they act on moving walls, as in steam and internal combustion engines. But when pressure is everywhere the same, without gradients and unconfined, it has no perceptible effect except as a source of gravity. The same may be said of a negative pressure; it may exist, but we cannot detect it except in the way it affects the behavior of the universe.

Our discussion on the Einstein equation showed that the left side of the equation represents curved and active spacetime and the right side represents "matter." Whereas the left side is crystal clear, the right side is murky because nobody knows exactly what "matter" comprises in this context. On occasions, very strange things have been placed on the right side. Customarily, the right side is kept simple and tidy with only the density of matter, the density of field energies, positive pressure, and things of the familiar world. But the universe is more than the familiar world and is not a star or a steam engine. A uniform cosmic stress of the nature of a negative pressure, having no effect on the structure of planets, stars, and galaxies, is quite unfamiliar and usually not considered. But, argued McCrea, the universe is perhaps governed by forces not directly manifest in laboratory experiments and astronomical systems, and we therefore cannot rely on normal experience to tell us what should be on the right side of the Einstein equation in cosmology.

We have seen how the cosmological term greatly enlarges the range of possible cosmological models; we find that a negative cosmic stress enlarges even further the range of possibilities. A convenient relativistic equation of state (an equation that relates pressure and density) is

$$P = (\gamma - 1)\rho c^2, \qquad [18.24]$$

where γ is a constant more general than the ratio of specific heats. Usually, in cosmology, $\gamma \geq 1$; for zero-pressure ("dust-filled") universes, $\gamma = 1$; for universes containing only thermal radiation, $\gamma = 4/3$.

Consider the effect of negative pressure ($\gamma < 1$) in an expanding universe. Such a universe in a state of tension releases energy as it expands. A piece of elastic – representing a one-dimensional universe – when stretched becomes warm because the work done by stretching releases thermal energy. This added energy has mass. A similar thing happens in a universe of negative pressure; during expansion, energy is released and this added energy might take various forms and even include newly created particles.

In a universe in which the tension equals the energy density ($\gamma = 0$), the equation of state is

$$P = -\rho c^2, \qquad [18.25]$$

and the energy released by expansion is sufficient to maintain a constant mass density. If the continually created energy is in the form of matter, we have the astonishing

situation of an expanding universe of constant mass density. Expansion creates matter and the density stays constant. This is McCrea's ingenious explanation of the continuous creation of matter in the Bondi–Gold–Hoyle steady-state universe. In 1951, McCrea wrote: "the single admission that the zero of absolute stress may be set elsewhere than is currently assumed on somewhat arbitrary grounds permits all of Hoyle's results to be derived within the system of general relativity theory. Also, this derivation gives the results an intelligible physical coherence." Following Alan Guth's suggestion in 1981, expansion at constant density is now referred to as inflation, and the cosmic tension in the inflationary universe proposed by Guth is attributed to the "false vacuum."

If we abandon the belief that pressure must always be positive, and accept the possibility of a cosmic tension, we are confronted with a bewildering array of new universes (Figure 18.15). The following are

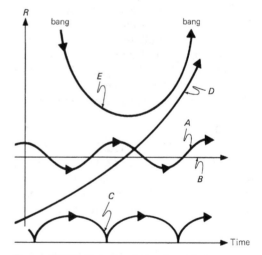

Figure 18.15. Universes in tension, where pressure is $P = (\gamma - 1)\rho c^2$, ρ is the mass density, and γ is a constant. (A) A closed, oscillating universe in which $0 < \gamma < 2/3$; (B) a closed, static, stable universe in which $0 < \gamma < 2/3$; (C) a closed, constant-density, oscillating universe in which $\gamma = 0$; (D) a flat, constant-density, steady-state universe in which $\gamma = 0$; and (E) all open and closed universes in which $\gamma < 0$, which expand and become not whimpers but big bangs.

examples. When the tension equals one-third the energy density,

$$P = -\tfrac{1}{3}\rho c^2, \qquad\qquad [18.26]$$

($\gamma = 2/3$) gravity ceases to have its usual effect and the universe is controlled only by the cosmological constant. When the tension is less than the energy density but greater than one-third the energy density ($2/3 > \gamma > 0$), or

$$\tfrac{1}{3}\rho c^2 < -P < \rho c^2, \qquad\qquad [18.27]$$

universes oscillate in size without attaining big bangs; their oscillations slowly diminish in amplitude and they become stable static universes. When tension equals the energy density ($\gamma = 0$), universes expand and contract at constant density. Furthermore, when the tension exceeds the energy density ($\gamma < 0$), we meet the incredible universes where density increases with expansion and decreases with contraction: whimpers expand to big bangs, and big bangs contract to whimpers.

WORLDS IN CONVULSION
The homogeneity riddle
Homogeneity – meaning all places are alike at the same time – is the most remarkable feature of the universe. Because the observed universe is remarkably isotropic, we feel compelled by the location principle to conclude that the universe is much the same everywhere. Widely separated regions, billions of light years apart, not only are similar in content but also are synchronized in their expansion. We would feel more comfortable with cosmic homogeneity – referred to as the cosmological principle – if only we could understand why it exists. The homogeneity riddle is discussed in Chapter 22 on inflation. Here we comment briefly on this subject.

When explaining anything we naturally look back to earlier times simply because causes precede effects. Can homogeneity be explained by looking back to a time when the universe was younger, and physical processes existed that created homogeneity out of inhomogeneity like the processes that create a calm sea after a storm?

The creation of calm in an initially stormy universe encounters a serious problem. The observable universe is roughly the Hubble sphere of radius $10h^{-1}$ billion light years, and light from things at greater distances has not reached us because the universe is not yet old enough. Regions of the universe cannot influence one another when they are separated more widely than a distance about equal to the Hubble length. As we look back in time, seeking the cause of homogeneity, the Hubble length shrinks, and regions now free to interact were once isolated from one another.

We saw in Chapter 14 that the rate of increase of the Hubble length L_H is

$$\frac{dL_H}{dt} = c(1 + q). \qquad\qquad [18.28]$$

Things at the Hubble distance, receding at velocity c, are overtaken by the boundary of the Hubble sphere receding at velocity $c(1 + q)$. We see that in a decelerating universe ($q > 0$), the Hubble sphere expands in the comoving frame and regions now able to interact inside the Hubble sphere were once outside the Hubble sphere and isolated from one another. If regions free to interact now were once isolated from one another, how can we hope to explain homogeneity by seeking its cause in the past? This is the homogeneity riddle (or horizon riddle). In the popular decelerating universes we are denied any kind of large-scale explanation that looks to causes acting in the past.

Attempts to solve the homogeneity riddle divide into two schools: the "chaos school" and the "antichaos school."

Chaos school
The chaos school follows mythological traditions and holds that initially the universe is amorphous (without form) and in a state of true chaos. Here chaos is used in the sense of total disorder, and not in the sense of apparent disorder, as found in deterministic nonlinear dynamical systems such as the weather. According to mythology, in the beginning when "heaven above

and earth below had not been formed" there existed indescribable chaos. By the operation of natural or supernatural causes there emerged out of chaos a state of order. The possibility that uniformity emerges from chaos has been investigated by Charles Misner of Maryland University.

The "chaoticists" seek to solve not only the problem of homogeneity but also other problems such as the origin of the cosmic background radiation and the initial conditions of galaxy formation. In this picture, the homogenizing processes, whatever they may be, release energy that heats the big bang, and the cosmic background radiation is the present-day evidence of this mechanism. From the initial inhomogeneity, small density fluctuations survive and later develope into galaxies.

Homogeneous but anisotropic universes have been studied to see whether the anisotropy decays into a state of isotropy. These "mixmaster universes" thrash backward and forward in giant convulsions; they expand in one direction while oscillating rapidly in the other two directions (imagine a cylindrical body oscillating in radius while expanding along its axis), and repeatedly, era after era, each era lasting longer than the previous era, the directions of expansion and oscillation interchange. The problem is how much the contained matter and radiation will dampen the convulsions by dissipative processes. It seemed at first that neutrinos in the early universe could abate the convulsions and create a state of isotropy, but investigations of a more general kind have shown that it is not possible to attain a high degree of isotropy from a preceding state of extreme anisotropy.

Inflation as a homogenizing mechanism conforms to the chaos school of thought. We can imagine a highly irregular surface representing primordial chaos. How, in this analogy, can we make the surface smooth? In the inflation picture we don't. Instead, we take a tiny region and distend (inflate) it to a size trillions of times larger. Its irregularities are stretched and smoothed into a vast flat surface.

Accelerated ($q < 0$) expansion in the very early universe caused by a state of tension solves the homogeneity riddle. Considerable inhomogeneity on large scales may exist in the extreme early universe before inflation. Because of causal interactions, a more-or-less homogeneous state exists on scales smaller than a Hubble length. During inflation, a Hubble sphere stays constant in size ($dL_H/dt = 0$), but its quasi-homogenized contents expand rapidly and form a homogeneous region vastly larger than a Hubble sphere. The expansion, in effect, stretches and smoothes out the original irregularity. Later, when inflation ceases and deceleration ($q > 0$) commences, the Hubble sphere begins to expand and overtake regions of the universe already in a homogeneous state. This neat solution of the homogeneity riddle was proposed in 1981 by Alan Guth, now at the Massachusetts Institute of Technology.

Antichaos

The antichaos school believes that the universe begins in a homogeneous state. This seems at first a simple-minded way of evading the issue, but actually it springs from a deep well of thought. We explain the state of individual things by seeking causes active in the past. But the universe is a unity, embracing space and time, and such a procedure is questionable and might even be inappropriate when trying to explain the design of the universe. The universe is not created in time any more than it is created in space – it contains time and space – and therefore we should not seek the cause of its design at a point in time any more than we would seek for it at a point in space. Homogeneity is perhaps fundamental and indispensable, and without it there might not be a universe. Furthermore, without it, we might not exist even if the universe did. This latter aspect of the problem, involving the anthropic principle, was considered by C. B. Collins and Stephen Hawking who wrote in 1973, "The fact that we have observed the universe to be isotropic is only a consequence of our existence." They showed that an inhomogeneous universe might not contain galaxies because of the unsuitability of the initial conditions, and hence stars and living creatures would not exist. There may exist many universes, some homogeneous and some inhomogeneous, and observers exist in only those that are homogeneous.

Inhomogeneous universes

Here we should mention a few of the many inhomogeneous universes that have been considered. Hierarchical universes, which have been prominent in the history of cosmology, have been discussed previously (Chapter 7). Other proposed universes, although homogeneous on scales less than a Hubble length, are inhomogeneous on larger scales. Widely separated unsynchronized regions, or "island universes," expand and contract independently. Remote regions far beyond the reach of observation are not necessarily synchronized with our region and may be contracting. Collapsed regions in a common spacetime form giant black holes that hide behind event horizons. Vast hierarchies of black holes may exist inside one another, each a world of internal activity seen in frozen stasis by the world outside.

The observed isotropy of the cosmic background radiation indicates a high degree of uniformity in our region, and all speculations on inhomogeneity are now limited to remote regions far beyond our observed region. The universe is isotropic to 1 part in 100 000, and therefore roughly these remote regions lie at least 100 000 Hubble distances away. If the universe is infinite in space, then at distances of thousands of trillions of light years, and perhaps even trillions of trillions of light years, there may exist unsynchronized regions, but we are not in a position to know.

KINEMATIC RELATIVITY
Milne's Stoic universe and his search for the explanation of gravity
Edward Milne of Oxford University, a famed astrophysicist and cosmologist,

turned a penetrating eye on general relativity and was not impressed. In 1948, two years before he died, he wrote in *Kinematic Relativity*: "Motion imposed in consequence of a geometry differing from the geometry commonly used in physics was a credible notion. Gravitation as a warping of space was a credible notion, though it gave not the least hint as to the nature or origin of gravitation; why the presence of matter should affect 'space' was left unexplained." Einstein's theory assumes that "geometry" and "matter" are linked together; Milne doubted the truth of that assumption,

Milne constructed his own theory of the universe, known as kinematic relativity, in which gravity is not included as an initial ingredient. With a supposedly small number of axioms, such as the cosmological principle and the rules of special relativity, he sought to create a picture of the universe that explained gravity and other laws of nature. He believed that the purpose of cosmology is to explain why things are as they are and not give elaborate descriptions of how things work. When measured by what he hoped to attain, his efforts failed; when measured by what he actually accomplished, however, his methods and insights succeeded in making considerable impact on cosmology.

Milne's picture of an expanding universe, when reduced to its simplest elements, is much easier to understand than general relativity. His universe consists of a spherical cloud of particles (galaxies) that expands within flat space, which is infinite and otherwise empty. It is the old Stoic cosmos, with center and edge, updated and made to conform to the rules of special relativity. Its expansion begins at a point in space; particles are shot out in all directions with velocities ranging from zero to close to that of light; and the edge of the cloud expands in space at a velocity close to that of light. About each particle the distribution and recession velocities of all other particles is isotropic in the frame of that particle. Owing to the relativity effect, most particles are crowded close to the edge of the cloud,

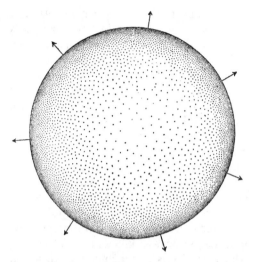

Figure 18.16. Milne's bounded universe expanding in space.

as shown in Figure 18.16. This finite and bounded universe, said Milne, has a large number of "particles in the field of view of any observer, merging towards the limit of visibility into a continuous background." In this descriptive framework of motions (hence the name kinematic relativity), where particles move freely unaffected by forces of any kind, Milne attempted to show that each particle manifests a behavior that simulates the effect of gravity. He wanted to explain gravity by starting with a cosmic framework that did not assume the existence of gravity. Few readers found his arguments convincing and the significance of his cosmological theory is still obscure.

Milne identified each particle with a galaxy. Because there are an "infinity of particles," his universe has an infinite mass. His finite and bounded universe can be transformed mathematically (by changing the intervals of space and time) into an infinite and unbounded universe. This new universe obeys the Robertson–Walker metric; it consists of expanding space, is homogeneous and isotropic, has negative curvature, and is uniformly populated with galaxies. This transformation of space and time makes the galaxies stationary in

expanding space, and the big bang no longer exists as a point in space.

Having transformed Milne's picture into a universe of dynamic, curved space, we are able to look at it more closely from the point of view of general relativity. In the Friedmann–Lemaître equations we make the cosmological constant Λ zero; also, because gravity on the cosmic scale was not required by Milne, we make the gravitational constant G zero. From Equations [18.19] and [18.20] we find

$$K = -H^2, \qquad [18.29]$$

$$q = 0, \qquad [18.30]$$

where $K = k/R^2$ and k is the curvature constant. Hence $k = -1$; thus space has negative curvature and is infinite in extent. We see that $\dot{R} = 1$, and therefore $R = t$, where t is the age of the universe. This universe expands at a constant rate ($H = t^{-1}$, $q = 0$), and the Hubble period equals the age of the universe. When Milne's universe is brought into the fold of general relativity, it loses its center and edge, has infinite space of negative curvature, and expands at a constant rate of zero deceleration.

CONTINUOUS CREATION

Ah Love! could thou and I with Fate conspire
To Grasp this sorry Scheme of Things entire,
Would not we shatter it to bits – and then
Re-mould it nearer to the Heart's Desire!
Edward FitzGerald (1809–1883), The Rubáiyát of Omar Khayyám

Creation theories
The steady-state expanding universe, proposed in 1948 by Herman Bondi and Thomas Gold, obeyed the perfect cosmological principle (all places are alike in space and time). An expanding universe in a steady state has infinite age. In 1948 this was an attractive feature of the model because of the time-scale difficulty besetting many evolutionary universes (the Hubble period seemed less than the age of the Solar System). An expanding universe that never changes requires a continuous creation of new matter to maintain a constant

density. "Hence," wrote Bondi and Gold, "there must be continuous creation of matter in space at a rate which is, however, far too low for direct observation." The new matter is created not out of radiation, as in the static MacMillan steady-state universe, but apparently out of nothing.

Spontaneous creation has been a recurrent theme throughout the history of science, and even in the twentieth century we find many examples. James Jeans, a famed astronomer, in *Astronomy and Cosmogony* (1929) surmised that the "centres of the nebulae are of the nature of 'singular points' at which matter is poured into our universe from some other, and entirely extraneous, spatial dimension, so that, to a denizen of our universe they appear as points at which matter is being continually created." Pascual Jordan of Germany developed in 1939 a scalar–tensor theory that modified general relativity so that matter is not conserved but created. He said, "The conjecture suggests itself that the cosmic creation of matter does not take place as a diffuse creation of protons, but by the sudden appearance of whole drops of matter." Jordan's "drops" were stars created in an embryonic form. Japanese mathematicians in Hiroshima at the same time developed a continuous creation theory for a de Sitter universe in which galactic embryos are spontaneously created.

Steady-state expanding universe

Nothing changes in the cosmic scenery of the expanding steady-state universe, and therefore the curvature K, the Hubble term H, the deceleration term q, and the density ρ must all stay constant. The curvature K is equal to k/R^2, and because the scaling factor R increases with expansion, the curvature remains constant only when k is zero, and space is flat and infinite in extent. Because the Hubble term $H = \dot{R}/R$ is constant, $\dot{R}$ is proportional to R, and the scaling factor grows exponentially as in the de Sitter universe. This ensures that the deceleration term has the constant value of -1. A constant density requires the creation of matter

at a rate of about 1 hydrogen atom per cubic meter every Hubble period, roughly equivalent to 1 galaxy per year per Hubble sphere.

The steady-state universe regenerates itself in one-third of a Hubble period. The average age of things that endure, such as nucleons and galaxies, is also one-third of a Hubble period. Some galaxies are young and have recently formed, others are very old, and the average age of all galaxies is one-third of a Hubble period. Our Galaxy, which is about 15 billion years old, is therefore about twice as old as the average galaxy in the steady-state universe. Most galaxies in a big bang universe have a similar age, about 15 billion years. A person believing in a cosmic steady state might feel concerned that our Galaxy is much older than the average galaxy, and yet coincidentally is the right age for a big bang universe.

Steady-state theories

Following a "discussion with Mr. T. Gold," Fred Hoyle showed how the theory of general relativity could be modified to allow for a continuous creation of matter. In the same year that Bondi and Gold put forward their steady-state theory, Hoyle used the scalar–tensor theory (shortly to be discussed) and found that the constant density of the universe and the Hubble term are related by the equation,

$$3H^2 = 8\pi G\rho. \qquad [18.31]$$

Hoyle's creation theory did not indicate the form in which matter is created. The theory breaks the law of conservation of matter – implicit in general relativity theory – by means of a mathematical device. William McCrea, as we saw earlier in this chapter, proposed that continuous creation is the result of cosmic tension. A cosmic tension, when equal to the energy density, maintains a state of constant density. McCrea's theory, like Hoyle's theory, does not explain why only matter, and not also antimatter, is created.

According to the original work of Bondi and Gold, creation is uniform everywhere, and newly created matter eventually condenses and forms new galaxies. McCrea in 1964 suggested an alternative theory in which the creation might not be uniform in space but more active in regions where matter already exists. "All matter," he proposed, "is the potential promoter of the creation of matter. All matter is normally in galaxies, and so the creation of fresh matter normally promotes the growth of galaxies. But occasionally a fragment of matter becomes detached from its galaxy. Any such fragment is a potential promoter of fresh creation; if it is successful as such, it is the embryo of a new galaxy."

End of the steady-state theory

Momentous discoveries have struck down the steady-state theory. The cosmic background radiation indicates that a big bang once existed, and the prevalence of quasars in the past indicates that the universe has evolved in a dramatic fashion. Desperate attempts have been made by Hoyle and Jayant Narlikar to salvage remnants of the theory. They invoke large-scale inhomogeneities and variations in the laws and constants of nature to preserve an eternal universe. The original elegant simplicity of the expanding steady-state idea, attractive to so many persons, has gone, and the theory is now of historical interest. In 1967, Dennis Sciama wrote: "I must add that for me the loss of the steady-state theory has been a cause of great sadness. The steady-state theory has a sweep and beauty that for some unaccountable reason the architect of the universe appears to have overlooked. The universe is in fact a botched job, but I suppose we shall have to make the best of it."

SCALAR–TENSOR THEORY

There was a Door to which I found no key:
There was a Veil past which I could not see.
Edward FitzGerald (1809–1883), The Rubáiyát of Omar Khayyám

The scalar–tensor theory of gravity was advanced in 1939 by Pascual Jordan of Germany. The idea is to lay in the Riemannian

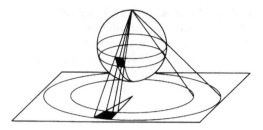

Figure 18.17. A conformal transformation changes space but preserves angles, as in this stereographic projection that maps a spherical surface onto a plane.

spacetime of general relativity a scalar field that varies from place to place. Gravity retains its character of dynamic curvature, but is modified by the scalar field in a special way known as a conformal transformation (Figures 18.17 and 18.18). The transformation consists of multiplying the spacetime interval by the scalar function, and space and time intervals are thus stretched or compressed equally. For example, the Minkowski metric

$$ds^2 = dt^2 - (dx^2 + dy^2 + dz^2),$$

is conformally transformed into

$$ds'^2 = dt'^2 - (dx'^2 + dy'^2 + dz'^2)$$
$$= F[dt^2 - (dx^2 + dy^2 + dz^2)]$$
$$= F\,ds^2,$$

where dt, dx, dy, dz, are transformed into dt', dx', dy', dz', and F is the scalar that varies in space and time. This transformation from spacetime intervals ds to intervals ds' is called conformal because it affects intervals of space and time similarly and angles and velocities remain unaltered. When the scalar F is constant in space and time, the process consists merely of a change in the conventional units of measurement. When everything in the universe is doubled in size, with the exception of a meter stick, all we need do is relabel the stick as half a meter, and nothing changes. Calling a centimeter a meter does not change the physical world. But when the scalar F varies from place to place in space and time, it changes relative sizes and durations, and thereby

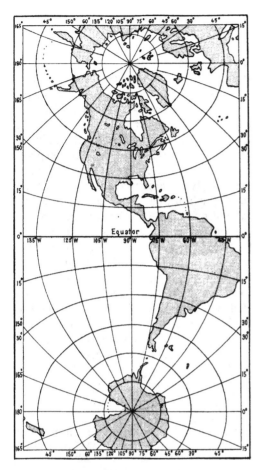

Figure 18.18. This Mercator map is a conformal transformation of the Western Hemisphere.

changes the physical properties of the universe.

A conformal transformation – known also as a units transformation – breaks the rigid constraints of general relativity and widens the range of physical theories. Let us take a universe in which atoms are everywhere alike and the gravitational constant G has a fixed universal value. We transform this normal universe into a new universe in which the units of measurement vary in time and from place to place. In this new universe we have no way of knowing that some places have been stretched and other places compressed. As we travel around with a ruler and a clock, a centimeter will still be a finger breadth and a second will still be a

slow heart beat. But what we notice is that electrons and protons have masses that vary, the gravitational constant has different values, and matter is being created or destroyed.

The scalar–tensor theory, as it is called, is like a black box that has two openings labeled IN and OUT. We insert a universe into the IN hole, and from the OUT hole comes a new and physically different universe. On the outside of this cosmic dream machine are a number of controls. Usually in cosmology we are not interested in making transformations that vary things from place to place, because we want to preserve homogeneity. The transformations of main interest are those that cause things everywhere to change in the same way in time. We are interested primarily in only three controls marked G, C, and M, and by their adjustment we determine the kind of universe that emerges from the dream machine. With only these knobs to adjust, the scalar field allows three things to happen that normally are forbidden. Adjustment of the G knob controls the way the gravitational constant varies in time; adjustment of the C knob controls creation and annihilation of matter; and adjustment of the M knob controls how particle masses vary in time. Our dream machine is thus able to manufacture multitudes of fantasy worlds from any given input world.

Jordan's world

Jordan's interest in a scalar–tensor theory was aroused by Paul Dirac's large-number hypothesis (Chapter 23) in which he proposed that G decreases in value as the universe expands. Jordan showed that G could be made to vary in the desired way by means of a scalar function. At first, the control knobs were not properly adjusted, and universes emerged in which everything had changed in an alarming way. Not only did G vary, as postulated by Dirac, but also C and M, and matter appeared or disappeared, and particle masses varied. Cosmologists are moderately tolerant of rival inventions provided that credulity is not taxed excessively. A variation in the gravitational constant, or a creation of matter, or a variation in subatomic masses is each by itself of tolerable interest; but patience is exhausted when two or all three happen simultaneously. Although Jordan pioneered the scalar–tensor theory, his work in cosmology has not made a lasting impact.

Hoyle–Narlikar world

Hoyle used the scalar–tensor theory to create matter in an expanding steady-state universe. Adjustment of the creation knob C insured that density remained always constant. A slip of the hand, however, could quite easily make the creation rate either too fast or too slow, in which case a steady-state situation would be lost.

The steady-state theory has been overthrown by unanticipated discoveries. Hoyle earned notoriety as its most active supporter, and with Narlikar has made modifications to the steady-state theory to bring it more into conformity with the discovery of the cosmic background radiation. One idea is that particle masses change with time. The G and C knobs are set to zero, and the M knob is adjusted in such a way that "the usual mysteries concerning the so-called origin of the universe begin now to dissolve," wrote Hoyle in 1975. The universe is assumed to be static, and atoms, human beings, and stars shrink in size slowly because of the growth in the mass of subatomic particles. The expanding universe of atoms of fixed size is transformed into a static universe of atoms of shrinking size. Atoms in the past were larger and therefore emitted longer wavelengths, thus explaining the redshift effect. The big bang, which Hoyle disliked, is banished and becomes a moment in time when all atomic masses happen to be close to zero. According to this picture, the cosmic background radiation is not a product of the big bang, but is starlight from an earlier phase of the universe scattered and thermalized by atoms of enormous size at the epoch of minimum mass. This "shrinking atom" universe, rejected by Eddington long before, retains the infinite age of the steady-state universe but abandons continuous creation.

Brans–Dicke world

Robert Dicke used the scalar–tensor theory as a basis for investigating Mach's principle. Ernst Mach believed that the inertial mass of a particle is the result of it "feeling" the presence of all other particles in the universe. Dicke argued that this principle explains why the gravitational mass m_{grav} of a particle, which is determined by its response to a gravitational field, is equal to the inertial mass m_{inert} of the particle, which is determined by its response to accelerated motion. Normally, we make these two masses equal and write them as m. But why are they equal? Dicke argued as follows. Let M be the gravitational mass and L the radius of the observable universe, and let GMm_{grav}/L be the gravitational energy of a single particle. According to Dicke, this energy equals $m_{inert}c^2$, and hence

$$\frac{m_{inert}}{m_{grav}} = \frac{GM}{Lc^2} = 1,$$

and the relation $GM/Lc^2 = 1$ serves as an expression of Mach's principle. The problem is to find a way in which Lc^2/M controls the value of G.

Carl Brans and Dicke used the scalar–tensor theory because it allows G to vary with expansion. By adjusting the G knob on the dream machine so that G-variation was small enough not to be in conflict with observation, and yet large enough to maintain $G = Lc^2/M$, they were able to claim that the universe obeys Mach's principle.

Radar observations of orbital motions in the Solar System and theoretical studies of helium synthesis in the early universe have since shown, however, that G-variation, if it exists, is extremely small. As a result, the Brans–Dicke universe has become practically indistinguishable from a universe in which G is constant.

The many universes

With the scalar–tensor theory we are able to create from any one universe a large number of physically distinct universes. One of the most intriguing aspects of the theory is that any scalar–tensor universe can be transformed into any other scalar–tensor universe. A Brans–Dicke universe of *G*-variation can be converted into a Hoyle–Narlikar universe of M-variation, and both can be converted into a universe of C-variation. They all interconvert by appropriate transformations, and all can be converted back to an ordinary general relativity universe by either abolishing the scalar function or absorbing it into the properties of matter.

REFLECTIONS

1   We are the music-makers
        And we are the dreamers of dreams,
    Wandering by lone sea-breakers,
        And sitting by desolate streams;
    World-losers and world-forsakers,
        On whom the pale moon gleams:
    Yet we are the movers and shakers
        Of the world forever it seems.
    Arthur O'Shaughnessy (1844–81), Ode

2   *Howard Robertson in 1933 wrote: "In giving this survey of cosmologies we are convinced that the underlying theory forms an integral part of the theory of general relativity, and that although the choice of a particular model may for the present be influenced by the predilection of the individual, we can hope that the future will reveal additional evidence to test its validity and to lead us to a satisfying solution." These words (in "Relativistic cosmology") reflect the views of most cosmologists who would also agree with Dennis Sciama: "A rigid theory has not yet been discovered. For instance, general relativity, which is the best theory of space, time and gravitation that has so far been proposed, is, as we shall see, consistent with an infinite number of different possibilities, or models, for the history of the Universe. Needless to say, not more than one of these models can be correct, so that the theory permits possibilities that are not realized in Nature. In other words, it is too wide. We can put this in another way. In the absence of a theory*

anything can happen. If we introduce a weak theory too many things can still happen. A strong enough theory has not yet been discovered" (Modern Cosmology, 1971).

•   *"A theory has only the alternatives of being right or wrong. A model has the third possibility: it may be right, but irrelevant" (Manfred Eigen, in* The Physicist's Conception of Nature, *edited by Jagdish Mehra, 1973).*

3   *"Much later, when I discussed cosmological problems with Einstein, he remarked that the introduction of the cosmological term was the biggest blunder he ever made in his life. But this 'blunder,' rejected by Einstein, is still sometimes used by cosmologists even today, and the cosmological constant denoted by the Greek letter Λ rears its ugly head again and again and again" (George Gamow,* My World Line, *1970). The Λ force is referred to by various names, such as the cosmological constant, cosmological term, cosmical constant or cosmical term.*

•   *William McCrea in a paper entitled "Cosmology today" (1970) wrote: "Lemaître was, I think, the first to seek to relate this problem [the formation of galaxies] to the general evolution of the universe as studied by relativistic cosmology. He considered that the time spent near the Einstein state in the Lemaître model was the time when galaxies and clusters of galaxies were formed out of gas-clouds. For in this phase there is a near balance between gravitational attraction and cosmical repulsion (represented by the L-term) that provides the sort of instability that leads to condensations."*

•   *"But here I shall provisionally retain Λ as an unknown parameter, both to show its influence on the models and to allow for the possibility that this theoretically allowable term may more legitimately arise in some future, more comprehensive field theory" (Howard Robertson, at a meeting of the American Association for the Advancement of Science, 1954).*

•   *"The tentative conclusion is reached that, if general relativity is to be treated as a self-contained theory, then the 'cosmical terms'* that contain the cosmical constant should be omitted. But if general relativity is only part of what is needed to construct a theoretical model of physical reality, then the cosmical terms ought to be retained as affording additional freedom in linking up with other parts of physical theory" (William McCrea, "The cosmic constant," 1971).

4   Imagine a large number of universes, some of which evolve while others are in a steady state, and suppose that all contain intelligent beings. What is the chance of living in an evolving universe? Each steady-state universe is of infinite duration and therefore contains infinitely more intelligent beings than an evolving universe that is habitable for only a finite period of time. Hence, if steady-state universes exist, the probability of occupying one is unity, and the probability of occupying an evolving universe is zero. But our universe is not in a steady state, and it follows that probably habitable steady-state universes do not exist.

5   Edward Milne (1896–1950) stressed: (i) the importance of the cosmological principle (a name he introduced), (ii) the distinction between the world map and the world picture, and (iii) the importance of operational methods of distance measurement, similar to those used in radar. Among his other ideas is that of dual time scales: an atomic time scale in which time is measured by atoms and a dynamic time scale in which time is measured by the motions of planets and other bodies. His dynamic intervals of time are the logarithm of the atomic intervals of time, and in dynamic time the gravitational constant G remains fixed, whereas in atomic time G increases. Although the idea of dual time scales is intriguing, few physicists have taken it seriously.

Eddington and Milne probed deeply, both asking searching questions, and it is perhaps true to say that little of science can survive penetrating and critical investigation. The roots of knowledge are buried in mystery, and when we dig too deep, too soon, the tree of knowledge falls down.

6   Fred Hoyle wrote in 1948: "By introducing continuous creation of matter into the field equations of general relativity a stationary universe showing expansion properties is obtained without recourse to a cosmical constant.... The following work is concerned with this aspect of the matter and arose from a discussion with Mr. T. Gold who remarked that through continuous creation of matter it might be possible to obtain an expanding universe in which the proper density of matter remained constant. This possibility seemed attractive, especially when taken in conjunction with aesthetic objections to the creation of the universe in the remote past. For it is against the spirit of scientific enquiry to regard observable effects as arising from 'causes unknown to science,' and this in principle is what creation-in-the-past implies" ("A new model for the expanding universe").

•   "It is the purpose of a scientific hypothesis to stick out its neck, that is, to be vulnerable. It is because the perfect cosmological principle is so extremely vulnerable that I regard it as a useful principle. It is something that could in practice be 'shot down' by experiment and observation far more easily than the ordinary cosmological principle, and I think you will agree with that" (Hermann Bondi, in Rival Theories of Cosmology, a 1959 BBC discussion).

•   "Therefore, when looking at the distant parts of the universe, we see them as they were a long time ago. Any meaningful comparison of these distant parts with the ones nearby presupposes that the laws of physics are the same in the two cases. Since the universe, by definition, includes the study of all observable phenomena, we may expect the laws of physics also to be somehow determined by the universe as a whole. If the state of the universe was once very much different from what it is now, what guarantee do we have that the laws of physics were the same in the past as they are now?" (Jayant Narlikar, "Steady state defended," 1973). Narlikar repeats the argument made by Bondi and Gold that only in a universe eternally the same can we be confident that the laws of nature are invariant. The argument contains, however, a flaw. It assumes that "the universe as a whole" exists only in space. This is a worm's eye view. A god's eye view sees an unchanging universe existing throughout spacetime. The unity of the universe, which embodies the laws of nature, is not a thing in space that evolves in time, but a thing that embraces all spacetime.

7 Cyclic or oscillating universes are not uncommon in mythology. The Hindu religion provides the best example, in which each cycle of the Hindu universe is a kalpa, or day of Brahma, that lasts 4320 million years. Vishnu, who controls the universe, has a life of a hundred "years," each of which contains 360 days of Brahma. After 36 000 cycles, lasting roughly 150 trillion real years, the world comes to an end and only the Absolute Spirit survives. After an indefinite period of time a new world and a new Vishnu emerge and the cyclic scheme is repeated. It is interesting that a virtual day of Brahma is not far short of a Hubble period in modern cosmology.

8 In Chapter 16 we derived the zero-pressure Friedmann–Lemaître equations:

$$\ddot{R} = -\frac{4\pi G\rho R}{3} + \frac{\Lambda R}{3},$$

$$\dot{R}^2 = \frac{8\pi G\rho R^2}{3} + \frac{\Lambda R^2}{3} - k,$$

$$\rho R^3 = \text{constant}.$$

With the substitution of the "observable" quantities,

curvature: $\qquad K = k/R^2$

Hubble term: $\qquad H = \dot{R}/R$

deceleration term: $\quad q = -\ddot{R}/RH^2$

the equations take the form

$$K = 4\pi G\rho - H^2(q+1),$$

$$\Lambda = 4\pi G\rho - 3qH^2,$$

used in the text (Equations 18.19 and 18.20).

The Einstein universe has no expansion or contraction, hence $H = 0$, and the Friedmann–Lemaître equations simplify to

(Equations 18.1 and 18.2)

$$K = 4\pi G\rho,$$

$$\Lambda = 4\pi G\rho.$$

Thus $K = \Lambda$, and the universe has positive curvature equal to the cosmological constant and consists of spherical space. The scaling factor is the radius of the universe:

$$R = (4\pi G\rho)^{-1/2}$$

(Equation 18.3).

The de Sitter universe has zero density $(\rho = 0)$ and space is flat $(K = 0)$. From the Friedmann–Lemaître equations we get

$$q = -1,$$

$$\Lambda = 3H^2$$

(Equations 18.6 and 18.7). This is an accelerating universe with a constant Hubble term $H = (\Lambda/3)^{1/2}$; hence

$$R = R_0\, e^{H(t-t_0)}, \qquad\qquad [18.32]$$

and $R = R_0$ at $t = t_0$, where t_0 is the present age and R_0 is the present value of the scaling factor.

9 In the Friedmann universes the cosmological constant is zero and hence

$$K = H^2(2q - 1),$$

$$4\pi G\rho = 3qH^2,$$

(Equations 18.8 and 18.9). The solution for $k = 1$ is the cycloid:

$$R = R_{\max} \sin^2\left(\frac{\phi}{2}\right), \qquad\qquad [18.33]$$

$$t = R_{\max}\left(\frac{\phi}{2} - \sin\phi\right), \qquad\qquad [18.34]$$

where ϕ is the "turning angle." The solution, for $k = -1$ is

$$R = R_{max} \sinh^2\left(\frac{\phi}{2}\right), \qquad\qquad [18.35]$$

$$t = R_{\max}\left(\sinh\phi - \frac{\phi}{2}\right). \qquad\qquad [18.36]$$

In the Einstein–de Sitter version, space is flat $(K = 0)$, therefore the deceleration term q

has the fixed value 0.5, and

$$8\pi G\rho = 3H^2. \qquad [18.37]$$

Because density ρ is proportional to $1/R^3$, we have $HR^{3/2}$ is constant, and hence

$$R = R_0\left(\frac{t}{t_0}\right)^{2/3}, \qquad [18.38]$$

and the age of the universe is $t_0 = 2/3H_0$ and $\rho t^2 = $ constant. The density of the Einstein–de Sitter universe has the critical value:

$$\rho_{\text{crit}} = \frac{3H^2}{8\pi G}, \qquad [18.39]$$

and therefore

$$\rho = 2q\rho_{\text{crit}}$$

(Equation 18.14).

10 *The density parameter Ω, nowadays in common use, expresses the density in terms of the critical density:*

$$\Omega = \frac{\rho}{\rho_{\text{crit}}}. \qquad [18.40]$$

In the Friedmann universes of zero pressure,

$$K = H^2(\Omega - 1),$$

$$8\pi G\rho = 3H^2\Omega,$$

$\rho R^3 = $ constant and $\Omega = 2q$. For $\Omega > 1$, the density is greater than the critical density and the universe is closed; for $\Omega \le 1$, the density is less than or equal to the critical density and the universe is open.

The Friedmann equations can be re-arranged in useful ways. One way is the following. We have

$$k = (RH)^2(\Omega - 1) = \text{constant},$$

$$R^3H^2\Omega = \text{constant}$$

(using $\rho R^3 = $ constant) and therefore

$$(RH)^2(\Omega - 1) = (R_0H_0)^2(\Omega_0 - 1), \quad [18.41]$$

$$R^3H^2\Omega = R_0^3H_0^2\Omega_0, \qquad [18.42]$$

where, as usual, the zero subscript denotes present values. On eliminating Ω and H in succession from these two equations, and

using $1 + z = R_0/R$, we find

$$H = H_0(1 + z)(1 + \Omega_0 z)^{1/2}, \qquad [18.43]$$

$$\Omega = \frac{\Omega_0(1 + z)}{1 + \Omega_0 z}, \qquad [18.44]$$

which show how these two parameters, important in observational cosmology, vary as functions of redshift z. The distance L in the world map to a body of redshift z (important in all measurements of H_0) can be calculated by inserting Equation [18.43] in Equation [15.36]. From Equation [18.43] we find in a closed universe that the maximum radius at $H = 0$ is given by

$$\frac{R_{\text{max}}}{R_0} = \frac{\Omega_0}{\Omega_0 - 1}, \qquad [18.45]$$

which is Equation [18.16] with $\Omega_0 = 2q_0$. Equation [18.44] is important in the flatness riddle (see Chapter 22). In the early universe, when z is large, we have from Equation [18.44] that Ω is very close to unity. A long-lived universe must therefore be extremely flat in its early stages. Later, we shall see how inflation explains how the universe became flat and long-lived.

11 *With a uniform pressure P included, the Friedmann–Lemaître equations become*

$$\ddot{R} = -\frac{4\pi G}{3}\left(\rho + \frac{3P}{c^2}\right)R + \frac{\Lambda R}{3},$$

$$\dot{R}^2 = \frac{8\pi G\rho R^2}{3} + \frac{\Lambda R^2}{3} - k,$$

$$\frac{d(\rho c^2 R^3)}{dt} + P\frac{dR^3}{dt} = 0.$$

With the relativistic equation of state $P = (\gamma - 1)\rho c^2$, γ constant, we find

$$K = 4\pi G\rho\gamma - H^2(q + 1), \qquad [18.46]$$

$$\Lambda = (3\gamma - 2)4\pi G\rho - 3qH^2, \qquad [18.47]$$

$$\rho R^{3\gamma} = \text{constant}. \qquad [18.48]$$

All universes discussed in this chapter, with the exception of MacMillan's steady-state model and Misner's mixmaster model, conform in one way or another to this set of general equations.

- When γ equals 1, we obtain the previous general results for zero-pressure:

$$K = 4\pi G\rho - H^2(q+1),$$

$$\Lambda = 4\pi G\rho - 3qH^2,$$

and $\rho R^3 =$ constant.
- When $\gamma = 4/3$, as in a radiation-dominated universe, we obtain

$$K = \frac{16\pi G\rho}{3} - H^2(q+1),\qquad[18.49]$$

$$\Lambda = 8\pi G\rho - 3qH^2.\qquad[18.50]$$

- The static Einstein universe of $H = 0$ is

$$K = 4\pi G\rho\gamma,\qquad[18.51]$$

$$\Lambda = (3\gamma - 2)4\pi G\rho,\qquad[18.52]$$

and the larger the value of γ, the smaller the radius R. Notice that when $\gamma = 2/3$, a static universe is possible with a cosmological term equal to zero.
- The Friedmann universes $(\Lambda = 0)$ containing pressure are given by

$$K = H^2\left(\frac{2q}{3\gamma - 2} - 1\right),\qquad[18.53]$$

$$(3\gamma - 2)4\pi G\rho = 3qH^2.\qquad[18.54]$$

The zero-pressure Friedmann universes of $\gamma = 1$ have already been discussed. In the radiation Friedmann universes of $\gamma = 4/3$, we obtain

$$K = H^2(q - 1),\qquad[18.55]$$

$$8\pi G\rho = 3qH^2.\qquad[18.56]$$

These universes are closed $(K > 0)$ when $q > 1$, and open $(K \le 0)$ when $q \le 1$. In the Einstein–de Sitter radiation universe $(K = 0)$, the deceleration term is $q = 1$, and because the density ρ varies as $1/R^4$, we find that R varies as $t^{1/2}$.

12 Adiabatic variations occur when energy is neither gained nor lost from a closed system. Consider a comoving volume V in which the net energy gained from other comoving volumes is zero. A small adiabatic energy change dE in V, corresponding to a small volume change dV, is

$$dE = -P\,dV,$$

where $E = \rho c^2 V$. When the pressure P is negative the energy increases because of the work done by the expansion. With an equation of state $P = (\gamma - 1)\rho c^2$, negative pressure (or cosmic tension) corresponds to γ having a value less than 1. Density ρ varies as $R^{-3\gamma}$, and when $\gamma = 0$, the density remains constant during expansion. Thus universes of $P = -\rho c^2$ expand and contract at constant energy density. During expansion energy (or matter) is continually created, and during contraction energy (or matter) is continually annihilated. In the $P = -\rho c^2$ universes in tension, the Friedmann–Lemaître equations become

$$K = -H^2(q + 1),\qquad[18.57]$$

$$\Lambda = -8\pi G\rho - 3qH^2.\qquad[18.58]$$

A flat $(K = 0)$, static $(H = 0)$, and stable universe is possible when Λ is attractive and equal to $-8\pi G\rho$. The steady-state and inflationary universes have constant H and q, and this is possible when $K = 0$ and $q = -1$. If we assume that the cosmological term Λ is zero (although nothing in steady-state theory requires this assumption) we obtain Hoyle's result $8\pi G\rho = 3H^2$ that he obtained from the scalar–tensor theory.

PROJECTS

1 Discuss the Einstein static universe. Will stars, pouring out radiation into space, nudge this unstable universe into expansion or collapse? Consider an Einstein universe in which gravity is repulsive (change G to $-G'$) and the Λ force is attractive (change Λ to $-\Lambda'$). Is it open or closed? Could life exist in such a universe?
2 "Whatever pushed the universe over the brink, why did it topple it on the expansion side of the abyss rather than on the other?" (Jagjit Singh, *Great Ideas and Theories in Modern Cosmology*, 1970). Is there an answer? Perhaps there are many Eddington universes and some expand and others collapse.
3 We live at a time in the history of the universe when stars are shining. In 1930, Eddington remarked that astronomers

must "count themselves as extraordinarily fortunate that they are just in time to observe this interesting but evanescent feature of the sky." In *The Measure of the Universe*, J. D. North wrote that Eddington appears "to rely on a Principle of the Improbability of Good Fortune." Discuss these remarks.

4 Suppose that in the past at a redshift of 9 the universe passed through an Einstein quasi-static state. What was its radius and density?

• Discuss the difference between static, steady-state, and static steady-state universes.

• Are you a banger or antibanger? Which class of universe appeals the most to you?

• Do you agree with Bishop Barnes about the possibility of understanding God's will in a finite universe?

5 Can you think of other ways to classify universes? For example, universes that are habitable and uninhabitable by organic life; or universes that contain only matter, or antimatter, or both matter and antimatter.

6 Throughout history people have lived in different universes (or thought they did). What kind of universe would you prefer to live in? Is there anything to stop you?

7 Discuss the following remarks made by Hermann Bondi and Thomas Gold in "The steady-state theory of the expanding universe" (1948). "The application of the laws of terrestrial physics to cosmology is examined critically. It is found that terrestrial physics can be used unambiguously only in a stationary homogeneous universe.... As the physical laws cannot be assumed to be independent of the structure of the universe, and conversely the structure of the universe depends upon the physical laws, it follows that there may be a stable position. We shall pursue the possibility that the universe is in such a stable, self-perpetuating state, without making any assumptions regarding the particular features which lead to this stability. We regard the reasons for pursuing this possibility as very compelling, for it is only in such a universe that there is any basis for the assumption that the laws of physics are

constant; and without such an assumption our knowledge, derived virtually at one instant of time, must be quite inadequate for an interpretation of the universe and the dependence of its laws on its structure, and hence inadequate for any extrapolation into the future or the past."

8 "A relevant question is whether at the present stage a proliferation of models in any way advances cosmology. Present observational data are inadequate to make a clear decision even when [the pressure is zero] and the cosmological constant is discarded as an irrelevant complication" (Harrison, "Classification of uniform cosmological models," 1967). Cosmology has always had a surfeit of heretical rival models – long may it remain so! What do you think?

9 Show that in the closed Friedmann universes:

$$R = R_{\max} \frac{\Omega - 1}{\Omega} = L_H \frac{1}{(\Omega - 1)^{1/2}}.$$

FURTHER READING

Bondi, H. *Cosmology*. Cambridge University Press, Cambridge, 1960.

Davies, P. C. W. *Other Worlds*. Dent, London, 1983.

Eddington, A. S. *The Expanding Universe*. Cambridge University Press, Cambridge, 1933.

Guth, A. H. and Steinhardt, P. J. "The inflationary universe." *Scientific American* (May 1984).

Sciama, D. W. *Modern Cosmology*. Cambridge University Press, Cambridge, 1971.

Trimble, V. "Cosmology: man's place in the universe." *American Scientist* (January–February, 1977).

Whitrow, G. J. *The Structure and Evolution of the Universe: An Introduction to Cosmology*. Hutchinson, London, 1959.

SOURCES

Barnes, E. W. *Scientific Theory and Religion: The World Described by Science and its Spiritual Interpretation*. Cambridge University Press, Cambridge, 1933.

Bondi, H., Bonnor, W. B., Lyttleton, R. A., and Whitrow, G. J. *Rival Theories of Cosmology: A Symposium and Discussion of Modern Theories of the Universe*. Oxford University Press, London, 1960.

Bondi, H. and Gold, T. "The steady-state theory of the expanding universe." *Monthly Notices of the Royal Astronomical Society* 108, 252 (1948).

Brans, C. and Dicke, R. H. "Mach's principle and a relativistic theory of gravitation." *Physical Review* 124, 125 (1961).

Collins, C. B. and Hawking, S. W. "Why is the universe isotropic?" *Astrophysical Journal* 143, 317 (1973).

Dicke, R. H. "Gravitational theory and observation." *Physics Today* (January 1967).

Dicke, R. H. *Gravitation and the Universe*. American Philosophical Society, Philadelphia, 1970.

Eddington, A. S. "On the instability of Einstein's spherical world." *Monthly Notices of the Royal Astronomical Society* 90, 668 (1930).

Einstein, A. "Cosmological considerations on the general theory of relativity," 1917. Reprinted in *The Principle of Relativity. A Collection of Original Memoirs on the Special and General Theory of Relativity*. H. A. Lorentz, A. Einstein, H. Minkowski, and H. Weyl. Dover Publications, New York, 1952.

Friedmann, A. "On the curvature of space." Translated by B. Doyle in *A Sourcebook in Astronomy and Astrophysics 1900–1975*. Edited by K. R. Lang and O. Gingerich. Harvard, Cambridge, Massachusetts, 1979.

Gamow, G. *My World Line: An Informal Autobiography*. Viking Press, New York, 1970.

Harrison, E. R. "Classification of uniform cosmological models." *Monthly Notices of the Royal Astronomical Society* 137, 69 (1967).

Hoyle, F. "A new model for the expanding universe." *Monthly Notices of the Royal Astronomical Society* 108, 372 (1948).

Jaki, S. L. *Science and Creation: From Eternal Cycles to an Oscillating Universe*. Science History Publications, New York, 1974.

Jeans, J. H. *Astronomy and Cosmogony*. Cambridge University Press, Cambridge, 1929. See p. 360.

Jordan, P. "Formation of the stars and development of the universe." *Nature* 164, 637 (1949).

Kaufmann, W. "The Hoyle–Narlikar cosmology." *Mercury* (May–June 1976).

Lemaître, G. "A homogeneous universe of constant mass and increasing radius accounting for the radial velocity of extra-galactic nebulae." *Monthly Notices of the Royal Astronomical Society* 91, 483 (1931).

Leslie, J. *Universes*. Routledge, London, 1989.

Mach, E. *The Science of Mechanics*. Open Court, Chicago, 1893.

McCrea, W. H. "Relativity theory and the creation of matter." *Proceedings of the Royal Society* 206, 562 (1951).

McCrea, W. H. "Continual creation." *Monthly Notices of the Royal Astronomical Society* 128, 335 (1964).

McCrea, W. H. "Cosmology today." *American Scientist* (September–October 1970).

McCrea, W. H. "The cosmic constant." *Quarterly Journal of the Royal Astronomical Society* 12, 140 (1971).

Mehra, J. Editor. *The Physicist's Conception of Nature*. Reidel, Dordrecht, Netherlands, 1973.

Milne, E. A. *Relativity, Gravitation, and World Structure*. Oxford University Press, Clarendon Press, Oxford, 1935.

Milne, E. A. *Kinematic Relativity*. Oxford University Press, Clarendon Press, Oxford, 1948.

Misner, C. W., Thorne, K. S., and Wheeler, J. A. *Gravitation*. Freeman, San Francisco, 1970.

Narlikar, J. V. "Steady state defended," in *Cosmology Now*. Edited by J. Laurie. British Broadcasting Corporation, London, 1973.

Narlikar, J. V. *Introduction to Cosmology*. Cambridge University Press, Cambridge, 1993.

North, J. D. *The Measure of the Universe: A History of Modern Cosmology*. Oxford University Press, Clarendon Press, Oxford, 1965.

Peebles, P. J. E. and Silk, J. "A cosmic book of phenomena." *Nature* 346, 233 (1990).

Robertson, H. P. "Relativistic cosmology." *Reviews of Modern Physics* 5, 62 (1933).

Schlegel, R. "Steady-state theory at Chicago." *American Journal of Physics* 26, 601 (1958). A discussion of MacMillan's steady-state theory.

Sciama, D. W. "Cosmology before and after quasars and the cosmic black-body radiation." *Scientific American* (September 1967).

Shu, F. *The Physical Universe*. University Science Books, Mill Valley, California, 1982.

Singh, J. *Great Ideas and Theories in Modern Cosmology*. Dover Publications, New York, 1970.

Tinsley, B. M. "The cosmological constant and cosmological change." *Physics Today* (June 1977).

Tolman, R. C. "Models of the physical universe." *Science* 75, 367 (1932).

Tolman, R. C. *Relativity Thermodynamics and Cosmology*. Oxford University Press, Clarendon Press, Oxford, 1934.

Tropp, E. A., Frenckel, V. Ya., Chernin, A. D. *Alexander A. Friedmann: The Man Who Made the Universe Expand*. Translated by A. Dron and M. Burov. Cambridge University Press, Cambridge, 1993.

Weinberg, S. *Gravitation and Cosmology: Principles and Applications of the General Theory of Relativity*. Wiley, New York, 1972.

Weinberg, S. "The cosmological constant problem." *Reviews of Modern Physics* 61, 1 (1989).

19 OBSERVATIONAL COSMOLOGY

For the history that I require and design, special care is to be taken that it be of wide range and made to the measure of the universe. For the world is not to be narrowed till it will go into the understanding (which has been done hitherto), but the understanding is to be expanded and opened till it can take in the image of the world.
Francis Bacon (1561–1626), Novum Organum

INTRODUCTION

"Now, what I want is Facts.... Facts alone are wanted in life."
Mr. Gradgrind in Hard Times *by Charles Dickens (1812–1870)*

Facts about the heavens

We begin on a philosophical note by quoting Arthur Eddington from his book *The Expanding Universe*: "For the reader resolved to eschew theory and admit only definite observational facts, all astronomical books are banned. There are no purely observational facts about the heavenly bodies. Astronomical measurements are, without exception, measurements of phenomena occurring in a terrestrial observatory or station; it is only by theory that they are translated into knowledge of a universe outside." Without books and theories our observations of the heavens lack content and significance.

We construct universes that are models of the true Universe. Our longing for absolute truth tempts us to believe that the current universe of our society is the Universe. Each society has its own universe (ours is the physical universe whose principles are discussed in Chapter 8), and each society interprets its observations in accord with the principles of that universe. The many historic universes – Babylonian, Egyptian, Platonic, Medieval, Newtonian, Victorian, to name a few – were all very different from one another, and all were strongly supported by the content and significance of the observations made by those who believed in them.

Stripped of most of our preconceptions, we observe – within the limits of our unaided senses – valleys, hills, rivers, buildings, and so forth, backed by a bright sky by day traveled by the Sun and a dark sky by night traveled by the Moon and points of light known as planets and stars. What in this picture is cosmologically significant? With the advance of knowledge and technology over thousands of years more and more of the seen world (and the worlds revealed by sensitive instruments) has become cosmologically significant. And now, in the physical universe of today, even a simple metal coin testifies to the existence of stars that died before the formation of the Solar System, and hence the universe is billions of years old, spans billions of light years in space, and contains billions of galaxies.

COSMOGRAPHY

Cosmography gives the "bare facts" about the universe. Our brief overview in this chapter divides into three parts: observations of local things, observations of things at intermediate distances, and observations of things at cosmic distances.

Local observations

Local observations of cosmological significance are restricted to the Solar System, the Galaxy, and the Local Group, and extend no farther than a few million light years. They help us to establish:

- the first steps in a "distance-ladder;"
- the distribution and density of matter;
- the ages of stars and the Galaxy, thus setting limits on the age of the universe;
- the abundances of the elements, particularly of the light elements hydrogen, deuterium, helium, and lithium, that depend on conditions in the early universe;
- the cosmic background radiation (CBR) that originates in the early universe and reveals the peculiar motion of the Galaxy;
- the cosmological parameters.

Local conditions yield information on the age of the universe t_0, the present baryonic density ρ_{b0}, and the value of the combination $\Omega_0 H_0^2$ of the density parameter and the Hubble term.

Intermediate-distance observations

Observations of things beyond the Local Group of galaxies at intermediate distances are confined mainly to the Local (or Virgo) Supercluster and extend to distances of a few hundred million light years. These observations explore only the sub-Hubble sphere and do not extend into the full Hubble flow. Nonetheless, they help to establish:

- the structure, distribution, and motion of galaxies;
- the more extended distance scales;
- the redshift–distance and redshift–velocity relations;
- the cosmological parameters.

These observations determine the age and baryon density of the universe, value of the Hubble term H_0, and value of the density parameter Ω_0.

Distant observations

Distant observations extend many billions of light years beyond the Local Supercluster to the limits of the observable universe. These deep-space observations, all of which are cosmologically significant, extend far outside the sub-Hubble sphere and help

to establish:

- cosmic distances: the redshift–distance relation ceases to be linear, and the linear velocity–distance law requires that what is seen in the world picture is projected onto the world map;
- evolutionary histories that allow us to make comparisons between nearby and distant astronomical systems;
- the cosmological parameters.

Our conclusion is that we are still a long way from a precise and secure knowledge of the values of the Hubble term H_0, density parameters Ω_{b0} and Ω_0, deceleration term q_0, cosmological term Λ, and the curvature constant k.

LOCAL OBSERVATIONS
Nearby distances

To discover the rate of expansion (the Hubble term H_0), its deceleration (the deceleration term q_0), and the global geometry (the curvature constant k) of the universe we must find the distances and redshifts of remote and faint systems. But distances are difficult to determine; fortunately redshifts are comparatively easy. We start with nearby astronomical systems in the sub-Hubble sphere in which peculiar motions dominate over the Hubble flow; then, stepwise, we construct overlapping distance-indicators (as in Figure 19.1) that eventually reach out beyond the sub-Hubble sphere into the fully developed Hubble flow where Hubble's linear redshift–distance law may be used. It is a ladder, or rather staircase, on which each step takes us more into the depths of space. Unfortunately, with each step, approximations accumulate and uncertainty grows.

The first steps on the distance-ladder are described in Chapter 5. We begin with the astronomical unit (the distance of the Earth from the Sun) and with the year (the orbital period of the Earth about the Sun). A body at distance 1 parsec (3.26 light years) subtends an angle of 1 arc second from two points separated 1 astronomical unit on a line perpendicular to the line of

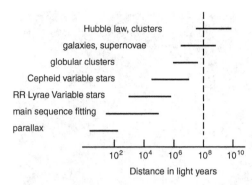

Figure 19.1. The distance-staircase, or distance-ladder. Different, overlapping methods of measuring distance must be used as the distance increases. The sub-Hubble boundary (inside which peculiar motions dominate over the Hubble flow) is shown arbitrarily at 100 million light years.

sight. The light year, which is perhaps a simpler unit in cosmology, is the distance traveled by light in 1 orbital period of the Earth.

Parallax measurements take us out beyond the Solar System to distances of 100 light years (0.03 arc seconds) and, with rising uncertainty, to distances of 300 light years. Recent statistical parallax measurements (the average of many stars in an open cluster) and the measurements from satellites above the Earth's atmosphere extend the range to 1000 light years (0.003 arc seconds). The Hyades (Figure 19.2), a loose aggregation of stars at distance 46 parsecs (150 light years), is the nearest and most important open cluster. Several methods of measuring distance, calibrated by the Hyades standard distance, now take over. The moving-cluster method (Chapter 5) includes the Hyades and other open clusters at greater distances. The important main-sequence fitting method (Chapter 5) measures the distances of star clusters – such as the Pleiades – relative to the Hyades and extends our distance measurements to the limits of the Galaxy. Correcting for extinction (interstellar absorption) of starlight becomes a major challenge.

The apparent brightness and the inverse-square law play a vital role in the astronomical problem of measuring distances. We saw in Chapter 5 that the apparent magnitude m and the absolute magnitude M of a luminous source are related by the distance modulus

$$\mu = m - M = 5\log(L/10), \qquad [19.1]$$

where L is the distance of the source in parsecs. The apparent magnitude m measures the observed brightness of the source – often in a specific spectral range – and the absolute magnitude M is the magnitude the source would appear to have if at distance of 10 parsecs. The distance L of a source can be found when its absolute magnitude M is known. The absolute magnitude is determined by

$$M = 4.7 - 2.5\log(\mathscr{L}/L_\odot), \qquad [19.2]$$

where the absolute magnitude of the Sun (of luminosity $\mathscr{L}_\odot$) is 4.7. In this chapter $\mathscr{L}$ denotes luminosity as distinct from L, which denotes distance. A sunlike star of $M = 5$ at 1 megaparsec distance has an apparent magnitude $m = 30$, far beyond the limits of ordinary ground-based telescopes and barely in reach of the Hubble Space Telescope. A galaxy of luminosity $10^{10}\mathscr{L}_\odot$ has an absolute magnitude $M = -20.3$, and an apparent magnitude $m = 19.7$ at distance 1000 megaparsecs.

Highly luminous RR Lyrae stars, and pulsating cepheids of known period–luminosity relation (Chapter 5), are the distance gauges to the outer reaches of the Galaxy, the globular clusters, and the Magellanic Clouds and other nearby galaxies. With various corrections for factors affecting apparent magnitude, such as different heavy-element compositions and extinction in the Galaxy and other galaxies, these stars extend our reach to the fringe of the Local Group.

Distribution and density of matter

The average density of most astronomical systems progressively decreases with increasing size. The density of matter in the disk of the Galaxy in the neighborhood of the Sun is around 1 solar mass per cubic parsec (equivalent to 10 hydrogen atoms per cubic centimeter), of which interstellar gas contributes about 10 percent. The total

Figure 19.2. The Hyades, a cluster of several hundred stars at 150 light years distance, is used for calibrating the distances of more distant clusters. (Yerkes Observatory photograph, Bruce 6-inch Telescope, 1906, courtesy Richard Dreiser.)

mass of the Galaxy inside the Sun's galactic orbit is about 200 billion solar masses (Chapter 6), half of which consists of luminous matter in the form of stars. The luminous matter of the Galaxy outside the Sun's orbit is much less, and therefore the luminous mass of the Galaxy is not much greater than 100 billion solar masses.

If the entire mass M of the Galaxy were concentrated inside a galactic orbit of radius r, the rotation velocity V_{rot}, given by $V_{rot} = (GM/r)^{1/2}$, would decrease with radius as $r^{-1/2}$. But the rotation velocity beyond the Sun's orbit, revealed by observations of 21-centimeter radio emission from hydrogen

gas, does not decrease but stays almost constant to a radial distance of more than 50 kiloparsecs, as shown in Figure 19.3. Other spiral galaxies also show similar constant rotation velocities. A rotation velocity that stays constant out to such large distances indicates that most of the mass of our Galaxy lies outside the Sun's orbit, perhaps in the halo, and the total mass may be as large as 10^{12} solar masses. According to this interpretation, the mass inside radius r increases with r, and therefore the density drops as $1/r^2$.

This raises the problem of dark matter, a problem that worsens as we go to larger

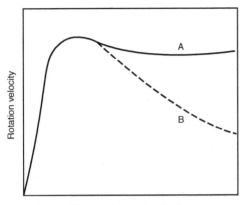

Figure 19.3. A schematic rotation curve of a spiral galaxy – such as the Galaxy – showing how the rotation velocity varies with radius. Beyond 30 000 light years the rotation velocity tends to be flat, as shown by curve A, indicating that considerable nonluminous mass exists in the outer regions of the Galaxy. If no mass existed beyond 30 000 light years, the rotation velocity would fall off in the way shown by the dashed curve B.

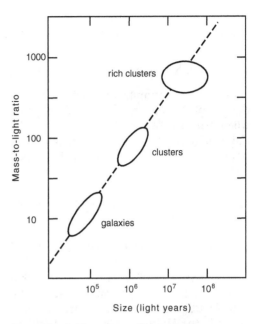

Figure 19.4. The mass-to-light ratio of astronomical systems increases with their size. In this schematic diagram, the mass M and luminosity L are measured in solar units.

systems. At least 90 percent of the matter in the universe responsible for the gravity holding together large astronomical systems is nonluminous. The existence of dark matter is quite certain; its true nature, however, is uncertain. Obvious candidates for various forms of dark matter are large molecular clouds in interstellar space, Jupiter-like brown dwarfs (too small to ignite hydrogen-burning), low-luminosity red dwarfs (just massive enough to burn hydrogen), white dwarfs, supernova remnants (neutron stars and black holes), and matter in less obvious forms, such as primordial black holes (Chapters 13 and 20). Large but as yet undetected populations of brown dwarfs, red dwarfs, and black holes, collectively referred to as MACHOS (massive compact halo objects), might be the answer to the dark-matter problem. Gravitational lensing, the bending of the light of background bodies by foreground bodies, offers hope of detecting the dark matter in the Galaxy, provided it exists in compact forms.

The mass-to-light ratio (Figure 19.4) is a useful tool in the study of dark matter. This ratio, denoted by $M/\mathscr{L}$, is the mass M of a body divided by its luminosity $\mathscr{L}$. Usually the mass is measured in solar-mass units and the luminosity is measured in solar-luminosity units. Hence the $M/\mathscr{L}$ ratio of the Sun is 1. In the Sun's neighborhood of the Galaxy, the ratio is approximately 3. The $M/\mathscr{L}$ value of the Galaxy as a whole is about 50.

Ages of the stars and the Galaxy

The universe is older than the Galaxy by perhaps one or two billion years, and the Galaxy is older than the oldest stars by perhaps less than one billion years. Dating the birth of the oldest stars helps us to fix an approximate lower limit on the age of the universe.

Stars have a finite main-sequence lifetime because of their limited supply of hydrogen. The smaller their mass, the lower their luminosity, and the greater becomes their luminous lifetime. Estimated ages of low-luminosity population II stars in globular clusters that are now turning off the main sequence is 15 billion years. This result shows that the age t_0 of the universe is perhaps at least

$$t_0 = 16 \text{ billion years.} \qquad [19.3]$$

Because of uncertainties in such estimates, the age of the universe lies plausibly somewhere between 14 and 18 billion years. The radioactive dating (nuclear cosmochronology) of the chemical elements made in stars and ejected into space indicates a similar result.

The Hubble time (the reciprocal of the Hubble term H_0) is

$$t_H = \frac{1}{H_0} = 9.8h^{-1} \text{ billion years} \qquad [19.4]$$

(see Equation 14.24), where $H_0 = 100h$ kilometers a second per megaparsec is the Hubble term in the velocity–distance law, and h is the Hubble coefficient thought to be between 0.5 and 1. As usual, the zero subscript indicates a present value. The age of a decelerating universe falls short of t_H (see Figure 19.5), and if the actual age is $t_0 = 16$ billion years, then clearly h must be less than $9.8/16 = 0.61$. The popular Einstein–de Sitter universe, in which the deceleration term q is permanently 0.5 (and the density parameter Ω is permanently 1),

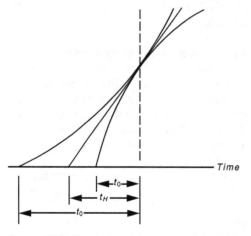

Figure 19.5. The age t_0 of the universe is less (greater) than the Hubble time $t_H = 1/H_0$ in a decelerating (accelerating) universe.

has an age

$$t_{\text{crit}} = \tfrac{2}{3}t_H = 6.5h^{-1} \text{ billion years,} \qquad [19.5]$$

and for this value to equal 16 billion years we want $h = 0.41$. In closed Friedmann universes of q_0 greater than 0.5, of age equal to 16 billion years, the Hubble coefficient h is smaller than 0.41. In open Friedmann universes of q_0 between 0 and 0.5, of age equal to 16 billion years, the Hubble coefficient h lies between 0.41 and 0.61. Many cosmologists feel that these low values of h are unacceptable and indicate perhaps the existence of a cosmological term Λ.

Abundances of the elements

The problem of the origin and age of the chemical elements has been mostly solved. Until early in the twentieth century, astronomers thought the composition of the Sun was similar to that of the Earth. But now we know that the Sun and other main sequence stars consist mostly of hydrogen and helium, and have comparatively only small amounts of all other elements. George Gamow in the late 1940s believed that elements heavier than hydrogen were primordially synthesized. More recent work has shown that only the light chemical elements – deuterium, helium, and lithium – were produced in this way when the universe was about 200 seconds old. All the stars on the main sequence, slowly converting hydrogen into helium, have added about 1 to 2 percent to the primordial helium abundance. The heavy chemical elements (e.g., carbon, oxygen, ... iron, nickel, ... thorium, uranium, ...) were synthesized in stars, and their present abundances are clues to the history of our Galaxy. The abundances of light elements, on the other hand, are clues to the history of the early universe (Chapter 20).

Fred Hoyle and Roger Tayler at Cambridge University in 1964 raised the problem of explaining the large helium abundance. The helium fraction of matter by mass is approximately $Y = 0.25$. They showed that a hot big bang can solve the problem. The predicted light-element abundances, according to more recent calculations (Chapter 20),

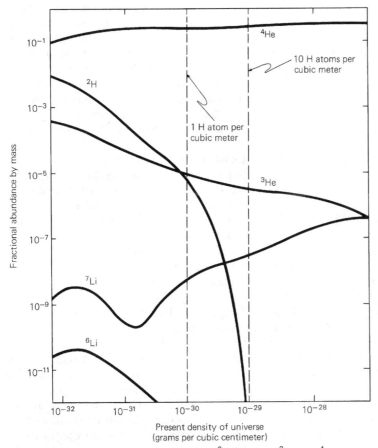

Figure 19.6. The abundances of deuterium (^{2}H), helium (^{3}He and ^{4}He), and lithium (^{6}Li and ^{7}Li) produced in the first 200 seconds of the early universe. The curves show how the mass fractions of these elements relate to the present-day average density of baryonic matter in the universe. Notice how the deuterium abundance is sensitive to the matter density: the more dense the matter, the more frequently deuterium nuclei collide with one another in the early universe and combine to form helium nuclei. Adapted from David Schramm and Robert Wagoner, "What can deuterium tell us?: *Physics Today* (1974).

are shown in Figure 19.6 for a range of values in the present average density of baryonic matter. Baryonic matter consists of atoms made of protons and neutrons. The calculations assume that the background radiation temperature is 2.73 kelvin, and only three species of leptons (electrons, muons, and tauons) exist. The abundance of deuterium is particularly sensitive to the density of baryonic matter.

A fraction of interstellar matter is recycled through stars (star formation, stellar winds, novas, and supernovas) and some primordial deuterium inevitably gets burned to helium. Therefore more deuterium was made in the early universe than is now observed. The problem of determining the original deuterium abundance is alleviated by using the combined abundance of ^{3}He + D (because deuterium burns to helium-3). The observed abundances of deuterium and helium-3 indicate a present baryon density

$$\rho_{b0} = 4 \times 10^{-31}$$
grams per cubic centimeter. [19.6]

If the universe contains only baryonic matter, then this is near enough the total density ρ_0. The critical density of the Einstein–de Sitter universe ($q_0 = 0.5$, or $\Omega_0 = 1$) is

$$\rho_{crit} = \frac{3H_0^2}{8\pi G} = 2 \times 10^{-29} h^2$$

grams per cubic centimeter. [19.7]

On using the density parameter for baryonic matter $\Omega_{b0} = \rho_0/\rho_{crit}$, we find

$$\rho_{b0} = 2 \times 10^{-29} \Omega_{b0} h^2, \qquad [19.8]$$

and hence, from Equations [19.6] and [19.8],

$$\Omega_{b0} h^2 = 0.02. \qquad [19.9]$$

If the universe contains only baryonic matter, then Ω_{b0} is near enough the total density parameter Ω_0. For a value of h between 0.5 and 1.0, the density parameter Ω_{b0} lies between 0.02 and 0.08. In this case, in a Friedmann universe of $\Omega_0 = 2q_0$, the deceleration term q_0 has a value between 0.01 and 0.04, and the universe is open, in agreement with the previous discussion on age.

The total density parameter Ω_0 includes radiation and all other sources of gravity. The present radiation density ρ_{r0} is negligibly small; but if the universe contains a significant amount of hitherto undetected nonbaryonic matter, such as heavy neutrinos or more exotic forms of weak interacting massive particles (known as WIMPS), then the total density ρ_0 of the universe consists of the sum of the baryonic (ρ_{b0}), nonbaryonic ($\rho_{?0}$), and radiation (ρ_{r0}) densities. The total density parameter is the sum

$$\Omega_0 = \Omega_{b0} + \Omega_{?0} + \Omega_{r0}, \qquad [19.10]$$

where $\Omega_{b0} = \rho_{b0}/\rho_{crit}$, $\Omega_{?0} = \rho_{?0}/\rho_{crit}$, and $\Omega_{r0} = \rho_{r0}/\rho_{crit}$.

In this book the term "dark" matter refers to nonluminous matter of all forms that comprise Ω_0; the term "missing" matter refers only to nonbaryonic forms that comprise $\Omega_{?0}$. Most of ρ_{b0} is dark matter. If $\Omega_0 = 1$, then most of the difference between ρ_{b0} and ρ_0 is not only dark but also missing. Dark matter undoubtedly exists, but does missing matter exist? If the universe passed through a period of inflation in the early universe – an idea that has grown in popularity in recent years – then most likely the density parameter Ω_0 has a value equal to 1. In that case the gap between ρ_{b0} and ρ_0 is large, and the universe is dominated by missing matter.

Cosmic background radiation

Since 1941, astronomers have found that some molecules in interstellar space are excited by low-energy radiation. Later, radio astronomers found a persistent noise or hiss in their receivers that at first had no obvious explanation. The molecular excitations and the radio noise were finally explained in the following way. Arno Penzias and Robert Wilson, of the Bell Telephone Laboratories, systematically eliminated all possible sources of noise from their radio receiver and horn-shaped antenna (Figure 19.7) at Holmdell in New Jersey, and finally arrived at an irreducible noise level having roughly a temperature of 3 kelvin. The mysterious noise seemed to have an extraterrestrial origin. Meanwhile, James Peebles, a young astrophysicist at Princeton University, was repeating the calculations made a decade or so earlier by Ralph Alpher and Robert Herman, and was exploring the idea of a hot early universe and a present-day vestigial cosmic background radiation. This work came to the attention of Penzias and Wilson, and in 1965, Robert Dicke, Peebles, and their colleagues at Princeton University identified the radio noise detected by Penzias and Wilson as the cosmic background radiation. Penzias and Wilson received in 1978 the Nobel prize in physics for their immensely important serendipitous discovery.

The cosmic background radiation (CBR), which decoupled from matter at redshift 1000, now has a temperature 2.728 kelvin determined by sensitive infrared detectors. (In this book the temperature is often stated less precisely, sometimes as 2.7, 2.73, or even 3 kelvin.) Observations

Figure 19.7. The 20-foot parabolic horn-shaped antenna of the Bell Telephone Laboratories at Holmdell, in New Jersey, used by Arno Penzias and Robert Wilson in their discovery of the cosmic microwave background radiation in 1965. (Courtesy of Robert Wilson. Copyright, 1978, with permission of Lucent Technologies.)

from the Cosmic Background Explorer (COBE) satellite have produced two surprises. First, as shown in Figure 19.8, the intensity follows the blackbody curve of thermal radiation with a deviation of only 1 part in 10^4. Second, after the subtraction of a large 24-hour anisotropy (i.e., maximum and minimum values occur in opposite directions of the sky), the radiation is surprisingly isotropic with only small amplitude anisotropies. From the large 24-hour anisotropy we know that the Galaxy is moving at 600 kilometers per second toward the constellation Leo. The small anisotropies, typically only a few parts in 10^5, which survive from the time of decoupling (at redshift 1000), are the imprint of density irregularities of matter that subsequently evolved into galaxies, clusters of galaxies, and superclusters (Figure 19.9).

The number density of photons in the cosmic background radiation is around 400 per cubic centimeter (Chapter 17); the number density of nucleons is ρ_{b0}/m_n, or around $1 \times 10^{-5} \Omega_{b0} h^2$; the photon–nucleon number ratio is hence

$$\eta = 4 \times 10^7 (\Omega_{b0} h^2)^{-1}, \qquad [19.11]$$

or 2×10^9, according to Equation [19.9]. To every nucleon there are 2 billion photons. This ratio, whose value cosmologists so far have not explained, is a measure of the baryon asymmetry and the specific entropy of the universe (see Chapter 20).

Cosmological parameters

We learn much about the universe, not by searching the far depths of space, but by searching the local depths of time. The oldest stars imply that the universe has an age of at least 16 billion years, creating an age problem for the Friedmann universes unless the Hubble coefficient h is less than about 0.6. From the abundances of the

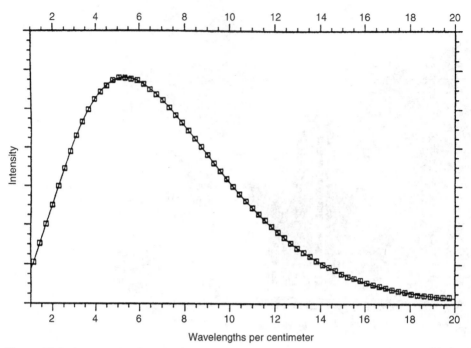

Intensity

Wavelengths per centimeter

Figure 19.8. The intensity of the cosmic microwave radiation at different frequencies corresponding to a temperature 2.728 kelvin determined by instruments on the Cosmic Background Explorer (COBE) satellite launched in November 1990. (Frequency is expressed as wavelengths per centimeter.) (Courtesy John Mather, Dale Fixsen, and Rick Shafer at NASA, Goddard Space Flight Center.)

COBE-DMR 4-year Sky Map

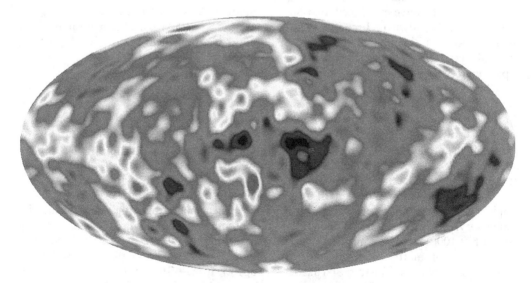

Figure 19.9. A full sky simulation of an anisotropy map based on four years of observations made with the Cosmic Background Explorer. The temperature difference between the black and white regions is 10^{-4} kelvin. (Reproduced courtesy of Gary Hinshaw at NASA, Goddard Space Flight Center.)

light elements made in the early universe we infer an average baryonic density of 4×10^{-31} grams per cubic centimeter (approximately 10 times more than the density of luminous matter). From this result we find quite generally that $\Omega_{b0}h^2 = 0.02$. Thus Ω_{b0} is small if h lies in the generally accepted range 0.5 to 1. An Einstein–de Sitter universe in which Ω is permanently equal to 1, which contains only baryonic matter, has a Hubble coefficient $h = 0.14$ that most astronomers would view as unacceptably low. Larger values of h require nonbaryonic additions referred to as missing matter. The word "missing" makes clear that nonbaryonic matter, if it exists, not only is dark – like most baryonic matter – but also at present is unknown. The cosmic background radiation – almost perfectly thermal (Figure 19.8) and almost perfectly isotropic (Figure 19.9) – is by far our best evidence that the universe is basically a homogeneous unity (Chapter 7), and its small amplitude anisotropies reveal the embryos of astronomical systems.

INTERMEDIATE-DISTANCE OBSERVATIONS
Structure, distribution, and motion of galaxies

The study of galaxies – their compositions, structures, motions, emissions, absorptions, masses, sizes, histories, and pathologies – provides us with standards and insights for comparison with more distant galaxies. By comparing local and distant galaxies, and allowing for the effects of evolution, we try to use galaxies as distance indicators.

The Local Group, dominated by two giant spiral galaxies (our Galaxy and M 31), is a swarm of galaxies, most of which are dwarf systems having typical speeds 100–200 kilometers per second. Beyond the Local Group lie hordes of nearby similar groups containing ellipticals (such as M 87), spirals (such as M 81 and M 101), irregulars (such as M 82), and numerous dwarf galaxies. These groups stretch away to the rich and irregular Virgo cluster of thousands of galaxies at a distance of 70 million light years. This vast complex of clusters, comprising more than 1 million galaxies in a cube of side 200 million light years, forms the Local Supercluster, known also as the Virgo supercluster. The Coma and Perseus clusters are vast concentrations of galaxies similar to the Virgo system.

Gerard de Vaucouleurs, in the 1950s, following the work of Vera Rubin on the systematic peculiar flow velocities of galaxies, proposed the existence of a local supercluster of galaxies that he called the Supergalaxy. Extragalactic surveys have since confirmed his proposal and shown furthermore that the universe consists of numerous superclusters connected by luminous bridges and separated by nonluminous voids. Superclusters form blocks or "plates" analogous to the plates in plate-tectonic theory of the Earth's surface. The cosmic blocks expand, but because each is partly held together by its own gravity, the spaces between the blocks slowly widen. Each block contains large internal systematic peculiar motions, adding to the difficulty of finding the true value of the Hubble term.

Intermediate distance scales

Often in astronomy we compute the distance of a source by comparing the received luminous flux with its emitted flux (as in Equations 19.1 and 19.2). This method, however, has the limitation that the emitted flux must be known. The source, whose distance we seek to determine, must be recognizably similar to nearer sources whose luminosity is already known.

In our first steps beyond the Galaxy we continue to use stars and star clusters as distance indicators, comparing them with nearer stars and star clusters of known distances. The bright RR Lyrae variable stars are used to a distance of 1 million light years, the pulsating cepheid stars to 10 million light years, the globular clusters of known size and absolute magnitude to 30 million light years, and supernovas (type Ia) of known luminosity to at least 100 million light years.

In the intermediate universe, we look out to distances extending several hundred million light years and look back in time several hundred million years. The evolutionary changes in stars, globular clusters, and galaxies on the time scale of 1 Galactic year (two hundred million years) are normally small and finding reliable standard candles is not the most serious difficulty. Distances measured in the world picture (on the backward lightcone) and in the world map (where the velocity–distance law holds true) are not greatly different.

The mass M and radius R of a gravitationally bound system are related by an expression known as the virial theorem:

$$GM = U^2R, \qquad [19.12]$$

where U is a characteristic internal velocity inferred either from the Doppler-shifted emissions and absorptions in different parts of the system, or simply from the Doppler-broadened lines of the whole system. The mass M includes all gravitational matter. The velocity U may be a rotational velocity in a spiral galaxy, or the relative velocity of two galaxies in orbit about each other, or the typical velocity of many galaxies in a cluster, and in each case M denotes the mass of the system. When mass is measured in this way, the mass-to-light ratio $M/\mathscr{L}$ tends to increase with the size of the system and the proportion of dark matter increases.

Let us select spiral galaxies that have similar $M/\mathscr{L}$ ratios. Their Doppler-broadened spectral lines of 21-centimeter radiation caused by disk rotation are proportional to their mass M according to the virial theorem. The mass M in turn is proportional to luminosity $\mathscr{L}$ according to the known $M/\mathscr{L}$ ratio. Hence there exists a correlation between line width and the luminosity (or absolute magnitude). Because of this "Tully–Fisher relation," first used by Brent Tully and Richard Fisher, distances of spiral galaxies as far away as 300 million light years can be determined from measurement of line widths and apparent magnitudes. One difficulty: the line widths depend on the inclination of the spiral galaxy to the line of sight and this inclination must be determined.

Each method of measuring extragalactic distances has its problems and uncertainties, and the whole subject often seems more an art than a science. Questions abound. One example: how do we explain the odd fact that there has not been a supernova in the Galaxy since 1604, whereas supernovas in other similar galaxies occur on the average once every 50 years?

The local universe, spanned by intermediate distance scales, is the sub-Hubble "sphere" (a realm not necessarily spherical) in which astronomical peculiar motions dominate over cosmic recession. In Chapter 14 we saw

$$L_{\text{sub-}H} = V_{\text{pec}}H_0^{-1}$$
$$= 10z_{\text{pec}}h^{-1} \text{ billion light years,}$$
$$[19.13]$$

where $L_{\text{sub-}H}$ is the sub-Hubble radius and V_{pec} is the maximum peculiar radial velocity corresponding to a redshift z_{pec}. For $V_{\text{pec}} = 9000$ kilometers a second, or $z_{\text{pec}} = 0.03$, the sub-Hubble distance $L_{\text{sub-}H}$ is $300h^{-1}$ million light years.

Redshifts

We can temporally sidestep the problem of finding distances by using redshift measurements in a redshift space of polar coordinates z, θ, ϕ (see Figure 19.10); alternatively, by using radial velocities in a radial velocity space of polar coordinates V, θ, ϕ, where

$$V = cz = 300\,000z$$
$$\text{kilometers per second.} \quad [19.14]$$

Redshifts and velocities must be corrected for the peculiar motion of the Earth.

When we know the Hubble term H_0, we may cautiously use Hubble's redshift–distance relation

$$cz = H_0L, \qquad [19.15]$$

(see Chapter 14) and the redshift z is proportional to distance L. An equivalent

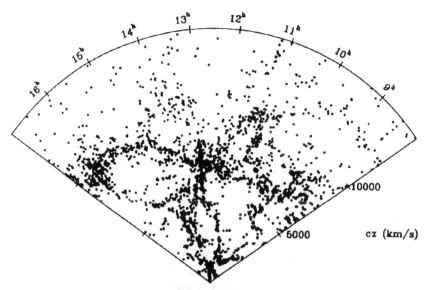

Figure 19.10. Center for Astrophysics redshift survey. This map contains 2529 galaxies of apparent magnitude m less than 15.6 in a slice of the north galactic cap between declination 26.5 and 44.5 degrees, displayed in right ascension. The radial coordinate is velocity in kilometers per second obtained from multiplying redshift by the speed of light. (Courtesy of John Huchra and Margaret Geller, with permission of the Harvard-Smithsonian Institute.)

relation is

$$L = zL_H = 10zh^{-1}$$
$$\text{billion light years,} \qquad [19.16]$$

where $L_H = c/H_0 = 10h^{-1}$ billion light years is the Hubble distance. Apart from their approximate nature, these relations neglect the peculiar velocity of the source.

A source of $z = 0.02$ lies at distance $200h^{-1}$ million light years and has a recession velocity 6000 kilometers per second. This is not an unreasonable estimate of the size of the sub-Hubble sphere. The Virgo cluster at $z = 0.004$, $V = 1150$ kilometers per second, lies inside the sub-Hubble sphere, whereas the Coma cluster at $z = 0.023$, $V = 7000$ kilometers per second, lies outside.

A common observational procedure uses Equation [19.1] in which the distance is given by Equation [19.15]:

$$m = M + 15 - 5\log h + 5\log cz, \quad [19.17]$$

where cz is the recession velocity, or

$$m = M + 42.4 - 5\log h + 5\log z, \quad [19.18]$$

where the speed of light c is 300 000 kilometers per second. When the absolute magnitude M is known, Equations [19.17] and [19.18] may be used to determine H_0 (in this case h) from the apparent magnitude m. This was how Hubble first determined H_0 (see Figure 14.3) and found that h was approximately 5. Note that when $\log z$ (or $\log cz$) is plotted against m, the curve has slope 0.2 because on intermediate distance scales we are not distinguishing between the world picture and the world map.

Cosmological parameters

Equations [19.13]–[19.18] may be used only for small values of z, less than, say, 0.1. The reasons: first, the linear Hubble redshift–distance relation is an approximation; second, the Fizeau–Doppler velocity–redshift relation is also an approximation; and third, we have failed to project the world picture (consisting of what is seen) onto the world map (where the fundamental velocity–distance law applies). We tread on

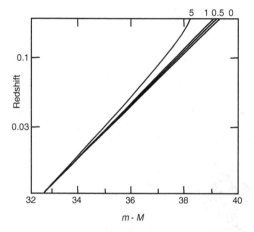

Figure 19.11. A set of curves according to Equation [19.29], showing redshift plotted against the distance modulus $m-M$ for $q = 5, 1, 0.5, 0$, assuming that $h = 1$. The curvature has no effect on the shape of these curves; similarly with Figure 19.12.

perilous ground when these approximate equations are used for larger redshifts.

Two rival schools emerged in the second half of the 20th century. The first, led by Allan Sandage and Gustav Tammann, held that H_0 is close to 50 kilometers a second per megaparsec (h equal to 0.5); the second, pioneered by Gerard de Vaucouleurs, held that H_0 is close to 100 kilometers a second per megaparsec (h equal to 1). Different distance indicators, methods of correction, and selection effects account for much of their different results.

Cosmologists since the time of Hubble have repeatedly been surprised by the complexity and extent of cosmic inhomogeneity. The sub-Hubble sphere – the realm of intermediate distances – has steadily grown larger as a result of advances in observational techniques. Great luminous arcs now stretch across large regions of the extragalactic sky as a tantalizing frosting on hidden worlds of dark matter. We have reached the stage in cosmology where the sub-Hubble sphere contains systematic streaming motions attaining velocities as large as and even larger than 10000 kilometers per second, corresponding to redshifts equal to

0.03 or greater. Unfortunately, at these velocities the linear redshift–distance relation begins to break down, just at the stage when we need it to determine a reliable value of H_0.

LARGE-DISTANCE OBSERVATIONS

Astronomers have an advantage over geologists in that they can directly observe the past.
Martin Rees, Perspectives in Astrophysical Cosmology.

Cosmic distances

The universe evolves and the things seen distributed at large distances in the world picture must be mapped into the world map; only in this way can they be assembled into a theoretically coherent pattern. The simplest mapping process employs no more than the deceleration term. Things, however, evolve and when seen in the distant past at large distances are unlike the nearby things as they are now. Cosmological evolution is complicated by astronomical evolution.

We continue to use, as in previous chapters, the simple geometric idea of distance as used in everyday life. This is the kind of distance we need in the world map. The world map consists of uniformly curved space. Geometric or tape-measured distances are used in the velocity–distance law and are fully defined by the Robertson–Walker metric (Chapter 14):

$$ds^2 = dt^2 - R^2$$
$$\times [dr^2 + S^2(d\theta^2 + \sin^2\theta\, d\phi^2)], \quad [19.19]$$

where r, θ, and ϕ are comoving coordinates, R is the scaling factor that varies with time, and S is a function of r that equals r in flat space $(k = 0)$, $\sin r$ in spherical space $(k = 1)$, and $\sinh r$ in hyperbolic space $(k = -1)$, where k is the curvature constant. The radial comoving distance at the time of observation t_0 to a source that emitted radiation at time t is

$$r = \int_t^{t_0} \frac{dt}{R} = \frac{1}{R_0} \int_0^z \frac{dz}{H}, \quad [19.20]$$

as shown in Equation [14.36], and the tape-measure distance $(L = R_0 r)$ in the world map to a source in the world picture of redshift z is

$$L = \int_0^z \frac{dz}{H}, \qquad [19.21]$$

where H is expressed as a function of z. Notice that R is measured in units of time, and distances are also measured in units of time (the light-travel time). By using Equation [19.21], sources seen in the world picture at redshift z are mapped into the world map at distance L, where their present velocity of recession is $H_0 L$. Equation [19.21] requires that we know how the Hubble term H varies as a function of z. When H is constant (i.e., $H = H_0$, $q_0 = -1$), L reduces exactly to the Hubble relation $L = z L_H$, and the recession velocity V reduces exactly to the Fizeau–Doppler relation $V = cz$, and is true for all values of z.

For small values of z we can expand H in the form:

$$H = H_0[1 + (1 + q_0)z + \cdots], \qquad [19.22]$$

and to this approximation, $S = r$, and flat and curved spaces are indistinguishable. According to Equation [19.21], the distance to a source of redshift z is

$$L = R_0 r = z L_H \left[1 - (1 + q_0)\frac{z}{2} - \cdots \right]. \qquad [19.23]$$

Notice that distance and redshift are no longer related by the linear Hubble approximation. Similarly, the velocity–redshift relation

$$V = H_0 L = cz \left[1 - (1 + q_0)\frac{z}{2} - \cdots \right], \qquad [19.24]$$

is no longer the linear Fizeau–Doppler approximation. Provided z is kept small, Equations [19.23] and [19.24] apply to all expanding uniform universes, and are independent of particular cosmological models (flat or curved, with or without a cosmological term).

Observers use various operational "distances." These so-called distances are physically useless until transformed into the geometric tape-measure distances sanctioned by the Robertson–Walker metric. The commonest of the operational distances is the "luminosity distance" L_{lum} defined by

$$F = \frac{\mathscr{L}}{4\pi L_{lum}^2}. \qquad [19.25]$$

where the luminous flux F (the radiant energy received per unit area per second) from a source of luminosity $\mathscr{L}$ (the total radiant energy emitted per second) is proportional to the inverse square of L_{lum}. The flux, according to the Robertson–Walker metric, is

$$F = \frac{\mathscr{L}}{4\pi (R_0 S)^2 (1 + z)^2}. \qquad [19.26]$$

The source of luminosity $\mathscr{L}$ is at distance $L = R_0 r$, and a sphere of this radius centered on the source has surface area $4\pi (R_0 S)^2$. Hence the observed flux is proportional to $\mathscr{L}/4\pi (R_0 S)^2$. Each photon emitted by the source has its energy reduced by $(1 + z)$; the number of photons emitted by the source in time interval dt equals the number crossing this spherical surface in time $dt_0 = (1 + z)\, dt$, which means the rate of reception of photons is further reduced by $(1 + z)$; hence the observed flux is reduced by $(1 + z)^2$. On comparing Equations [19.25] and [19.26], we find

$$L_{lum} = (1 + z) R_0 S = (1 + z) L \frac{S}{r}, \qquad [19.27]$$

where L is the geometric distance. In flat, static space, the luminosity distance is $L_{lum} = L$. Calculation shows with Equation [19.23] that approximately, for small z,

$$L_{lum} = z L_H \left[1 + (1 - q_0)\frac{z}{2} - \cdots \right]. \qquad [19.28]$$

When this distance is used in the magnitude–distance Equation [19.1], we find

$$m = M + 42.4 - 5 \log h + 5 \log z + 1.09(1 - q_0)z, \qquad [19.29]$$

where $1.09 = 2.5/\log_e 10$. This relation between the distance modulus $m - M$ and the redshift, shown in Figure 19.11, should

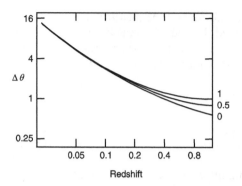

Figure 19.12. Curves showing how the angular diameter $\Delta\theta$ of clusters of galaxies varies with redshift according to Equation [19.33]. The angular diameter $\Delta\theta$ is in minutes of arc.

be compared with Equation [19.17] in which no distinction is made between the world picture and the world map.

A second operational distance is the "apparent-size distance" L_{dia}. In flat, static space a source of diameter D at distance L subtends an angle $\Delta\theta = D/L$ in radians. In general, let

$$\Delta\theta = \frac{D}{L_{\text{dia}}}. \qquad [19.30]$$

According to the Robertson–Walker metric, this angle is expressed as $\Delta\theta = D/RS$. The circumference of the sky in the world picture at comoving distance r is $2\pi RS$; it subtends a total angle 2π (or 360 arc degrees) and hence the fraction $D/2\pi RS$ equals $\Delta\theta/2\pi$. Because $R_0/R = (1+z)$, we find

$$\Delta\theta = (1+z)\frac{D}{R_0 S}, \qquad [19.31]$$

and by comparing this result with Equation [19.30] we find that the distance by apparent size is

$$L_{\text{dia}} = \frac{R_0 S}{1+z} = \frac{L}{1+z}\frac{S}{r}. \qquad [19.32]$$

In flat and static space $L_{\text{dia}} = L$. Using Equation [19.22], we find for small redshifts

$$\Delta\theta = \frac{D}{zL_H}\left[1 + (3+q_0)\frac{z}{2} + \cdots\right] \qquad [19.33]$$

as shown in Figure 19.12. Unfortunately for this method, most galaxies (and clusters of galaxies) at large distances appear smaller than they actually are. This is because their outer regions, being fainter than their inner regions, tend not to be seen. Corrections must be made for this and other effects and the results have a large aura of uncertainty.

Our small redshift relations become inadequate in surveys far beyond the sub-Hubble sphere, and more precise relations must be used. Such relations, as shown in the Reflections, are found in the context of specific models, such as the Friedmann models. Our explorations of the universe of long ago and far away have not reached a consensus on the appropriate model.

Evolutionary effects

The evolution of galaxies and their clusters over long periods of time is a fascinating study still in its infancy. Cosmologists try to select galaxies of predetermined luminosity that will serve as standard candles and act as distance indicators. Sometimes the brightest galaxies in rich clusters are chosen. The aim is to determine the Hubble term H_0 and the deceleration term q_0 from the observations of their apparent magnitudes and redshifts. Unavoidably, however, these astronomical standard candles evolve and reliable estimates depend on knowing how their absolute magnitudes vary over long periods of time.

Evolution means change in time. The lookback time $t_0 - t$ as a function of redshift is:

$$t_0 - t = \int_t^{t_0} dt = \int_0^z \frac{dz}{(1+z)H}. \qquad [19.34]$$

We see luminous systems as they were long ago, and to compare them with present-day systems of a similar nature we need to know how they have evolved over the lookback period $t_0 - t$. By using the approximation for H in Equation [19.22], we find, to second order in z, that

$$t_0 - t = zt_H\left[1 - z\left(1 + \frac{q_0}{2}\right) - \cdots\right], \qquad [19.35]$$

where $t_H = 1/H_0$ is the Hubble time. If we limit ourselves to $z \leq 0.1$, then the maximum allowed lookback time is little more than 1 billion years and the evolutionary change in most galaxies is not large. The lookback time for larger redshifts must be determined in the context of specific models.

Do galaxies get brighter or fainter with age? One might expect them to get fainter, yet theory and observation suggests that bright galaxies at the centers of rich clusters are swallowing smaller galaxies and growing larger and brighter in the process. These changes are sufficient to mask the effect of the deceleration term – astronomical evolution dominates over cosmological evolution – and the once popular method of seeking to determine H_0 and q_0 by means of the magnitude–redshift relation now seems less practical.

Solid state detectors, a hundred times more sensitive than photographic plates, reveal numerous faint galaxies, as many as 100 per square arcminute, so distant that their light has been in transit during most of the lifetime of the universe. Quasars blaze as bright beacons at cosmological distances but their wide differences in luminosity make them unsuitable as distance indicators. Presumably most quasars and other active galactic nuclei originate during the early stages of galactic evolution. The large redshifts of many quasars (the large redshifts, such as PC 1247 + 3406 at $z = 4.89$) show that galaxies had already formed when the universe was perhaps no more than 10 percent of its present age. Quasars were most abundant in the past at redshifts between 3 and 4, and their numbers have since greatly declined. Absorption lines at different redshifts in their spectra suggest that large clouds of intergalactic gas may exist that have not condensed to form galaxies.

How do clusters of galaxies evolve? The x-ray emission from intergalactic gas in rich clusters has increased in time, and this observation supports the view that large clusters are probably not as old as the galaxies.

Cosmological parameters

The magnitude–redshift relation (Equation 19.29), with various modifications and different selection procedures made by astronomers, has been the workhorse of modern observational cosmology since the time of Edwin Hubble. The aim has been to determine from observations the value of H_0 at smaller redshifts and the value of q_0 at larger redshifts. But observations are confusing at smaller redshifts because of large sub-Hubble systematic motions and at larger redshifts because of evolutionary effects. Over the whole subject looms the possibility that inexact relations are used beyond their range of validity. For example, Hubble's linear redshift–distance relation breaks down at modest redshifts and a more precise relation must be formulated in the context of a specific model.

In summary, we are still a long way from a secure knowledge of the values of the Hubble term H_0, the density parameters Ω_{b0} and Ω_0, the deceleration term q_0, the cosmological term Λ, and the curvature constant k.

IS THE UNIVERSE OPEN OR CLOSED?

Is space curved or flat on the cosmic scale? Cosmologists have sought the answer to this question since 1900, when Karl Schwarzschild first attempted to measure the curvature of space. We notice that none of the approximate expressions deriving from the expansion of the Hubble term in Equation [19.22] contain the curvature constant k. At small redshifts, to second order, the geometric distance L defined by the Robertson–Walker metric is independent of curvature. Our observations of distant regions tell us how fast the universe expands (the Hubble term H_0) and how fast it decelerates (the deceleration term q_0), but not the curvature of space. Only in the context of a particular model or family of models, such as the Friedmann models, can we discover from the values of H_0 and q_0 the answer to our question. No consensus exists, however, on what is the correct model and the problem seems no nearer solution than in Schwarzschild's day.

Debates on whether the universe is open or closed still continue. Almost all cosmologists assume that general relativity is valid on the cosmic scale (although it has never been proved), and many cosmologists assume that the Λ force is zero and the universe is of the Friedmann type. With these assumptions we have three ways to determine if a Friedmann universe is open or closed. First, by comparing the observed age of the oldest stars with the critical age t_{crit}; second, by comparing the observed density of the universe with the critical density ρ_{crit}; and third, by comparing the observed value of q_0 with the critical value 0.5. When restricted to this limited class of models, most of the evidence indicates that our universe is open and therefore of infinite extent in space. If, however, the Λ force exists, then this conclusion does not necessarily apply.

REFLECTIONS

1 *"But suppose that redshifts are velocity-shifts, and that the nebulae are receding. Then the law of redshifts furnishes information, not concerning the contents or properties of internebular space, but concerning the expansion of the universe as a whole. There seems to be no a priori necessity for a linear law of expansion, a strict proportionality between redshifts and distance. Indeed, the general theory indicates that the law must depart from linearity. If our sample is sufficiently large we may, perhaps, observe the departure and determine its trend. The information would immediately restrict our actual world to a particular family of possible worlds"* (Edwin Hubble, The Observational Approach to Cosmology, 1937).

2 *"The dominating feature of recent observational work has undoubtedly been the revision of the distance scale, and with it of Hubble's constant, by Baade and Sandage. It is not easy to appreciate now the extent to which for more than fifteen years all work in cosmology was affected and indeed oppressed by the short value of T (1.8 × 10⁹ years) so confidently claimed to have been established observationally"* (Hermann Bondi, Cosmology, 1960). Here T is the reciprocal of the Hubble term.

3 *A candle flame at distance 1 kilometer is as bright as the first magnitude $(m = 1)$ stars Aldebaran and Altair. The pole star Polaris is second magnitude $(m = 2)$ and similar to a candle flame at distance $\sqrt{2.5} = 1.6$ kilometers. The higher the magnitude the fainter the apparent brightness of the source. A candle flame at 10 kilometers is 100 times fainter and therefore like a sixth magnitude $(m = 6)$ star. Sources brighter than zero magnitude have negative magnitudes; for example, Sirius $(m = -1.6)$, Venus $(m = -4)$, and the full Moon $(m = -12.5)$ are respectively 10, 100, and 250 000 times brighter than a star of first magnitude. From the rule: brightness is proportional to $2.5^{(1-m)}$, and from the inverse-square rule, brightness is proportional to $1/L^2$, where L is the distance of the source, we find that L^2 is proportional $2.5^{(m-1)}$. But $2.5^5 = 100$, and therefore L is proportional to $10^{(m-1)/5}$. Astronomers define the absolute magnitude M as that of a first magnitude star at distance 10 parsecs. Hence*

$$\frac{L}{10} = 10^{(m-M)/5},$$

where L is measured in parsecs. Taking the logarithm of both sides, we get (Equation 19.1):

$$m - M = 5\log(L/10).$$

The Sun has an absolute visual magnitude of $M = 4.8$, which means that at 10 parsecs it is a star of apparent magnitude $m = 4.8$. The visual magnitude m_{vis} seen by the eye is more than the bolometric magnitude m_{bol} that includes all forms of radiation. The bolometric absolute magnitude of the Sun is not 4.8 but 4.7. Because $\mathcal{L}_\odot$ is the total luminosity of the Sun, the absolute magnitude of any source of luminosity $\mathcal{L}$ is

$$M = 4.7 - 2.5\log(\mathcal{L}/\mathcal{L}_\odot)$$

(Equation 19.2). The luminosity of the Galaxy is around $10^{10} \mathscr{L}_\odot$, corresponding to an absolute magnitude of $M = -20.3$.

4 *Imagine a universe containing uniformly distributed sources all having identical luminosities. In a Euclidean static space the number N observed increases with distance L as L^3, the observed flux F from each decreases with distance as $1/L^2$, and therefore*

$$F = \text{constant} \times \frac{1}{N^{2/3}}.$$

Both the flux F from a source and the number N of sources out to the observed source are in principle observable quantities and the results do not depend on measurements of redshift. If the observed flux decreases faster or slower than the number N raised to the inverse two-thirds power, either the sources are not of similar luminosities, or the sources evolve, or space is not static, or the geometry of the universe is not Euclidean. Astronomers once hoped that the two-thirds rule would provide a simple test of the geometry of the universe. But in the real world the problems of selection of similar sources and the effects of evolution make impractical the determination of the curvature of space by this method.

• *Radioastronomers in the 1950s, notably Martin Ryle and his colleagues, discovered the first real evidence that the universe is evolving. At that time very few radio sources had known redshifts determined from their optical counterparts. Instead, radioastronomers used number-counts. They found that the number per unit volume of faint and presumably very distant radio sources greatly exceeded the number per unit volume of bright and presumably less distant radio sources. This was persuasive evidence of the greater abundance of radio sources in the past than today, thus disproving the basic tenet of the steady-state theory. The discovery by optical astronomers of the greater abundance of quasars in the past provided further evidence that the universe is evolving.*

• *"Our nearest bright quasar, 3C 273, is 500 Mpc away. At [a time corresponding to redshift] $z = 2$–2.5, the nearest quasar would have been 30 times closer and as bright as a 4th-magnitude star. (It is an anti-anthropic irony that the best time to be an astronomer was at that early era, before the Earth had formed)"* (Martin Rees, Perspectives in Astrophysical Cosmology, *1995*).

5 *In "Discovery of the cosmic microwave background" (1990), Robert Wilson writes, "Arno and I of course were very happy to have any sort of an answer to our dilemma. Any reasonable explanation would have probably made us happy. In fact, I do not think that either of us took the cosmology very seriously at first. We had been used to the idea of steady-state cosmology; I had come from Caltech and had been there during many of Fred Hoyle's visits. Philosophically, I liked the steady-state cosmology. So I thought that we should report our result as a simple measurement: the measurement might be true after the cosmology was no longer true!"*

• *In* The First Three Minutes *(1977), Steven Weinberg writes: "The detection of the cosmic microwave radiation background in 1965 was one of the most important scientific discoveries of the twentieth century. Why did it have to be made by accident? Or to put it another way, why was there no systematic search for the radiation, years before 1965?" The technology existed but nobody had in mind the need to make a serious search. Weinberg suggests three reasons. First, the success of stellar nucleosynthesis seemed to eliminate all need for primordial nucleosynthesis and to give credence to the steady-state theory. Second, communication between theorists and experimenters had failed; theorists did not know that 3 kelvin radiation could be detected and experimenters (such as Penzias and Wilson) did not know that it might exist. Third, before 1965 physicists had difficulty taking cosmology seriously, and many of them regarded anything to do with the early universe as fanciful speculation.*

6 *"Gerard de Vaucouleurs on the one hand, and Allan Sandage and Gustav Tammann on the other, arrived at estimates of the size of the universe, as measured by the Hubble constant, differing from each other by a factor of two. Moreover, when I asked the protagonists* what was the range outside which they could not imagine the Hubble constant lying, these ranges did not even overlap" (Michael Rowan-Robinson, The Cosmological Distance Ladder, 1985).

7 Dark matter in the universe is not a new idea. In 1907, Edward Fournier d'Albe in Two New Worlds pointed out that the darkness of the night sky (Chapter 24) could be explained if, to every bright star, there are 10^{12} dark stars. For other reasons astronomers realized that this was far too much dark matter. On the basis of his studies of the motions of stars in the Sun's neighborhood, the Dutch astronomer Johannes Kapteyn wrote in 1922, "we have the means of estimating the mass of dark matter in the Universe." Also in 1922, James Jeans wrote: "There must be about three dark stars in the universe to every bright star." (See Virginia Trimble, "Dark matter" in The Encyclopedia of Cosmology.)

8 "Ideally, the question of whether the universe is open or closed is a nineteenth-century Riemannian problem that should be resolved by geometrical surveying using nothing more than the Robertson–Walker metric, and it should not be entangled in the ambiguities of twentieth-century Einsteinian dynamics. This means the nonlinearity of the redshift–distance relation becomes the central issue. The first-order term in the redshift–distance relation determines the Hubble constant, the second-order and higher nonlinear terms determine the deceleration constant, and the third-order and higher nonlinear terms determine the curvature constant. Unfortunately, we are still at the beginning of this cosmographical project, debating the correct value of the Hubble constant. Because of astronomical evolutionary effects, we have virtually given up hope at present of measuring directly the deceleration constant. This leaves us stranded on the shores of the unknown a long way from reaching the curvature constant by direct surveying. In the last sixty or more years we have heard repeatedly the question: is the universe open or closed? and have heard numerous times that it is closed and numerous times that it is

open. Occasionally, we need to remind ourselves that in cosmology almost nothing is certain" (Harrison, in a review of Is the Universe Open or Closed? by Peter Coles and George F. R. Ellis).

9 In the Robertson–Walker metric (Equation 19.19), the term S, important in cosmological observations, can be expanded:

$$k = 0: \quad S = r,$$

$$k = 1: \quad S = \sin r = r - \frac{r^3}{6} + \frac{r^5}{120} - \cdots,$$

$$k = -1: \quad S = \sinh r = r + \frac{r^3}{6} + \frac{r^5}{120} + \cdots.$$

To second order in z,

$$r = \frac{z}{R_0 H_0}\left[1 - \frac{z}{2}(1 + q_0) - \cdots\right], \quad [19.36]$$

$$S = \frac{z}{R_0 H_0}\left[1 - \frac{z}{2}(1 + q_0) - \cdots\right], \quad [19.37]$$

and S is equal to r and both are independent of the curvature constant k. Second-order redshift corrections to the linear Hubble ($cz = H_0 L$) and Fizeau–Doppler ($V = cz$) relations are insufficient to determine by direct observation whether the universe is open or closed. Only indirectly, in the context of an assumed model, can we determine the curvature of the universe from measurements of H_0 and q_0.

10 It was shown in Chapter 18, for the Friedmann universes (cosmological constant equal to zero), that

$$H = H_0(1 + z)(1 + \Omega_0 z)^{1/2}, \quad [19.38]$$

$$R_0 H_0 = [k/(\Omega_0 - 1)]^{1/2}, \quad [19.39]$$

where we have used the density parameter $\Omega_0 = 2q_0$. In these models the comoving coordinate distance r to a source of redshift z is

$$r = \frac{Z}{R_0 H_0} = Z\left(\frac{\Omega_0 - 1}{k}\right)^{1/2} \quad [19.40]$$

and $L = Z/H_0$, where

$$Z = \int_0^z \frac{dz}{(1 + z)(1 + \Omega_0 z)^{1/2}}. \quad [19.41]$$

The tape-measure distance in the world map and the velocity–distance law are, respectively,

$$L = ZL_H,$$ [19.42]

$$V = cZ,$$ [19.43]

correct for all values of redshift z. When z is small,

$$Z = z$$ (to first order)

$$Z = z\left[1 - \frac{z}{2}\left(1 + \frac{\Omega_0}{2}\right)\right]$$ (to second order). [19.44]

The evaluation of the integral Z in Equation [19.41] is best expressed in the form

$$Y = \frac{2}{\Omega_0^2(1+z)}[\Omega_0 z - \Omega_0 + 2$$
$$+ (\Omega_0 - 2)(1 + \Omega_0 z)^{1/2}],$$ [19.45]

where

$$\frac{Y}{Z} = \frac{S}{r}$$ [19.46]

and

$$S = \frac{Y}{R_0 H_0} = Y\left(\frac{\Omega_0 - 1}{k}\right)^{1/2}.$$ [19.47]

The curvature term $k = 1$ applies when $\Omega_0 > 1$ and $S = \sin r$; $k = 0$ applies when $\Omega_0 = 1$ and $S = r$; and $k = -1$ applies when $\Omega_0 < 1$ and $S = \sinh r$.

The luminosity distance according to Equation [19.27] is

$$L_{\text{lum}} = (1 + z)YL_H,$$ [19.48]

which, when used in the magnitude–distance relation, gives a more precise result than Equation [19.29] in the context of the Friedmann models. Also, the apparent-size distance is

$$L_{\text{dia}} = \frac{YL_H}{(1+z)},$$ [19.49]

instead of Equation [19.32].

PROJECTS

1 Discuss the following: "To begin with, there is the fundamental problem of why the scientific procedure I have outlined should be possible at all! Why is it, for instance, an astronomer who studies the sky tonight can go through certain operations on a piece of paper, or in a computing machine, and then say with the utmost precision what some other astronomer will see if he studies the sky in, say, a thousand years hence? If you want an answer I will give you one. We have no idea!" (W. H. McCrea, in Cosmology Now. Editor L. John).

2 Peter Sheuer, a radioastronomer at Cambridge, England, said in 1963 "there are only two and a half facts in cosmology." The first is the sky is dark at night (Chapter 24); the second is the recession of the galaxies (Chapter 14); and the half fact is that the contents of the universe are evolving. Are there now other "facts" we can add?

3 "Jupiter's moons are invisible to the naked eye and therefore can have no influence on the Earth, and therefore would be useless, and therefore do not exist" (Francisco Sizzi, 1610). Most of the universe is unobserved. The Sizzi principle – if you can't see it, it either doesn't matter or it doesn't exist – in various ways is still widely practiced and perhaps always will be. Discuss Sizzi's extreme positivist philosophy.

4 Consider the thought: We live in a "block-tectonic" universe (as compared with living on the plate-tectonic Earth's surface). Systematic motions within each block dominate over the Hubble flow, and to determine the true Hubble flow we must reach out far beyond our local block.

5 At a press conference on April 24, 1992, presenting the recent results from NASA's Cosmic Background Explorer (COBE) satellite, George Smoot said of the cosmic background radiation, "If you're religious, it's like seeing God." This remark created front-page news in newspapers around the world with headlines such as, "the theory of creation," and "grand unification of religion and science." What do you think of Smoot's remark?

FURTHER READING

Audouze, J. "Ages of the universe," in *Physical Cosmology*. Edited by R. Balian, J. Audouze, and D. N. Schramm. North Holland, Amsterdam, 1979.

Bernstein, J. *Three Degrees Above Zero: The Bell Labs in the Information Age*. Charles Scribner's Sons, New York, 1984.

Bolte, M. and Hogan, C. J. "Conflict over the age of the universe." *Nature* 376, 399 (1995).

Chown M. *Afterglow of Creation*. Arrow Books, London, 1993.

Gott, J. R., Gunn, J. E., Schramm, D. N., and Tinsley, B. M. "Will the universe expand forever?" *Scientific American* (March 1976).

Hodge, P. W. "The cosmic distance scale." *American Scientist* 72, 472 (September–October 1984).

Longair, M. *Our Evolving Universe*. Cambridge University Press, Cambridge, 1996.

McCrea, W. H. "The cosmical constant." *Quarterly Journal of the Royal Astronomical Society* 12, 140 (1971).

Miley, G. K. and Chambers, C. K. "The most distant radio galaxies." *Scientific American*, 54 (June 1993).

Pasachoff, J. M. *The Farthest Things in the Universe*. Cambridge University Press, Cambridge, 1994.

Rees, M. J. "Understanding the high-redshift universe: progress, hype, and prospects." *Quarterly Journal of the Royal Astronomical Society* 34, 279 (1993).

Rees, M. J. *Perspectives in Astrophysical Cosmology*. Cambridge University Press, Cambridge, 1995.

Rowan-Robinson, M. *Ripples in the Cosmos: A View Behind the Scenes of the New Cosmology*. Freeman, New York, 1993.

Sandage, A. R. "Cosmology: A search for two numbers." *Physics Today* (February 1970).

Schramm, D. N. and Wagoner, R. V. "What can deuterium tell us?" *Physics Today* (December 1974).

Sciama, D. W. "The renaissance of observational cosmology." *Revista Nuovo Cimento*, numero special 1, 371 (1969).

Smoot, G. and Davidson, K. *Wrinkles in Time*. William Morrow, New York, 1993.

Tayler, R. J. *Hidden Matter*. Horwood, Chichester, England, 1991.

Turner, E. L. "Gravitational lenses." *Scientific American* (July 1988).

Tyson, A. "Mapping dark matter with gravitational lenses." *Physics Today* (June 1992).

Van Helden, A. *Measuring the Universe: Cosmic Dimensions From Aristarchus to Halley*. University of Chicago Press, Chicago, 1985.

Wilson, R. W. "Discovery of the cosmic microwave background" in *Modern Cosmology in Retrospect*. Edited by B. Bertotti et al. Cambridge University Press, Cambridge, 1990.

SOURCES

Bergh, S. van den. "The extragalactic distance scale," in *Stars and Stellar Systems*. Edited by A. Sandage et al. University of Chicago Press, Chicago, 1975.

Blandford, R. D. "Gravitational lenses." *Quarterly Journal of the Royal Astronomical Society* 31, 305 (1990).

Boesgaard, A. M. and Steigman, G. "Big bang nucleosynthesis: Theories and observations." *Annual Review of Astronomy and Astrophysics* 319, 23 (1985).

Bondi, H. *Cosmology*. Cambridge University Press, Cambridge, 1960.

Burbidge, E. M., Burbidge, G. R., Fowler, W. A. and Hoyle, F. "Synthesis of the elements in stars." *Reviews of Modern Physics* 29, 547 (1957).

Coles P. and Ellis, G. F. R. *Is the Universe Open or Closed? The Density of Matter in the Universe*. Cambridge University Press, Cambridge, 1997.

Coles, P. and Lucchin, F. *Cosmology: The Origin and Evolution of Cosmic Structure*. Wiley, Chichester, 1995.

Eddington, A. S. *The Expanding Universe*. Cambridge University Press, Cambridge, 1933.

Fournier d'Albe, E. *Two New Worlds*. Longmans, Green, London, 1907.

Fowler, W. A. "The age of the universe." *Quarterly Journal of the Royal Astronomical Society* 28, 87 (1987).

Fukugita, M., Hogan C. J., and Peebles, P. J. E. "The history of the galaxies." *Nature* 381, 489 (1996).

Harrison, E. R. "The redshift-distance and velocity–distance laws." *Astrophysical Journal* 403, 28 (1993).

Harrison, E. R. "Is the universe open or closed?" *American Journal of Physics* 403, 28 (1998).

Hodge, P. W. "The extragalactic distance scale." *Annual Reviews of Astronomy and Astrophysics* 19, 357 (1981).

Hoyle, F. and Tayler, R. J. "The mystery of the cosmic helium abundance." *Nature* 203, 1108 (1964).

Hubble, E. *The Observational Approach to Cosmology*. Clarendon Press, Oxford, 1937.

Jeans, J. H. "Motions of stars in a Kapeteyn-universe." *Monthly Notices of the Royal Astronomical Society* 82, 122 (1922).

Longair, M. S. Editor. *Confrontation of Cosmological Theories With Observational Facts*. Reidel, Dordrecht, Holland, 1974.

Longair, M. S. "Observational cosmology." *Reports on Progress in Physics* 34, No. 12 (1971).

McCrea, W. H. "The problem of the galaxies," in *Cosmology Now*. Editor L. John. British Broadcasting Corporation, London, 1973.

Narlikar, J. V. *Introduction to Cosmology*. Second Edition. Cambridge University Press, Cambridge, 1995.

Partridge, R. B. *3K: The Cosmic Microwave Background Radiation*. Cambridge University Press, Cambridge, 1995.

Peebles, P. J. E. *Principles of Physical Cosmology*. Princeton University Press, Princeton, 1993.

Penzias, A. A. and Wilson, R. W. "A measurement of excess antenna temperature at 4080 Mc/s." *Astrophysical Journal* 142, 419 (1965).

Robertson, H. P. "Relativistic cosmology." *Reviews of Modern Physics* 5, 62 (1933).

Rowan-Robinson, M. *The Cosmological Distance Ladder*. Freeman, New York, 1985.

Rubin, V. C. and Coyne, S. J. Editors. *Large-Scale Motions in the Universe*. Princeton University Press, Princeton, 1988.

Sandage, A. R. "Observational tests of world models." *Annual Reviews of Astronomy and Astrophysics* 26, 561 (1981).

Smith, R. C. *Observational Astrophysics*. Cambridge University Press, Cambridge, 1995.

Tammann, G. A. "The Hubble constant and the deceleration parameter," in *Confrontation of Cosmological Theories with Observational Data*. Edited by M. S. Longair. Reidel, Dordrecht, Netherlands, 1974.

Trimble, V. "The origin and abundance of the chemical elements." *Reviews of Modern Physics* 47, 877 (1975).

Trimble, V. "Dark matter," in *The Encyclopedia of Cosmology*. Garland, New York, 1993.

Weinberg, S. *The First Three Minutes: A Modern View of the Origin of the Universe*. Basic Books, New York, 1977.

Part III

The living throne, the sapphire-blaze,
Where angels tremble while they gaze,
He saw; but blasted with excess of light,
Closed his eyes in endless night.
John Milton (1608–1674), Progress of Poesy

THE PRIMEVAL ATOM

The universe expands, and naturally we conclude that in the past the universe was in a more condensed state than at present. If we journeyed back in time we would expect to see the universe get steadily denser. Ultimately, we would arrive at the very high-density state popularly called the "big bang." This conclusion seems unavoidable. It might be a mistake, however, to forget entirely the many debates among cosmologists concerning the reality of a big bang beginning. Eddington was firmly against the idea of a universe that begins in a dense state, and many persons – particularly those who were drawn to science by Eddington's popular works – have felt disinclined to set his views aside lightly. The steady-state theory of an expanding universe, proposed in the late 1940s, attracted many who were united in their dislike of the big bang idea, and even now, as the 20th century closes, a few cosmologists continue to think that a big bang interpretation of the observations is mistaken.

What do we mean by the expression "big bang?" The actual singularity of maximum density at the origin of time? Or an early period in cosmic history? If the latter, how long a period? Like most persons, we shall use the expression "big bang" loosely in both senses, sometimes meaning the beginning, and sometimes the early universe up to the time when ordinary matter became the dominant constituent at the cosmic age of roughly 100 000 years. It must be

admitted that the ambiguous expression "big bang" is more sensational than sensible. It serves as a useful label when distinguishing between a big bang universe and some other world system, such as the steady-state universe.

Alexander Friedmann in 1922 led the way to the idea of an initial singular state of very high, if not infinitely high, density. But the idea made little impact until some years later when Georges Lemaître investigated and championed a dense origin that he referred to as the "primeval atom." Soon, other expressions were in use, such as: "singular state" and "big squeeze." Lemaître suggested that the dense early universe resembles a large radioactive atomic nucleus. This cosmic nucleus, or primeval atom, explodes and forms fragments that evolve into galaxies. In his later years he wrote: "These considerations, besides providing a natural beginning, supply what can be called an inaccessible beginning. I mean a beginning which cannot be reached, even by thought, but which can be approached in some asymptotic manner." Lemaître was a priest and some cosmologists viewed the primeval atom with reservation and regarded Lemaître's theory as an amalgam of science and theology.

The steady-state theory emerged in the late 1940s and rose in popularity in the 1950s. From earlier chapters we recall that the big bang universe was plagued by the age problem. The Hubble time (the reciprocal of the Hubble term), which is a rough indication of the age of the universe, was observed to be less than the age of the Earth. The infinitely old steady-state universe avoided this problem. But new observations by Walter Baade in 1952 and Allan Sandage in 1958 solved, or at least greatly alleviated, the big bang age problem.

In the late 1940s, a small group of scientists, the "Gamow group," began to make important developments in our understanding of the early universe. The principal members of this group were the charismatic George Gamow of George Washington University and the young scientists Ralph Alpher and Robert Herman of Johns Hopkins Applied Physics Laboratory. The dense early universe was called the big squeeze until Fred Hoyle in 1949 coined

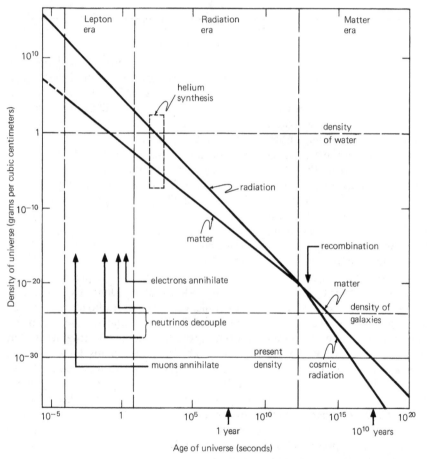

Figure 20.1. Standard model of the early universe showing how density (in grams per cubic centimeter) varies with time (in seconds). The total density varies as the inverse of the square of time. An increase in age by a factor 10 reduces the density by a factor 1/100. In the matter era, density is due almost entirely to matter and is proportional to T^3, where T is the temperature of the cosmic radiation; hence T is proportional to $t^{-2/3}$, where t is the age of the universe. In the radiation era the density is due almost entirely to thermal radiation and is proportional to T^4; hence T is proportional to $t^{-1/2}$. Note that in the matter era the radiation density is proportional to $t^{-8/3}$ (and decreases faster than the density of matter), and in the radiation era the matter density is proportional to $t^{-3/2}$ (and decreases slower than the density of radiation).

the popular title big bang. Gamow referred to the dense matter of the very early universe as *ylem*, pronounced "eye-lem" (derived from Greek *hylo*, meaning original substance, as in ethyl and methyl). The name ylem is not widely used. Perhaps a better word is *elem* pronounced "ee-lem," as in "element," and having the same root.

Planck units of time

1	beginning of big bang
10^5	monopoles, inflation
10^{10}	quarks, leptons, gluons
10^{15}	
10^{20}	
10^{25}	
10^{30}	
10^{35}	birth of hadrons
10^{40}	hadrons, leptons, photons
10^{45}	photons dominate
10^{50}	
10^{55}	end of big bang
	birth of atoms
10^{60}	birth of galaxies,
	stars, life
10^{65}	death of stars

Figure 20.2. Evolution of the universe in logarithmic time. The younger the universe the faster that events occur, and the older the universe the slower that they occur, and cosmic history seems best told in logarithmic time. Each order-of-magnitude step increases the age of the universe tenfold. The big bang (early universe), measured from the Planck epoch, extends 55 orders of magnitude; the subsequent matter-dominated era of stars and galaxies, from the end of the big bang until the present time, extends only 5.5 orders of magnitude. Reckoned in this way, more than ninety percent of cosmic history occurred in the big bang.

The Gamow group studied the physics of the early universe and made many important developments. Their major breakthrough, in 1948, was the realization that the early universe was not only very dense but also very hot. This led to two important conclusions. First, there exists a radiation era in the early universe during which the density of radiation greatly exceeds the density of matter. Second, this radiation, cooled by expansion, survives and bathes the whole universe in the afterglow of the early universe. Gamow and his colleagues estimated that the afterglow, now named the cosmic background radiation (CBR) has a present temperature somewhere between 5 and 50 degrees kelvin.

Almost all cosmologists now firmly believe that the discovery of the cosmic background radiation establishes beyond all doubt that we live in a big bang universe and confirms the prediction by the Gamow group that the early universe was very dense and hot. The bang–antibang controversy has been set to rest, and the high-density, high-temperature beginning of the universe is a fact as secure as any in cosmology.

This chapter outlines the standard model of the early universe (Figures 20.1 and 20.2), and takes us back to a time when the universe was only 100 microseconds old. Theories on what happened before 100 microseconds are touched on in more detail in later chapters (Chapter 22 on inflation and Chapter 25 on creation).

THE LAST FIFTEEN BILLION YEARS
A safari in time

We think the age of the universe lies somewhere between 10 and 20 billion years. For convenience we assume 15 billion years. As a rough guide we use the rule that the average mass density of the universe is inversely proportional to the square of the age of the universe (the Einstein–de Sitter universe). Thus, when the universe is 1/10 its present age, the average density is 100 times greater. Also, for simplicity, we assume that the present density is roughly 1 hydrogen atom

Table 20.1 *Energies*

Energy (electron volts)	Temperature (kelvin)	Age of universe (seconds)
1 eV	10^4	10^{12}
1 MeV (10^6 eV)	10^{10}	1
1 GeV (10^9 eV)	10^{13}	10^{-6}
1 TeV (10^{12} eV)	10^{16}	10^{-12}

M = mega; G = giga; T = tera.

per cubic meter. At age 15 million years, at redshift 100, the universe has a density 1 million times greater, or 1 hydrogen atom per cubic centimeter. A typical average density within galaxies is also 1 hydrogen atom per cubic centimeter, and hence we know that galaxies in their present form do not exist at a time earlier than 15 million years at redshifts greater than 100. We occasionally use the relations between particle energy, temperature, and cosmic time, as shown in Table 20.1.

We shall travel back in time and attempt to reconstruct the history of the universe. We must, like all time-traveling spectators, speak of the past (or the future) in the present tense. As our time machine counts down the billions of years, we see the galaxies moving slowly closer together. We see the birth and death of countless stars, and witness the origin of life on Earth and possibly the origin and even death of life in myriads of planetary systems throughout the Galaxy. When the Galaxy is about 5 billion years younger, we see the birth of the Solar System.

We continue on our safari into the distant past, watching the galaxies getting younger and drifting closer together. Presumably, when the galaxies were young, long before the birth of the Sun, lifeforms of many kinds arose and evolved, and it is a matter of absorbing interest to speculate on their fate and the possibility that many still exist in highly advanced states of development.

When the universe is about 1 billion years old (at roughly redshift 10) we see the galaxies swelling into gigantic orbs of gas.

Then, between 1 billion and 100 million years (between redshift 10 and 30) lies the heroic age of galaxy formation. We pause and watch immense and slowly turning orbs of primordial gas lit by constellations of newborn stars in their central regions. Gas descends between these stars and settles into either the rotating disks of giant spiral galaxies or the bright nuclei of giant elliptical galaxies.

Earlier still, we enter the "dark ages." Little or nothing is known of this starless prenatal era of the galaxies. Perhaps swirling gas is illuminated with flickers of light from shock waves; perhaps primordial black holes of all sizes forged in the big bang are surrounded by regions of glowing gas; and perhaps things not yet imagined bestrew the dark universe. At age 15 million years, the universe is comfortably warm at the room temperature 300 kelvin. Earlier still, at age 1.5 million years (redshift 450), the universe is filled with a red glow. The density is now 100 hydrogen atoms per cubic centimeter (still much better than a laboratory vacuum) and the temperature is 1200 kelvin.

At an age of approximately 500 thousand years (redshift 1000), the universe is flooded with brilliant yellow light, as bright as in a furnace at a temperature 3000 kelvin. We at last stand at the threshold of the early universe – the big bang. Behind us, back in the dark future, lies a universe dominated by matter, before us, in the bright past, lies a universe dominated by radiation. From this epoch, at the end of the big bang, descends the cosmic background radiation, cooled by expansion, that now has redshift 1000 and a temperature of 3 kelvin.

We have journeyed back to a time when the universe is less than 1 million years old, and are about to enter a period of cosmic history known as the early universe. As we enter, two things happen that greatly alter the cosmic scenery.

THE FIRST MILLION YEARS
Decoupling epoch
The first thing that happens as we enter the early universe (or big bang) in our time

machine is the disappearance of atoms. From now on, as we journey back toward the frontier of time, the rising temperature is too high for atoms to exist, and they are dissociated into their positive nuclei and negative electrons. In this plasma state, the nuclei and electrons move freely and independently. The transformation is not abrupt, but slow, and lasts about 10 000 years.

The ionization of the atoms – mostly hydrogen and helium – releases a flood of electrons that constantly scatter the radiation in all directions. Light no longer travels freely and the universe glows like a luminous mist. This explains why nowadays we cannot

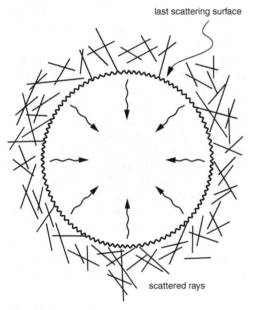

last scattering surface

scattered rays

Figure 20.3. We look out in space and back in time to the early universe. We see the big bang in all directions at a distance (measured in time) of roughly 10^{10} years. Light travels freely after the decoupling epoch, which is why we observe it as the cosmic background radiation. Before the decoupling epoch light rays were constantly scattered by free electrons. The decoupling epoch, as shown here, is the "last scattering surface." The density perturbations of matter that evolved into galaxies are shown as irregularities in the last scattering surface. The temperature irregularities that accompany the density irregularities are observed in the cosmic background radiation (see Figure 19.9).

look back beyond the decoupling epoch. We see only the light emitted by this luminous mist in the last moments of the early universe. This light, the most ancient in the universe, is the cosmic background radiation that has traveled from the last moments of the big bang, as shown in Figure 20.3. Before the decoupling epoch, radiation is continually scattered by free electrons. At the end of the early universe, at the decoupling epoch, gas has cooled to 3000 kelvin and most electrons and atomic nuclei have recombined. The radiation now decouples from weakly ionized matter and is free to travel unimpeded. In later ages, when gas is heated and ionized in intergalactic and interstellar space, its density is too low to have much effect on the cosmic background radiation.

From the decoupling epoch (known also as the recombination epoch because atomic nuclei and electrons recombine to form neutral atoms) comes directly the cosmic background radiation that we observe.

Epoch of equal densities

The second thing that happens at the end of the the early universe is the changeover in the relative mass densities of matter and radiation. Radiation has energy, and therefore has mass. (Energy = mass $\times c^2$, where c is the speed of light.) At the epoch of equal densities, 1 cubic centimeter of matter has the same mass as 1 cubic centimeter of radiation. As we look back into the past, we see the density of matter steadily increasing as $(1 + z)^3$, where z is the cosmic redshift. Because the present density is around 1 hydrogen atom per cubic meter, the density of matter at a redshift z equal to 1000 is 1 billion times greater and equal to 1000 hydrogen atoms per cubic centimeter. The density of radiation, however, increases faster, as $(1 + z)^4$, and at $z = 1000$ its density is 1 trillion times greater than at present. The cosmic background radiation of 2.73 kelvin has a density roughly 1/4000 that of matter; and hence, at temperature 10 000 kelvin (redshift about 4000) and at age approximately 100 000 years, the radiation and matter densities are equal. The epoch of equal densities occurs close to the decoupling epoch and both mark the end of the early universe.

As a rough guide in the early universe we shall use the relations

$$\rho t^2 = 1 \times 10^6,$$ [20.1]

$$tT^2 = 1 \times 10^{20},$$ [20.2]

between density ρ, time t, and temperature T, where the density is in grams per cubic centimeter (water has a density of 1 in these units), time is measured in seconds, and temperature in kelvin. For example, at temperature 10 billion kelvin, the universe is 1 second old and 1 million times the density of water.

Radiation era

In the radiation era the universe consists predominantly of radiation. This era of early cosmic history begins when the expanding universe is 1 second old, lasts for 100 thousand years, and during this period of time the radiation is more dense than matter. The radiation era begins at age 1 second and temperature 10 billion degrees kelvin, when the radiation density is 1 million grams per cubic centimeter (on Earth 1 ton per thimbleful) and the density of matter is 1 tenth of a gram per cubic centimeter. The radiation era ends at age about 100 000 years and temperature a few thousand degrees, when radiation and matter have similar densities, equivalent to about 1 thousand hydrogen atoms per cubic centimeter. As we go back in time, the density of radiation overtakes and then exceeds the density of matter because the radiation density increases as T^4 (where T is temperature), whereas the matter density increases more slowly as T^3.

With trepidation we have ventured into the big bang, and behold! we find ourselves in a silent world of radiant splendor. As we travel farther back in time, plunging deeper into an incandescent early universe, the intensity soars and the radiation becomes brighter than the central furnaces of the hottest stars. Although matter during this stage of our journey is of relatively low density, it nonetheless is important because free electrons incessantly scatter the energetic rays of light and the radiation behaves like a dense fluid.

Origin of helium

An event of great importance in the radiation era is the production of primordial helium. About 25 percent of all matter (in the form of free protons and electrons) is transformed into helium nuclei in the first few hundred seconds of the radiation era. For 10 billion years the stars have worked industriously at converting hydrogen into helium and have succeeded in transforming only about 2 percent of all hydrogen in the universe. Yet 10 times as much hydrogen is burned into helium in only 200 or so seconds in the early radiation era. The universe at this time is like an enormous hydrogen bomb and the energy released by the fusion of hydrogen into helium is immense. But the energy already existing in the radiation is vastly greater, and the thermonuclear detonation of the universe raises its temperature by less than 10 degrees and increases the radiation energy by a few parts in 100 million.

The amount of helium produced can be estimated in the following way. At the beginning of the radiation era there exists a dense sea of radiation and a low-density gas consisting of protons, neutrons, and electrons. Because of earlier conditions (to be discussed shortly) there are two neutrons to every 10 protons. The neutrons and protons continually collide together and readily combine to form deuterons, which are the nuclei of atoms of heavy hydrogen known as deuterium. But the deuterons are readily dissociated back into neutrons and protons by the intense radiation. Thus, when the universe is only a few seconds old, the neutrons and protons combine continually to form deuterons, and the deuterons dissociate continually back into neutrons and protons. After 100 seconds, the temperature has dropped to 1 billion degrees and the radiation is then insufficiently energetic to break apart the deuterons. Neutrons and protons now form deuterons that can survive.

Meanwhile, the free neutrons have been slowly decaying into protons and electrons. This is because neutrons, when free and not bound in atomic nuclei, are unstable and decay in about 1000 seconds. By the time neutrons are able to combine permanently with protons, their abundance has decreased, and of every 16 nucleons, two are neutrons and 14 are protons. The two neutrons combine with two protons, and the result is two deuterons and 12 protons.

At high temperature, deuterons rapidly combine to produce helium nuclei. At 1 billion degrees, the temperature is insufficient to dissociate the deuterons, but sufficient for them to combine quickly and form helium nuclei. Two deuterons combine to form 1 helium nucleus, thus yielding 1 helium nucleus and 12 protons for every 16 nucleons. The whole process is over in roughly 200 seconds, and in that time 25 percent (by mass) of all matter is converted into helium (four out of every 16 nucleons form a helium nucleus) and the remainder consists predominantly of hydrogen nuclei (protons). Slight amounts of deuterium, helium-3 (helium nucleus of two protons and one neutron), and lithium are also produced.

The abundance of helium in the universe is not a very sensitive indicator of the density of matter, and therefore not a good indicator of how much matter in the form of nucleons (baryonic matter) exists in the universe today. As our rough estimate has shown, the helium abundance is mainly controlled by temperature and depends on the initial neutron–proton ratio. Not all deuterons succeed in combining to form helium nuclei; a small fraction survives and accounts for the deuterium that now exists. Between one hundredth and one thousandth of 1 percent of all hydrogen is at present in the form of deuterium (heavy hydrogen). The amount of deuterium that survives (and is not burned into helium) in the early universe is very sensitive to the density of matter, as shown in Figure 19.6. When the density is

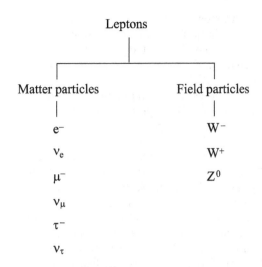

Figure 20.4. Leptons are the weak interacting particles. They consist of matter particles (called fermions) and field particles (called bosons). The matter leptons on the left comprise electrons, muons, tauons, and their neutrinos; the field particles on the right comprise the negative W^-, positive W^+, and zero charged Z^0 bosons.

low, fewer deuterons collide with one another and a larger fraction survives; and when the density is high, more deuterons collide and a smaller fraction survives. By determining the fraction of hydrogen now in the form of deuterium (and also the abundance of other light elements formed in the early universe), we can estimate the present average density of matter in the universe. The results (Chapter 19) indicate that the density of baryonic matter consisting of nucleons is about 10 percent of the critical Einstein–de Sitter density.

THE FIRST SECOND
The lepton era

Immediately preceding the radiation era exists what is sometimes called the lepton era. Leptons are light particles – hence the name – such as electrons, positrons (positive electrons), and neutrinos. The lepton era begins when the universe is 100 microseconds (one ten-thousandth of a second) old, when the temperature is 1 trillion kelvin, and the density is 100 trillion grams per cubic centimeter, and lasts until the beginning of the radiation era when the universe is 1 second old, has a temperature 10 billion kelvin, and a density 1 million grams per cubic centimeter.

In the lepton era the temperature is high enough for the production of electron pairs. Electron pairs (i.e., electrons and positrons) are continually created and annihilated, and an incessant interchange of energy occurs among photons, electron pairs, and neutrinos. Because of thermal equilibrium there are roughly equal numbers of photons, electrons (both positive and negative), and electron neutrinos and antineutrinos. Buried in this dense photon–lepton ferment are the nucleons (protons and neutrons), and to each nucleon there exists very roughly 1 billion photons, 1 billion electron pairs, and 1 billion neutrino pairs. Each nucleon continually collides with leptons; a neutron captures a positron and becomes a proton, a proton captures an electron and becomes a neutron, and in each case, neutrinos are emitted and absorbed. Very nearly half the nucleons are neutrons and half are protons.

A neutron is 1 seventh of 1 percent heavier than a proton, and in a free state it decays into a proton, an electron, and a neutrino in a time of roughly 1000 seconds. Slightly more energy is needed to create a neutron than a proton, and because of this small energy difference, each nucleon in the lepton era tends to be a proton for a slightly longer time than a neutron. For this reason there are slightly more protons than neutrons. Initially, at very high temperature, the difference is small. For example, at a temperature of 100 billion degrees, to every 10 000 neutrons there are 10 001 protons. At the end of the lepton era, the temperature has dropped enough for the difference in neutron and proton masses to become important. Electrons now have scarcely enough energy to convert protons into neutrons, whereas the conversion of neutrons into protons by positron absorption is much easier. As a result, two neutrons to every 10 protons exist at the end of the lepton era at the beginning of the radiation era.

Disappearance of muons and electrons

So far, we have encountered only electrons and their neutrinos; but other leptons, known as muons, that possess their own kind of neutrinos, await us. All leptons are shown in Figure 20.4. The muon is similar to the electron, but 207 times heavier; it is unstable and decays in 1 millionth of a second into an electron and neutrinos. The antiparticle of the electron is the positron, or positive electron, and the antiparticle of the muon is the antimuon, or positive muon. The temperature rises as we travel back in time, and as it approaches and reaches 10^{10} kelvin, there occurs copious pair-production of electrons and their neutrinos, but no pair-production of muons because of their greater mass. Not until the temperature approaches and reaches 10^{12} (1 trillion) kelvin is the thermal energy sufficient to start copious pair-production of muons and their neutrinos.

When the universe is only 100 microseconds old and the temperature is about 1 trillion degrees, the whole universe is flooded with photons, electrons and their neutrinos, muons and their neutrinos, and also particles such as pions that are relics of an earlier era. The universe expands, the temperature declines, the negative and positive muons annihilate each other, and soon vanish from the scene. At high temperature, the muons are created at the rate they annihilate each other, but below 1 trillion kelvin, annihilation occurs faster than creation. The energy released by the disappearance of the muon population is shared among the surviving species of particles: mostly photons, muon neutrinos (which are not annihilated), and electrons and their neutrinos.

When, as a result of expansion, the temperature drops to 10 billion kelvin, electron pairs annihilate faster than they are created. Below a temperature of 5 billion kelvin, the electron hordes vanish, and their energy is inherited by the photons.

Decoupling of neutrinos

The events of interest in the lepton era are the decoupling of the muon neutrinos at the beginning (100 microseconds) of the era and the decoupling of the electron neutrinos at the end (1 second) of the era.

Neutrinos are uncharged and weakly interacting particles that travel at the speed of light. There are three kinds: electron neutrinos, muon neutrinos, and tauon neutrinos (the tauon has not been previously mentioned), corresponding to the three kinds of leptons: electrons, muons, and tauons. Each kind of lepton, with its own neutrino, has its antiparticles. Neutrinos normally interact weakly with matter, and only when their energy and density are very high can they significantly participate in the give-and-take of particle reactions in thermal equilibrium. Tauons are about 3500 times heavier than electrons, and owing to their large mass, they and their antiparticles annihilate much earlier than the lepton era, and only their neutrinos survive.

In the early lepton era, as the universe expands and cools, the muons and anti-muons begin to annihilate each other faster than they are created. The muons begin to disappear; the muon neutrinos and anti-neutrinos, however, stay around because they cannot easily annihilate one another. Owing to the high thermal energy per particle, the muon neutrinos continue to interact with the muons until almost all muons have vanished. To every four photons there now exists a total of nine neutrinos of all kinds. With the decline of the muon dynasty, hordes of muon neutrinos are released to wander forever freely through the universe, interacting with virtually nothing.

After the decoupling of the muon neutrinos, the temperature continues to decline, and eventually the multitudes of electrons and their antiparticles (positrons) also begin to disappear. Up to this point, the electrons and their neutrinos have interacted intimately. But now, with the decline of the electron population, the electron neutrinos and antineutrinos quickly decouple and join the muon neutrinos.

Neutrinos from the early universe have traveled freely at the speed of light and steadily lost energy because of the expansion of the universe. They still exist and form a cosmic background of neutrinos. The red-shift of the decoupled muon neutrinos is approximately 4×10^{11} (10^{12} divided by 2.73 kelvin) and of the electron neutrinos is approximately 4×10^{9} (10^{10} divided by 2.73 kelvin).

The temperature of the neutrino background is less than the 2.73 kelvin of the photons of the cosmic background radiation. The reason for the difference in temperature is as follows. Muon neutrinos decouple after most muons (with which they interact) have vanished. Later, the electron neutrinos decouple, but because the temperature is now lower and they therefore interact more weakly, they decouple before most electrons (with which they interact) have vanished. The energy released by the annihilation of the muons is shared by all surviving particles: muon neutrinos, electrons and their neutrinos, and photons. Although muon neutrinos interact with nothing, their temperature declines (because of expansion) in step with the temperature of the photons and other particles. After the electron neutrinos decouple, the electrons annihilate and their energy is then inherited by the photons. None of this released energy goes to the decoupled muon and electron neutrinos. The neutrino cosmic background radiation has a temperature 70 percent of the photon background temperature, or approximately 2 kelvin at the present epoch.

There are four kinds of neutrinos (electron and muon neutrinos and their anti-neutrinos), and to every eight photons of the cosmic background radiation there are three of each kind still with us. Hence, to every eight photons there are 12 assorted neutrinos. Thus in each cubic centimeter of space there are now roughly 400 background photons and 600 background neutrinos, making a total of 1000 particles. Roughly, with the tauon neutrinos, the total is 1300.

The interaction of the background neutrinos with matter is exceedingly weak. Presumably, in the future they will be detected, and we shall then have a means of probing the first second of the universe.

THE FIRST HUNDRED MICROSECONDS

Why go further?

Our journey back through the radiation and lepton eras has brought us to a time when the universe is 100 microseconds old. The temperature has steadily risen and reached 1 trillion kelvin; and the density has soared to 10^{14} grams per cubic centimeter (equivalent on Earth to 100 million tons per thimbleful). It is a tribute to modern physics that we can, in broad outline, trace the history of the universe back to this early epoch. What we have studied so far is known as the standard model of the early universe. Considerable uncertainty exists concerning what happens earlier in the first ten-thousandth of a second.

We stand so close to the beginning of time that it seems absurd to want to continue the journey farther. Yet everything happens with such rapidity that perhaps more of cosmic history occurs in the first 100 microseconds than has since occurred in the subsequent 10 billion years. Before us in the extreme early universe lies a blur of intense action. We are not sure exactly what happens, simply because we do not yet know enough physics. Inspired by the words of Pascal, "if our view be arrested there, let our imagination pass beyond," we examine some ideas concerning the earliest moments of the universe. The possibility of inflation in the extreme early universe with its many consequences is explored in Chapter 22.

The hadron era

On traveling farther back in time, we leave the lepton era and enter what might be called the hadron era. Hadrons (from *hadro* meaning stout or strong), in addition to their weak and electromagnetic interactions, interact among one another with a force

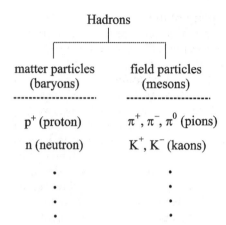

Figure 20.5. Hadrons are the strongly interacting particles. Like leptons they consist of matter particles (fermions) that are the baryons, and field particles (bosons) that are the mesons. The matter particles on the left comprise nucleons (neutrons and protons) and many short-lived particles (such as the lambdas, sigmas, xis, and omegas); the field particles on the right comprise mesons (such as the pions and kaons). Unlike the leptons, which are structureless and irreducible, the hadrons possess complex structure and are composed of quarks.

known as the strong interaction. Familiar hadrons (Figure 20.5) are the nucleons consisting of protons and neutrons. The nuclei of atoms contain also particles called pions that are hadrons with masses 270 times that of electrons. The pions are the field particles of the strong force; they skip to and fro among the nucleons and hold an atomic nucleus together despite the electrical repulsion of its protons. In the hadron era, the universe swarms with hadrons because the temperature is high enough for the creation of pions, nucleons and antinucleons, and also other hadrons and their antiparticles. Leptons and photons also exist, but the universe is dominated by the presence of a dense sea of hadrons.

We pause in the hadron era for a moment's thought. Around us exists a dense conglomeration of photons, leptons and antileptons, hadrons and antihadrons. Particles and their antiparticles continually annihilate each other and the energy released is continually recycled to create fresh particles and antiparticles. But as the temperature drops, annihilation overtakes creation, and toward the end of the brief hadron era all matter and antimatter have practically vanished. The lowest mass hadrons – the pions – although still abundant, must soon suffer the same fate and disappear in the opening act of the lepton era. This raises an interesting question in cosmology: Why does matter still exist today? Why was all the matter not annihilated with antimatter long ago at the end of the hadron era? Antimatter has apparently gone, but a small fraction of the matter survives. The simplest answer is that the early universe contained slightly more matter than antimatter, and the slight excess of matter over antimatter has survived and now constitutes the present material universe. The question of why the early universe favors matter very slightly more than antimatter cannot easily be answered. The question of how much more matter than antimatter is more easily answered.

It is possible to calculate the excess of matter over antimatter in the hadron era. We find, as the following argument shows, that the excess is about 1 part in a billion. In Chapter 17 it was shown that most of the entropy of the universe resides in the 2.73 degree cosmic background radiation. We should amend this remark and say that the entropy is mostly in the cosmic background radiation and the cosmic background neutrinos. For our purpose we may assume that entropy in the universe is measured by the number – roughly 1000 per cubic centimeter or 1 billion per cubic meter – of cosmic photons and neutrinos. Also, in each cubic meter we have on the average 1 nucleon. In the early universe the entropy in a comoving volume is much the same as it is now in that same comoving volume. The total number of particles of all kinds sharing that entropy has not greatly changed. The particles that flood the universe may alter their nature, from hadrons in the hadron era, to leptons in the lepton era, to photons in the radiation era, but their total

number remains approximately constant. If N particles occupy a comoving volume in the hadron era, there are now approximately N particles in the same comoving volume today. (The number fluctuates due to spin and statistics of different particle populations, but the fluctuation in this discussion can be neglected.) The N particles of the dense hadron era have become, in the greatly expanded universe, N_n nucleons and $N - N_n$ photons and neutrinos. In this discussion we neglect all other particles, such as electrons, which do not greatly affect the argument. At present there are approximately 1 billion cosmic photons and neutrinos to each nucleon, or $(N - N_n)/N_n = 10^9$. Because N_n is so small compared with N, we can say that N nucleons and antinucleons in the hadron era have become the N photons and neutrinos that now exist. The neglect of other particles in the late stages of the hadron era, such as the pions and leptons, again involves only relatively small errors.

To see what happens, let N^+ be the number of nucleons and N^- be the number of antinucleons in a comoving volume in the hadron era. The total number of nucleons of both kinds is given by

$$N = N^+ + N^-.$$

The excess of nucleons over antinucleons can be written as

$$\Delta N = N^+ - N^-,$$

and this difference is the conserved baryon number in the comoving volume. In a universe that has no preference for either matter or antimatter, $N^+ = N^-$, and the baryon number ΔN is zero. The fractional difference of matter and antimatter in the hadron era is expressed in the form

$$\frac{\Delta N}{N} = \frac{N^+ - N^-}{N^+ + N^-}, \qquad [20.3]$$

and this is the ratio of the baryon number (the number difference between baryons and antibaryons) and the total number of baryons (the sum of the baryons and antibaryons).

The original N hadrons have annihilated, with the exception of the small excess ΔN of baryons, and are now N photons and neutrinos. The surviving ΔN baryons are the present N_n nucleons ($\Delta N = N_n$); and because $N_n/N = 10^{-9}$, we have $\Delta N/N = 10^{-9}$. Hence the fractional difference of matter and antimatter in the hadron era is the present ratio of the number of nucleons and the number of photons plus neutrinos.

The universe, according to this argument, has been transformed in a spectacular manner. Once, long ago in its earliest moments, the universe consisted of intermingled matter and antimatter of almost equal amounts, and the slight difference, about 1 part per billion, was of very little significance. But this small difference has survived and now constitutes all the galaxies, stars, planets, and living creatures. Without this difference, the universe would consist only of photons and neutrinos, and we would not be here marveling on this fortunate circumstance. All the immense energy of the hadron era, released by the annihilation of matter and antimatter, has passed into the cosmic ocean of photons and neutrinos that have since been cooled by expansion. The cosmic radiation consisting of 3-degree photons and 2-degree neutrinos, which at present appears to be so unimportant, represents the unimaginable energy that once existed in the very early universe, whereas the matter that we now prize so highly is the result of a small, freakish difference that was once of no apparent importance.

Cosmology and particle physics

Our understanding of the early universe between 100 microseconds and 100 thousand years is moderately secure. After this period, with the emergence of atoms and astronomical systems, the universe, although governed by familiar laws and forces, loses simplicity and its complexity often defies understanding. Further back in time, before 100 microseconds, the universe is governed by laws and forces that get progressively more unfamiliar, and the universe again loses simplicity and defies our understanding with its complexity.

Forces of nature

All the richness and diversity of the world around us stems from the interplay of four forces. In order of increasing strength, they are the gravitational force, weak force, electromagnetic force, and strong force. These four forces, separately and in combination, have shaped the history of the universe since the hadron era.

Perhaps in the beginning, possibly at the Planck epoch, a single superforce dominates the universe. The Planck particle energy, characteristic of this epoch, is 10^{19} GeV (1 GeV is 1 billion electron volts and 10^{19} GeV is 10 billion billion billion electron volts); this enormous particle energy (each particle has a mass 10^{-5} grams) corresponds to a temperature T of 10^{32} kelvin, a cosmic age t of 10^{-44} seconds, and a density ρ of 10^{94} grams per cubic centimeter. At this epoch, or thereabouts, the superforce splits into two forces: gravity, with which we are familiar, and the hyperweak force, as shown in Figure 20.6. The hyperweak force makes little or no distinction between matter and antimatter. According to current versions of the grand unified theory, after the expanding universe has cooled to a temperature 10^{28} kelvin (corresponding to a particle energy 10^{15} GeV), the hyperweak force splits into two separate forces: the strong force and the electroweak force. According to current theory, this grand-unified transition marks the onset of an inflation period that lasts from 10^{-36} to about 10^{-34} seconds. After inflation, the universe is ruled by gravity, the electroweak force, and the strong force. Later, at temperature 10^{15} kelvin and age 10^{-10} seconds, the electroweak force splits into the familiar weak and electromagnetic forces. Thus, just before the hadron era, the original single superforce has become the four forces we know today.

High-energy particle accelerators on Earth in the late 20th century have attained energies of hundreds of billions of electron volts, and no doubt in the future will attain

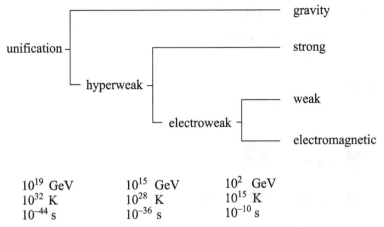

Figure 20.6. The forces of nature. The low-energy world we live in is ruled by the four forces shown on the right. As we move left, toward higher energies, the forces converge and perhaps, at the Planck epoch, unite into a single superforce. First, the electromagnetic and weak forces unite to form the electroweak force at about 100 GeV. (1 GeV $= 10^9$ electron volts), corresponding to a temperature 10^{15} kelvin at cosmic time 10^{-10} seconds. Above this energy the hadrons dissolve into a mixture of strongly interacting quarks, gluons, and electroweak particles called a quark plasma. According to grand unified theory, at the much higher energy of about 10^{15} GeV, corresponding to a temperature 10^{28} kelvin at cosmic age 10^{-36} seconds, the electroweak and strong forces unite to form the hyperweak force; the electroweak field particles (W^+, W^-, Z^0) and the strong field particles (colored gluons) become the hyperweak gluons that ignore the difference between leptons and quarks. Possibly, at the highest energy, the hyperweak force unites with gravity to form a superweak force, but no credible theory exists at present. The term gluon is used generally for all field particles. Thus the low-energy gluons are photons of the electromagnetic field, Ws and Zs of the weak field, pions of the strong field, and gravitons of the gravitational field.

even higher energies of thousands of billions of electron volts. Experiments in high-energy particle physics support the theory that strongly interacting particles (baryons and mesons), which are collectively known as hadrons, are composed of basic particles known as quarks, and that also the electromagnetic and weak forces jointly form an electroweak force at energies above 100 GeV. But accelerators on Earth can never attain the enormous Planck energy of 10 billion billion billion electron volts, or even the grand-unification energy of 1 million billion billion electron volts. Perhaps the only natural place for these enormous energies is in the very early universe, which unfortunately is inaccessible to direct observation. But undoubtedly what happens in the early universe greatly affects the design

of the universe of today. The universe is itself a laboratory in which theories on the nature of matter at extremely high energies are tested and judged by their effect on the world around us. For example, the absence of an abundance of the magnetic monopoles predicted by grand unified theory was initially thought to be a fatal defect in the theory until the discovery of inflation.

The quark era

The quark theory, proposed independently by Murray Gell-Mann and George Zweig, both at the California Institute of Technology, simplifies the bewildering complexity of the hadron family of particles. According to the proposed theory, hadrons consist of structureless basic particles called quarks. Two kinds of hadrons exist: baryons and

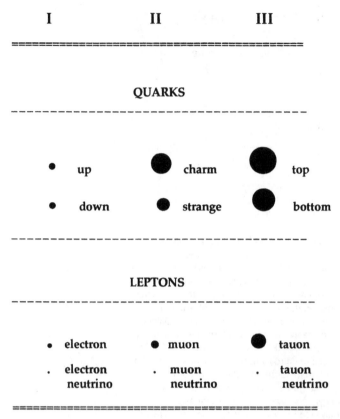

I   II   III

QUARKS

- up   ● charm   ● top
- down   ● strange   ● bottom

LEPTONS

- electron   ● muon   ● tauon
. electron   . muon   . tauon
  neutrino   neutrino   neutrino

Figure 20.7. This shows the three lepton–quark families of the grand unified theory. Size indicates mass (or energy). Family I consists of the electron and its neutrino, and the up and down quarks of nuclear matter. The ordinary low-energy world around us is composed solely of these fundamental matter particles. Family II (the muon and its neutrino plus the charm and strange quarks), and family III (the tauon and its neutrino plus the top and bottom quarks) consist of particles that compose the worlds of high energy.

mesons. An early version of the theory proposed three different quarks (or quark flavors), each with its own antiquark. Baryons (hadrons of half-integer spin), which include nucleons, consist of quarks arranged in groups of three. Mesons (hadrons of integer spin and zero baryon number), which include the pions, consist of quarks and antiquarks arranged in groups of two. Quarks are electrically charged, but their charge is fractional, either $\frac{2}{3}$ or $-\frac{1}{3}$ of the elementary charge on the proton.

More recently, the quark flavors have increased to six (thus paralleling the six leptons), and are distinguished by the names up, down, charm, strange, top, and bottom, denoted by the symbols u, d, c, s, t, and b. These six basic states are called flavors. Each flavor has three distinct colors, thus making 18 distinct kinds of quark, or 36 when antiquarks are included. Quarks, like leptons, have no internal structure, as far as we know, and arguments of symmetry suggest that the number of quark flavors should equal the number of leptons.

In our world of comparatively low energy the particles that mediate the weak force are the W bosons; the particles that mediate the electromagnetic force are the photons; and the particles that mediate the strong force are the mesons. At higher energies the Z bosons mediate the electroweak force, and at even higher energies the gluons mediate the hyperweak force. At a time less than 100 microseconds, the universe consists of quarks and leptons, and their mediating forces consist of gluons. The strong force that mediates between quarks consists of colored gluons, the force that mediates between leptons and electric charges in general consists of electroweak gluons (photons, W, and Z mesons), and the gravitational force consists of gluons called gravitons.

No quark in an isolated state has been discovered. The force acting between quarks is unlike all other forces in nature; it does not decrease in strength as the separating distance increases. If we energetically try to separate two quarks, the energy expended in pulling them apart creates a new quark–antiquark pair, and instead of separating the original quarks, we have created two additional quarks. It is like trying to isolate the ends of a piece of string: when the string is pulled too hard, it snaps into two pieces, creating two new ends. In this analogy, the piece of string is a hadron and its ends are quarks. Alternatively, it is like trying to isolate the north and south poles of a bar magnet: by breaking the magnet into two pieces, we create two magnets, each with its opposite poles. In this analogy, the magnet is a hadron and its poles are quarks.

As we travel back in time and enter the hadron era we encounter a dense sea of pions at a temperature of about 1 trillion degrees and a density of about 100 trillion grams a cubic centimeter (similar to the density of atomic nuclei). Then, still earlier in the hadron era, the universe is filled with pair-created nucleons, antinucleons, and heavier hadrons. Closely packed hadrons overlap one another and the quarks of neighboring hadrons are close together. Earlier still, at higher densities, the hadrons dissolve, their boundaries melt away, and the quarks break free, and the universe is a dense sea of quarks. Structureless particles – quarks, leptons, and gluons – all in thermal equilibrium, are continually annihilated and created.

In our back-to-the-past journey we have entered the hadron era and found it short-lived. It soon gives way to an earlier era – the quark era – consisting of structureless particles. In this new era, particles densely overlap (remember particles are waves), and quarks have energies too high for them to recombine and form into hadrons.

GRAND UNIFIED ERA
Quantum cosmology

What lies before us in the grand unified era, before the quark era, when the universe is younger than 10^{-36} seconds, is veiled from view by theoretical and conceptual problems. But undeterred, like seafarers of old who voyaged across unknown and hazardous seas, we push on, seeking the frontier of time. How far can we travel back in time? Can we in thought travel back to a cosmic singularity at "zero time" when density is infinitely great? In an unreal world without quantum phenomena the answer is yes, but in the real world the answer is no.

When the universe is 1 hundred-million-trillion-trillion-trillionth (or 10^{-44}) of a second old, at the Planck epoch, our journey ultimately comes to a halt before a totally impenetrable wall: the Planck barrier. The density, although not infinite, has the enormous Planck value 10^{94} grams per cubic centimeter. The corresponding Planck temperature is 10^{32} kelvin.

Before us lies the mysterious realm of quantum cosmology, of which almost nothing is at present known. The quantum fluctuations of spacetime, on the scale of the Planck length (10^{-34} centimeters) and Planck time (10^{-44} seconds), are now of cosmic magnitude. Space and time are scrambled inextricably and discontinuously (Figure 20.8). John Wheeler at Princeton University, who explored this subject, visualizes spacetime under these conditions as a chaotic foam. The energy density is so immense that even the virtual particles of spacetime itself – quantized black holes of

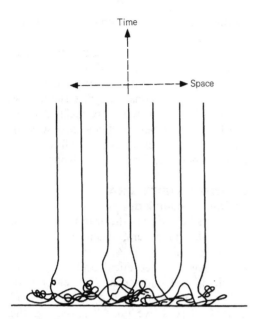

Figure 20.8. Perhaps the universe begins in a chaotic state. Imagine a large number of strings hanging vertically. At floor level the strings are coiled, tangled, and knotted together in a dense layer. In this analogy the strings are world lines along which intervals of individual time are measured. High above the floor exists an orderly sequence of events as in ordinary spacetime. Close to the floor the world lines are jumbled into closed and open loops and no common sequential property exists that can be identified with time. Is this the way the universe begins?

Planck mass 10^{-5} gram – are real. Spacetime is a foam of quantized black holes, and space and time no longer exist in the sense that we normally understand. There is no "now" and "then," and no "here" and "there," for everywhere is torn into discontinuities. We cannot go farther because an orderly historical sequence of events, which hitherto has unfolded during our journey backward in time, no longer exists. Here in the realm of quantum cosmology lie perhaps secrets that foretell the design and architecture of the physical universe.

REFLECTIONS

1 *"Yet there* are *new things to discover, if we have the courage and dedication (and*

money!) to press onwards. Our dream is nothing else than the disproof *of the standard model and its replacement by a new and better theory. We continue, as we have always done, to search for a deeper understanding of nature's mystery: to learn what matter is, how it behaves at the most fundamental level, and how the laws we discover can explain the birth of the universe in the primordial big bang"* (Sheldon Glashow, The Charm of Physics).

2 *The big bang is frequently portrayed as an explosion. An explosion occurs at a point in space, whereas the big bang embraces all of space. In an explosion, gas is driven outward by a steep pressure gradient, and a large pressure difference exists between the center and edge of the expanding gas. In the universe, there is no pressure gradient and the pressure is everywhere the same, and no center and edge exist. The expression "exploding big bang," although vivid, is a metaphor that conveys the wrong idea. George McVittie, a British cosmologist, wrote in 1974: "a 'bang' suggests that sound waves are emitted and a noise is heard. But again the equations defining the model universe show that no such sound waves are produced. In semipopular expositions of cosmology, terms such as 'the big bang hypothesis' or 'the big bang theory' are to be found. If these expressions have any meaning at all, they must be disguised ways of referring to the singular state found in the model universes of general relativity. For all these reasons it is unfortunate that the term 'big bang,' so casually introduced by Hoyle, has acquired the vogue which it has achieved"* ("Distance and large redshifts").

• *"The 'early universe' conveys more meaning [than big bang] in that it refers, in a general sense, to an early period of cosmic history. Perhaps cosmologists should follow geologists and use terms such as:* proto-cosmos *(from* proto *meaning first), for the extreme early universe of the first few Planck periods; and* paleocosmos *(from* paleo, *meaning early), for the very early universe of, say, the first 100 microseconds. With this terminology we can use the convenient* adjectives protocosmic and paleocosmic" (Edward Harrison, "A century of changing perspectives in cosmology," 1992).

- In an article "Needed: a better name for the big bang" (Sky and Telescope, August 1993), Timothy Ferris wrote: "Aesthetically speaking, 'Big Bang' sounds inappropriately bellicose. The dictionary definitions of bang include 'a sudden loud noise or thump' and 'an earsplitting, explosive noise.'" In an accompanying article entitled "The big bang challenge," the editors opened a competition soliciting suggestions from readers for a better name. At the close of the competition one month later, 13 099 entries had been received from 41 countries. A handful of entries came from cosmologists; most, however, came from "kindergartners, octogenarians, prison inmates, physicians, and many others from all walks of life" (Sky and Telescope, March 1994). No entry was judged a worthy successor of the egregious expression big bang. Although Hoyle did not submit an entry, he was nonetheless deemed the winner. The competition was widely viewed in the press as an attempt to eliminate sexist language from astronomy, and Henry Allen of the Washington Post labeled the editors of Sky and Telescope "cosmic correctness police."

3 Two revolutions have overtaken cosmology in the 20th century. In the first half of the century, the discovery of the expansion of the universe established that the early universe is very dense. In the second half of the century, the discovery of the cosmic background radiation established that the early universe is also very hot.

- George Ellis (at the University of Cape Town) in "Innovation, resistance and change: the transition to the expanding universe," distinguishes five major conceptual viewpoints in 20th-century cosmology. These viewpoints, referred to as paradigms, are:

(i) Unchanging models. Everybody, including Einstein, assumes that the universe as a whole is static and unevolving.
(ii) Evolving models. A major sea change in thought, prompted by observations associated mainly with Hubble's name, leads to the popular concept of an evolving universe.
(iii) The hot big bang. Inspired by Gamow, physicists in growing numbers begin to take a serious interest in cosmology.
(iv) Horizons and causal limits. Wolfgang Rindler's clarification of cosmological horizons (Chapter 21) ushered in a style of thought that eventually resulted in the now widely accepted but not unchallenged inflationary universe.
(v) Threshold of classical models. The study of the initial conditions of the universe developed into quantum cosmology.

A paradigm is a fixed attitude of mind; it consists of a set of related concepts widely shared in society. The history of science consists of the rise and fall of paradigms.

4 In 1948, George Gamow and his colleagues Ralph Alpher and Robert Herman were attracted by the thought that all elements heavier than hydrogen are synthesized in the big bang. "We conclude first of all," Gamow wrote, "that the relative abundance of various atomic species (which are found to be essentially the same all over the observed region of the universe) must represent the most ancient archaeological document pertaining to the history of the universe. These abundances must have been established during the earliest stages of expansion when the temperature of the primordial matter was still sufficiently high to permit nuclear transformations to run through the entire range of chemical elements" (George Gamow, "The evolution of the universe," 1948). That year Alpher and Herman wrote, "The temperature in the universe at the present time is found to be about 5 K." This predicted result is remarkably close to the value discovered by Penzias and Wilson in 1965. In their calculations they assumed that all matter in the big bang is initially in the form of neutrons, and nuclear synthesis begins at a temperature 1 billion kelvin (high enough for many nuclear reactions, but not too high to dissociate deuterons). With these assumed initial conditions they found that 50 percent of all matter was converted into helium.

Gamow and his colleagues were able to explain the origin of helium – a remarkable achievement – but their hope of explaining the origin of all other heavier elements was shattered by an insuperable difficulty. Elements much heavier than helium, such as carbon, nitrogen, and oxygen, are not produced abundantly because of the absence of stable nuclei of atomic weights 5 and 8. "The trouble lies in the fact that the nucleus of mass 5, which would be the next stepping stone, is not available. Due to some peculiar interplay of nuclear forces, neither a single proton nor a single neutron can be rigidly attached to the helium nucleus, so that the next stable nucleus is that of mass 6 (the lighter isotope of lithium), which contains two extra particles. On the other hand, under the assumed physical conditions, the probability that two particles will be captured simultaneously by a helium nucleus is negligibly small, and the building-up process seems to be stopped short at that point" (George Gamow, The Creation of the Universe, *1952). The answer to Gamow's problem was discovered by Fred Hoyle. Two helium nuclei come together, stay together in a brief state of resonance, and before the resonance decays and the helium nuclei separate, a third helium nucleus has time to join the two and the three together make a stable carbon nucleus. The three-particle collision in this case succeeds because it is really two two-particle collisions occurring one after the other. This reaction is possible if the temperature and density are both very high. But in the big bang, the temperature and density continually drop, and by the time helium has been synthesized, the physical conditions are no longer favorable for the production of carbon. In the interior of stars, however, temperature and density steadily rise as the star evolves, and eventually, after hydrogen has been converted into helium, the physical conditions necessary for the transformation of helium into carbon, and then into heavier elements, are attained. This is why we now believe that most elements heavier than helium are produced in stars.*

5 *A proton (p) and a neutron (n) combine to form a deuteron (d):*

$$p + n \rightarrow d,$$

and in this way deuterium is created in the early radiation era. At first, the deuterons are dissociated back into protons and neutrons by energetic rays of radiation:

$$d \rightarrow p + n,$$

but when the temperature has dropped to the neighborhood of 1 billion kelvin, the deuterons are no longer easily dissociated and they freely combine to form helium nuclei. Two deuterons combine to form 1 helium nucleus:

$$d + d \rightarrow \alpha,$$

where α denotes an alpha particle that is the nucleus of the helium atom. With 14 protons to every two neutrons (see below), we obtain finally 12 protons and one helium nucleus:

14 protons + 2 neutrons

$\rightarrow$ 12 protons + 2 deuterons

$\rightarrow$ 12 protons + 1 helium nucleus.

Of the initial 16 nucleons, four have united to form the helium nucleus, and $\frac{1}{4}$, or 25 percent, of all nucleons (and therefore 25 percent of the mass) is in the form of helium, and 75 percent in the form of hydrogen.

Leptons (electrons, muons, tauons, neutrinos, and all their antiparticles), and lepton numbers are shown in Table 20.2. The total electron lepton number is always conserved, as in

$$p + e^- = n + \nu_e,$$

where the electron lepton number of each side is +1. Similarly with muons and tauons. Neutrons and protons, with energetic leptons, interconvert, as shown:

$$n + e^+ \rightarrow p + \bar{\nu}_e, \quad n + \nu_e \rightarrow p + e^-,$$

$$p + e^- \rightarrow n + \nu_e, \quad p + \bar{\nu}_e \rightarrow n + e^+.$$

Note the conservation of electric charge and electron lepton number. Similar reactions

Table 20.2 *Leptons (charge, lepton number)*

Leptons			Antileptons	
e^-	electron $(-1,+1)$		e^+	positron $(+1,-1)$
ν_e	electron neutrino $(0,+1)$		$\bar{\nu}_e$	electron antineutrino $(0,-1)$
μ^-	muon $(-1,+1)$		μ^+	antimuon $(+1,-1)$
ν_μ	muon neutrino $(0,+1)$		$\bar{\nu}_\mu$	muon antineutrino $(0,-1)$
τ^-	tauon $(-1,+1)$		τ^+	antitauon $(+1,-1)$
ν_τ	tauon neutrino $(0,+1)$		$\bar{\nu}_\tau$	tauon antineutrino $(0,-1)$

exist for the muon and tauon leptons. When electron pairs begin to vanish at the end of the lepton era, and the neutrinos decouple, the neutron and proton abundances are in the ratio of two neutrons to every 10 protons. At this stage the neutrons begin to decay slowly according to

$$n \rightarrow p + e^- + \bar{\nu}_e,$$

at a rate of about 10 percent every 100 seconds. Helium production starts when the temperature has dropped to 1 billion degrees, and by then there remains approximately 2 neutrons to every 14 protons.

6 At the beginning of this century the only known subatomic particle was the electron; since then, decade by decade, slowly at first, new particles have been discovered. To cope with the motley crowd of more and more particles, various symmetries, classifications, and laws of conservation were devised to create a state of reasonable order. High-energy particle accelerators discovered hundreds of different particles, mostly excited hadrons of very short lifetime, uncovering a surprising complexity in the subatomic realm. For a time this state of affairs was not unlike that in the nineteenth century when chemists and physicists sought to find order in the chemical-atomic realm. Science often advances, so it seems, by decomposing an intricate system into an activity of components, each component itself a system of simpler components. The world decomposes into molecules that decompose into atoms that decompose into subatomic particles, disclosing worlds of subatomic complexity that can be understood only by further decomposition

into even smaller particles. These particles are the quarks, leptons, and gluons that constitute matter and have no internal structure.

Some important hadrons are shown in Table 20.3. Hadrons fall into two classes: baryons and mesons. Each baryon has a baryon number $+1$, and its antiparticle a baryon number -1, and the baryon number is conserved in all particle interactions. Thus, when a proton and an antiproton are created or annihilated, their combined baryon number is zero.

Baryons have half-integer spin, such as $\frac{1}{2}, \frac{3}{2}$, and $\frac{5}{2}$ times h, and belong to a broad class of particles named fermions. The spin of nucleons, for example, is $\frac{1}{2}h$. Mesons have zero baryon number; also they have integer spin, such as 0, 1, and 2 times h, and belong to a broad class of particles named bosons.

The fractional electric charges and the fractional baryon numbers of the quarks are displayed in Table 20.4. The situation is actually more complicated than shown because

Table 20.3 *Some hadrons and their quark compositions*

Hadrons			Quark compositions
baryons	p	proton	uud
	n	neutron	udd
	Λ^0	lambda	uds
mesons	π^+	positive pion	u$\bar{\text{d}}$
	π^0	neutral pion	u$\bar{\text{u}}$ + d$\bar{\text{d}}$
	π^-	negative pion	d$\bar{\text{u}}$
	K^+	positive kaon	u$\bar{\text{s}}$
	K^-	negative kaon	s$\bar{\text{u}}$

Table 20.4 *Quarks*

Flavor		Charge	Baryon number
u	up	$\frac{2}{3}$	$\frac{1}{3}$
d	down	$-\frac{1}{3}$	$\frac{1}{3}$
c	charm	$\frac{2}{3}$	$\frac{1}{3}$
s	strange	$-\frac{1}{3}$	$\frac{1}{3}$
t	truth	$\frac{2}{3}$	$\frac{1}{3}$
b	beauty	$-\frac{1}{3}$	$\frac{1}{3}$

each species has three colors, and each has its antiquark. Examples of how hadrons are constructed from quarks are shown in Table 20.3. Each quark (u, d, etc.) has a baryon number $\frac{1}{3}$ and each antiquark ($\bar{u}$, $\bar{d}$, etc.) a baryon number $-\frac{1}{3}$. Note that the quark composition always gives a baryon number $+1$ for a baryon (-1 for an antibaryon) and zero for a meson.

7   Sheldon Glashow (Harvard University), Steven Weinberg (then at Harvard University), and Abdus Salam (at the Center for Theoretical Physics, Trieste) developed in the early 1970s the electroweak theory that unifies the weak and electromagnetic forces. The W bosons of the weak interaction theory became allied to the photons (also bosons) of the electromagnetic theory. At energies above 100 GeV, the bosons of the electromagnetic and weak fields assume similar properties. At energies below 100 GeV, their unified symmetry breaks into the lesser symmetries of the electromagnetic and weak forces. After the success of the electroweak theory, Glashow and Howard Georgi (Harvard) took the next step and developed the hyperweak force that unifies the electroweak and strong forces. The leptons of the electroweak field become allied to the quarks of the strong field with X bosons, and at sufficiently high energy are indistinguishable and interchangeable. Three families of quarks imply three families of leptons. From studies of the primordial helium abundance, James Gunn (Princeton University), David Schramm (University of Chicago), and Gary Steigman (State University of Iowa) determined that the allowed number of lepton families cannot

exceed three, (e, ν_e), (μ, ν_μ), and (τ, ν_τ), thus implying three families of quarks, (u, d), (c, s), and (t, b), in confirmation of the theory.

Can we be sure that quarks and leptons are the ultimate constituents of matter? Altogether, there are 36 quarks (six flavors, each possessing three colors, thus making a total of 36 with their antiquarks). Is it possible that quarks and leptons are not the ultimate elementary states of matter but are themselves composite structures of yet more elementary entities? Are particles like Chinese boxes, ever enclosing particles of a more fundamental nature? "We do not know how much further we shall have to probe into subatomic phenomena before we reach an end to novelties, if indeed that will ever happen. Nor do we know if we as individuals, and as a species, are capable of scientific investigations to the point where this happens. These are questions for future humans to answer" (Gerald Feinberg, What Is the World Made Of? 1977).

8   Particles possess characteristics, such as mass, charge, and spin, and they combine to make composite particles having internal structure. When we look at particles on the basis of their spin, we find two distinct kinds: fermions (half-integral spin) and bosons (integral spin). Fermions are the particles of matter, and bosons are the particles of the forces that interact with matter. Thus the world consists of matter particles and force particles. As an illustration, electrons (fermions) are matter particles, whereas photons (bosons) are particles of the electromagnetic force that interacts with matter. Similarly, nucleons (fermions) are matter particles, whereas mesons (bosons) are particles of the strong force that interacts with nucleons. In the theory of supersymmetry, a mirror reality exists in which bosons are matter particles and fermions are force particles.

All particles of matter (fermions) consist of leptons (which are light, such as electrons) and baryons (which are heavy, such as nucleons). The leptons, which are few in variety, have no internal structure, whereas the baryons, which are numerous in variety, have complex internal structure. But baryons consist of quarks, which have no internal structure, and simplicity is restored by the idea that matter is composed not only of structureless leptons but also structureless quarks. Basically, there are six kinds of leptons (electrons, muons, tauons, with their neutrinos) and six kinds of quark (up, down, strange, charm, top, and bottom).

9   Arthur Schuster wrote in 1898: "Surely something is wanted in our conception of the universe. We know positive and negative electricity, north and south magnetism, and why not some extra-terrestrial matter related to terrestrial matter as the source is to the sink, gravitating towards its own kind, but driven away from the substances of which the solar system is composed. Worlds may have formed from this stuff, with elements and compounds possessing identical properties with our own, indistinguishable in fact from them until they are brought into each other's vicinity. If there is negative electricity, why not negative gold, as yellow and valuable as our own, with the same boiling point and identical spectral lines; different only in so far that if brought down to us it would rise up into space with . . . acceleration?" ("Potential matter: a holiday dream"). We know from experiments with antiparticles that antimatter does not possess negative gravity and is attracted to matter in the same way as ordinary matter. Schuster apparently was the first to speculate on the idea of antimatter. He concluded his essay with the words: "Whether such thoughts are ridiculed as the inspirations of madness, or allowed to be the serious possibilities of a future science, they add renewed interest to the careful examination of the incipient worlds which our telescopes have revealed to us. Astronomy, the oldest and most juvenile of the sciences, may still have some surprises in store. May antimatter be commended to its care!"

•   The idea that the universe is baryon symmetric (meaning that it contains matter and antimatter in equal amounts) seems attractive. To avoid complete annihilation, some segregation must occur so that isolated

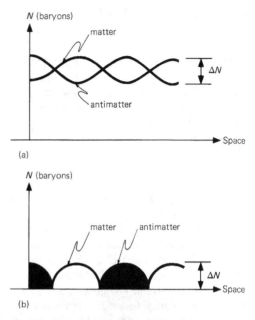

Figure 20.9. In (a) the densities of matter and antimatter vary relative to each other, and their average densities are supposed equal. After annihilation, only their difference survives, as shown in (b), and isolated regions exist of matter and antimatter.

regions of matter and antimatter can survive mutual annihilation. Segregation mechanisms have been considered but so far none has been accepted as satisfactory. The simplest idea is shown in Figure 20.9, in which the densities of matter and antimatter in the hadron era (or earlier) vary in space relative to each other. (In the absence of variations, the densities are supposed everywhere equal.) After annihilation has ceased in the hadron era, there remain isolated pockets of matter and antimatter that evolve and become galaxies and antigalaxies. This proposal by itself fails to explain why the initial fluctuations occurred. Furthermore, we have no evidence of the existence of antigalaxies or other astronomical systems of antimatter.

•   To create particle pairs abundantly in a very hot gas, the existing particles – such as photons – must have thermal energies of $2mc^2$ or more. A typical thermal energy is kT, where k is the Boltzmann constant, and nucleons and antinucleons of mass m_n are pair-created abundantly when the temperature T equals or exceeds $2m_{\mathrm{n}}c^2/k$, about 10^{13} kelvin.

Grand unified theories give, in principle, a credible explanation of why matter is favored cosmologically more than antimatter. The slight differences in the decay schemes of matter and antimatter are not important in a state of perfect thermodynamic equilibrium. Owing to the rapid expansion of the universe, however, thermodynamic equilibrium is disturbed and the matter and antimatter decay schemes are affected slightly differently. We may also appeal either to the theistic or the anthropic principles. If the one-part-in-a-billion difference in the matter and antimatter densities had not existed in the very early universe, the universe would be starless and lifeless and we would not be here discussing the subject. It may be said a supreme being designed our universe for habitation by life and arranged that matter and antimatter are unequally favored. Alternatively, it may be said, a universe is self-aware (contains conscious observers) when such a difference exists; we are here in this particular universe – one of many hypothetical universes – because its initial difference in the abundances of matter and antimatter is just right.

10 Galaxy formation (the subject of how galaxies originate) deals with some of the most perplexing problems in cosmology. At the end of the 20th century, after more than fifty years of research, cosmologists still do not understand how galaxies originate and form in an expanding universe. Conceivably, small condensations such as globular clusters or dwarf galaxies aggregate to form galaxies, galaxies aggregate to form clusters, then clusters to form superclusters, and so on, to perhaps even larger systems. Also conceivably, large condensations, such as superclusters, come first and fragment into clusters, clusters fragment into galaxies, then galaxies into dwarf galaxies and globular clusters, and so on, to perhaps smaller systems? The first is the bottom-up theory, and the second is the top-down theory. A combination of both seems most probable.

An attractive possibility is that the irregularities that ultimately develop into galaxies originate quantum mechanically in the extreme early universe. Thermal fluctuations are much too small, whereas quantum fluctuations look more promising. After the decoupling epoch the irregularities evolve and eventually form into galaxies when the universe is one or more billion years old. It is commonly assumed that the irregularities at the time of decoupling consist mainly of small-amplitude variations in the density of matter. The ratio of the density variation $d\rho$ and the average density ρ, or $d\rho/\rho$, called the density contrast, grows after decoupling and is proportional to the scale factor R. The present value of the scale factor has increased 1000 fold since the decoupling epoch, and if we suppose that the contrast density on the largest scales is now between 0.01 and 0.1, the contrast density was between 10^{-5} and 10^{-4} at the decoupling epoch.

A temperature variation dT, accompanying the density variation $d\rho$ at the time of decoupling, should have a magnitude dT/T not greatly different from $d\rho/\rho$. After decoupling, the temperature variation dT/T stays constant, unlike the contrast density $d\rho/\rho$ that grows. From the observed variations of dT/T of order 10^{-5} on large scales in the cosmic background radiation, we infer the existence of density variations of this magnitude at the decoupling epoch.

11 When a star in an advanced state of evolution collapses, it terminates its collapse as a white dwarf, or a neutron star, or a black hole. A black hole is the ultimate state of collapse. To an external observer the black hole is frozen in a perpetual state of collapse, whereas to an internal observer who falls with the star, the collapse continues and ends in a singularity of maximum density. Such singularities, predicted by the laws of nature, are themselves beyond the reach of the known natural laws.

Singularity theorems have been developed by Roger Penrose, Stephen Hawking, and George Ellis. Penrose showed that a singularity is inevitable whenever a self-gravitating *region is enclosed within a trapped surface. Inside a trapped surface (see Chapter 13), all light rays are dragged inward faster than they can escape outward. Rotation and non-spherical collapse cannot avert a singularity when the collapsing region is surrounded by a trapped surface. It has also been shown that when certain well-defined conditions are satisfied – for example, density and pressure must both be positive – a collapsing universe terminates unavoidably in a singular state. This singular state, according to classical theory, occurs at infinite density. But according to modern physics an infinite density is unlikely, and probably the singular state occurs at the finite Planck density.*

PROJECTS

1 Discuss: "I would bet odds of 10 to 1 on the validity of the general 'hot big-bang concept' as a description of our universe since it was around 1 second old at a temperature of 10^{10} K (or ~1 MeV). Some people are even more confident. In a memorable lecture at the International Astronomical Union back in 1982, Zel'dovich claimed that the big bang was 'as certain as that the Earth goes around the Sun.' He must even then have known his compatriot Landau's dictum that cosmologists are 'often in error but never in doubt'!" (Martin Rees, *Perspectives in Astrophysical Cosmology*).

2 Consider: "Many people would argue that it makes no physical sense to talk about half-an-hour that took place ten billion years ago. To answer that criticism, let us consider a site, somewhere in Nevada where an atomic bomb was set off several years ago. The site is still hot with long-lived fission products, and it took only about one microsecond for the nuclear explosion to produce all the fission products" (Paul Vogel, Amherst College, 1978). See a discussion on the same theme by George Gamow in *The Creation of the Universe* (1952, pp. 63–64).

3 If the universe everywhere were baryon symmetric, and matter and antimatter were totally annihilated in the hadron era, the universe would contain almost nothing but radiation, and the cosmic background radiation temperature 2.73 kelvin would now be slightly higher by about one billionth of a degree. But for reasons still not entirely clear, matter exceeded antimatter by one part in a billion in the hadron era. If the matter excess were now converted into thermal radiation, the temperature would be, not 2.73 kelvin, but roughly 20 degrees kelvin. In both cases the universe contains only radiation. Try and explain the difference in temperature.

4 Think about reductionism (the belief that things are explained in terms of an activity of component things) and holism (the belief that the whole is more than a summation of its components, particularly in living creatures).

5 Discuss the matter–antimatter difference in the early universe from the viewpoint of the anthropic and theistic principles (and any other viewpoint you wish).

6 "'Do we do fundamental physics to explain the world about us?' is a question that is often asked. The answer is NO! The world about us was explained 50 years ago or so. Since then, we have understood why the sky is blue and why copper is red. That's elementary quantum mechanics. It's too late to explain how the work-a-day world works. It's been done. The leftovers are things like neutrinos, muons, and K mesons – things that have been known for half a century, and have no practical application, and probably never will: little mysteries like how the universe began and how it will end." (Sheldon Glashow, *The Charm of Physics*). Contrast this challenging statement with: "Science is a fast-moving subject, and the scientific conception of the cosmos has changed radically over the last few centuries. It may well change again. New laws of physics no doubt await our discovery, new concepts and ideas that could remold the entire intellectual framework on which our present judgments about creation, evolution, and cosmic collapse are based. The role of man as an intellectual observer and as an active force for restructuring the world through technology could easily shift in perspective in the coming centuries" (Paul Davies, *The Runaway Universe*).

FURTHER READING

Alpher, R. A. "Large numbers, cosmology and Gamow." *American Scientist* (January–February 1973).

Alpher, R. A. and Herman, R. "Reflections on 'big bang' cosmology," in *Cosmology, Fusion and Other Matters: George Gamow Memorial Volume*. Editor F. Reines. Colorado Associated University Press, Boulder, 1972.

Barrow, J. and Silk, J. "The structure of the early universe." *Scientific American* (April 1980).

Barrow, J. and Silk, J. *The Left Hand of Darkness*. Basic Books, New York, 1983.

Brush, S. G. "How cosmology became a science." *Scientific American* (August 1992).

Burbidge, G. "Was there really a big bang?" *Nature* 233, 36 (September 3, 1971).

Chown, M. *Afterglow of Creation: From the Fireball to the Discovery of Cosmic Ripples*. Arrow Books, London, 1993.

Davies, P. C. W. *The Forces of Nature*. 2nd edition. Cambridge University Press, Cambridge, 1986.

Drell, S. D. "When is a particle?" *American Journal of Physics* 46, 597 (January 1976).

Feinberg, G. *What Is the World Made Of? Atoms, Leptons, Quarks, and Other Tantalizing Particles*. Doubleday, Anchor Press, New York, 1977.

Fritzsch, H. *Quarks: The Stuff of Matter*. Basic Books, New York, 1983.

Gamow, G. *The Creation of the Universe*. Viking Press, New York, 1952.

Gribbin, J. *In Search of the Big Bang: Quantum Physics and Cosmology*. Basic Books, New York, 1986.

Lemaître, G. *The Primeval Atom*. Van Nostrand, New York, 1951.

Overbye, D. *Lonely Hearts of the Cosmos: The Scientific Quest for the Secret of the Universe*. Harper Collins, New York, 1991.

Pasachoff, J. and Fowler, W. "Deuterium in the universe." *Scientific American* (May 1974).

Peebles, P. J. E., Schramm, D. N., Turner, E. L., and Kron, R. G. "The case for the relativistic hot big bang cosmology." *Nature* 769, 352 (August 29, 1991).

Rees, M. J. *Perspectives in Astrophysical Cosmology*. Cambridge University Press, Cambridge, 1995.

Rees, M. J. *Before the Beginning: Our Universe and Others*. Addison Wesley Longman, Reading, Massachusetts, 1997.

Schramm, D. N. "The early universe and high-energy physics." *Physics Today* (April 1983).

Schramm, D. N. and Wagoner, R. V. "What can deuterium tell us?" *Physics Today* (December 1974).

Schwarz, C. *A Tour of the Subatomic Zoo: A Guide to Particle Physics*. American Institute of Physics, New York, 1992.

Silk, J. *The Big Bang*. Second Edition. W. H. Freeman, San Francisco, 1989.

Weinberg, S. *The First Three Minutes: A Modern View of the Origin of the Universe*. Basic Books, New York, 1977.

SOURCES

Alpher, R. A. and Herman, R. C. "Evolution of the universe." *Nature* 162, 774 (November 13, 1948).

Alpher, R. A. and Herman, R. C. "Early work on 'big-bang' cosmology and the cosmic background radiation," in *Modern Cosmology in Retrospect*. Editors B. Bertotti et al. Cambridge University Press, Cambridge, 1990.

Alpher, R. A., Follin, J. W., and Herman, R. C. "Physical conditions in the initial stages of the expanding universe." *Physical Review* 92, 1347 (1953).

Arp, H. C., Burbidge, G., Hoyle, F., Narlikar, J. V., and Wickramsinghe, N. C. "The extragalactic universe: An alternative view." *Nature* 807, 346 (August 30, 1990).

Davies, P. C. W. *The Runaway Universe*. Dent, London, 1978.

Dicke, R. H., Peebles, P. J. E., Roll, P. G., and Wilkinson, D. T. "Cosmic black-body radiation." *Astrophysical Journal* 142, 414 (1965).

Ellis, G. F. R. "Singularities in general relativity." *Comments on Astrophysics and Space Physics* 8, 1 (1978).

Ellis, G. F. R. "Innovation, resistance and change: the transition to the expanding universe," in *Modern Cosmology in Retrospect*. Editors B. Bertotti et al. Cambridge University Press, Cambridge, 1990.

Gamow, G. "The evolution of the universe." *Nature* 162, 680 (October 30, 1948).

Gamow, G. *The Creation of the Universe*. Viking Press, New York, 1952.

Glashow, S. L. *The Charm of Physics*. Simon and Schuster, New York, 1991.

Harrison, E. R. "The early universe." *Physics Today* (June 1968).

Harrison, E. R. "A century of changing perspectives in cosmology." *Quarterly Journal Royal Astronomical Society* 33, 335 (1992).

Kolb, W. E. and Turner, M. S. *The Early Universe*. Addison-Wesley, New York, 1990.

Lemaître, G. "The beginning of the world from the point of view of quantum theory." *Nature* 127, 706 (May 9, 1931).

Lemaître, G. *The Primeval Atom*. Van Nostrand, New York, 1951.

McVittie, G. C. "Distance and large redshifts." *Quarterly Journal of the Royal Astronomical Society* 15, 246 (1974).

Peebles, P. J. E. "Primordial helium abundance and the primordial fireball." *Astrophysical Journal* 146, 542 (1966).

Peebles, P. J. E. *Principles of Physical Cosmology*. University Press, Princeton, New Jersey, 1993.

Penrose, R. "Singularities and time-asymmetry," in *Einstein Centenary Volume*. Editors S. W. Hawking and W. Israel. Cambridge University Press, Cambridge, 1979.

Schuster, A. "Potential matter: A holiday dream." *Nature* 58, 367 (August 18, 1898).

Tayler, R. J. "Neutrinos in the universe." *Quarterly Journal of the Royal Astronomical Society* 93, 22 (1981).

Wilson, R. W. "Discovery of the cosmic microwave background," in *Modern Cosmology in Retrospect*. Editors B. Bertotti et al. Cambridge University Press, Cambridge, 1990.

HORIZONS IN THE UNIVERSE

I am a part of all that I have met;
Yet all experience is an arch wherethro'
Gleams that untravelled world, whose margin fades
For ever and for ever when I move.
Tennyson (1809–92), Ulysses

WHAT ARE COSMOLOGICAL HORIZONS?

Horizons

We look out in space and back in time and do not see the galaxies stretching away endlessly to an infinite distance in an infinite past. Instead, we look out a finite distance and see only things within the "observable universe." Like the sea-watching folk in Robert Frost's poem, we "cannot look out far" and "cannot look in deep."

The observable universe is normally only a portion of the whole universe. We are at the center of our observable universe; its distant boundary acts as a cosmic horizon beyond which lie things that cannot be observed. Observers in other galaxies are located at the centers of their observable universes that are also bounded by horizons. A person on a ship far from land, who sees the sea stretching away to a horizon, is at the center of an "observable sea." People on other ships are at the centers of their own observable seas that are bounded by horizons. Despite this analogy the horizons of the universe are not as simple as the horizons of the sea.

Particle and event horizons

The subject of cosmic horizons was confusing until Wolfgang Rindler cleared up the muddle in 1956. He showed that in discussing the observable and unobservable we must distinguish between two kinds of observables: things that endure in time and things that have only momentary existence.

World lines in spacetime represent things such as particles and galaxies that endure; they occupy at each instant in time a place in space. Points in spacetime represent events or brief happenings, such as the flash of a firefly or the explosion of a supernova, that occupy a place in space and only a moment in time. World lines are in effect strings of events. In this chapter, the events of main interest emit light, and the world lines of interest, other than the observer's, are of luminous bodies, such as galaxies.

To discover what is observable and what is unobservable we must specify the nature of the things observed. If they are particles or galaxies that endure and have world lines, we discover one kind of answer; if they are events that occur briefly, we discover another kind of answer. For example, if a person is asked, "Have you met Mr. X?" the answer could be quite different to that in response to, "Did you see Mr. X at his wedding?" The first question asks if a world line has been observed at some time or other, and the second asks if an event was observed that occurred at a particular time.

There are thus two types of horizon, a world line horizon and an event horizon, and both are important. Rindler referred to the world line horizon as a "particle" horizon, and because this latter term is now widely adopted we shall continue to use it. It must be understood, however, that the word "particle" in this case means world line and represents anything that endures.

In the following we first define particle and event horizons. With the help of illustrations we try to make clear their significance, first in a static universe, and then in an expanding universe. Our discussion concerns only horizons in universes that are isotropic and homogeneous (i.e., all directions are alike and all places are alike at each instant in cosmic time).

A particle horizon is the surface of a sphere in space that has the observer at the center. This horizon divides the whole of space into two regions: the region inside the horizon contains all galaxies that are visible, and the region outside contains all galaxies that are not visible. Thus the particle horizon is a spherical surface in space that encloses the observable universe. The horizon at sea is of this type; it is a frontier that divides all things into two groups: those inside that are visible and the rest outside that are not visible.

The event horizon divides all events into two groups: those visible at some time or other and those that are never visible. An observer sees events on the backward lightcone. The event horizon is therefore not a surface in space but a null surface in spacetime (in this case the backward lightcone) separating the events that can be observed at some time from the events that can never be observed. The event horizon is not quite so obvious as the particle horizon with its sea-horizon analogy, but this need not cause concern; the next section will help to clarify this obscure subject.

HORIZONS IN STATIC UNIVERSES

The two types of horizon are most easily demonstrated in an infinite and static universe. We forget for a moment that the universe is expanding and suppose that we live in a static universe that contains uniformly distributed galaxies.

Particle horizon

We suppose that the galaxies have been luminous for 10 billion years. Either this hypothetical universe was created 10 billion years ago with luminous galaxies or galaxies

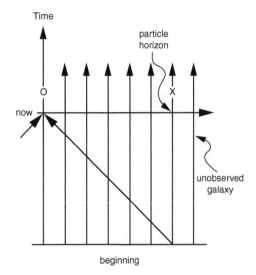

Figure 21.1. This diagram represents a static universe that has a beginning and consists of uniformly distributed galaxies. We are the observer O who looks out now and sees the world lines of luminous galaxies intersecting our backward lightcone. World line X is at the particle horizon. Galaxies having world lines beyond X cannot be seen.

became luminous 10 billion years ago in a preexisting dark universe. The situation is shown in Figure 21.1. The universe consists of world lines of luminous galaxies that commence at a "beginning."

The world line labeled O represents our Galaxy from which we observe the universe. From O, at the instant "now," we look out in space and back in time and see the other galaxies on our backward lightcone. We see galaxies because their world lines intersect our lightcone, and we see each at some instant in its lifetime. All galaxies have been shining for 10 billion years and it is therefore possible to look out and see them stretching away to a distance of 10 billion light years. Galaxies at greater distances cannot be seen because we look back either to a time when the universe was created or to a time before galaxies were born.

A particle horizon divides all luminous sources into those observed and those not observed. Hence, in a static universe the

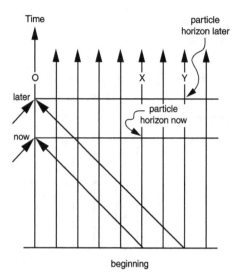

Figure 21.2. At the moment "now" we – the observer O – see no farther than the world line X. Subsequently, at the moment labeled "later," we see beyond X to world line Y. The particle horizon thus recedes in a static universe and the observable universe, bounded by the particle horizon, expands.

particle horizon is at the distance indicated by world line X, and in this example it lies at distance 10 billion light years. Galaxies at distances less than 10 billion light years are visible and lie inside the observable universe, and galaxies at distances greater than 10 billion light years are not visible and lie outside the observable universe.

We wait a period of time – say 1 billion years – and repeat our observations. At the instant "later," shown in Figure 21.2, when galaxies have been shining for 11 billion years, we see them stretching away to a distance of 11 billion light years. The particle horizon has receded to a distance of 11 billion light years. Thus the particle horizon moves outward away from the observer at the speed of light, and although the universe is static, the observable universe actually expands. This is important. In all uniform (i.e., homogeneous and isotropic) universes, static and nonstatic, expanding and contracting, the particle horizon moves outward at the speed of light relative to the galaxies. As time passes we always see more and more of the universe.

Event horizon

We turn now to the event horizon and ask whether in the static universe events exist that can never be seen at any time by an observer. If such events exist, we can divide the universe into two parts: one that contains all the events observable from the observer's world line O; and the other that contains the remaining events unobservable from O. The surface separating the two parts is the event horizon for observer O. ("Observers" in this chapter are immortal; they are born with the universe and die with the universe.)

If the universe is eternal and galaxies shine forever, no event horizon exists. O's lightcone advances up O's world line, and any pointlike event in spacetime will eventually lie on the lightcone and be visible. Hence, in an eternal static universe, in which galaxies are forever luminous, there exists no event horizon and every event in the universe at some time or other is observed by every observer.

An event horizon exists in a universe that has an "end." Either the whole universe terminates, or the galaxies cease to shine and the universe becomes dark. As a result, all world lines of luminous galaxies come to an end, as in Figure 21.3. Figure 21.4 shows clearly that such a universe has an event horizon: it is O's lightcone at the last

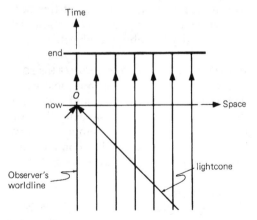

Figure 21.3. A static universe that has an ending. The time labeled "end" is the observer's last moment of observation.

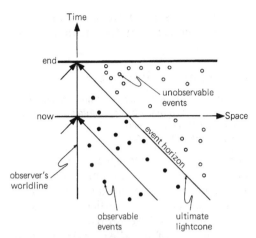

Figure 21.4. The event horizon is the observer's backward lightcone at the moment when the universe ends. Inside this ultimate lightcone are the events that can be observed at some time, and outside are the events that can never be observed.

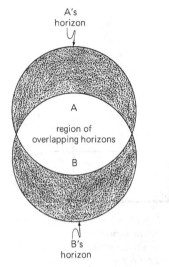

Figure 21.5. Albert (A) and Bertha (B) have overlapping horizons, but each can apparently see things that the other cannot. By communicating with each other can they enlarge their individual horizons into a joint horizon? If they can, then their individual horizons are not true information horizons.

possible moment. Inside the event horizon are the events that have been seen, and outside are the events that can never be seen. The lightcone cannot advance farther into the future and all events outside this ultimate lightcone remain unseen.

The static universe serves to illustrate moderately well the nature of cosmic horizons. From it we learn that beyond the particle horizon are world lines (particles, stars, galaxies) that cannot at the time of observation be seen at any stage in their existence, and beyond the event horizon are events (happenings of short duration) that cannot be seen at any time in the observer's existence.

Before proceeding to nonstatic universes, we must discuss the horizon riddle and the horizon problem.

THE HORIZON RIDDLE

Consider two widely separated observers, A (for Albert) and B (for Bertha). We suppose they can see each other. Each has a horizon such that A cannot see things beyond his horizon and B cannot see things beyond her horizon. Each sees things the other cannot see, as illustrated in Figure 21.5. We ask: Can B communicate to A information that

extends A's knowledge of things beyond his horizon? If so, then a third observer C may communicate to B information that extends her horizon, which can then be communicated to A. Hence, an unlimited sequence of observers B, C, D, E, ... may extend A's knowledge of the universe to indefinite limits. According to this argument A has no true horizon. This is the horizon riddle.

The riddle arises from our experience with horizons on the surface of the Earth. If A and B are on ships at sea, within sight of each other, they each see the sea stretching away to the horizon. A sees things that B cannot see, and similarly, B sees things that A cannot see. By flag signals or by radio they can keep each other informed of things not directly visible. By communication, A and B share information and succeed in extending their horizons. A pre-twentieth-century admiral had a horizon that embraced his entire fleet.

When we speak of things that are seen or not seen we usually have in mind those that endure and are represented by world lines.

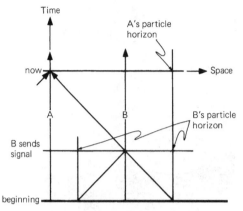

Figure 21.6. Proof that Bertha cannot help Albert to see beyond his horizon (and similarly Albert cannot help Bertha to see beyond her horizon). The horizons in this case are particle horizons: B communicates information to A by sending it at the speed of light on A's backward lightcone; but when B sends the information, her horizon extends no farther than A's horizon, and she cannot see farther than A.

Thus the horizon riddle applies to the particle horizon of the universe. We consider the particle horizon in a static universe (Figure 21.6) and show that the riddle has a simple solution. We have supposed that luminous galaxies originated 10 billion years ago and the particle horizon is therefore at distance 10 billion light years. Observers A and B see each other and have overlapping horizons. Suppose A and B are separated by a distance of 6 billion light years. B sends out information that travels at the speed of light and takes 6 billion years to reach A. Hence A receives from B information that was sent 6 billion years ago when the universe was 4 billion years old. But B's particle horizon in the past at the time when the information was sent was only 4 billion light years distant. Thus B's horizon at that time did not extend beyond A's present horizon. With this argument, and the help of Figure 21.6, we see that neither B nor any other observer can extend A's particle horizon. The particle horizon is a true information horizon and no information can be obtained from other observers concerning what lies beyond. Although we

have used the static universe, the argument applies quite generally to particle horizons in all universes.

THE HORIZON PROBLEM

While to deny the existence of an unseen kingdom is bad, to pretend that we know more about it than its bare existence is no better.
Samuel Butler (1835–1902)

For many years cosmologists have debated a subject referred to as the horizon problem. The problem exists in all static and expanding universes that have particle horizons. As an illustration of the problem, consider a static universe of age t_0. An observer cannot see farther than the particle horizon at distance ct_0, where c is the speed of light, or distance t_0 in units of light-travel time. Figure 21.7 shows the observer as a dot at

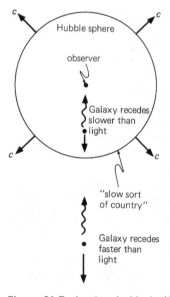

Figure 21.7. A galaxy inside the Hubble sphere recedes from us (the observer) at subluminal velocity, and the light emitted by the galaxy in our direction is able to approach us. A galaxy outside the Hubble sphere recedes from us at superluminal velocity, and the light emitted by the galaxy in our direction is unable to approach us and actually recedes. The edge of the Hubble sphere is the country of the Red Queen – the photon horizon – where the recession velocity of the galaxies is transluminal and light emitted in our direction stands still.

the center of a sphere of radius ct_0. The edge of the sphere is the particle horizon that encloses the observable universe; outside the sphere lies the unobservable universe from which light has not yet reached the observer. The sphere expands as the universe ages and its edge – the particle horizon – recedes from the observer at the speed of light. In the course of time more and more of the universe becomes visible. Thus when the universe was 1 year old, the horizon was at distance 1 light year. When the universe is 10 billion years old, the horizon is at distance 10 billion light years. In the future when the universe is 20 billion years old the horizon will be at distance 20 billion light years. In all universes, static and nonstatic, the particle horizon sweeps past the galaxies at the speed of light and the observer progressively sees more and more of the universe.

Suppose observer O sees A in one direction at distance L and B in the opposite direction also at distance L. How large must L be in order that A and B are unaware of each other's existence at the time when they are seen by O? In a static universe, the answer is $L = \frac{1}{3}ct_0$, as shown in Figure 21.8. More generally, in static and nonstatic universes, A and B cannot see each other when L is greater than $\frac{1}{3}L_P$, where L_P is the distance to the particle horizon. When A and B are each farther away than one-third the particle horizon distance L_P, they see

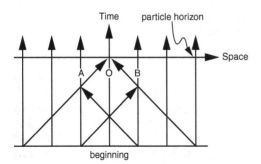

Figure 21.8. The observer sees A and B at equal distances L in opposite directions. When L is greater than one-third the distance to the particle horizon, A and B are unaware of each other's existence.

the observer but cannot see each other. They do not know that each other exists. To make clear the nature of the problem, imagine that A and B have similar genetic coding. If they do not know that each other exists – are outside each other's horizon – and previously had no history of interaction, how can we explain why they are genetically alike? Stated more generally, why should galaxies, stars, chemical elements, and subatomic particles exist in similar states when their horizons do not overlap?

The particle horizon is important because it determines not only the maximum distance an observer can see, but also the maximum distance between things that are able to communicate and affect one another. It determines the range of causal interactions. A body observed at distance $L = \frac{1}{2}L_P$, and now inside our horizon, was outside at a cosmic age earlier than $\frac{1}{2}t_0$. Normally we look back into the past for causes that explain the way things are now. But how can we explain the way things are now on the scale of 10 billion light years by causes that existed when the universe was less than 10 billion years old? This is the horizon problem.

The horizon problem has no known scientific solution in a static universe. A possible scientific solution in an expanding universe requires a period of accelerated expansion in the early universe. Alan Guth introduced the idea of accelerated expansion as a serious possibility in 1981 and called it inflation, and his inflationary model is discussed in Chapter 22.

HUBBLE SPHERES

Static universes serve to illustrate the fundamental nature of cosmological horizons but are not very realistic. First, in a preambling manner we discuss a few basic properties of expanding universes.

According to the velocity–distance law the recession velocities V of the galaxies increase linearly with distance L:

$$V = HL, \qquad [21.1]$$

where H is the Hubble term and L is the sort of distance one would obtain with a tape measure stretched on a curved surface. The value of H at the present epoch is $H_0 = 100h$ km per second per megaparsec and the coefficient h lies perhaps between 0.5 and 1. When L is doubled the recession velocity V is also doubled. Distances are measured in the world map that covers homogeneous space at a common instant in cosmic time. In a homogeneous universe that stays homogeneous during expansion, the velocity–distance law is necessarily linear in the world map (i.e., velocity must increase in strict proportion to distance). Thus if the universe consists of infinite space, at infinite distance the recession velocity is infinitely large. The velocity–distance law $V = H_0 L$ tells us how fast a galaxy at distance L recedes at the present time.

The recession velocity equals the velocity of light at the Hubble distance

$$L_H = \frac{c}{H_0}, \qquad [21.2]$$

and at the present time

$$L_H = 9.8 \times 10^9 h^{-1} \text{ light years.} \qquad [21.3]$$

The Hubble distance lies somewhere between 10 and 20 billion light years, and for illustration we assume an intermediate value 15 billion light years. From Equations [21.1] and [21.2] we have

$$\frac{V}{c} = \frac{L}{L_H}, \qquad [21.4]$$

and this shows clearly that galaxies at distance L greater than L_H recede faster than the velocity of light. According to the expanding space paradigm, galaxies are stationary in space and recede from one another because of the expansion of intergalactic space. We are at the center of our Hubble sphere, a sphere whose present radius is 15 billion light years. Inside this Hubble sphere are the galaxies that recede slower than light velocity, and outside are those that recede faster than light velocity. The Hubble sphere must not be confused with the observable universe. The observable universe is bounded by the particle horizon, and if the Hubble sphere and observable universe were the same, the observable universe would be infinitely large in a static universe ($H_0 = 0$, $L_H = \infty$). But static universes of finite age have particle horizons at finite distance, and therefore the Hubble sphere cannot be the observable universe.

RECEPTION AND EMISSION DISTANCES

We look out and see galaxies of various redshifts and must be careful about assuming that distance increases always with redshift. Reception distances (measured in the world map) increase always with redshift. The larger the redshift, the greater the distance at the time of reception. But we cannot see galaxies at their present distances; we see them in the past at the time when they emitted the light now seen, and their emission distances do not continually increase with redshift in a big bang universe.

Each receding galaxy has two distances: the distance at the time of reception and the distance at the time of emission. The first is the reception distance measured in the world map and denoted by L, and the second is the emission distance and denoted by L_{emit} (see Figure 15.6). These two distances have the simple redshift relation

$$\frac{L}{L_{\text{emit}}} = 1 + z, \qquad [21.5]$$

where z is the redshift of the galaxy. The reception distance is always greater than the emission distance in an expanding universe.

The emission distances of galaxies increase at low redshifts and decrease at high redshifts. Faint galaxies of large redshifts seemingly far away were actually nearer to us at the time of emission than bright galaxies of small redshifts. This odd state of affairs occurs in all expanding universes in which the deceleration term q is greater than -1. It does not occur in the de Sitter and steady-state universes in which q is equal to -1.

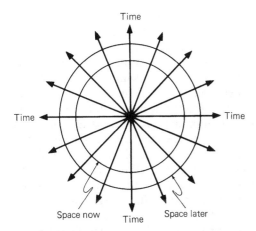

Figure 21.9. A big bang with world lines diverging in all directions. Do not let this diagram mislead you into thinking that the big bang occurs at a point in space. Time is measured along the world lines, and space is represented by any spherically curved surface perpendicular to the world lines.

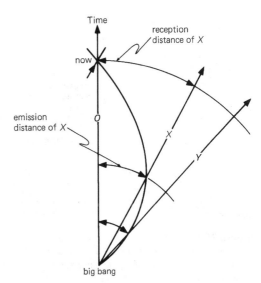

Figure 21.10. O's lightcone curves back into the big bang, and for this reason we are able to observe the cosmic background radiation that has traveled freely since it decoupled in the early universe. This diagram shows the reception and emission distances of galaxy X. Although galaxy Y has a greater reception distance, its emission distance is smaller than that of X. Thus Y, which is now farther away than X, was closer to us than X at the time of emission of the light we now see.

A spacetime diagram convenient for our immediate purpose is shown in Figure 21.9, in which world lines diverge radially in all directions from a big bang. Space is represented by spherical surfaces perpendicular to the world lines, and time is measured along the radial world lines. World lines are galaxies fixed in space, and as time advances, they recede from one another because of the expansion of space. An expanding spherical balloon is a helpful analogy. Galaxies are points marked on the surface of the balloon and as the balloon inflates the "galaxies" recede from one another. The two-dimensional surface of the balloon represents our three-dimensional space. The radial direction in which the balloon expands represents time and should not be confused with the third dimension of space.

Any world line – it does not matter which – is chosen as the observer and labeled O, as in Figure 21.10. At any instant in time – call it "now" – the observer's lightcone stretches out and back and intersects other world lines such as X and Y. Because of the expansion of space, the lightcone does not stretch out straight as in a static universe, but contracts

back into the big bang. All world lines and all backward lightcones converge into the big bang. The observer, by looking in any direction, looks back into the big bang, and the light the observer receives from the big bang is the cosmic background radiation. The luminous events on the backward lightcone are redshifted, and the closer they are to the big bang, the larger is their redshift. Thus redshift increases steadily as we proceed along the lightcone toward the big bang.

Figure 21.11 shows the emission and reception distances of two galaxies X and Y. X's emission distance is measured in space at the time X emitted the light that O now sees, and X's reception distance is measured in space at the time its light reaches O. Similarly with Y. As shown, X's reception distance is smaller than Y's and we may say X is now nearer than Y. Also X's redshift is smaller than Y's redshift.

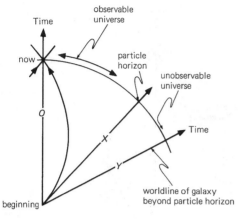

Figure 21.11. At the instant labeled "now" the particle horizon is at worldline X. In a big bang universe, all galaxies at the particle horizon have infinite redshift.

Redshift always increases with reception distance. As we move back along the light-cone, the emission distance at first increases, then reaches a maximum, and thereafter decreases. In Figure 21.10, the emission distance of Y is less than that of X, even though Y is now farther away and has the greater redshift. The maximum emission distance L_{max} in the Einstein–de Sitter universe is 8/27 of the Hubble distance L_H, or almost 5 billion light years, and has redshift of 1.25. Galaxies at distance L_{max} at the time of emission had a recession velocity equal to the velocity of light; they were at the edge of the observer's Hubble sphere at the time of emission. Notice in Figure 21.10 that when Y emitted light toward O, the backward lightcone was diverging away from O's world line. The light rays leaving Y and moving toward O were at first dragged away from O by the expansion of space, and then, on reaching the maximum distance, began to approach O.

THE PHOTON HORIZON IN COSMOLOGY

Country of the Red Queen

A galaxy outside the Hubble sphere emits a ray of light in our direction, as shown in Figure 21.7. Although the ray hurries toward us, it actually recedes; it travels through space at the speed of light, but the space through which it travels recedes from us faster than light. As Eddington in 1933 wrote: "Light is like a runner on an expanding track with the winning-post receding faster than he can run." All light rays emitted in our direction within the Hubble sphere approach us, whereas all light rays emitted outside the Hubble sphere recede from us. At the edge of the Hubble sphere, the light rays traveling toward us stand still; they hurry toward us at the same velocity that expanding space carries them away. "Now, here, you see, it takes all the running you can do, to keep in the same place," said the Red Queen to Alice.

All galaxies inside the Hubble sphere recede subluminally (slower than light) and all galaxies outside recede superluminally (faster than light). The edge of the Hubble sphere is what might be called the photon horizon, a curious sort of horizon, not of particles but of photons. Light rays moving toward us inside the photon horizon approach us, all light rays outside must recede. The photon horizon, where recession is transluminal, is the country of the Red Queen.

Galaxies outside the Hubble sphere recede superluminally, and their light rays recede, but this does not mean that galaxies and their events outside the photon horizon are permanently hidden from the observer's view. If that were so, the photon horizon would also be an event horizon. In most universes the Hubble term H is not constant. In a decelerating universe, in which the Hubble term decreases with time, the Hubble distance $L_H = c/H$ increases, and the Hubble sphere expands in the comoving frame. Hence, it expands faster than the universe and the edge of the Hubble sphere – the photon horizon – overtakes the receding galaxies. Light rays outside the Hubble sphere moving toward us may therefore eventually be overtaken by the photon horizon; they will then be inside the Hubble sphere and will at last start approaching us. Eddington's runner sees the winning-post receding, but he must keep running and not give up; the expanding track is slowing down and eventually the winning-post will be reached.

How fast does the Hubble sphere expand? Its radius is $L_H = c/H$, and it expands at velocity $U_H = dL_H/dt$, and this can be shown to be (Equation 14.29)

$$U_H = c(1 + q),\qquad\qquad [21.6]$$

where q is the deceleration term. This is the recession velocity of the photon horizon. Galaxies at the photon horizon recede at the velocity of light c, whereas the horizon itself recedes at the velocity $c(1 + q)$. In a decelerating universe, such as a Friedmann model, the deceleration term q is positive and the Hubble sphere expands faster than the universe. Thus the photon horizon overtakes the galaxies and sweeps past them at relative velocity cq. The deceleration term in the Einstein–de Sitter universe, for example, has a value 0.5, and hence $U_H = 1.5c$, and the photon horizon sweeps past the galaxies at $0.5c$.

THE PARTICLE HORIZON

The receding particle horizon

We recall that beyond the particle horizon are galaxies that at the present time cannot be observed at any stage in their evolution. Their world lines do not intersect the observer's backward lightcone. Inside the particle horizon are the galaxies whose world lines do intersect the backward light-cone, and they comprise the observable universe.

The spacetime diagrams shown in Figures 21.11 and 21.12 illustrate the nature of the particle horizon in an expanding universe. In Figure 21.11 we see how space at the moment "now" divides into two regions: the nearer contains all world lines intersecting the observer's backward lightcone; the farther contains all world lines not intersecting the backward lightcone. The first region is the observable universe, the second region is the unobservable universe, and the particle horizon separates the two. The distance of the particle horizon is measured in the observer's world map (the space at

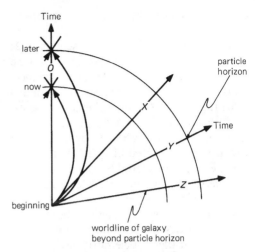

Figure 21.12. At the instant labeled "later" the particle horizon has receded to world line Y. Notice the distance of the particle horizon is always a reception distance, and the particle horizon always overtakes the galaxies and always the fraction of the universe observed increases.

time "now") and is the reception distance of galaxies of infinite redshift. The redshift is infinite because L_{emit} in Equation [21.5] is zero.

We consider a later instant, labeled "later" in Figure 21.12, when the universe is more expanded. Clearly, the observer's lightcone intersects more world lines and the particle horizon is at a greater distance. Thus the particle horizon recedes and the observable part of the universe expands faster than the universe itself. We have seen that in a static universe the particle horizon sweeps out past the galaxies at the speed of light, and it can be shown quite generally in all universes, static and non-static, that the particle horizon sweeps past the galaxies at the speed of light. The observed fraction of the universe always increases. The particle horizon at distance L_P recedes at velocity $U_P = dL_P/dt$, and it can be shown

$$U_P = c + H_0 L_P.\qquad\qquad [21.7]$$

At the particle horizon the galaxies recede at velocity $V_P = H_0 L_P$, and the particle horizon itself recedes at $U_P = c + V_P$; hence the horizon overtakes the galaxies at the speed of light c.

The Einstein–de Sitter universe illustrates what happens. We find that the particle horizon is at twice the Hubble distance. Thus the observable universe has a radius twice that of the Hubble sphere. The recession velocity of the galaxies at the particle horizon, according to the velocity–distance law, is twice the speed of light. Because the particle horizon overtakes the galaxies at the speed of light, the particle horizon in the Einstein–de Sitter model recedes at three times the speed of light. The redshift of the galaxies at the photon horizon is $z = 3$, and at the particle horizon is infinite. Notice that galaxies at the photon horizon recede at velocity c and yet have finite cosmological redshift. (The redshift for the special relativity Doppler effect would be infinite, demonstrating once again that cosmological redshifts are not a Doppler effect.) In universes of constant positive deceleration q, the distances of the particle and photon horizons have the ratio $L_P/L_H = 1/q$. In the radiation-dominated early universe, $q = 1$, and the Hubble sphere and observable universe have the same size; the photon and the particle horizons are coincident and both are often referred to as the "horizon," although they have distinctly differently properties. Generally, when q is not constant, comoving bodies can be inside and outside the Hubble sphere at different times. But not so for the observable universe; once inside, always inside. Horizons are like membranes; the photon horizon acts as a two-way membrane (comoving bodies can cross in both directions depending on the value of q), and the particle horizon acts like a one-way membrane (comoving bodies always move in and never out).

Universes without particle horizons

Some universes, such as the Milne, de Sitter, and steady-state universes, lack particle horizons. In these universes, all world lines intersect an observer's backward lightcone and all galaxies in the universe are visible at some stage in their evolution. To show

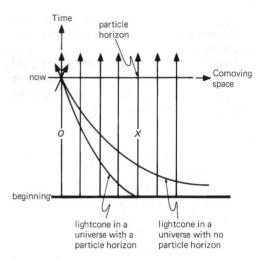

Figure 21.13. A spacetime diagram of comoving space (in which all world lines are parallel) and cosmic time. Some universes have particle horizons and in their case the lightcone stretches out and back to the beginning at a finite comoving distance indicated by the world line X. In universes without particle horizons, the lightcone stretches out to an unlimited distance and intersects all world lines.

why such universes are possible we use a different type of spacetime diagram. This diagram, shown in Figure 21.13, depicts comoving rather than ordinary space coordinates. All comoving bodies are separated from one another by constant comoving distances, and in this new diagram all world lines are parallel. But light rays are not straight lines. The backward lightcone does not diverge as in the static universe but flares out. In universes with particle horizons, such as the Friedmann versions, the lightcone extends back to the beginning of the universe at finite comoving distance and a particle horizon exists at world line X. In other universes, however, such as the Milne model (in which the scale factor $R = t$ and $H = 1/t$, $q = 0$), the lightcone reaches the beginning of time $t = 0$ at an infinite comoving distance and there is no particle horizon. The observable universe fills the entire actual universe and all galaxies are in principle visible. The de Sitter and steady-state universes are of this kind, but are more complicated and will be considered when we discuss event horizons.

CONFORMAL DIAGRAMS

The time has come to introduce the reader to a powerful tool. Quite simply, we transform the units of space and time so that the spacetime diagram looks the same as for a static and flat universe. Everything that we have learned about horizons in the simple static Euclidean universe then applies to all universes, static and nonstatic, flat and curved. This mathematical tool is known as a conformal transformation. In our case it is conformal because the spacetime diagram resembles (conforms to) the Minkowski diagram of special relativity and leaves unchanged spatial angles.

In Figure 21.13, comoving space takes the place of ordinary space. This kind of diagram has the advantage that all world lines of comoving bodies are parallel to one another. They look like the world lines of stationary bodies in a static universe. But light rays are not straight and the lightcone spreads out awkwardly, as in Figure 21.13, and the horizons are not obvious. We have already altered the intervals of space from ordinary to comoving space; we now in addition alter the intervals of time so that light rays are straight and the lightcone is conical, as in the static universe. We then have a spacetime diagram of conformal space and conformal time (Figure 21.14) that looks like the ordinary spacetime diagram of a static universe of Euclidean geometry. Because we know how to handle horizons in a static universe, we now know how to handle horizons in nonstatic universes.

The spacetime interval between two events close together is given by the line element

(spacetime interval)^2

$$= \text{(time interval)}^2 - \text{(space interval)}^2,$$

$$[21.8]$$

in which space intervals are measured in light-travel time. We have already changed space intervals into $R \times$ intervals of comoving (conformal) distance, and this suggests

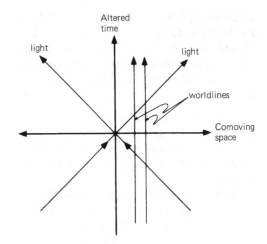

Figure 21.14. A spacetime diagram of conformal coordinates consisting of comoving space and altered time. When we straighten out the lightcone by altering the intervals of time, the spacetime diagram of a nonstatic universe looks like that of a static universe. This coordinate transformation allows us to study horizons in nonstatic universes just as easily as in static universes.

we change time intervals to $R \times$ intervals of conformal time. In the new conformal line element the spacetime interval becomes

(spacetime int.)^2

$$= R^2[\text{(conf. time int.)}^2$$
$$- \text{(conf. space int.)}^2], \qquad [21.9]$$

with obvious abbreviations. All light rays follow paths – technically called null geodesics – on which spacetime intervals are zero. Thus if a light ray travels 1 light second in 1 second, the spacetime interval is zero, and from Equation [21.9] the equation for the lightcone is

(conf. space int.)^2 = (conf. time int.)^2,

and hence

conf. space int. = ±conf. time int., [21.10]

where the plus sign is for the forward lightcone into the future and the minus sign is for the backward lightcone into the past. This last relation (Equation 21.10) between intervals of conformal space and time is for a spacetime, shown in Figure 21.14, in which world lines are parallel and light rays are straight as in a static universe. The advantage of this kind of diagram is that it allows us to treat horizons in the same way as for a static universe. Four possibilities must be considered:

(a) The universe has a beginning and an ending in conformal time, as in Figure 21.15. The closed Friedmann universe that begins and ends with big bangs belongs to this class.

(b) The universe has a beginning but no ending in conformal time, as in Figure 21.16. The Einstein–de Sitter universe and the Friedmann universe of negative curvature, which begin with a big bang and expand forever, belong to this class.

(c) The universe has an ending but no beginning in conformal time, as in Figure 21.17. The de Sitter and steady-state universes belong to this class.

(d) The universe has no beginning and no ending in conformal time, as in Figure 21.18. The Einstein static, the Eddington–Lemaître, and the Milne universes are members of this class.

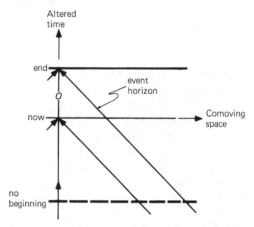

Figure 21.17. A conformal spacetime diagram in which altered time has an ending but no beginning. Only an event horizon exists.

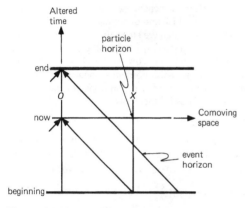

Figure 21.15. A conformal spacetime diagram in which altered time has a beginning and an ending. The world line X is at the particle horizon. Notice the existence of an event horizon.

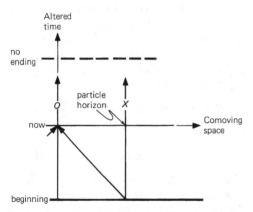

Figure 21.16. A conformal spacetime diagram in which altered time has a beginning but no ending. The particle horizon is at world line X, and no event horizon exists.

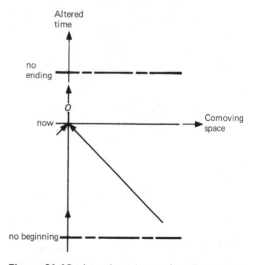

Figure 21.18. A conformal spacetime diagram in which altered time has no beginning and no ending. There are no particle and event horizons.

We can now see which universes have particle horizons. The necessary condition for a particle horizon is that conformal time has a beginning, as in class (a) shown in Figure 21.15, and in class (b) shown in Figure 21.16. The observer's lightcone stretches back and terminates at the lower boundary where the universe begins. When conformal time has no beginning, hence no lower boundary, as in class (c) shown in Figure 21.17, and in class (d) shown in Figure 21.18, the lightcone stretches back without limit and intersects all world lines in the universe. In these universes there are no particle horizons. Note that a beginning in conformal time does not necessarily mean a beginning in ordinary time.

By constructing spacetime diagrams with coordinates that are conformal with those of a static universe, we find that particle horizons exist when conformal time has a beginning. As in ordinary time in the static universe, the observer's lightcone advances in conformal time into the future and the particle horizon recedes. Once a galaxy is inside the particle horizon, and part of the observable universe, it remains always

inside the particle horizon, as seen in Figure 21.19.

EVENT HORIZONS
The ultimate lightcone
Inside the observer's event horizon exist events that can be observed at some time or other, and outside exist events that can never be observed. With our new spacetime diagrams of conformal time and conformal (or comoving) space, event horizons are easy to understand. Let us consider an event located somewhere in spacetime (Figure 21.20). The observer's lightcone advances up O's world line, sweeping through spacetime, and the event eventually lies on the lightcone and the observer sees it. In this way, in the course of time, all events are disclosed to the observer.

But this is not so when conformal time has an ending, as in class (a) shown in Figure 21.15, and in class (c) shown in Figure 21.17. In these classes there exist events that can never be seen, as in Figure 21.20: O's lightcone cannot advance beyond the upper limit of conformal time, and the events that never lie on O's lightcone will never be observed.

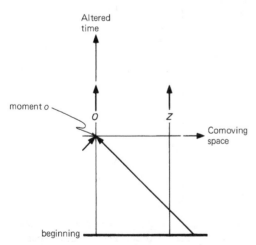

Figure 21.19. As the moment of observation *o* advances up the observer's world line O, the particle horizon recedes. Once a world line, such as Z, lies within the particle horizon, it remains inside forever. This means that a galaxy inside the observable universe will remain inside and always observable as long as it exists.

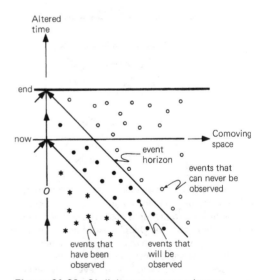

Figure 21.20. O's lightcone cannot advance beyond the end of altered time. Hence there are events that can never be observed by O, and the ultimate lightcone is the event horizon.

The necessary condition for an event horizon is that conformal time has an ending. The event horizon in cosmology is thus nothing more than the observer's ultimate lightcone at the end of conformal time. All the events inside the event horizon (the ultimate lightcone) are at some time observed, and all events outside are never observed. Note that an end in conformal time does not necessarily mean an end in ordinary time.

Blueshifts and redshifts at event horizons
Universes that end in big bangs, such as the closed Friedmann universe, have event horizons. An observer (world line O) in a collapsing universe receives signals from event a (Figure 21.21) at redshift

$$z + 1 = R_0/R, \qquad [21.11]$$

where R_0 is the value of the scaling factor at the time of reception and R the value at the time of emission. Because the universe is collapsing, R_0 is less than R, and the redshift is negative. A negative redshift means that light is blueshifted toward the blue end of the visible spectrum. At the last moment of observation R_0 is zero. Hence the redshift is -1 and the blueshift is maximum. Everything seen close to the event horizon happens rapidly, and at the last possible moment, at the event horizon, happens infinitely rapidly.

There are universes that expand forever and yet have endings in conformal time. The de Sitter and steady-state universes are of this kind and therefore have event horizons. But R_0 is now not zero but infinity. At the event horizon all events seen have infinite redshift and happen infinitely slowly.

An alternative way of looking at the de Sitter and steady-state universes considers a spacetime diagram of conformal (comoving) space and ordinary cosmic time, as in Figure 21.22. The observer O at moment o sees event a. As the moment of observation advances into the unlimited future, the lightcone moves upward and approaches more and more slowly but

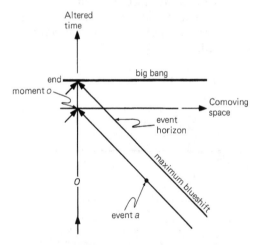

Figure 21.21. Collapsing universes that terminate in big bangs have an end in altered time (and in cosmic time), and therefore possess event horizons. At moment o, observer O looks back into the past and sees all events, such as a on A's world line, blueshifted. As the moment o of observation advances and approaches the end, events are seen with increasing blueshift, and at the last possible moment, all events on the observer's backward lightcone are seen to happen infinitely rapidly. In this case, the event horizon has maximum blueshift.

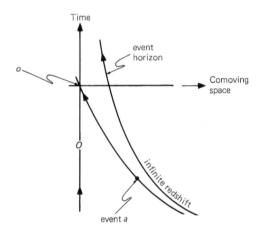

Figure 21.22. This shows the de Sitter universe in a spacetime diagram of comoving space and cosmic time. The event horizon is as shown. As the moment o now advances into the infinite future, the observer's backward lightcone approaches the event horizon more and more closely. Events close to this horizon have large redshifts because of expansion, and at the horizon all events have infinite redshift.

never reaches the event horizon. The event horizon is the observer's lightcone in the infinite future. Events outside this horizon can never be observed.

Exponentially expanding universes

Admittedly, our comments so far have not greatly clarified what happens in the de Sitter, steady-state, and other exponentially expanding universes, such as the inflationary universe. We first note that the Hubble term is fixed and the Hubble sphere has constant radius. At the edge of the Hubble sphere – the photon horizon – the recession velocity equals the velocity of light, and calculation shows that in these universes the redshift at the photon horizon is infinite. The Hubble sphere is the observable universe. Why is the boundary of the Hubble sphere an event horizon and not a particle horizon? The observed galaxies are carried farther and farther away from the observer and become progressively more redshifted. What happens to these galaxies – do they eventually cross the edge of the Hubble sphere and disappear from view? These questions, which perplexed many cosmologists in the past, can be answered with the help of Figure 21.23.

The boundary of the Hubble sphere has become a true horizon. It is not a particle horizon because at any instant all galaxies in the universe are visible to an observer. Any galaxy now beyond the Hubble sphere, no matter how far away, had in the past a part of its world line inside the Hubble sphere and is therefore observable at some stage in its history. All galaxies recede and move out of the Hubble sphere, yet the observer never sees them crossing the Hubble boundary.

Exponentially expanding universes have spacetime diagrams in comoving space and conformal time of the kind shown in Figure 21.17. We see that these universes have event horizons but no particle horizons. In the case of these universes, however, it is more convenient to use a diagram of ordinary space and ordinary time, as in Figure 21.23. The Hubble sphere is of constant

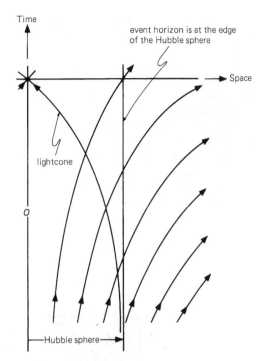

Figure 21.23. The de Sitter universe displayed in ordinary space and ordinary time. The Hubble sphere has constant radius about the observer's world line O. All world lines diverge away from O, and every world line intersects the edge of the Hubble sphere and at some time is inside the Hubble sphere. The observer's lightcone curves back, as shown, and approaches but never crosses the edge of the Hubble sphere. The observer sees all world lines and there is no particle horizon. Events outside the Hubble sphere are never observed and the edge of the Hubble sphere is the event horizon, and is the observer's ultimate backward lightcone.

radius, as shown, and all world lines diverge away from the observer's world line O. The lightcone approaches asymptotically the edge of the Hubble sphere, intersecting all world lines in the universe, and never extends beyond the Hubble sphere. No particle horizon exists because all world lines intersect the observer's lightcone. The edge of the Hubble sphere is an event horizon because it is the observer's ultimate lightcone, and all events outside the Hubble sphere can never be observed.

Figure 21.23 makes clear that the farther the observer looks out in space and back in time the closer the galaxies approach the edge of the Hubble sphere and the greater becomes their redshift. But the observer never sees the galaxies disappearing across the edge. At the edge of the Hubble sphere are crowded an infinite number of infinitely redshifted galaxies.

Galaxies, of course, do not shine forever, and as luminous sources their world lines are therefore of finite length. This is something for the reader to think about and make suitable amendments where necessary in what has previously been said. World lines of finite length, as in Figure 21.24,

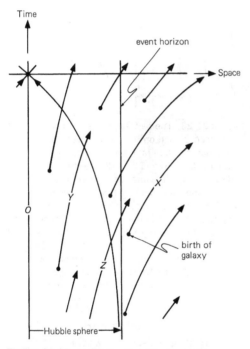

Figure 21.24. In the steady-state expanding universe, galaxies are continually created so as to maintain a constant average density. Galaxies are born, and in this diagram die when they cease to be luminous. Most galaxies, such as X, are born outside the Hubble sphere and are never seen by O; some galaxies, such as Y, are born inside the Hubble sphere and may die before reaching the Hubble edge; and other galaxies, such as Z, cross the Hubble edge while still luminous. As in the de Sitter universe, the number of galaxies of infinite redshift at the event horizon is infinitely great. Thus there are an infinite number of galaxies crowded at the edge of the Hubble sphere.

were of particular interest in the case of the steady-state universe. This universe expands in the same way as the de Sitter universe and has an event horizon at the edge of the Hubble sphere. New galaxies continually form everywhere, maintaining a constant average density of matter, and it is therefore not true to say that all galaxies are observable in a continuous creation steady-state universe. Galaxies do not originate in the infinite past inside the observer's Hubble sphere, but in the finite past and mostly outside the Hubble sphere. World line X in Figure 21.24 is an example of a galaxy never seen by the observer. A galaxy formed inside the Hubble sphere may die and become nonluminous before it reaches the Hubble boundary, as indicated by world line Y. The number of luminous galaxies crossing the Hubble edge, however, having world lines such as Z, is still infinite. At the event horizon, where the redshift is infinite and light rays emitted in our direction stand still, there exists an infinite number of galaxies.

REFLECTIONS

1 *Cosmological horizons were investigated in 1956 by Wolfgang Rindler in a classic paper ("Visual horizons in world-models") in which he wrote, "A horizon is here defined as a frontier between things observable and things unobservable," and he distinguished between two kinds of "things," events and world lines (particles), thus leading to two kinds of horizon: the event horizon and the particle horizon.*

2 *The rational method explains present conditions as the result of past conditions: "things are as they are because things were as they were." This method becomes embarrassing when initial conditions must be arranged in special and even improbable ways to explain present conditions. We are left wondering what explains the special initial conditions. Perhaps this is true of all rational universes?*

3 *"We are unable to obtain a model of the universe without some specifically cosmological assumptions that are completely unverifiable" (George Ellis, "Cosmology and verifiability," 1975). The problem is*

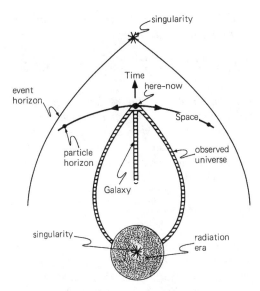

Figure 21.25. The observed universe consists of only those events that lie on the observer's backward lightcone. A small region about the observer's world line contains events not on the lightcone whose existence can be inferred from the immediate environment. The history of the Galaxy, the Solar System, the Earth, and the human race is confined to this region. All the rest of spacetime contains events that at present are unobserved. If there is an event horizon, then beyond this horizon lie all the events that can never be observed.

that we observe isotropy, which we cannot explain, and we assume homogeneity, which we cannot verify. We observe only those events that lie on our backward lightcone, as in Figure 21.25, and the rest of spacetime – except for a small region about our world line – is unobserved. All our knowledge of the universe is limited to a small region surrounding the Earth's world line and around the backward lightcone. The theory of inflation (Chapter 22) changes this picture, but so far inflation is itself an unverified theory.

4 *How can the properties of the universe be explained by causes in the past when interactions over large distances could not exist? The horizon problem became more urgent with the discovery of the smoothness and isotropy of the cosmic background radiation that decoupled at the end of the early universe (Chapter 20) when the age of the universe*

was a few hundred thousand years. The difference in the radiation from opposite sides of the sky is less than 1 part in 100 000. Yet the emitting regions at the time of decoupling were far apart and outside each other's particle horizons. If these emitting regions do not "know" that each other exists, how can they be in identical states? The beauty of inflation is that it solves this problem.

5 *Consider two visible bodies at equal distances in opposite directions from us, as shown by world lines A and B in Figure 21.8. We see these bodies, but can they see each other? Let T be the time (in units of conformal time) that light takes to travel to us from A and B. The time that light takes to travel from A to B, or from B to A, is obviously 2T. Hence when the universe is older than 3T, we not only see A and B, but they also see each other at the time they emit the light we now see. If the universe is younger than 3T, and older than T, we see A and B, but they cannot yet see each other. There is thus a maximum distance beyond which the observed bodies A and B do not know that each other exists. By examining Figure 21.8, we see that this maximum distance is one-third the distance to the particle horizon. The answer to our question is that bodies equidistant in opposite directions, farther away from us than one-third the distance to the particle horizon, cannot see each other. In a matter-dominated Einstein–de Sitter universe, this distance is $\frac{1}{3}L_P = \frac{2}{3}L_H$ at redshift $z = 1.25$.*

This highlights the apparently insoluble problem of understanding why the universe is homogeneous (all places are alike). Regions visible to us in opposite directions at large redshifts have not had time to influence each other and are unaware of each other's existence. Yet they exist in identical states. How can things be exactly similar when they lie outside one another's horizons? This is the horizon problem.

6 *Let $dr = dL/R$ be an interval of comoving distance and $d\tau = dt/R$ an interval of conformal time, as in Chapter 14. Thus $dL = R\,dr$ is an interval of proper (or tape-measure) distance and $dt = R\,d\tau$ is an interval of cosmic* time. Equation [21.10] states $dr = -d\tau$ on the backward lightcone, and on integrating we find

$$r = \tau_0 - \tau, \qquad [21.12]$$

where

$$\tau = \int_0^t \frac{dt}{R}, \qquad [21.13]$$

is the conformal time of an event on O's backward lightcone measured from the beginning to time t, and

$$\tau_0 = \int_0^{t_0} \frac{dt}{R}, \qquad [21.14]$$

is the conformal time at the moment of observation, also measured from the beginning of t. Alternatively,

$$\tau_0 - \tau = \frac{1}{R_0} \int_0^z \frac{dz}{H}, \qquad [21.15]$$

in terms of redshift (Equations 14.36 and 19.21), where H is a function of z.

The particle horizon corresponds to $\tau = 0$, and hence $r_P = \tau_0$, and the proper distance to the particle horizon is $R_0\tau_0$, or

$$L_P = R_0\tau_0, \qquad [21.16]$$

and τ_0 is found from Equation [21.14], or

$$\tau_0 = \frac{1}{R_0} \int_0^\infty \frac{dz}{H}. \qquad [21.17]$$

7 Assume that the scale factor R varies as t^n, where t is the age of the universe and n a constant number. In these power-law big bang universes: $H = n/t$, hence n is a positive number in an expanding universe; $q = (1-n)/n$, hence n is less than 1 in a decelerating universe. The Hubble distance, where galaxies recede at the velocity of light c, is

$$L_H = ct_0/n = ct_0(1+q). \qquad [21.18]$$

The Hubble sphere itself expands at velocity $U_H = dL_H/dt$, or

$$U_H = c/n = c(1+q), \qquad [21.19]$$

and the edge of the Hubble sphere, or photon horizon, overtakes and sweeps past the galaxies when n is less than unity. The particle horizon is at the distance

$$L_P = \frac{nL_H}{1-n} = \frac{L_H}{q} \qquad [21.20]$$

and this distance – the radius of the observable universe – is greater than, equal to, or less than the Hubble distance when n is greater than, equal to, or less than 0.5. The observable universe expands at velocity $U_P = dL_P/dt$, or

$$U_P = c + V_P = \frac{c}{1-n} = c\left(1+\frac{1}{q}\right), \qquad [21.21]$$

and equals the recession velocity of the galaxies $V_P = HL_P$ at the particle horizon plus the velocity of light c; the particle horizon always overtakes the galaxies at the velocity of light. In the matter-dominated Einstein–de Sitter universe of $n = 2/3$, we have $L_H = 1.5ct_0$, $U_H = 1.5c$, $L_P = 3ct_0$, and $U_P = 3c$; and in the radiation-dominated version of this universe of $n = 0.5$, we have $L_H = 2ct_0$, $L_P = 2ct_0$, and $U_P = 2c$. When n is equal to or greater than unity, there is no particle horizon, and the observer sees all luminous objects in the universe. Thus Milne's universe of $n = 1$ lacks particle and event horizons, and he regarded the absence of horizons in his universe as a distinct advantage.

The maximum distance of the lightcone from an observer's world line is found to be

$$L_{max} = n^{1/(1-n)}L_H, \qquad [21.22]$$

and the redshift of sources at this maximum distance is

$$z \text{ (at } L_{max}) = n^{-n/(1-n)} - 1, \qquad [21.23]$$

and this gives $L_{max} = 8L_H/27$ and $z = 1.25$ for $n = 2/3$, and $L_{max} = L_H/4$ and $z = 1$ for $n = 0.5$. The recession velocity of sources at maximum emission distance, at the time of emission, is always equal to the velocity of light; these sources at maximum distance are therefore at the edge of the observer's Hubble sphere at the time they emitted the light that is now seen.

PROJECTS

1 If light traveled with infinite speed, the world map and the world picture would be identical. What would happen to the horizons?

2 Find when the observable universe and the Hubble sphere are the same in size. Discuss the behavior of the Hubble sphere and the observable universe in the Dirac universe of $n = 1/3$.

3 As time passes, the observable universe contains more and more galaxies; is this true also in a collapsing universe?

4 Discuss the maximum proper (tape-measure) distance of an observed world line in an expanding big bang universe. How is it possible at the time of emission that a source of redshift $z = 2$ is nearer than a source of $z = 1$?

5 Can you think of cosmic horizons that might exist because of the observer's forward lightcone? Such horizons determine the observer's ability to influence future events and particles elsewhere in the universe.

6 Most physical scientists like to use formulas rather than words. As distinct from most humanists they also like to use lots of diagrams. Is there a reason for this fondness of diagrams?

SOURCES

Centrella, J. C. "Visual horizons in cosmological models: A study." Senior honors thesis, Department of Physics and Astronomy, University of Massachusetts, 1975. Much of this chapter is based on Joan Centrella's honors thesis.

Eddington, A. S. *The Expanding Universe.* Cambridge University Press, Cambridge, 1933.

Ellis, G. F. R. "Cosmology and verifiability." *Quarterly Journal of the Royal Astronomical Society* 16, 245 (1975).

Ellis, G. F. R. and Williams, R. F. *Flat and Curved Space-Times.* Clarendon Press, Oxford, 1988.

Guth, A. H. and Steinhardt, P. J. "The inflationary Universe." *Scientific American* (May 1984).

Harrison, E. R. "Hubble spheres and particle horizons." *Astrophysical Journal* 383, 60 (1991).

Hawking, S. W. and Ellis, G. F. R. *The Large Scale Structure of Space-Time.* Cambridge University Press, Cambridge, 1973.

Macallum, M. A. H. In *The Origin and Evolution of the Galaxies*, page 9. Edited by B. J. T. Jones and J. E. Jones. Reidel, Dordrecht, 1983.

North, J. D. *The Measure of the Universe.* Clarendon Press, Oxford, 1965.

Penrose, R. In *Relativity Groups and Topology*, page 565. Edited by C. M. de Wit and J. A. Wheeler. Benjamin, New York, 1964.

Rindler, W. "Visual horizons in world-models." *Monthly Notices of the Royal Astronomical Society* 116, 662 (1956).

22 INFLATION

Alice laughed. "There's no use trying," she said: "one can't believe impossible things."

"I dare say you haven't had much practice," said the Queen. "When I was your age, I always did it for half-an-hour a day. Why, sometimes I've believed as many as six impossible things before breakfast."

Lewis Carroll (Charles Dodgson, 1832–1898), Alice in Wonderland

PERFECT SYMMETRY

According to current thinking the evolution of the early universe consists of a series of transitions to states of progressively lower symmetry, starting at an initial state of utmost symmetry. Perhaps, in the beginning, the harmonious unity of all forces falls apart into two derivative forces: the gravitational force and the hyperweak force. At first, both are of equal strength. The strange, grand unified era of the extreme early universe has begun and the universe consists of what might aptly be called "elem" (after ylem, introduced by George Gamow). Little or no distinction exists between matter and antimatter, or between quarks and leptons.

We use again the rough rule-of-thumb relations of the standard model:

$$\rho t^2 = 10^6, \qquad [22.1]$$

$$tT^2 = 10^{20}, \qquad [22.2]$$

where ρ represents the density of the universe in grams per cubic centimeter, t the age of the universe in seconds, and T the temperature in kelvin. As we saw in Chapter 20, at age 1 second, the density is 10^6 grams per cubic centimeter and the temperature is 10^{10} kelvin.

At the Planck epoch, 10^{-44} seconds, the cosmic density is of order 10^{94} grams per cubic centimeter and the temperature is 10^{32} kelvin, equivalent to a particle energy of 10^{19} GeV, where 1 GeV is 1 billion electron volts, and 1 electron volt equals the thermal energy of a particle at 10 000 kelvin. Possibly, at the Planck epoch, spacetime consists of a dense foam of real (not virtual) quantum fluctuations on length scales of 10^{-33} centimeters and time scales of 10^{-43} seconds. (A 10^{19} GeV quantum of energy, equivalent to the Planck mass of 10^{-5} grams, is sufficient energy to light a 100 watt lamp for almost a year.)

According to current ideas, grand unification in the extreme early universe lasts for a very short time. Because of expansion, the temperature falls from the Planck value and at time 10^{-36} seconds reaches a critical value 10^{28} kelvin, equivalent to 10^{15} GeV. The stage is set for the second act of symmetry-breaking: the breakup of grand unification. The hyperweak force falls apart into two new and different forces: the electroweak and strong forces of the quark–lepton era. This important phase transition initiates inflation and ushers in the distinction between matter and antimatter, and between quarks and leptons.

THE MONOPOLE PROBLEM

James Clerk Maxwell (1831–1879), who unified electricity and magnetism, was guided by Michael Faraday's ideas and the results of his experiments with electric and magnetic fields. Maxwell's electromagnetic equations have since served as a model for all field equations.

The electromagnetic equations reflect a strange asymmetry in nature: we find free electric charges but not free magnetic poles. The motions of electric charges, such as negative electrons, constitute the electric currents that generate magnetic fields. But magnetic poles exist only in pairs, forming the north and south poles of dipole magnets, and never in a free state as magnetic particles. If magnetic particles, called monopoles, actually existed, their independent motions would constitute magnetic currents that generate electric fields. In this way, perfect symmetry would exist between electricity and magnetism. North (positive) and south (negative) monopoles would attract each other in the same way as positive and negative electric charges. Nothing known in nature forbids the existence of monopoles, and many experimenters have searched for them, but so far none has been found.

According to grand unified theories, monopoles are created at extremely high energy (10^{15} GeV or even higher) and have a mass at least 10^{-9} gram. They exist abundantly in the extreme early universe. Their north and south poles are the particle and antiparticle forms of monopoles, and they annihilate each other just like all other particle–antiparticle pairs. Otherwise they are stable and do not decay into particles of lesser mass. Because of their small cross-section (their size is 10^{-29} centimeters) and the rapidity of expansion of the universe, annihilation fails to deplete significantly the monopole population. Monopoles should still exist and be as abundant as the photons of the cosmic background radiation.

Grand unified theory restores the lost symmetry of electromagnetism in the extreme early universe but makes more urgent the question of why monopoles are absent from the world around us.

DISCOVERY OF INFLATION

Grand unification greatly changes the landscape of the early universe. Sydney Coleman of Harvard University proposed the idea of a phase transition from a state dominated by the hyperweak force to a state of lower energy consisting of quarks and leptons dominated by the strong force and electroweak force. This phase transition is much like the transition from water to ice as temperature drops. Normally, the water–ice phase transition occurs at the freezing point of zero celsius, or 273 kelvin. Generally, when temperature drops, undisturbed pure water supercools to a temperature lower than freezing point before transforming into ice. Similarly, in the early universe, as the temperature drops, the transition to quarks and leptons fails to occur at the instant the temperature reaches the critical value 10^{28} kelvin. Instead of transforming abruptly into a quark–lepton mix, the elem supercools and becomes what Coleman called the "false vacuum." The false vacuum is the lowest possible energy state available to the hyperweak force. When the transition finally and spontaneously occurs, the false vacuum releases its immense latent energy, restoring the temperature back almost to the grand unified value and creating quarks, leptons, and gluons.

Alan Guth in 1980, while at Stanford University, investigated the monopole problem and studied also the properties of the false vacuum. He realized that the false vacuum existed in an extraordinary state of negative pressure that caused the expansion of the universe to accelerate. Others had already noticed much the same effect. Guth was the first to realize that accelerated expansion (which he called inflation) solved not only the monopole problem but also the flatness and horizon problems. Any universe having a period of accelerated expansion in its infancy is now known as an inflationary universe.

COSMIC TENSION

Negative pressure is just another name for tension. A stretched piece of elastic in a state of tension serves as an analogy of what happens in the inflation era of the extreme early universe. The act of stretching the piece of elastic performs work, and energy in the form of heat is generated. The greater the tension, the greater the energy generated.

Let an endless length of elastic represent a one-dimensional universe. As this universe in tension expands, its density (mass per unit length) decreases. But the energy generated contributes mass (equal to energy divided by c^2), and density does not decrease quite as much as one might suppose. The greatest possible tension exists when the energy generated by expansion maintains a constant mass density.

Much the same happens in an expanding three-dimensional universe in tension: the energy generated contributes to the existing mass and the density does not decrease quite as fast as one might suppose. The greatest possible tension exists when the energy generated by expansion maintains a constant density. This is the remarkable thing about inflation; it occurs at maximum (or almost maximum) tension, and the energy created maintains (or almost maintains) a constant density.

The total energy density equals the mass density ρ multiplied by the square of the speed of light c, or ρc^2, and maximum tension equals the total energy density. But tension is negative pressure, hence

$$P = -\rho c^2, \qquad [22.3]$$

where P denotes pressure. This is the simplest version of the equation of state of the false vacuum during inflation. The universe expands and both density and pressure remain unchanged. The equation that relates energy and pressure is the first law of thermodynamics, and in its constant entropy form,

$$\frac{\mathrm{d}E}{\mathrm{d}t} + P\frac{\mathrm{d}V}{\mathrm{d}t} = 0, \qquad [22.4]$$

where $E = \rho c^2 V$ is the total energy in a comoving volume V. On inserting the relation $P = -\rho c^2$ into this equation, we find that the density remains constant independent of V.

INFLATION
How it works
Figure 22.1 shows how the extreme early universe expands as in the Lemaître universe. A pre-inflation period extends from the Planck epoch at 10^{-44} seconds (at

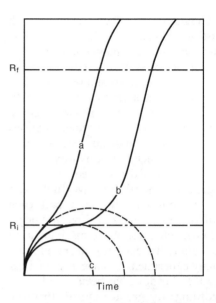

Time

Figure 22.1. Curve (a): the universe in the pre-inflation era decelerates, then at age 10^{-36} seconds at R_i it accelerates rapidly in the inflation era, and finally at age roughly 10^{-34} seconds at R_f it again decelerates in the post-inflation era. Curve (b): the universe in the pre-inflation era barely reaches inflation and enters into a hesitation period before inflation. Curve (c): the expanding universe fails to reach inflation and collapses and has a lifetime less than 10^{-36} seconds. In some theories it is assumed that inflation begins close to the Planck density at particle energy 10^{19} GeV; here it is assumed that inflation begins at particle energy 10^{15} GeV.

density 10^{94} grams per cubic centimeter and temperature 10^{32} kelvin) to the grand unified epoch at 10^{-36} seconds (density 10^{78} grams per cubic centimeter and temperature 10^{28} kelvin), and during this period pressure stays positive and expansion decelerates as in a normal Friedmann universe. Delay in the phase transition to quarks and leptons at the grand unified temperature creates the false vacuum and causes supercooling. (Figures 22.2 and 22.3 illustrate transitions to states of lower symmetry and lower energy.) During the transition the universe is thrown into tension and the density stays more or less constant and expansion accelerates exponentially. The accelerated expansion lasts for a brief period from age 10^{-36} to approximately 10^{-34}

(a)

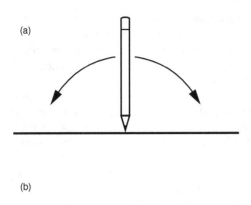

(b)

Figure 22.2. Spontaneous symmetry-breaking occurs because energy in the symmetrical state is higher than in the asymmetrical state. An upright pencil (a) standing on its point is in an example of a symmetrical state. The pencil is unstable and falls over in a random direction, as in (b), and by breaking its state of symmetry it attains a state of lower energy. This illustrates the grand unified phase transition in the original inflation theory.

(a)

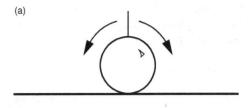

(b)

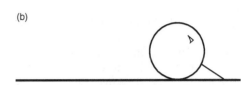

Figure 22.3. A smooth ball on a smooth surface is another example of a state of symmetry. The ball rolls freely to different positions on the surface and its symmetry remains unbroken and in its lowest state of energy. Now push a large pin into the ball with the pin in an upright position, as shown in (a). The ball, as in the case of the upright pencil in Figure 22.2, is in a symmetrical but unstable state. The ball rolls over, as in (b), to a less symmetrical state of lower energy. The symmetry-breaking transition occurs slowly because the ball rolls much slower than the pencil falls, thus illustrating the grand unified phase transition in the revised version of the inflation theory.

seconds, and in this brief period the universe inflates enormously.

We can understand why the expansion accelerates during a period of negative pressure by looking at the cosmological equations

$$3qH^2 = 4\pi G\left(\rho + \frac{3P}{c^2}\right) - \Lambda, \qquad [22.5]$$

$$H^2 = \frac{8\pi G\rho}{3} + \tfrac{1}{3}\Lambda - K, \qquad [22.6]$$

where $H = \dot{R}/R$ is the Hubble term, $q = -\ddot{R}/RH^2$ the deceleration term, R the scale factor, Λ the cosmological constant, $K = k/R^2$ the curvature, and $k = 1, 0,$ or -1 the curvature constant. Both Λ and K are comparatively small in the early universe and can be ignored. The first of the above two equations (Equation 22.5) shows that when the pressure P is negative and less than $-\rho c^2/3$, gravity in effect becomes repulsive, just as if the universal gravitation constant G had changed sign from positive to negative. The maximum negative pressure $P = -\rho c^2$ of Equation [22.3], when inserted in Equation [22.5], yields

$$3qH^2 = -8\pi G\rho. \qquad [22.7]$$

Because H^2 and the density ρ are positive, the deceleration term q is negative, and the expansion of the universe accelerates. The second equation (Equation 22.6) states

$$3H^2 = 8\pi G\rho, \qquad [22.8]$$

and by comparing Equations [22.7] and [22.8], we see that $q = -1$. Hence H is constant, thus making the density ρ constant, and the scale factor R expands exponentially during inflation:

$$R = R_i\, e^{H(t - t_i)}, \qquad [22.9]$$

where R_i is the initial value of the scale factor at the beginning of inflation at time t_i.

The inflation factor

During inflation the Hubble time $1/H$ is constant and roughly equal in value to the cosmic age $t_i = 10^{-36}$ seconds at the beginning of inflation. Let t_f denote the age at the end of inflation. The term

$$\eta = H(t_f - t_i) = \frac{t_f}{t_i} - 1,$$

measures the duration of inflation in Hubble periods, and the initial R_i and the final R_f values of the scale factor are related by

$$\frac{R_f}{R_i} = e^{\eta} = \text{inflation factor.} \qquad [22.10]$$

When $\eta = 10$, the inflation factor equals 2×10^4; when $\eta = 50$, it equals 5×10^{21}; and when $\eta = 100$, it equals 3×10^{43}. (A proton inflated by $\eta = 92$ would fill the present observable universe.) Estimates vary and usually η lies between 50 and 100. Inflation lasts for a period of time ηH^{-1} during which the universe expands enormously. Grand unified theories cannot predict with

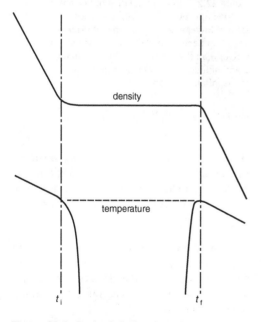

density

temperature

t_i t_f

Figure 22.4. During inflation, from time t_i at approximately 10^{-36} seconds to t_f at 10^{-34} seconds, the density of the universe stays constant at about 10^{78} grams per cubic centimeter, but the temperature drops to a very low value. Initially, the temperature T_i is approximately 10^{28} kelvin, and the release of latent energy at the end of inflation restores the temperature T_f close to the initial value T_i.

confidence the amount of inflation. Conveniently, most conditions at the end of inflation, such as density, reheat temperature, and rate of expansion, are independent of the inflation factor.

During inflation the density stays constant and the temperature plunges, as shown in Figure 22.4. At the end of inflation the latent energy locked in the false vacuum breaks free and reheats the universe to approximately its temperature at the beginning of inflation and populates the universe with a dense sea of quarks, leptons, and gluons in thermal equilibrium. The highly inflated universe now embarks on its second stage of decelerated expansion.

The matter–antimatter difference

In the universe at present the baryon number (number of baryons minus the number of antibaryons) in a comoving volume is strictly conserved. Higher-mass baryons decay into lower-mass baryons, and neutrons decay into protons. But protons are the lowest-mass baryons and cannot decay into anything that conserves baryon number. Hence they are stable. But conservation of baryon number is absent in the preinflation world, a world dominated by the hyperweak force that ignores the difference between matter and antimatter. Probably, in the world of today, the hyperweak force still exists in a vanishing vestigial form, and theory indicates that protons decay extremely slowly into lower-mass particles (e.g. photons, positrons, neutrinos) in a time longer than 10^{32} years.

Quarks and leptons emerge at the end of inflation when the hyperweak force is in rapid decline. In a previous chapter (Chapter 20), we saw that in the early universe matter is favored over antimatter by about 1 part in 10^9. Why this bias toward matter when matter and antimatter are symmetrical? A straight and simple answer is that particles and antiparticles are not exact mirrors of each other in an expanding universe. This slight asymmetry, exploited by a lack of perfect thermodynamic equilibrium (because of expansion), tips the balance very slightly in favor of matter. The small difference in the abundances of matter and antimatter emerges after inflation at the beginning of the quark–lepton era and amounts to 1 part in 10^9. This equals roughly the ratio of the number of baryons and the number of photons in the universe of today.

INFLATION SOLVES THE MONOPOLE PROBLEM

And don't forget inflation also solves the problem of a flat automobile tire.
David Van Blerkom, 1993

The magnetic monopoles discussed above exist in the pre-inflation period of the early universe. If the grand unified phase transition happens abruptly, with no supercooling and no inflation, and quarks and leptons form immediately, the monopoles would nowadays be as abundant as cosmic background photons. The inflationary universe reveals why monopoles are now rare and difficult to find.

Grand unified theories are still in an exploratory state, and guesswork and analogy have so far failed to determine the duration of inflation. We can easily calculate, however, the minimum inflation needed to solve the monopole puzzle.

Monopoles are massive and stable; they annihilate each other, but otherwise do not decay into particles of lesser mass. Instead, inflation sweeps them far apart and makes them a dilute ingredient of the false vacuum. At the end of inflation, when reheating floods the universe with newborn particles, the monopoles are widely separated from one another.

The cores of monopoles have a radius 10^{-29} centimeters (normally they are surrounded by clouds of virtual particles and tend to be larger), and this core size is comparable to the average separation of monopoles at temperature $T_G = 10^{28}$ kelvin, where T_G is the grand unified temperature at which inflation begins. In a noninflationary scenario, the universe expands roughly by a factor $T_G/T_0 = 3 \times 10^{27}$, where $T_0 = 3$ kelvin is the present temperature of the cosmic background radiation. If we multiply the initial separation by this factor, we find that the present monopole separation is a fraction of a millimeter and therefore similar to the separation of photons in the cosmic background radiation. In the absence of inflation, monopoles – contrary to observation – become by far the dominate constituent of the universe because of their huge mass.

Inflation saves the universe from monopole domination. The inflation factor e^η and the subsequent expansion factor T_f/T_0 give a total expansion of $e^\eta T_f/T_0$. The reheat temperature T_f at the end of inflation equals roughly the initial grand unified temperature. Hence the total expansion factor is $e^\eta T_G/T_0$, or e^η times that in a universe without inflation. Calculation shows that when η equals 60, each galaxy contains on the average 1 monopole; and when η equals 67, the entire Hubble sphere contains 1 monopole.

Inflation neatly solves the monopole problem. It solves other problems, as we shall now see.

INFLATION SOLVES THE FLATNESS PROBLEM

The flatness problem addresses a subtle issue discussed in 1979 by Robert Dicke and James Peebles of Princeton University. We have noticed previously that the curvature of space, denoted by K, is negligible in the early universe and have used the Einstein–de Sitter model (in which $K = 0$) as a convenient model. In fact, the rule-of-thumb relations of Equations [22.1] and [22.2] are for the flat Einstein–de Sitter universe. The flatness problem is this: why in the beginning is the universe almost perfectly flat? Why is curvature K so very small? If curvature is not small, and is positive, the universe long ago would have collapsed and ended, or if negative, the universe would now be almost entirely devoid of matter. The anthropic principle provides one answer: stars and living creatures cannot exist if the curvature is not extremely small in the beginning. Here, as in other applications, the anthropic principle does little more than remind us of the poverty of science.

Guth in 1981 gave a scientific answer. He pointed out that the enormous and rapid expansion during inflation literally flattens the universe. Inflation of a wrinkled balloon illustrates what happens: as the balloon inflates, its surface stretches and its wrinkles get flatter.

The curvature K of homogeneous and isotropic space is k/R^2, where k is the fixed curvature constant. Why the universe begins with a particular value of k, we do not know. With $k = 0$, the curvature K is zero, space is permanently flat and open, and the geometry is Euclidean; with $k = 1$, the curvature is positive, space is curved and closed, and the geometry is spherical; and with $k = -1$, the curvature is negative, space is curved and open, and the geometry is hyperbolic. Clearly, when $k = 0$, the flatness problem does not arise, only the question of why the curvature has the particular value $K = 0$ in an apparently unlimited range of negative and positive values. If $k = 1$, the curvature is $K = 1/R^2$, and if $k = -1$, then $K = -1/R^2$, and in both geometries the magnitude of the curvature during inflation decreases by the huge amount $e^{-2\eta}$:

$$K_f = K_i\, e^{-2\eta},$$

where K_f and K_i are the final and initial values of curvature in the inflation era. The final value of the curvature is much less than the initial value. For example, if $\eta = 67$, sufficient to sweep all monopoles out of the visible universe, inflation reduces the curvature by the factor 10^{-58}. When inflation ceases, the universe has a large energy density (the same as at the beginning of inflation), a large rate of expansion (the same as at the beginning of inflation), and a greatly reduced curvature.

The cosmological equation (Equation 22.6)

$$H^2 = \frac{8\pi G\rho}{3} + \Lambda - K,$$

shows what happens. The cosmological term Λ is constant and, for our present purpose, plays an unimportant role in the early universe. During inflation, the density remains constant and the curvature rapidly decreases. Soon after the beginning of inflation, the curvature becomes negligible and the above equation reduces to $3H^2 = 8\pi G\rho$. As we have seen, the Hubble term $H = \dot{R}/R$ is constant and the scale factor increases exponentially as e^{Ht}, where $H = (8\pi G\rho/3)^{1/2}$.

The density of a flat universe ($K = 0$) in which the cosmological constant is zero ($\Lambda = 0$) is the critical density

$$\rho_{\text{crit}} = \frac{3H^2}{8\pi G}. \qquad [22.11]$$

With the density parameter

$$\Omega = \frac{\rho}{\rho_{\text{crit}}}, \qquad [22.12]$$

Equation [22.6] converts to

$$K = H^2(\Omega - 1), \qquad [22.13]$$

and we see that Ω for positive curvature is greater than 1 and for negative curvature is less than 1. The density parameter equals 1 in the permanently flat Einstein–de Sitter universe of $K = 0$. At the end of inflation the curvature K_f is a very small negative or positive quantity, and hence Ω must be less or more than unity by only a very small amount.

Let K_0, H_0, Ω_0, and T_0 be the present values of the curvature, Hubble term, density parameter, and cosmic background radiation temperature, respectively, and let K, H, Ω, and T be their values at some other time. In the noninflationary universe, we have $K_0/K = (R/R_0)^2 = (T_0/T)^2$, and in the Einstein–de Sitter early universe, $H_0/H = t/t_0 = (R/R_0)^2$, and therefore $H_0/H = (T_0/T)^2$. Hence in the early universe, the density parameter is

$$\Omega = 1 + (T_0/T)^2(\Omega_0 - 1). \qquad [22.14]$$

With $T = 10^{28}$ and $T_0 = 3$ kelvin, we find $\Omega = 1 + 10^{-55}(\Omega_0 - 1)$. The observed average density of the universe is not greatly larger or greatly smaller than the critical density, and we can safely say that Ω_0 lies somewhere between 0.01 and 10. This leads us to the conclusion that the value of Ω in the very early universe was astonishingly close to unity, between 1 plus 10^{-54} and 1 minus 10^{-55} in the present example. Fine tuning in the initial conditions is necessary if the universe is to attain its present great age and be a fit place for the evolution of life.

Consider now the situation in the case of inflation. During inflation the Hubble term H is constant and hence $K_i/(\Omega_i - 1) = K_f/(\Omega_f - 1)$ according to Equation [22.13], where K_i and Ω_i are initial values and K_f and Ω_f are final values. We also have $K_f = K_i e^{-2\eta}$, and hence

$$\Omega_i = 1 + e^{2\eta}(\Omega_f - 1).$$

Combining this result with Equation [22.14], using $\Omega = \Omega_f$, $T = T_f = T_i = T_G$, we find

$$\Omega_i = 1 + (e^{\eta} T_0/T_G)^2(\Omega_0 - 1), \qquad [22.15]$$

where T_G is the grand unified temperature at which inflation begins. Ideally, e^{η} is equal or close to T_G/T_0. Then $e^{\eta} = 3 \times 10^{27}$, or $\eta = 63$, and $\Omega_i = \Omega_0$. This value of η solves the monopole and flatness problems and does not require that Ω_0 is close to unity. We consider two possibilities: either "large inflation" or "small inflation." Large inflation of e^{η} greater than T_G/T_0 solves the monopole and flatness problems but drives the density parameter Ω_0 close to unity, contrary to most observations. Small inflation of e^{η} less than T_G/T_0 may, if not too small, solve the monopole problem but fails to solve the flatness problem and the early universe is inexplicably flat. Either the universe is extremely flat now (Ω_0 very close to unity), or the universe is extremely flat in the beginning (Ω_i very close to unity). Ideal inflation (e^{η} close to T_G/T_0) requires fine tuning to avoid the difficulties of large and small inflation.

INFLATION SOLVES THE HORIZON PROBLEM

The horizon problem (see Chapter 21) concerns both static and expanding universes of finite age. We consider first a static universe of age t_0. An observer cannot see farther than a distance ct_0, where c is the speed of light. We imagine the observer as a dot at the center of a circle (actually a sphere) of radius ct_0. Inside the circle lies the observable universe, and outside lies the unobservable universe from which light has not yet reached the observer. The circle is the horizon (more technically, the particle horizon) of the observable universe. The horizon recedes from the observer at the speed of light, and as the universe ages, more and more of it becomes visible. When the universe is 1 year old, the horizon is at distance 1 light year; and when it is 10 billion years old, the horizon is at distance 10 billion light years. Although the universe is static, its visible region – the observable universe – expands at the speed of light and in the course of time the observer sees more and more of it. Far back in time, when the horizon was very near, each tiny part of the universe was isolated inside its horizon. The horizon is important because it determines not only the maximum distance an observer sees, but also the maximum distance between things able to interact and influence one another.

The situation is much the same in an expanding universe of age t_0. Space itself now expands and light travels at speed c in expanding space. Hence the maximum distance traveled by light in the cosmic lifetime t_0 is greater than ct_0. In the matter-dominated Einstein–de Sitter universe, for example, light travels $3ct_0$ in the cosmic lifetime and this is the horizon distance. Always the horizon recedes at light speed relative to comoving bodies in expanding space. In the course of time the observable universe contains more and more, and once things are inside the horizon, they are always inside and can never leave.

Habitually we look back into the past for causes that explain the way things are now. In cosmology this creates a problem. How can the way things are now on the scale of billions of light years be explained by causes existing when the universe was much less than billions of years old? For example, in a static universe, an observed body at distance L less than ct_0 was outside the horizon and unobserved at a time t when ct was less than L.

The horizon problem became even more puzzling with the discovery of the amazing isotropy of the cosmic background radiation that decoupled at the end of the early universe. The difference in the radiation intensity from opposite sides of the sky is less than 1 part in 100 000. Yet the emitting regions were far apart and outside each other's horizon at the time of decoupling. If the emitting regions had never interacted and did not "know" each other existed, how could they be in identical states? In the noninflationary universe the horizon problem has no known solution other than appeal to the anthropic or theistic principles.

The scientific solution to the horizon problem requires that the universe passes through an early period of accelerated expansion. In the pre-inflation period of expansion, things interact, equilibrate, and become homogenized in tiny causally interconnected regions on the scale of ct, where t is the age of the universe. During the inflation period, each tiny homogenized region is stretched superluminally to enormous size, and in the post-inflation period is vastly larger than a Hubble sphere (Figure 22.5). Subtle aspects of the horizon problem are dealt with more fully in Chapter 21. Homogenizing occurs initially in tiny causally interconnected regions the size of a Hubble sphere (which in the early universe is the same as the particle horizon), and inflation stretches these tiny regions to enormous size.

The velocity–distance law states that bodies at distance L recede at velocity cL/L_H, where L_H is the radius of the Hubble sphere. Galaxies inside the Hubble sphere recede subluminally (slower than light) and galaxies outside recede superluminally (faster than light). The Hubble sphere itself expands at velocity

$$\frac{dL_H}{dt} = c(1 + q), \qquad [22.16]$$

where q is the deceleration term. Comoving galaxies at the Hubble distance recede at the velocity of light c, and the Hubble sphere

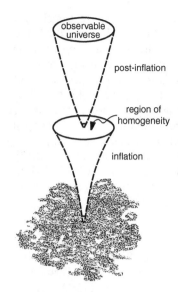

Figure 22.5. Tiny homogenized regions in the preinflationary era are the size of a Hubble sphere of radius 10^{-26} centimeters. Such regions rapidly expand in the inflationary era and become enormously large. In the post-inflation era these enormous regions continue to expand with positive deceleration. A small part of one of these enormous expanding regions, initially about 1 centimeter radius, is now our present observable universe of 10 billion light years radius.

expands at $1 + q$ times c. In a decelerating Friedmann universe, the Hubble sphere expands faster than the universe and its boundary overtakes the galaxies at velocity cq. A comoving body outside the Hubble sphere recedes superluminally, but the Hubble sphere expands faster than the universe, and eventually it overtakes the body and the body then lies inside the Hubble sphere and recedes subluminally. On the other hand, in an accelerating universe of q negative, the Hubble sphere expands slower than the universe and bodies move out of the Hubble sphere. A body, at first inside the Hubble sphere and receding subluminally, moves eventually out of the Hubble sphere, and then recedes superluminally.

The horizon problem is a causal problem. How at time t can observed bodies separated by distance L less than ct have causally interacted at an earlier time when L was greater than ct? The cosmic background radiation seen in opposite directions of the sky decoupled at time $t = 3 \times 10^5$ years in regions separated at that time by distance $L = 3 \times 10^7$ light years. How did those relatively small and widely separated regions equilibrate their temperatures? The solution requires that the universe passed through a period of accelerated expansion (inflation), and an initially self-interacting small region became distended into a large homogeneous and isotropic region, part of which is our present observable universe.

NONLUMINOUS MATTER

There are more things in heaven and earth, Horatio,
Than are dreamed of in your philosophy.
Shakespeare, Hamlet

Dark baryonic matter

The stars in the heavens shine and their light reveals a universe of galaxies. Starlight shows us how luminous matter is distributed in the universe. But not all stars are luminous like the Sun and not all matter is in stars. Dark matter undoubtedly exists; it betrays itself in various ways, mainly by its gravitational effect on the motions of luminous matter. The motions of galaxies in clusters reveal in a striking way that most of their matter is nonluminous. The galaxies move much too fast relative to one another to be confined in their clusters by the gravitational attraction of visible matter alone. It seems, as with icebergs, that invisible matter greatly outweighs visible matter. At every level in the astronomical hierarchy we find evidence of dark matter, and usually the larger the system the greater is the fraction of dark matter.

Twentieth-century astronomers have long speculated on the nature of the dark matter governing the motions of stars and galaxies. Various candidates have been suggested: gas, dust, low-luminosity stars, primordial black holes, stellar black holes, supermassive black holes, unborn galaxies, and changes in the laws of physics. All matter in known and unknown forms affects the expansion of the universe, and this is why

determination of the unseen matter is important in cosmology.

The critical density ρ_{crit} of the Einstein–de Sitter universe serves as a benchmark. The density of all matter is measured in terms of the critical density by means of the density parameter Ω_0:

$$\rho_0 = \Omega_0 \rho_{\mathrm{crit}}$$

$$= 1.9 \times 10^{-29} \Omega_0 h^2$$

grams per cubic centimeter   [22.17]

where $H_0 = 100h$ kilometers per second per megaparsec. A Friedmann universe is closed when the density is greater than the critical density and Ω_0 exceeds 1, and open when the density is equal to or less than the critical density and Ω_0 is equal to or less than 1.

The cosmic background radiation is only 0.0001 (or 1×10^{-4}) of the critical density. All forms of luminous matter amount approximately to 0.01 of the critical density, but the actual amount is not known precisely because of the difficult nature of the observations and because of uncertainty in the value of h. We can be more on target by using the parameter $\Omega_0 h^2$ instead of Ω_0, but many uncertainties in the measurements still remain, and in this discussion we shall occasionally assume that $h = 0.5$ (the critical density is then 4.7×10^{-30} grams per cubic centimeter) and stay with the density parameter Ω_0. Very approximately, all the unseen dark matter, which holds together galaxies and clusters of galaxies, raises the value of the density parameter Ω_0 from 0.01 to a value between 0.1 and 0.2.

The amount of hydrogen converted into helium, deuterium, and other light elements in the early universe by nucleosynthesis depends, among other things, on the photon–baryon number ratio. The value of this ratio is determined by observations of the cosmic background radiation temperature and the total number of baryons of visible and invisible matter. Studies in primordial nucleosynthesis show that Ω_0 is of order 0.1. Hence the dark matter inferred by astronomers (for which Ω_0 is also of order 0.1) consists mostly of baryons –

protons and neutrons – and is the same as the matter in planets and stars. Any evidence suggesting that Ω_0 is much greater than 0.1 must mean the additional dark matter cannot be baryonic (not if we accept that h is of order unity) and must be radically different from ordinary matter composing planets and stars.

Dark nonbaryonic matter

Thus far the facts are compelling: dark matter exists and its gravitational attraction holds together galaxies and clusters of galaxies; it outweighs luminous matter by a factor of roughly 10, raising the density parameter Ω_0 roughly from 0.01 to 0.1.

But if we accept that $\Omega_0 = 0.1$, in agreement with the observations and with primordial nucleosynthesis, we are in conflict with inflation theory that argues persuasively that the density parameter is close to unity. Thus we have a wide gap of "missing matter" between $\Omega_0 = 0.1$ and $\Omega_0 = 1$ that must be filled with nonbaryonic matter. Baryonic matter (protons and neutrons) cannot fill the gap because of the constraints set by primordial nucleosynthesis. Moreover, missing matter seems to be more smoothly distributed than ordinary matter and avoids concentration in galaxies and clusters. Hence, two kinds of nonluminous matter exist: dark baryonic matter taking Ω_0 from 0.01 to 0.1, and dark nonbaryonic "missing matter" taking Ω_0 from 0.1 to 1. All missing matter is dark but not all dark matter is missing.

Many suggestions have been made concerning the nature of the missing dark matter. Before embarking on flights of fancy, the reader should bear in mind that the astronomical evidence for a universe dominated by exotic forms of matter is slim, and the laboratory evidence for the various proposed candidates is equally slim. Effective inflation, unless finely tuned, mandates the missing matter, yet we do not know what form it takes and so far have no evidence that it actually exists. These are still early days, however, and undoubtedly many discoveries still await us in the future.

Cosmic neutrinos have been popular candidates for the missing dark matter. They are as abundant in the universe as the photons of the cosmic background radiation, and as far as we know, they have no rest mass and move at the speed of light. They are a form of hot (or relativistic) dark matter (HDM), and their contribution to the density is small and roughly the same as that of the cosmic background radiation. If, however, neutrinos possess a small rest mass, small enough not to be ruled out by existing experimental limits, their combined mass could outweigh all other forms of matter. The neutrino mass for achieving critical density, expressed in electron volts, lies between 10 and 100 eV, depending on the theory and kind of neutrino. Neutrinos would be relativistic (kinetic energy greater than rest energy) in the early universe, and later would become nonrelativistic because of kinetic energy lost by expansion. They would start as hot dark matter and later become cold dark matter (CDM).

The missing particles, in whatever form, cannot possess electric charges like electrons because we would see the light emitted by their interactions with one another and with ordinary matter. They interact gravitationally, at least on the cosmic scale, and a large class of such particles is known as WIMPS (weakly interacting massive particles). Many exotic forms have been proposed, forming combinations of HDM and CDM designed to meet inflation demands and explain the origin of galaxies, clusters of galaxies, and superclusters.

THE ORIGIN OF GALAXIES

The smoothness of the cosmic background radiation tells us that long ago at the decoupling epoch the universe was much less irregular than at present. Farther back in time the universe was probably even less irregular; but never perfectly smooth, because variations of some kind must have existed that later grew and developed into galaxies and their clusters.

Cosmologists since the 1930s have tried to understand how galaxies form in an expanding universe. Let $\rho + d\rho$ be the density at any point, where ρ is the average density and $d\rho$ is a small-amplitude variation. Generally, both ρ and $d\rho$ decrease in time in an expanding universe, but the contrast density, the ratio $d\rho/\rho$, increases. The spatial variation $d\rho$ in density consists of a spectrum of wavelengths, each wavelength having its own amplitude. Instead of a spectrum of wavelengths of different amplitudes, we can for convenience imagine separate regions of different sizes (wavelengths) having different density perturbations (amplitudes). Thus, here is a perturbed region of a certain size and density, and here is another perturbed region of a different size and density. We see that in addition to a spectrum of wavelength, we need also a distribution function specifying the relative number of perturbed regions at each wavelength.

All regions of different sizes expand with the universe and it helps if we think in terms of their comoving sizes (or comoving wavelengths) $\bar{\lambda}$, where the actual expanding wavelength is $\lambda = R\bar{\lambda}$. In a power-law spectrum, the contrast density $d\rho/\rho$ varies as $\bar{\lambda}^{-a}$:

$$\frac{d\rho}{\rho} = \left(\frac{d\rho_i}{\rho_i}\right)\left(\frac{\bar{\lambda}}{\bar{\lambda}_i}\right)^{-a},$$

where i denotes the initial value and a is a constant. A spectrum, such as this, which stays constant and does not change during expansion, is said to be scale-invariant. The physical processes causing the perturbations determine the kind of spectrum and distribution function. A not uncommon distribution has a bell-shaped (Gaussian) curve.

The universe originally is much larger than a Hubble sphere. The Hubble sphere expands faster than the universe; thus the baryons in a Hubble sphere at 10^{-4} seconds have the same mass as the Earth; at 1 second they have the same mass as the Sun; and at 10^8 seconds they have the same mass as a large galaxy. All disturbed regions destined to become galaxies and clusters of galaxies are smaller than a Hubble sphere during

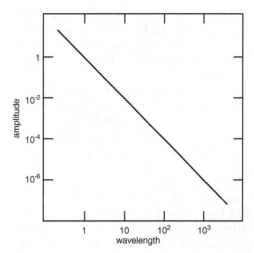

Figure 22.6. A spectrum in arbitrary units showing how the amplitude of the initial contrast density fluctuations, which develop into galaxies and clusters of galaxies, might depend on the wavelength.

their initial stages in the extreme early universe.

Calculation shows that in the early universe the contrast density of regions larger than the Hubble sphere grows linearly in time. Thus, for a power-law spectrum, $d\rho/\rho$ is proportional to $t/\bar{\lambda}^a$, or $t(R/\lambda)^a$ in terms of real wavelengths. The amplitude of the contrast density increases with time and decreases with wavelength. The spectrum $a = 2$ (Figure 22.6) has rather interesting properties. It varies as $t(R/\lambda)^2$, and because R increases as $t^{1/2}$ in the early universe, the spectrum varies as $(t/\lambda)^2$. Each disturbed region, when its size equals ct, or a multiple of ct, has a similar contrast density. Because $L_H = 2ct$, each perturbed region has the same density contrast when its size equals a Hubble sphere.

After the Hubble sphere has expanded and become larger than a perturbed region, its contrast density ceases to grow during the radiation era. Perturbations less than a galactic mass and smaller than a Hubble sphere tend to decay in the radiation era because of photon diffusion, as first shown by Joseph Silk of the University of California. Thus all galactic-mass perturbations that are overtaken by the Hubble sphere in the radiation era have similar amplitudes and also similar amplitudes at the end of the radiation era when radiation and matter decouple.

The surviving density perturbations begin to grow again after emerging from the radiation era. To explain the origin of galaxies and of clusters of galaxies we require typically a contrast density amplitude of roughly 10^{-4} at the decoupling epoch. The contrast density is 10^{-4} as each wavelength crosses into a Hubble sphere and later emerges from the radiation era. After decoupling the contrast density grows and at first is proportional to R, and if, for argument's sake, we suppose on very large scales, in uncondensed forms, it is now 0.1 then roughly

$$\left(\frac{d\rho}{\rho}\right)_{dec} = \frac{0.1}{1 + z_{dec}}. \qquad [22.18]$$

According to this crude argument we find that at the decoupling epoch the required contrast density is $(d\rho/\rho)_{dec} = 10^{-4}$ for $z_{dec} = 10^3$.

The perturbed regions of interest in the study of galaxy formation are much larger than a Hubble sphere in the early universe. This raises the question: how can a region in the early universe have a coherent density perturbation when its parts are flying from one another faster than light? They cannot communicate and do not even know that one another exists. Either we must suppose that the variations in density are imprinted in the universe from the beginning or we must devise some neat idea such as the inflation theory.

Density fluctuations may originate at a phase transition when matter spontaneously transforms from one phase to another. The grand unified phase transition is a promising possibility, and quantum fluctuations of the fields that drive inflation also produce density irregularities. Some studies find that the density irregularities on scales less than a Hubble length obey an $a = 2$ spectral law. These perturbed regions expand with the universe and when larger than a Hubble length are stretched superluminally into vastly larger regions. The characteristics of the spectrum in this process stay essentially unchanged.

REFLECTIONS

1 *"Inflation was being talked about in the economic sense, and certainly what the universe did was inflate, so it seemed like a natural word"* (*Alan Guth quoted in* Lonely Hearts of the Cosmos *by Dennis Overbye*).

● *"The great Russian physicist Lev Landau said cosmologists are often wrong but never in doubt. That may have been true once, but after having years of observations refuse to agree with theories, there's no shortage of doubt in the field"* (*Robert Mathews, "Cosmologists meet to face their fears"*).

● *"Cosmology has much in common with archaeology: just as it is hard to be certain how Stonehenge was erected given only the current state of the stones, so there is a variety of plausible theories of the origin of the galaxies and their present-day large-scale clustering. The cure for the uncertainty is a time machine, and astronomers are now starting to amass data on the distribution of distant galaxies, seen when the universe was young"* (*John Peacock, "Lumps in the early universe"*).

2 *Consider the following. In a universe of age t, two comoving immortals – A for Albert and B for Bertha (Figure 22.7) – are separated by a distance L that is less than ct and small enough that A and B are free to communicate with each other and learn a common language. How the ratio L/ct subsequently changes in an expanding universe depends on the deceleration term q. In a decelerating universe ($q > 0$), the Hubble sphere expands faster than the universe and A and B remain in communication. They recede from each other subluminally and the ratio L/ct steadily decreases. In an accelerating universe ($q < 0$), the Hubble sphere expands slower than the universe and A and B soon lose contact. They then recede from each other superluminally and the ratio L/ct increases and becomes much greater than unity. Consequently A and B have only fading memories of*

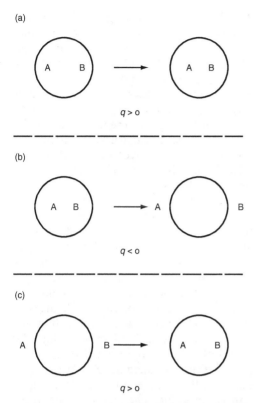

(a)

$q > o$

(b)

$q < o$

(c)

$q > o$

Figure 22.7. (a) A and B are initially separated by distance L less than a Hubble distance. The Hubble distance equals roughly ct, where t is the age of the universe. A and B recede from each other at less than the speed of light and are free to communicate and learn a common language. In a decelerating pre-inflationary era, L/ct decreases and A and B continue to communicate. (b) Deceleration ceases and the expansion accelerates. In the inflationary era that follows, L/ct increases, and when B is outside the Hubble sphere of A (and A is outside the Hubble sphere of B) they recede faster than the speed of light, and A and B are no longer able to communicate with each other. (c) Acceleration ceases and the expansion decelerates. In the post-inflationary era that follows, L/ct again decreases and eventually A and B recede from each other slower than the speed of light. Once more they are free to communicate and to their surprise they discover that they share a common language.

each other. Imagine that much later the acceleration changes into deceleration, as happens in the inflationary universe. The Hubble sphere starts to expand faster than the universe and eventually the ratio L/ct is once

again less than unity. Albert and Bertha are again able to communicate. They find with surprise that they speak the same language, and wonder how this is possible in a decelerating universe.

3 *Opinions vary, but most workers on inflation theory agree that a grand unified phase transition involves supercooling and some form of false vacuum. As in the supercooled water–ice transition, small isolated crystals eventually form at random points and grow rapidly and coalesce. Guth's original ideas conformed to this analogy. First comes supercooling, followed by spontaneous transitions at random points in the false vacuum. These quark–lepton crystals grow rapidly and finally coalesce. Further study, however, showed that such coalescence cannot occur. Once the crystals form they cease to inflate. But the surrounding false vacuum continues to inflate. Although individual crystals grow in size, they never coalesce because they recede from one another faster than the speed of light. Each crystal forms an isolated world containing a trillion quarks, leptons, and gluons, and because each particle has an energy 10^{15} GeV, equivalent to a mass of 1 billionth of a gram, the total mass of each crystal amounts to roughly a few kilograms. Further study (by Andrei Linde at the Lebedev Institute in Moscow, and Andreas Albrecht and Paul Steinhardt at the University of Pennsylvania) revealed a possible solution to the difficulty. The supercooled false vacuum inflates and the transition to a quark–lepton state occurs as before in random regions, each region the size of a Hubble sphere. But the transition in these regions is slow and inflation continues during the process (see Figures 22.2 and 22.3). Each tiny region inflates into a vast world that at the end of inflation fills with quarks, leptons, and gluons. The size of this inflated world depends on the inflation factor.*

A phase transition from vapor to liquid provides another analogy. Saturated vapor supercools and condenses into droplets. In an expanding cloud chamber, for example, the temperature drops rapidly and supercooled vapor condenses into droplets at nucleation sites. In the inflation picture, these tiny bubble worlds balloon into vast isolated worlds. The initial trillion particles occupying a Hubble sphere, multiplied by $e^{3\eta}$ (the cube of the inflation factor), become the final number of particles in the inflated bubble world. The total number of photons, neutrinos, and other particles in the Hubble sphere of today is of order 10^{90} (Chapter 23), and hence the inflation exponent η must exceed 60 if today's Hubble sphere forms part of a larger inflated bubble world.

4 *The following calculations illustrate the importance of initial flatness. Let the density parameter be Ω_i at time t_i in the early universe. We find that at a later time t, when the density parameter is Ω, that*

$$\frac{t}{t_i} = \frac{1 - \Omega}{1 - \Omega_i}. \qquad [22.19]$$

What is the condition, at age $t_i = 1$ second, that the universe can attain an age of at least 10^{12} seconds (the decoupling epoch) and have a density parameter of $\Omega = 0.5$? If $\Omega_i = 1 - \varepsilon$, then Equation [22.19] shows that ε is equal to or less than 0.5×10^{-12}. From the fact the universe was so flat at age 1 second we conclude that it was even flatter at earlier times. Suppose that a radiation-dominated closed universe begins at the Planck epoch $t^* = 10^{-44}$ seconds, what is the initial value of the density parameter Ω^* for this universe to have a total lifetime t_0? We find

$$\frac{t_0}{t^*} = \frac{2\Omega^{*1/2}}{\Omega^{*1/2} - 1}, \qquad [22.20]$$

and with $\Omega^* = 1 + \varepsilon$, with ε small, we find $t_0/t^* = 4/\varepsilon$. Thus, for a lifetime greater than 1 second, ε must be less than 4×10^{-44}. From these two illustrations we see that the universe must initially be exceedingly flat to last for even a short period of time.

PROJECTS

1 A constant density universe expands exponentially and the Hubble term H remains constant. Why?

2 Discuss the following statement: The outcast steady-state theory of Bondi, Gold, and Hoyle has come in from the cold and now receives shelter in the big bang theory that it so strongly opposed.

3 The prevailing scientific method explains present conditions in terms of initial conditions. On finding that the initial conditions must be finely tuned for the existence of observers, we get involved in debates on metaphysical issues such as the anthropic principle. Why was the universe initially so flat? – because otherwise we would not be here marveling over its remarkable initial flatness. But now comes the idea of inflation, and we find that the original flatness of the universe does not have to be finely tuned. It can be almost anything, provided the primordial universe lasts long enough to attain inflationary conditions. Is it conceivable, in view of this discovery, that one day we shall explain all "fine tuning" by means of physical, not metaphysical, arguments? Can we imagine a universe that begins with all its fundamental constants unfixed in value, and subsequently these constants become fixed at values determined by natural processes? One day we might have a grandiose quantum cosmology in which the sum over the histories of all virtual universes, each weighted by an "experiential probability" factor, determines the actual observed universe. The experiential probability is large for a universe billions of years old, spanning billions of light years, and containing billions of galaxies, and this might explain why the fundamental constants have their observed values. Is this yet another and more subtle way of stating the anthropic principle?

FURTHER READING

Dodd, J. E. *The Ideas of Particle Physics: An Introduction for Scientists.* Cambridge University Press, Cambridge, 1984.

Ford, K. W. "Magnetic monopoles." *Scientific American* (December 1983).

Glashow, S. L. *The Charm of Physics.* Simon and Schuster, New York, 1991.

Gribbin, J. *In Search of the Big Bang: Quantum Physics and Cosmology*. Heinemann, London, 1986.

Gribbin, J. and Rees, M. *Cosmic Coincidences: Dark Matter, Mankind, and Anthropic Cosmology*. Bantam Books, New York, 1989.

Guth, A. H. *The Inflationary Universe: The Quest for a New Theory of Cosmic Origins*, Addison-Wesley, New York, 1997.

Guth, A. H. and Steinhardt, P. J. "The inflationary universe." *Scientific American* (May 1984).

Overbye, D. *Lonely Hearts of the Cosmos: The Scientific Quest for the Secret of the Universe*. Harper, New York, 1991.

Pagels, H. R. *Perfect Symmetry: The Search for the Beginning of Time*. Simon and Schuster, New York, 1985.

Riordan, M. and Schramm, D. N. *The Shadows of Creation: Dark Matter and the Structure of the Universe*. W. H. Freeman, New York, 1990.

Rowan-Robertson, M. *Ripples in the Cosmos: A View Behind the Scenes of the New Cosmology*. W. H. Freeman, New York, 1993.

Rubin, V. C. "Weighing the universe: dark matter and missing mass," in *Bubbles, Voids, and Bumps in Time: The New Cosmology*. Editor, J. Cornell. Cambridge University Press, Cambridge, 1989.

Schramm, D. N. "The early universe and high-energy physics." *Physics Today* (April 1983).

Silk, J. *The Big Bang*. Revised edition. Freeman, New York, 1989.

Weinberg, S. *Dreams of a Final Theory*. Pantheon Books, New York, 1992.

SOURCES

Bernstein, J. and Feinberg, G. Editors. *Cosmological Constants: Papers in Modern Cosmology*. Columbia University Press, New York, 1986.

Blau, S. K. and Guth, A. H. "Inflationary cosmology," in *300 Years of Gravitation*. Editors, S. W. Hawking and W. Israel. Cambridge University Press, Cambridge, 1987.

Chuang, I., Durrer, R., Turok, N., and Yurke, B. "Cosmology in the laboratory: Defect dynamics in liquid crystals." *Science* 251, 1336 (1991).

Coles, P. and Ellis, G. F. R. *Is the Universe Open or Closed: The Density of Matter in the Universe*. Cambridge University Press, Cambridge, 1997.

Coles, P. and Lucchin, F. *Cosmology: The Origin and Evolution of Cosmic Structure*. Wiley and Sons, Chichester, 1995.

Dicke, R. H. and Peebles, P. J. E. "The big bang cosmology – enigmas and nostrums," in *General Relativity: An Einstein Centenary Survey*. Editors, S. W. Hawking and W. Israel. Cambridge University Press, Cambridge, 1979.

Fourner d'Albe, E. E. *Two New Worlds*. Longmans and Green, London, 1907.

Guth, A. H. "Inflationary universe: A possible solution to the horizon and flatness problems." *Physical Review* D23, 347 (1981).

Kazanas, D. "Dynamics of the universe and spontaneous symmetry breaking." *Astrophysical Journal* 241, L59 (1980).

Kolb, E. W. and Turner, M. S. *The Early Universe*. Addison-Wesley, New York, 1990.

Linde, A. "Particle Physics and Inflationary Cosmology." *Physics Today* (September 1987).

Mathews, R. "Cosmologists meet to face their fears." *Science* (5 November 1993).

Partridge, R. B. *3K: The Cosmic Microwave Background Radiation*. Cambridge University Press, Cambridge, 1995.

Peacock, J. "Lumps in the early universe." *Nature* (8 July 1993).

Preskill, J. "Monopoles in particle physics and cosmology," in *Inner Space, Outer Space*. Editors, E. W. Kolb, M. S. Turner, D. Lindley, K. Olive, and D. Seckel. University of Chicago Press, Chicago, 1989.

Rindler, W. "Visual horizons in world models." *Monthly Notices of the Royal Astronomical Society* 116, 662 (1956).

Trimble, V. "Existence and nature of dark matter in the universe." *Annual Reviews of Astronomy and Astrophysics* 25, 425 (1988).

THE COSMIC NUMBERS

Do you believe then that the sciences would ever have arisen and become great if there had not beforehand been magicians, alchemists, astrologers, and wizards who thirsted and hungered after secret and forbidden powers?
Friedrich Nietzsche (1844–1900), The Will to Power

CONSTANTS OF NATURE

Natural units

We measure distances in units such as meters and light years, intervals of time in units such as seconds and years, and masses in units such as grams and kilograms. There is nothing sacred about these units, which are determined by our history, environment, and physiology. If we communicate with beings in another planetary system and inform them that something has a size of so many meters, an age of so many seconds, and a mass of so many kilograms, they will not understand because their units of measurement are undoubtedly different. But they will understand if we say the size is so many times that of a hydrogen atom, the age is so many times that of a certain atomic period, and the mass is so many times that of a hydrogen atom, simply because their atoms are the same as ours (if they were not, it would be an incoherent universe, incomprehensible, and we might not be able to communicate with them). The basic uniformity of the universe provides us all with the same set of natural units of measurement.

The only objects that appear exactly the same everywhere are atoms and their constituent particles. A natural unit of mass is that of the nucleon equal approximately to that of the hydrogen atom. A natural unit of length is the size of a subatomic particle, such as a nucleon, a unit known as the fermi in honor of Enrico Fermi, and is equal to 1×10^{-13} centimeters. A natural unit of time is the period required by light to travel a distance of 1 fermi, and Richard Tolman suggested the name jiffy for the unit equal to 1×10^{-23} seconds. In round numbers, a human being has a mass of 10^{29} nucleons, a height of 10^{15} fermis, and a lifetime of 10^{32} jiffies. With such natural units from the subatomic world we appreciate the lavish scale on which the universe is constructed. A planet such as the Earth has a mass of 10^{52} nucleons, a star such as the Sun a mass of 10^{57} nucleons, and a galaxy such as our own a mass of 10^{68} nucleons.

Cosmic numbers dealing with the scale and structure of the universe have attracted much attention in past decades. If the observable universe is measured in natural units we find that it has approximately a mass of 10^{80} nucleons, a size of 10^{40} fermis, and an age 10^{40} jiffies. Cosmic numbers such as these have a seductive fascination.

A remarkable contribution to the subject of cosmic numbers was made more than 2000 years ago by Archimedes, one of the greatest scientists of the ancient world. Archimedes invented a numerical system called the "naming of numbers," and in a work entitled *The Sand-Reckoner* he used this system to calculate the total number of grains of sand the universe could contain. In effect, he estimated the volume of the universe as it was then known, using a grain of sand as his unit of volume. He found that the universe had a volume of 10^{63} grains of sand. What is remarkable to us is that this volume of sand, as estimated by Archimedes, had a mass equal to our present estimates of the mass of the observable universe (see Reflections).

Constants of nature

Generally, each fundamental constant indicates the involvement of a particular branch of physics:

G: gravity
c: relativity
h: quantum mechanics
m_n, m_e, e: subatomic particles.

The gravitational constant G appears in calculations whenever gravity is involved. The speed of light c is always associated with relativity and the propagation of light. Planck's constant h, the cornerstone of quantum mechanics, is associated with the wavelike nature of particles and the corpuscular properties of radiation. Associated with the subatomic particles are the mass m_n of the nucleon, the mass m_e of the electron, and the electric charge e that is positive for a proton and negative for an electron. Nucleons (i.e., protons and neutrons) have a mass 1836 times the mass of the electron, and in each gram of matter there are almost a trillion trillion (10^{24}) nucleons. One coulomb of electric charge contains nearly 10 million trillion (10^{19}) elementary e charges, and this is the number of electrons that flow each second through the filament of a 100-watt electric light bulb (the current is assumed to be 1 ampere).

The constants of nature can be arranged to form natural numbers (often referred to as dimensionless numbers) that are independent of human units of measurement. We start by giving two examples. The first is the ratio of the nucleon and electron masses

$$\frac{m_n}{m_e} = 1836, \qquad [23.1]$$

and is obviously a natural number independent of whether we measure masses in grams or kilograms or other units. The second example is the Sommerfeld fine structure constant denoted by α:

$$\alpha = \frac{2\pi e^2}{hc} = \frac{1}{137}. \qquad [23.2]$$

The fine structure constant appears whenever radiation interacts with particles, and the combination of c, h, and e indicates a wavelike (h) interaction between particles (e) and light (c). The value $1/137$ is a natural dimensionless number, used by terrestrial and extraterrestrial scientists alike, and is independent of our units of measurement. Planck's constant h frequently occurs in the form $h/2\pi$, and for convenience we write $\hbar = h/2\pi$, and the fine structure constant is hence $\alpha = e^2/\hbar c$.

A characteristic size of atoms is the radius a_0 of the hydrogen atom:

$$a_0 = \frac{\hbar^2}{m_e e^2} = 0.5 \times 10^{-8} \text{ centimeters}, \qquad [23.3]$$

referred to as the Bohr orbit radius. The absence of G and c indicates that gravity and relativity are not of primary importance in atomic structure. A characteristic wavelike size of an electron, traveling close to the speed of light, is the electron Compton length,

$$\lambda_e = \frac{\hbar}{m_e c} = 4 \times 10^{-11} \text{ centimeters}. \qquad [23.4]$$

We notice that $\lambda_e = \alpha a_0$, and the electron Compton length is $1/137$ times smaller than the size of a hydrogen atom. The wavelength of an electron moving at speed V is $\hbar/m_e V$. An electron in a hydrogen atom has a typical speed $V = \alpha c$ (when its wavelength equals the circumference of the atom) equal to $1/137$ times the speed of light. The classical electron radius a is the size of an electron calculated prior to the introduction of quantum mechanics. It is obtained by assuming that the energy $m_e c^2$ of the electron is in the form of electrical energy e^2/a, thus yielding a radius a expressed by

$$a = \frac{e^2}{m_e c^2} = 3 \times 10^{-13} \text{ centimeters}, \qquad [23.5]$$

equal to 3 fermis. Electrons are wavelike and do not behave in a commonsense corpuscular way like billiard balls, and the classical electron radius is not a meaningful measurement of electron size. It is interesting, nonetheless, that the classical electron radius a is characteristic of the size of a nucleon (and of many subatomic particles having internal structure), and we use it as a convenient subatomic unit of length. We see that $a = \alpha\lambda_e = \alpha^2 a_0$, and hence the proton – the nucleus of the hydrogen atom – is $\alpha^2 = 1/18\,769$ times the size of the hydrogen atom.

So far we have found two natural numbers (1836 and 1/137) that involve relativity, quantum mechanics, and the properties of subatomic particles. We now construct a third number that involves the gravitational constant G. The electrical and gravitational forces between a proton and an electron are both attractive and proportional to the inverse square of their separating distance r. The electrical force is e^2/r^2, the gravitational force is Gm_nm_e/r^2: and the ratio of these two forces is the large natural number

$$\frac{e^2}{Gm_nm_e} = 0.2 \times 10^{40}. \qquad [23.6]$$

Electrical forces between neighboring charged particles are vastly stronger than their gravitational attractions. On the atomic scale, electrical forces dominate. On much larger scales the positive and negative charges of many particles tend to neutralize one another and electrical forces become comparatively weak. Gravity cannot be neutralized and the larger the number of particles in a system the stronger becomes the combined gravitational force. Although gravity is weak and its effect is negligible among a few particles, in large systems of many particles gravity becomes strong and can overwhelm all other forces.

If a nucleon were a nonrotating black hole it would have a Schwarzschild radius of $2Gm_n/c^2$. The coefficient 2 is not important and we shall say that $a_g = Gm_n/c^2$ is the gravitational size of a nucleon. If the nucleon had a radius a_g, then gravity would be of primary importance in determining its size. The actual radius is a, and we see that

$$\frac{a}{a_g} = \frac{e^2}{Gm_nm_e} = 0.2 \times 10^{40}. \qquad [23.7]$$

The nucleon, small by ordinary standards, is vastly larger than if collapsed to form a black hole. Thus gravity is not important in subatomic structure. Often the nucleon Compton length $\lambda_n = \hbar/m_nc$ (equal to 2×10^{-14} centimeters) is a convenient measure of the size of a nucleon, and we find

$$\frac{\lambda_n}{a_g} = \frac{\hbar c}{Gm_n^2} = 1.5 \times 10^{38}. \qquad [23.8]$$

This large number shows that quantum mechanical and not gravitational forces are of dominant importance in the structure of particles.

Cluster hypothesis

The constants of nature yield two groups of natural numbers. The first group consists of the relatively small numbers that are clustered around unity:

$$\frac{m_e}{m_n}, \quad \frac{e^2}{\hbar c}, \quad \frac{\hbar c}{e^2}, \quad \frac{m_n}{m_e}, \qquad [23.9]$$

and these numbers are 1/1836, 1/137, 137, and 1836, respectively. The second, group consists of the relatively large numbers that are clustered about 10^{40}:

$$\frac{e^2}{Gm_n^2}, \quad \frac{\hbar c}{Gm_n^2}, \quad \frac{e^2}{Gm_nm_e},$$
$$\frac{\hbar c}{Gm_nm_e}, \quad \frac{e^2}{Gm_e^2}, \quad \frac{\hbar c}{Gm_e^2}. \qquad [23.10]$$

These numbers are 1/1836, 137/1836, 1, 137, 1836, and 137×1836 multiplied by 0.2×10^{40}, respectively. Each group consists of numbers covering a range that is small in comparison with the wide separation of the two groups, as shown in Figure 23.1. We shall refer to the first group of numbers as the unity group and the second group of numbers as the N_1 group.

The clustering of natural numbers into two groups of relatively small spread is sufficiently remarkable for us to postulate a cluster hypothesis. This tentative hypothesis states that all natural numbers compounded from the fundamental constants of nature are members of either the unity or N_1 group. Other numbers, such as $N_1^{1/2}$ or N_1^2,

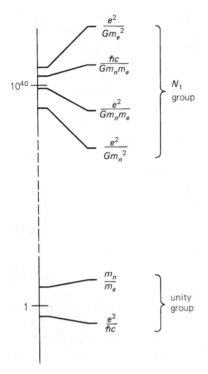

Figure 23.1. Sample numbers in the unity and N_1 groups of numbers.

may be obtained from N_1 and cannot be regarded as basic in the same sense as N_1. We have no theory to support the cluster hypothesis, but presumably the explanation must have something to do with the basic design of the universe. Life would not exist in a universe in which the constants were arranged differently.

Time variations of the natural constants
On several occasions scientists have considered the possibility that the constants of nature, either singly or in combination, change with time as the universe evolves. To these scientists it has seemed reasonable that if the constants are related in some unknown Machian manner to the properties of the universe as a whole, then they should change their values as the universe expands and evolves. Let us see if this idea contradicts the cluster hypothesis.

A not infrequent suggestion is that the universal gravitational constant G decreases in value as the universe expands. The unity group does not contain G, and therefore remains unaffected by G-variation. All members of the N_1 group are inversely proportional to G, and therefore G-variation merely moves the N_1 group without altering its tight clustering. Hence G-variation does not contradict the cluster hypothesis and all natural numbers retain their membership in two widely separated groups. Some consequences of G-variation are mentioned later in connection with the Dirac universe.

Some years ago it was suggested that the elementary charge e changes its value slowly with time. This idea, no longer popular, was proposed as a way of explaining the redshifts of distant galaxies. If e was smaller in the past, then atoms were larger (a_0 was bigger) and their emitted wavelengths were longer, thus making distant sources appear to be receding because of their observed redshifts. This was the shrinking-atom hypothesis advocated by many who disliked the idea of an expanding universe. We notice that when e varies the numbers containing e^2 will wander outside the unity and N_1 groups and these groups will become dispersed. Hence e-variation contradicts the cluster hypothesis. Furthermore, if e varies, then the value of the fine structure constant $\alpha = e^2/\hbar c$ also changes. Many details in the spectra of atoms are sensitive to the value of the fine structure constant, and by studying the atomic emissions of distant quasars we learn that α has not changed over long periods of time. The abundance of the isotopes of many terrestrial elements is also sensitive to the value of α, and therefore we know that α cannot have varied much during the lifetime of the Earth. Also the structure of stars is sensitive to the value of α, and changes by very small amounts eliminate the possibility of long-lived luminous stars.

We are left with the possibility that subatomic particle masses change with time. Comparison of atomic and molecular spectral emissions of extragalactic sources show that the electron mass does not vary. We know also that the combination $\alpha m_n/m_e$

cannot change much, if at all, because the redshifts of distant galaxies are the same when measured in optical and 21-centimeter radiation. Hence the ratio of the nucleon and electron masses is constant because α, as shown by other methods, is constant. Thus m_n remains unchanged over long periods of time. Cosmologists now firmly believe the fundamental constants are truly constant and have values that are everywhere the same throughout space and time.

The evidence supports the conclusion that all natural numbers derived from the constants of nature reside in two groups that are tightly clustered. The only fundamental constant that can vary without disturbing the clustering is the gravitational constant G.

Number sequences

Max Planck in 1913 showed that the constants G, c, and $\hbar$ can be combined to create natural units of length, time, and mass. The Planck length is

$$a^* = \left(\frac{G\hbar}{c^3} \right)^{1/2}$$

$$= 2 \times 10^{-33} \text{ centimeters,} \qquad [23.11]$$

and the Planck mass is

$$m^* = \left(\frac{\hbar c}{G} \right)^{1/2} = 2 \times 10^{-5} \text{ grams.} \quad [23.12]$$

The Planck unit of time t^* is obtained by dividing the Planck length a^* by the speed of light c, and is equal to 10^{-43} seconds. The Planck units can be found in the following way. We suppose that a particle exists of mass m^* that has a gravitational length equal to its Compton length. Its gravitational length is $a^* = Gm^*/c^2$, and its Compton length is $a^* = \hbar/m^*c$; on equating the two we obtain the Planck units. Such a particle has the properties of a black hole of quantum mechanical size and in effect is a quantum black hole. The creation of a quantum black hole requires an enormous energy (10^{19} GeV) and these particles have not yet been observed. Particles of Planck mass are

possibly the basic constituents of spacetime. If spacetime could be examined on the scale of 10^{-33} centimeters we might see a dense foam of virtual quantum black holes, popping into and out of existence on a time scale of 10^{-43} seconds. Conceivably, in the beginning, the universe consisted of an extremely dense sea of real quantum black holes (see Chapter 20). On this matter we cannot be certain because we lack a quantum theory of gravity.

On comparing quantum black holes with nucleons we find

$$\frac{\lambda_n}{a^*} = \left(\frac{\hbar c}{Gm_n^2} \right)^{1/2} = N_1^{1/2}, \qquad [23.13]$$

$$\frac{m^*}{m_n} = \left(\frac{\hbar c}{Gm_n^2} \right)^{1/2} = N_1^{1/2}. \qquad [23.14]$$

The size λ_n of a nucleon is roughly 10^{20} times larger than the size a^* of a Planck particle, and the mass m^* of a Planck particle is roughly 10^{20} times larger than the mass m_n of a nucleon.

A star is held together by gravity and is supported by the pressure resulting from the thermal motions of its individual particles. Unlike a planet, a star consists of high-temperature gas and its constituent particles are freely moving protons and electrons. These charged particles, as they rush around, have close encounters and are strongly affected by their electrical forces. Normal stars are dominated by gravity on the large scale and by electrical forces on the small scale. Despite their wide range of sizes and luminosities, stars have only a relatively small range of masses. Those only 1/10 the mass of the Sun are not hot enough to count as luminous hydrogen-burning stars, and those much more than 10 times the mass of the Sun are rare and burn their hydrogen rapidly. Because the structure of a star is governed by gravitational and electrical forces, a reasonable guess would be that the number of particles in a star depends in some way on the number N_1 that expresses the relative strengths of electrical and gravitational forces. Let N_{star} be the number of nucleons in a typical star. It can be shown (see Reflections) that

$$N_{\text{star}} = \left(\frac{\hbar c}{Gm_{\text{n}}^2}\right)^{3/2} = 2 \times 10^{57}. \quad [23.15]$$

Thus $N_{\text{star}} = N_1^{3/2}$, and the typical mass of a star is given by

$$M_{\text{star}} = m_{\text{n}} N_{\text{star}} = 4 \times 10^{33} \text{ grams}.$$

This result, surprising in view of the roughness of the calculation, is twice the mass of the Sun.

We have derived two large numbers $N_1^{1/2}$ and $N_1^{3/2}$, and by interchanging e^2 and $\hbar c$, and m_{n} and m_{e}, we can construct other numbers that belong to the $N_1^{1/2}$ and $N_1^{3/2}$ groups. For example, a stellar mass has a gravitational length GM/c^2, and this is approximately its radius in the form of a black hole, which is $N_1^{3/2}$ times the gravitational length of a nucleon, or N_1 times a Planck length, or $N_1^{1/2}$ times the size of a nucleon.

The constants of nature form natural numbers in the unity and N_1 groups, and we have found other derivative groups of physical significance, such as $N_1^{1/2}$ and $N_1^{3/2}$. Breakdown in the cluster hypothesis would destroy the "magic-number" sequence: $1, N_1^{1/2}, N_1^{3/2}, \ldots$.

THE COSMIC CONNECTION
A striking coincidence

The observable universe extends out to a distance of about 10 billion light years. This distance is the Hubble length denoted by L_H, and when expressed in units of the nucleon size a, we find it has the value

$$\frac{L_H}{a} = N_2 = 3 \times 10^{40}. \quad [23.16]$$

Alternatively, it is 3×10^{38} electron Compton lengths or 5×10^{41} nucleon Compton lengths. Roughly speaking, the observable universe has a size 10^{40} fermis, and the Hubble period L_H/c (a measure of the age of the universe) is 10^{40} jiffies, where 1 jiffy (10^{-23} seconds) is the approximate time taken by light to travel a distance of 1 fermi. Thus the characteristic size and age of the universe, when measured in natural

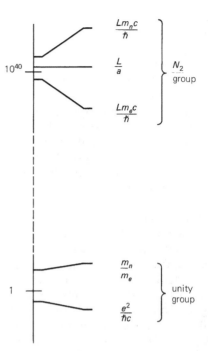

Figure 23.2. The unity and N_2 groups of numbers.

or subatomic units, are represented by a group of numbers clustered in the neighborhood of 10^{40}. We shall refer to these numbers as the N_2 group (Figure 23.2).

The approximate coincidence

$$N_1 = N_2 \quad [23.17]$$

is striking. Both the N_1 and N_2 groups have large values and their approximate equality suggests the possibility of a concealed relation between the constants of nature and the size of the universe. A fortuitous coincidence cannot be ruled out, but seems unlikely with numbers so large.

Large-number hypothesis

In 1937, the famed physicist Paul Dirac (1902–1984) postulated the large-number hypothesis. This hypothesis, in Dirac's words, states: "Any two of the very large dimensionless numbers occurring in Nature are connected by a simple mathematical relation, in which the coefficients are of order unity." The large numbers referred to by Dirac are the members of the N_1 and N_2 groups. Dirac's speculative hypothesis implies that a physical relation exists between the N_1 and N_2 groups and that both groups remain permanently equal in value to each other.

On the one hand we have the cluster hypothesis stating that the natural dimensionless numbers fall into the unity and N_1 groups, and on the other hand we have the large-number hypothesis stating that the N_1 and N_2 groups are physically related and permanently equal. We clearly have a cosmic-number problem.

The cosmic-number problem

The N_1 group of numbers stays constant in an expanding universe, whereas the N_2 group of numbers measures the size of the universe in fermis or the age in jiffies and increases in time as the universe expands and ages. Thus at the Planck epoch, $N_2 = 10^{-20}$, and when the universe is 1 second old, $N_2 = 10^{20}$. In the early universe N_1 had its present value 10^{40}. The present equality of N_1 and N_2 is therefore puzzling. Are the two groups now equal by chance, or are they, in accordance with Dirac's large number hypothesis, permanently equal and related in an unknown way? Before considering how the problem can be solved, we show that the large number hypothesis

leads to an impressive array of magic numbers.

MAGIC NUMBERS

Because of the N_1 and N_2 equality, there exists a number sequence 1, $N^{1/2}$, N, $N^{3/2}$, N^2, ..., where N stands for either N_1 or N_2.

Deliberately we have not assigned precise values to either N_1 or N_2 because each stands for a group of numbers clustered around 10^{40}. Let N represent any number somewhere in the range 10^{38} to 10^{42} that comes from either the N_1 or N_2 groups. Thus $N_1 = N_2 = N$ and the equality sign in this case represents only approximate equality. We find that there exists a sequence of magic numbers

$$\ldots N^{-2}, N^{-3/2}, N^{-1}, N^{-1/2}, 1,$$

$$N^{1/2}, N, N^{3/2}, N^2, \ldots.$$

The numbers preceding 1 are obtained by simply inverting those that follow 1, and we need consider only the numbers that are unity and greater. We have a sequence of numbers of the kind 1, 10^{20}, 10^{40}, 10^{60}, 10^{80}, ..., compounded from N_1 and N_2. Some examples of how these numbers may be derived are found later in this chapter, and more information is in Table 23.1

Table 23.1 *Cosmic numbers*

1	$N^{1/2}$	N	$N^{3/2}$	N^2
N_1 N_2	$\dfrac{a^*}{a_g} = N_1^{1/2}$	$\dfrac{a}{a_g} = N_1$	$\dfrac{R_S}{a_g} = N_1^{3/2}$	$\dfrac{L_H}{a_g} = N_1 N_2$
	$\dfrac{a}{a^*} = N_1^{1/2}$	$\dfrac{R_S}{a^*} = N_1$	$\dfrac{L_H}{a^*} = N_1 N_2^{1/2}$	$\dfrac{M_H}{m_n} = N_1 N_2$
	$\dfrac{R_S}{a} = N_1^{1/2}$	$\dfrac{L_H}{a} = N_2$	$\dfrac{M_\odot}{m_n} = N_1^{3/2}$	$\dfrac{\rho^*}{\rho_n} = N_1^2$
	$\dfrac{L_H}{R_S} = \dfrac{N_2}{N_1^{1/2}}$	$\dfrac{M_\odot}{m^*} = N_1$	$\dfrac{M_H}{m^*} = N_1 N_2^{1/2}$	$\dfrac{\rho_j}{\rho_{crit}} = N_2^2$
	$\dfrac{m^*}{m_n} = N_1^{1/2}$	$\dfrac{\rho^*}{\rho_j} = N_1$		
	$\dfrac{M_H}{M_\odot} = \dfrac{N_2}{N_1^{1/2}}$	$\dfrac{\rho_j}{\rho_n} = N_1$		
		$\dfrac{\rho_n}{\rho_{crit}} = \dfrac{N_2^2}{N_1}$		

Table 23.2 *Symbols and quantities*

Nucleon gravitational length	$a_g = \dfrac{Gm_n}{c^2} = 1.24 \times 10^{-42}$ cm	
Planck length	$a^* = \left(\dfrac{G\hbar}{c^3}\right)^{1/2} = 1.61 \times 10^{-33}$ cm	
Classical electron radius	$a = \dfrac{e^2}{m_e c^2} = 2.82 \times 10^{-13}$ cm	
Gravitational radius of Sun	$R_S = \dfrac{GM_\odot}{c^2} = 1.5 \times 10^5$ cm	
Hubble length	$L_H = \dfrac{c}{H} = 9.25 \times 10^{27}\, h^{-1}$ cm	
Planck time	$t^* = \left(\dfrac{G\hbar}{c^5}\right)^{1/2} = 5.39 \times 10^{-44}$ sec	
Jiffy time	$j = \dfrac{e^2}{m_e c^3} = 9.40 \times 10^{-24}$ sec	
Hubble time	$t_H = H^{-1} = 3.09 \times 10^{17}\, h^{-1}$ sec	
Electron mass	$m_e = 9.11 \times 10^{-28}$ g	
Nucleon mass	$m_n = 1.66 \times 10^{-24}$ g	
Planck mass	$m^* = \left(\dfrac{\hbar c}{G}\right)^{1/2} = 2.18 \times 10^{-5}$ g	
Solar mass	$M_\odot = 1.99 \times 10^{33}$ g	
Hubble mass	$M_H = \dfrac{4\pi \rho_{univ} L_H^3}{3} = 2 \times 10^{55}\, \Omega h^{-1}$ g	
Planck density	$\rho^* = \dfrac{c^5}{G^2 \hbar} = 5.12 \times 10^{93}$ g cm^{-3}	
Jiffy density	$\rho_j = \dfrac{3}{8\pi G j^2} = 2 \times 10^{52}$ g cm^{-3}	
Nucleon density	$\rho_n = \dfrac{3m_n}{4\pi a^3} = 1.77 \times 10^{13}$ g cm^{-3}	
Density of universe	$\rho_{univ} = \dfrac{3\Omega H^2}{8\pi G} = 1.88 \times 10^{-29}\, \Omega h^2$ g cm^{-3}	

(Table 23.2 identifies unfamiliar symbols). The equality sign "=" in the text and tables of this chapter indicates "approximately equal to."

Unity group

The fine structure constant and the ratio of the nucleon and electron masses are numbers that belong to the unity group. Also in this same group we have $(N_1/N_2)^{\pm 1}$, of which a representative sample is

$$\frac{N_1}{N_2} = \frac{e^4}{GL_H m_n m_e^2 c^2} = 0.4. \qquad [23.18]$$

Other members in the unity group are obtained by interchanging the particle masses m_n and m_e and also using the fine structure constant. John Stewart of Princeton University Observatory in 1931 drew attention to the importance of unity numbers of the N_1/N_2 kind. He proposed that elementary particles have a characteristic mass

$$\text{particle mass} = \left(\frac{e^4}{GL_H c^2}\right)^{1/3}. \qquad [23.19]$$

Stewart suggested that subatomic particle masses are related in a fundamental way to the size of the universe. Variation of particle masses, however, has been disproved.

A unity number of a different kind is found as follows. The expansion of the universe according to accepted theory is governed by gravity, and when numerical coefficients and other numbers such as π are ignored, this is expressed by the approximate relation $G\rho = H^2$, where ρ is the average density of the universe. The Friedmann universes, particularly the Einstein–de Sitter model, are examples. Because the Hubble length L_H is c/H, we have

$$G\rho L_H^2/c^2 = 1. \qquad [23.20]$$

The mass of the observable universe is approximately $M = \rho L_H^3$, and hence, from Equation [23.20],

$$GM/L_H c^2 = 1. \qquad [23.21]$$

This says that the mass of the observable universe has a gravitational length GM/c^2 equal to the Hubble length L_H. Some scientists, particularly Dennis Sciama of Oxford University and Robert Dicke of Princeton University, have proposed that this last result might hold rigorously and is not an approximate relation. Their argument, based on Mach's principle, is that the value of the gravitational constant G is determined by the distribution of matter in the universe. If this interpretation of Mach's principle is correct, then it seems plausible, they argue, that G equals $L_H c^2/M$ at every instant, and therefore it must vary with expansion.

The $N^{1/2}$ group
We have already encountered numbers of the 1, $N^{1/2}$, and N kind. In addition, the radius of a black hole of stellar mass is roughly $R_S = GM_{star}/c^2$, and because M_{star} equals $m_n N_1^{3/2}$, we easily find

$$\frac{L_H}{R_S} = \frac{N_2}{N_1^{1/2}} = N^{1/2}. \qquad [23.22]$$

Also, because the mass of the observable universe is $M_{univ} = L_H c^2/G$,

$$\frac{M_{univ}}{M_{star}} = \frac{N_2}{N_1^{1/2}} = N^{1/2}. \qquad [23.23]$$

Put in words, the observable universe is 10^{20} times the size of a black hole of stellar mass, and the mass of the observable universe is 10^{20} times that of a star.

The N group
The N_1 and N_2 numbers have already been discussed. Of their combinations that yield N, we single out the ratio of the density of a nucleon to the density of the universe. If we continue to neglect small numbers such as π, the density of a nucleon is m_n/a^3, and the density of the universe is M_{univ}/L_H^3, from which we get

$$\frac{\text{density of nucleon}}{\text{density of universe}} = \frac{N_2^2}{N_1} = N. \qquad [23.24]$$

The density of a nucleon is therefore roughly 10^{40} times the density of the universe.

The $N^{3/2}$ group
The mass of a star is $N_1^{3/2}$ times the mass of a nucleon. We can also show from the previous results that the mass of the observable universe, measured in Planck units of mass, is

$$\frac{M_{univ}}{m^*} = N_1^{1/2} N_2 = N^{3/2}, \qquad [23.25]$$

and the size of the observable universe, measured in Planck units of length, is given by

$$\frac{L_H}{a^*} = N_1 N_2^{1/2} = N^{3/2}. \qquad [23.26]$$

The N^2 group
The total number of nucleons in the observable universe is the famous Eddington number $N_E = N^2$. This number is found by dividing the mass of the observable universe by the nucleon mass:

$$N_E = N_1 N_2 = N^2. \qquad [23.27]$$

Eddington attached great importance to this result. Because the Eddington universe is closed he was able to argue that N^2, equal roughly to 10^{80}, is the actual number of nucleons in a finite universe. Generally, the Eddington number amounts to no more than a rough estimate of the nucleons in the observable universe of roughly the Hubble radius. Archimedes' 10^{63} grains of sand contains 10^{80} nucleons, and by a remarkable coincidence Archimedes' number equals Eddington's number.

Other numbers

The above numbers serve as examples of various possible combinations of N_1 and N_2 having physical significance. There seem to be no numbers of the $N_2^{1/2}$ and $N_2^{3/2}$ kind, and no obvious combinations that yield $N^{5/2}$ and $N^{7/2}$. Notice that in the Hubble sphere there are N_1^3 fermi cells, or $(N_2 N_1^{1/2})^3 = N^{9/2} = 10^{180}$ Planck cells. Each Planck cell is a spacetime fluctuation lasting $1/N_1^{1/2}$ jiffies, and in the Hubble 4-volume (the Hubble sphere lasting a Hubble period) there exist a total of $(N_2 N_1^{1/2})^4 = N^6 = 10^{240}$ spacetime fluctuations or virtual quantum black holes. Is 10^{240} the largest possible finite number of physical significance?

SOLVING THE COSMIC CONNECTION

How can we make N_1 and N_2 permanently equal and thereby preserve the magic-number sequence? There are various solutions.

The Eddington solution

Eddington was the first to insist that the proper cosmic yardstick is not the Hubble length L_H, which continually changes in an expanding universe, but $c/\Lambda^{1/2}$, determined by the cosmological constant Λ. Hence N_2 is not equal to L_H/a, but rather,

$$N_2 = \frac{c}{a\Lambda^{1/2}}. \qquad [23.28]$$

The Eddington and Lemaître universes contain the cosmological constant, and therefore, according to this interpretation, they have constant values of N_2. In both universes, N_1 is roughly equal to N_2, thus satisfying Dirac's large-number hypothesis. Eddington's solution of the cosmic-number problem is attractive, although it has not been widely accepted, mainly because of

our uncertainty concerning the reality of the Λ force.

The Dirac solution

Dirac suggested that N_1 increases with time and in this way remains permanently equal to N_2. The only way N_1 can change is by slow variation of one or more of the constants of nature. The cluster hypothesis eliminates all such variations except G-variation, and Dirac assumed that the gravitational constant G decreases slowly in an expanding universe. We have seen that N_1 is proportional to $1/G$, and N_2 is proportional to L_H; in the Dirac universe $N_1 = N_2$ exists because GL_H is constant.

From the unity group of numbers we have (Equation 23.20) the relation $G\rho L_H^2 = c^2$, where ρ is the density of the universe. Because GL_H is constant, ρL_H is also constant in the Dirac universe. But density ρ varies as $1/R^3$, where R is the scale factor, and because the Hubble distance $L_H = c/H$ and the Hubble term equals $\dot{R}/R$, we have that $\dot{R}R^2$ is constant. Hence the scale factor R varies as $t^{1/3}$, where t is the age of the universe. Therefore

age of Dirac universe

$$= \tfrac{1}{3} \times \text{Hubble period}, \qquad [23.29]$$

and for a Hubble period of 15 billion years this gives an age of 5 billion years. Also G is proportional to $1/t$, and the gravitational constant has an infinitely large value at the beginning of expansion.

The Dirac universe is only slightly older than the Solar System. This young age of the universe is a little deceptive, however, because as we go back in time the gravitational constant increases and all self-gravitating systems speed up in their evolution. The distance of the Earth from the Sun changes as $1/G$ and the length of the year changes as $1/G^2$; thus 4 billion years ago the Sun was 1/5 its present distance and the year was 1/25 its present value. Also, the luminosity of main sequence stars varies as G^7; two and a half billion years ago the Sun was $2^7 = 128$ times more luminous than at present, and only half a billion years ago the oceans on Earth were at boiling point owing to the brightness and closeness of the Sun. Variations in G cannot affect atomic and molecular processes or the biochemistry of life. Self-gravitating systems evolve rapidly in the Dirac universe, but life evolves no more rapidly than in a universe of constant G. From the fossil record we know that algae existed on Earth more than 3 billion years ago. But 3 billion years ago in the Dirac universe the Earth's surface was scorched by the intensely bright Sun and life was impossible. This strongly suggests, as first shown by Edward Teller, that if G varies, it cannot vary in the way proposed by Dirac. We have made N_1 permanently equal to N_2 at the cost of eliminating the creatures who notice the numerical coincidence.

Other suggestions have been advanced in which the gravitational constant varies at a much slower rate than suggested by Dirac. These theories involve modifications of general relativity discussed in Chapter 17. Although a slower variation of G in time is not in discord with the existence of life on Earth, radar observations of the motions of planets have made even a very slow variation seem unlikely.

The static solution
In a static finite universe, N_2 measures the size of the universe, which is constant, and hence $N_1 = N_2$ is a permanent characteristic. This solution, however, comes at the cost of explaining the extragalactic recession redshifts with a new theory such as the tired-light theory. But unconventional theories have not gained wide support.

The steady-state solution
In the de Sitter and the continuous creation steady-state universe nothing ever changes on the cosmic scale. The Hubble term H is constant and the Hubble length $L_H = c/H$ never changes. Hence the ratio $L_H/a = N_2$ is unchanging and the equality $N_1 = N_2$ is permanently secure. Almost certainly we do not live in a de Sitter-like universe, and

observations have firmly ruled out the steady-state universe.

The cosmic yardstick solution
The Hubble length – or the size of the observable universe – is a convenient cosmic yardstick (see Figure 23.3). But it has the undesirable property of changing with time. We might therefore ask: Is it possible that we have failed to identify the correct cosmic yardstick when we determine the value of N_2? The cosmological constant Λ serves as a fixed measure of the scale of the universe $(N_2 = c/\Lambda^{1/2})$ and possibly Eddington was right when he argued that Λ is the natural standard that determines the sizes and masses of subatomic particles. We might then use the anthropic principle to show that life exists only in those universes in which the value of Λ is such that N_1 and N_2 are equal.

There is yet another way we might approach this subject. Why choose as characteristic of the size of the universe the

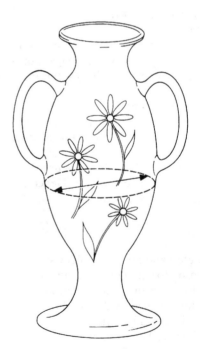

Figure 23.3. What is the diameter of this vase? The question illustrates the difficulty of using the Hubble length as a cosmic yardstick.

Hubble length that has different values at different times? A diameter of a vase does not necessarily inform us of its size. We should perhaps try to rid ourselves of pre-relativity ideas and take a four-dimensional view and ask ourselves: What cosmic scale is invariant?

Expanding universes spanning infinite space and infinite time (not in a steady state) generally do not possess an invariant cosmic scale. By using Eddington's reasoning we could argue that such universes have nothing that determines the sizes and masses of subatomic particles; hence there is no natural value for N_2, and such universes probably cannot contain life. Expanding universes spanning finite space but infinite time also do not have an invariant geometric scale and perhaps also suffer similar limitation.

But expanding universes spanning finite time do contain an invariant spacetime scale. A universe that expands and collapses has an invariant cosmic length equal to its lifetime. In such a universe there exists a fixed fundamental scale that can determine the sizes and masses of subatomic particles. Expanding universes that span finite space and finite time are the most attractive because in all directions of spacetime there exists roughly the same maximum cosmic length. The closed Friedmann universe is an example; it has a lifetime of πR_{max}, and a maximum spatial scale determined by R_{max}, and the natural and only cosmic yardstick is R_{max} (see Figure 23.4). In this case we can write $N_2 = R_{max}/a$; and $N_1 = N_2$ is then a constant relation throughout spacetime.

The Dicke solution

If N_2 changes with time in our expanding universe, we should at least try to explain why the coincidence $N_1 = N_2$ exists at the present time (Figure 23.5). Robert Dicke of Princeton University proposed in 1961 an ingenious explanation that initiated the anthropic principle in its modern form. In the distant past, when the universe was young, the number N_2 was small but nobody was around to notice its discordant value. In the distant future, when the universe is old and the stars are dead, N_2 will be large and again nobody will be around to notice the difference. Dicke argued that N_2 is approximately equal to N_1 during the period of cosmic history when stars shine and organic life exists.

Organic life as we understand it cannot begin until the first generation of stars has evolved and produced elements, such as carbon, oxygen, and nitrogen, that are essential for all biological structures. These and other

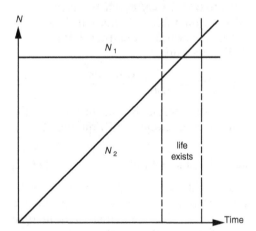

Figure 23.5. N_1 remains constant and N_2 increases in time. Life exists after the first stars have evolved and produced the elements necessary for planetary systems and organic life and ceases when all sunlike stars have died. During this zoicosmic era, N_1 and N_2 are in approximate agreement, as first shown by Robert Dicke.

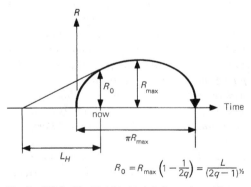

$$R_0 = R_{max}\left(1 - \frac{1}{2q}\right) = \frac{L}{(2q-1)^{1/2}}$$

Figure 23.4. The Hubble length L_H varies with time. In a closed universe, R_{max} is the natural cosmic length because it does not change.

elements are ejected into interstellar space by dying stars and later incorporated into other stars and their planetary systems. Therefore life does not begin until the first generation of stars has died.

Main sequence stars have a typical lifetime given by the equation

$$\text{stellar lifetime} = \frac{1}{1000} \times \frac{M_{\text{star}}c^2}{\text{luminosity}}.$$
[23.30]

The total energy of a star is its mass M_{star} multiplied by c^2, of which only 1 percent is released by the nuclear reactions that convert hydrogen into helium and heavier elements; also only about 10 percent of the hydrogen is converted. The released energy, about one thousandth of $M_{\text{star}}c^2$, divided by the luminosity gives the lifetime of the star.

Light, as it streams away from the surface of a star, pushes on the atomic particles in the star's atmosphere. The maximum possible luminosity that a star can have is when the radiation pressure, pushing outward at the surface, equals the gravitational pull inward. A star of greater luminosity blows itself away. This maximum luminosity, known as the Eddington luminosity, is equal to $GM_{\text{star}}m_{\text{n}}c/a^2$, where a is the classical electron radius. A normal star has a luminosity typically one-thousandth of the Eddington value:

$$\text{luminosity} = \frac{1}{1000} \times \frac{GM_{\text{star}}m_{\text{n}}c}{a^2}.$$
[23.31]

When this result is substituted in the stellar lifetime expression (Equation 23.30), we find

$$\text{stellar lifetime} = N_1 \text{ jiffies},$$
[23.32]

where 1 jiffy $= a/c = 10^{-23}$ seconds. Dicke argued that life cannot begin until the age of the universe is at least equal to the lifetime of the first generation of stars:

age of universe = stellar lifetime.

The present age of the universe is 10 or more times greater, but a factor of 10 is unimportant and lies well within the spread of the unity and the N groups of numbers. Because

the age of the universe is N_2 jiffies, we have

$$N_1 = N_2,$$
[23.33]

as a necessary condition for the existence of organic life. Eventually, N_2 will become much larger than N_1, but all stars will then have died and life will have perished. The cosmic numbers N_1 and N_2 are in approximate agreement while intelligent life exists.

This is Dicke's anthropic argument. A universe is fit for inhabitation by organic life while the values of N_1 and N_2 are in approximate agreement. This argument explains not only why these two numbers are equal, but also helps us to understand why they are so large. We can exist only in a universe that is generously endowed with the eons of time necessary for stellar and biological evolution. Hence N_1 must be large, and the existence of life requires a universe billions of years old, spanning billions of light years of space, and containing billions of galaxies.

REFLECTIONS

1 *"All the systems of units which have hitherto been employed, including the so called absolute c.g.s. [centimeter-gram-second] system, owe their origin to the coincidences of accidental circumstances, inasmuch as the choice of the units lying at the base of every system has been made, not according to general points of view that would necessarily retain their importance for all places and all times, but essentially with reference to the special needs of our terrestrial civilization.... These quantities [the Planck units of length, time, and mass] therefore must be found always the same when measured by the most widely differing intelligences according to the most widely differing methods"* (Max Planck, The Theory of Heat Radiation, *1913, based on lectures at the University of Berlin in 1906–7).*
2 *"From the intrinsic evidence of his creation, the Great Architect of the Universe now begins to appear as a pure mathematician"* (James Jeans, The Mysterious Universe, *1930). Mathematicians are thought*

ANY PHYSICS TOMORROW?

Figure 23.6. Cover of *Physics Today*, January 1949. (With permission of the American Institute of Physics.)

to be cold and remote individuals. God as a creator, architect, or designer are common attributions, but God as a mathematician caused a considerable stir in theological circles in the 1930s.

3 *"Finally, I repeat my personal conviction that the cosmical constant Λ is connected with the relation between electromagnetic and gravitational units, and that sooner or later a theory giving an accurate value of Λ will be forthcoming"* (Arthur Eddington, *"The expansion of the universe"*).

• *"I believe there are 15,747,724,136,275, 002,577,605,653,961,181,555,468,044,717, 914,527,116,709,366,231,425,076,185,631, 031,296 protons in the universe, and the same number of electrons."* (Arthur Eddington, The Philosophy of Physical Science).

4 *"It is proposed that all the very large dimensionless numbers which can be constructed from the important natural constants of cosmology and atomic theory are con-*

nected by simple mathematical relations involving coefficients of the order of magnitude unity. The main consequences of this assumption are investigated and it is found that a satisfactory theory of cosmology can be built up from it"* (Paul A. M. Dirac, "A new basis for cosmology").

5 *"There are several amusing relationships between the different scales. For example, the size of a planet is the geometric mean of the size of the universe and the size of an atom; the mass of man is the geometric mean of the mass of a planet and the mass of a proton. Such relationships, as well as the basic dependence on α and α_G [where $1/\alpha_G = N_1$] from which they derive, might be regarded as coincidences if one did not appreciate that they can be deduced from known physical theory"* (Bernard Carr and Martin Rees, "The anthropic principle and the structure of the physical world").

6 *"There are some, King Gelon, who think that the number of sand is infinite in multitude; and I mean by the sand not only that which exists about Syracuse and the rest of Sicily but also that which is found in every region whether inhabited or uninhabited. Again there are some who, without regarding it as infinite, yet think that no number has been named which is great enough to exceed its multitude"* (Archimedes [287–212 BC], introductory words of The Sand-Reckoner). In the third century BC the Greeks had a numerical system with which they were able to count moderately easily up to a myriad, where a myriad is 10 000. This system could be extended to a myriad myriads or a 100 million, but became awkward and cumbersome for larger numbers. Archimedes introduced a new system, called the "naming of numbers," which greatly extended the range of numbers beyond a myriad myriads. In this new system, the number expressed as p units of the qth order and the rth period is given by

$$\text{number} = pM^{2[(q-1)+(r-1)M^2]}, \qquad [23.34]$$

where $M = 10^4$, and p, q, and r are integers in the range of 1 to M^2. The largest number in Archimedes' system is M^2 units of the M^2

order and M^2 period, and hence

largest number $= M^{2M^4}$

$$= 10^{8 \times 10^{16}}, \qquad [23.35]$$

which is 1 followed by 80 000 trillion zeros. The large numbers in modern cosmology are 10 followed by 40 or 80 zeros, and it is clear that Archimedes' system is adequate for almost all purposes.

Archimedes then discussed the size of the universe. "Now you are aware that 'universe' is the name given by most astronomers to the sphere whose centre is the centre of the Earth and whose radius is equal to the straight line between the centre of the Sun and the centre of the Earth. This is the common account, as you have heard from astronomers. But Aristarchus of Samos brought out a book consisting of some hypotheses, in which the premises lead to the result that the universe is many times greater than that now so called. His hypotheses are that the fixed stars and the Sun remain unmoved, that the Earth revolves about the Sun in the circumference of a circle, the Sun lying in the middle of the orbit, and that the sphere of the fixed stars, situated about the same centre as the Sun, is so great...." There is some ambiguity concerning Aristarchus's estimate of the size of the universe, and Archimedes assumed that he meant that the size of the Sun's orbit is the geometric mean between the size of the universe and the size of the Earth:

$$\frac{\text{size of universe}}{\text{size of Sun's orbit}} = \frac{\text{size of Sun's orbit}}{\text{size of Earth}}.$$
$$[23.36]$$

Expressed in modern units, the Aristarchean heliocentric universe had a radius of 1 light year. Archimedes found that this universe, if filled completely with sand, would contain a number of grains of sand equal to one thousand myriad units of the eighth order and first period:

number of grains of sand

$$= 10^7 \times 10^{8(8-1)} = 10^{63}. \qquad [23.37]$$

The modern universe, which is 10^{10} times larger than the Aristarchean universe, could therefore contain 10^{30} times as many grains of sand. But the modern universe has a density 10^{-30} times that of the density of sand, and by coincidence the Aristarchean universe, when filled with sand, contains matter of the same mass as the modern observable universe.

Archimedes estimated that a poppy seed has a diameter of one-tenth of a finger breadth and a volume equal to that of one myriad grains of sand. If we suppose that a finger breadth is 1 centimetre, and the density of sand is 3 grams per cubic centimetre, then each grain of sand is rather small, of mass 2×10^{-7} grams. On dividing this result by the mass of a nucleon, we see that each grain of sand contains 10^{17} nucleons. Thus the Aristarchean universe, filled with sand, contains $10^{63+17} = 10^{80}$ nucleons. Archimedes knew nothing about nucleons, and yet by pure chance his number of grains of sand is equal to the Eddington number of nucleons.

"I conceive that these things, King Gelon, will appear incredible to the great majority of people who have not studied mathematics, but that to those who are conversant therewith and have given thought to the question of the distances and sizes of the Earth, the Sun, and Moon and the whole universe, the proof will carry conviction. And it was for this reason that I thought the subject would be not inappropriate for your consideration" (closing words of The Sand-Reckoner).

Most of Archimedes' work is lost. He perished in the sack of Syracuse. According to Plutarch (AD 46–127): "For it chanced that he was by himself, working out some problem with the aid of a diagram, and having fixed his thoughts and his eyes upon the matter of his study, he was not aware of the incursion of the Romans, or of the capture of the city. Suddenly a soldier came upon him and ordered him to go with him to Marcellus. This Archimedes refused to do until he had worked out his problem, whereupon the soldier flew into a passion and drew his sword and slew him."

7 *A googol stands for the number 10^{100} and was made famous by Edward Kasner and invented by his nine-year-old-son. Black holes of galactic mass have an evaporation lifetime of approximately one googol years. A googolplex is 10^{googol} (one followed by a googol of zeros). Archimedes' largest number (1 followed by 8×10^{16} zeros) greatly exceeds a googol but is very much less than a googolplex. A time period of a googolplex years, figuratively speaking, is still only a blink of an eye in a universe that expands forever. A similar remark applies to a $10^{googolplex}$ (one followed by a googolplex of zeros and obviously a googoogolplex).*

8 *A characteristic energy of a particle in a star is the energy required by a nucleon to escape from the surface to infinity. If R is the radius of a star of mass M_{star}, this energy is roughly $GM_{star}m_n/R$. In the deep interior of the star, protons and electrons rush around, each with energy of this magnitude, and repeatedly have close encounters with one another. If d is the distance of closest approach, the kinetic energy is equal to e^2/d, and hence*

$$\frac{e^2}{d} = \frac{GM_{star}m_n}{R}. \qquad [23.38]$$

The number of nucleons in the star is $N_{star} = M_{star}/m_n$, and from Equation [23.38] we find

$$N_{star} = \frac{R}{d} \times \frac{e^2}{Gm_n^2}.$$

Now let l be the mean separating distance between neighboring particles, such that $N_{star} = (R/l)^3$, and we find

$$N_{star} = \left(\frac{l}{d} \frac{e^2}{Gm_n^2} \right)^{3/2}$$

$$= 10^{54} \times \left(\frac{l}{d} \right)^{3/2}. \qquad [23.39]$$

When l and d are almost equal, the particles are squeezed together, as in a solid or liquid, and have little freedom to move. A body of 10^{54} nucleons has a mass 1/1000 that of the Sun, and is therefore a large planet such as Jupiter consisting mainly of light elements (hydrogen and helium) in a condensed-matter state.

Stars are luminous, hot, and gaseous, not cold like Jupiter. In their deep interiors, the radiation, which slowly diffuses out, is very important in the determination of stellar structure. We find that the wavelength of the radiation in their central regions is approximately equal to the separating distance l between particles. Or, expressed differently, the number of photons in a star is roughly equal to the number of electrons. Photons have an average energy $\hbar c/l$, equal to the average energy e^2/d of the particles. This gives the approximate relation $l/d = 1/\alpha$ that defines a luminous main sequence star, where $\alpha = e^2/\hbar c$ is the fine structure constant. Neighboring particles deep inside a hot star are thus separated by an average distance 137 times their distance of closest approach. The number of nucleons (equal to the number of photons) in a star is therefore

$$N_{star} = \left(\frac{\hbar c}{Gm_n^2} \right)^{3/2} = N_1^{3/2}, \qquad [23.40]$$

and $N_{star} = 10^{57}$. The mass is $M_{star} = m_n N_{star}$, and is of order 10^{33} grams. This result was first derived by Pascual Jordan in 1939 by different arguments.

9 *"I certainly wouldn't give up attempts to make the anthropic principle unnecessary by finding a theoretical basis for the values of all the constants. It's worth trying, and we have to assume that we shall succeed, otherwise we shall surely fail" (Steven Weinberg, BBC broadcast, 1984). John Gribbin and Martin Rees (Cosmic Coincidences) remark: "So perhaps it is best, if they are to retain their scientific motivation, that theoretical physicists should not take the strong anthropic principle, the idea that the universe is tailor-made for man, too seriously. If there is a unique 'theory of everything,' then there is certainly a sense in which the laws of physics could not have been otherwise. We would then have to accept it as genuinely coincidental, or even providential, that the constants determined by high-energy physics happen to lie in the narrowly restricted range that allows complexity and* *consciousness to evolve in the low-energy world we inhabit."*

PROJECTS

1 In his article "Any physics tomorrow?" (1949), George Gamow wrote: "If and when all the laws governing physical phenomena are finally discovered and all the empirical constants occurring in these laws are finally expressed through the four independent basic constants, we will be able to say that physical science has reached its end, that no excitement is left in further explorations, and that all that remains to a physicist is either tedious work on minor details of the self-educational study and adoration of the magnificence of the completed system. At that stage physical science will enter from the epoch of Columbus and Magellan into the epoch of the *National Geographic Magazine....*" What do you think?

2 Do you believe that one day we will have a physical theory that determines the values of the fundamental constants and shows why they necessarily have their observed values?

3 Discuss the anthropic principle and the cosmic numbers. Why is life unlikely when N_1 and N_2 are grossly unequal?

FURTHER READING

Barrow, J. D. and Tipler, F. J. *The Anthropic Cosmological Principle.* Clarendon Press, Oxford, 1986.

Davies, P. C. W. *The Accidental Universe.* Cambridge University Press, Cambridge, 1982.

Davies, P. C. W. *The Cosmic Blueprint.* Heinemann, London, 1987.

Gribbin, J. and Rees, M. *Cosmic Coincidences: Dark Matter, Mankind, and Anthropic Cosmology.* Bantam Books, New York, 1989. (Particularly good.)

SOURCES

Carr, B. J. and Rees, M. J. "The anthropic principle and the structure of the physical world." *Nature* 278, 605 (April 12, 1979).

Carter, B. "Large number coincidences and the anthropic principle in cosmology," in *Confrontation of Cosmological Theories with Observational Data.* Editor M. S. Longair. Reidel, Dordrecht, Netherlands, 1974.

Dicke, R. H. "Dirac's cosmology and Mach's principle." *Nature* 192, 440 (November 4, 1961).

Dirac, P. A. M. "The cosmological constants." *Nature* 139, 323 (February 20, 1937).

Dirac, P. A. M. "A new basis for cosmology." *Proceedings of the Royal Society*, A165, 199 (1938).

Dyson, F. J. "The fundamental constants and their time variation," in *Aspects of Quantum Theory.* Editors A. Salam and E. P. Wigner. Cambridge University Press, Cambridge, 1972.

Eddington, A. S. "The expansion of the universe." *Monthly Notices of the Royal Astronomical Society* 91, 412 (1931).

Eddington, A. S. *The Philosophy of Physical Science.* Cambridge University Press, Cambridge, 1939.

Gamow, G. "Any physics tomorrow?" *Physics Today* (January 1949).

Harrison, E. R. "The cosmic numbers." *Physics Today* (December 1972).

Heath, T. L. *The Works of Archimedes.* Cambridge University Press, Cambridge, 1897. Reprint: Dover Publications, New York, 1953. Contains *The Sand-Reckoner.*

Jeans, J. *The Mysterious Universe.* Cambridge University Press, Cambridge, 1930. Reprint: Macmillan, London, 1937.

Jordan, P. "Formation of the stars and development of the universe." *Nature* 164, 637 (October 15, 1949).

Planck, M. *The Theory of Heat Radiation.* Dover Publications, New York, 1959.

Stewart, J. Q. "Nebular red shift and universal constants." *Physical Review* 38, 2071 (1931).

Teller, E. "On the change of physical constants." *Physical Review* 73, 80 (1948).

Whittaker, E. *From Euclid to Eddington: A Study of Conceptions of the External World.* Cambridge University Press, Cambridge, 1958. Reprint: Dover, New York.

24 DARKNESS AT NIGHT

Deep into the darkness peering, long I stood there,
wondering, fearing,
Doubting, dreaming dreams no mortal ever
dared to dream before.
Edgar Allan Poe, The Raven (1845)

THE GREAT RIDDLE
An inferno of stars

There is a simple and important experiment in cosmology that almost everybody can perform. It consists of gazing at the night sky and noting its state of darkness. When we ask, why is the sky dark at night? (Figure 24.1) the natural response is the Sun is shining on the other side of the Earth and starlight is weaker than sunlight. It takes an unusual mind to realize that the relative weakness of starlight is of cosmological significance, and such a person was the astronomer Johannes Kepler, imperial mathematician to the emperor of the Holy Roman Empire.

In a forest (Figure 24.2), a line of sight in any horizontal direction must eventually intercept a tree trunk, and the distant view consists of a background of trees. Similarly, on looking away from Earth at night, we see a "forest" of stars (Figure 24.3). If the stars stretch away endlessly, a line of sight must eventually intercept the surface of a distant star. The distant view of the universe should consist of a continuous background of bright stars with no separating dark gaps.

We know that the universe consists of clusters of stars, galaxies, and clusters of galaxies, and has astronomical structure more complex than the uniform distribution of stars often supposed in previous centuries. The universe expands and may consist of non-Euclidean geometry. Even so, we shall continue to think of the riddle of darkness in its original form, and later show that

developments in cosmology have made little difference to the riddle.

In a universe of infinite extent, populated everywhere with bright stars, the entire sky should be covered by stars with no separating dark gaps. Hence, when all stars are bright like the Sun, the entire sky at every

Figure 24.1. A line of sight in an endless universe populated with luminous stars. If the universe has no end in space, and the stars stretch away endlessly, every line of sight must eventually intercept the surface of a star. Why then are there dark gaps between the stars? (*Darkness at Night*, with permission of Harvard University Press.)

Figure 24.2. A forest of trees. The trees stretch away forming a continuous background and there are no gaps. (*Darkness at Night*, with permission of Harvard University Press.)

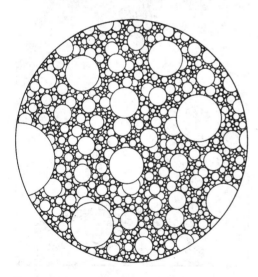

Figure 24.3. A forest of stars. The stars stretch away forming a continuous background and there are no gaps. Otto von Guericke, mayor of Magdeburg in the seventeenth century, may have been the first to use the forest analogy. (With permission, E. R. Harrison, *American Journal of Physics* 45, 120, 1977.)

point should blaze with a brilliance equal to the Sun's disk. The sky is 180 000 times larger than the Sun's disk, and starlight falling on Earth should be 180 000 times more intense than sunlight. In the midst of this inferno of intense light, life would cease in seconds, the atmosphere and oceans boil away in minutes, and the Earth turn to vapor in days. Fortunately, the sky at night is dark. What then is wrong with the forest analogy?.

The riddle begins

Kepler was not the first to discover the riddle. The epic poem *The Nature of the Universe* by Lucretius in 55 BC, discovered in 1417, had awakened the idea of an infinite universe. This new and exciting idea influenced the thoughts of many, including Thomas Digges, Giordano Bruno, and William Gilbert.

In 1576, only 33 years after the death of Copernicus, Thomas Digges took the first step. He dismantled the Aristotelian sphere of fixed stars and dispersed the stars – infinite in number – throughout infinite space (Chapter 8): "This orb of starres fixed infinitely up," wrote Digges in the *Perfit Description of the Cælestiall Orbes*, "extendeth hit self in altitude sphericallye." Digges main contribution was to point out that although the stars are infinitely numerous, yet only a finite number can be seen because "the greatest part rest by reason of their wonderfull distance invisible unto us." With these words he originated the riddle of the dark night sky. He was the first to realize that dark gaps between the stars call for an explanation. His solution, however, which seemed at the time obvious and acceptable, is incorrect.

Kepler terrified by infinity

Kepler believed in the Copernican heliocentric theory and was excited by Galileo's astronomical observations with the newly discovered telescopes. In company with Copernicus and in keeping with Aristotelian cosmology, he believed that the starry universe is finite and bounded. Digges had torn away the outer edge and transformed the sphere of fixed stars into an infinite universe of stars. But Kepler was terrified by such an idea and vehemently rejected it.

In 1610, Kepler received a copy of Galileo's small book *The Starry Messenger*. After only a few days he dashed off a long letter to Galileo, and a month later this letter was published as a short book entitled *Conversation with the Starry Messenger*. In this book can be found Kepler's most potent argument against the idea of an infinite universe. "You do not hesitate to declare," he said, "that there are visible over 10 000 stars. The more there are, and the more crowded they are, the stronger becomes my argument against the infinity of the universe." For if the universe stretched away endlessly, with stars like the Sun swarming everywhere, then the whole "celestial vault would be as luminous as the sun." It was clear that "this world of ours does not belong to an undifferentiated swarm of countless others."

According to Kepler, the universe is not like an endless forest; instead, it is like a finite clump of trees in which we looked out between the tree trunks to a dark enclosing wall. He did not use the forest analogy. Nonetheless, he realized that in an endless universe the stars would collectively outshine the Sun and flood the heavens with light far more intense than we actually observe.

The infinite stellar universe encountered in Kepler's *Conversation* its most devastating criticism. The choice was clear: either a cosmic edge and a dark night sky, or no cosmic edge and a blazing sky. Astronomers who followed disliked the idea of a cosmic edge, and over the centuries have sought for the solution of the dark night-sky riddle.

TWO INTERPRETATIONS

There are two alternative interpretations of the darkness of the night sky, as shown in Figures 24.4 and 24.5. Either the sky is covered with overlapping stars or is not covered with stars.

According to the first interpretation (*interpretation* A), the sky is actually covered by

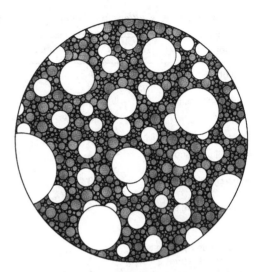

Figure 24.4. Interpretation A: the covered sky. Stars indeed cover the entire sky, but the most distant stars for some reason cannot be seen. The riddle becomes: Why is the starlight missing? (*Darkness at Night*, with permission of Harvard University Press.)

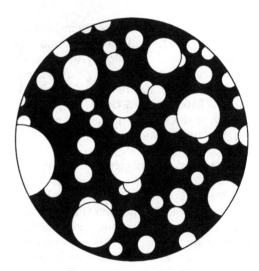

Figure 24.5. Interpretation B: the uncovered sky. The dark gaps are indeed empty of stars. The riddle becomes: Why are stars missing? (From *Darkness at Night*, with permission of Harvard University Press.)

stars with no gaps in between, and the riddle in effect asks: what has happened to the missing starlight that never reaches us on Earth? Most studies of the riddle have assumed that this is the correct interpretation, as

Table 24.1 *Proposed solutions of the riddle of night-sky darkness*

Solution	Interpretation	Author	Date
Starlight is too feeble	A	Digges	1576
Dark cosmic wall	B	Kepler	1610
Stoic finite cosmos	B	Guericke	1672
		Clerke	1890
		Shapley	1917
Geometric effect	A	Halley	1720
Interstellar obscuration	A	Chéseaux	1744
		Olbers	1823
Hierarchical structure	B	Herschel	1848
		Proctor	1870
Cosmic age too short	B	Poe	1848
		Mädler	1861
		Kelvin	1901
Obscuration by dark stars	A	Fournier d'Albe	1907
Static steady state[†]	A	MacMillan	1922
Redshift	A	Bondi	1955
Fill-up time too large	B	Harrison	1964

A: The covered-sky missing-starlight interpretation.
B: The uncovered-sky missing-stars interpretation.
[†] Discussed in Chapter 18.

shown in Table 24.1, in which some, not all, of the proposed solutions of the riddle of cosmic darkness are listed. (All presently known scientific solutions are given in *Darkness at Night: A Riddle of the Universe* by the author.) Thomas Digges chose interpretation A when he supposed that the light from very distant stars was too feeble to be seen by the eye.

In the second interpretation (*interpretation B*), the sky is not covered with stars and the dark gaps are real. The riddle in effect asks: what has happened to the missing stars? Kepler chose this interpretation when he argued that the dark gaps between stars was evidence of a finite universe bounded by a dark enclosing wall.

HALLEY'S SHELLS
Newton's infinite universe
Kepler's *Conversation with the Starry Messenger* was widely read and many no doubt were teased by the conflict of the infinite universe and the darkness of the night sky. The conflict became acute, tantamount to a paradox, with the rise of the Cartesian and Newtonian infinite world systems. These world systems of Euclidean geometry provided a basis for calculation. Nearby stars are few in number, yet each gives a large contribution of starlight; distant stars are numerous, but each gives only a small contribution of starlight.

No doubt Newton was aware of the riddle, but was more concerned with a similar problem in the theory of gravity. Both the gravitational pull and the light from a star decrease as the inverse square of the star's distance. The number of stars at any assumed distance, however, increases as the square of the distance, thus compensating for the loss of light and gravitational pull of each star. We should be pulled in different directions by large gravitational forces and also receive large quantities of light from all directions. Newton resolved the gravity problem, sometimes referred to as the "gravity paradox," by assuming that the infinite universe is homogeneous (the same at all places), and hence equal forces pull from all directions and cancel out one another's pull (Chapter 16). Starlight from different directions, however, cannot cancel in this way, but instead adds up, and reason is brought into direct conflict with observation.

Edmund Halley

The first person to attempt to discuss in a mathematical fashion the problem of the dark night sky was Edmund Halley. In 1720 he published two short papers on the infinite universe. In the first, he wrote: "Another Argument I have heard urged, that if the number of Fixt Stars were more than finite, the whole superficies of their apparent Sphere would be luminous."

Halley introduced the idea of concentric shells of equal thickness (Figure 24.6). We occupy a position in space and add up the contributions of light received from all stars. To do this we construct a series of imaginary concentric spheres, of increasing radius, with the center at the point in space

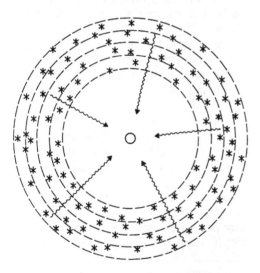

Figure 24.6. Imaginary concentric spheres of increasing radius (with the observer at the center) form a series of shells of constant thickness. When the shells are large, the number of stars in each shell increases as the square of the radius of the shell. The light received by the observer at the center from each star in a shell decreases as the square of the radius of the shell. These two effects – the number of stars increasing and their light decreasing – compensate each other and all shells contribute equal amounts of light.

that we occupy. Let the radius of each successive sphere increase by a fixed amount, as shown, so that the spaces between the spherical surfaces form shells of equal thickness. The volume of each shell (surface area × thickness) increases as the square of the radius. If we assume that stars are uniformly distributed, their number in each shell increases also as the square of the radius.

But the light received by us at the center from any single star is inversely proportional to the square of its distance. Consequently, when the number of stars in a shell is multiplied by the amount of light received from each of these stars, we obtain a quantity of light that is fixed and independent of the radius of the shell. All shells contribute equal quantities of light. The total amount of light reaching us is the quantity of light from one shell multiplied by the number of shells.

A universe of stars stretching away endlessly contains an infinite number of shells. Each shell contributes a finite quantity of light and hence, according to this argument, at our chosen point in space there should exist an infinite amount of light. Our chosen point can be anywhere and therefore at all points in space light is infinitely intense. This conclusion is of course absurd and the error in the argument can be spotted almost immediately. In a forest we do not see all the trees of the forest. Tree trunks obstruct our view of more distant trees, and a line of sight extends to a background consisting of a fusion of trees that lies not very far away. Stars also have a finite size and tend to obstruct our view of more distant stars, and therefore a line of sight extends to a continuous background of stars that lies at finite distance (Figure 24.7).

Halley's solution to the problem was that "the more remote Stars, and those far short of the remotest, vanish even in the nicest Telescopes, by reason of their extreme minuteness; so that, tho' it were true, that some such Stars are in such a place, yet their Beams, aided by any help yet known, are not sufficient to move our Sense; after

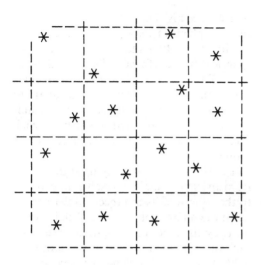

Figure 24.7. A forest in which each tree occupies an average area A. If each tree has a width w the lookout limit in the forest is A/w. Let each star occupy an average volume V. If the cross-section of each star is s, the lookout limit in the universe is V/s. (Edward Hamson with permission, *American Journal of Physics*.)

the same manner as a small Telescopical fixt Star is by no means perceivable to the naked Eye." Halley tried to explain the darkness of the night sky by a geometric argument that the remotest stars are invisible because of their "extreme minuteness." Digges almost 150 years previously had similarly argued that the light from distant stars is too feeble to be detected by the eye.

The light emitted by a single atom is far too feeble to be detected by the eye. Yet the collective light from many atoms, as in a candle flame, is easily seen. The arguments made by Digges and Halley cannot be true because both neglected the collective light of many stars. In the forest analogy, their solution assumes that distant trees are invisible and we see only foreground trees.

Jean-Philippe Loys de Chéseaux

A few years later, in 1744, the gifted Swiss astronomer-mathematician Jean-Philippe Loys de Chéseaux turned his attention to the problem of the darkness of the night sky. He first showed that the whole sky is 180 000 times larger than the apparent disk

of the Sun, and if all stars are sunlike, then the starlight falling on the Earth's surface from a background of stars would be 180 000 times more intense than sunlight. Chéseaux calculated the distance to the background of stars and found (when expressed in modern units) a value of 3000 trillion light years. He then showed that the number of visible stars covering the sky was 10 billion trillion trillion trillion (1 followed by 46 zeros). More than 150 years were to pass before these results were improved by Lord Kelvin. Chéseaux attributed the darkness of the night sky to interstellar absorption. The corresponding analogy is that of a foggy forest in which distant trees are obscured from view by fog and only foreground trees are seen clearly.

Heinrich Olbers

Almost eighty years later, in 1823, Heinrich Olbers, a renowned physician and astronomer in Bremen, presented similar arguments, but without Chéseaux's insightful calculations. He also said that starlight is absorbed while traveling between the stars. The solution proposed by Chéseaux and Olbers adopts interpretation A. Where is the missing starlight? The answer according to Chéseaux and Olbers is that it has been absorbed by interstellar matter. This solution fails, as shown by John Herschel in 1848, because the absorbing matter heats up and soon emits as much radiation as it absorbs.

Olbers' principal contribution was the line-of-sight argument. In an endless universe populated with stars of finite size that stretch away without limit, every line of sight must intercept the surface of a star. Hence there can be no dark gaps, and if all stars are like the Sun, every part of the sky should blaze with light as bright as the Sun's disk. This way of explaining the problem of darkness has the advantage that it does not require the assumption that stars are distributed uniformly, as with Halley's shells. If all the trees in a forest were clustered into clumps, and the clumps clustered into woods, and the woods clustered into larger woods, we would still be surrounded by a continuous background of trees. Similarly, if all stars form clusters, which form galaxies, which form clusters of galaxies, the sky would still be covered by a continuous background of stars. (An exception exists in certain hierarchical arrangements, as we shall see.)

The puzzling darkness of the night sky, bringing theory into direct conflict with observation, is now known as "Olbers' paradox." The intriguing riddle has been discussed in recent decades by numerous authors, notably Hermann Bondi, most of whom thought the riddle was discovered by Olbers, and were unaware of the earlier work by Digges, Kepler, Halley, Chéseaux, and others. Also they were unaware that Lord Kelvin at the beginning of the twentieth century had given the definitive answer to the riddle.

BRIGHT-SKY UNIVERSES

We continue to adopt for the time being the Newtonian picture of an infinite universe, populated more or less uniformly with sun-like stars, in an attempt to understand the riddle of cosmic darkness without the complications of galaxies, expansion, and other more recent discoveries.

The distance to the background

Let A represent an area in a forest that contains on the average one tree, and let w be the typical width of a tree trunk at eye level. The distance seen in the forest is the *lookout limit*, determined by the relation

$$\text{lookout limit} = \frac{A}{w}, \qquad [24.1]$$

and the total number of trees visible from any point is therefore found to be

$$\text{number of visible trees} = \frac{\pi A}{w^2}. \qquad [24.2]$$

As an example, if the average distance between trees is 5 meters, the area A occupied by a single tree is 25 square meters. If the diameter w of a tree trunk at eye level is typically 0.5 meter, the lookout limit in the

forest is 50 meters, and the number of visible trees is 314.

Stars, like tree trunks, have a certain size and tend to block our view of more distant stars. The lookout limit in the universe, as in a forest, is easily calculated. Let V be the volume of space that contains on the average one star, and let s be the typical cross-sectional area of a star (equal to π times the square of the star's radius). The expression for the distance seen in the universe is

$$\text{lookout limit} = \frac{V}{s}, \qquad [24.3]$$

and the total number of stars visible from any point in space is given by the relation

$$\text{number of stars visible} = \frac{4\pi V^2}{3s^3}. \qquad [24.4]$$

Although the number of stars is infinite in a universe of infinite extent, only a finite number can be seen from any single point, and these visible stars cover the sky and prevent us from seeing the rest that lie beyond.

We now realize that the light reaching us comes only from stars within the lookout limit, and all light from stars further away is intercepted by the nearer stars and never reaches us. We must therefore add up only the contributions of light from successive shells out to a distance equal to the lookout limit. Hence the light reaching us is of finite and not infinite intensity.

A bright-sky universe

We have found that a line of sight in every direction terminates at the surface of a star, and the sky is covered with stars with no empty spaces between them. Of course, if considerable absorption occurs, as Chéseaux and Olbers said, then most lines of sight terminate on dust grains and other absorbing particles. But because the absorbing particles heat up and then emit as much radiation as they absorb, the sky will be as bright as with no absorbing particles. It is as if we were enclosed within a spherical surface, of radius equal to the lookout limit, that has a temperature equal to the surface temperature of the stars. According to this argument, we live inside a furnace that has incandescent walls. Wherever we stand in space we are surrounded by an unbroken wall of incandescent stars. Everything is bathed in a flood of intense light, and the temperature everywhere is the same as that at the surfaces of stars. Because sunlike stars have a surface temperature of about 6000 kelvin, this is the temperature everywhere in space according to this argument.

A seemingly logical argument, leading to a conclusion in contradiction with reality, constitutes the riddle of cosmic darkness.

Absorption of starlight in interstellar space cannot avert a bright sky. It is as ineffective as putting an absorbing gas into a furnace in the hope that it will keep the objects inside cool. The gas quickly heats up to the same temperature as the furnace and nothing is gained. Whatever is put in a bright-sky universe to shield us from the blinding rays of zillions of stars rapidly heats up and becomes part of the inferno.

The cosmic-edge solution

There is available an obvious resolution of the riddle. All we need do is restore the cosmic edge of antiquity and place it at a distance much less than the lookout limit. If the radius of a finite and bounded universe is less than the lookout limit, it has insufficient stars to cover the sky and the sky is dark at night. This was Kepler's solution of the problem. But a spatially bounded universe, as in the Aristotelian world system, is nowadays unacceptable because space cannot terminate abruptly at a wall-like cosmic edge.

Wall-like edges went out of fashion in the High and Late Middle Ages, but the popularity of cliff-like edges waxed and waned and finally collapsed in the early decades of the twentieth century. A one-island universe, or a Stoic cosmos floating in an infinite void of empty space, was frequently proposed as a solution of the riddle: We stand, in effect, inside a clump of trees and look out through the trees to a treeless plain beyond. Agnes Clerke, astronomer

and historian, echoed a widespread view when she wrote in 1890 in *The System of the Stars* that the entire stellar content of the universe composed one "all-embracing" scheme – the Milky Way – "all-embracing, that is to say, so far as our capacities of knowledge extend. With the infinite possibilities beyond, science has no concern." She rejected the infinite starry universe because it would create a bright night sky: "for from innumerable stars a limitless sum-total of radiations should be derived, by which darkness would be banished from our skies; and the 'intense inane,' glowing with the mingled beams of suns individually indistinguishable, would bewilder our feeble senses with its monotonous splendour. This laying bare, so to speak, of the empyrean would be the simple and certain result of the continuance of sidereal objects comparable with that prevailing in our neighborhood."

Harlow Shapley, a famous American astronomer, as recently as 1917 wrote: "Either the extent of the star-populated space is finite or 'the heavens would be a blazing glory of light' … since the heavens are not a blazing glory, and since space absorption is of little moment throughout the distances concerned in our galactic system, it follows that the defined stellar system is finite." But not a shred of evidence now remains in support the one-island or Stoic universe.

Spherical space

Traditional arguments have led us to the conclusion that an unbounded and infinite universe has a bright sky. What about an unbounded universe of finite size?

Many have suggested that the night sky is dark because the universe is finite in size and has unbounded spherical space. It seems not unreasonable that the sky would be dark in such a universe when the distance to the antipode is less than the lookout limit. The universe would contain too few stars to cover the sky. This is similar to Kepler's solution but avoids the objectionable cosmic edge. Unfortunately, it fails to solve the riddle.

The surface of a globe is unbounded, yet of finite area, and can be taken as representing a spatially finite and unbounded universe. We saw in previous chapters that in a universe of positive curvature, having spherical geometry, that by traveling in any direction we eventually arrive back at our starting point. Rays of light circumnavigate such a universe and continue to go around and around until absorbed.

We imagine that the spherical surface of a planet is covered with trees. At any place on the surface we stand in a finite but endless forest. We must suppose in this analogy that light rays travel parallel to the surface of the planet at eye level. We look out into an endless forest of trees. If the lookout limit is more than half the circumference of the planet (greater than the distance to the antipode), we see all trees, also the backs as well as the fronts of some trees by looking in opposite directions. When the lookout limit is more than the circumference, we see the fronts and the backs of all trees. Because the forest is endless, we must always see a continuous background of trees at the lookout limit. When, for example, the lookout limit is 100 times the circumference of the planet, we see each tree repeated 100 times.

The same thing happens in a universe of finite and unbounded space. We see a continuous background of stars formed by repeated circumnavigations of light rays. The sky is covered with stars just as in an infinite universe. A finite but unbounded universe therefore fails to solve the paradox.

Hierarchical solutions
Solutions of the riddle using hierarchical astronomy (Chapter 4) were introduced by John Herschel and Richard Proctor in the nineteenth century. This approach was later adopted by Fournier d'Albe in England and Carl Charlier in Sweden in the early twentieth century. Kant's idea of a hierarchy of clusters of increasing size was adopted by Charlier, whose work received wide publicity. The argument goes as follows. In a hierarchy of stars, clusters of progressively

larger size have progressively lower average density, and the lookout limit therefore progressively increases. By arranging that the density of the clusters decreases sufficiently rapidly with increasing size, the lookout limit can be made indefinitely large. In this way the sky at night becomes dark. Fournier d'Albe put forward an alternative hierarchy in which the visible universe is only one of a series of universes of increasing size, arranged in such a way that the solar systems in one universe are the atoms in the next larger universe. He showed that with such an arrangement the sky at night is dark.

A hierarchical resolution, however, is not very satisfactory. On all scales the universe is anisotropic, contrary to optical observations and the isotropy of the cosmic background radiation. Furthermore, a hierarchical resolution of the paradox is quite unnecessary, as we shall see.

THE PARADOX RESOLVED
A more realistic universe
So far in this discussion we have thought in terms of an infinite, static, Newtonian universe populated uniformly with stars that shine forever. We must now ask whether the riddle of cosmic darkness is valid in a more realistic universe.

We have seen that absorption by dust and gas is of no help. The gathering together of different kinds of stars into galaxies, and of galaxies into clusters, is also of no help. Ordinary clustering, as in a hierarchy of only a few levels of clustering, merely alters the value of the lookout limit while the sky continues to blaze with light. When trees in a forest are clumped together into groups, and groups into woods, our line of sight in all horizontal directions still terminates at tree trunks. The kind of hierarchy needed to solve the riddle consists of an unlimited number of levels of clustering. We have seen that a finite but unbounded space also fails to solve the riddle, because each line of sight stretches around and around the universe until it eventually intercepts the surface of a star. Expansion of the universe has not yet been considered. But first, we must turn to other matters.

Energy considerations

The conclusion that the night sky should be as bright as the Sun derives from pre-twentieth century science. That something is seriously amiss is shown by the following energy argument. The average density of matter of all kinds in the visible universe is about equal to the mass of 1 hydrogen atom per cubic meter. We imagine that all matter in the universe is annihilated and converted directly into thermal radiation. Mass and energy are equivalent and calculation shows that the thermal radiation everywhere has a temperature of 20 kelvin. This is very much less than the surface temperature of stars. Thus the fearsome furnace of Olbers' paradox has at most a temperature of only 20 kelvin and can never be 6000 kelvin. We are forced to conclude that the universe does not contain enough energy to create a bright sky.

If our universe contained 10 billion times more matter than it does now in the form of stars, and if all this matter were annihilated and converted totally into thermal radiation, the temperature everywhere would equal that at the surface of the Sun. But stars do not convert all their mass into radiation with 100 percent efficiency. Sunlike stars burn hydrogen into helium and during their entire luminous lifetimes convert only approximately 0.1 percent of their mass into starlight. Instead of increasing the number of stars by a factor of 10^{10} (10 billion), we need a factor of 10^{13} (10 trillion) to create a bright-sky universe. On energy grounds, bright skies can in principle exist, but only in universes at least 10 trillion times more dense than our own. But can stars exist in such an inferno of light? Almost certainly not.

By using energy considerations we have shown that in our universe a bright sky cannot exist. The traditional arguments that deduce a bright night sky are therefore wrong, and the old riddle collapses in the face of modern science. Let us try to track down what is actually wrong with the traditional argument.

Lookback limit is greater than the luminous lifetime

We continue to suppose for convenience that all stars are similar to the Sun. With an average cosmic density of 1 hydrogen atom per cubic meter (or 10^{-24} grams per cubic meter), and with this matter all lumped into stars, we find that the lookout limit in the universe is 10^{23} (100 billion trillion) light years. This is larger than Chéseaux's result because he knew nothing about galaxies and assumed that stars everywhere are distributed as in the neighborhood of the Sun. Most of the starlight contributing to a bright sky comes from remote regions at immense distances. The number of visible stars covering the entire sky has the enormous value 10^{60}, or 1 trillion trillion trillion trillion trillion. These exorbitant numbers of the lookout limit and the number of visible stars provide the essential clue we need to solve the riddle.

Light travels at finite speed, and when we look out in space we also look back in time (Figure 24.8). A lookout limit in space of 10^{23} light years corresponds to a *lookback limit* in time of 10^{23} years. Hence the most distant stars that contribute to a bright sky

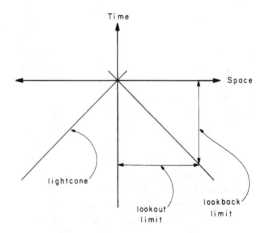

Figure 24.8. Spacetime diagram showing the lookout limit in space and the lookback limit in time.

were shining 10^{23} years ago when they emitted the light now seen. In a homogeneous universe, distant stars are similar to nearby stars, and because nearby stars are still shining, distant stars must also still be shining. Therefore distant stars have been shining continuously for at least 100 billion trillion years. But this is impossible and would require that in its lifetime each star radiates into space energy having a mass 10 billion times the mass of the star.

A rough-and-ready luminous lifetime for sunlike stars is 10^{10} years. This typical luminous lifetime is short compared with the lookback limit. A bright sky was obtained by adding up the contributions of light from successive shells of stars out to the lookout limit in space. We now realize that beyond a distance of 10^{10} light years we are looking back to a time before the stars became luminous (see Figure 24.9). The stars at distances greater than 10^{10} light years are now shining, the same as nearby stars, but their light has not yet reached us.

All shells of visible stars contribute equal quantities of light, and the total amount of light reaching us from stars out to a distance of 10^{10} light years is therefore only $10^{10}/10^{23} = 10^{-13}$ (1 ten-trillionth) of the amount required to create a bright sky. The number of stars visible is hence only 10^{21} and not the 10^{60} that is needed to cover the entire sky.

This then is why the sky is dark at night in the infinite, static Newtonian universe. It is dark because the luminous lifetime of stars is very much less than the lookback limit in time. The riddle is solved more or less in the historical context in which it was discovered. According to the forest analogy, we stand in a clump of trees, ringed with successive zones of progressively younger trees, and we look out beyond the farthest seedlings to a treeless plain. A hierarchical distribution of clustered stars is unnecessary, for it merely increases the lookout limit, which is already quite large enough to ensure a dark night sky.

Suppose that luminous stars are not all created at the same moment. For example,

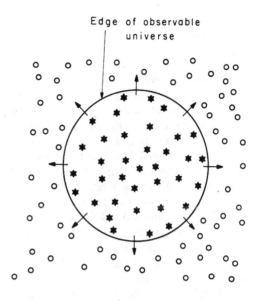

Figure 24.9. Why the sky at night is dark in a static universe. We look out and see luminous stars surrounding us out to a maximum distance determined by the luminous lifetime of stars – roughly 10 billion light years. At greater distances we look back to a time before the stars were luminous. Although the stars are stationary, the outer boundary of the sphere of visible stars in a static universe expands at the speed of light. If we wait long enough, the stars around us will begin to die out. Thereafter we will be surrounded by an expanding dark sphere of dead stars, and beyond the dark sphere will lie an expanding shell of luminous stars that has a constant thickness of 10 billion light years, and beyond this shell of visible stars will lie a dark universe of stars yet to be seen.

imagine that of all the uniformly distributed stars only 10 percent are luminous at any one moment. The lookout limit to luminous stars is now 10 times greater and the corresponding lookback limit in time is 10^{24} years. When these stars begin to die after 10^{10} years, the next 10 percent become luminous, and then the next, and so on, thus giving an overall luminous lifetime for all stars of 10^{11} years. The total amount of light is still only $10^{11}/10^{24} = 10^{-13}$ of that required for a bright sky. Switching stars on sequentially over many generations fails to increase the brightness of the night sky.

Bright-sky universes

It is not difficult to design hypothetical universes with bright skies. All that is needed is a lookback limit less than the luminous lifetime of stars. If the luminous lifetime remains unchanged, we must reduce the lookout limit by increasing the number of stars. With a lookback limit of 10^{10} years, equal to the luminous lifetime, we must, according to Equation [24.3], reduce by a factor 10^{-13} the average volume V occupied by each star. Equation [24.4] then tells us that in a bright sky universe the sky is covered with 10^{34} stars.

A bright sky can be created by abandoning homogeneity. Let us arrange that all stars are luminous on an observer's backward lightcone. That is, in a static universe, homogeneity in the world picture but not in the world map. The more distant a star, the earlier it starts shining. Starlight converges on the observer from all directions and creates in a region about the observer an incandescently bright sky. In this way it is possible to form a bright sky when the lookout limit is larger than the luminous lifetime. The observer, now roasted in the glare of focused starlight, is not in the least privileged by occupying the cosmic center.

"THE GOLDEN WALLS OF THE UNIVERSE"

For more than four hundred years astronomers have proposed various solutions of the dark night-sky riddle. With few exceptions their solutions were off target. Not until Lord Kelvin (William Thomson, 1824–1907) at the beginning of the twentieth century at age 77 cast his eagle eye on the riddle had anyone performed the correct calculations showing why the sky is dark at night. Oddly enough, the first person to come close to suggesting the correct solution was the American poet and writer Edgar Allan Poe (1809–1849).

In his imaginative essay *Eureka*, published in 1848 two years before he died at age 40, Poe wrote of the "golden walls of the universe" formed from a "myriad of shining bodies that mere number has appeared to blend into unity." Moreover, "Were the succession of stars endless, then the background of the sky would present us a uniform luminosity, like that displayed by the Galaxy – since there could be absolutely no point in all that background at which would not exist a star. The only mode, therefore, in which, under such a state of affairs, we could comprehend the voids which our telescopes find in innumerable directions, would be by supposing the distance of the invisible background so immense that no ray from it has yet been able to reach us at all." Thus Poe's solution adopts interpretation B, and the sky is not covered with stars because the lookout limit (the "invisible background") is too far away for light to have yet reached us. In other words, the lookback limit is greater than the time that stars have been shining. The finite speed of light and the finite age of luminous stars have come together for the first time specifically to solve the riddle of cosmic darkness. Poe, however, did not believe in an infinite universe, and he added the words, "That this *may* be so, who shall venture to deny? I maintain, simply, that we have not even the shadow of a reason for believing that it *is* so." He discarded the correct solution in favor of a finite cosmos of stars. Even so, his prescient vision on this and other subjects discussed in *Eureka* is remarkable.

The astronomer Johann von Mädler also had the right idea. In 1861, in his *Popular Astronomy*, he wrote: "The velocity of light is finite, a *finite* time has passed from the beginning of Creation until our day, and we, therefore can only perceive the heavenly bodies out to the distance that light has traveled during that finite amount of time." The sky at night is dark, he said, because light from very distant stars has not yet reached us, and the absorption of starlight proposed by Olbers is unnecessary.

THE CELEBRATED HYPOTHESIS

The definitive solution was derived by Lord Kelvin in 1901 in an article "On ether and gravitational matter through infinite space." First he showed that the stars in the Galaxy are insufficient in number to cover the entire sky. He found that the fraction of the sky covered with stellar disks was less than one trillionth, and said, "This exceedingly small value will help us to test an old and celebrated hypothesis that if we could see far enough into space the whole sky would be seen occupied with discs of stars all perhaps of the same brightens as our sun."

Kelvin showed that the fraction of the sky covered by stars (assumed to be Sunlike) is related to the relative brightness of the sky by

sky-cover fraction

$$= \frac{\text{brightness of starlit sky}}{\text{brightness of Sun's disk}}. \quad [24.5]$$

It is remarkable that in the four hundred year history of the riddle, Kelvin stands out as the only person to draw this simple conclusion and show in a quantitative manner the connection between the sky-cover fraction and the brightness of the starlit sky.

Kelvin showed that even if the stars stretched away "in a great sphere," and were not confined to a small sphere (the Galaxy), the sky would still be dark. The supposition of uniform density, he said, is arbitrary and "we ought in the greater sphere to assume the density much smaller than in the smaller sphere." By assuming that stars everywhere are distributed as in the solar neighborhood, he calculated that the lookout limit – the distance to a continuous background of stars – was 3000 trillion light years, in fact similar to Chéseaux's previous result of which he was unaware. Hence, argued Kelvin, light from the most distant visible stars must travel for 3000 trillion years. But this travel time was much greater than the luminous lifetime of stars.

Kelvin had devoted considerable thought to the source of energy radiated by the Sun, and had come to believe that the source is the slow gravitational contraction of the Sun. This was before the discovery of nuclear energy and meant, as Kelvin had shown, that the Sun and similar stars have ages between 10 and 100 million years. Kelvin wrote, "if all the stars through our vast sphere commenced shining at the same time ... at no one instant would light be reaching the earth from more than an excessively small proportion of all the stars." His calculations showed

sky-cover fraction

$$= \frac{\text{luminous lifetime of stars}}{\text{lookback limit}}, \quad [24.6]$$

and therefore we find from Equations [24.5] and [24.6]

$$\frac{\text{brightness of starlit sky}}{\text{brightness of Sun's disk}}$$

$$= \frac{\text{luminous lifetime of stars}}{\text{lookback limit}}. \quad [24.7]$$

According to the data available to Kelvin, the fraction of the sky covered by stars out to a distance of 100 million light years was 3×10^{-8}. With more modern data (typical luminous lifetime of 10^{10} years, lookback limit of 10^{23} years) we obtain 10^{-13} (1 ten-trillionth) for the sky-cover fraction, and the average brightness of the night sky at any point due to starlight is 1 ten-trillionth of the brightness at any point of the Sun's disk.

Perhaps the most remarkable aspect of "Olbers' paradox" is not its discovery several hundred years before Olbers, but that Kelvin's definitive work was totally overlooked and never once referred to in numerous subsequent discussions by scientists and historians.

EXPANSION AND DARKNESS
Expansion fails to solve the riddle

A popular belief in recent decades holds that "Olbers' paradox" is solved by the expansion of the universe. Starlight from distant regions of the universe is weakened by the cosmological redshift, and it was said that this redshift explains why the sky is dark at night. According to this argument, the act of gazing at the night sky and noting its state of darkness provides immediate proof that the universe is expanding. If the universe were static, it has been said, the sky would blaze with light and we could not exist. The expansion redshift keeps the universe cool and habitable. The redshift solution adopts interpretation A: it assumes that the sky is actually covered with stars and concludes that they are invisible because of their large redshift.

Little thought is now needed to realize that something is seriously wrong with the redshift solution. When John Herschel and Richard Proctor proposed that hierarchy solved the riddle, they accepted the argument that every line of sight should intercept a star and the sky should blaze with starlight. Similarly, advocates of the redshift solution also accepted the traditional argument. But we have seen that the static Newtonian universe has a dark night sky and the universe does not contain enough energy to make a bright night sky. Hierarchy and redshift cannot solve a riddle that is already solved; they merely make the night sky darker than in the uniform and static universe. The redshift solution, if it were correct, would mean that the sky is actually covered with an enormous number of stars that we cannot see because their light is weakened by expansion. Yet we have found that it is impossible for the sky to be covered by stars because the luminous lifetime of stars is much less than the lookback limit.

The cosmic box

We turn now for a moment to a more powerful method of solving the riddle (see Figure 24.10). This method follows from the discussion in Chapter 17. Each star can be thought of as occupying an average volume V of the universe. We imagine that an average star is surrounded by perfectly reflecting walls forming a box of volume V. Light emitted

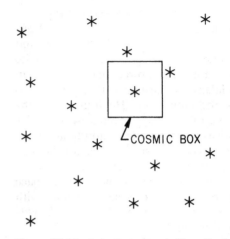

Figure 24.10. A star is surrounded by reflecting walls that form a cosmic box of the same volume as the average volume V occupied by an average star. The conditions for filling this cosmic box with radiation from a single average star are the same as the conditions for filling the universe with radiation from all stars. The box is full when the star receives as much radiation as it emits.

by this star, instead of streaming away into endless space, bounces from wall to wall and remains trapped inside the box. It is intuitively obvious that the radiation inside the box is the same as the radiation outside. The star in the box retains its radiation in its vicinity, and the stars outside the box mingle their radiation; otherwise there is no difference.

In a bright-sky universe, as visualized in Olbers' paradox, the stars emit as much radiation as they receive. Radiation fills space up to the level where it equals that at the surface of stars. The temperature at every point in space is the same as at the surfaces of stars. The time required to fill all space with radiation from all stars up to this level equals the time required by an average star to fill the box with its own radiation. This *fill-up time* is easily calculated. A ray of light in the box travels to and fro between the reflecting walls and is finally intercepted and absorbed by the star itself. The average distance traveled by a ray of light is the lookout limit. The corresponding average time a ray travels is the lookback limit and this is the fill-up time of the box.

fill-up time = lookback limit.   [24.8]

This is true for the box and also the whole universe. After that, the star absorbs as much radiation as it emits, and the box is filled with radiation in equilibrium with the star.

The average star must shine for 10^{23} years to fill the box with radiation. But a star is capable of shining for only a small fraction of this time, typically 10^{10} years. The radiation level remains low because the luminous lifetime of stars is much less than the fill-up time. The same applies to the universe. Stars do not contain enough energy to fill the universe with radiation up to the level at the surface of stars.

An expanding cosmic box

We consider now an expanding universe and imagine an average star inside a cosmic box that expands with the universe. The box has a comoving volume V, and V is an average volume occupied by a star. Light rays in the box are repeatedly reflected by the receding walls and receive at each reflection a small Fizeau–Doppler redshift. These repeated small redshifts are the same as the cosmological expansion redshift, as shown in Chapter 17, and the radiation inside the box at any instant is exactly the same as the radiation outside the box in the expanding universe.

If the sky at night is dark because the universe is expanding, then also the radiation in the box is feeble because the box is expanding. Take two boxes, one expanding and the other static, and let both contain identical stars that have been shining for identical periods of time (Figure 24.11). Calculation shows that radiation in the expanding box is not much weaker than radiation in the static box at the instant when they have equal volumes, as shown in Figure 24.11. Expansion reduces the intensity of radiation in a decelerating universe to a level generally not less than 50 percent of that in a static universe. The effect of expansion cannot be the cause of a dark night sky because in a

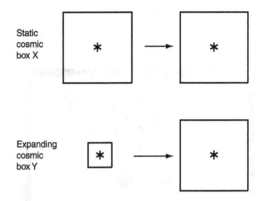

Figure 24.11. The redshift solution, once popular, adopts interpretation A and assumes that the sky is covered with stars and that most stars cannot be seen because of their extreme redshift. Imagine two identical stars that shine for the same period of time and are enclosed in separate cosmic boxes X and Y, as shown on the left side of the diagram. Box X is static. Box Y, which is initially small, expands and finally has the same volume as X, as shown on the right of the diagram. Thus on the right side we have two boxes of identical volume containing identical stars that have been shining for identical periods of time. Calculation shows that for continuous decelerating expansion the radiation level in Y is generally never less than half that in X, no matter how small Y is initially. This demonstrates that the effect of expansion on starlight cannot explain the darkness of the sky at night.

bright-sky universe the light must be reduced to a level one ten-trillionth of that in the static universe.

Starlight is too feeble to fill the dark universe

Darkness of the night sky is due not to absorption of starlight, not to hierarchical clustering of stars, not to the finiteness of the universe, not to expansion of the universe, and not to many other proposed causes. The explanation is quite simple and can be stated in various equivalent ways. Because of the finite luminous age of stars and the finite speed of light, the number of visible stars is too few to cover the entire sky; most stars needed to cover the sky are so far away that their light has not reached us; the light-travel time from the most

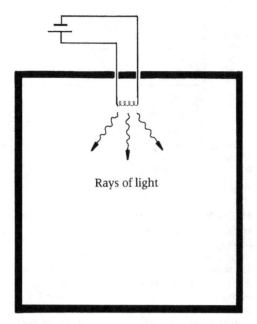

Rays of light

Figure 24.12. A large box of perfectly reflecting walls and containing a small source of radiation such as a flashlight filament connected to a flashlight cell. The filament is the only object in the box that absorbs radiation. Emitted rays travel to and fro between the walls, traveling on the average a distance equal to the lookout limit, and are eventually absorbed by the filament.

distant stars is greater than their luminous lifetime; the luminous lifetime of stars is shorter than the time needed to fill space with radiation to the temperature it has at the surfaces of stars; and stars do not contain enough energy to fill space with radiation of this temperature. Why is the sky dark at night? Because starlight is too feeble to fill the dark universe.

The cosmic background radiation
Out of doors at night we look up at the night sky. Between the stars we look out immense distances in space and far back in time before the formation of the galaxies and their firstborn stars. Our sight extends to the limit of the visible universe at the frontier of the big bang. In all directions we look back to the creation of the universe and see the big bang covering the sky. Twentieth-century cosmology has solved the old riddle of darkness at night and shown that the night sky is covered not by stars but by the big bang mercifully veiled from view by expansion of the universe.

REFLECTIONS
1 *Otto von Guericke (1602–1686), mayor of Magdeburg, constructed the first air pump and performed various experiments with large evacuated vessels. He also may have originated the forest analogy. In* New Magdeburg Experiments on Void Space *(1672), he wrote: "Although many stars cannot be seen we should not form the opinion that they do not exist. A forest does not end where individual trees cannot be seen any farther." He believed in a finite starry cosmos immersed in an infinite void (as did Newton in his first years at Cambridge), and thus explained the darkness of the night sky.*

2 *The historian Stanley Jaki, in his book* The Paradox of Olbers' Paradox, *writes: "This constitutes, in effect, the most paradoxical aspect of Olbers' paradox. In this sense, Olbers' paradox is not Olbers', nor is it Halley's. It is the paradox of the unscientific habits of scientific workers and writers. For it is no small matter that some scientists can be shockingly careless when it comes to the presentation of a detail of scientific history." Jaki writes that almost all commentators on Olbers' paradox did not bother to read the original works but merely repeated one another. We might also add that of the many writers who have discussed the paradox very few performed any calculations to support their views.*

3 *In* The New Star *(1606), Johannes Kepler (1571–1630) expressed his thoughts concerning the infinite universe: "This very cogitation carries with it I don't know what secret, hidden horror; indeed one finds oneself wandering in this immensity to which are denied limits and centre and therefore also all determinate places." (Alexandre Koyré,* From the Closed World to the Infinite Universe, *1958.)*

"Suppose that we took 1000 fixed stars, none of them larger than 1' (yet the majority

of them in the catalogues are larger). If these were all merged in a single round surface they would equal (and even surpass) the diameter of the sun. If the little disks of 10 000 stars are fused into one, how much will their visible light exceed that of the disk of the sun? If this is true, and if they are suns having the same nature as our sun, why do not these suns collectively outdistance our sun in brilliance? ... Hence it is quite clear that ... this world of ours does not belong to an undifferentiated swarm of countless others" (Kepler, 1610). See Edward Rosen, Kepler's Conversation with Galileo's Sidereal Messenger (1965). In his haste, Kepler mistranslated Galileo's Starry Message into Starry Messenger, and according to Rosen, "He thereby unintentionally supplied a powerful weapon to the deadliest enemies of Galileo, whom he would never have deliberately injured in the slightest way."

Koyré in From the Closed World to the Infinite Universe quotes Kepler: "The explanation of this fact is easy: whereas the planets shine by the reflected light of the sun, the fixed stars shine by their own, like the sun. But if so, are they really suns as Bruno has asserted? By no means. The very number of the new stars discovered by Galileo proves that the fixed stars, generally speaking, are much smaller than the sun, and that there is in the whole world not a single one which in dimensions, as well as luminosity, can be equal to our sun. Indeed, if our sun were not incommensurably brighter than the fixed stars, or these so much less bright than it, the celestial vault would be as luminous as the sun."

4 Diffraction (the deflection of light passing through an aperture or by a sharp edge) increases with wavelength, thus accounting for the prismatic colors in haloes about brightly lit points and edges. The amount of diffraction decreases as the size of the aperture increases. The human eye cannot resolve objects much smaller than about 1 minute of arc, and telescopes generally cannot resolve angles much smaller than about 1 second of arc. (A golf ball at a distance of 9 kilometers subtends an angle of 1 second.) Nearby stars subtend geometric angles of roughly 1 millisecond, and stars farther away subtend smaller angles. Because of diffraction, stars visible in the telescope subtend apparent angles of about 1 second. Hence we need only 7×10^{11} stars to cover the whole sky with diffraction-limited stellar disks, and not the immense number calculated by Chéseaux. But, on reflection, it should become clear that diffraction is irrelevant. Olbers' paradox is based on geometric angles subtended by stars, and the deflection and diffraction of starlight cannot affect the average intensity of starlight.

5 Edmund Halley (1656–1742), a friend of Newton, is best known for Halley's Comet that he observed in 1682 and predicted would return in 1758. It has since returned again in 1835, 1910, and 1986. Halley became the Astronomer Royal in 1720, and in that year he published two short papers relating to the darkness of the night sky entitled "Of the infinity of the sphere of fix'd stars" and "Of the number, order, and light of the fix'd stars." His explanation of the darkness of the night sky is not altogether clear and seems to combine Digges's argument (the light from individual stars is too feeble to be seen by the eye) and a geometric argument (light decreases faster than the inverse square of distance). An account of his work is found in the Journal Book of the Royal Society: "The other objection against an infinite number of stars is from the small quantity of light which they all give, whereas were there an infinite number it would seem to be much more. To this Dr Halley replies that light is not divisible in infinitum and that consequently when stars are at very remote distances their light diminishes in a greater proportion than according to the common rule, and at last becomes entirely insensible even to the largest telescopes" (see Michael Hoskin, "Dark skies and fixed stars").

6 Jean Philippe Loys de Chéseaux (1718–1751) in 1744 wrote: "The enormous difference which we find between this conclusion and actual experience shows either that the sphere of fixed stars is not infinite but

actually incomparably smaller than the finite extension I have supposed, or that the force [flux] of light decreases faster than the inverse square of distance. This latter supposition is quite plausible; it requires only that starry space is filled with a fluid capable of intercepting light very slightly" (Treatise on Comets). *According to Chéseaux, "Even if this fluid were 330,000,000,000,000,000 times more transparent or thinner than water," it would reduce starlight to "1 part in 430,000,000 the amount of light from the sun."*

7   *Heinrich Olbers (1758–1840) solved the riddle with these words: "Because the celestial vault has not at all points the brightness of the Sun, must we reject the infinity of the stellar system? Must we restrict this system of stars to one small portion of limitless space? Not at all. In our inference drawn from the hypothesis that an infinite number of fixed stars exists, we have assumed that space throughout the whole universe is absolutely transparent, and that light, consisting of parallel rays, remains unimpaired as it propagates great distances from luminous bodies. This absolute transparency of space, however, is not only undemonstrated but also highly improbable"* (Heinrich Olbers, On the Transparency of Space, *1823 translated in* E. R. Harrison, Darkness at Night). *In this paper, Olbers refers to Halley but not to Chéseaux. Olbers possessed among his books a copy of Chéseaux's* Treatise on Comets *in which he had at some time made marginal notes, presumably some years previously. The evidence, such as the considerable difference in their treatments, indicates that Olbers probably had forgotten Chéseaux's work and the fact that Chéseaux had proposed interstellar absorption as a solution. Olbers' unique contribution was the line-of-sight argument, which culminated in the realization by Kelvin that the sky-cover fraction and the radiation level are related.*

8   *"Light, it is true, is easily disposed of. Once absorbed, it is extinct forever, and will trouble us no more. But with radiant heat the case is otherwise. This, though absorbed, remains still effective in heating the absorbing medium, which must either increase in temperature, the process increasing,* ad infinitum, *or, in its turn becoming radiant, give out from every point at every instant as much heat as it receives"* (John Herschel, Edinburgh Review, *1848). It was by no means clear in Herschel's day that heat and light are different though interchangeable forms of energy. Bondi, who reawakened interest in the riddle of the dark night sky, wrote in* Cosmology *(1960): "What happens to the energy absorbed by the gas? It clearly must heat the gas until it reaches such a temperature that it radiates as much as it receives, and hence it will not reduce the average density of radiation." This argument assumes that the absorbing medium heats up in a time less than the luminous lifetime of stars. Edward Fournier d'Albe, a scientist-engineer who transmitted television pictures from London in the early 1920s, suggested in* Two New Worlds *(1907) several solutions. Concerning the absorption solution he wrote: "If a hot star is something altogether exceptional – a freak happening in a billion times – then the average temperature of an infinite universe will be quite comfortable." His billion was what Americans call a trillion. A trillion dark absorbing stars to every bright emitting star creates a dark sky and acts as an absorbing medium that will never heat up.*

9   *The possibility of a hierarchical solution was suggested by John Herschel (*Edinburgh Review, *1848), who presumably had in mind a system similar to what Kant had imagined in the previous century: "Nothing is easier than to imagine modes of systematic arrangement of the stars in space," wrote Herschel. In a letter in 1869 to Richard Proctor (a writer of popular books on astronomy), Herschel explained: "One of the arguments advanced in favor of the spatial extinction of light was that, if there is not such extinction, the whole heavens ought to be one blaze of solar light – admitting the universe to be infinite, because it was contended that there then could be no direction in space in which the visual ray would not encounter a star (i.e., a sun). This argument is fallacious, for it is* *easy to imagine a constitution of a universe literally infinite which would allow of any amount of such directions of penetration as not to encounter a star. Granting that it consists of systems subdivided according to the law that every higher order of bodies in it should be immensely more distant from the centre than those of the next inferior order – this would happen."*

In Other Worlds Than Ours *(1870), Proctor wrote: "it is worth noticing that ... if we adopt the belief in an infinite succession of orders of systems; that is, first satellite-systems, then planetary-systems, then star-systems, then systems of star-systems, then systems of systems of star-systems, and so on to infinity; ... we no longer have as a conclusion that the whole heavens should be lighted up with stellar (that is solar) splendor; even though, in this view of the subject, there are in reality an infinite number of stars, just as in the view according to which the sidereal system extends without interruption to infinity."*

Carl Charlier, a Swedish astronomer, strongly believed in a hierarchical universe. In "How an infinite world may be built up" (1922), he derived the conditions for a dark night sky. By making the lookout limit always larger than the size of a cluster, we can find the conditions for a dark sky. Let us call systems of stars (galaxies) the first level, systems of systems of stars (clusters of galaxies) the second level, and so on; and let us consider systems of the nth level, where n is a number in the range from 1 to infinity. A system of the nth level has radius R_n and contains N_n systems of the next lower level of radius R_{n-1}. This nth system becomes transparent when R_n is greater than $\sqrt{N_n}$ times R_{n-1}:

$$R_n > \sqrt{N_n} R_{n-1}. \qquad [24.9]$$

Let systems at all levels be transparent in this manner. Thus if a cluster contains 1600 galaxies, its radius must exceed 40 times the radius of a galaxy; similarly, if a supercluster contains 900 clusters, its radius must exceed 30 times the radius of a cluster; and so on, for clusters of higher and higher order. The total number of stars in a cluster of the nth

level is

$$N(n) = N_1 N_2 N_3 \cdots N_n. \qquad [24.10]$$

Because, from Equation [24.9], $N_1 < (R_1/R_0)^2$, $N_2 < (R_2/R_1)^2$, $N_3 < (R_3/R_2)^2$, and so on, we find from Equation [24.10] that the total number of stars N_n in the nth system is

$$N(n) < (R_n/R_1)^2. \qquad [24.11]$$

The average density of stars in the nth system is therefore

$$\text{number density} = \frac{3N(n)}{4\pi R_n^3}$$
$$< \frac{3}{4\pi R_1^2 R_n}, \qquad [24.12]$$

from Equation [24.11]. Thus the average density decreases as $1/R_n$. We see that as R_n goes to infinity (as it must in an infinite hierarchical universe) the average density of stars goes to zero. Although the universe contains an infinite number of stars, their average number per unit volume goes to zero in a universe having an infinitely large lookout distance.

The most unrealistic assumption in this solution of the riddle of darkness is that stars are reservoirs of unlimited energy and shine for a time greater than R_n/c, which is eternity when R_n goes to infinity.

A hierarchy of stars, as proposed by Charlier, is a fractal arrangement (see Chapter 4). As before, let $N(n)$ be the number of stars occupying a volume of radius R_n. The fractal dimension D is defined by $N(n) = (R_n/R_0)^D$, where R_0 in this case is the radius of a typical star. A general theorem states that fractal arrangements occupying a space of d dimensions need not intersect when the sum of their fractal dimensions is less than d. A ray of light, or a line of sight, has a fractal dimension of 1. Thus a line of sight in a fractal forest of $d = 2$ need not intersect a tree when $D + 1$ is less than 2, or D is less than 1. A line of sight in a fractal universe of $d = 3$ need not intersect a star, and the sky is dark, when $D + 1$ is less than 3, or D is *less than 2, in agreement with Equation [24.11].*

10 *"Then, indeed amid unfathomable abysses will be glaring unimaginable suns. But all this will be merely a climactic magnificence foreboding the great End. Of this End the new genesis described can be but a partial postponement. While undergoing consolidation, the clusters themselves, with a speed prodigiously accumulative, have been rushing towards their own general centre – and now, with a million-fold electric velocity, commensurate only with their material grandeur and their spiritual passion for oneness, majestic remnants of the tribe of Stars flash, at length, into a common embrace. The inevitable catastrophe is at hand. . . . Are we not, indeed, more than justified in entertaining a belief – let us say, rather, in indulging a hope – that the processes we have here ventured to contemplate will be renewed forever, and forever, and forever; a novel Universe swelling into existence, then subsiding into nothingness, at every throb of the Heart Divine"* (Edgar Allan Poe, Eureka). *This remarkable essay anticipated the expansion, collapse, and possible oscillation of the universe.*

11 *Lord Kelvin (William Thomson 1824–1907) solved the riddle of darkness in a paper "On ether and gravitational matter through infinite space" (1901) that for unknown reasons was later omitted from all bibliographies of Kelvin's works and was therefore totally overlooked in subsequent discussions of the riddle. Kelvin often said "paradoxes have no place in science." He took the view that paradoxes are in ourselves and not the external world. It is ironic that he was the first quantitatively to solve with utmost lucidity a riddle that later, when his work lay forgotten, became a confusion of unsubstantiated assertions known as Olbers' paradox.*

Kelvin argued as follows. The volume of a shell of radius q and thickness dq is $4\pi q^2\, dq$, and if there are n stars per unit volume, the number of stars in the shell is $4\pi nq^2\, dq$. Let each star have a cross-section of s. The stars cover an area $4\pi nsq^2\, dq$ in the shell. If we

divide this area by the surface area $4\pi q^2$ of the shell, we obtain the sky-cover fraction of the shell: $d\alpha = ns\, dq$, or $d\alpha = dq/\lambda$, where $\lambda = 1/ns$ is the lookout limit. Thus the sky-cover fraction in a sphere of stars of radius r^* is found by integrating $d\alpha = dq/\lambda$ from $q = 0$ to $q = r^*$:

$$\alpha = \frac{r^*}{\lambda}. \qquad [24.13]$$

If $t^* = r^*/c$, and $\tau = \lambda/c$ is the lookback limit, we obtain

$$\alpha = \frac{t^*}{\tau}. \qquad [24.14]$$

If t^* is the luminous lifetime of stars, the sky-cover fraction equals the luminous lifetime divided by the lookback limit, which is Kelvin's result (Equation 24.6).

Let each star have luminosity L. The stars in the shell radiate energy at the rate $4\pi nLq^2\, dq$, and on dividing by $4\pi q^2 c$, where c is the speed of light, we obtain $nL\, dq/c$ as the contribution du from the shell to the radiation energy density at the center. The radiation density at the surface of a star is $u^* = L/sc$, and hence $du = u^*\, dq/\lambda$. On integrating as before, we find $u = u^* r^*/\lambda$, or

$$\alpha = \frac{u}{u^*}. \qquad [24.15]$$

This also is Kelvin's result (Equation 24.5). From Equations [24.14] and [24.15] we obtain

$$\frac{u}{u^*} = \frac{t^*}{\tau},$$

as shown in the text (Equation 24.7).

12 *The general solution of Olbers' paradox recognizes that*

(i) light propagates at finite speed;
(ii) either the universe is of finite age or stars have a finite luminous lifetime.

The night sky is dark when

(iii) the average separation of stars is such that the lookback limit is greater than

the age of the universe or the luminous lifetime of stars, whichever is less.

Statement (i) means that when we look out in space we also look back in time. Statement (ii) means that an unimpeded line of sight extends back to either the birth of luminous stars or the birth of the universe. According to statement (iii), if all stars are sunlike, the night sky is bright when the age of the universe or the lifetime of luminous stars, whichever is smaller, equals $10^{14}L^3$ years, and the number of stars covering the sky is $10^{42}L^6$, where L is the average separating distance between stars measured in light years.

13 "I cannot resist commenting on what seems to me an inhibition in the minds of astronomers that must surely have affected the development of cosmology from the seventeenth to the nineteenth century. The inhibition was a marked reluctance to recognize or acknowledge the principle that when we look out far in space we also look back far in time. This inhibition has received scant recognition in the history of science and may, among its many effects, have delayed the general solution to the riddle of cosmic darkness. The heavens gave visual evidence that observed stars at distances of tens of thousands of light years have existed for tens of thousands of years. Other worlds are seen as they were long ago. This evidence showing that stars originated at least tens of thousands of years ago controverted scriptural testimony on the age of the heavens.... We can easily imagine the astonishment of the public if astronomers had openly said that when we look out in space we look back thousands and perhaps millions of years to a time when the heavens were created!" (E. R. Harrison, "Olbers' paradox in recent times").

14 Let us imagine we have constructed a large box with perfectly reflecting walls (Figure 24.12). Inside this box we place a source of light such as the filament of a flashlight bulb connected to a supply of electrical energy. The filament has a cross-sectional area of about 1 square millimeter, and we assume that the box is a large cube with sides measuring 1 kilometer. When the filament is switched on, the emitted rays of light travel on the average a distance $V/s = 10^{17}$ centimeters, or 0.1 light year, before interception by the filament. After 0.1 year, or roughly 5 weeks, the filament absorbs as much radiation as it emits and the box is filled with radiation in equilibrium with the source. But suppose the filament is connected to a supply of limited electrical energy, such as a flashlight cell, which is capable of maintaining a bright filament for only 5 hours. The luminous lifetime of the filament is now much less than the fill-up time of the box, and the radiation level remains low for the same reason the sky is dark at night.

15 The radiation energy in a box of volume V is given by

$$\frac{d}{dt}(uV^{4/3}) = \frac{V^{4/3}}{\tau}(u^* - u), \qquad [24.16]$$

where u^* is the radiation level at the surface of the sources, $\tau = 1/nsc$ is the fill-up time, and n is the number per unit volume of the luminous sources, each of cross-sectional area s. This equation can be integrated (with $nV = $ constant) when we know how the volume V changes with time. First, we assume that V is constant, and the box represents a typical region in a static universe. We find that

$$u = u^*(1 - e^{-t/\tau}), \qquad [24.17]$$

where t is the time the luminous sources have been shining. When t is small compared with the fill-up time τ, then

$$u = u^*\frac{t}{\tau}, \qquad [24.18]$$

in agreement with Kelvin's Equation [24.15]. Second, let us assume that the size of the box (i.e., $V^{1/3}$) is proportional to t^n (where n is a constant), and that u is small compared with u^*. We find

$$u = \frac{u^*}{1 + n}\frac{t}{\tau}, \qquad [24.19]$$

and by comparing the last two equations we

see that the radiation in the expanding box is reduced by a factor $1/(1+n)$. In an expanding, decelerating universe we have $0 < n < 1$, and therefore expansion cannot greatly reduce the radiation level when the sources continuously emit during expansion. (The radiation from the big bang is greatly redshifted because it originated long ago and was not continuously emitted during the lifetime of the universe.) If t, τ, and u^* are the same in the static and expanding boxes, then both boxes contain the same number of photons, and therefore the average redshift of all photons in the expanding box is $\langle z \rangle = n$. In an expanding, decelerating universe the average redshift is less than 1.

16 "The reason for the cosmological significance of such a simple fact as the darkness of the night sky is that this is one of the phenomena that depend critically on circumstances far away" (Hermann Bondi, Cosmology, 1960). Concerning the riddle, Bondi wrote, "Since ... the rigour of the deductive argument appears to be unimpeachable, we must conclude that some of Olbers' assumptions are wrong. The assumptions may be restated here as:

(i)   The average density of stars and their average luminosity do not vary throughout space.
(ii) The same quantities do not vary with time.
(iii) There are no large systematic movements of the stars.
(iv) Space is Euclidean.
(v)   The known laws of physics apply."

Olbers explicitly stated only assumption (i). Bondi suggested that the paradox can be resolved if assumption (iii) is dropped. "If distant stars are receding rapidly the light emitted by them will appear reddened on reception and hence will have lost part of its energy. If the recession velocity of distant stars is great enough the loss of energy may be sufficient to reduce the radiation density to the observed level." Calculation shows, however, that redshift by itself is not sufficient. We have seen in the text (also Figures 24.11 and 24.12) that the redshift at most

reduces the radiant energy by only a small amount (less than 50 percent). The assumption that must be dropped is (ii), because stars cannot shine at a constant rate for a time equal to the lookback time.

Bondi's redshift solution of the riddle of cosmic darkness is correct in the steady-state expanding universe that he and Thomas Gold and Fred Hoyle had strongly advocated since 1948. In the steady-state universe, the energy levels u and u^*, the fill-up time τ, and the Hubble term $H = \dot{V}/3V$ are all constant, and Equation [24.16] therefore gives

$$u = \frac{u^*}{4H\tau + 1}. \qquad [24.20]$$

Roughly, $1/H$ is 10^{10} years and τ is 10^{23} years, and therefore we find u/u^* is about

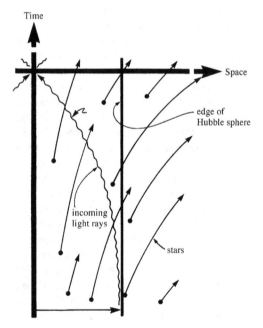

Figure 24.13. In the steady-state universe the Hubble length is constant. This spacetime diagram shows the observer's backward lightcone stretching back and gradually approaching the edge of the Hubble sphere. World lines are shown diverging out of the Hubble sphere. An infinite number of world lines intersect the backward lightcone and in the steady-state universe the sky is actually covered with stars, in agreement with interpretation B, and most stars are highly redshifted.

10⁻¹³, which is much the same as in a big bang universe. In the steady-state universe, interpretation A applies, and the sky is indeed covered with stars and is dark because of the redshift effect. Figure 24.13 shows that in this universe the observer's backward lightcone stretches back and gradually approaches the edge of the Hubble sphere. An infinite number of world lines diverge out of the Hubble sphere and intersect the backward lightcone and the sky is covered with stars. Poe's golden walls actually exist but are redshifted into invisibility.

PROJECTS

1 Discuss the following: John Arbuthnot, Alexander Pope, Jonathan Swift, and other members of the Scriblerus Club, under the name Martinus Scriblerus, asked in 1741, "How long a new star was lighted up before its appearance to the inhabitants of our earth?" Also in 1741, Edward Young wrote in *Night Thoughts on Life*:

> How distant some of the nocturnal suns!
> So distant, says the sage, 'twere not absurd
> To doubt that beams set out at Nature's birth
> Had yet arrived at this so foreign world,
> Though nothing half as rapid as their flight.

2 What is the average distance traveled by an arrow in a forest before it strikes a tree? When we speak of a background of trees at the lookout limit it must be understood that the lookout limit is an average sort of distance (much like the mean free path of particles in kinetic gas theory).

3 Assume a static universe of uniformly distributed sunlike stars in Euclidean space. If the lookout limit is 10^{12} light years, what is the average separating distance between stars. If the luminous lifetime is 10^{10} years, how many can be seen and what is the sky-cover fraction?

4 Show with diagrams and explanations the solutions corresponding to the following analogies. A forest surrounded by a dark wall; a forest of finite size; a misty forest; a forest covering the surface of a planet; a forest in which the trees get younger with distance; an infinite forest consisting of copses forming small woods, which are clustered to form larger woods, which are clustered to form even larger woods, and so on.

5 The discovery that Lord Kelvin had solved the riddle, and related the sky-cover fraction to the radiation level was submitted by the author to a journal of the history of astronomy. The Editor rejected the paper ("Kelvin on an old and celebrated hypothesis," subsequently published in *Nature*) on the grounds that because scientists had been unaware of Kelvin's work, it was of no relevance in the history of astronomy. Discuss this interpretation of the history of science.

FURTHER READING

Bondi, H. *Cosmology*. Cambridge University Press, Cambridge, 1960.

Charlier, C. V. L. "How an infinite world may be built up." *Arkiv förn Matamatik, Astronomi och Fysik* 16, no. 22 (1922).

Chéseaux, J. P. Loys de. *Traité de Comète*. Bousequet, Lausanne, 1744. Translated in E. R. Harrison, *Darkness at Night*.

Clerke, A. M. *The System of the Stars*. Longmans, Green, London, 1890.

Digges, T. "A perfect description of the caelestill orbes." Reproduced in E. R. Harrison, *Darkness at Night*.

Fournier d'Albe, E. E. *Two New Worlds*. Longmans, Green, London, 1907.

Halley, E. "Of the infinity of the sphere of fix'd stars." *Philosophical Transactions* 31, 22 (1720–1721). "Of the number, order, and light of the fix'd stars." *Philosophical Transactions* 31, 24 (1720–1721). Reproduced in E. R. Harrison, *Darkness at Night*.

Harrison, E. R. "The dark night sky paradox." *American Journal of Physics* 45, 119 (1977).

Harrison, E. R. "The dark night sky riddle: A paradox that resisted solution." *Science* 226, 941 (1984).

Harrison, E. R. "Kelvin on an old and celebrated hypothesis." *Nature* 322, 417 (1986).

Harrison, E. R. *Darkness at Night: A Riddle of the Universe*. Harvard University Press, Cambridge, 1987.

Harrison, E. R. "Olbers' paradox in recent times." In *Modern Cosmology in Retrospect*. Editors B. Bertotti et al. Cambridge University Press, Cambridge, 1990.

Herschel, J. F. W. "Essays from the Edinburgh and Quarterly Reviews." Longman, Brown, Longman and Roberts, 1857.

Hoskin, M. "Dark skies and fixed stars: A Christmas lecture." *Journal of the British Astronomical Association* 83, 4 (1973).

Jaki, S. L. *The Paradox of Olbers' Paradox.* Herder, New York, 1969.

Kelvin. See Thomson, W.

Kepler, J. See Koyré, A. and Rosen, E.

Kerby-Miller, C. Editor. *Memoirs of the Extraordinary Works and Discoveries of Martinus Scriblerus, Written in Collaboration by the Members of the Scriblerus Club.* Volume 1, page 167. Russell and Russell, New York, 1966.

Koyré, A. *From the Closed World to the Infinite Universe.* Johns Hopkins University Press, Baltimore, 1957. Reprint: Harper Torchbooks, New York, 1958.

Mädler, J. See Tipler, F. J.

Olbers, H. W. M. "Ueber die Durchsichtigkeit des Weltraumes." In *Astronomisches Jahrbuch für das Jahr 1826.* Editor J. E. Bode. Späthen, Berlin, 1823. Translated in E. R. Harrison, *Darkness at Night.*

Poe, E. A. "Eureka: An essay on the material and spiritual universe." See *The Science Fiction of Edgar Allan Poe.* Editor H. Beaver. Penguin Books, New York, 1976.

Proctor, R. A. *Other Worlds than Ours: The Plurality of Worlds Studied Under the Light of Recent Scientific Researches.* Longmans, Green, London, 1870.

Rosen, E. *Kepler's Conversation with Galileo's Sidereal Messenger.* Johnson Reprint Corp., New York, 1965.

Thomson, W. (Lord Kelvin). "On ether and gravitational matter through infinite space." *Philosophical Magazine* 2, 161 (1901).

Tipler, F. J. "Johann Mädler's resolution of Olbers' paradox." *Quarterly Journal of the Royal Astronomical Society* 29, 313 (1988).

Young, E. "Night Thoughts on Life" in *Edward Young: The Complete Works of Poetry and Prose.* Editor J. Nichols. Olms Verlagsbuchhandlung, Hildesheim, 1968.

25 CREATION OF THE UNIVERSE

Eer time and place were, time and place were not,
When primitive *Nothing* something streight begot,
Then all proceeded from the great united – What?
John Wilmot (1647–1680), Upon nothing

COSMOGENESIS I

First, some definitions: Cosmogeny (soft g) is the study of the origin (cosmogenesis) of the universe in the sense of how it comes into being. Cosmothanatology is the study of the end (cosmothanatos) of the universe in the sense of how it ceases to be. Cosmogony (hard g) is the study of the early stages in the history of the universe, and eschatology is the study of the final stages. Comparing the history of the universe with the life of a person, we have

cosmogenesis	birth
cosmogony	infancy
eschatology	senility
cosmothanatos	death

Whereas cosmogony (evolution of the early universe and the formation of structure) and eschatology (evolution of the dying universe and the dissolution of structure) are long-established subjects of scientific inquiry, cosmogeny and cosmothanatology are branches of cosmology that traditionally have been more philosophical, theological, and mythological than scientific. But this distinction should not diminish our scientific interest in cosmogenesis and cosmothanatos, for both subjects have much to gain from the discipline of scientific inquiry. Here we consider some topics in cosmogeny, cosmothanatology, and eschatology. Cosmogony (infancy) has already been considered in previous chapters.

We begin by turning to the creation myths that offer insight into the views of earlier societies on the birth and death of the universe.

CREATION MYTHS

Generally, neolithic cosmology made little or no distinction between the organic and inorganic kingdoms, and all animate and inanimate things were created together in a cosmic womb. In the myths of later ages, the living and nonliving things tended to be distinguished, and creation occurred as a sequential process, often as a twofold act, in which the living and nonliving worlds were created separately. Some of the creation myths in recorded history are as follows.

Sumerian

The Sumerian epic of creation, *Enuma Elish*, tells that in the beginning, "when Heaven above and Earth below had not been formed," there existed the primal Apsu – an encircling watery abyss – and Tiamat, a female being. From Apsu and Tiamat arose a dynasty of more than 600 gods and goddesses who controlled the various realms of existence, and the accounts of their intrigues and wars, and of catastrophes (such as the Deluge), serve as a basis in the study of the mythology of classical antiquity.

Egyptian

Before the creation, according to the myths of ancient Egypt (see Figure 25.1), there dwelt in Nun – the primal oceanic abyss – "a spirit, still formless that bore within itself

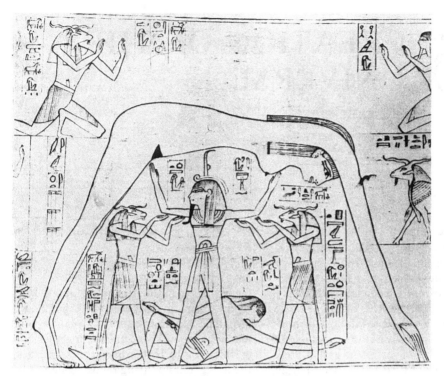

Figure 25.1. Shu, the Egyptian god of the atmosphere, raises his daughter Nut, the sky goddess, above the recumbent body of his son Geb, the Earth god.

the sum of all existence," called Atum, whose name signifies "to be complete." Then Atum, in the manifestation of Atum-Ra, created the gods and goddesses, all the living creatures, and the worlds they inhabit. Atum-Ra became personified as Ra the Sun god, and thereafter the gods and goddesses abounded in profusion: no less than 740 are listed in the tomb of Thutmosis III, who lived in the fifteenth century BC.

Indian

With Indian myths we are confronted by an imaginative riot of Vedic, Hindu, Buddhist, and Jainic deities. In the early myths of the Rig Veda it was said: "Who verily knows and can declare whence came this creation? He, the first origin of this creation, whose eye in highest heaven controls this world, whether he did or did not form it all, he verily knows it, or perhaps he knows it not." Later, according to the Hindu law of Manu: "All was darkness, without form,

beyond reason and perception, as if wholly asleep. Then the self-existent Lord became manifest, making all discernible with his power, unfolding the universe in the form of its elements, and scattering the shades of darkness." The primal undifferentiated state evolved into elements, these elements in their most subtle form combined to create living creatures, and then assumed the grosser states of the nonliving world. The Wheel of Time turned, the Sun rose and set, the Moon waxed and waned, the seasons came and went, the king died and lived again, birth and death alternated in endless incarnations, nation triumphed over nation, catastrophe followed catastrophe, wheels turned within wheels, cycles enfolded cycles, yuga followed yuga, and maha yuga followed maha yuga. Yet the Days of Brahma, though seemingly endless, were numbered, and the great gods and goddesses, creating and destroying worlds, were themselves doomed to die, tied to the relentless Wheel of Time. (A yuga on the average lasted 1 080 000 years, and a maha yuga – a day in the life of Brahma – lasted 4 320 000 years.)

Stoic and Mayan

Belief in cyclic time and periodic cosmic birth and death flourished in the Mediterranean world and was a principal part of the Stoic philosophy with its message of fortitude in adversity and the inevitability of fate. But the Mayans, more than all others, were obsessed with the idea of cyclic time. They believed that their calendric calculations sustained the universe and ensured the repeated return of the time-carrying gods; errors in their calculations would cause their whirligig cosmos to collapse into ruin.

Chinese

The Chinese created more practical myths. Heaven was conceived as a well-organized bureaucracy in which the gods and goddesses devoted their time to compiling registers, making reports to one another, and issuing directives. Later, in the Confucian scriptures, we find the elements of ether, fire, air, water, and earth (common to several mythologies after the rise of Greek science), each possessing its own degree of subtlety and each having correspondence with one of the five notes of harmony, the five flavors, and the five colors. The masculine qualities of light, warmth, and dryness, associated with the Sun, were called the yang; whereas the feminine qualities of shade, shelter, and moisture, associated with the Moon, were called the yin (see Figure 25.2). The convolutions of yang and yin generated order, sense, and the way of all things.

Greek

In the beginning, according to Greek myths, there existed four primal beings. First came Chaos (the abyss), then Gaea (the Earth), Tartarus (the lower world), followed by Eros (the unifying spirit of love). Hosts of gods and goddesses were created by the

Figure 25.2. The convoluted yin and yang.

four primal beings, by their matings, and by each alone. The genealogical charts of the various deities cover hundreds of pages. A significant early event was the begetting of Uranus the sky god by Gaea; then, from the mating of Uranus with Gaea, arose the Titans who were the first terrestrial rulers.

Rise of science and ethical religions

The invocation of cosmogenic beings in the creation myths makes the origin of the universe an intelligible event at the cost of raising legitimate questions lacking intelligible answers. Questions such as: What universe did the cosmogenic beings occupy? What created them and their universe? Why, how, and out of what was the universe created? What in the scheme of things distinguishes between the living and the non-living? Unfortunately, other than appeals to mysticism and faith, the myths rarely provide clear answers.

Ionian philosopher-scientists in the sixth century BC took the extraordinary step of dispensing with gods and goddesses as controlling forces. They introduced the idea of innate forces and natural laws, disentangled the sequences of cause and effect in the natural world, and formulated the rudiments of the scientific method. To this day science inherits their curiosity and incredulity.

It is interesting that simultaneously (2500 years ago) with the rise of Greek science there emerged around the world the enlightened ethical religions of Confucianism, Buddhism, and Zoroastrianism that associated moral behavior with religion, and condemned the common religious practice of human sacrifice. The teachings of Zoroastrianism of rewards in heaven and punishments in hell in an afterlife world, with elaborate angelogy and demonology, led to Mithraism, Manichaeism, Neoplatonism, and other mystery religions of the Middle East that shaped the ethical ideals of the Judaic, Christian, and Islamic worlds.

GENESIS
Mosaic chronology
According to the Mosaic chronology, common to the Judaic, Christian, and Islamic scriptures, the universe originated a few thousand years ago. (The word Mosaic pertains to the name Moses.) From available biblical sources, Dante Alighieri, author of *The Divine Comedy,* estimated that Adam and Eve were created in 5198 BC; Johannes Kepler, a famed astronomer, in his book *Mysterious Cosmography*, on the basis of scriptural texts and astronomical records, set the year of creation at 3877 BC; and James Ussher, an Irish bishop, set the date of creation at 4004 BC. Nowadays we view these estimates as amusing and fail to appreciate that in their day they were serious and honest attempts at cosmochronology. Isaac Newton, greatest of all scientists, in his posthumously published book *The Chronology of Ancient Kingdoms Amended*, indicated that creation occurred in 3988 BC.

On the basis of biblical texts, many persons – Jews, Christians, and Arabs – still believe that the universe, or at least the known world, was created by a supreme being in the recent geological past. Believers often do not realize, or do not want to realize, that this view of creation is now untenable in the light of modern knowledge. From the time of Newton to the present day the estimated age of the universe has increased roughly by a factor of 100 every 100 years. Nowadays, the estimated age of the universe is somewhere between 10 and 20 billion years.

Omphalos
Mark Twain in *Letters from the Earth* said that Earthlings who believe in the Mosaic

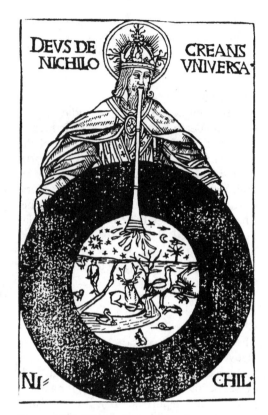

Figure 25.3. The creation of the universe from nothing (ex nihilo), depicted in *Libelius de nichilo* (Paris, 1510) by Charles de Bouelles, an early French humanist. This illustration shows God inflating ("inspiring") the embryonic universe by breathing into it. (S. K. Hetherington. *The Cosmographical Glass: Renaissance Diagrams of the Universe.*)

chronology are unable to explain the light they receive from stars more distant than a few thousand light years. In fact, after the sixth day when the heavens were created, "not a single star winked in that black vault" until light reached the Earth 4.3 years later from the Alpha Centauri system.

But Mark Twain underestimated the imaginative ingenuity of Earthlings. There is nothing logically wrong with the unfalsifiable idea that the universe originated a few thousand years ago, or even yesterday, provided we accept a highly intricate set of initial conditions. If we accept the belief that a supreme being created the universe, why not also accept the belief that the universe was created with starlight in transit from the stars?

The zoologist Philip Gosse firmly believed that the world was created roughly 6000 years ago in six days, and he accepted the geological evidence pointing to a world of immense age. In his book *Omphalos* (a Greek word meaning navel), published in 1867, he wrote that Adam, the archetype male, was created with a navel and hence carried the vestiges of a birth that had never occurred. If God saw fit to create Adam with a navel, wrote Gosse, surely God also saw fit to create a world complete with the vestiges of past geological eras that also had never occurred. In one supreme cosmogenic act that happened only a few thousand years ago, the universe was outfitted with a billion-year-old fictitious history. The impact of rain drops etched in sedimentary clays, footprints and teethmarks from beasts who roamed the Earth millions of years ago, ancient fossils still buried deep underground, light in transit from the distant stars and galaxies, and all the interlocking and elaborate complexity of a self-consistent universe functioning according to natural laws was created in the recent past.

Philip Gosse was excited by his discovery of how to reconcile Genesis and science, and expected praise for *Omphalos*. But he died a disappointed man, for his book was coldly received by the public, and ridiculed by critics who saw in it the implication that God is a joker and the universe a hoax. Many persons unthinkingly still have ideas similar to those proposed by Gosse.

Eugene Wigner in his book *Symmetries and Reflections* wrote: "The world is very complicated and it is clearly impossible for the human mind to understand it completely. Man has therefore devised an artifice that permits the complicated nature of the world to be blamed on something which is called accidental, and thus permits him to abstract a domain in which simple laws can be found. The complications are called the initial conditions, the domain of regularities is called the laws of nature." The initial conditions are accidental (similar to the Aristotelian god-given "accidentals") and subsequent conditions are the result of these earlier conditions evolving in accordance with the laws of nature.

Science advances by extending and generalizing the laws of nature, and by reducing and simplifying the initial conditions. Progressively, the "accidental" content of nature becomes more understood and is thereby transferred to the "domain of regularities." At each step in the advance of science, more of the accidental is explained and shown to be the natural consequence of preceding simpler conditions. A world of complex initial conditions and simple laws historically evolves into a world of simple initial conditions and complex laws. Similarly, modern cosmology seeks to explain the universe with the simplest initial conditions and the most general laws. Clearly, *Omphalos* runs counter to the scientific method.

COSMOGENESIS II
Two kinds of creation

Some scientists have speculated that new matter is created in the universe. "The type of conjecture that presents itself, somewhat insistently, is that the centers of the nebulae are of the nature of 'singular points,' at which matter is poured into our universe," wrote James Jeans in 1929, "so that to a denizen in our universe they appear as points at which matter is continually created." In the steady-state theory (1948) of Hermann Bondi, Thomas Gold, and Fred Hoyle, matter is continuously created everywhere. Matter is equivalent to energy, hence the creation of new matter violates the law of conserved energy. Instead of the conservation of energy, the steady-state cosmologists proposed what seemed a more fundamental law: the conservation of the present state of the universe. In the ensuing controversy between big-bangers and steady-staters, it was frequently claimed that we have a choice between the instant creation of a big bang or the continuous creation of a steady state, a choice between creation all at once or creation little by little. Debaters assumed that both kinds of creation, instant and continuous, are on equal footing. But this is a false assumption.

There are two kinds of creation: creation of the universe and creation in the universe. On the one hand, we have creation (as in cosmogenesis) of the whole universe complete with space and time; on the other, we have creation of things in the space and time of an already existing universe. In the big bang universe, everything including space and time is created; in the steady-state universe, matter is created in the space and time of a universe already created. Failure to distinguish between the two violates the containment principle (Chapter 9). The steady-state theory employs creation in the magical sense that at a certain place in space at a certain instant in time there is nothing, and at the same place a moment later there is something. But the creation of the universe has not this meaning, unless we revert to the old belief that time and space are metaphysical and extend beyond the physical universe; in that case, creation of a universe is in principle the same as the creation of a hazel nut. But in fact uncontained creation (cosmogenesis) is totally unlike contained creation. Cosmogenesis involves the creation of space and time, and this is what makes it so difficult to understand.

Kalam universe

In the tenth and eleventh centuries, the scholars of the ilm al-kalam, a religious school of Arab philosopher-theologians known as the mutakallimum, developed the idea of atomic time (see Chapter 9). They sought to demonstrate the total dependence of all physical and mental phenomena on the will of the supreme being (the "sole agent"). They denied the existence of natural laws and natural forces, and believed in an inert world in which all is shaped into form and jerked into motion by the will of the sole agent. According to the kalem time-atom theory, the universe is repeatedly created by the sole agent, and in each creation (each an atom of time) a new and slightly different universe is occupied by human beings with slightly different memories designed to link the time atoms into a coherent sequence. The kalem theory of atomic time anticipated the theory of occasionalism advanced later by René Descartes and other philosophers as a solution to the mind–body problem.

Occasionalism and parallelism

How can the physical human body interact with the nonphysical human mind? Two seventeenth-century theories proposed solutions that depend on different kinds of creation. According to the first theory, proposed by Descartes and known as occasionalism, the universe is created not once, as commonly supposed, but repeatedly, and in each creation the universe exists in a slightly different and more evolved form. God constantly recreates the mental and physical worlds such that readjusted minds and bodies exist in coordinated states. According to the second theory, proposed by Gottfried Leibniz and known as parallelism, a perfect self-running universe is created in the beginning in "pre-established harmony," and the mental and physical worlds ever since have run in parallel in perfect synchronism. The physical and mental worlds can be compared to two clocks: in occasionalism, the clocks require repeated adjustment to maintain their coincidence; in parallelism, the clocks, perfectly constructed, never require adjustment and run always in unison.

COSMOGENESIS III

How did the universe begin? Such a question seems impossible to answer in a comprehensible manner. Can it possibly be that the question is meaningless? Or are we perhaps asking the wrong question? The universe is a unified four-dimensional continuum; why should it have a "beginning" at an instant in time any more than a beginning at a point in space? The universe contains space and time, but is itself spaceless and timeless, and words such as "begin" and "end" seem inappropriate.

The universe itself has no location in space and time, and the creation of the universe of space and time is a spaceless and timeless act. Because we cannot say at one moment there exists nothing and at the next moment there exists the universe, what can cosmogenesis mean? At least the following general remark may be made. Some cosmologists, for example, Arthur Eddington, have strongly disliked the idea of a big bang because of its mythological implications and have pushed the beginning of the universe out of sight back into an infinite past. Not a few cosmologists, ancient and modern, have thought that this trick solves the cosmogenic problem. We are concerned here with the creation of the whole universe, including its space and time. An infinite span of physical time has to be created the same as a finite span of physical time. A universe of infinite age, such as the steady-state universe, is created the same as a universe of finite age. This applies also to space; an open universe of infinite space, such as the steady-state universe, is created the same as a universe of finite space, such as a closed big bang universe. The problem of cosmogenesis confronts all universes, with and without big bangs, infinite or finite in space and time.

It is an attractive thought that in the "beginning" the universe begins in a state of utmost symmetry, formless yet potential of many forms. Perhaps time was without direction, and time and space were without distinction. Perhaps the first symmetry-breaking transition was the birth of space-time from an embryonic manifold, which itself emerged from "the great united – What?"

Space and time are the most basic elements of the physical universe, and we must realize that it is created neither at a place in space nor at a moment in time (unless that place and moment are in the space and time of another universe occupied by the creating agent). We may say the universe began – in the sense of evolving – at the earliest moment in its time, but cannot say that it was created at that moment, or at any other moment in its time. Creation of the universe involves the creation of space and time including everything in space and time. The physical universe, if created, is created in one stupendous spaceless and timeless act.

Can we in modern cosmology accept the popular belief that an initial state (the big bang) is created and the rest of cosmic history then automatically unfolds, act by act? No, because cosmogenesis then ceases to be timeless and becomes a process in time. We are trapped in a maze of creation myths when we persist in thinking the universe was created in the big bang, or in some other initial state. Most cosmogenic theories are of the mythic kind, consisting of the creation of only an initial state, implying the pre-existence not only of time, but also space and the laws of nature.

Undoubtedly, it is a great advantage if a universe has a simple initial state from which all subsequent structure unfolds within the space and time of that universe. By virtue of its laws, the complexity of its evolved state is implicit in the simplicity of its initial state. The steady-state universe lacks this advantage, for it does not evolve from a simple initial state, and all its complexity remains a permanent feature throughout time. Thus the creation of a big bang universe is a simpler problem than the creation of a steady-state universe; moreover, a steady-state universe not only is created, but also contains creation. We have a cosmogenic Ockham's razor: the simpler the initial conditions and more general the laws, the fewer the assumptions needed in the creation theory.

In modern cosmogenesis, the creation of our universe is a timeless event and cannot be viewed as an act of becoming in our time. We may not say, for example, "the universe was created in the big bang." Cosmogenesis becomes empty of rational meaning when we speak of the creation of a universe in the time frame of that universe.

FITNESS OF THE UNIVERSE

Why is the universe so favorable in numerous ways to the existence of life? Throughout history, mythology and theology have urged the idea of a universe designed for the benefit of life. In the twentieth century an increasing number of contributions from science have made more persuasive the case for cosmic design at a fundamental level. The design of the universe is fixed by the physical constants and the laws of physics. In a universe containing luminous stars and chemical elements essential for organic life, the physical constants are necessarily precisely adjusted (or finely tuned). Slight deviations from their observed values result in a starless and lifeless universe. The laws, including the forces (strong, electromagnetic, weak, and gravitational, and their unified electroweak and hyperweak forms), reflect in some degree the symmetries of spacetime and its relativistic decomposition into three-dimensional space and one-dimensional time. The constants, such as c the speed of light, e the charge and m_e the mass of the electron, h Planck's constant, G the gravitation constant, determine the values of the coupling constants of the forces. They provide also a set of absolute units (the Planck units) of distance, time, mass. The constants, when combined, yield dimensionless numbers (Chapter 23) that occupy two widely separated numerical groups: a "unity-group" consisting, for example, of the ratio m_p/m_e of the proton and electron masses, and the fine structure constant e^2/hc; and a "large-number group" consisting, for example, of the ratio $e^2/Gm_e m_p$ of the electric and gravitational forces between an electron and a proton of order 10^{40}. The observable universe is also 10^{40} times larger than a characteristic size of an elementary particle. This coincidence between two very large and unrelated numbers is all the more striking because the electrical–gravitational force ratio stays constant at 10^{40}, but the scale of the universe in subatomic units steadily increases, beginning at 10^{-20}, reaching unity when the universe is 10^{-23} seconds old, and now is

10^{40}. Robert Dicke showed that the present coincidence is the natural consequence of stellar evolution. Not until the first stars have evolved and synthesized and expelled into interstellar space elements, such as carbon, oxygen, and silicon, can planets form and life evolve. By that time both numbers are in approximate agreement. Thus began the anthropic principle in its modern scientific form. A vast universe, billions of years old, spanning billions of light years, teeming with galaxies is a precondition for the existence of organic life, and such a universe requires precisely adjusted values of the fundamental constants.

The fundamental constants in various combinations determine a host of conditions that make life possible. The size of a terrestrial planet is the geometric mean of the sizes of the universe and an atom; the mass of a multicellular organism is the geometric mean of the masses of a planet and a proton. More striking are the dramatic changes in the heavens produced by small changes in the values of the physical constants. When G is slightly reduced, stars cease to be luminous; when slightly increased, they burn too quickly and their luminous lifetimes are too short for biological evolution. The nuclear binding energy of deuterium (first stepping stone in nucleosynthesis) is only slightly greater than the neutron–proton mass difference; a small decrease in the strong force (or small increase in e) creates a universe without elements, other than hydrogen, vital to life; a small increase in the strong force (or decrease in e) creates a universe lacking hydrogen and luminous stars. The triple-alpha reaction (Chapter 5) that converts helium into carbon in stars depends on a ^{4}He–^{8}Be–^{12}C resonance so critical that Hoyle could predict its energy level with precision. If the energy level were slightly higher or lower, carbon and most other elements would not exist.

These are a few examples of fine tuning. An intricate network of interlocking critical relations leaves the investigator puzzled by what appears to be evidence that the universe has been designed at its most fundamental level to make it compatible with the existence of organic life.

Science advances by explaining what previously was viewed as accidental and irreducible. Possibly in the future, the physical constants will be explained self-consistently and will not be inexplicable. This development must come as a result of recognizing that fitness is separate from creation and is thereby open to rational inquiry.

Of the many current theories of cosmogenesis, few attempt to explain fitness, perhaps because fitness has not always been sharply distinguished from creation; perhaps because we lack a theory that explains why the basic constants have their observed values; and perhaps because, by courtesy of the anthropic principle, if enough universes are created in a scatter-shot manner, by chance one might be fit and that universe is ours.

Theories that attempt to explain why the universe is the way it is are not under an obligation to explain also why the universe exists. The problem of fitness is sufficiently onerous without burdening it with the problem of creation. Atomic physicists study the structure of atoms and are not required at the same time to explain why atoms exist.

FITNESS AND CREATION

The question "why does the universe exist?" is not the same as "why is the universe the way it is?" Nor are the answers necessarily the same. The first question concerns the creation of the universe and the second concerns the fitness of the universe (the fine tuning of the fundamental constants of physics and the compatibility of the universe with the existence of life). Most theories of creation and fitness fall into four classes: theistic theories concerned with both creation and fitness, anthropic theories concerned primarily with fitness, spontaneous creation theories concerned primarily with creation, and natural selection theories concerned with both creation and fitness.

THEISTIC THEORIES

The "Lord ... who created it not as a formless waste but as a place to be lived in."
Isaiah 45:18.

Theistic theories are as old as the human race. The common belief in the Judaic-Christian-Islamic world is that a supreme being (God) created the universe and designed it specifically for inhabitation by life. Why does the universe exist? Because God created it. Why is the universe the way it is? Because God created it intentionally that way. These joint answers to the questions of creation and fitness form the basis of the theistic principle. The usual failure to distinguish between creation and fitness may be because traditional theological arguments roll both into a single subject.

When logically examined, as done by Saint Augustine of Hippo (354–430), theistic theories of cosmogeny lead to the realization that everything in the created universe is predetermined. Whatever happens is fated to happen in accordance with the cosmic blueprint. Subsequent acts of creation, such as the creation of life, are superfluous, for everything has been timelessly created and there is no need to create again what has already been created. Moreover, the perfect work of an omnipotent supreme being may not be altered by human beings, who must act in accord with the grand design. Saint Augustine, the architect of deterministic theism (the belief that God created a universe of perfect design), sought to reconcile Platonic reason with Christian dogma, and argued in the *Confessions* that free will is incompatible with God's unalterable design. Adam and Eve in the beginning had freedom of will, but because of willful failure to conform to the grand design, free will was withdrawn from them and their descendents. All who disagreed with this deterministic doctrine, such as the British monk Pelagius, were condemned as heretics.

If God created one universe, why not many? and why not all perfect for their purpose? Giordano Bruno's argument (1584) advocating the creation of not one but a multitude of stars, can be extended to the creation of a multitude of universes: "Thus is the excellence of God magnified and the greatness of his kingdom made manifest; he is glorified not in one, but in countless suns; not in a single earth, but in a thousand, I say, in an infinity of worlds." In the modern scheme of cosmic plenitude, God creates a multitude of universes.

David Hume in his posthumously published *Dialogue Concerning Natural Religion* (1779) separated creation and fitness by arguing that many universes "might have been botched and bungled throughout an eternity ere this system was struck out; much labour lost, many fruitless trials made, and a slow but continual improvement carried out during infinite ages in the art of world-making." Bruno's principle of cosmogenic plenitude conjures up the prospect of many universes, and Hume's introduction of fitness opens up the possibility of many barren universes, unfit for inhabitation by life.

ANTHROPIC THEORIES

Anthropic theories originated in antiquity with thoughts such as "man is the measure of all things" (Protagoras). The anthropic principle states in its weak form that "what we can expect to observe must be restricted by the conditions necessary for our presence as observers" (Brandon Carter), and its controversial strong form that the universe exists and is the way it is because we exist. The strong form of the anthropic principle is a tangle of issues open to endless debate and we shall stay with the less controversial weak form.

Physics lacks a theory that explains why the constants of nature have their observed values. Lacking a scientific explanation, we suppose the constants are distributed with random values in all possible combinations throughout an array of universes, and we occupy the universe, or one of a subset of universes, that is compatible with the existence of life. Unlike the theistic principle, in which design is intentional, in the anthropic principle, design is fortuitous and at the cost of requiring the creation of many starless and lifeless universes. The virtue of the anthropic principle is that it treats cosmic design as a separate subject. Freed from the mysteries of creation, fitness becomes a rational subject open to scientific inquiry.

SPONTANEOUS CREATION THEORIES

Various cosmogenic theories of spontaneous creation have emerged in recent years. One group of theories proposes that the universe is spontaneously created either from nothing, or from a quantum fluctuation borrowed from the created universe. In effect, the universe self-creates without extracosmic midwifery. Edward Tryon in 1973, in support of such theories, argued that the total cosmic energy (potential plus kinetic) is zero, and the universe is therefore a zero-energy creation. Design was omitted from the argument.

Another group of theories proposes that universes spontaneously create other universes. Spacetime seethes with foam-like quantum fluctuations in the form of semi-closed microscopic worlds of virtual reality. At the Planck density sufficient energy exists to make real these expanding and collapsing microscopic bubble worlds. On rare occasions a bubble world, instead of collapsing, expands to the point where it reaches inflation, and then transforms into an independent universe. Thus a parent universe spontaneously creates embryonic ("baby") universes that inflate into an adult universe, and in turn give birth to more universes. We can genealogically link these bubble worlds, but we cannot draw pictures showing them sequentially creating one another. In all such schemes of quantum-mechanical creation, the fitness of our universe is generally explained by the anthropic principle: in an ensemble of many universes, our universe is fortuitously fit for inhabitation by life.

Joe Rosen's creation theory provides an illustration of the tenseless nature of cosmogenesis. Universes X and Y create each other: X quantum-mechanically creates Y, and Y quantum-mechanically creates X. In Rosen's theory, a quantum fluctuation in X in X's time creates Y, and a quantum fluctuation in Y in Y's time creates X. In this cosmogenic scheme, X and Y form a self-creating pair. The scheme can readily be extended to a self-creating plurality of universes. Does this really solve the cosmogenic problem?

NATURAL SELECTION THEORIES
Black hole generation of universes
We have a cosmic picture of universes spontaneously creating universes. Lee Smolin of Syracuse University in 1992 proposed a natural selection theory in which offspring universes are created in black-hole singularities, and the most prolific universes are those with the greatest number of black holes. The idea is that the constants of nature are slightly different in each universe, and the variations that favor the formation of black holes will become the most abundant. Smolin argued that black holes imply stars, which imply planetary systems, which imply living creatures. Thus universes finely-tuned to maximize their black hole population will contain living creatures. In this way creation and fitness are explained by a single theory.

A difficulty with Smolin's theory is its neglect of the possibility that most black holes are primordial, and their natural selection would favor universes tuned for their maximum production, and such universes need not contain stars and living creatures. John Gribbin (*In the Beginning*) discusses Smolin's theory of universes reproducing universes via black-hole singularities, and remarks: "Cosmologists are now having to learn to think like biologists and ecologists, and to develop their ideas not within the context of a single, unique, universe, but in the context of an evolving population of universes. Each universe starts from its own big bang, but all universes are interconnected in complex ways by black hole 'umbilical cords,' and closely related universes share

the 'genetic' influence of a similar set of physical laws."

Creation of universes by intelligent life
Not impossibly, our universe is created by intelligent beings in another universe who are millions and perhaps billions of years more advanced than human beings. These beings occupy a universe that is compatible with the existence of intelligent life and is therefore similar to our own. This suggests a natural selection theory in which intelligent life in parent universes creates offspring universes, and in offspring universes fit for inhabitation, new life evolves to a high level of intelligence and creates further universes. Universes unfit for inhabitation lack intelligent life and cannot reproduce. Plausibly, offspring universes have properties similar to their parent universes – apart from small random genetic variations in the constants of physics – and the universes most hospitable to intelligent life are naturally selected by their ability to reproduce.

Alan Guth has discussed how it might be possible to create a universe. "Now we have the mathematical tools that allow us to seriously discuss the prospects for creating a universe in your basement" (quoted by M. Mitchell Waldrop, 1987). If roughly 10 kilograms of particles at energy 10^{15} GeV is compressed to the size of a black hole, the interior has the right conditions for inflation, and will create a detached bubble world (Figure 25.4) that inflates into a universe containing billions of galaxies. If human beings can have such wild ideas, might not more intelligent beings know how to make these ideas come true? "The birth of a new universe may be the result of a planned project to test whether cosmogeny is practical, or to create universes more hospitable to intelligent life, or to transfer, if possible, from one universe to another. The ultimate aim in the evolution of intelligence is conceivably the creation of universes that nurture intelligence" (Harrison, "The natural selection of universes containing intelligent life," 1995).

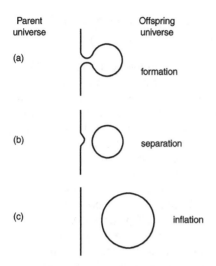

Parent universe — Offspring universe

(a) — formation

(b) — separation

(c) — inflation

Figure 25.4. Creation of an offspring universe from a parent universe. In (a) a bubble world (or microuniverse) forms either from a fluctuation or from a do-it-yourself project in the laboratory. In (b) the bubble world separates (its umbilical cord evaporates by Hawking radiation), and then in (c) it inflates and becomes a new universe. In a do-it-yourself project the trick is to assemble a bubble world ripe for inflation.

ESCHATOLOGY
Cosmic doom

The *Götterdämmerung* (*Twilight of the Gods*) in Norse myths is an apocalyptic vision that illustrates the eschatology and cosmothanatology found in many myths. From the beginning, the world is doomed and the gods are destined to die. The end is foreshadowed by oath-breaking, baleful omens, and titanic warfare among gods and men. In the final carnage of Doomsday, the Sun becomes swollen and blood red, and the Earth, frozen in the grip of a paralyzing winter, sinks back into the abyss. Out of the cosmic wreckage, goes the tale, emerges a new universe, ruled by better gods, inhabited by better mortals.

Into the far future

Let us explore the future of the physical universe, venturing farther than any time traveler of science fiction has dared. As our time machine clicks off the billions of years we see the galaxies drifting farther and farther apart. In about 5 billion years, the Sun swells into a red giant and engulfs the inner planets. The Earth has long since lost its atmosphere and oceans in the intense red glare and life has either perished or fled to other places.

The stars fade like guttering candles and are snuffed out one by one. Out in the depths of space the great celestial cities, the galaxies, cluttered with the memorabilia of ages, are slowly dying. Tens of billions of years pass in the growing darkness. Occasional flickers of light pierce the fall of cosmic night and flurries of activity delay the end of a universe destined to become a galactic graveyard.

The far future offers two possibilities: In the first, the universe ceases to expand, collapses, and dies in a state similar to the initial big bang. In the second, the universe expands forever and lifeless galaxies voyage in frozen darkness to the end of time. Or so it was once thought. The picture presented by this second possibility has changed in recent years. We now know that dead galaxies cannot endure forever.

The big bang returns

According to the first possibility, at some time in the future, perhaps in 40 or 50 billion years, the universe ceases to expand and commences to collapse. Dead and dying galaxies halt their headlong flight and begin to approach one another. The cosmic background radiation, which has cooled to little more than 1 degree kelvin, begins to get warmer. At first, only the nearest galaxies are seen approaching and therefore blueshifted; the rest at greater distances, seen as they were in the past when the universe was still expanding, are redshifted. As time passes, more and more galaxies are blueshifted.

Slowly the clusters of galaxies approach one another, then overlap, and melt away; first superclusters, then the great clusters, and finally all other clusters. Then the galaxies slowly approach one another. When the temperature of the cosmic background radiation reaches 100 kelvin, the galaxies begin to overlap and melt away. Following the dissolution of the galaxies, the universe now consists mostly of stars and their clusters. Most stars are old and are dark dwarfs, faded white dwarfs, neutron stars, and black holes.

A remarkable thing now happens. The stars, moving freely with random peculiar motions, begin to accelerate in the collapsing universe, just like the particles in a gas that is slowly heated by compression. Stars rushing at high speed occasionally collide and erupt in brilliant explosions of light. But most stars never collide and continue to accelerate to relativistic speeds. Streaking through the interstellar tumult like comets, they leave behind long trails of incandescent gas. The uproar rises to a crescendo as the universe grows denser and hotter, and then, in the throes of unimaginable devastation, a new radiation era dawns.

After a lifespan of 100 or so billion years, the big bang returns and the universe reverts to the state it had originally. What follows then we do not know. It may not even be meaningful to inquire what happens in this singular state, because time may have lost its sequential orderliness.

Eternal darkness and emptiness

We turn to the second possibility consisting of everlasting expansion. In trillions of years the galaxies become dark and dead. Yet they cannot remain eternally in this dormant state. Given enough time – and eternity has always more than enough – all systems must lapse into the lowest possible energy states and the entropy must rise to the maximum possible value. Eternity has no bottlenecks and barriers. The nuclei of ordinary hydrogen gas at room temperature, for example, fuse and transform into helium nuclei in 10^{20} years. A hundred million million million years sounds a long time, but is less than the proverbial blink of an eye compared with eternity.

Our Galaxy and the great galaxy in Andromeda are in orbit about each other and are slowly spiraling together because of the gravitational waves emitted by their orbital motions. In 10^{30} years, ignoring all other dissipative processes, the two galaxies will merge together and become one large system. Also, clusters of galaxies emit gravitational radiation because of the orbital motions of their members, and in 10^{35} years will have contracted and become supergalaxies. An additional process, in some cases very important, helps to explain why clusters of all kinds and sizes slowly contract. The fastest moving members occasionally escape from the cluster, and this evaporative process "cools" the cluster and it contracts slowly.

Stars in binary systems also spiral together owing to orbital gravitational radiation, and in about 10^{30} years most will have amalgamated. Galaxies and star clusters will slowly shrink owing to orbital gravitational radiation and the evaporation of their fastest members. In a time less than 10^{40} years most stars in galaxies will have aggregated into supermassive black holes.

According to grand unified theories of particle physics, protons are probably not permanently stable particles. On a time scale of 10^{10} years – the present age of the universe – the law of baryon conservation is reliable. But on a time scale approaching 10^{35} years, protons decay into lower mass particles such as photons, positrons, and neutrinos. Electrons and positrons tend to annihilate and become photons. The hyperweak force, which ruled in the extreme early universe and ignored the law of baryon conservation, although now exceedingly feeble, is still alive and active. In the beginning, at very high energies, hyperweak interactions occurred on a time scale 10^{-40} seconds; now, at very low energy, they occur on a time scale 10^{40} seconds (10^{33} years). In the future, when the cosmic background radiation has cooled to 10^{-16} kelvin, stars and all forms of matter not engulfed by black holes will have dissolved into radiation. We are left with a universe consisting of black holes of various masses immersed in a cosmic bath of weak radiation.

Stephen Hawking at Cambridge University showed in 1974 that black holes emit thermal radiation from their event horizons and have a temperature inversely proportional to their mass. Because of "Hawking radiation" (Chapter 13), which consists mostly of photons and neutrinos, the mass of the black hole steadily decreases. As the mass decreases, the temperature rises, and luminosity, which is proportional to the inverse square of the mass, also rises. Calculations show that black holes of solar mass radiate away their entire mass in a lifetime of 10^{66} years, and superholes of galactic mass last 10^{100} years. Ultimately, when the cosmic background radiation has cooled to 1 trillion-trillion-trillion-trillion-trillionth (10^{-60}) of a kelvin, the universe consists only of photons, neutrinos, gravitational waves, and some electron pairs too far apart to annihilate. All else has vanished. Containing no matter or antimatter, only radiation, the universe now has zero baryon number and maximum entropy – for all eternity. *Sic transit gloria mundi*!

REFLECTIONS

1 Before all Time, all Matter, Forme, and Place;
  God all in all, and all in God it was:
  Immutable, immortall, infinite,
  Incomprehensible, all spirit, all light,
  All Majestie, all selfe Omnipotent,
  Invisible, impassive, excellent,
  Pure, wise, just, good, God raign'd alone at rest,
  Himselfe alone selfes Pallace, hoast and guest.
  Devine Weekes, *Du Bartas, 1605*

2 *Until the eighteenth century it was widely believed that the Earth was only thousands of years old. Mounting evidence in geology and paleontology (the study of fossils), however, indicated a much greater age. Compromise doctrines were proposed in which the Earth had been periodically visited by catastrophes, such as life-destroying deluges, and natural and supernatural laws had alternated in their control over the Earth. Fossils were the relics of less perfect forms of life that had been wiped out in preparation for the creation of more perfect forms.*

• *James Hutton, a Scottish farmer and physician, opposed these views and advanced in 1785 the theory that nature behaves in an* orderly and uniform way, and the formation and erosion of mountains are continuous processes that have acted steadily for an indefinitely long period of time. He concluded that there is "no vestige of a beginning – no prospect of an end." This was the uniformitarian principle that in the early Victorian era was impressively argued by Charles Lyell. The whole landscape, he declared, is continually modified by natural processes, and yet the terrestrial surface remains essentially unchanged. The catastrophists believed in periodic bursts of divine intervention, whereas the uniformitarians believed in a self-running steady-state world controlled only by natural laws. The controversy between the advocates of these rival beliefs raged until the middle of the nineteenth century and was much more heated than the parallel controversy between the big-bangers and steady-staters of this century.*

• *Then came the Darwinian revolution. Genesis, biblical testimony, and a whole outlook on life went on trial. "It is hopeless trying to understand the nineteenth century with its fulminations from pulpit and platform without realizing that numerous persons were struggling to save their imperiled world pictures that gave meaning and purpose to life on Earth; nor can we hope to understand the furor of the scenes enacted in the present [twentieth] century unless we realize that many societies were struggling, and still are, to find new beliefs, often with dismal and tragic results." (Harrison,* The Masks of the Universe*).*

• *In the late nineteenth century, physicists under the leadership of Lord Kelvin attacked the uniformitarians. Calculations on tidal effects and terrestrial heat losses showed that the Earth could not be as old as the geologists supposed. The calculated age of the Sun seemed irrefutable. At first, Kelvin assumed that the Sun derived the energy for its luminosity from the infall of meteoroids; later, he adopted Hermann Helmholtz's suggestion that the Sun obtains its energy by slow contraction, thus releasing gravitational energy, and estimated an age for the Sun of 20 million years. Because light from the Sun is essential for life on Earth, this implied that* rock strata containing fossils could not be older than 20 million years. Insistence by the physicists on this short time span created dismay in the Earth and life sciences, and attempts were made to compress geological history and biological evolution into Kelvin's chronology. In the early years of this century, the physicists redeemed themselves by developing radioactive dating methods, and were able to show that the age of the Earth is measured in billions – not millions – of years. In 1904, in "Radiation and emanation of radium," Ernest Rutherford wrote, "The discovery of the radioactive elements, which in their disintegration liberate enormous amounts of energy, thus increases the possible limit of the duration of life on this planet, and allows the time claimed by the geologist and biologist for the purpose of evolution."

3 "Thousands of books were published in the nineteenth century, most of them in England, attempting to harmonize geology and Genesis. In this dreary and pathetic literature, one book stands out from all others as so delightful and fantastic that it deserves special mention. It was called Omphalos (the Greek word for navel), and was written by zoologist Philip Gosse . . . . Not the least of its remarkable virtues is that although it won not a single convert, it presented a theory so logically perfect, and so in accord with geological facts that no amount of scientific evidence will ever be able to refute it." (Martin Gardner, Fads and Fallacies in the Name of Science, 1957.) "Never was a book cast upon the waters with greater anticipation of success than was this curious, this obstinate, this fantastic volume," wrote Edmund Gosse, the poet and son of Philip. "He offered it with a glowing gesture to atheists and Christians alike . . . . But, alas! atheists and Christians alike looked at it and laughed, and threw it away . . . even Charles Kingsley, from whom my father had expected the most instant appreciation, wrote that he could not 'believe that God has written on the rocks one enormous and superfluous lie.' "

4 "If the world had begun with a single quantum, the notion of space and time would altogether fail to have any meaning at the beginning; they would only begin to have sensible meaning when the original quantum had been divided into a sufficient number of quanta. If this suggestion is correct, the beginning of the world happened a little before the beginning of space and time" (Georges Lemaître, "The beginning of the world from the point of view of quantum theory," 1931). The single quantum referred to is Lemaître's primeval atom. The suggestion that time and space have no meaning in the beginning is interesting, but the remark that the world began a little before the beginning of time is surely without meaning.

• In The Primeval Atom (1951), Lemaître later wrote the memorable words: "the atom world broke up into fragments, each fragment into still smaller pieces. Assuming for the sake of simplicity that this fragmentation occurred in equal pieces, we find that two hundred and sixty successive fragmentations were needed in order to reach the present pulverization of matter into our poor little atoms, which are almost too small to be broken further. The evolution of the world can be compared to a display of fireworks that has just ended: some few red wisps, ashes, and smoke. Standing on a cooled cinder, we see the slow fading of the suns, and we try to recall the vanished brilliance of the origin of the worlds."

5 Olaf Stapledon, in his imaginative The Star Maker (1937), described how the Star Maker designs and creates universes of increasing complexity. In each creation intelligence is interwoven in a different and highly intricate fashion. "In vain my fatigued, my tortured attention, strained to follow the increasingly subtle creations which, according to my dream, the Star Maker conceived. Cosmos after cosmos issued from his fervent imagination, each one with a distinctive spirit infinitely diversified, each in its fullest attainment more awakened than the last; but each one less comprehensible to me. . . . Sometimes the Star Maker fashioned a cosmos which was without any single, objective, physical nature. Its creatures were wholly without influence on one another; but under the direct stimulation of the Star Maker each creature conceived an illusory but reliable and useful physical world of its own, and peopled it with figments of its imagination. These subjective worlds the mathematical genius of the Star Maker correlated in a manner that was perfectly systematic." This cosmos is reminiscent of Gottfried Leibniz's theory of monads.... "But at the close of his maturity he willed to create as fully as possible, to call forth the full potentiality of his medium, to fashion worlds of increasing subtlety, and of increasing harmonious diversity. As his purpose became clearer, it seemed also to include the will to create universes each of which might contain some unique achievement of awareness and expression. For the creatures' achievement of perception and of will was seemingly the instrument by which the Star Maker himself, cosmos by cosmos, woke into keener lucidity. Sometimes the Star Maker flung off creations which were in effect groups of many linked universes, wholly distinct physical systems of very different kinds, yet related by the fact that the creatures lived their lives successively in universe after universe, assuming in each habitat an indigenous physical form, but bearing with them in their transmigrations faint and easily misinterpreted memories of earlier existences. In another way also, this principle of transmigration was sometimes used. Even creations that were not thus systematically linked might contain creatures that mentally echoed in some vague but haunting manner the experience or temperament of their counterparts in some other cosmos. In some creations each being had sensory perception of the whole physical cosmos from many spatial points of view, or even from every possible point of view. In the latter case, of course, the perception of every mind was identical in spatial range, but it varied from mind to mind in respect of penetration or insight. This depended on the mental calibre and disposition of particular minds.... Sometimes these beings had not only omnipresent perception but omnipresent volition. They could take action in every region of space, though with varying precision and vigour according to their mental calibre. In a manner they were disembodied spirits, striving over the physical cosmos like chess-players, or like Greek gods

over the Trojan Plain." Stapledon omits the possibility of universes in which creatures are omnipresent in time. "In one inconceivably complex cosmos, whenever a creature was faced with several possible courses of action, it took them all, thereby creating many temporal dimensions and distinct histories of the cosmos. Since in every evolutionary sequence of the cosmos there were very many creatures and each was constantly faced with many possible courses, and the combination of all their courses were innumerable, an infinity of distinct universes exfoliated from every moment of every temporal sequence in this cosmos.... At length, so my dream, my myth, declared, the Star Maker created his ultimate and most subtle cosmos, for which all others were but tentative preparations. Of this final creation I can say only that it embraced within its own organic texture the essence of all its predecessors; and far more besides. It was like the last movement of a symphony, which may embrace, by the significance of its themes, the essence of the earlier movements; and far more besides.... I strained my fainting intelligence to capture something of the form of the ultimate cosmos. With mingled admiration and protest I haltingly glimpsed the final subtleties of world and flesh and spirit, and of the community of those most diverse and individual beings, awakened to full self-knowledge and mutual insight. But as I strove to hear more inwardly into that music of concrete spirits in countless worlds, I caught echoes not merely of joys unspeakable, but of griefs inconsolable."

6 One of Stapledon's ideas on cosmogeny has emerged in modern cosmology in the many-worlds interpretation of quantum mechanics. Electrons in excited atoms make transitions; each transition is to one of several potential states; the actual transition that occurs cannot be predicted, only its probability. Hugh Everett of Princeton University in 1957 proposed that at each transition the universe splits so that all potential states are realized, each in a separate real universe. Thus from each atom exfoliates many new universes each time a transition occurs, and

these new universes are identical except that in each the atom is in a different final state. An object such as a rain drop or a human being, consisting at any moment of many quantum transitions, is continually splitting this universe into many universes, in each of which the object is in a slightly different state and follows a different history.

• "Pure chance and free choice are unwelcome guests in any rational and orderly scheme. What is not mandatory is forbidden. Imagine you are walking in a wood and come to a place where the path divides into alternate routes. There is no reason to take one path more than the other and you are perfectly free to choose whichever you please. But freedom of choice is an illusion in a rational universe, for all is determined by the inviolable laws of the universe. Liberty to do this or that as you wish, go here or there as you will is intolerable, for it contradicts the rational determinism of the universe you live in. Hence you take both paths" (Harrison, The Masks of the Universe). Both paths are taken but in different universes. We preserve the ultimate rationality of the scheme of things at the price of creating new universes. This is the many-worlds interpretation of the freedom of will.

• All rational universes, theological and scientific, are deterministic. Given a sufficient set of initial conditions and the regulating laws of that universe, the outcome is predictable. This is true not only in the physical universe, but also in the Augustinian universe in which all events are fully determined by divine laws and initial conditions. Independent action by human beings, contrary to the laws, is forbidden in both universes. Alternatively, all choices are realized in a multiuniverse, and the laws are fully deterministic in each universe.

• The British monk Pelagius (late fourth and early fifth centuries) strongly disagreed with fatalistic arguments. His protests against the decadence of life in Rome were parried by excuses that human weakness is preordained, and being inevitable, is hence forgivable. Pelagius taught that a predestinate universe threw on God the blame for the evil that properly belonged to human beings. How can members of the Church, he said, talk about sin and its redemption when "if it is necessary, it is not a sin." Men and women, he declared, if they choose, are free to live untainted by sin. He was condemned as a heretic and forced to flee from Rome.

• "Freedom of will is the conviction that we as individuals have some control over our lives and are not the sport of merciless fate and the helpless victims of inflexible laws. All scientific, philosophical, and theological theories that explain how things work are in conflict with our personal awareness of free will. As Dr. Johnson said, 'All theory is against the freedom of the will, all experience for it'" (Harrison, The Masks of the Universe). In this work I argued that free will is a property of the Universe, which contains us, but is not a property of a rational universe, which does not contain us. I also suggested that the mind–matter problem stems from the Universe–universe dichotomy. The mind belongs to the Universe, the brain belongs to the physical universe.

7 When a star in an advanced state of evolution falls inward, it terminates its collapse either as a white dwarf, a neutron star, or a black hole. A black hole is the ultimate state of collapse: to an observer, far from the star, it is frozen in a perpetual state of collapse; to an internal observer, who falls with the star, the collapse continues and ends in a singularity of maximum possible physical density. Such collapse singularities, predicted by the laws of nature, are beyond the reach of known laws of nature and are not understood. Singularity theorems have been developed by Roger Penrose, Stephen Hawking, and George Ellis. Ignoring all forces other than gravity, Penrose first showed that a singularity is inevitable whenever a collapsing region is enclosed in a "trapped surface." Inside a trapped surface (see Chapter 13), all light rays are dragged inward faster than they can escape outward. Rotation and nonspherical collapse cannot avert a singularity if the collapsing region is surrounded by a trapped surface. It has been shown that if certain general conditions are satisfied (density and pressure are

both positive, for example) a collapsing universe terminates in the singular state. Possibly, the singular state occurs at the Planck density.

8   *The three essential elements of a theory of natural selection of universes are:*

(i)   a population of self-reproducing universes;

(ii) "genetic" variations in the values of the natural constants and other fundamental parameters;

(iii) a process that selects for reproduction only universes that are inhabitable by organic life.

In biology the criteria of "fitness" are adaptability to the environment and the reproductive fertility of members of a population. In the natural selection by intelligent life the criteria of fitness is the inhabitability of the universe and the intelligence of the inhabitants.

9   *Is it possible to have a cosmogenic theory in which only universes exist that contain consciousness, and these universes are necessarily finely tuned for the support of conscious organisms? This is an advanced form of the strong anthropic principle in which consciousness guarantees existence. What is consciousness? We do not know, but each of us knows that it exists. Conceivably, in future theories of the physical world, the conscious acts of observing and conceptualizing may play a central role in the structure of the physical world.*

PROJECTS

1   Discuss: "I am encouraged to believe that the origin and properties of our universe may be explicable within the framework of conventional science." Edward Tryon, 1973.

2   The theistic principle asserts that the universe is the way it is in order that we exist. Discuss this principle.

3   Can you prove that the universe was not created on the day you were born and will not end on the day that you die?

4   If universes create universes, what created the multiuniverse that consists of a population of reproducing universes? Have we merely enlarged the scope of cosmogenesis and must now explain the origin of an indefinitely large number of universes? Notice that this question is similar in some respects to asking: "If God created the universe, who created God?"

5   Are you an Augustinian believer or a Pelagian heretic? Explain.

6   If a Day in the life of Brahma is 4 320 000 years, and the life of Brahma is 100 Years, what is the lifetime of the Hindu universe? Compare the answer with the modern estimate based on the Hubble term.

7   In his article "Eschatology" (in *The Encyclopedia of Time*), Charles de Apollo writes, "Wherever minorities lived under social and religious oppression, and whenever plagues and wars erupted, eschatological forecasters arose, motivated by zealous piety and the overbearing desire for a better future." Predictions of cosmic doom are frequent, even nowadays. As I write, today's newspaper reports that a religious group in Massachusetts claims to have found biblical evidence predicting the end of the universe on a certain day in one month's time. Give reasons why eschatological and cosmothanatological predictions are common.

8   Thanatophobia is fear of death. Can you describe any period in history when cosmothanatophobia was not uncommon? Thanatophilia is a longing for death; can we say that prophets of doom suffer from cosmothanatophilia?

9   Read and enjoy:

> I saw eternity the other night
> Like a great ring of pure and endless light.
> All calm, as it was bright:
> And round beneath it, Time in hours, days, years,
> Driven by the spheres
> Like a vast shadow moved; in which the world
> And all her train were hurled.
> The World, *Henry Vaughan (1622–95)*

FURTHER READING

Arrhenius, S. *Worlds in the Making*. Harper, London, 1908.

Campbell, J. *The Masks of God: Creative Mythology*. Penguin Books, New York, 1976.

Davies, P. C. W. *The Last Three Minutes*. Basic Books, New York, 1994.

Dawkins, R. *The Blind Watchmaker*. Penguin Books, London, 1988.

Enteman, W. F. *The Problem of Free Will: Selected Readings*. Scribner's Sons, New York, 1967.

Fabian, A. C. Editor. *Origins: The Darwin College Lectures*. Cambridge University Press, 1988.

Gribbin, J. *The Omega Point*. Heinemann, London, 1987.

Gribbin, J. *In the Beginning: After COBE and Before the Big Bang*. Little Brown, New York, 1993.

Guth, A. H. "Starting the universe: The big bang and cosmic inflation," in *Bubbles, Voids, and Bumps in Time: The New Cosmology*. Editor, J. Cornell. Cambridge University Press, Cambridge, 1989.

Hawking, S. *Black Holes and Baby Universes and Other Essays*. Bantam, New York, 1993.

Rees, M. J. "Origin of the universe," in *Origins: The Darwin College Lectures*. Editor, A. C. Fabian. Cambridge University Press, Cambridge, 1988.

Rees, M. J. *Perspectives in Astrophysical Cosmology*. Cambridge University Press, Cambridge, 1995.

Rees, M. J. *Before the Beginning: Our Universe and Others*. Addison-Wesley Longman, Reading, Massachusetts, 1997.

Sproul, B. C. *Primal Myths: Creating the World*. Harper and Row, New York, 1979.

Tryon, E. P. "What made the world?" *New Scientist*, p. 14 (8 March 1984).

Waldrop, M. M. "Do-it-yourself universes." *Science*, p. 845 (20 February 1987).

Weinberg, S. "The decay of the proton." *Scientific American* (June 1981).

SOURCES

Borges, J. L. "The garden of forking paths," in *Labyrinths: Selected Stories and Other Writings*. Editor, J. E. Irby. New Directions, Norfolk, Connecticut, 1962.

Carter, B. in *Confrontations of Cosmological Theories with Observational Data*. Editor, M. S. Longair. Reidel, Dordrecht, 1974.

Eddington, A. S. "The end of the world from the standpoint of mathematical physics." *Nature*, Supplement, p. 447 (March 24, 1931).

Eliade, M. *Myths of Creation and of Origin: From Primitives to Zen*. Collins, London, 1967.

Eliade, M. *The Myth of the Eternal Return, or Cosmos and History*. Princeton University Press, Princeton, New Jersey, 1971.

Ellis, G. F. R. *Before the Beginning: Cosmology Explained*. Boyers-Bowerdean, London, 1993.

Friedrich, O. *The End of the World: A History*. Coward, McCann, Geoghegan, New York, 1982.

Gardner, M. *Fads and Fallacies in the Name of Science*. Dover Publications, New York, 1957.

Gosse, P. See Gardner, M.

Haber, F. C. *The Age of the World: Moses to Darwin*. Johns Hopkins University Press, Baltimore, 1959.

Hardin, G. *Nature and Man's Fate*. Rinehart, New York, 1959.

Harrison, E. R. *The Masks of the Universe*. Macmillan, New York, 1985.

Harrison, E. R. "Atomicity of time," and "Cosmology," in *Encyclopedia of Time*. Editor, S. L. Macey. Garland Publishing, New York, 1994.

Harrison, E. R. "The natural selection of universes containing intelligent life." *Quarterly Journal of the Royal Astronomical Society* 36, 193 (1995).

Harrison, E. R. "Creation and fitness of the universe." *Astronomy and Geophysics* 39, 27 (1998).

Heninger, S. K. *The Cosmographical Glass: Renaissance Diagrams of the Universe*. The Huntington Library, San Marino, California, 1977.

Islam, J. N. "Possible ultimate fate of the universe." *Quarterly Journal of the Royal Astronomical Society* 18, 3 (1977).

Jeans, J. *Astronomy and Cosmogony*. Cambridge University Press, Cambridge, 1929.

Kasner, E. and Newman, J. *Mathematics and the Imagination*. Simon and Schuster, New York, 1952.

Lemaître, G. "The beginning of the world from the point of view of quantum theory." *Nature* 127, 706 (1931).

Lemaître, G. *The Primeval Atom*. Van Nostrand, New York, 1951.

Linde, A. "Particle physics and inflationary cosmology." *Physics Today* (September 1987).

Linde, A. *Inflation and Quantum Cosmology*. Academic Press, Boston, 1990.

Maclaglan, D. *Creation Myths*. Thames and Hudson, London, 1977.

Maimonides, M. *The Guide for the Perplexed*. Dover, New York, 1927. Part 1, Chapters 72–76.

Nelkin, D. *The Creation Controversy: Science or Scripture in the Schools*. Norton, New York, 1982.

Newton, I. *The Chronology of Ancient Kingdoms Amended*. London, 1728.

North, J. D. "Chronology and the age of the world," in *Cosmology, History, and Theology*. Editors, W. Yourgrau and A. D. Breck. Plenum Press, New York, 1977.

Numbers, R. L. "Creationism in 20th-century America." *Science* 218, 538 (1982).

Osterbrock, D. E. and Raven, P. H. Editors. *Origins and Extinctions*. Yale University Press, New Haven, 1988.

Peacock, A. "Cosmos and creation," in *Cosmology, History, and Theology*. Editors, W. Yourgrau and A. D. Breck. Plenum Press, New York, 1977.

Poe, E. A. "Eureka: A Prose Poem," in *The Science Fiction of Edgar Allan Poe*. Editor, H. Beaver. Penguin Books, Harmondsworth, 1976.

Rees, M. J. "The collapse of the universe: An eschatological study." *Observatory* 89, 193 (1969).

Russell, R. J., Murphy, N., and Isham, C. J. Editors. *Quantum Cosmology and the Laws of Nature: Scientific Perspectives on Divine Action*. Center for Theology and the Natural Sciences, Berkeley, 1993.

Rutherford, E. "Radiation and emanation of radium," 1904, in *The Collected Works of Lord Rutherford*, volume 1, page 650. Allen and Unwin, London, 1962–1965.

Schmithals, W. "Eschatology," in *Dictionary in the History of Ideas: Studies of Selected Pivotal Ideas*. Editor, P. P. Wiener. Volume 2, page 154. Charles Scribner, New York, 1973.

Smolin, L. "Did the universe evolve?" *Classical and Quantum Gravity* 9, 173 (1992).

Smolin, L. *The Life of the Cosmos*. Oxford University Press, Oxford, 1997.

Stapledon, O. *Last and First Men and Star Maker*. Dover Publications, New York, 1968.

Tryon, E. P. "Is the universe a vacuum fluctuation?" *Nature* 396, 246 (1973).

Twain, M. (S. L. Clemens), *The Science Fiction of Mark Twain*. Archer Books, New York, 1984.

Whittaker, E. T. *The Beginning and End of the World*. Oxford University Press, Clarendon Press, Oxford, 1942.

Wigner, E. *Symmetries and Reflections*. Indiana University Press, Bloomington, 1967.

LIFE IN THE UNIVERSE

Life, like a dome of many-coloured glass,
Stains the white radiance of eternity.
Percy Bysshe Shelley (1792–1822), Adonais

ORIGIN OF LIFE ON EARTH

How did life originate on Earth? There are various theories, most of which fall into the four classes: special creation theories, spontaneous creation theories, panspermia theories, and biochemical theories.

Special creation theories

The belief that life originated as a supernatural event is the metaphysical theory of special creation. It has numerous mythic variations. Most recorded myths distinguish between nonliving and living things. Often the nonliving world comes first, the living world follows, and creation is thus a twofold act. Catastrophe theories of the eighteenth and nineteenth centuries elaborated on such beliefs and proposed that many acts of creation had occurred in the past. After a catastrophe had destroyed the terrestrial environment, newly created life arose in more evolved forms, and evolution occurred supernaturally, not naturally. Organic life even in its rudest forms was thought to be composed of substances fundamentally different from those of nonliving things. To this day we speak of organic and inorganic chemistry, although this distinction is now a matter of convenience only, and organic chemistry deals mostly with the numerous compounds containing carbon atoms. It came as a shock when the chemist Friedrich Wöhler in 1828 first made urea (a simple organic substance) from inorganic chemicals. Subsequent developments showed that chemicals are interchangeable between inorganic and organic things, thereby unifying the living and nonliving worlds at the atomic and molecular levels. We no longer believe that living things are made of nonmaterial substances requiring special creation in a material world. Unification at the atomic level has placed severe constraints on special creation theories.

Spontaneous creation theories

For ages it was believed that living things can be created spontaneously in a magic manner. "Even now multitudes of animals are formed out of the earth with the aid of showers and the sun's genial warmth," wrote the Roman poet Lucretius. Until recent times, spontaneous creation was widely regarded as part of the cyclic fecundity of nature in a world animated by supernatural agents.

Spontaneous creation in the form proposed by Aristotle was later widely accepted, even by men such as Descartes and Newton. Worms in the earth, eels in the mud, maggots in apples, flies around waste matter, and many other lower links in the chain of being were thought to be spontaneously created in warm, moist, shady places. In Judaic-Christian-Islamic societies, spontaneous creation was a vestigial creative urge surviving from the original act of special creation. It was believed that witches and wizards had the power to influence this creative urge in its demonic forms.

William Harvey, a physician who discovered in 1628 the circulation of the blood, speculated that all forms of life, seemingly spontaneously created, might actually be born from seeds and eggs too small to be seen by the eye. The Italian physician Francesco Redi tested this hypothesis in 1668 and found, using meat isolated in flasks or covered with gauze, that maggots are born from eggs laid by flies. But a new difficulty soon arose: Van Leeuwenhoek of Holland, with carefully made microscopes, discovered a teeming world of small organisms. Many persons thought this new world of microscopic life was maintained by spontaneous generation and was perhaps the missing link between the living and nonliving worlds.

The deathblow was delivered by Louis Pasteur in 1884, who demonstrated beyond doubt that microorganisms are not spontaneously created. Using filtered air and isolating nutrient fluids from contact with the atmosphere, he showed that the generation of microorganic life could be avoided. When exposed to the atmosphere the nutrient fluids became cloudy and swarmed with microscopic life. The conclusion was clear: These lower forms of life are conveyed by the atmosphere and breed wherever they have access to a nutrient medium.

Panspermia theories

Special and spontaneous creation seemed natural and even necessary in a world where life, according to the Mosaic chronology, had existed for only thousands of years.

The Epicurean atomic philosophy, revived in the Late Middle Ages, opened up the dizzy prospect of life existing throughout infinite space over indefinitely long periods of time. Inevitably, in this cosmic context, the idea emerged that life in primitive form propagates like pollen from planet to planet and flourishes wherever it finds a hospitable environment. In the panspermia theory, proposed early in the twentieth century by the Swedish chemist Svante Arrhenius, spores or other minute organisms in a dormant state travel through interstellar space propelled by the pressure of starlight. These organisms originate on planets and then are wafted through the heavens, ready to initiate life wherever possible. In a more recent version of this theory, proposed by Francis Crick and Leslie Orgel, and called directed panspermia, advanced civilizations direct automated space vehicles loaded with organic genetic material to planetary systems potentially favorable to the evolution of life. According to this theory, we on Earth have kinship with the races of the skies.

Actually, panspermia is only a transport theory. It does not explain the origin of life. Life is not created afresh each time it ascends the chain of being, and the basic problem of creation is left unsolved.

Some astronomers have said that minute organisms, however hardy, are incapable of surviving the hazards of long periods of interstellar travel, and stellar winds, ultraviolet radiation, and cosmic rays will destroy them. This argument is not entirely convincing. Stars originate from condensations of gas and dust in cool and dense interstellar clouds rich in organic and inorganic molecules. Not impossibly, in these clouds, complex molecules and even cells, attached perhaps to the surfaces of dust grains, survive for long periods of time. Almost certainly we have not heard the last word on the panspermia theory.

Biochemical theories

The physical manifestation of life consists of molecules obeying the laws of physics. The molecules form intricate cellular structures that also obey the laws of physics. Hence we are free to investigate in a scientific manner how such structures might have originated in the physical world. This is the biochemical theory of the origin of life. With an abundance of prebiotic (before life) organic chemicals on the surface of the early Earth, and a medley of exotic conditions, we cannot reject out of hand the possibility that self-replicating molecular systems formed naturally. Once such systems originate there seems no reason why they should not evolve into more complex systems by means of natural selection. Charles Darwin wrote in 1871 in a letter,

"But if (and oh! what a big if!) we could conceive in some warm little pond, with all sorts of ammonia and phosphoric salts, light, heat, electricity, &c present, that a protein compound was chemically formed ready to undergo still more complex changes." The biochemical origin of life was investigated by John Haldane and Alexander Oparin; it has been developed by many scientists, and is now widely accepted. Haldane, in "Origin of life" (1954), wrote, "Critics will say that a self-reproducing machine is still a machine, and there is an absolute gulf between any possible activity of such a machine and the most elementary feeling or desire, let alone human consciousness. Of such critics I ask, 'Do you think that your idea or perception of a stone is like a stone?' "

THE EXUBERANT EARTH

Life began on Earth almost four billion years ago. The terrestrial environment was probably rich in organic chemicals – a veritable organic Garden of Eden – and conditions were favorable for the biochemical origin of life. How the first steps were taken is still an uncertain subject.

The Earth trembled incessantly with earthquakes and its surface was cratered repeatedly with the infall of meteors of all sizes. The atmosphere of hydrogen, nitrogen, ammonia, methane, and other gases, including traces of oxygen, was wracked by titanic storms and fed by numerous volcanic plumes. The dark sky, torn by continual lightning, glowed red from the reflected glare of lava flows. Volcanic ash and torrential rain fell continually on the smoldering land and steaming seas. Ultraviolet radiation, electrical discharges, radioactivity, shockwaves, and all the razzmatazz of an exuberant Earth established an immense biochemical industry producing myriads of organic compounds in vast quantities. These compounds, including amino acids and nucleotides, concentrated in the warm and shallow seas.

Possibly mountains were not very high because the Earth's crust was thin; possibly less water covered the Earth's surface (because it had not yet all come from the interior in volcanic gases and other forms) and deep oceans were a thing of the future. But shallow seas, scattered everywhere, covered much of the Earth's surface. Each was a pool of "primeval broth," boiled, cooled, diluted, shaken, decanted, and occasionally thrown into the sky by some violent event. Trillions of biochemical experiments were performed every moment under all possible conditions. In the seas, in rock fissures, deep underground, and on the surfaces of dust particles, molecules of exotic form had their hour and were then incinerated, buried under ash, and enfolded within the Earth. Irresistibly, the wizardry of biochemistry advanced, becoming more ingenious with the passage of millions of years. Nucleotides joined together to form chainlike molecules of numerous, often freakish, codings that controlled the assembly of amino acids into proteins of novel design.

At some stage – and why not? – a molecule became self-replicating and was thus able to multiply. In this way, an entire sea might be dominated by a species of replicating molecule. Possibly each sea found its own solution to the challenge of replication; and volcanic eruptions, inundations, and strong winds enabled the seas to exchange genetic codings and compete with one another.

Probably more than once the whole biotic enterprise was destroyed. The scarred face of a visiting planetesimal, tens of kilometers in diameter, loomed in the sky and minutes later an unimaginable explosion devastated the entire terrestrial surface. After millions of years, when volcanic fury had created a new atmosphere and new seas, it would all begin again.

The bombardment era (during which the planets and moons swept up the debris left over from the formation of the Solar System) ended about four billion years ago. During the next half billion years, in a way not yet understood, cells originated in their simplest form. The growing dilution of the seas and the decreasing abundance of organic chemicals perhaps favored the survival of replicating molecular structures retaining their own enriched environment. The invention of the membrane, which enclosed an enriched environment, transferred the individuality of life from the seas to the cells, and life in recognizable form began.

THE EVOLUTION OF LIFE

The building blocks of organisms are relatively simple molecules known as amino acids and nucleotides that are made from atoms of hydrogen, carbon, nitrogen, oxygen, and other elements in lesser amounts. The amino acids join together to form the long chainlike molecules of the numerous proteins; the nucleotides join together to form the long chainlike molecules of the nucleic acids DNA and RNA. The DNA molecules have the shape of a double helix and contain the genetic coding of the entire organism.

The word cell was introduced in 1665 by Robert Hooke and the cell theory (stating that all organisms are constructed from cells) was advanced in the nineteenth century. Unicellular (single-celled) organisms vary greatly in size, from small bacteria hundreds of times the size of a hydrogen atom to large ostrich eggs. Multicellular organisms contain numerous interacting cells having various specialized functions; the human body consists of about 10^{14} cells, and on the average each cell consists of 10^{14} atoms. The two basic kinds of cell are the prokaryotes, which were the first to evolve, and the more complex eukaryotes, which evolved later. Eukaryote cells have their DNA confined within a small region of the cell known as the nucleus; prokaryote cells, however, lack nuclei and their DNA is dispersed in the cell. Of the five kingdoms of life, the Monera (bacteria, blue-green algae, etc.) consist of organisms composed from prokaryotes; the remaining kingdoms of Protista (diatoms, amoeba, seaweed, slime molds, etc.), Fungi (toadstools, mushrooms, etc.), Plantae (mosses, trees, etc.), and Animalia (worms, insects, mammals,

etc.) consist of organisms composed mostly of eukaryotes.

There exist two functionally different kinds of organism: autotrophs and heterotrophs. The autotrophs are self-nourished; most are plants; their food supply consists mainly of inorganic chemicals, and usually their energy comes from photosynthesis in which carbon dioxide and water are converted by sunlight into sugars:

carbon dioxide + water + sunlight

$$= \text{sugars} + \text{oxygen}, \qquad [26.1]$$

and energy is stored in the sugars. The heterotrophs are other-nourished; most of them are animals; their food supply consists of organic substances, and their energy comes from the inverse process:

sugars + oxygen = carbon dioxide + water

$$+ \text{energy}, \qquad [26.2]$$

involving respiration. Photosynthetic autotrophs create the sugars, discharge oxygen into the atmosphere, and are consumed as food by heterotrophs that convert the oxygen back to carbon dioxide. Many elaborations of this simplified picture exist.

Symbiosis is the living together of dissimilar organisms for mutual benefit (parasitism is when the association is not to mutual benefit). The dependence of flowering plants on insects, which feed on and pollinate flowers, is an example of communal symbiosis. In host–guest symbiosis the host can be a heterotroph and the guest an autotroph, as in the case of lichens, where the host is a fungus (providing moisture and minerals) and the guest is an alga (providing food by photosynthesis).

Some of the components in eukaryote cells (such as the mitochondria and chloroplasts) possess their own genetic material. Not impossibly the eukaryotes evolved long ago by incorporating useful prokaryotes in a host–guest symbiotic relationship.

How cells originated is unknown. The first cells were perhaps primitive heterotrophs of bacterial form that depended on an abundant supply of organic molecules.

Table 26.1 *Geological time scales*

Time (millions of years before present)	Era	Period	Epoch	Events	
		Quaternary	Recent Pleistocene	Homo sapiens	
			(*pleisto* = most)		
5			Pliocene (*plio* = more)		5
10	Cenozoic (*ceno* = recent)	Tertiary			10
			Miocene (*mio* = less)	hominids primates	
			Oligocene (*oligo* = few)	grass, flowering plants	
50			Eocene (*eos* = dawn)	mammals	50
			Paleocene	dinosaurs disappear	
100	Mesozoic (*meso* = middle)	Cretaceous			100
		Jurassic		birds dinosaurs	
		Triassic		reptiles	
		Permian		fish, amphib-	
	Paleozoic (*paleo* = ancient)	Carboniferous to Cambrian	Devonian Silurian Ordovician	ians, insects vertebrates forests invertebrates metazoans ozone layer	
500					500
1000		Precambrian era or Proterozoic era (*protero* or *proto* = earliest)		oldest photo-synthetic plants oldest fossil cells oldest rocks origin of Earth	1000
5000					5000
		Pre-Solar era		origin of Galaxy	

Simple autotrophs, similar to the photosynthetic blue-green algae, presumably came later as the supply of organic chemicals declined. According to the fossil record, these autotrophs first appeared approximately 3.5 billion years ago. Perhaps the hard-pressed heterotrophs, which earlier had evolved in a world of plenty, survived by forming host–guest symbiotic relations that gave many autotrophs protection in exchange for their special services. Thus some cells attached themselves to the heterotrophs as threadlike flagella, and the combined organism became mobile and able to search for food; others became resident inside the heterotrophs as mitochondria and chloroplasts. In such ways, groups of prokaryotes combined to form sophisticated cells more capable of surviving. How sex and cell division originated are still mysteries, and the elaboration and perfection of cell-division may have occupied most of the Proterozoic era that ended less than a billion years ago.

Toward the end of the Proterozoic era the highest forms of life were still unicellular. Suddenly – or so it seems – under the protection of the newly formed ozone layer in the upper atmosphere, which filtered out ultraviolet radiation from the Sun, these single-celled organisms formed a profusion of multicellular creatures. Various invertebrates (creatures without backbones) appeared in a hundred million years or so, and numerous vertebrates and plants followed. Atmospheric oxygen continued to increase throughout the Paleozoic and Mesozoic eras, attaining its present level at about the beginning of the Cenozoic era. The drifting land masses, driven by large-scale motions in the Earth's mantle, formed and reformed, creating a changing pattern of continents and oceans. The forests, inland seas, marshes, and rivers swarmed with life: fish, insects, amphibia, and reptiles. Then in the Triassic period came the dinosaurs. Of these diverse creatures, some species were small and fleet-footed, some were able to fly, some were scaled and armored, and some were colossal in size.

The dinosaurs flourished in the Jurassic and Cretaceous periods and undoubtedly were the most vigorous and intelligent creatures of those times. By the end of the Cretaceous period they had almost vanished (perhaps because of the mass extinctions caused by meteoritic impacts and other environmental catastrophes), leaving the birds as their descendants.

Lush grasses, brightly colored flowering plants, and the foliage of deciduous trees transformed the world. Mammals arose and flourished in the early Cenozoic era, and about 20 million years ago, in the forests and on the savannas, the first primates appeared. Intelligence, crudely estimated from measurements of fossil skulls, has advanced rapidly in the last few million years, and in the recent 10 or so thousand years, since the retreat of the northern glaciers, agrarian societies, city states, and advanced industrial cultures have arisen.

NATURAL SELECTION

Astronomers speak of the evolution of the Sun and often seem to be the only scientists who use "evolution" in its correct sense. Evolution has nothing to do with progress, and means unrolling, as in scrolling down a computer screen, and applies to "an orderly succession in a long train of events." Debates on creation versus evolution rarely use the word evolution in its correct sense.

Jean Baptiste de Lamarck, a French naturalist who popularized the word "biology," in 1809 elaborated on the old idea that skills and other aptitudes acquired by individuals are inherited by their descendants. According to this common-sense belief, evolution is both self-directed and goal-oriented. Parents who learn to read and write have children who, even if they do not live with their parents, are motivated to learn to read and write. This belief, called Lamarckism, is still popular and accepted as obviously true. But Lamarckism fails to explain the way that life evolves. Skills acquired by the efforts of one generation are not inherited genetically by the next generation; at most, all that is inherited is the capacity to acquire skills.

Natural selection, a new theory of evolution, was independently proposed in 1858 by Charles Darwin and Alfred Wallace. Darwin presented a year later a detailed argument in his book *The Origin of Species*, an argument on which he had pondered for years. Darwin and Wallace, as naturalists, were intrigued by the adaptations of animal and plant species to different environments. Each hit on the idea of natural selection after reading *An Essay on Population* by Thomas Malthus, written in 1798.

The theory of natural selection brings together two streams of thought. The first is that the individuals of an interbreeding population are never exactly alike but vary from one another by small differences. The second is that the growth of a population is continually checked by environmental constraints. From this confluence of thought comes the new theory: those genetic differences that aid individuals to survive and reproduce are shared increasingly among members of the interbreeding population, and the population thus evolves (changes) by selective reproduction.

In all species there are advantageous and disadvantageous variations among the members, and by the continual elimination of the latter, a species evolves and becomes better adapted to the environment. The giraffe's long neck is not the result of striving to reach greater heights, as supposed by Lamarck, but the result of natural selection that has nothing to do with conscious desire. Giraffes have lived in competition with species that browsed on lower vegetation, and successive increases in height over many generations have been advantageous to the survival of giraffes with long necks. The natural selection process applies to populations of reproducing members, and though it may produce many strange twists and turns, it is as inexorable as any other law of the physical world. Because the environment is generally in a state of change, existing adaptations tend always to be out of date, and evolution is a dynamic process

of trying to catch up with whatever favors survival.

How fortunate, we exclaim, that natural selection preserves advantageous variations. Only the fittest survive! But let us not forget that advantageous is defined by what survives. The "survival of the fittest" is a phrase that must be used with care, for many peculiar creatures have evolved and then later become extinct, such as the Irish Elk with its immense antlers that in the end were more a burden than an advantage. If we use survival of the favored instead, it must be understood that the favoring is done by the environment – a complex yet witless environment that continually changes. Natural selection is never teleological (directed toward final goals) and cannot guarantee progress of any desired kind. Each step in biological evolution is directed by what survives and reproduces the most under the reigning circumstances; many species and many variations in a species are eliminated that under other circumstances might later have proved superior to those that actually survived.

Darwin's theory did not explain the origin of individual variations in a species. We now know that genetic variations are the fundamental cause of individual differences. Mutations in the genetic code are natural changes that occur inevitably in molecular systems of many interacting atoms. At first it was thought that interbreeding would blend together all variations and produce similar individuals; for example, a tall man and a short woman would have children of intermediate height. But botanical experiments by Gregor Mendel, an Austrian monk, demonstrated in 1865 that variations do not blend together and disappear with interbreeding. Mendel planted seeds of tall and short pea plants and noticed, after crossbreeding their peas, that the new plants were not of intermediate height. He found that tall and short plants were produced in numbers that have a fixed ratio. Mendel's experiments passed unnoticed at the time and did not attract attention until the beginning of the twentieth century.

INTELLIGENT LIFE

Why human beings developed large brains hundreds of thousands of years ago is not fully understood. Early human beings lived in primitive social groups and seemingly had no urgent need for large brains equal to those of modern human beings. They made and used tools, and this helps to explain their adept hands, but not their large brains. Apes use tools and yet do not have such large brains, and it seems doubtful that tool-making by itself accounts for the large human brain. The conditions promoting large brains must have been of a kind that naturally selected intelligence because of its survival value. We must ask: how did thin-skinned primates defend themselves against beasts with thick skins, sharp claws, and long tusks and horns? A good guess is that their survival was the result of organized and cooperative action made possible by the development of speech. The breakthrough to large brains came when our remote ancestors learned how to talk. Communication with a large vocabulary of symbolic sounds and the representation of the external world with acoustic imagery became a highly effective way of surviving. Organized methods of defense and attack turned human beings into formidable opponents. Maybe "man is a talking animal" is nearer the truth than "man is a tool-making animal."

In food-hunting and food-gathering societies the young were taught the language and initiated into the laws and myths, and the old were cherished as guardians of the cultural heritage. The society survived by the vigor of its youth and the knowledge of its elders. In the hundreds of thousands of years of the unrecorded past there were surely great singers, great artists, great story-tellers, great thinkers, and great leaders, some perhaps surpassing those known in the short span of recorded history.

To grasp the situation let us suppose that the evolution of life on Earth is compressed into the time span of a single day. Every 24 seconds of the day is equivalent to a million years of biological evolution. For 18 hours unicellular organisms thrive; at 6:00 pm in the evening, multicellular creatures appear; the mammals arrive at 11:00 pm, and hominoids emerge a minute or so before 12:00 pm; and human intelligence blossoms within the last few seconds. In this compressed picture of biological history, we cannot help but wonder what human intelligence will be like a few seconds after midnight. Will it have changed to something beyond the reach of our imagination or will it have vanished like a match flaring briefly in the long night?

WHAT IS LIFE?

Science explains the world around us by decomposing it into an activity of smaller and smaller parts. In this way the world is reduced to molecules, then atoms, then subatomic particles and their interactions. When this reductionist method of explanation is applied to living things, organisms reduce to cells, then molecules, atoms, and their subatomic particles. Thus living things reduce to the same basic constituents as nonliving things. All recognizable properties of life and mind are lost in the process, and yet remain potential in particles and their fields. Many thoughtful persons oppose this reductionist philosophy and believe life and mind cannot be fully explained as a dance of atoms and waves.

Vitalism is the theory that life is something essentially nonphysical that permeates and animates the physical world. This vital essence is the psyche according to Aristotle or *élan vital* according to the Irish-French philosopher Henri Bergson. In Bergson's theory of creative evolution, the world of material things is orderly and deterministic, whereas the *élan vital* – the breath of life – is creative and free. "The vitalist principle," he wrote, "may indeed not explain much, but it is at least a sort of label affixed to our ignorance, so as to remind us of this occasionally, while mechanism invites us to ignore that ignorance." Pierre Teilhard de Chardin, a Jesuit paleontologist-vitalist, believed that mind is interwoven in the physical universe and progresses to an "omega" state where individuality becomes submerged in a cosmic social mind.

Vitalism introduces nonphysical agents into the physical universe and hence violates the containment principle (Chapter 9). Many biologists oppose vitalism and regard it as an attempt to enliven the physical universe with an inlay of magical properties, a sort of magic universe animated by unseen spirits. One might indeed ask how a vital force (a sort of psychic efflorescence of material things) can explain the mind that is conceiving the vital force. On the other hand, one might also ask the reductionist (who claims that life and mind consist of an activity of physical things) where in the atoms and waves is the conceiver of a world of atoms and waves? Vitalists and reductionists believe themselves to be adequately portrayed in their own imagery.

Life, viewed objectively, seems sufficiently explained in terms of organic structures and their functions. Viewed subjectively, however, its inner world of experience seems inadequately explained by its own concepts of the physical world. No instrument in the laboratory can detect the existence of consciousness and yet each of us knows that consciousness exists.

LIFE BEYOND THE EARTH

Since classical antiquity people have believed in the existence of life beyond the Earth. The gods and goddesses lived in the skies long before the medieval universe adopted them, and modern science fiction is often no more than a resurrection of mythological imagery in a framework of pseudoscience. From a cosmological viewpoint it seems highly improbable that Earth is the only place in the universe where life exists. It would be preposterous to suppose that our Galaxy, out of billions of galaxies, is the only one in the universe where life has arisen. Finding ourselves in a vast universe, we feel impelled to believe that we are not alone, and in this sense the ancient myths are credible.

Speculation about the existence of life in other galaxies is interesting but at present far beyond all means of verification. Speculation about life elsewhere in our Galaxy, on which astronomy has much more to say than cosmology, is a possibility not beyond all means of verification. This has awakened interest in recent decades and directed attention to the problem of communicating with intelligent life elsewhere in the Galaxy. From the marriage of astronomy and biology comes the subject of exobiology.

Biologists are acutely aware of the hazards of evolution. To many biologists it is not in the least obvious that multicellular organisms must arise, evolve, and become intelligent. Their reservations concerning life elsewhere have been overshadowed by the optimism of astronomers. The latter have taken their cue from the Epicureans who long ago visualized a universe of endless worlds all teeming with life. Christiaan Huygens in *The Celestial Worlds Discover'd* (Figure 26.1) wrote: "Why should not every one of these stars or suns have as great a retinue as our sun, of planets, with their moons, to wait upon them?" These planets, he said, "must have their plants and animals, nay and their rational creatures too, and these as great admirers and diligent observers of the heavens as ourselves" (published posthumously in 1698). From this optimistic outlook in astronomy we see our Galaxy plenitudinously strewn with sunlike stars, each formed in much the same way as the Sun, and each encircled with planets, many of them earthlike. Intelligent beings live on Earth, and therefore intelligent beings must also live in other planetary systems. A technological civilization exists here on Earth; why not other such civilizations elsewhere? Surely we should try to communicate with them? If we do not try we shall never know. The debate on the existence in the Galaxy of extraterrestrial technological civilizations (or ETCs) has prompted astronomical searches and inspired several interesting discussions.

Extraterrestrial technological civilizations
Do ETCs exist? A common argument is as follows. Let N be the number of stars in

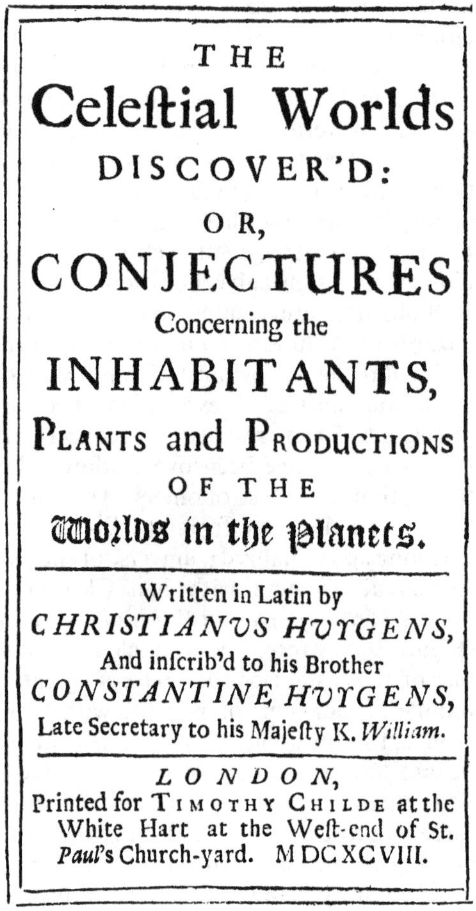

THE
Celeſtial Worlds
DISCOVER'D:
OR,
CONJECTURES
Concerning the
INHABITANTS,
PLANTS and PRODUCTIONS
OF THE
𝔚𝔬𝔯𝔩𝔡𝔰 𝔦𝔫 𝔱𝔥𝔢 𝔭𝔩𝔞𝔫𝔢𝔱𝔰.

Written in Latin by
CHRISTIANVS HUYGENS,
And inſcrib'd to his Brother
CONSTANTINE HUYGENS,
Late Secretary to his Majeſty K. *William.*

LONDON,
Printed for TIMOTHY CHILDE at the
White Hart at the Weſt-end of St.
Paul's Church-yard.　M DC XC VIII.

Figure 26.1. Title page of *The Celestial Worlds Discover'd* (1698) by Christiaan Huygens.

the Galaxy. The total number of extraterrestrial technological civilizations that have arisen in the lifetime of the Galaxy is expressed by the formula:

$$\text{number of ETCs} = N \times A \times B, \qquad [26.3]$$

where A is a fraction determined by astronomical considerations and B is a fraction determined by biological considerations. The fraction A can be written:

$$A = p_1 \times p_2 \times p_3. \qquad [26.4]$$

Here p_1 is the fraction of all stars in the Galaxy similar to the Sun, stars that are not too blue and not too red, not too luminous and not too under-luminous, and not members of close binary systems. A reasonable estimate is that p_1 has a value 0.1. The second term, p_2, is the fraction of these sun-like stars that have earthlike planets; it has been supposed in some arguments that p_2 has a value of unity, but we shall be conservative and assign to it a value 0.1. The third term, p_3, is the fraction of such planets occupying a habitable zone, not too close (like Venus) and not too far (like Mars) from the parent star; for this we shall also assume a value 0.1. In the Galaxy there are approximately 100 billion stars ($N = 10^{11}$), and therefore:

$$\text{number of ETCs} = 10^8 \times B. \qquad [26.5]$$

This rough estimate gives 100 million planets in the Galaxy where conditions are similar to those on Earth and where life might have originated.

Our real difficulties begin when we attempt to estimate a value for the biological fraction B, which can be written:

$$B = p_4 \times p_5 \times p_6 \times p_7. \qquad [26.6]$$

In this expression, p_4 is the probability that life originates in a unicellular form; p_5 is the probability that life evolves into multicellular organisms, such as mammals; p_6 is the probability that such organisms develop intelligence equal to or greater than that of human beings; and p_7 is the probability that intelligent life develops an advanced technological civilization.

Optimistic studies assume that $p_4 = 0.1$, $p_5 = p_6 = p_7 = 1$, and therefore $B = 0.1$. According to this argument of low credibility the number of extraterrestrial technological civilizations is 10 million. These civilizations have existed at different times in the history of the Galaxy, and the number existing at any moment, including the present, is given by

number of ETCs at any moment

$$= \text{total number of ETCs} \times \frac{t}{T}, \qquad [26.7]$$

where t is the average lifetime of such a civilization and T is the age of the Galaxy (approximately 10 billion years). Let us continue in this optimistic vein and assume that technological civilizations endure on the average for 1 million years. The number existing at any moment, including the present, is therefore 1000. A simple calculation then shows that in the disk of the Galaxy the average separating distance is roughly 1000 light years. Technological civilizations lasting for a million years have therefore plenty of time to communicate with other existing technical civilizations.

A pessimistic and perhaps more credible view of the value of B is found as follows. We again assume $p_5 = 0.1$ is the probability that life originates on an earthlike planet in a habitable zone, although we have no clue as to how far this might be wrong. Also we have no guarantee that life will evolve, even over billions of years, into multicellular organisms and in recognition of the hazards and traps involved we shall assign to p_5 a value of 0.1. Nothing compels us to conclude that advanced intelligence is inevitable for numerous species on Earth have survived long periods of time without it. The environment must affect the right species in the right way at the right time so that natural selection favors the development of large brains. The probability of this happening could be extremely small, but not to overdo the pessimism we shall assume that $p_6 = 0.1$. Finally, we must ask, what is the probability that intelligent life develops science and its handmaiden of advanced technology? Science and the scientific method were not discovered by the numerous cultures of Africa, America, China, India, Japan, or almost all other places, and its discovery was by no means a simple and straightforward set of events. Science made its first hesitant steps in the Hellenic world because of a few incredulous individuals who were ridiculed by their contemporaries, and was later developed in Europe in the face of organized hostility. Science arose because of accidental and improbable circumstances that existed in Greece and later in Europe. The probability p_7 of intelligent life constructing an effective scientific view of the universe is perhaps no greater than 0.1. With $p_4 =$

$p_5 = p_6 = p_7 = 0.1$, we have $B = 10^{-4}$, and the number of technological civilizations that have arisen in the lifetime of the Galaxy is therefore 10 000. For the Galaxy to contain at least one ETC at any one time, each must last for at least 1 million years. But a pessimistic (realistic?) estimate of the expected lifetime of an advanced technological civilization might be only a few centuries. Let us be generous and estimate a thousand years. This suggests that we, an advanced technological civilization, are alone in the Galaxy, and after our demise the next will occur somewhere in about 1 million years.

Galactic colonization

Estimates of the number of technological civilizations existing in the lifetime of the Galaxy range from the high value 10 million to the low value 10 thousand, and the low value is probably nearer the truth. Possibly culturally developed societies, lacking an advanced technology, are more numerous than those with advanced technology (they endure for longer periods of time without destruction of themselves and their biospheres), and hence many could exist in the Galaxy at any one time. We must, however, stress the importance of advanced science and technology in this discussion, not just because of the enhanced prospect of communication, but because the attainment of advanced science and technology, as in our own society, marks a critical stage in biological evolution. Development of advanced science not only creates hazards, but also opens up vistas and avenues of exploration that are denied nonscientific civilizations.

One very important avenue of exploration can already be seen. Human beings are approaching the stage in their evolution where they will probably be able to redesign themselves genetically. The natural selection of Darwinian evolution will be replaced, at least in some aspects of human evolution, by the self-directed evolution of Lamarckism.

Undoubtedly, the development of science creates new and serious hazards, and the chances are that most technological civilizations are short-lived. When numerous persons each possess the power to devastate a planetary environment and destroy their own species with doomsday weapons, technologically advanced civilizations cannot be expected to last longer than a few centuries. Perhaps only 1 in every 10 survives the first thousand years, and all others either self-destroy or revert to an earlier low-technological state. Of the pessimistically estimated 10 000 technological civilizations that arise in the lifetime of the Galaxy, it is possible that only 1000 survive for longer than 1000 years.

We must consider what happens to those technological civilizations that survive and do not self-destroy in the first thousand years of their existence.

A thousand years should be sufficient to develop interstellar space travel. Fusion power and other technologies still beyond our present reach will enable a technologically advanced civilization to construct large space vehicles that can travel at, say, one-thousandth the speed of light. A journey of 10 light years distance, from one planetary system to another, will last 10 000 years. This is not unthinkable with a large space vehicle having its own biosphere containing a social unit of millions of individuals. A halt at their destination might last no more than 10 000 years before the embarkation of one or more spaceships on the next interstellar journey. In this way, step by step, in a growing wave of space colonies, life could diffuse outward from the home planet at a rate of 10 light years every 20 000 or so years. Given this rate of diffusive migration, the entire Galaxy will be colonized in 100 or so million years – a period equal to 1 percent of the age of the Galaxy. Despite many challenges and setbacks on different fronts, the growth and magnitude of such an enterprise ensures that it will survive and continue to spread. This not entirely implausible picture leads to the conclusion that the Galaxy is perhaps now colonized by highly intelligent forms of life that originated from about 1000 technological civilizations.

Confronted with this intriguing scenario, we feel impelled to ask why we are not aware of the existence of these other forms of intelligent life. "Where are they?" Surely they should rally to our aid and welfare and show us how to solve our most pressing problems?

Galactic selection

We are the outcome of the inexorable processes of natural selection. To natural selection can be attributed the present fitness of the human body and brain. When, however, an intelligent species takes control of its planetary environment, the evolutionary game of adaptation changes and new rules determine what is fit and unfit.

Not impossibly the survival of ETCs depends on a "galactic selection" law, a law that states: "Destructively aggressive intelligent forms of life cannot colonize the Galaxy." Conceivably it operates in two modes: the first is automatic, and the second is judicial.

When a species becomes highly intelligent and develops a technologically advanced civilization, the environment that previously directed natural selection falls under the control of the members of the species. Many previous checks and balances are overridden, creating an unstable situation in which irresponsible behavior leads to disastrous consequences. A technologically advanced society composed of irresponsible and destructive members has little chance of surviving. At this stage, galactic selection takes over automatically. Intelligent and destructive forms of life, prone to warfare, are unlikely to survive for very long, particularly if they devastate their biospheres. They are unlikely to survive long enough to attain command of interstellar travel on a significant scale. Sealed in their biospheres they are like virulent organisms in planetary test tubes. Before embarking on interstellar voyages of galactic conquest they either self-destruct or revert to much lower technological conditions. Intelligent creatures who colonize the Galaxy are probably peaceful, not aggressive, and do not build oppressive galactic empires of subject races.

Galactic selection as described is natural and automatic. It is a fail-safe sort of natural selection law ensuring that irresponsible behavior and destructive aggression is self-terminating. Let us suppose that occasionally, owing to unusual circumstances, an aggressive technological civilization breaks free from its planetary test tube. Galactic selection might then operate in its judicial mode. The highly intelligent creatures who have colonized the Galaxy, reckoning up the woeful cost of an interstellar race of vandals, may find that they have only one recourse: the deliberate termination of the aggressive civilization. Possibly they themselves will not watch and wait for a premature breakaway, but will place automated devices in the neighborhood of solar systems where life is awakening into an advanced intelligent state. These monitoring machines will read the signs of technological advancement, and if excessive aggression occurs, they will await automatic self-termination. When, on rare occasions, aggression continues unarrested, and there are signs of preparation for long voyages of interstellar travel, the machines might then follow their programmed instructions and proceed to effect termination. How this might be done is a matter of more than academic interest to the human race in the next thousand years.

Where are they?

Mythology has accustomed us to the belief that the gods and angels are intimately concerned with the affairs of humanity. The probable existence of extraterrestrial life of advanced intelligence in the Galaxy is puzzling, because we find it difficult to understand why it ignores us. It offends our self-esteem by ignoring us.

The explanation may actually be quite simple. Galactic selection restricts all forms of direct contact with technological civilizations that are still confined to their planets. Such civilizations cannot be encouraged or aided to quit their planets prematurely. They must demonstrate their fitness to mingle with alien creatures, and self-destruction is the perfect way of demonstrating unfitness.

Astronomy and biology, while stressing different viewpoints, lead to the conclusion that advanced technological civilizations may have colonized the Galaxy. Because of galactic selection, alien intelligent creatures are probably more angelic than demonic.

EPILOGUE

Our revels now are ended. These our actors,
As I foretold you, were all spirits and
Are melted into air, into thin air;
And, like the baseless fabric of this vision,
The cloud-capped towers, the gorgeous palaces,
The solemn temples, the great globe itself,
Yea, all that it inherit, shall dissolve
And, like this insubstantial pageant faded,
Leave not a rack behind. We are such stuff
As dreams are made on, and our little life
Is rounded with a sleep.
The Tempest, *Shakespeare's last play*

REFLECTIONS

1 *"I have no doubt that in reality the future will be vastly more surprising than anything I can imagine. Now my own suspicion is that the universe is not only queerer than we suppose, but queerer than we can suppose"* *(John Haldane,* Possible Worlds and Other Papers*).*

2 *What is life? This question concerns us all and is of paramount importance in cosmology. With the ancients, we divide the world into living and nonliving things, but have no widely accepted meaning of the word "life." Organisms are composed of cells that are composed of molecules that are composed of atoms, and it is not clear at what level of complexity life first emerges. The cell is a miracle of the physical world and required billions of years to evolve; can we say that it is nonliving and claim that life must exist only in complex multicellular organisms?*

Living organisms feed, grow, move, reproduce, and behave in response to their environments. Many nonliving things exhibit similar properties. An automobile moves and consumes food; a crystal grows; a candle flame needs nourishment, reacts to its environment,

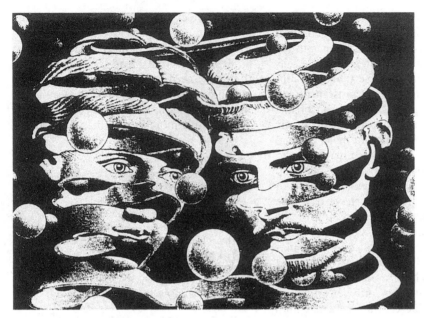

Figure 26.2. Life and the universe: *Bond of Union* (1956), by M. C. Escher. (Courtesy of the Collection Haags Gemeentemuseum, The Hague.) Is life accidental or essential in the scheme of things?

and self-reproduces with sometimes alarming consequences. Computers play chess with each other and are taking control of more and more of the routine functions of society. With so many nonliving things mimicking the characteristics ascribed to living organisms, it is difficult to know exactly what defines life. Are reproduction and natural selection the hallmarks of life? According to biochemistry, reproduction is a natural possibility in highly organized chemical systems. According to biology, natural selection operates automatically and consists of successive adaptations to a continually changing world. The nonliving physical world, it seems, has an astonishing power for creating organized complexity, and nothing of a physical nature sets life apart from the rest of the physical world.

When we search within ourselves for the meaning of life, we find that life is essentially psychic and consists of thoughts, emotions, and all that contributes to a state of self-awareness. But we are not sure what psychic means any more than what life means.

Many persons believe the word psychic denotes a nonphysical realm that interfaces with the physical realm; others hold that its explanation lies in the physical world.

● *"I have been saying that modern science broke down the barriers that separated the heavens and the earth, and thus it unified the universe. And this is true. But, as I have said, too, it did this by substituting for our world of quality and sense perception, the world in which we live, and love, and die, another world – a world of quantity, of reified geometry, a world in which, though there is a place for everything, there is no place for man. Thus the world of science – the real world – became estranged and utterly divorced from the world of life, which science has been unable to explain – not even to explain away by calling it 'subjective'. This is the tragedy of the modern mind which 'solved the riddle of the universe,' but only to replace it by another riddle: the riddle of itself"* (Alexander Koyré, Newtonian Studies).

3 *On two occasions at least, it was thought the vital force of life had been discovered.*

Luigi Galvani, an Italian anatomist, discovered accidentally in 1786 that the amputated hind legs of a frog would kick convulsively when in contact with a source of electricity. The legs also twitched when in contact with two different metals. Galvani thought that he had found the vital force of life and referred to it as "animal electricity." Nowadays we say he or she is galvanized into action. Robert Brown, a Scottish botanist, discovered in 1827 that minute particles of pollen, suspended in water, have continual irregular motion. He thought that this ceaseless jittery behavior revealed the activity of a vital force. But the Brownian motion of small particles is due to the random kicks of atoms and provides visible evidence of the atomic nature of matter.

4 "It goes against Nature, in a large field to grow only one shaft of wheat, and in the infinite universe only one living world" (Metrodorus, a student of Democritus, 400 BC).

• "Nothing in the universe is unique and alone, and therefore in other regions there must be other earths inhabited by different tribes of men and breeds of beasts" (Lucretius, De Rerum Natura).

• "In the cosmos there must be an infinite number of suns, with planets, with life around them" (Giordano Bruno, Infinite Universe and Worlds, 1584).

• "The atmospheres of celestial bodies as well as whirling cosmic nebulae can be regarded as the timeless sanctuary of animate forms, the eternal plantations of organic germs" (Justus von Liebig, Letters on Chemistry, 1861).

• "Yet whenever I see a frog's eye warily ogling the shoreward landscape, I always think inconsequentially of those twiddling mechanical eyes that mankind manipulates nightly from a thousand observatories. Someday ... we are going to see something not to our liking, some looming shape outside there across the great pond of space. Whenever I catch a frog's eye ... I stand quite still and try hard not to move or lift a hand since it would only frighten him. And standing thus it finally comes to me that this is the most enormous extension of vision of which life is capable: the projection of itself into other lives" (Loren Eiseley, The Immense Journey).

5 "Who would deny that such bodies, floating everywhere in the universal space, do not leave behind them the germs of life wherever the planetary conditions are already suitable to promote creation?" (Herman von Helmholtz, Formation of Planetary Systems, 1884). Helmholtz thought that microscopic organisms had been brought to Earth by meteorites.

• "In this manner life may have been transplanted for eternal ages from solar system to solar system and from planet to planet of the same system. But as among the billions of grains of pollen which the wind carries away from a large tree – a fir tree, for instance – only one may on an average give birth to a new tree, thus of the billions, or perhaps trillions, of germs which the radiation pressure drives out into space, only one may really bring life to a foreign planet on which life had not yet arisen and become the originator of living things on that planet.... Finally, we perceive that according to this version of the theory of panspermia, all organic beings in the whole universe should be related to one another" (Svante Arrhenius, Worlds in the Making).

6 "I say the power of the population is indefinitely greater than the power of the earth to produce subsistence for man ... the population, when unchecked, increases in geometrical ratio, subsistence only increases in an arithmetical ratio" (Thomas Malthus, An Essay on Population, 1798).

• "If variations useful to any organic being ever do occur, assuredly individuals thus characterized will have the best chance of being preserved in the struggle for life; and from the strong principle of inheritance, these will tend to produce offspring similarly characterized. This principle of preservation, I have called, for the sake of brevity, Natural Selection" (Charles Darwin [1809–1882], The Origin of Species, 1859).

• "There is grandeur in this view of life, with its several powers, having been originally breathed into a few forms or into one; and that, while this planet has gone cycling on according to the fixed laws of gravity, from so simple a beginning endless forms most beautiful and most wonderful have been, and are being, evolved" (final sentence of Darwin's The Origin of Species).

• "These checks – war, disease, famine and the like – must, it occurred to me, act on the animals as well as man. Then I thought of the enormously rapid multiplication of animals, causing these checks to be much more effective in them than in the case of man; and while pondering vaguely on this fact there suddenly flashed upon me the idea of the survival of the fittest – that the individuals removed by these checks must be on the whole inferior to those that survived" (Alfred Russel Wallace [1823–1913]).

7 "One afternoon, thinking about these facts [concerning the stability of the biosphere], the thought came that such constancy requires the existence of an active control system" (James Lovelock, "Hands up for the Gaia hypothesis," 1990). The idea of a self-regulating, self-sustaining biosphere is referred to by Lovelock as the Gaia hypothesis. Biologically, Gaia denotes a complex system of interdependent states. Advocates of the Gaia hypothesis, however, usually have much more in mind and ascribe living, even mystical, properties to Gaia; in this sense the Gaia hypothesis is a variation on vitalism.

8 "How, then, was an organ [the brain] developed so far beyond the needs of its possessor [primitive man]? Natural selection could only have endowed the savage with a brain a little superior to that of an ape, whereas he actually possesses one but little inferior to that of the average member of our learned societies" (Alfred Wallace). "Savages" have languages as intricate as our own, and Wallace fails to realize that linguistic communication implies large brains. Natural selection favors the survival of social groups of communicating individuals, and communicating individuals have large brains.

• Language "is not merely a reproducing instrument for voicing ideas but rather is itself the shaper of ideas, the program and guide for the individual's mental activity, for his analysis of impressions, for his synthesis of his mental stock in trade.... We dissect nature along lines laid down by our native languages" (Benjamin Lee Whorf, Language, Thought, and Reality, 1956). Whorf, who worked in an insurance office in Hartford, Connecticut, found many striking differences between the European and Hopi languages. He wrote: "We cut up and organize the spread and flow of events as we do, largely because, through our mother tongue, we are parties to an agreement to do so, not because nature itself is segmented in exactly that way for all to see. Languages differ not only in how they build their sentences but also in how they break down nature to secure the elements to put in those sentences." Edward Sapir, Yale University, had earlier written: "Human beings do not live in the objective world alone, nor alone in the world of social activity as ordinarily understood, but are very much at the mercy of the particular language which has become the medium of expression for their society.... The worlds in which different societies live are distinct worlds, not merely the same world with different labels attached." Language is thus the framework that contains and shapes our thoughts. Our languages and brain size are the two things that distinguish us from all other animals on this planet. Possibly both evolved together and are the consequence of each other.

9 Is science automatically guaranteed when life becomes intelligent? Not necessarily. See "Wrong Number?" (1979), by Robert Wesson, who says: "The odds that another creature like Homo sapiens, with individual and social drives and capacities, would attain an electronic civilization may be compared to those of winning a lottery. ... In sum, the likelihood of an intelligent creature attaining an electronic or higher civilization may be much less than is often assumed and the number of such civilizations to be expected in our Galaxy is therefore correspondingly reduced." We have assumed in the text that the probability is as high as 0.1 because nontechnological civilizations can exist for tens,

perhaps hundreds, of thousands of years during which science might accidentally begin to flourish.

- "Perhaps we should first attempt reciprocal communication with non-human organisms here on earth – say with ... a termite queen-mother who represents the highest natural societal organization known on this planet. Foolish suggestions, yes, but they suggest the difficulty and probable impossibility of interplanetary communication" (Harlow Shapley, View from a Distant Star, 1963).

- "If the intelligence of these creatures were sufficiently superior to ours, they would choose to have little if any contact with us" (Brookings Institution, The Implications of a Discovery of Extraterrestrial Life, 1961).

- "The probability of success is difficult to estimate; but if we never search, the chance of success is zero" (Giuseppe Cocconi and Philip Morrison, "Search for interstellar communications," 1959).

- "The Galaxy may contain an assortment of civilizations at various levels of development ... and we would need to ask why we are unaware of each. Alternatively, as Bracewell suggests, there may be a 'galactic club' of interacting civilizations. Or, more likely in my view, the civilization that is number one exercises control and enforces the rules. They may keep us separate from our neighbors to prevent unfavorable or disastrous interactions" (John Ball, "Extraterrestrial intelligence: Where is everybody?"). Various suggestions have been made to explain why extraterrestrial intelligent life, assuming it exists, ignores human beings. One interesting suggestion is the zoo hypothesis, according to which we are creatures in a planetary zoo and are continually observed. Unknown to us, we are objects of interest to extraterrestrial visitors who are either scientists studying life on Earth or, like members of the public who visit zoos, are just curious sightseers.

PROJECTS

1 Michael Hart in "An explanation for the absence of Extraterrestrials on Earth" (1975) argues: "If ... there were intelligent beings elsewhere in our Galaxy, then they would eventually have achieved space travel, and would have explored and colonized the Galaxy, as we have explored and colonized the Earth. However, they are not here, therefore they do not exist." He classifies possible explanations for the apparent absence of extraterrestrials as follows:

(a) because of physical, astronomical, biological, and engineering difficulties;
(b) because they choose not to visit us for social, political, or lack-of-interest reasons;
(c) because they want to visit us but have not yet had sufficient time to reach us;
(d) because they have visited the Earth in the past, but we do not know it.

Hart finds that all explanations are inadequate and concludes that we are truly alone in the Galaxy. What are your views?

2 Formulate your own views on the origin and evolution of life on Earth. If the paradigm is survival of the fittest, who judges what is fit and unfit? If the answer is past environments, then, because the environment changes, life in general consists always of out-of-date adaptations.

3 Is aggression an essential characteristic of intelligent life? Many individuals believe the answer is yes, and think that adversarial relations are the essence of a vibrant society. What are the pros and cons in this debate?

4 Is human evolution becoming self-directed by means of social and biological engineering, with goals set by our present values? Is this good or bad?

5 Consider the following gruesome situation. A devilish assassin travels back in time and eliminates at birth 100 of the greatest artists in recorded history. As a result, the world of today becomes artistically a duller place. But nothing else greatly changes. The assassin then travels back in time and eliminates at birth 100 of the greatest writers and poets. The world of today is again made a duller place. Even without Shakespeare, the world we live in is not greatly altered. Finally, the assassin travels back and eliminates at birth 100 of the greatest scientists in recorded history. When the assassin returns to the world of today he discovers that this last excursion has wiped out at least 90 percent of the world's present population. Most people remaining live in slave- and serf-powered societies, ravaged by plagues and wars, ruled by tyrants who claim to be either gods, or appointed by gods, or inspired by gods. This provocative argument serves as a subject for debate. An alternative argument is made in *The Promise of the Coming Dark Age* by L. S. Stavrianos, who thinks a return to barbarism offers the promise of a new renaissance.

6 Should we seek to make contact with intelligent life outside the Solar System?

7 Will space travel solve the population problem and allow us always to have as many offspring as we please?

8 Why is intelligence not automatically guaranteed by evolution? See *This View of Life* by George Gaylord Simpson, who argues that the probability of "humanoid" creatures is exceedingly small. In the text, I assume that this probability is as large as 0.1, which is not impossible if we consider the billions of years (not just the millions of years of the recent past) in which planetary life has had a chance of becoming intelligent.

9 Imagine you are a wise, compassionate, angelic and powerful being in the Galaxy. What would you do if an intelligent but destructively aggressive form of life on a planet evolves to the point where it is in a state of imminent self-destruction? What would you do if it survives self-destruction and is bent on the conquest of other worlds? The possible existence of powerful angelic beings and the role they might play is left out of most discussions on extraterrestrial life.

10 Consider the Shakespearean view of cosmology. The universes are baseless fabrics we as playwrights weave to give sense and substance to our experiences. The universes are masks fitted on the face of the Universe. Perhaps behind the masks there is no face! Only ourselves weaving the baseless fabrics in which we portray ourselves as actors in the cosmic drama.

FURTHER READING

Ashpole, E. *The Search for Extraterrestrial Intelligence*. Blandford, London, 1989.

Ball, J. A. "Extraterrestrial intelligence: Where is everybody?" *Icanus* 19, 347 (1973).

Beadle, G. and Beadle, M. *The Language of Life: An Introduction to the Science of Genetics*. Doubleday, Garden City, New York, 1960.

Beck, L. W. "Extraterrestrial intelligent life," in *Extraterrestrials: Science and Alien Intelligence*. Editor, E. Regis. Cambridge University Press, Cambridge, 1985.

Blakemore, C. *Mechanics of the Mind*. Cambridge University Press, Cambridge, 1977.

Bracewell, R. N. *The Galactic Club: Intelligent Life in Outer Space*. Freeman, San Francisco, 1974.

Crick, F. *Of Molecules and Men*. University of Washington Press, Seattle, 1966.

Crowe, M. J. *The Extraterrestrial Life Debate, 1750–1900*. Cambridge University Press, Cambridge, 1986.

Dick, S. J. *Plurality of Worlds: The Origins of the Extraterrestrial Life Debate from Democritus to Kant*. Cambridge University Press, Cambridge, 1982.

Dick, S. J. *The Biological Universe: The Twentieth-Century Extraterrestrial Life Debate and the Limits of Science*. Cambridge University Press, Cambridge, 1996.

Goldsmith, D. *The Quest for Extraterrestrial Life*. University Science Books, Mill Valley, California, 1980.

Harrison, A. A. *After Contact: The Human Response to Extraterrestrial Life*. Plenum, New York, 1997.

Huang, S. "Life in space and humanity on the Earth." *American Scientist* (June 1965).

Napier, J. *The Roots of Mankind*. Harper and Row, New York, 1973.

Newman, W. I. and Sagan, C. *Interstellar Migration and the Human Experience*. Editors, B. R. Finney, et al. University of California, Berkeley, 1985.

O'Neill, G. K. "The colonization of space." *Physics Today* (September 1974).

Orgel, L. E. *The Origins of Life: Molecules and Natural Selection*. John Wiley, New York, 1973.

Ponnamperuma, C. *The Origins of Life*. Dutton, New York, 1972.

Sagan, C. and Drake, F. "The search for extraterrestrial intelligence." *Scientific American* (May 1975).

Schrödinger, E. *What Is Life?* Cambridge University Press, Cambridge, 1946.

Schrödinger, E. *Mind and Matter*. Cambridge University Press, Cambridge, 1958.

Shapley, H. *Of Stars and Men: The Human Response to an Expanding Universe*. Beacon Press, Boston, 1958.

Shklovsky, I. S. and Sagan, C. *Intelligent Life in the Universe*. Holden-Day, New York, 1966.

Sneath, P. H. A. *Planets and Life*. Thames and Hudson, London, 1970.

Wald, G. "Life and mind in the universe." *International Journal of Quantum Chemistry, Quantum Biology Symposium* No. 11 (1984).

Wilkinson, D. *Alone in the Universe*. Monarch, Crowborough, England, 1997.

SOURCES

Arrhenius, S. *Worlds in the Making*. Harper and Brothers, New York, 1908.

Brookings Institution. *The Implications of a Discovery of Extraterrestrial Life*. Prepared for NASA, March 24, 1961, 87th Congress. U.S. Government Printing Office, Washington, D.C.

Bruno, G. See Singer, D. W.

Cameron, A. G. W. Editor. *Interstellar Communication: A Collection of Reprints and Original Contributions*. Benjamin, New York, 1963.

Cocconi, G. and Morrison, P. "Search for interstellar communications." *Nature* 184, 844 (September 19, 1959).

Darwin, C. *The Origin of Species by Means of Natural Selection; or, the Preservation of Favoured Races in the Struggle for Life*. 1859. Reprint: Penguin Books, Harmondsworth, Middlesex, 1968.

Frankfort, H., Frankfort, H. A., Wilson, J. A., and Jacobsen, T. *Before Philosophy*. Penguin Books, London, 1949. First published as *The Intellectual Adventure of Ancient Man*. University of Chicago Press, Chicago, 1946.

Gillie, O. *The Living Cell*. Funk and Wagnalls, New York, 1971.

Gould, S. J. *Ever Since Darwin: Reflections in Natural History*. Norton, New York, 1977.

Haldane, J. B. S. *Possible Worlds and Other Papers*. Chatto and Windus, London, 1927.

Haldane, J. B. S. "Origin of life," in *New Biology*, Number 16. Penguin, London, 1954.

Hart, M. H. "An explanation for the absence of extraterrestrials on Earth." *Quarterly Journal of the Royal Astronomical Society* 16, 128 (1975).

Heilbroner, R. *An Inquiry into the Human Prospect*. Norton, New York, 1975.

Hoerner, S. von. "Population explosion and interstellar expansion." *Journal British Interplanetary Society* 28, 691 (1975).

Holland, H. D. *The Chemical Evolution of the Atmosphere and Oceans*. Princeton University Press, Princeton, 1984.

Koyré, A. *Newtonian Studies*. Chapman and Hall, London, 1965.

Lovelock, J. E. "Hands up for the Gaia hypothesis." *Nature* 344, 100 (1990).

Maynard Smith, J. and Szathamáry, E. *The Major Transitions in Evolution*. Freeman, Oxford, 1995.

Mayr, E. "Darwin and natural selection." *American Scientist* (May–June 1977).

Murray, B., Gulkis, S., and Edelson, R. E. "Extraterrestrial intelligence: An observational approach." *Science* 199, 485 (1978).

Oparin, A. I. *The Origin of Life*. Second edition. Dover Publications, New York, 1953.

Ori, J., Miller, S. L., Ponnamperuma, C., and Young, R. S. *Cosmochemical Evolution and the Origins of Life*. Reidel, Dordrecht, Netherlands, 1974.

Papagiannis, M. "Are we all alone, or could they be in the asteroid belt?" *Quarterly Journal of the Royal Astronomical Society* 19, 277 (1978).

Ponnamperuma, C. and Cameron, A. G. W. *Interstellar Communication: Scientific Perspectives*. Houghton Mifflin, Boston, 1974.

Shapley, H. *View from a Distant Star: Man's Future in the Universe*. Basic Books, New York, 1963.

Simpson, G. G. *This View of Life: The World of an Evolutionist*. Harcourt, Brace and World, New York, 1963.

Singer, D. W. *Giordano Bruno: His Life and Thoughts, With an Annotated Translation of His Work on The Infinite Universe and Worlds*. Schumann, New York, 1950.

Stavrianos, L. S. *The Promise of the Coming Dark Age*. W. H. Freeman, San Francisco, 1976.

Teilhard de Chardin, P. *The Phenomenon of Man*. Harper and Row, New York, 1961.

Wesson, R. G. "Wrong number? A skeptic argues against the likelihood of advanced extra-terrestrial civilizations." *Natural History* (March 1979).

Whorf, B. L. *Language, Thought, and Reality: Selected Writings.* Editor, J. B. Carroll. M.I.T. Press, Cambridge, Massachusetts, 1956.

APPENDIX
FUNDAMENTAL
QUANTITIES

BASIC UNITS

speed of light

gravitation constant

Planck constant

Boltzmann constant

electron charge

fine structure constant

Hubble term

density term

$c = 3.00 \times 10^{10} \, \mathrm{cm \, sec^{-1}}$

$G = 6.67 \times 10^{-8} \, \mathrm{cm^3 \, g^{-1} \, sec^{-2}}$

$h = 6.63 \times 10^{-27} \, \mathrm{cm^2 \, g \, sec^{-1}}$

$\hbar = h/2\pi = 1.05 \times 10^{-27} \, \mathrm{cm^2 \, g \, sec^{-1}}$

$k_B = 1.38 \times 10^{-16} \, \mathrm{erg \, K^{-1}}$

$e = 4.80 \times 10^{-10} \, \mathrm{esu}$

$ = 1.60 \times 10^{-19} \, \mathrm{coulomb}$

$\alpha = e^2/\hbar c = 7.30 \times 10^{-3}$

$H_0 = 100h \, \mathrm{km \, sec^{-1} \, Mpc^{-1}}$

$\Omega = \rho_{\mathrm{univ}}/\rho_{\mathrm{crit}}$

UNITS OF LENGTH

Planck length

classical electron radius

Bohr radius

Earth radius

Sun's radius

astronomical unit

light year

parsec

Hubble length

$a^* = (G\hbar/c^3)^{1/2} = 1.61 \times 10^{-33} \, \mathrm{cm}$

$a = e^2/m_e c^2 = 2.82 \times 10^{-13} \, \mathrm{cm}$

$a_0 = \hbar^2/m_e c^2 = 0.53 \times 10^{-8} \, \mathrm{cm}$

$R_\oplus = 6.34 \times 10^8 \, \mathrm{cm}$

$R_\odot = 6.96 \times 10^{10} \, \mathrm{cm}$

$\mathrm{AU} = 1.50 \times 10^{13} \, \mathrm{cm}$

$\mathrm{ly} = 9.46 \times 10^{17} \, \mathrm{cm}$

$\mathrm{pc} = 3.09 \times 10^{18} \, \mathrm{cm}$

$L_H = c/H_0 = 3000h^{-1} \, \mathrm{Mpc}$

$ = 9.78 \times 10^9 h^{-1} \, \mathrm{ly} = 9.25 \times 10^{28} h^{-1} \, \mathrm{cm}$

UNITS OF TIME

Planck unit

jiffy unit

day

year

Hubble time

$t^* = (G\hbar/c^5)^{1/2} = 5.38 \times 10^{-44} \, \mathrm{sec}$

$j = a/c = 9.40 \times 10^{-24} \, \mathrm{sec}$

$\mathrm{d} = 86\,400 \, \mathrm{sec}$

$\mathrm{y} = 3.16 \times 10^7 \, \mathrm{sec}$

$t_H = 1/H_0 = 3.08 \times 10^{17} h^{-1} \, \mathrm{sec}$

UNITS OF MASS

Planck mass

electron mass

nucleon mass

mass of Earth

$m^* = (\hbar c/G)^{1/2} = 2.18 \times 10^{-5} \, \mathrm{g}$

$m_e = 9.11 \times 10^{-28} \, \mathrm{g}$

$m_n = 1.66 \times 10^{-24} \, \mathrm{g}$

$M_\oplus = 5.98 \times 10^{27} \, \mathrm{g}$

mass of Sun	$M_\odot = 1.99 \times 10^{33}\,\text{g}$
mass of Hubble sphere	$M_H = \dfrac{4\pi\rho_{\text{univ}}L_H^3}{3} = 3 \times 10^{22}\,\Omega h^{-1}\,M_\odot$

UNITS OF DENSITY

Planck density	$\rho^* = \dfrac{c^5}{G^2\hbar} = 5.12 \times 10^{93}\,\text{g}\,\text{cm}^{-3}$
jiffy density	$\rho_{\text{j}} = \dfrac{3}{8\pi G j^2} = 2 \times 10^{52}\,\text{g}\,\text{cm}^{-3}$
nucleon density	$\rho_{\text{n}} = \dfrac{3m_{\text{n}}}{4\pi a^3} = 2 \times 10^{13}\,\text{g}\,\text{cm}^{-3}$
critical density	$\rho_{\text{crit}} = \dfrac{3H^2}{8\pi G} = 1.88 \times 10^{-29}h^2\,\text{g}\,\text{cm}^{-3}$
density of universe	$\rho_{\text{univ}} = \Omega\rho_{\text{crit}} = \dfrac{3\Omega H^2}{8\pi G} = 1.88 \times 10^{-29}\,\Omega h^2\,\text{g}\,\text{cm}^{-3}$

UNITS OF ENERGY

Planck mass	$m^*c^2 = 1.2 \times 10^{19}\,\text{GeV} = 1.4 \times 10^{32}\,\text{K}$
nucleon mass	$m_{\text{n}}c^2 = 938\,\text{MeV} = 1.1 \times 10^{13}\,\text{K}$
electron mass	$m_{\text{e}}c^2 = 0.51\,\text{MeV} = 1.1 \times 10^{10}\,\text{K}$
degree kelvin	$1\,\text{K} = 8.62 \times 10^{-5}\,\text{eV}$
electron volt	$1\,\text{eV} = 1.60 \times 10^{-12}\,\text{erg} = 1.16 \times 10^4\,\text{K}$

Units: cm: centimeter; sec: second; g: gram; K: kelvin; eV: electron volt (energy of electron falling through a potential difference of 1 volt); MeV: $10^6\,$eV; GeV: $10^9\,$eV.

[†] 1 coulomb is the electric charge flowing for 1 second in an electric current of 1 ampere. In each second 6×10^{18} electrons flow in a current of 1 ampere. This is approximately the number of electrons flowing each second through the filament of a 100 watt bulb.

INDEX

The first page only is cited when an entry covers more than one page.

CPSIA information can be obtained
at www.ICGtesting.com
Printed in the USA
LVHW060518010921
696659LV00004B/294

9 780521 661485